Selected Titles in This Series

93 **Nikolai K. Nikolski,** Operators, functions, and systems: An easy reading. Volume 2: Model operators and systems, 2002

92 **Nikolai K. Nikolski,** Operators, functions, and systems: An easy reading. Volume 1: Hardy, Hankel, and Toeplitz, 2002

91 **Richard Montgomery,** A tour of subriemannian geometries, their geodesics and applications, 2002

90 **Christian Gérard and Izabella Łaba,** Multiparticle quantum scattering in constant magnetic fields, 2002

89 **Michel Ledoux,** The concentration of measure phenomenon, 2001

88 **Edward Frenkel and David Ben-Zvi,** Vertex algebras and algebraic curves, 2001

87 **Bruno Poizat,** Stable groups, 2001

86 **Stanley N. Burris,** Number theoretic density and logical limit laws, 2001

85 **V. A. Kozlov, V. G. Maz'ya, and J. Rossmann,** Spectral problems associated with corner singularities of solutions to elliptic equations, 2001

84 **László Fuchs and Luigi Salce,** Modules over non-Noetherian domains, 2001

83 **Sigurdur Helgason,** Groups and geometric analysis: Integral geometry, invariant differential operators, and spherical functions, 2000

82 **Goro Shimura,** Arithmeticity in the theory of automorphic forms, 2000

81 **Michael E. Taylor,** Tools for PDE: Pseudodifferential operators, paradifferential operators, and layer potentials, 2000

80 **Lindsay N. Childs,** Taming wild extensions: Hopf algebras and local Galois module theory, 2000

79 **Joseph A. Cima and William T. Ross,** The backward shift on the Hardy space, 2000

78 **Boris A. Kupershmidt,** KP or mKP: Noncommutative mathematics of Lagrangian, Hamiltonian, and integrable systems, 2000

77 **Fumio Hiai and Dénes Petz,** The semicircle law, free random variables and entropy, 2000

76 **Frederick P. Gardiner and Nikola Lakic,** Quasiconformal Teichmüller theory, 2000

75 **Greg Hjorth,** Classification and orbit equivalence relations, 2000

74 **Daniel W. Stroock,** An introduction to the analysis of paths on a Riemannian manifold, 2000

73 **John Locker,** Spectral theory of non-self-adjoint two-point differential operators, 2000

72 **Gerald Teschl,** Jacobi operators and completely integrable nonlinear lattices, 1999

71 **Lajos Pukánszky,** Characters of connected Lie groups, 1999

70 **Carmen Chicone and Yuri Latushkin,** Evolution semigroups in dynamical systems and differential equations, 1999

69 **C. T. C. Wall (A. A. Ranicki, Editor),** Surgery on compact manifolds, second edition, 1999

68 **David A. Cox and Sheldon Katz,** Mirror symmetry and algebraic geometry, 1999

67 **A. Borel and N. Wallach,** Continuous cohomology, discrete subgroups, and representations of reductive groups, second edition, 2000

66 **Yu. Ilyashenko and Weigu Li,** Nonlocal bifurcations, 1999

65 **Carl Faith,** Rings and things and a fine array of twentieth century associative algebra, 1999

64 **Rene A. Carmona and Boris Rozovskii, Editors,** Stochastic partial differential equations: Six perspectives, 1999

63 **Mark Hovey,** Model categories, 1999

62 **Vladimir I. Bogachev,** Gaussian measures, 1998

(Continued in the back of this publication)

Mathematical
Surveys
and
Monographs

Volume 93

Operators, Functions, and Systems: An Easy Reading

Volume 2: Model Operators and Systems

Nikolai K. Nikolski

Translated by
Andreas Hartmann

American Mathematical Society

Based on a series of lecture notes, in French,
Université de Bordeaux, 1991–1995.
Translated by ANDREAS HARTMANN and greatly revised by the author

2000 *Mathematics Subject Classification*. Primary 47–02, 30–02, 93–02,
30D55, 47B35, 47A45, 93B05, 93C05.

Abstract. The book joins four formally distant topics of analysis and its applications: Volume 1 contains 1) Hardy classes of holomorphic functions, 2) Spectral theory of Hankel and Toeplitz operators, and Volume 2 contains 3) Function models for linear operators on a Hilbert space and free interpolation, and 4) Infinite-dimensional system theory and signal processing. Beyond standard topics covered by these titles, it includes elements of maximal functions, Littlewood-Paley techniques, the Riemann zeta-function, Wiener filtering (all in Volume 1), as well as moment problems, reproducing kernel spaces, Schatten-von Neumann ideals, stationary processes, best rational approximations, similarity theory, and controllability with the least control operators (in Volume 2).

Library of Congress Cataloging-in-Publication Data

Nikol′skii, N. K. (Nikolai Kapitonovich)

Operators, functions, and systems : an easy reading / Nikolai K. Nikolski ; [translated by Andreas Hartmann and greatly revised by the author].

p. cm. — (Mathematical surveys and monographs, ISSN 0076-5376 ; v. 92)

Contents: v. 1. Hardy, Hankel, and Toeplitz

ISBN 0-8218-1083-9 (v. 1 : alk. paper)

1. Operator theory. 2. Harmonic analysis. 3. Control theory. I. Title. II. Mathematical surveys and monographs ; no. 92.

QA329.N55 2002
515′.724—dc21 2001053556

∞ The paper used in this book is acid-free and falls within the guidelines
established to ensure permanence and durability.
Visit the AMS home page at URL: http://www.ams.org/

10 9 8 7 6 5 4 3 2 1 07 06 05 04 03 02

A Few Words about the Book

WHAT THIS BOOK IS ABOUT

The book represents a mixture of harmonic and complex analysis with operator theory. The interplay between these disciplines is one of the most significant features of the second half of Twentieth century mathematics. It gave rise to several jewels of analysis, such as the theory of singular integral operators, Toeplitz operators, mathematical scattering theory, Sz.-Nagy-Foiaş model theory, the L. de Branges proof of the Bieberbach conjecture, as well as solving the principal interpolation problems in complex analysis and discovering the structural properties of function spaces (from Besov to Bergman).

The principal ingredients of the book are clear from the Contents and Subject Index, and indeed a simple list of key words tells more than long explanations. Without reproducing these lists nor the introductions to the four parts A, B, C, and D of the book, I would like give an abridged list of my favorite subjects, ordered by their appearance in the book:

Hardy classes
The Hilbert transformation
Weighted polynomial approximation
Cyclicity phenomena
Maximal and Littlewood-Paley functions
The Marcinkiewicz weak type interpolation
Wiener filtering theory
Riemann ζ function
Hankel operators: spectral theory, Peller's theory, moment problems
Reproducing kernel Hilbert spaces
Schatten-von Neumann operator ideals
Toeplitz operators
The operator corona problem
Spectral theory of normal operators
Sz.-Nagy-Foiaş function model
Von Neumann inequalities
Carleson and generalized free interpolations
Theory of spectral multiplicities
Elements of semigroup theory
Classical control theory of dynamical systems
Bases of exponentials on intervals of the real line
Elements of the H^∞ control theory

Style

I have tried to follow the logic of the above subjects as I understand it. As a consequence, this book is neither a function theory monograph, nor an operator theory manual. It is a treatise on operator-based function theory, or, if you prefer, function-based operator theory. As in my previous book "Treatise on the shift operator" (Springer, 1986) I have in mind a picture close to mathematical reality, where the most interesting and important facts take part of several disciplines simultaneously. This is why the way in which things proceed in this book is sometimes different from the appoved didactic style of presentation, when, first of all, background materials should be developed (even if you will need it 300 pages later...), then you go to the next preparatory level, and so on.

Here, new concepts and auxiliary materials appear when they are needed to continue the main theme. This theme is developed as theory of functions on the circle group and of operators acting on them, starting with the basic shift operator, then passing to stationary filtering, and Hankel and Toeplitz operators as compressions of the multiplication operators. Next, we arrive at the model theory for Hilbert space operators as (advanced) compressions of the same shift operator, and, finally, all this machinery is brought together to control dynamical systems. Therefore, taken as a style to telling mathematics, this is more a passion or a tale of mental intrigue than a rationally arranged catalog of facts.

It is also worth mentioning that this book has its origins in four courses I gave in 1992-1996 to graduate students in the University of Bordeaux, France. Although the courses were considerably extended when preparing this book, the text, perhaps, preserves the flavor of interaction with the audience: sometimes I repeat some notions or ideas already stated some tens (or hundreds...) of pages earlier to remind the reader of something what he may have forgotten from the last course.

Background

As it is clear from the preceding lines, the book can be read by anyone having a standard analysis background: Lebesgue measure, L^p spaces, elements of Fourier series and Fourier transforms (the Plancherel theorem), elementary holomorphic functions, Stone-Weierstrass theorem, Hilbert and Banach spaces, reflexivity, the Hahn-Banach theorem, compactness, and so on.

Formal structure

Parts A and B form the first volume of the book, and parts C and D form the second. Formally speaking, parts A, B, C and D are (reasonably) independent of each other in the sense that, for example, I may employ in part B some results of parts A, C, or D, but in the same way that I use (rarely) results from some exterior basic monographs.

The Parts are divided into chapters; there are 25 in the book. All chapters but one contain two special sections: *Exercises and Further Results*, and *Notes and Remarks*. These are important and inseparable parts of the book. To illustrate, the book contains 1428 propositions conventionally called theorems, lemmas, corollaries, and exercises. For *Exercises and Further Results*, the proofs are called "hints", and while they are shorter they still contain all the principal ingredients to understand the proof. All exercises were tested by a team of volunteer readers whose names are listed below (there were no casualties...). Some (rare) facts included in exercises and not proved are marked by asterisk $*$.

Sections *Notes and Remarks* usually contain surveys of the rest of the theory presented in the main body of the corresponding chapter.

Reference $A.\alpha.\beta.\gamma$ means subsection $\alpha.\beta.\gamma$ of Part A; $\alpha.\beta$ means section β of chapter α of the Part where you are; $\alpha.\beta.\gamma(3)$ means point (3) of subsection $\alpha.\beta.\gamma$, etc. Sign $\square$ indicates the end of a proof or a reasoning.

THE READER AND THE AUTHOR

As it is clear from the subtitle of the book, I expect that some readers are novices, graduate or undergraduate students, possessing the needed knowledge indicated above. It is also supposed that some readers are experts. Well, I shall be rewarded if there is at least one. In this case, and also anticipating the inevitable reproaches as to why I selected such and such subjects and not others, I permit myself to quote (in my translation from Russian) a great philologist, an expert of texts as such.

"The answers books give us are to questions that are not exactly the same as the author set before himself, but to those that we are able to raise ourselves... The books encircle us like mirrors, in which we see only our own reflexion; the reason why it is, perhaps, not everywhere the same is because all these mirrors are curved, each in its own way."

M.L. Gasparov
"Philology as morality"

WHAT IS MISSED

Of course, I do not intend to list here the rest of mathematics but just to mention explicitly some border subjects that could have been included but were not. These are (without any ordering) *extremal problems* of complex analysis (starting from results of S.Ya. Khavinson and H. Shapiro in the 1960's); the *problems of harmonic analysis-synthesis* (from L. Schwartz and B. Malgrange of the 1950's); *invariant subspaces*, from the existence problem in a Hilbert space, up to (more important) classification problems for concrete operators (including descriptions of *closed ideals* in algebras of holomorphic functions); *singular integral models* for hypo- and semi-normal operators; *scattering*; *univalent functions* via quasi-orthogonal decompositions; *realization theory*; *operator valued constrained interpolation*, and some other themes. I have no better way to excuse these omissions than to follow E. Beckenbach and R. Bellman who quoted the following verses (for a similar purpose):

Oh, the little more, and how much it is!
And the little less, and what worlds away!

R. Browning (Saul.st.39)

ACKNOWLEDGEMENTS

Several people read preliminary versions of the book and gave me their opinions, both mathematical and technical, especially by testing numerous exercises. For this I am greatly indebted to E. Abakumov, A. Aleksandrov, A. Baranov, A. Borichev, M. Gamal, V. Kapustin, S. Kislyakov, S. Shimorin, V. Vasyunin, P. Vitse. Also, I appreciate the time spent by my colleages A. Borichev, A. Hartmann and S. Kupin in helping me verify the reference list.

It has happened that a rather big part of this book, being lecture notes of my Bordeaux courses, was written in French, excepting some more recent additions. This part of the text was carefully translated into English by Andreas Hartmann, whose work essentially surpassed the simple translation including several constructive mathematical criticisms. I am very grateful to him for this work, as well as to Th.V. Pedersen for reading the translation and to D. Sherman for occasional advice.

My own function for completing preliminary lecture notes up to a self-contained book largely surpassed in time all predicted limits. During all these long months, my wife Ludmila steadily bore this somewhat rash enterprise, and all in all what is done bears the mark of her support.

I am also grateful to the publisher, the American Mathematical Society, for including the book in this series, and for having much patience during entire period of my work and enough flexibility at the moment when it became clear that the result would double the predicted size.

July 23, 2001
Gradignan

Contents

Volume 1: Hardy, Hankel and Toeplitz

Part C

Model Operators and Free Interpolation

Part C

Model Operators and Free Interpolation

Foreword to Part C

Part C of the book represents a quick introduction to function model techniques, to elements of spectral theory in the language of the function model and to applications to free interpolation. The function model of a Hilbert space contraction consists in writing it as a compression of the shift operator $S : f \longrightarrow zf$ to a coinvariant subspace of the (vector valued) Hardy space H^2. The main advantage of this writing is that it makes the function calculus of the operator very explicit, and, as a consequence, allows one to interpret all the Hardy classes techniques in terms of spectral theory. In Chapter 3 below, we give an example of such a relationship between these theories by considering the problem of free interpolation and generalized free interpolation by Hardy spaces functions as a problem of unconditionally convergent spectral decompositions of a suitable model operator. But, first, we construct (Chapter 1) and study (Chapter 2) the model itself.

To place the function model in a natural historical framework, we recall some other models of linear operators or matrices.

From a general point of view, a model of an operator $T : X \to X$ is another operator $M : K \to K$ which is in a certain sense equivalent to the first one. One can distinguish three types of equivalence.

(1) M is *similar* to T if $VT = MV$ for an isomorphism $V : X \to K$,

(2) M is *unitarily equivalent* to T if $VT = MV$ for some unitary operator $V : X \to K$, and X, K are Hilbert spaces,

(3) M is *quasi-similar* to T if $VT = MV$, $TU = UM$ for two operators U, V for which U, U^*, V, V^* are injections.

The subsequent list of examples gives some classical models.

EXAMPLES. **(1) The Schur model**. Every operator $T : X \to X$ on a finite dimensional space X, $\dim X = n < \infty$, is similar (unitarily equivalent if X is a Hilbert space) to a triangular matrix $\mathbb{C}^n \to \mathbb{C}^n$

$$\begin{pmatrix} a_{11} & a_{12} & \dots & a_{1n} \\ 0 & a_{22} & \dots & a_{2n} \\ \dots & \dots & \dots & \dots \\ 0 & 0 & \dots & a_{nn} \end{pmatrix}.$$

This is equivalent to say that there is a chain $\{0\} = E_0 \subset E_1 \subset E_2 \subset \cdots \subset E_n = X$ of invariant subspaces, or a *Schur basis* of T: $e_i \in E_i \ominus E_{i-1}$, $1 \le i \le n$.

(2) The Jordan model. Let $T : X \to X$, $\dim X = n < \infty$. Then T is similar to an operator of the following type

$$M = \sum_{k=1}^{m} \oplus \left(\lambda_k I_{d(k)} + J_{d(k)}\right)$$

where $\cup_{k=1}^{m}\{\lambda_k\} = \sigma(T)$ is the spectrum of T and J_d denotes the d-dimensional Jordan block $J_d : \mathbb{C}^d \to \mathbb{C}^d$,

$$J_d \simeq \begin{pmatrix} 0 & 0 & \dots & 0 & 0 \\ 1 & 0 & \dots & 0 & 0 \\ 0 & 1 & \dots & 0 & 0 \\ \dots & \dots & \dots & \dots & \dots \\ 0 & 0 & \dots & 1 & 0 \end{pmatrix}.$$

We have $\sum_{k=1}^{m} d(k) = n$, and the triplet $(m, \{\lambda_k\}_1^m, \{d(k)\}_1^m)$ is a complete invariant of T with respect to similarity, that is, T_1 and T_2 are similar if and only if

$$(m_1, \{\lambda_k^1\}_1^m, \{d_1(k)\}_1^m) = (m_2, \{\lambda_k^2\}_1^m, \{d_2(k)\}_1^m),$$

where $m = m_1 = m_2$.

(3) The von Neumann model. Recall that an operator $U : H \to H$ acting on a Hilbert space is called unitary if $(Ux, Uy) = (x, y)$ for every $x, y \in H$ and $UH = H$. In other words if and only if $UU^* = U^*U = I$, where U^* is the usual adjoint operator $(Ux, y) = (x, U^*y)$. The von Neumann model is given by the spectral theorem in the following form (von Neumann), see Subsection 1.5.1 for proofs.

THEOREM (Spectral Theorem). *Let $U : H \to H$ be unitary on a Hilbert space H. Then there is*

(i) a finite Borel measure μ on $\mathbb{T} = \{\zeta \in \mathbb{C} : |\zeta| = 1\}$,

(ii) an orthogonal projection valued (weak) Borel function $\zeta \longmapsto P(\zeta)$,

$$P(\zeta) : H \to H, \quad \zeta \in \mathbb{T},$$

such that the operator U is unitarily equivalent to the multiplication operator $f(\zeta) \longmapsto \zeta f(\zeta)$, $\zeta \in \mathbb{T}$, acting on the space

$$\int_{\mathbb{T}} \oplus H(\zeta)\, d\mu(\zeta),$$

where $H(\zeta) = P(\zeta)H$ and

$$\int_{\mathbb{T}} \oplus H(\zeta)\, d\mu(\zeta) = \left\{f \in L^2(H, \mu) : f(\zeta) \in H(\zeta) \text{ a.e. } \mu\right\}.$$

Here $L^2(H, \mu)$ is the space of H valued, measurable and square integrable functions (see A.3.11, Volume 1).

If we require that $\dim H(\zeta) = d_U(\zeta) \neq 0$ a.e. μ, then the equivalence class $[\mu] = \{\nu : \nu \sim \mu\}$ of the measure μ and the function $d_U(\cdot)$ are unitary invariants of U. Moreover they form a complete system of unitary invariants: $[\mu_U] = [\mu_V]$ and $d_U = d_V$ a.e. if and only if U is unitarily equivalent to V.

THEOREM. *The assertions of the previous theorem are true for* normal operators $N : H \to H$ *(*$N^*N = NN^*$ *or* $\|Nx\| = \|N^*x\|$ *for every* $x \in H$*) with the only difference that the measure* μ *is now carried by the spectrum of* N, $\sigma(N) = \operatorname{supp} \mu$ *(which is a compact subset in* $\mathbb{C}$*).*

(4) The Friedrichs model. This model applies to (smooth) weak perturbations of self-adjoint operators: $D + K$. So D will be assumed to be selfadjoint with continuous spectrum and K a "smooth" perturbation of D (for example, of the Hilbert–Schmidt class). Then one can show that in the spectral von Neumann model of D the operator $D + K$ has the following representation

$$f(x) \longmapsto xf(x) + \int_a^b N(x, y)f(y)dy,$$

where N is a kernel verifying certain supplementary conditions.

The model considered in this part of the book is of a different nature. So far we have seen models that were based on decompositions of an operator T in simpler operators (through sums, series or integrals). Here we will interpret the operator T as a part of another, complicated operator. Even if this seems to complicate our task of describing the action of T we obtain a very profound analytic structure that opens the door to most of those refined techniques of one-variable complex analysis which was developed in Parts A and B of Volume 1.

This model was discovered by M. Livshic at Kharkov in 1946, and was developed in the 50s and 60s by M. Naimark, M. Brodski and M. Krein.

In the sequel it was transformed by B. Sz.-Nagy et C. Foiaş (in the 60s) into a more customary form by representing it in the Hardy space H^2 (instead of its Fourier transform on the half line as in the Livshic model). Even if these latter differences are rather technical, they contributed largely to the development of the theory and its applications.

Below, we shorten to WOT the mention of the weak operator topology on the space $L(E, F)$ of bounded linear operators from E to F defined by the family of seminorms $A \longrightarrow |(Ax, y)|$, where $x \in E$, $y \in F^*$. The similar short form, SOT, is used for the strong operator topology defined by the seminorms $A \longrightarrow \|Ax\|$, where $x \in E$.

Returning to the contents of this Part C of the book we give one-line descriptions of its chapters.

Chapter 1 contains principal model and dilation constructions.
Chapter 2 contains elements of the model based spectral theory.
Chapter 3 contains applications to unconditional bases and spectral decompositions.

CHAPTER 1

The Basic Function Model

This chapter contains the construction and basic properties of the free function model and some of its "coordinate realizations". Namely, we start by proving the existence of the minimal unitary dilation U of a given Hilbert space contraction T (Section 1.1). Then we make a half-step towards the von Neumann model for U (see the Foreword above), that is, we define two "functional mappings" π, π_* which "intertwine" U with the basic unitary operator $f(e^{it}) \longmapsto e^{it}f(e^{it})$ and which are isometric embeddings of vector valued L^2 spaces on $\mathbb{T}$ into the space of the unitary dilation. The triple $\{U, \pi, \pi_*\}$ represents the coordinate free function model of T (Section 1.3). It is essentially uniquely defined by the so-called characteristic function of T which is an H^∞ operator valued function defined by $\Theta_T = \pi_*^*\pi$ (Section 1.2).

In Section 1.3 we write down two coordinate transcriptions of the model (the B. Sz.-Nagy–C. Foiaş's transcription, and B. Pavlov's transcription) by making particular choices of the functional embeddings. Section 1.4 contains the dissipative counterpart of the theory that allows us (via the Cayley transform) to treat operators A with $\operatorname{Im} A = (A - A^*)/2i \geq \mathbb{O}$. There are also some concrete examples of integral and differential operators.

Section 1.5, Exercises and Further Results, contains many various facts, both of classic operator theory and model theory. In particular, we give 1) a complete treatment of the von Neumann spectral theorem for normal operators and the Hilbert–Schmidt theorem on compact normal operators; 2) elements of C^*-algebras (including the Gelfand–Naimark theorem and the Fuglede–Putnam commutator theorem); 3) intertwining properties between the model operator M_Θ and "the residual part" of its minimal unitary dilation; 4) an intensive account of the problems of similarity and quasi-similarity to a unitary operator (using both model and harmonic analysis approaches), including their interplays with the left invertibility "corona problem" from Section B.9.2 (Volume 1); 5) the $C_{\alpha\beta}$ classification of contractions; and 6) a collection of facts about simultaneous unitary dilations of several contractions and multi-operator von Neumann inequalities (including the Parrott and Crabb–Davie counterexamples). The chapter ends with a survey of the state of the dilation based operator theory (Section 1.6).

1.1. Unitary Dilations

We begin with general dilations, then study those of them which are unitary. The section culminates in the general form of minimal unitary dilations.

1.1.1. DEFINITION. Let $T : H \to H$ be an operator acting on a Hilbert space H, and let $U : \mathcal{H} \to \mathcal{H}$ be acting on another Hilbert space $\mathcal{H} \supset H$. The operator

U is called a *dilation* of T if

$$T^n h = P_H U^n h, \quad h \in H, n \geq 0,$$

where P_H is the orthogonal projection of $\mathcal{H}$ onto H. The space $\mathcal{H}$ is called a *dilation space*.

The dilation property clearly implies that $q(T)h = P_H q(U)h$ for every $h \in H$ and every complex polynomial $q \in \mathcal{P}ol_+ = \mathcal{L}\text{in}(z^n : n \geq 0)$. A useful and obvious generalization of this notion is the *dilation of an operator algebra* $A \subset L(H)$. This is an algebra $B \subset L(\mathcal{H})$, $H \subset \mathcal{H}$, such that the natural compression $b \longmapsto P_H b|H$, $b \in B$, is a surjective homomorphism of B onto A.

The following lemma gives a description of the structure of the dilation space.

1.1.2. LEMMA (D. Sarason, 1965). *An operator $U : \mathcal{H} \to \mathcal{H}$ is a dilation of its compression $P_H U|H =: T$ if and only if $\mathcal{H}$ decomposes in the following way*

$$\mathcal{H} = G_* \oplus H \oplus G, \tag{1.1.1}$$

where $UG \subset G$ and $U^* G_* \subset G_*$ (in particular, the latter means that $H \oplus G$ is U-invariant).

PROOF. For the necessity, let U be a dilation of T, and set

$$G_1 = \text{span}(U^n H : n \geq 0),$$

the linear closed span of $U^n h$, $n \geq 0$, $h \in H$. Let $G = G_1 \ominus H$ be the orthogonal complement of H in G_1 and let $G_* = \mathcal{H} \ominus G_1$. By construction $UG_1 \subset G_1$. Observe that in general if $UE \subset E$ for some $E \subset \mathcal{H}$, then $U^* E^\perp \subset E^\perp$ — and vice versa. As $G_* = G_1^\perp = \mathcal{H} \ominus G_1$ we get $U^* G_* \subset G_*$.

In order to show that $UG \subset G$, let now $g \in G \subset G_1$ and $h \in H$. By definition of G_1 there are elements $h_{k,n} \in H$ such that

$$g = \lim_n \sum_{k \geq 0} U^k h_{k,n}.$$

This implies that

$$\begin{aligned}
(Ug, h) &= \lim_n \Big(U \sum_{k \geq 0} U^k h_{k,n}, h\Big) = \lim_n \Big(P_H U \sum_{k \geq 0} U^k h_{k,n}, h\Big) \\
&= \lim_n \Big(T \sum_{k \geq 0} T^k h_{k,n}, h\Big) = \lim_n \Big(\sum_{k \geq 0} T^k h_{k,n}, T^* h\Big) \\
&= \lim_n \Big(\sum_{k \geq 0} U^k h_{k,n}, T^* h\Big) = (g, T^* h).
\end{aligned}$$

Since $g \perp H$ and $T^* h \in H$, the scalar product vanishes. Thus $(Ug, h) = 0$, and as $h \in H$ was arbitrary, we have $Ug \in H^\perp$. By construction G_1 is invariant for U and hence $Ug \in G_1$. This yields $Ug \in G$ and $UG \subset G$.

For the sufficiency, first note that (1.1.1) and $UG \subset G$, $U^* G_* \subset G_*$ imply that $P_H U^n P_G = \mathbb{O}$ and $P_{G_1} U^n P_{G_1} = U^n P_{G_1}$ for every $n \geq 0$, where $G_1 = H \oplus G$ (cf. also (1.1.2) and (1.1.3) below). Thus, by induction, we get for $h \in H$ and $n \geq 0$

$$\begin{aligned}
T^{n+1} h &= TT^n h = P_H U P_H U^n h = P_H U (P_{G_1} - P_G) U^n h \\
&= P_H U (1 - P_G) U^n h = P_H U^{n+1} h.
\end{aligned}$$

The last equality is a consequence of the fact that $UG \subset G \perp H$, and hence $P_H U P_G (U^n h) = \mathbb{O}$. $\square$

1.1.3. REMARK. The sufficiency part of this proof can be illustrated in a more transparent way through matrix representations. In fact, suppose that we have decomposition (1.1.1) and $UG \subset G$, $U(H \oplus G) \subset H \oplus G$ (that is $U^*G_* \subset G_*$), then U may be rewritten as a triangular matrix acting on $G_* \oplus H \oplus G$:

$$U = \begin{pmatrix} P_{G_*}U|G_* & \mathbb{O} & \mathbb{O} \\ * & P_H U|H & \mathbb{O} \\ * & * & U|G \end{pmatrix}. \tag{1.1.2}$$

As $P_H U|H = T$, successive matrix multiplications show that

$$U^n = \begin{pmatrix} * & \mathbb{O} & \mathbb{O} \\ * & T^n & \mathbb{O} \\ * & * & * \end{pmatrix}. \tag{1.1.3}$$

Hence $T^n = P_H U^n|H$, $n \geq 0$. □

1.1.4. DEFINITION. A dilation $U : \mathcal{H} \to \mathcal{H}$ of an operator $T : H \longrightarrow H$ is called *unitary* if $U : \mathcal{H} \to \mathcal{H}$ is a unitary operator on $\mathcal{H}$ ($UU^* = U^*U = 1$). The dilation is *minimal* if the minimal subspace of $\mathcal{H}$ containing H and being invariant with respect to U and U^* (that is, a *reducing* subspace) coincides with $\mathcal{H}$:

$$\mathcal{H} = \operatorname{span}(U^n H : n \in \mathbb{Z}).$$

Clearly, if T has a unitary dilation, then $\|T\| \leq 1$.

1.1.5. COROLLARY. *The decomposition $\mathcal{H} = G_* \oplus H \oplus G$ of a minimal unitary dilation is unique.*

Indeed, suppose that $\mathcal{H} = \tilde{G}_* \oplus H \oplus \tilde{G}$ is another decomposition of the dilation space with the same invariance properties. By definition of $G_1 = H \oplus G$, and because of the minimality of the decomposition $G_* \oplus H \oplus G$, we have $H \oplus G \subset H \oplus \tilde{G}$, from which we deduce that $G \subset \tilde{G}$.

Pick $g \in \tilde{G} \ominus G = \tilde{G} \cap (G_* \oplus H) = \tilde{G} \cap G_*$. Then $U^n g \in \tilde{G}$, for every $n \geq 0$ and $U^n g \in G_*$ for every $n < 0$, and hence $U^n g \perp H$ for every $n \in \mathbb{Z}$. By definition of minimality we get $g = 0$, thus $G = \tilde{G}$. Necessarily we then also have $G_* = \tilde{G}_*$. □

1.1.6. COROLLARY. *A minimal unitary dilation has the following matrix representation*

$$U = \begin{pmatrix} P_{G_*}U|G_* & \mathbb{O} & \mathbb{O} \\ A & T & \mathbb{O} \\ C & B & U|G \end{pmatrix}$$

for suitable A, B, C. □

The aim of the rest of this section is to determine the structure of this matrix.

1.1.7. The structure of the minimal unitary dilation. Let U be the minimal unitary dilation of a contraction T. Then the restrictions $U|G$ and $U^*|G_*$ are *pure isometries* (that is, without unitary parts): if $G' \subset G$ such that $UG' = G'$, then $H \perp G \supset U^n G'$ for every $n \in \mathbb{Z}$. Hence $U^n H \perp G'$, $n \in \mathbb{Z}$, which in view of the minimality shows that $G' = \{0\}$. Clearly, the same arguments work for the case of $U^*|G_*$.

Now, we can apply the Wold–Kolmogorov lemma A.1.5.1 to the isometries $U|G$ and $U^*|G_*$. The following corollary, which is an immediate consequence of this lemma, is the first step towards a clarification of the intrinsic structure of the minimal unitary dilation, see Lemma 1.1.14 below.

1.1.8. COROLLARY. *If $U : \mathcal{H} \longrightarrow \mathcal{H}$ is the minimal unitary dilation of a contraction $T : H \longrightarrow H$ and $\mathcal{H} = G_* \oplus H \oplus G$, then*

$$\begin{aligned} G &= \sum_{n\geq 0} \oplus U^n E, \\ G_* &= \sum_{n\geq 0} \oplus U^{*n} E_*, \end{aligned}$$

where $E = G \ominus UG$ and $E_* = G_* \ominus U^* G_*$. □

1.1.9. The function model of a pure isometry, and vector valued Hardy spaces. Let us transfer the representation of the previous corollary,

$$G = \sum_{n\geq 0} \oplus U^n E = \Big\{ \sum_{n\geq 0} U^n e_n : e_n \in E, \sum_{n\geq 0} \|e_n\|^2 < \infty \Big\},$$

which lives on an abstract Hilbert space, to a subspace of the space L^2 (see also A.1.6.3, Volume 1).

Consider the operator

$$V : \sum_{n\geq 0} U^n e_n \longmapsto f(\zeta) = \sum_{n\geq 0} e_n \zeta^n, \quad |\zeta| = 1,$$

where f is now an element of the space $L^2(\mathbb{T}, E)$ of all E valued square-summable functions (with respect to the Lebesgue measure m on $\mathbb{T}$). We have

$$\int_{\mathbb{T}} \|f(\zeta)\|_E^2 \, dm(\zeta) = \sum_{n\in\mathbb{Z}} \|\widehat{f}(n)\|^2,$$

where $\widehat{f}(n)$, $n \in \mathbb{Z}$, are the Fourier coefficients of f:

$$\widehat{f}(n) = \int_{\mathbb{T}} f(\zeta)\overline{\zeta}^n dm(\zeta), \quad n \in \mathbb{Z}.$$

As in the scalar case, we may define the *Hardy space* $H^2(\mathbb{T}, E)$ as the subspace of $L^2(\mathbb{T}, E)$ of all functions with vanishing negative Fourier coefficients (see also A.1.6.3 and, especially, Section A.3.11, Volume 1):

$$H^2(\mathbb{T}, E) = H^2(E) = \{f \in L^2(\mathbb{T}, E) : \widehat{f}(n) = 0, n < 0\}.$$

Clearly, V is a unitary surjection from G onto $H^2(E)$ satisfying

$$VUV^{-1}(z^n e) = VUU^n e = VU^{n+1} e = z^{n+1} e$$

for all $n \geq 0$ and $e \in E$. Hence the operator

$$VUV^{-1} = S_E : f(z) \rightarrow zf(z)$$

is nothing but the multiplication by the independent variable z in the space $H^2(E)$.

1.1.10. COROLLARY. *With the above notation, the restrictions $U|G$ and $U^*|G_*$ of the minimal unitary dilation U of a contraction T and its adjoint U^* are unitarily equivalent to the* shift operators *(multiplication by z) S_E and S_{E_*} on $H^2(E)$ and $H^2(E_*)$, respectively.* □

1.1.11. Defect operators, partial isometries, and polar decompositions. Let $T : H \to H$ be a contraction, $\|T\| \le 1$. Then $((I - T^*T)x, x) = \|x\|^2 - \|Tx\|^2 \ge 0$ for every $x \in H$, and hence $I - T^*T \ge \mathbb{O}$. For such an operator the non-negative square root $(I - T^*T)^{1/2}$ exists, see B.1.4.6 (Volume 1). Here is another proof of this fact. The operator $A = I - T^*T$ is a self-adjoint positive operator. Its von Neumann model is given by $A : f(t) \longmapsto tf(t)$ for every f,

$$f \in \int_0^{\|A\|} \oplus H(t)\, d\mu(t),$$

see C.1.5.1(k). Now $A^{1/2}f = t^{1/2}f(t)$ is the desired square root. □

The operator $D_T = (I - T^*T)^{1/2}$ is called the *defect operator* of T. Clearly, for every $x \in H$ we have $\|Tx\|^2 + \|D_T x\|^2 = \|x\|^2$. The spaces $\mathcal{D}_T, \mathcal{D}_{T^*}$ defined by

$$\begin{aligned} \mathcal{D}_T &= \operatorname{clos} D_T H, \\ \mathcal{D}_{T^*} &= \operatorname{clos} D_{T^*} H \end{aligned}$$

are called the *defect spaces* of the contraction T.

Recall that an operator $V : H \to K$ is a *partial isometry* if $H = H_i \oplus H_0$ where $V : H_i \to H$ is an isometry and $V|H_0 \equiv \mathbb{O}$. The space H_i is then called the *initial space* of V and its range $VH_i = VH$ the *final space* of V.

It is not hard to see that the adjoint operator V^* is a partial isometry if and only if this is the case for V. Then VH_i is the corresponding initial space of V^* and H_i its final space.

The following two lemmas were already mentioned in Section B.1.5 (Volume 1), but we recall them for the reader's convenience. The first one contains the classical *polar decomposition* of a Hilbert space operator, and the second gives an intertwining property of defect operators.

1.1.12. LEMMA. *Let $M : H \to K$ be a bounded operator acting between two Hilbert spaces H and K, and let $|M| = (M^*M)^{1/2}$. Then there is a partial isometry $V : H \to K$ such that $M = V|M|$. Moreover, V is unique if we require* $\operatorname{Ker} V = \operatorname{Ker}|M| = \operatorname{Ker} M$.

For the proof we refer to B.1.5.5 (Volume 1).

1.1.13. LEMMA. *The operators T and T^* intertwine the defect operators:*

$$\begin{aligned} D_{T^*}T &= TD_T, \\ D_T T^* &= T^* D_{T^*}. \end{aligned}$$

PROOF. Clearly $TD_T^2 = T(I - T^*T) = (I - TT^*)T = D_{T^*}^2 T$, and by induction we get $TD_T^{2n} = D_{T^*}^{2n}T$, $n \ge 0$. Let $(p_n)_{n\ge1}$ be a sequence of polynomials tending uniformly to $\sqrt{x}$ on $[0, 1]$, then

$$Tp_n(D_T^2) = p_n(D_{T^*}^2)T,$$

which implies $TD_T = D_{T^*}T$. See also B.1.5.2(d), Volume 1. □

Now, we are in a position to prove the following main lemma disclosing the intrinsic structure of unitary dilations. For comments see Subsection 1.1.15 below.

1.1.14. LEMMA. *Let $T : H \longrightarrow H$ be a contraction and U its minimal unitary dilation on $\mathcal{H} = G_* \oplus H \oplus G$. Then*

$$U = \begin{pmatrix} P_{G_*}U|G_* & \mathbb{O} & \mathbb{O} \\ D_{T^*}V_*^* & T & \mathbb{O} \\ -VT^*V_*^* & VD_T & U|G \end{pmatrix},$$

where V, V_* are partial isomtries of H into G, G_* with respectively $\mathcal{D}_T$ and $\mathcal{D}_{T^*}$ as initial spaces and $E = G \ominus UG$ and $E_* = G_* \ominus U^*G_*$ as final spaces.

PROOF. We will use $U^*U = UU^* = I$ and the representations

$$U = \begin{pmatrix} U_{G_*} & \mathbb{O} & \mathbb{O} \\ A & T & \mathbb{O} \\ C & B & U_G \end{pmatrix}$$

$$U^* = \begin{pmatrix} U_{G_*}^* & A^* & C^* \\ \mathbb{O} & T^* & B^* \\ \mathbb{O} & \mathbb{O} & U_G^* \end{pmatrix},$$

where $U_G = U|G$ and $U_{G_*} = P_{G_*}U|G_*$. Hence

$$UU^* = \begin{pmatrix} * & * & * \\ AU_{G_*}^* & AA^* + TT^* & * \\ CU_{G_*}^* & CA^* + BT^* & * \end{pmatrix} = I$$

and

$$U^*U = \begin{pmatrix} * & * & * \\ * & T^*T + B^*B & B^*U_G \\ * & * & * \end{pmatrix} = I.$$

We obtain the following system of equations

$$\begin{array}{ll} \text{(a)} & CA^* + BT^* = \mathbb{O}, \\ \text{(b)} & AA^* + TT^* = I, \quad T^*T + B^*B = I, \\ \text{(c)} & AU^*|G_* = \mathbb{O}, \quad CU^*|G_* = \mathbb{O}, \quad B^*U|G = \mathbb{O}. \end{array}$$

Let us start with (c). Since $G_* = \sum_{k \geq 0} \oplus U^{*n}E_*$, we have $U^*G_* = G_* \ominus E_*$, and hence

$$A|(G_* \ominus E_*) \equiv \mathbb{O}. \tag{1.1.4}$$

Analogously for $B^* : G \longrightarrow H$:

$$B^*|(G \ominus E) \equiv \mathbb{O}. \tag{1.1.5}$$

Using (b) we get $AA^* = I - TT^* : H \to H$, that is, $AA^* = D_{T^*}^2$, where $A^* : H \to G_*$. By Lemma 1.1.12 there is a partial isometry $V_* : H \to G_*$ such that $A^* = V_*D_{T^*}$ and $\operatorname{Ker} V_* = \operatorname{Ker} D_{T^*}$. Moreover, $\operatorname{Range} V_*^* = (\operatorname{Ker} V_*)^\perp = \mathcal{D}_{T^*}$. Applying again Lemma 1.1.12, we likewise get a partial isometry $V : H \to G$ such that $B = VD_T$ and $\operatorname{Ker} V = \operatorname{Ker} D_T$. As before we deduce that $\operatorname{Range} V^* = \mathcal{D}_T$. By (1.1.5) we have $G \ominus E \subset \operatorname{Ker}\ V^*$ and hence $E \supset \operatorname{Range} V$. The same arguments applied to (1.1.4) show that $E_* \supset \operatorname{Range} V_*$. These inclusions are actually equalities. Indeed, let, for instance, $y \in E_* \ominus V_*H \subset G_*$. Then $Ay = D_{T^*}V_*^*y = \mathbb{O}$. On the other hand, as $U^*G_* \perp E_*$ by definition, we have $P_{G_*}U|_{E_*} = \mathbb{O}$. We deduce that

$$Uy = \begin{pmatrix} P_{G_*}U|E_* & * & * \\ A & * & * \\ C & B & * \end{pmatrix} \begin{pmatrix} y \\ 0 \\ 0 \end{pmatrix} = \begin{pmatrix} 0 \\ 0 \\ Cy \end{pmatrix} \in G$$

(here $C : G_* \to G$). The invariance of G implies that $U^n y \in G$, $n \geq 1$. By assumption, we also have $U^{*n} y \in G_*$, $n \geq 0$. Consequently $\mathcal{E} := \mathrm{span}(U^n y : n \in \mathbb{Z}) \subset G_* \oplus G$. Clearly $U^n \mathcal{E} = \mathcal{E}$, $n \in \mathbb{Z}$, and obviously $U^n \mathcal{E} \perp H$, $n \in \mathbb{Z}$, that is, $\mathcal{E} \perp U^n H$, $n \in \mathbb{Z}$. The minimality of U now implies that $\mathcal{E} = \{0\}$, and hence $E_* = V_* H$.

In the same way one verifies that $VH = E$. Hence V has the initial space $(\mathrm{Ker}\, V)^\perp = \mathrm{Range}\, V^* = \mathcal{D}_T$ and the final space E; likewise V_* has initial space $\mathcal{D}_{T^*}$ and final space E_*.

It remains to determine C. To this end we use the identity (a), where we replace A and B by the expressions given above. First note that as V_* is a partial isometry, we have

$$V_*^* V_* = \begin{cases} 1 & \text{on } \mathcal{D}_{T^*} \\ \mathbb{O} & \text{on } \mathcal{D}_{T^*}^\perp. \end{cases}$$

Hence

$$V_*^* V_* D_{T^*} = D_{T^*}.$$

Now by Lemma 1.1.13 we get

$$\begin{aligned} \mathbb{O} &= CA^* + BT^* = CV_* D_{T^*} + VD_T T^* = CV_* D_{T^*} + VT^* D_{T^*} \\ &= (C + VT^* V_*^*) V_* D_{T^*}. \end{aligned} \tag{1.1.6}$$

We have already seen that $\mathrm{Range}\, V_* = V_* H = E_*$, and it is easy to verify that $\mathrm{clos}(\mathrm{Range}\, V_* D_{T^*}) = \mathrm{Range}\, V_*$. Thus $\mathrm{clos}(\mathrm{Range}\, V_* D_{T^*}) = E_*$, and we obtain from (1.1.6) that

$$C | E_* = -VT^* V_*^* | E_*.$$

By (c) we have $C | U^* G_* = \mathbb{O}$ and also $V_*^* = \mathbb{O}$ on $(V_* D_{T^*})^\perp = E_*^\perp = U^* G_*$, which implies

$$C = -VT^* V_*^*.$$

This justifies the claimed matrix structure of U.

To prove the uniqueness of the representation, observe that

a) $S_E \simeq S_{E'}$ (unitary equivalence) if $\dim E = \dim E'$ (here $\dim E = \dim \mathcal{D}_T$ depends only on T),

b) $S_{E_*} \simeq S_{E'_*}$ if $\dim E_* = \dim E'_*$ (here $\dim E_* = \dim \mathcal{D}_{T^*}$),

c) it is also easy to see that if we replace the partial isometries V and V_* by V' and V'_* with the same initial spaces $\mathcal{D}_T$ and $\mathcal{D}_{T^*}$, then we get a new operator U' unitarily equivalent to the former one.

The three points a)–c) imply the required uniqueness. □

1.1.15. Comments. 1) In order to better understand the matrix structure of U, observe that the essential part of U, that is,

$$\alpha = \begin{pmatrix} D_{T^*} V_*^* & T \\ -VT^* V_*^* & VD_T \end{pmatrix},$$

is actually an isometry. Indeed, for $x \in \mathcal{D}_{T^*}$ (the final space of V_*^*) and $y \in H$, we have

$$\|D_{T^*} x + Ty\|^2 + \| - T^* x + D_T y\|^2 = \|x\|^2 + \|y\|^2,$$

which implies that α defines an isometry on the space $E_* \oplus H$ (note that E_* is the initial space of V_*^*).

In view of Corollary 1.1.8, we then get the following orthogonal decomposition of $\mathcal{H}$,

$$\mathcal{H} = \underbrace{\cdots \oplus U^{*n} E_* \oplus \cdots \oplus E_*}_{G_* = H^2_-(E_*)} \oplus H \oplus \underbrace{E \oplus UE \oplus \ldots}_{G = H^2(E)},$$

and hence U can be represented in this decomposition as a bi-infinite matrix,

$$\begin{pmatrix}
. & . & . & . & . & . & . & . & . & . & . & . & . \\
\ldots & \mathbb{O} & \mathbb{O} & \mathbb{O} & & & & & & & & & \\
\ldots & I & \mathbb{O} & \mathbb{O} & & & & & & & & & \\
\ldots & \mathbb{O} & I & \mathbb{O} & & & & & & & & & \\
. & . & . & . & . & . & . & . & . & . & . & . & . \\
 & & & \mathbb{O} & I & \mathbb{O} & & & & & & & \\
 & & & & \mathbb{O} & a & T & & & & & & \\
 & & & & & c & b & \mathbb{O} & & & & & \\
 & & & & & & \mathbb{O} & I & \mathbb{O} & & & & \\
. & . & . & . & . & . & . & . & . & . & . & . & . \\
 & & & & & & & & \mathbb{O} & I & \mathbb{O} & . & \ldots \\
 & & & & & & & & & \mathbb{O} & I & \mathbb{O} & \ldots \\
. & . & . & . & . & . & . & . & . & . & . & . & \ldots
\end{pmatrix},$$

where

$$\begin{pmatrix} a & T \\ c & b \end{pmatrix} = \begin{pmatrix} D_{T^*} V_*^* | E_* & T \\ -VT^* V_*^* | E_* & V D_T \end{pmatrix}.$$

Here, we in fact have realized $U|G$ and $U^*|G_*$ as shift operators on $H^2(E)$ and $H^2(E_*)$, respectively (cf. also Corollary 1.1.10).

2) It is not hard to see that the uniqueness part of the theorem can be generalized in the following way. If T and T' are two unitarily equivalent contractions, that is, $XT = T'X$ for a unitary mapping X, and U, U' their respective minimal unitary dilations, then the operator V given by

$$V\Big(\sum U^n h_n\Big) = \sum U'^n X h_n$$

(a priori for finitely supported sequences $(h_n)_{n \in \mathbb{Z}} \subset H$) is well-defined and isometric on the linear hull $\mathcal{L}\mathrm{in}(U^n H : n \in \mathbb{Z})$. It can thus be extended to a unitary mapping $V : \mathcal{H} \to \mathcal{H}'$ for which $VU = U'V$ and $V|H \equiv X$.

It is easily seen that the arguments of the above proof are reversible, and hence the following converse of Lemma 1.1.14 holds.

1.1.16. THEOREM. *Let $\mathcal{U}$, $\mathcal{U}_*$ be two pure isometries with defect spaces of the same dimension as $\mathcal{D}_T$ and $\mathcal{D}_{T^*}$, respectively:*

$$\dim E = \dim \mathcal{D}_T,$$
$$\dim E_* = \dim \mathcal{D}_{T^*},$$

where $E^\perp = \mathrm{Range}\ \mathcal{U}$, $E_*^\perp = \mathrm{Range}\ \mathcal{U}_*$. Let further $V : H \to G$ and $V_* : H \to G_*$ be partial isometries with respective initial spaces $\mathcal{D}_T$ and $\mathcal{D}_{T^*}$ and final spaces E

and E_*. Then the matrix

$$U = \begin{pmatrix} P_{G_*}\mathcal{U}_* & \mathbb{O} & \mathbb{O} \\ D_{T^*}V_*^* & T & \mathbb{O} \\ -VT^*V_*^* & VD_T & \mathcal{U} \end{pmatrix}$$

defines a minimal unitary dilation of T. □

1.1.17. EXAMPLE. Here, we determine the minimal unitary dilation of an isometry $V : H \to H$. Let $H = H_u \oplus H_0$ be its Kolmogorov–Wold decomposition (see A.1.5.1, Volume 1), where $V|H_u$ is unitary and $V|H_0$ is a pure isometry. Hence, with $E = H \ominus VH$

$$H_0 = \sum_{n\geq 0} \oplus V^n E.$$

On H_u, the operator V is already unitary, so we consider the nonunitary part H_0. On this part V acts as a right-shift (cf. Corollary 1.1.10, $l^2(\mathbb{N}, E) \simeq H^2(E)$), and in order to turn it into a unitary application it suffices to adjoin the "negatively indexed" part $G_* = l^2(\mathbb{Z}_-, E) = \sum_{n\leq -1} \oplus E \simeq H^2_-(E)$. For $h \in H = H_u \oplus H_0$ set $h = x_u + x^0$ and $x^0 = (x_0, x_1, \dots) \in \sum_{n\geq 0} \oplus V^n E = H_0$ (the latter space is identified with $l^2(\mathbb{N})$, cf. Corollary 1.1.10). Now define an operator

$$U : G_* \oplus H_u \oplus H_0 \longrightarrow G_* \oplus H_u \oplus H_0$$

by

$$(\dots, x_{-n}, \dots, x_{-1}) \oplus x_u \oplus x^0 \longmapsto (\dots, x_{-n-1}, \dots, x_{-2}) \oplus Vx_u \oplus (x_{-1}, x_0, \dots).$$

It is clear that U is a minimal unitary dilation of V.

In particular, for the shift operator $V : f \longmapsto zf$, $f \in H^2(E)$, which is completely nonunitary, the minimal unitary dilation is given by the operator of multiplication by z on $L^2(E)$: $z : L^2(E) \longrightarrow L^2(E)$ (the two-sided shift).

1.1.18. REMARK. In addition, if we do not mind about a kind of minimality, Theorem 1.1.16 can provide a formula for a unitary dilation defined on a space $\mathbf{H}$ that is *the same* for all contractions T acting on a given Hilbert space H. Indeed, we can set

$$\mathbf{H} = l^2(H) \oplus H \oplus l^2(H),$$

where $l^2(H) = \{x = (x_k)_{k\geq 0} : x_k \in H, \sum_{k\geq 0} \|x_k\|^2 < \infty\}$, identifying H with $\{0\} \oplus H \oplus \{0\}$, and next choose

$\mathcal{U} = \mathcal{U}_*^* = S$ (the shift operator on $l^2(H)$),
$Vh = (h, 0, 0, ...)$ for $h \in H$, and
$V_*^* x = x_0$ for $x \in l^2(H)$.

Now, it is clear that the operator

$$U = U_T : \mathbf{H} \longrightarrow \mathbf{H}$$

defined by the matrix formula of Theorem 1.1.16 is a unitary dilation of T.

1.2. Functional Embeddings and the Characteristic Function

Here we introduce the main objects of the function model — functional embeddings and the characteristic function of a completely nonunitary contraction. The latter is expressed in terms of the defect operators and the resolvent of the contraction in question.

1.2.1. Functional embeddings of a given contraction. Let $T : H \to H$ be a contraction, let $U : \mathcal{H} \to \mathcal{H}$ be its minimal unitary dilation on $\mathcal{H} = G_* \oplus H \oplus G$, and let $G \ominus UG$, $G_* \ominus U^* G_*$ be the final spaces of V and V_*. In order to get an explicit description of the dilation, we will use the von Neumann model for the unitary mappings

$$\begin{aligned} U|\operatorname{span}(U^n(G \ominus UG) : n \in \mathbb{Z}) &= U|\sum_{n\in\mathbb{Z}} \oplus U^n(G \ominus UG) \\ U|\operatorname{span}(U^n(G_* \ominus U^* G_*) : n \in \mathbb{Z}) &= U|\sum_{n\in\mathbb{Z}} \oplus U^n(G_* \ominus U^* G_*). \end{aligned}$$

These models are simply the multiplication operators $f(z) \longmapsto zf(z)$ on the spaces $L^2(G \ominus UG)$, respectively $L^2(G_* \ominus U^* G_*)$, or on $L^2(E)$ and $L^2(E_*)$, where E, E_* are such that

$$\dim E = \dim G \ominus UG, \quad \dim E_* = \dim G_* \ominus U^* G_*.$$

Passing to the structure of the unitary dilation, we begin with two unitary maps

$$v : E \to G \ominus UG, \quad v_* : E_* \to G_* \ominus U^* G_*.$$

Next, we define two unitary applications:

$$\pi : L^2(E) \to \sum_{n\in\mathbb{Z}} \oplus U^n(G \ominus UG)$$

$$\pi\left(\sum_n z^n e_n\right) = \sum_n U^n v e_n$$

and

$$\pi_* : L^2(E_*) \to \sum_{n\in\mathbb{Z}} \oplus U^n(G_* \ominus U^* G_*)$$

$$\pi_*\left(\sum_n z^n e_n^*\right) = \sum_n U^{n+1} v_* e_n^*$$

(for technical reasons that will become clear below we use U^{n+1} rather than U^n). Finally set

$$\Pi = (\pi, \pi_*) : L^2(E) \oplus L^2(E_*) \to \mathcal{H}$$

$$\Pi(f, g) = \pi f + \pi_* g.$$

The operators π, π_* are called *functional embeddings.*

1.2.2. Basic properties of the functional embeddings. (1) Clearly we have

$$\begin{aligned} \pi^* \pi &= I_{L^2(E)}, \\ \pi_*^* \pi_* &= I_{L^2(E_*)}, \end{aligned}$$

(that is, π and π_* are isometries).

(2) We have $\pi z = U\pi$ and $\pi_* z = U\pi_*$. Hence

$$\Pi z = U\Pi.$$

(3) Since we have chosen U^{n+1} instead of U^n in the definition of π_* we get

$$G = \pi H^2(E) \perp \pi_* H^2_-(E_*) = G_*.$$

(4) If we consider π and π_* as operators from $L^2(E)$ and, respectively, $L^2(E_*)$ into $\mathcal{H}$, we may compose

$$\pi_*^* \pi : L^2(E) \to L^2(E_*)$$

to obtain an application commuting with z:

$$\pi_*^* \pi z = z \pi_*^* \pi$$

(cf. (2)). By (3), we have $\pi_*^* \pi H^2(E) \perp H^2_-(E_*)$, and hence

$$\pi_*^* \pi H^2(E) \subset H^2(E_*).$$

In order to identify the operator $\pi_*^* \pi$ with an analytic function, we need the following lemma, which is an operator valued generalization of the Wiener lemma A.7.2.1.

1.2.3. LEMMA. *Let $A : L^2(E) \to L^2(E_*)$ be a linear bounded operator such that $Az = zA$. Then*

(1) There is a unique function $a \in L^\infty(E \to E_*)$ such that

$$Af(\zeta) = a(\zeta) f(\zeta), \;\; f \in L^2(E).$$

Moereover $\|A\| = \|A|H^2(E)\| = \|a\|_\infty = \operatorname{ess\,sup}_{\zeta \in \mathbb{T}} \|a(\zeta)\|$.

(2) $AH^2(E) \subset H^2(E_*)$ if and only if $a \in H^\infty(E \to E_*) = L^\infty(E \to E_*) \cap H^2(E \to E_*) = \{\varphi \in L^\infty(E \to E_*) : \widehat{\varphi}(k) = 0, k < 0\}$.

PROOF. (1) To begin with, we give an intuitive idea of the construction of a. Just "set" $a(\zeta)e = (Ae)(\zeta)$, where $\zeta \in \mathbb{T}$ and $e \in E$. Then

$$\begin{aligned}(Ap)(\zeta) &= \left(A\Big(\sum_n e_n z^n\Big)\right)(\zeta) = \Big(\sum_n z^n A e_n\Big)(\zeta) = \sum_n \zeta^n a(\zeta) e_n \\ &= a(\zeta)\Big(\sum_n \zeta^n e_n\Big) = a(\zeta) p(\zeta)\end{aligned}$$

for every E valued trigonometric polynomial $p = \sum_n e_n z^n$ and for a.e. $\zeta \in \mathbb{T}$.

Unfortunately, as the values $(Ae)(\zeta)$ exist only on a set depending on $e \in E$, this formula can be regarded only as a hint for the proof. Below, we use some properties of vector valued L^2 and H^2 spaces, for which we refer to Section A.3.11 (Volume 1).

For a more accurate definition, let $E_k \subset E_{k+1}$, $\dim E_k = k < \infty$, be such that $\operatorname{clos}(\bigcup_n E_n) = E$. Set $A_k = A|L^2(E_k)$ and choose an orthonormal basis $\{e_1, e_2, \ldots, e_k\}$ of E_k. Then the functions $(Ae_j)(\zeta) \in E_*$, $1 \le j \le k$, are well defined almost everywhere on $\mathbb{T}$. For an arbitrary vector

$$e = \sum_1^k b_j e_j \in E_k, \quad b_j \in \mathbb{C},$$

we get

$$(Ae)(\zeta) = \sum_1^k b_j(Ae_j)(\zeta) \in E_*.$$

We thus obtain an operator valued function $a_k(\zeta) : E_k \to E_*$, well defined a.e. on $\mathbb{T}$, by

$$a_k(\zeta)e = (Ae)(\zeta), \quad e \in E_k.$$

Clearly, for arbitrary $c_n \in E_k$ we have by linearity

$$\begin{aligned}\left(A\left(\sum_{\text{finite}} z^n c_n\right)\right)(\zeta) &= \sum_{\text{finite}} (z^n Ac_n)(\zeta) = \sum_{\text{finite}} \zeta^n a_k(\zeta)c_n = \\ &= a_k(\zeta)\left(\sum_{\text{finite}} \zeta^n c_n\right)\end{aligned}$$

a.e. on $\mathbb{T}$. Therefore, for every E_k valued polynomial f and, passing to the limit, for every $f \in L^2(E_k)$, we get

$$(Af)(\zeta) = a_k(\zeta)f(\zeta)$$

a.e. on $\mathbb{T}$. Clearly, such a function a_k is unique a.e., and the norm of the restriction $A_k = A|L^2(E_k)$ is controlled by that of A: $\|A_k\| \le \|A\|$. We claim that

$$\big\|A|H^2(E_k)\big\| = \|A_k\| = \text{ess sup}_{\mathbb{T}} \|a_k(\zeta) : E_k \to E_*\|.$$

Indeed, both inequalities "$\le$" are obvious. Conversely, taking $f = p \cdot e$, where $e \in E$, we get

$$\left(\int_{\mathbb{T}} |p(\zeta)|^2 \|a_k(\zeta)e\|^2 \, dm(\zeta)\right)^{1/2} \le \|A|H^2(E_k)\| \cdot \|p\|_2 \|e\|$$

first for $p \in \mathcal{P}ol_+$, and hence for every $p \in \mathcal{P}ol$ (since $z^n p \in \mathcal{P}ol_+$ for a suitable n). Lemma A.7.2.1 implies that ess $\sup_{\zeta\in\mathbb{T}} \|a_k(\zeta)e\| \le \|A|H^2(E_k)\| \cdot \|e\|$ for every $e \in E_k$. Hence

$$\|a|E_k\|_\infty \le \|A|H^2(E_k)\|,$$

and the claim follows.

Now by definition $A_{k+1}|L^2(E_k) = A_k$, and hence $a_{k+1}(\zeta)|E_k = a_k(\zeta)$ a.e. on $\mathbb{T}$. By the above arguments we have $\|A_k\| = \|a_k\|_\infty \le \|A\|$, and consequently,

$$\|a_k(\zeta)\|_{E_k \to E_*} \le \|A\|$$

a.e. on $\mathbb{T}$ for arbitrary $k \ge 1$. Since $a_{k+1}(\zeta)|E_k = a_k(\zeta)$, we may define an operator $a(\zeta)$ on $\bigcup_{k\ge1} E_k$ by setting $a(\zeta)x = a_k(\zeta)x$ for $x \in \bigcup_{k\ge1} E_k$ (for a suitable k). We thus get an application $a(\zeta)$ defined on a dense subset $\bigcup_{k\ge1} E_k$ of E. Moreover, $\|a(\zeta)e\| \le \|A\| \cdot \|e\|$ for almost every $\zeta \in \mathbb{T}$ and every $e \in \bigcup_{k\ge1} E_k$.

Consequently, we can extend $a(\zeta)$ for almost every $\zeta \in \mathbb{T}$ to an operator on E, and this operator clearly verifies

$$(Af)(\zeta) = a(\zeta)f(\zeta)$$

for almost every $\zeta \in \mathbb{T}$ and every $f \in \bigcup_{k\ge1} L^2(E_k)$. Note that both sides of this equality are continuous operators that coincide on a dense subset, which proves the first assertion.

It is clear that the norm equalities follow from the proof.

(2) Suppose that $aH^2(E) \subset H^2(E_*)$. Then, in particular, $ae \in H^2(E_*)$ for every $e \in E$, and hence $\widehat{(ae)}(n) = 0$ for $n < 0$. As $\widehat{(ae)}(n) = \widehat{a}(n)e$ for $e \in E$, we get $\widehat{a}(n) = \mathbb{O}$, $n < 0$, and $a \in H^\infty(E \to E_*)$. The converse is obvious. $\square$

Using Property 1.2.2(4) and Lemma 1.2.3(2) we can now introduce the following notion.

1.2.4. Definition. The function $\Theta \in H^\infty(E \to E_*)$ verifying $\Theta f = \pi_*^* \pi f$, $f \in L^2(E)$ is called the *characteristic function* of T.

Notation: $\Theta = \Theta_T$.

1.2.5. Corollary. *(1) We have*

$$\Pi^*\Pi = \begin{pmatrix} \pi^* \\ \pi_*^* \end{pmatrix} \begin{pmatrix} \pi & \pi_* \end{pmatrix} = \begin{pmatrix} I & \Theta^* \\ \Theta & I \end{pmatrix} : L^2(E) \oplus L^2(E_*) \to L^2(E) \oplus L^2(E_*).$$

(2) If we replace the unitary maps v, v_* by other unitary maps v', v'_*, we get a new characteristic function Θ' that is equivalent to Θ in the following sense $u_*\Theta(\zeta) \equiv \Theta'(\zeta)u$, where $u = v'^* v$, $u_* = v'^*_* v_*$ which are unitary constants *(cf. 1.2.1 above).* $\square$

The following commutative diagram illustrates the action of the functional embeddings

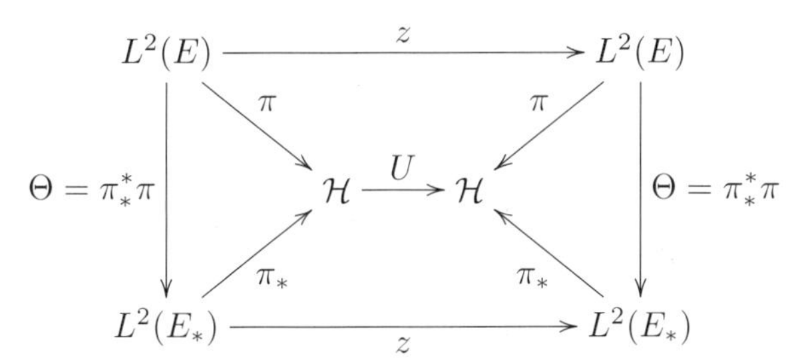

We will use the characteristic function for the construction of a function model of T. As Θ_T does not contain any information about the unitary part of T, we first have to separate this part.

1.2.6. Lemma (H. Langer, 1961). *(1) Let $T : H \to H$ be a contraction, $\|T\| \le 1$. Then there is a unique decomposition*

$$H = H_0 \oplus H_u$$

such that $TH_u \subset H_u$, $TH_0 \subset H_0$ and $T|H_u$ is unitary, whereas $T|H_0$ is a so-called completely nonunitary (c.n.u.) *operator, that is, $T|H_0$ is not unitary on any of its invariant subspaces.*

(2) If Π is the functional embedding of the minimal unitary dilation, then

$$H_u = (\operatorname{Range}\Pi)^\perp.$$

(3) T is completely nonunitary if and only if

$$\operatorname{clos}(\operatorname{Range}\Pi) = \mathcal{H}.$$

Proof. We will prove that the decomposition is actually given by $H = H_0 \oplus (\operatorname{Range}\Pi)^\perp$, where Π is the functional embedding of the minimal unitary dilation of T. The rest will then be clear.

Let U be the minimal unitary dilation of T and set $F = (\operatorname{Range}\Pi)^{\perp}$. As $\operatorname{Range}\Pi$ is invariant with respect to U and U^{-1}, the space F is reducing for U. Moreover, $F \subset H$, since $G, G_* \subset \operatorname{Range}\Pi$. On the other hand, $T = P_H U|H$ and $UF \subset F$, $U^*F \subset F$, which implies that $TF \subset F$ and also that

$$T^*F = P_H U^* F = U^* F \subset F.$$

The subspace F is thus reducing for T, and the restriction $T|F = U|F$ is unitary. This shows that $F \subset H_u$.

Suppose now that $H_1 \subset H$ is an invariant subspace of T, $TH_1 \subset H_1$, such that $T|H_1$ is unitary. Then for every $n \geq 0$ and $h \in H_1$ we have

$$\|h\| = \|T^n h\| = \|P_H U^n h\| \leq \|U^n h\| = \|h\|. \tag{1.2.1}$$

Hence $U^n h = P_H U^n h \in H$ for all $n \geq 0$. Replacing T by T^* in (1.2.1) we also get $U^{*n} h = U^{-n} h \in H$ for all $n \geq 0$. This shows that $U^n H_1 \subset H$, $n \in \mathbb{Z}$. Thus $U^n H_1 \perp G, G_*$, $n \in \mathbb{Z}$, which is equivalent to $H_1 \perp U^n G, U^n G_*$, $n \in \mathbb{Z}$, and finally $H_1 \perp \operatorname{Range}\Pi$. We conclude that $H_1 \subset F$, and thus $H_u = F = (\operatorname{Range}\Pi)^{\perp}$. □

1.2.7. The characteristic function as unitary invariant. We will now show that two c.n.u. contractions are unitarily equivalent if their characteristic functions are equivalent. Let us define *equivalence* of two functional embeddings. Consider the functional embeddings

$$\begin{aligned} \Pi : L^2(E) \oplus L^2(E_*) &\to \mathcal{H}, \\ \Pi' : L^2(E') \oplus L^2(E'_*) &\to \mathcal{H}'. \end{aligned}$$

We say that Π and Π' are *equivalent* if there are unitary applications $u : E \to E'$, $u_* : E_* \to E'_*$, $V : \mathcal{H} \to \mathcal{H}'$ such that

$$\Pi' W = V \Pi,$$

where

$$W = \begin{pmatrix} u & \mathbb{O} \\ \mathbb{O} & u_* \end{pmatrix}.$$

This means that the equivalence of two functional embeddings is a special unitary equivalence.

1.2.8. THEOREM. *Let T, T' be two c.n.u. contractions on H and H' respectively with corresponding minimal unitary dilations U, U' and functional embeddings Π, Π'. Let further Θ and Θ' be the characteristic functions of T and T', respectively. The following assertions are equivalent.*

(1) T and T' are unitarily equivalent.

(2) Π and Π' are equivalent.

(3) Θ and Θ' are equivalent (that is, $\Theta' u = u_* \Theta$ for some unitary constants u, u_*).

PROOF. (1)$\Rightarrow$ (2). Suppose that $XT = T'X$, where X is unitary. Define

$$V\Big(\sum_n U^n h_n\Big) = \sum_n U'^n X h_n$$

with $h_n \in H$. Then, V is an isometry:

$$\begin{aligned}(U'^n X h_n, U'^k X h_k) &= (X h_n, U'^{k-n} X h_k) = (X h_n, P_{H'} U'^{k-n} X h_k)\\ &= (X h_n, T'^{k-n} X h_k) = (X h_n, X T^{k-n} h_k)\\ &= (h_n, T^{k-n} h_k) = (h_n, P_H U^{k-n} h_k)\\ &= (h_n, U^{k-n} h_k) = (U^n h_n, U^k h_k).\end{aligned}$$

Moreover, V has dense range since the dilations are supposed to be minimal. Thus V is unitary. Set $u = V|E$, $u_* = V|E_*$ and

$$W = \begin{pmatrix} u & \mathbb{O} \\ \mathbb{O} & u_* \end{pmatrix}.$$

It is not hard to see that $\pi' u = V\pi$ and $\pi'_* u_* = V\pi_*$, hence $\Pi' W = V\Pi$.

(2)$\Rightarrow$ (3). It is clear that if $\Pi' W = V\Pi$, and hence $W^*\Pi'^* = \Pi^* V^*$, then $W^*\Pi'^*\Pi' W = \Pi^*\Pi$. Consequently $\Theta = u_*^*\Theta' u$.

(3)$\Rightarrow$ (2). Using $\Theta' u = u_*\Theta$ and Corollary 1.2.5 we get

$$(\Pi'^*\Pi')W = W(\Pi^*\Pi), \tag{1.2.2}$$

where as before

$$W = \begin{pmatrix} u & \mathbb{O} \\ \mathbb{O} & u_* \end{pmatrix}$$

is a unitary operator on $L^2(E) \oplus L^2(E_*)$. Define an operator $V : \mathcal{H} \to \mathcal{H}'$ by $V(\Pi F) = \Pi'(WF)$ for $F \in L^2(E) \oplus L^2(E_*)$. The reader may now verify that, because of (1.2.2), V is unitary, which implies (2).

(2)$\Rightarrow$ (1). Let $\Pi' W = V\Pi$. As Π intertwines the operators U and z: $U\Pi = \Pi z$ (and the same is true for Π': $\Pi' z = U'\Pi'$, see 1.2.2(1)), we get

$$VU\Pi = V\Pi z = \Pi' W z = \Pi' z W = U'\Pi' W = U' V \Pi$$

(note that W is a constant and as such it commutes with the operator of multiplication by z). The range of Π is dense in $\mathcal{H}$, so U and U' are unitarily equivalent: $VU = U'V$. Using again $V\Pi = \Pi' W$ we get

$$\begin{aligned} V\pi H^2(E) &= \pi' H^2(E'),\\ V\pi_* H^2(E_*) &= \pi'_* H^2(E'_*).\end{aligned}$$

Moreover, $G = \pi H^2(E) \equiv \mathrm{span}(U^n(G \ominus UG) : n \geq 0)$, $G' = \pi' H^2(E')$, and the same is true for the terms with index "$*$". Hence $VG = G'$ and $VG_* = G'_*$. Since $H = \mathcal{H} \ominus (G \oplus G_*)$ and $H' = \mathcal{H}' \ominus (G' \oplus G'_*)$, we get $VH = H'$. Setting now

$$X = V|H$$

we obtain a unitary operator $X : H \to H'$ verifying

$$\begin{aligned} T'X &= P_{H'} U' X = P_{H'} U' V|H = P_{H'} V U|H = V V^* P_{H'} V U|H = V P_H U|H\\ &= XT.\end{aligned}$$

□

1.2.9. An explicit formula for the characteristic function. We will use the same notation as in the preceding subsections. In particular, let

$$U : \mathcal{H} = G_* \oplus H \oplus G \to \mathcal{H}$$

be the minimal unitary dilation of a completely nonunitary contraction T,

$$P_H U|H = T$$

with defect spaces

$$E = G \ominus UG, \quad E_* = G_* \ominus U^* G_*.$$

The applications

$$\pi : L^2(E) \to \mathcal{H}, \quad \pi\left(\sum_n a_n z^n\right) = \sum_n U^n a_n, \quad a_n \in E,$$

$$\pi_* : L^2(E_*) \to \mathcal{H}, \quad \pi_*\left(\sum_n b_n z^n\right) = \sum_n U^{n+1} b_n$$

are the functional embeddings, and we let

$$V : \mathcal{D}_T \longleftrightarrow E, \quad V_* : \mathcal{D}_{T^*} \longleftrightarrow E_*$$

be the unitary identifications of the corresponding subspaces. By Lemma 1.2.6(3) we have

$$\operatorname{span}(\operatorname{Range}\pi, \operatorname{Range}\pi_*) = \mathcal{H}.$$

1.2.10. THEOREM. *With the above notation we have*

$$\Theta_T(z)h = V_*(-T + zD_{T^*}(I - zT^*)^{-1}D_T)V^*h, \quad h \in E.$$

PROOF. To begin with, we determine the adjoint operator $\pi_*^* : \mathcal{H} \to L^2(E_*)$. Let $x \in \mathcal{H}$. Then

$$\begin{aligned}\left(\pi_*^* x, \sum_n a_n z^n\right)_{L^2(E_*)} &= \left(x, \sum_n U^{n+1} a_n\right)_{\mathcal{H}} = \sum_n (x, U^{n+1} a_n)_{\mathcal{H}} \\ &= \sum_n (U^{*n+1} x, a_n)_{\mathcal{H}} = \sum_n (P_{E_*} U^{*n+1} x, a_n)_{\mathcal{H}} \\ &= \left(\sum_n z^n P_{E_*} U^{*n+1} x, \sum_k z^k a_k\right)_{L^2(E_*)}\end{aligned}$$

for every finite family $\{a_n\}_n \subset E_*$. Consequently

$$\pi_*^* x = \sum_n z^n P_{E_*} U^{*n+1} x, \quad x \in \mathcal{H}.$$

Note that for $n < 0$ and $h \in E$ we have $P_{E_*} U^{*(n+1)} h = 0$. Remark also that by construction of π we have $\pi h = h$ for $h \in E$. Hence

$$\pi_*^* \pi h = \sum_{n \geq 0} z^n P_{E_*} U^{*(n+1)} h, \quad h \in E,$$

and $\pi_*^* \pi : L^2(E) \to L^2(E_*)$. Identify now $h \in E$ with

$$h = \begin{pmatrix} 0 \\ 0 \\ h \end{pmatrix} \in \begin{pmatrix} G_* \\ H \\ G \end{pmatrix}.$$

Recall that

$$U^* = \begin{pmatrix} U^*|G_* & V_*D_{T^*} & -V_*TV^* \\ \mathbb{O} & T^* & D_TV^* \\ \mathbb{O} & \mathbb{O} & P_GU^*|G \end{pmatrix},$$

and

$$U^* \begin{pmatrix} 0 \\ 0 \\ h \end{pmatrix} = \begin{pmatrix} -V_*TV^*h \\ D_TV^*h \\ P_GU^*h \end{pmatrix} = \begin{pmatrix} -V_*TV^*h \\ D_TV^*h \\ 0 \end{pmatrix},$$

where we have used the fact that $E \perp UG$ and hence $U^*h \in U^*E \perp G = \text{span}\,(U^nE : n \geq 0)$. Applying again U^* to this vector we get

$$U^{*2} \begin{pmatrix} 0 \\ 0 \\ h \end{pmatrix} = U^* \begin{pmatrix} -V_*TV^*h \\ D_TV^*h \\ 0 \end{pmatrix} = \begin{pmatrix} \{U^*G_*\} + V_*D_{T^*}D_TV^*h \\ T^*D_TV^*h \\ 0 \end{pmatrix},$$

where $\{U^*G_*\}$ denotes a vector in U^*G_*. Next,

$$U^{*3} \begin{pmatrix} 0 \\ 0 \\ h \end{pmatrix} = U^* \begin{pmatrix} \{U^*G_*\} + V_*D_{T^*}D_TV^*h \\ T^*D_TV^*h \\ 0 \end{pmatrix} = \begin{pmatrix} \{U^*G_*\} + V_*D_{T^*}T^*D_TV^*h \\ T^{*2}D_TV^*h \\ 0 \end{pmatrix},$$

where $\{U^*G_*\}$ is another vector in U^*G_*. By induction we obtain

$$U^{*(n+1)} \begin{pmatrix} 0 \\ 0 \\ h \end{pmatrix} = \begin{pmatrix} \{U^*G_*\} + V_*D_{T^*}T^{*(n-1)}D_TV^*h \\ T^{*n}D_TV^*h \\ \mathbb{O} \end{pmatrix}$$

for $n > 0$. Hence

$$P_{E_*}U^{*(n+1)} \begin{pmatrix} 0 \\ 0 \\ h \end{pmatrix} = \begin{cases} \begin{pmatrix} -V_*TV^*h \\ 0 \\ 0 \end{pmatrix}, & n = 0 \\ \begin{pmatrix} V_*D_{T^*}T^{*(n-1)}D_TV^*h \\ 0 \\ 0 \end{pmatrix}, & n > 0. \end{cases}$$

This gives

$$\begin{aligned} \pi_*^*\pi h &= \sum_{n\geq 0} z^n P_{E_*}U^{*(n+1)}h = \begin{pmatrix} \sum_{n\geq 1} z^n V_*D_{T^*}T^{*(n-1)}D_TV^*h - V_*TV^*h \\ 0 \\ 0 \end{pmatrix} \\ &= -V_*TV^*h + V_*z\sum_{k\geq 0} z^k D_{T^*}T^{*k}D_TV^*h \\ &= -V_*TV^*h + V_*zD_{T^*}\left(I - zT^*\right)^{-1}D_TV^*h \\ &= V_*\left(-T + zD_{T^*}\left(I - zT^*\right)^{-1}D_T\right)V^*h \end{aligned}$$

for every $h \in E$. □

1.2.11. REMARKS. 1) Recall that $T\mathcal{D}_T \subset \mathcal{D}_{T^*}$ (using $TD_T = D_{T^*}T$), and that $V^* : E \to \mathcal{D}_T$, $V_* : \mathcal{D}_{T^*} \to E_*$ are unitary mappings identifying E with $\mathcal{D}_T$ and E_* with $\mathcal{D}_{T^*}$. Choosing simply $E = \mathcal{D}_T$, $E_* = \mathcal{D}_{T^*}$, we get

$$\Theta_T = -T + zD_{T^*}(I - zT^*)^{-1}D_T.$$

2) If we multiply Θ_T on the right by D_T we get

$$\Theta_T(z)D_T = D_{T^*}(I - zT^*)^{-1} \cdot (zI - T).$$

Similarly, multiplying from the left by D_{T^*} we obtain

$$D_{T^*}\Theta_T(z) = (zI - T)(I - zT^*)^{-1}D_T|\mathcal{D}_T.$$

1.2.12. COROLLARY. *For every $h \in \mathcal{D}_T$, we have $\|\Theta_T(0)h\| = \|-Th\|$, and hence, if $h \neq 0$, we get*

$$\|\Theta_T(0)h\| < \|h\|.$$

Indeed, since $\|Th\|^2+\|D_Th\|^2 = \|h\|^2$ for arbitrary $h \in H$, we have $\|\Theta_T(0)h\| = \|Th\| = \|h\|$ if and only if $D_Th = 0$. This means that $h \in \operatorname{Ker} D_T$, that is, $h \perp \operatorname{clos}\operatorname{Range} D_T = \mathcal{D}_T$. □

1.2.13. DEFINITION. A contractive function $\Theta(z) : E \to E_*$, $\Theta \in H^\infty(E \longrightarrow E_*)$, is called *pure* if $\|\Theta(0)h\| < \|h\|$ for every $h \in H$, $h \neq 0$.

The characteristic function of a contraction is thus always pure. The reader may verify that for every function $\Theta \in H^\infty(E \to E_*)$, $\|\Theta(z)\| \leq 1$, $z \in \mathbb{D}$, there are unique decompositions $E = E_0 \oplus E_1$ and $E_* = E_{*_0} \oplus E_{*_1}$ such that $\Theta(z)|E_1$ is a unitary constant mapping from E_1 onto E_{*_1} (i.e., $UE_1 = E_{*_1}$, $\Theta(z)|E_1 \equiv U$, $\|Uh\| = \|h\|$ for every $h \in E_1$) and $\Theta(z)|E_0$ is a pure contraction $E_0 \longrightarrow E_{*_0}$. The latter function is called the *pure part of* Θ.

1.2.14. Abstract functional embeddings. Here we solve the inverse problem, that is, construct a Hilbert space contraction for given functional embeddings. By an *abstract functional embedding* we mean a (linear) mapping

$$\Pi = (\pi, \pi_*) : L^2(E) \oplus L^2(E_*) \longrightarrow \mathcal{H}$$

satisfying the following properties.

(1) The restrictions $\pi = \Pi|L^2(E)$, $\pi_* = \Pi|L^2(E_*)$ are isometries;
(2) $\pi H^2(E) \perp \pi_* H^2_-(E_*)$;
(3) $\operatorname{Range}\Pi$ is dense in $\mathcal{H}$;
(4) $\pi_*^*\pi = \Theta$ is a purely contractive $H^\infty(E \longrightarrow E_*)$ function.

The functional embedding constructed in Subsection 1.2.1 for a given c.n.u. contraction T and its minimal unitary dilation will be denoted by Π_T.

1.2.15. THEOREM. *Let Π be an abstract functional embedding. There exists a c.n.u. contraction M such that Π is equivalent to Π_M. Every two such contractions are unitarily equivalent.*

Moreover, M can be chosen of the form

$$M = P_H U|H,$$

where

$$H = \mathcal{H} \ominus \left(\pi_* H^2_-(E_*) \oplus \pi H^2(E)\right)$$

and $U : \mathcal{H} \longrightarrow H$ defined by the relation $U\Pi = \Pi z$, is a minimal unitary dilation of M.

PROOF. The uniqueness part follows from Theorem 1.2.8. We show the existence.

Using the notation of Subsection 1.2.14 we set $G = \pi H^2(E)$ and $G_* = \pi_* H^2_-(E_*)$, and define, relying on (2) of 1.2.14, the subspace $H \subset \mathcal{H}$ by the equality

$$\mathcal{H} = G_* \oplus H \oplus G.$$

It is easy to check that the equality $U(\Pi F) = \Pi(zF)$ defines an isometric operator from Range Π to $\mathcal{H}$, which thus extends to a unitary operator $U : \mathcal{H} \longrightarrow \mathcal{H}$ (see condition (3) of 1.2.14). Clearly, $UG \subset G$ and $U^*G_* \subset G_*$, so U is a unitary dilation of $M := P_H U|H$, where P_H stands for the orthogonal projection onto the subspace $H \subset \mathcal{H}$ (see Lemma 1.1.2). If we can show that U is a *minimal* unitary dilation of M, then we will get $\Pi = \Pi_M$, where Π_M is as in Subsection 1.2.1 (with $v = \pi|E$, $v_* = \pi_*|E_*$).

To prove the minimality property, we first show that $\pi E \subset \operatorname{clos}(H + UH)$. Indeed, let $e \in E$. Since $P_H = I - P_G - P_{G_*}$, and UP_GU^* is the projection onto UG and $UG \perp \pi e$, we have

$$UP_HU^*\pi e = \pi e - UP_GU^*\pi e - UP_{G_*}U^*\pi e = \pi e - UP_{G_*}U^*\pi e.$$

On the other hand, $UG_* = G_* \oplus U\pi_*E_*$ and $\pi e \perp G_*$ imply that $UP_{G_*}U^*\pi e \in U\pi_*E_*$, which is orthogonal to G_*. Hence, $P_{G_*}(UP_{G_*}U^*\pi e) = 0$, and therefore

$$UP_{G_*}U^*\pi e = P_H(UP_{G_*}U^*\pi e) + P_G(UP_{G_*}U^*\pi e).$$

Since $P_G = \pi P_+ \pi^*$, $P_{G_*} = \pi_* P_- \pi_*^*$, where $P_\pm$ are the Riesz projections on L^2 spaces, we get

$$\begin{aligned} P_G(UP_{G_*}U^*\pi e) &= \pi P_+\pi^* U\pi_* P_- \pi_*^* \pi \bar{z} e = \pi P_+ \pi^* U \pi_* P_- \Theta \bar{z} e \\ &= \pi P_+ \pi^* U \pi_* \bar{z} \Theta(0) e = \pi P_+ \pi^* \pi_* \Theta(0) e \\ &= \pi P_+ \Theta^* \Theta(0) e = \pi \Theta(0)^* \Theta(0) e. \end{aligned}$$

Therefore,

$$UP_HU^*\pi e = \pi e - P_H(UP_{G_*}U^*\pi e) - \pi\Theta(0)^*\Theta(0)e,$$
$$\pi e - \pi\Theta(0)^*\Theta(0)e = UP_HU^*\pi e + P_H(UP_{G_*}U^*)\pi e,$$

which belong to $UH + H$. Since Θ is pure, Range$(I - \Theta(0)^*\Theta(0))$ is dense in E, which yields the inclusion $\pi E \subset \operatorname{clos}(H + UH)$.

The latter inclusion obviously implies that

$$G = \operatorname{span}(U^n \pi E : n \geq 0) \subset \operatorname{span}(U^n H : n \geq 0).$$

Similarly, replacing U, G, G_*, π, π_* and πe by U^*, G_*, G, π_*, π and $U^*\pi_* e_*$, respectively, we prove that

$$U^*\pi_*e_* - U^*\pi_*\Theta(0)\Theta(0)^*e_* = U^*P_HU(U^*\pi_*e_*) + P_H(U^*P_GU(U^*\pi_*e_*)),$$

which yields $U^*\pi_*E_* \subset \operatorname{clos}(U^*H + H)$, and hence

$$G_* = \operatorname{span}(U^{*n}\pi_*E_* : n > 0) \subset \operatorname{span}(U^{*n}H : n \geq 0).$$

Finally, $H, G, G_* \subset \operatorname{span}(U^nH : n \in \mathbb{Z})$, which means that U is a minimal dilation. The theorem follows. □

We finish this section with the following corollary of the Langer lemma 1.2.5.

1.2.16. COROLLARY. *Let U be the minimal unitary dilation of a c.n.u. contraction $T : H \longrightarrow H$. Then the spectral measure of U is absolutely continuous with respest to Lebesgue measure m, and therefore, $\lim_n (T^n x, y) = \lim_n (T^{*n} x, y) = 0$ for every $x, y \in H$.*

Indeed,

$$(T^n x, y) = (U^n x, y) = \int_{\mathbb{T}} \zeta^n (x(\zeta), y(\zeta))\, d\mu(\zeta)$$

for $n \geq 0$, where we used the von Neumann spectral representation of U (see the introduction to this chapter, as well as 1.5.1(k) below). Consequently, the second assertion follows from the first one and the Riemann-Lebesgue lemma. For the first assertion, see explanations in Subsection 1.3.3 below. □

1.3. The Function Model and Its Transcriptions

Now, we are in a position to define the coordinate free function model and to prove the main result of the theory (Theorem 1.3.2 below). We also briefly consider different coordinate realizations (transcriptions) of the model and give some simple examples.

1.3.1. Coordinate free definition. Let $T : H \to H$ be a c.n.u. contraction and let $\Theta_T(z) : \mathcal{D}_T \to \mathcal{D}_{T^*}$ be its characteristic function

$$\Theta_T(z) = -T + zD_T(I - zT^*)^{-1} D_T | \mathcal{D}_T.$$

The following points represent the steps towards the construction of the function model of a given completely nonunitary contraction.

(1) Let $\mathcal{H}, E, E_*$ be such that

$$\begin{aligned} \dim E &= \operatorname{rank} D_T = \partial_T, \\ \dim E_* &= \operatorname{rank} D_{T^*} = \partial_{T^*}, \end{aligned}$$

where ∂_T and ∂_{T^*} are the *defect indices*. Let further

$$\Pi = (\pi, \pi_*) : L^2(E) \oplus L^2(E_*) \to \mathcal{H}$$

be such that π, π_* are isometries and such that $\pi_*^* \pi$ is equivalent to Θ_T, and $\operatorname{clos} \operatorname{Range} \Pi = \mathcal{H}$. The existence of such a pair π, π_* is proved in Section 1.2. As in Theorem 1.2.14, we define a unitary operator $U : \mathcal{H} \to \mathcal{H}$ by

$$\Pi z = U \Pi.$$

(2) The space

$$\mathcal{K}_\Theta = \left(\pi_* H^2_-(E_*) \oplus \pi H^2(E)\right)^\perp, \tag{1.3.1}$$

where $\Theta = \Theta_T$, is called the *model space* of T and the operator

$$M_\Theta = P_\Theta U | \mathcal{K}_\Theta,$$

where P_Θ is the orthogonal projection onto $\mathcal{K}_\Theta$, the *model operator*.

We have

$$\begin{aligned} P_\Theta &= I - P_{\pi_* H^2_-(E_*)} - P_{\pi H^2(E)} \\ &= I - \pi_* P_- \pi_*^* - \pi P_+ \pi^*, \end{aligned}$$

where $P_\pm$ are the Riesz projections $L^2 \to H^2_\pm$ acting on vector valued L^2 spaces.

We call the 5-tuple $\{\mathcal{H}, \pi, \pi_*, \mathcal{K}_\Theta, M_\Theta\}$ the *function model* of T. The following theorem is the principal consequence of the previous construction.

1.3.2. THEOREM. *Given a completely nonunitary contraction T and a function Θ which is equivalent to the characteristic function Θ_T, then the operator M_Θ defined above is unitarily equivalent to T, and U is a minimal unitary dilation of M_Θ.*

PROOF. This is straightforward from Theorem 1.2.15. □

1.3.3. Introducing spectral coordinates. The von Neumann spectral representation of the minimal unitary dilation U of a c.n.u. contraction can be interpreted as a coordinate writing of this operator:

$$U : f(\zeta) \longmapsto \zeta f(\zeta), \quad f \in \int_{\mathbb{T}} \oplus H(\zeta)\, d\mu(\zeta).$$

Let us first recall some facts about the *spectral multiplicity* d_U of U defined by $d_U(\zeta) := \dim H(\zeta)$. For the complete statement, the notation, and the proof see Subsection 1.5.1 below. Observe that the Lebesgue measure is absolutely continuous with respect to μ: $m << \mu$ (that is, $\mu\sigma = 0$ implies $m\sigma = 0$). Indeed, as $\pi L^2(E), \pi_* L^2(E_*) \subset \mathcal{H}$, there are restrictions of U that are unitarily equivalent to the multiplication by z on $L^2(E)$ or $L^2(E_*)$ (namely, $U|\pi L^2(E)$ or $U|\pi_* L^2(E_*)$). Conversely, as $\operatorname{span}(\pi L^2(E), \pi_* L^2(E_*)) = \mathcal{H}$, we also have $\mu << m$. Hence $\mu \simeq m$ and we may assume that $\mu = m$. Now, since the embeddings π, π_* are isometries, we have

$$\max(\dim E, \dim E_*) \le d_U(\zeta) \le \dim E + \dim E_*,$$

that is,

$$\max(\operatorname{rank} D_T, \operatorname{rank} D_{T^*}) \le d_U(\zeta) \le \operatorname{rank} D_T + \operatorname{rank} D_{T^*}$$

a.e. on $\mathbb{T}$. Now, we need a generalized (weighted) form of the von Neumann model. As we have already mentioned, we may assume that $m = \mu$, but for more flexibility we add an operator valued weight. Namely, let $\mathcal{E}$ be an auxiliary space such that $\dim \mathcal{E} \ge \dim \mathcal{D}_T + \dim \mathcal{D}_{T^*}$, and let $W(\zeta) : \mathcal{E} \to \mathcal{E}$ be a nonnegative function, $(W(\zeta)x, x) \ge 0$ for every $x \in \mathcal{E}$, with $\operatorname{rank} W(\zeta) = d_U(\zeta)$ for almost every $\zeta \in \mathbb{T}$. Set

$$L^2(\mathcal{E}, W) = \left\{ f : \mathbb{T} \longrightarrow \mathcal{E} : f \text{ measurable}, \int_{\mathbb{T}} (Wf, f)\, dm < \infty \right\},$$

and define a semi-scalar product by

$$(f, g)_W = \int_{\mathbb{T}} (Wf, g)\, dm.$$

Now we may identify the operator U with the multiplication by the independent variable on $L^2(\mathcal{E}, W)$:

$$U : f(\zeta) \longmapsto \zeta f(\zeta), \quad f \in L^2(\mathcal{E}, W).$$

In fact, in order to obtain a Hilbert space based on this construction, we should factorize the space (to obtain a definite inner product) and complete it. We will avoid these unpleasant operations using the identity $\int_{\mathbb{T}} (Wf, f)\, dm = \int_{\mathbb{T}} \|W^{1/2} f\|^2\, dm$.

Thus, the mapping $f \longmapsto W^{1/2}f$ is an isometry of $L^2(\mathcal{E}, W)$ onto the closure of $W^{1/2}L^2(\mathcal{E})$:

$$L^2(\mathcal{E}, W) \to \operatorname{clos}_{L^2(\mathcal{E})} W^{1/2}L^2(\mathcal{E}).$$

The dimension condition can be satisfied by choosing $\mathcal{E} = E \oplus E_*$, where $\dim E = \partial_T$, $\dim E_* = \partial_{T^*}$.

1.3.4. The transcription equations. In order to realize the plan outlined above, we consider a functional embedding

$$\Pi : L^2(E \oplus E_*) \to \mathcal{H} = L^2(E \oplus E_*, W)$$

such that

$$\operatorname{clos} \operatorname{Range} \Pi = L^2(\mathcal{E}, W),$$

and

$$\pi^*\pi = I, \quad \pi_*^*\pi_* = I, \quad \pi_*^*\pi = \Theta_T. \tag{1.3.2}$$

These *transcription equations* can be rewritten in matrix form as follows

$$\Pi^*\Pi = \begin{pmatrix} I & \Theta^* \\ \Theta & I \end{pmatrix}.$$

Suppose that W, given by

$$W = \begin{pmatrix} w_1 & w_2 \\ w_3 & w_4 \end{pmatrix},$$

is bounded and nonnegative, $W \geq \mathbb{O}$ (in particular, $w_1(\zeta) = w_1(\zeta)^* \geq \mathbb{O}$, $w_4(\zeta) = w_4(\zeta)^* \geq \mathbb{O}$ and $w_2(\zeta)^* = w_3(\zeta)$ a.e. on $\mathbb{T}$). Assume for simplicity that, say, π_* is the identity mapping

$$\pi_* = \begin{pmatrix} \mathbb{O} \\ I \end{pmatrix},$$

and set

$$\pi = \begin{pmatrix} x \\ y \end{pmatrix}.$$

Then

$$\Pi = (\pi, \pi_*) = \begin{pmatrix} x & \mathbb{O} \\ y & I \end{pmatrix},$$

so we can rewrite (1.3.2) in the following form (note that the adjoints π^*, π_*^* are defined with respect to the scalar product in $L^2(E \oplus E_*, W)$):

$$\begin{aligned} x^*(w_1 - w_3^*w_3)x &= \Delta^2, \\ w_4 &= I, \\ y &= \Theta - w_3 x, \end{aligned} \tag{1.3.3}$$

where $\Delta(\zeta) = (I - \Theta_T(\zeta)^*\Theta_T(\zeta))^{1/2} = D_{\Theta_T(\zeta)}$ for a.e. $\zeta \in \mathbb{T}$, that is, $\Delta(\zeta)$ is the defect operator of the contraction $\Theta_T(\zeta) : E \to E_*$.

1.3.5. The principal B. Sz.-Nagy–C. Foiaş transcription. Let W be a diagonal weight $w_2 \equiv w_3 \equiv 0$:

$$W = \begin{pmatrix} w_1 & \mathbb{O} \\ \mathbb{O} & w_4 \end{pmatrix},$$

so (1.3.3) reduces to

$$x^* w_1 x = \Delta^2, \quad w_4 \equiv I, \quad y = \Theta,$$

and hence

$$W = \begin{pmatrix} w_1 & \mathbb{O} \\ \mathbb{O} & I \end{pmatrix}.$$

The density condition $\operatorname{clos} \operatorname{Range} \Pi = L^2(E \oplus E_*, W)$ is then equivalent to

$$\operatorname{clos} \operatorname{Range} x = L^2(E, w_1).$$

Set for instance $x = \Delta$, $w_1(\zeta) = P_{\operatorname{Range} \Delta(\zeta)}$, $\zeta \in \mathbb{T}$, $w_1(\zeta) : E \to E$. Then

$$W = \begin{pmatrix} P_{\overline{\Delta(\zeta)E}} & \mathbb{O} \\ \mathbb{O} & I \end{pmatrix}$$

$$\pi_* = \begin{pmatrix} \mathbb{O} \\ I \end{pmatrix}, \quad \pi = \begin{pmatrix} x \\ y \end{pmatrix} = \begin{pmatrix} \Delta \\ \Theta \end{pmatrix}.$$

Moreover

$$\begin{aligned} \mathcal{K}_\Theta &= \left(\begin{pmatrix} L^2(E) \\ L^2(E_*) \end{pmatrix}, W \right) \ominus \left(\pi_* H^2_-(E_*) \oplus \pi H^2(E) \right) \\ &= \begin{pmatrix} L^2(E, P_{\overline{\Delta(\zeta)E}}) \\ L^2(E_*) \end{pmatrix} \ominus \left[\begin{pmatrix} \mathbb{O} \\ H^2_-(E_*) \end{pmatrix} \oplus \begin{pmatrix} \Delta \\ \Theta \end{pmatrix} H^2(E) \right] \\ &= \begin{pmatrix} \operatorname{clos} \Delta L^2(E) \\ H^2(E_*) \end{pmatrix} \ominus \begin{pmatrix} \Delta \\ \Theta \end{pmatrix} H^2(E). \end{aligned}$$

Usually the places of E and E_* are interchanged:

$$\mathcal{K}_\Theta = \begin{pmatrix} H^2(E_*) \\ \operatorname{clos} \Delta L^2(E) \end{pmatrix} \ominus \begin{pmatrix} \Theta \\ \Delta \end{pmatrix} H^2(E),$$

$$M_\Theta \begin{pmatrix} f \\ g \end{pmatrix} = P_\Theta z \begin{pmatrix} f \\ g \end{pmatrix}.$$

This representation is called the *B.Sz-Nagy and C. Foiaş transcription* of the model.

1.3.6. COROLLARY. *For every pure contractive function $\Theta \in H^\infty(E \longrightarrow E_*)$ there exist functional embeddings π, π_* satisfying conditions (1)-(3) of Subsection 1.2.14 and the equation $\pi_*^* \pi = \Theta$.*

Indeed, we can take $\mathcal{H}, \pi, \pi_*$ defined above. □

1.3.7. REMARKS. We leave it to the reader to check the following facts.

(1)

$$P_\Theta = I - \pi P_+ \pi^* - \pi_* P_- \pi_*^* = \begin{pmatrix} P_+ - \Theta P_+ \Theta^* & -\Theta P_+ \Delta \\ -\Delta P_+ \Theta^* & I - \Delta P_+ \Delta \end{pmatrix}.$$

(2)

$$M_\Theta^* \begin{pmatrix} f \\ g \end{pmatrix} = \begin{pmatrix} P_+ \bar{z} f \\ \bar{z} g \end{pmatrix}.$$

[*Hint*:

$$z\begin{pmatrix}\Theta\\ \Delta\end{pmatrix}H^2 \subset \begin{pmatrix}\Theta\\ \Delta\end{pmatrix}H^2,$$

and hence

$$P_+\overline{z}\left(\begin{pmatrix}\Theta\\ \Delta\end{pmatrix}H^2\right)^\perp \subset \left(\begin{pmatrix}\Theta\\ \Delta\end{pmatrix}H^2\right)^\perp. \quad]$$

(3) Let $\Theta(z) = \Theta_u(z) \oplus \Theta_0(z) : E_1 \oplus E_0 \longrightarrow E_{*1} \oplus E_{*0}$ be the decomposition mentioned in the comments to Definition 1.2.13, where Θ_u is a unitary constant and Θ_0 is a purely contractive H^∞ function. Then Θ_{M_Θ} is equivalent to Θ_0. (See Theorem 1.2.15.)

1.3.8. The symmetric B. Pavlov transcription. This transcription requires that both $\pi : L^2(E) \longrightarrow L^2(E \oplus E_*)$ and $\pi_* : L^2(E_*) \longrightarrow L^2(E \oplus E_*)$ are natural identical embeddings. With the notation of Subsection 1.3.4 this means that $x = I$, $y = 0$, and hence $w_3 = \Theta$ and $w_1 = I$. Therefore $W = W_\Theta$, where

$$W_\Theta = \begin{pmatrix} I & \Theta^* \\ \Theta & I \end{pmatrix}, \quad \mathcal{H} = L^2(E \oplus E_*, W_\Theta),$$
$$G = \begin{pmatrix} H^2(E) \\ \mathbb{O} \end{pmatrix}, \quad G_* = \begin{pmatrix} \mathbb{O} \\ H^2_-(E_*) \end{pmatrix}.$$

One can check the following formula for the model operator

$$M_\Theta \begin{pmatrix} f \\ g \end{pmatrix} = \begin{pmatrix} zf - (z(\Theta^* g + f))\widehat{\ }(0) \\ zg \end{pmatrix}.$$

In this model, it is easy to follow the evolutions of U^n and U^{*n} (the model itself originates in scattering theory), but the model space $\mathcal{K}_\Theta$ is complicated. Moreover, it may happen that elements of $\mathcal{H}$ are no longer pairs of L^2 functions, because the weight W_Θ, in general, degenerates, so one needs to factorize and to complete the above written pre-Hilbert space.

1.3.9. Some special cases. (1) Let $\Theta(\zeta) : E \to E_*$ be isometric a.e. on $\mathbb{T}$, $\Theta \in H^\infty(E \to E_*)$. Then Θ is called an *inner* (or *left inner*) function. If Θ is inner, then $\Theta(\zeta)^*\Theta(\zeta) \equiv I_E$, $\zeta \in \mathbb{T}$, hence $\Delta \equiv \mathbb{O}$, and the model is reduced to

$$\begin{aligned} \mathcal{K}_\Theta &= H^2(E_*) \ominus \Theta H^2(E), \\ M_\Theta f &= P_\Theta z f, \quad f \in \mathcal{K}_\Theta, \\ P_\Theta &= P_+ - \Theta P_+ \Theta^* = I - \Theta P_+ \Theta^*. \end{aligned}$$

In case $\Theta(\zeta)$ is unitary almost everywhere on $\mathbb{T}$ (Θ is then a *two-sided inner function*), we obtain

$$P_\Theta = \Theta\Theta^* - \Theta P_+ \Theta^* = \Theta P_- \Theta^*.$$

Moreover, in this case $\partial_T = \partial_{T^*}$. The characteristic function of M_Θ coincides with the pure part of Θ (see 1.2.13 above).

(2) If Θ is scalar valued (i.e. $\partial_T = \partial_{T^*} = 1$), we may identify E and E_* with $\mathbb{C}$. The spaces $L^2(E), L^2(E_*), H^2(E)$ and $H^2(E_*)$ are then reduced to the classical

scalar spaces L^2, H^2, and

$$\mathcal{K}_\Theta = \begin{pmatrix} H^2 \\ \operatorname{clos} \Delta L^2 \end{pmatrix} \ominus \begin{pmatrix} \Theta \\ \Delta \end{pmatrix} H^2,$$

where Θ is an H^∞ function, $|\Theta(z)| \leq 1$, and

$$\Delta = (1 - |\Theta|^2)^{1/2}.$$

If moreover Θ is inner, $\Theta \in H^\infty$, $|\Theta(\zeta)| = 1$ a.e. on $\mathbb{T}$, and hence two-sided inner, we get

$$\begin{aligned} \mathcal{K}_\Theta &= H^2 \ominus \Theta H^2, \\ M_\Theta f &= P_\Theta z f, \\ P_\Theta &= \Theta P_- \overline{\Theta}. \end{aligned}$$

(3) Let $\Theta = z^{n+1}$, $n \geq 0$. In this case, $\mathcal{K}_\Theta$ is nothing else than the space of polynomials of degree $\leq n$. In fact, $f \perp z^{n+1} H^2$ if and only if $f \in \mathcal{L}\mathrm{in}(1, \dots, z^n)$. In $\mathcal{K}_\Theta$, the family $\mathcal{B} = \{z^k\}_0^n$ is an orthonormal basis, and we can give an explicit formula for the orthogonal projection P_Θ:

$$P_\Theta \left(\sum_{k=0}^{\infty} \widehat{f}(k) z^k \right) = \sum_{k=0}^{n} \widehat{f}(k) z^k.$$

The matrix of the model operator $M_\Theta f = P_\Theta z f$, $f \in \mathcal{K}_\Theta$, with respect to the basis $\mathcal{B}$ is given by a *Jordan block*

$$M_\Theta \simeq \begin{pmatrix} 0 & 0 & \dots & 0 & 0 \\ 1 & 0 & \dots & 0 & 0 \\ 0 & 1 & \dots & 0 & 0 \\ \cdot & \cdot & \cdot & \cdot & \cdot \\ 0 & 0 & \dots & 1 & 0 \end{pmatrix}.$$

(4) Let

$$\Theta = B = \prod_{k \geq 1} b_{\lambda_k}$$

be a (finite or infinite) Blaschke product with zero set $\{\lambda_k\}_{k\geq 1} \subset \mathbb{D}$ satisfying the Blaschke condition: $\sum_{k \geq 1} (1 - |\lambda_k|) < \infty$ (see Section A.3.7, Volume 1). Here, as before, $b_\lambda = \frac{\lambda - z}{1 - \overline{\lambda} z} \cdot \frac{|\lambda|}{\lambda}$.

1.3.10. Lemma. *If $\lambda_i \neq \lambda_j$, $i \neq j$, then*

$$\mathcal{K}_B = \operatorname{span} \left(\frac{1}{1 - \overline{\lambda_k} z} : k \geq 1 \right).$$

Proof. Let $k_\lambda(z) = \frac{1}{1 - \overline{\lambda} z}$, $\lambda \in \mathbb{D}$, be the *reproducing kernel* of H^2:

$$(f, k_\lambda) = \left(f, \frac{1}{1 - \overline{\lambda} z} \right)_{H^2} = f(\lambda), \quad |\lambda| < 1.$$

Clearly,

$$k_{\lambda_l} \perp BH^2, \quad l \geq 1.$$

Hence

$$\mathcal{L} = \operatorname{span} (k_{\lambda_l} : l \geq 1) \subset \mathcal{K}_B.$$

Suppose now that $f \perp \mathcal{L}$. Then $f(\lambda_l) = 0$, $l \geq 1$. Hence, by Section A.3.7, $g := f/B$ is in H^2, and $f = Bg \in BH^2$. We deduce that $\mathcal{L}^\perp \subset BH^2 = \mathcal{K}_B^\perp$, and finally $\mathcal{L} \supset \mathcal{K}_B$. □

1.3.11. LEMMA. *Let Θ be an inner function. Then $f \in \mathcal{K}_\Theta$ is an eigenfunction of M_Θ^*, $M_\Theta^* f = \overline{\lambda} f$, if and only if $\Theta(\lambda) = 0$ and*

$$f = \frac{c}{1 - \overline{\lambda} z},$$

where $c \in \mathbb{C}$. (See also B.3.1.2(3), Volume 1.)

PROOF. Recall that for $f \in \mathcal{K}_\Theta$ we have

$$M_\Theta^* f = P_\Theta \overline{z} f = P_+ \overline{z} f = z^* f$$

(note that $z\Theta H^2 \subset \Theta H^2$ and hence $z^* \mathcal{K}_\Theta \subset \mathcal{K}_\Theta$). We deduce that

$$M_\Theta^* f = \overline{\lambda} f$$

for $f \in \mathcal{K}_\Theta$ if and only if $z^* f = \frac{f - f(0)}{z} = \overline{\lambda} f$, and hence if and only if

$$f = c \frac{1}{1 - \overline{\lambda} z}.$$

Now, $\overline{\lambda}$ is an eigenvalue of M_Θ^* if and only if

$$k_\lambda = \frac{1}{1 - \overline{\lambda} z} \in \mathcal{K}_\Theta,$$

that is, $k_\lambda \perp \Theta H^2$, which is equivalent to $\Theta(\lambda) = 0$. □

1.3.12. COROLLARY. *If $\Theta = B$ is a Blaschke product with simple zeros $(\lambda_k)_{k \geq 1}$, then the eigenvectors k_{λ_l}, $l \geq 1$, of M_B^* span $\mathcal{K}_B$:*

$$\operatorname{span}\left(\frac{1}{1 - \overline{\lambda_l} z} : l \geq 1 \right) = \mathcal{K}_B.$$

1.3.13. REMARK. The same reasoning as in Lemma 1.3.11 gives a description of the eigenvectors of the adjoint M_Θ^* of the model operator (in its Sz.-Nagy–Foiaş form) for a general purely contractive function $\Theta \in H^\infty(E \longrightarrow E_*)$.

Indeed, using 1.3.7(2) we obtain that a pair $F = (f, g)$ is an eigenvector,

$$M_\Theta^* F = \overline{\lambda} F,$$

if and only if $g = 0$ and $f = e_* (1 - \overline{\lambda} z)^{-1}$, where $e_* \in E_*$ is such that $\Theta(\lambda)^* e_* = 0$.

The completeness of the eigenvectors of a model operator M_Θ^* as described in Corollary 1.3.12 is often considered as a definition of *Blaschke products for operator valued functions.*

Similarly, one can show that

$$\operatorname{Ker}(\lambda I - M_\Theta) = \left\{ F = (\lambda - z)^{-1} (\Theta e, \Delta e)^{col} : e \in E, \Theta(\lambda) e = 0 \right\}.$$

1.4. Models for Certain Accretive and Dissipative Operators

In this section we consider two examples of integral and differential operators including them in the framework of model theory. Section 1.5 below contains some more examples of these types. The operators considered are not contractions but have positive real, or respectively, imaginary part. This kind of operators can be treated in model theory after a suitable rational transformation known as the Cayley transform. Later on, we apply the calculations of this section to a kind of spectral analysis of the operators in question.

We start by giving the prerequisites on the Cayley transform.

1.4.1. The Cayley transform. Here we consider bounded Hilbert space operators, say $V : H \longrightarrow H$, whose quadratic forms (Vx, x) take values in the upper half-plane $\mathbb{C}_+ = \{\zeta \in \mathbb{C} : \mathrm{Im}(\zeta) \geq 0\}$ (*dissipative operators*) or, respectively, in the right half-plane $\mathbb{C}^+ = \{\zeta \in \mathbb{C} : \mathrm{Re}(\zeta) \geq 0\}$ (*accretive operators*). In other words, an operator is dissipative if

$$\mathrm{Im}(V) = (V - V^*)/2i \geq \mathbb{O},$$

and accretive if

$$\mathrm{Re}(V) = (V + V^*)/2 \geq \mathbb{O}.$$

Operators of these types become contractions when we apply a conformal mapping $C_+ : \mathbb{C}_+ \longrightarrow \mathbb{D}$ (respectively, $C^+ : \mathbb{C}^+ \longrightarrow \mathbb{D}$).

Our goal is to compute the characteristic function and the function model of $T = C_+(V)$, respectively, $T = C^+(V)$, in terms of the initial operator V. To this end, we need to know the resolvent of T (which can be expressed in terms of the resolvent of V), the defect operators D_T, D_{T^*}, and the Langer decomposition $T = T_u \oplus T_0$ of T into the unitary and completely nonunitary parts (see Lemma 1.2.6 above). The following theorem shows that, in the language of dissipative operators V, this decomposition amounts to the decomposition $V = V_s \oplus V_0$ where V_s is a selfadjoint operator ($V_s^* = V_s$) and V_0 is completely nonselfadjoint, that is, has no selfadjoint restrictions to its (nonzero) reducing, or simply invariant, subspaces. For an accretive V this corresponds to a similar decomposition for iV.

1.4.2. THEOREM. *(1) A bounded operator V acting on a Hilbert space H is accretive if and only if its* Cayley transform

$$T = C^+(V) = (I - V)(I + V)^{-1}$$

is a well defined contraction with $-1 \notin \sigma(T)$.

Moreover, $C^+(V^*) = C^+(V)^*$, and iV is selfadjoint if and only if $C^+(V)$ is unitary. For an accretive operator V, a (closed) subspace $E \subset H$ is V-invariant if and only if it is $C^+(V)$-invariant, and iV is completely nonselfadjoint if and only if $C^+(V)$ is completely nonunitary.

(2) A bounded operator A is dissipative if and only if its Cayley transform

$$T = C_+(A) = (iI - A)(iI + A)^{-1}$$

is a well-defined contraction and $-1 \notin \sigma(T)$. A dissipative operator A is completely nonselfadjoint if and only if $C_+(A)$ is completely nonunitary.

PROOF. We first verify that for an accretive operator V the sum $I + V$ is invertible. Indeed,

$$\|(I + V)y\|^2 = \|y\|^2 + \|Vy\|^2 + 2\,\mathrm{Re}(Vy, y) \geq \|y\|^2$$

for every $y \in H$, and the same estimate holds for $\|(I+V)^*y\|$. Therefore $(I+V)$ is invertible and hence T is well-defined. Moreover, $T = 2(I+V)^{-1} - I$, and hence $T + I = 2(I+V)^{-1}$, so $-1 \notin \sigma(T)$.

In order to see that T is a contraction, we use the following equivalences

$$\begin{aligned}
&\|(I-V)(I+V)^{-1}x\| \le \|x\|, \quad \forall x \in H,\\
&\quad \Leftrightarrow \|(I-V)y\|^2 \le \|(I+V)y\|^2, \quad \forall y \in H\\
&\quad \Leftrightarrow \|y\|^2 + \|Vy\|^2 - 2\operatorname{Re}(y, Vy) \le \|y\|^2 + \|Vy\|^2 + 2\operatorname{Re}(y, Vy), \quad y \in H\\
&\quad \Leftrightarrow 2\operatorname{Re}(y, Vy) \ge 0, \quad y \in H\\
&\quad \Leftrightarrow \operatorname{Re} V \ge \mathbb{O}.
\end{aligned}$$

Clearly, this also shows that if T is a contraction then V is accretive.

Further, the equality $C^+(V^*) = C^+(V)^*$ and the equivalence $(V^* = -V) \Leftrightarrow (C^+(V)$ is unitary) are obvious.

Let $E \in \operatorname{Lat} V$. From Lemma 2.1.10 (or Exercise 1.5.1(b), see below) we know that the set $\rho(V) = \{\lambda \in \mathbb{C} \setminus \sigma(V) : (\lambda I - V)^{-1}E \subset E\}$ is the union of a number of connected components of $\mathbb{C}\setminus\sigma(V)$, and that $\rho(V)$ contains the unbounded component of $\mathbb{C} \setminus \sigma(V)$. Since $(-\mathbb{C}^+) \subset \mathbb{C} \setminus \sigma(V)$ (see the arguments above), we get $(I+V)^{-1}E \subset E$. This implies that $E \in \operatorname{Lat} T$, $T = C^+(V)$.

Therefore, if $E \in \operatorname{Lat} V$ and $V + V^* = \mathbb{O}$ on E, we get $E \in \operatorname{Lat} V^*$, and hence $E \in \operatorname{Lat} T$ and $E \in \operatorname{Lat} T^*$. Now, Lemma 1.4.3 (see below) implies that $D_{T|E} = D_T|E = \mathbb{O}$ and $D_{(T|E)^*} = D_{T^*}|E = \mathbb{O}$, and so $T|E$ is unitary.

The converse can be proved analogously by using the relation $I+V = 2(T+I)^{-1}$ (see above).

(2) The proof for the dissipative case is similar. □

The following lemma gives the defect operator of the Cayley transform of an accretive operator.

1.4.3. LEMMA. *Let V be an accretive bounded operator and let $T = (I-V)(I+V)^{-1}$ be its Cayley transform. Then*

$$\begin{aligned}
D_T^2 &= (I+V^*)^{-1}2(V^*+V)(I+V)^{-1},\\
D_{T^*}^2 &= (I+V)^{-1}2(V^*+V)(I+V^*)^{-1}.
\end{aligned}$$

Similarly, for the Cayley transform $T = C_+(A)$ of a bounded dissipative operator A, we have

$$\begin{aligned}
D_T^2 &= (iI-A^*)^{-1}2i(A-A^*)(iI+A)^{-1},\\
D_{T^*}^2 &= (iI+A)^{-1}2i(A-A^*)(iI+A^*)^{-1}.
\end{aligned}$$

PROOF.

$$\begin{aligned}
D_T^2 &= I - T^*T = I - (I+V^*)^{-1}(I-V^*)(I-V)(I+V)^{-1}\\
&= (I+V^*)^{-1}[(I+V^*)(I+V) - (I-V^*)(I-V)](I+V)^{-1}\\
&= (I+V^*)^{-1}\cdot 2(V+V^*)(I+V)^{-1}.
\end{aligned}$$

□

1.4.4. Examples of accretive operators. (1) Let

$$(Vf)(x) = \int_0^x K(x,t)f(t)dt, \quad x \in (0,a),$$

$f \in L^2(0,a)$, be a bounded *Volterra type operator*. Then

$$(V^*f)(x) = \int_0^a \overline{K(t,x)}\chi(t,x)f(t)dt,$$

where

$$\chi(x,t) = \begin{cases} 1, & t < x \\ 0, & t > x. \end{cases}$$

Hence

$$\big((V+V^*)f\big)(x) = \int_0^a k(x,t)f(t)dt$$

with

$$k(x,t) = K(x,t)\chi(x,t) + \overline{K(t,x)}\chi(t,x) = \begin{cases} K(x,t), & x > t \\ \overline{K(t,x)}, & t > x \end{cases}.$$

If the function k is a *positive definite kernel*, that is, if

$$\int_0^a \int_0^a k(x,t)f(t)\overline{f(x)}\,dtdx \geq 0$$

for every $f \in L^2(0,a)$, then $2\operatorname{Re} V = V + V^* \geq \mathbb{O}$.

(2) **A special case**. Let $K(x,t) = N(x-t)$ for some real valued function N. Then $k(x,t) = N(|x-t|)$. Hence, $V + V^* \geq \mathbb{O}$ if and only if

$$\int_0^a \int_0^a N(|x-t|)f(t)\overline{f(x)}\,dtdx \geq 0$$

for every $f \in L^2(0,a)$. In this case the function N is called *positive definite.*

We may add that if k is a degenerated kernel of rank n, that is, if

$$k(x,t) = \sum_{i=1}^n \varphi_i(x)\overline{\psi_i(t)},$$

then $\operatorname{rank}(V+V^*) \leq n$. It is clear that in this case the kernel $k(x,t)$ is positive definite if $\psi_i \equiv \varphi_i$, $1 \leq i \leq n$.

1.4.5. The classical Volterra operator V**.** Define the *Volterra operator* V on $L^2(0,a)$ by

$$Vf(x) = \int_0^x f(t)dt, \quad 0 < x < a.$$

Then

$$V^*f(x) = \int_0^a \chi(t,x)f(t)dt = \int_x^a f(t)dt.$$

Consequently,

$$(2\operatorname{Re} Vf)(x) = \big((V+V^*)f\big)(x) = \left(\int_0^x f(t)\,dt + \int_x^a f(t)\,dt\right) = \int_0^a f(t)\,dt;$$

in other words, $2(\operatorname{Re} V)f = (f,1)1$. Hence

$$(2\operatorname{Re} Vf, f) = \left|\int_0^a f(t)\,dt\right|^2 \geq 0$$

for every $f \in L^2(0,a)$, which means that $\operatorname{Re} V$ is positive, that is, V is accretive, and moreover $\operatorname{rank}(\operatorname{Re} V) = 1$.

The aim of what follows is to find the characteristic function of the Cayley transform of the Volterra operator. To this end we need some auxiliary results.

1.4.6. LEMMA. *For $z \in \mathbb{C}$ and $f \in L^2(0,a)$, we have*

$$(I - zV^*)^{-1} f(x) = f(x) + z \int_x^a e^{z(t-x)} f(t)\, dt.$$

PROOF. By induction, we get

$$V^{*n} f(x) = \frac{1}{(n-1)!} \int_x^a (t-x)^{n-1} f(t)\, dt, \quad n \geq 1,$$

and hence, the spectral radius of V is 0: $\|V^{*n} f\|_2 \leq \frac{a^n}{(n-1)!} \|f\|_2$ which gives $\lim_n \|V^{*n}\|^{1/n} = 0$. Moreover,

$$\begin{aligned} (I - zV^*)^{-1} f(x) &= \sum_{n \geq 0} z^n V^{*n} f(x) = f(x) + z \sum_{n \geq 1} z^{n-1} \int_x^a \frac{(t-x)^{n-1}}{(n-1)!} f(t)\, dt \\ &= f(x) + z \int_x^a e^{z(t-x)} f(t)\, dt. \end{aligned}$$

□

In the same way one can show that

$$(I - zV)^{-1} f(x) = f(x) + z \int_0^x e^{z(x-t)} f(t)\, dt,$$

so we get the following corollary.

1.4.7. COROLLARY. *We have*

$$\begin{aligned} (I + V^*)^{-1} f(x) &= f(x) - \int_x^a e^{x-t} f(t)\, dt, \\ (I + V)^{-1} f(x) &= f(x) - \int_0^x e^{t-x} f(t)\, dt \end{aligned}$$

for $f \in L^2(0,a)$ and $x \in (0,a)$. □

1.4.8. COROLLARY. *The operator iV is completely nonselfadjoint, and hence $C^+(V)$ is a c.n.u. contraction.*

Indeed, if $E \in \operatorname{Lat} V$ such that $\operatorname{Re} V|E = 0$, and $f \in E$, then $(2 \operatorname{Re} V) V^n f = (V^{n+1} f)(a) = 0$ for $n \geq 0$. The computations of Lemma 1.4.6 and the previous corollary show that $\int_0^a (a-t)^n f(t) dt = 0$, $n \geq 0$. The Weierstrass completeness theorem yields $f = 0$, and hence $E = \{0\}$. The rest is contained in Theorem 1.4.2.
□

1.4.9. LEMMA. *The defect operator of $T = C^+(V)$ is given by*

$$D_T = \lambda_1 (\cdot, e^x) e^x,$$

where $\lambda_1 = \sqrt{2} e^{-a} / \|e^x\|$.

PROOF. We have

$$\begin{aligned} D_T^2 &= (I+V^*)^{-1}2(V^*+V)(I+V)^{-1} = (I+V^*)^{-1}\cdot 2(\cdot,1)1(I+V)^{-1} \\ &= 2\big((I+V)^{-1}\cdot,1\big)(I+V^*)^{-1}1 = 2\big(\cdot,(I+V^*)^{-1}1\big)(I+V^*)^{-1}1. \end{aligned}$$

As

$$(I+V^*)^{-1}1(x) = 1 - \int_x^a e^{x-t}\,dt = 1 - e^x\int_x^a e^{-t}\,dt = 1 - 1 + e^{x-a} = e^{x-a},$$

we get

$$D_T^2 f(x) = 2(f,e^{x-a})e^{x-a} = 2e^{-2a}(f,e^x)e^x,$$

and

$$D_T f(x) = \lambda_1(f,e^x)e^x,$$

where λ_1 will be determined by the following: for every $f\in L^2(0,a)$ we have

$$\begin{aligned} D_T^2 f(x) &= D_T(\lambda_1(f,e^x)e^x) = \lambda_1(\lambda_1(f,e^x)e^x,e^x)e^x = \lambda_1^2(f,e^x)\|e^x\|^2 e^x \\ &= 2e^{-2a}(f,e^x)e^x, \end{aligned}$$

and hence $\lambda_1^2\|e^x\|^2 = 2e^{-2a}$. As λ_1 is positive, we deduce that $\lambda_1 = \sqrt{2}e^{-a}/\|e^x\|$. □

1.4.10. LEMMA. *The defect operator of $T^* = C^+(V^*)$ is given by*

$$D_{T^*} = \lambda_2(\cdot,e^{-x})e^{-x},$$

where $\lambda_2 = \sqrt{2}/\|e^{-x}\|$.

PROOF. As before, we have

$$\begin{aligned} D_{T^*}^2 f &= (I+V)^{-1}2(V+V^*)(I+V^*)^{-1}f = 2\big((I+V^*)^{-1}f,1\big)(I+V)^{-1}1 \\ &= 2\big(f,(I+V)^{-1}1\big)(I+V)^{-1}1, \end{aligned}$$

where

$$(I+V)^{-1}1(x) = 1 - \int_0^x e^{t-x}\,dt = 1 - e^{-x}(e^x-1) = e^{-x}.$$

Hence

$$D_{T^*}^2 f(x) = 2(f,e^{-x})e^{-x}.$$

Then $D_{T^*} = \lambda_2(\cdot,e^{-x})e^{-x}$, where $\lambda_2^2\|e^{-x}\|^2 = 2$, which completes the proof. □

1.4.11. The characteristic function and the function model of $C^+(V)$. Now, we are in a position to compute the characteristic function $\Theta_T : \mathcal{D}_T \longrightarrow \mathcal{D}_{T^*}$. First, we observe that by the preceding two lemmas the defect spaces are one-dimensional, $\mathcal{D}_T = \mathcal{L}(e^x)$, $\mathcal{D}_{T^*} = \mathcal{L}(e^{-x})$, and we have

$$\Theta_T = -T + zD_{T^*}(I-zT^*)^{-1}D_T|\mathcal{L}(e^x).$$

Hence Θ_T can be represented with respect to the normalized one-element bases $\{e^x/\|e^x\|\}$ and $\{e^{-x}/\|e^{-x}\|\}$ in the follwoing form

$$\Theta_T(z)\frac{e^x}{\|e^x\|} = \theta(z)\frac{e^{-x}}{\|e^{-x}\|},$$

or

$$\Theta_T(z)e^x = \frac{\|e^x\|}{\|e^{-x}\|}\theta(z)e^{-x}.$$

(Here $(\theta(z))$ is the associated (1×1)-matrix.) By Corollary 1.4.7 and the definition of V we obtain

$$Te^x = C^+(V)e^x = (I - V)(I + V)^{-1}e^x = e^{-x},$$

and Lemma 1.4.9 gives

$$D_T e^x = \lambda_1(e^x, e^x)e^x = \alpha_1 e^x,$$

where $\alpha_1 = \|e^x\|^2\lambda_1$. It remains to compute the term $zD_{T^*}(1 - zT^*)^{-1}e^x$. As $T^* = (I - V^*)(I + V^*)^{-1}$, we get

$$\begin{aligned}(I - zT^*)^{-1} &= \big(I - z(I - V^*)(I + V^*)^{-1}\big)^{-1}\\ &= \big((I + V^*)^{-1}(I + V^* - z(I - V^*))\big)^{-1}\\ &= (I - z + V^* + zV^*)^{-1}(I + V^*)\\ &= (1 - z)^{-1}\left[I + \frac{1+z}{1-z}V^*\right]^{-1}(I + V^*).\end{aligned}$$

A straightforward calculation shows that $(I + V^*)e^x = e^a$. Hence by Lemma 1.4.6 (with z replaced by $-\frac{1+z}{1-z}$) we have

$$\begin{aligned}(I - zT^*)^{-1}e^x &= (1 - z)^{-1}\left[I + \frac{1+z}{1-z}V^*\right]^{-1}(I + V^*)e^x\\ &= (1 - z)^{-1}\left[I + \frac{1+z}{1-z}V^*\right]^{-1}e^a\\ &= (1 - z)^{-1}\left[e^a + \frac{z+1}{z-1}\int_x^a e^{\frac{z+1}{z-1}(t-x)}e^a\,dt\right]\\ &= (1 - z)^{-1}e^a e^{\frac{z+1}{z-1}(a-x)}.\end{aligned}$$

Let us apply $D_{T^*} = \lambda_2(\cdot, e^{-x})e^{-x}$, $\lambda_2 = \sqrt{2}/\|e^{-x}\|$, to this function:

$$\begin{aligned}D_{T^*}(I - zT^*)^{-1}e^x &= \lambda_2 e^{-x}\Big((1 - z)^{-1}e^a e^{\frac{z+1}{z-1}(a-x)}, e^{-x}\Big)\\ &= \lambda_2 e^{-x}(1 - z)^{-1}e^{\frac{2az}{z-1}}\Big(e^{-\frac{z+1}{z-1}x}, e^{-x}\Big).\end{aligned}$$

Next,

$$\Big(e^{-\frac{z+1}{z-1}x}, e^{-x}\Big) = \int_0^a e^{-\frac{z+1}{z-1}x - x}\,dx = \int_0^a e^{-\frac{2z}{z-1}x}\,dx = -\frac{z-1}{2z}e^{-\frac{2az}{z-1}} + \frac{z-1}{2z}.$$

Consequently,

$$\begin{aligned}D_{T^*}(I - zT^*)^{-1}e^x &= \lambda_2 e^{-x}(1 - z)^{-1}e^{\frac{2az}{z-1}}\left(\frac{z-1}{2z} - \frac{z-1}{2z}e^{-\frac{2az}{z-1}}\right)\\ &= \lambda_2 e^{-x}\left(\frac{1}{2z} - \frac{1}{2z}e^{\frac{2az}{z-1}}\right).\end{aligned}$$

Finally,

$$\begin{aligned}\Theta_T(z)e^x &= \big(-T + zD_{T^*}(I - zT^*)^{-1}D_T\big)e^x\\ &= e^{-x}\left(-1 + z\alpha_1\lambda_2\left(\frac{1}{2z} - \frac{1}{2z}e^{\frac{2az}{z-1}}\right)\right).\end{aligned}$$

We have $\alpha_1\lambda_2 = 2e^{-a}\frac{\|e^x\|}{\|e^{-x}\|}$. An explicit calculation gives

$$\|e^x\|^2 = \frac{e^{2a}}{2}\left(1-e^{-2a}\right), \quad \|e^{-x}\|^2 = \frac{1}{2}\left(1-e^{-2a}\right).$$

Hence $\alpha_1\lambda_2 = 2$ and

$$\Theta_T(z)e^x = e^x(-1+1-e^{\frac{2az}{z-1}}) = -e^a\theta(z)e^{-x}.$$

On the other hand,

$$\Theta_T(z)e^x = \frac{\|e^x\|}{\|e^{-x}\|}\theta(z)e^{-x} = -e^a\theta(z)e^{-x},$$

from which we deduce that

$$\theta(z) = -e^{a\frac{z+1}{z-1}}.$$

Therefore, we have proved the following theorem.

1.4.12. THEOREM. *The characteristic function of the Cayley transform* $T = C^+(V) = (I-V)(I+V)^{-1}$ *of the Volterra operator* $V : L^2(0,a) \longrightarrow L^2(0,a)$,

$$Vf(x) = \int_0^x f(t)dt,$$

is equivalent to the scalar function θ:

$$\theta(z) = e^{a\frac{z+1}{z-1}}.$$

As the operator $T = C^+(V)$ *is a c.n.u. contraction, it is unitarily equivalent to its function model* M_θ *on the space* $\mathcal{K}_\theta = H^2 \ominus \theta H^2$.

□

1.4.13. Dissipative Sturm–Liouville operators. Here we consider the differential operator ℓ_h defined by the differential expression

$$\ell_h y = -y'' + qy,$$

where q is a real valued function, and by the boundary condition $y'(0) = hy(0)$ with a complex number h. The operator ℓ_h is defined on the domain

$$\mathrm{Dom}(\ell_h) = \{y \in W^2_{2,loc}(\mathbb{R}_+) : \ell_h y \in L^2(\mathbb{R}_+), y'(0) = hy(0)\}$$

as an operator $\ell_h : \mathrm{Dom}(\ell_h) \longrightarrow L^2(\mathbb{R}_+)$ taking $y \longmapsto \ell_h y$. We will use some general facts about the second-order differential operators ℓ_h that can be found in textbooks, see for instance M. Reed and B. Simon [**RS**] or F. Atkinson [**At**]. In particular,

$$\ell_h^* = \ell_{\bar{h}}.$$

Moreover, the operator ℓ_h is a rank one perturbation of the operator ℓ_∞ defined by the same differential expression ℓ_h and the boundary condition $y(0) = 0$. The operator ℓ_∞ is self-adjoint, and the operator ℓ_h is dissipative if and only if $\mathrm{Im}(h) > 0$.

Our aim is to compute the characteristic function of the Cayley transform $C_+(\ell_h)$, which enables us to include the study of ℓ_h into the model theory.

To this end, we consider two solutions φ_ζ and ψ_ζ of the equation $\ell_h y = \zeta y$ satisfying the boundary conditions

$$\varphi_\zeta(0) = 0,\ \varphi'_\zeta(0) = 1 \quad \text{and} \quad \psi_\zeta(0) = -1,\ \psi'_\zeta(0) = 0.$$

It is well known that for every ζ with $\mathrm{Re}(\zeta) \neq 0$ there exists a unique L^2 solution y_ζ of the equation $\ell_h y = \zeta y$ representable as a linear combination of φ_ζ and ψ_ζ:

$$y_\zeta = \psi_\zeta + m(\zeta)\varphi_\zeta.$$

The function $\zeta \longmapsto m(\zeta)$ determined by this condition for all $\zeta \neq \overline{\zeta}$ is called the *Weyl function.* We compute the characteristic function of $C_+(\ell_h)$ in terms of the Weyl function m.

An important remark is that now we cannot use the formulas from Lemma 1.4.3 for the defect operators, because the operator $A = \ell_h$ is unbounded, its domain is different from the domain of the adjoint, and the imaginary part is not well defined. We overcome these difficulties as follows.

First, we note that if $T = C_+(A) = (iI - A)(iI + A)^{-1}$, then $T^* = (iI + A^*)(iI - A^*)^{-1}$. Let $f \in \mathrm{Dom}\, A$ and $x = (A + iI)f$. Then $Tx = (iI - A)f$, whence

$$\begin{aligned} D_T^2 x &= (A + iI)f - T^*(iI - A)f \\ &= (A + iI)f - (A^* + iI)(A^* - iI)^{-1}(A - iI)f. \end{aligned}$$

If $g = (iI - A^*)^{-1}(iI - A)f$, then

$$(A - iI)f = (A^* - iI)g,$$

or

$$\ell_h(f - g) = i(f - g),$$

which implies that $f - g = cy_i$, and

$$D_T^2 x = (A + iI)f - (A^* + iI)g = (\ell_h + iI)(f - g) = 2icy_i.$$

Thus, we can see that the defect subspace $\mathcal{D}_T$ of the Cayley transform $T = C_+(\ell_h)$ is the one dimensional subspace generated by the solution $y_\zeta = y_i$ with $\zeta = i$,

$$\mathcal{D}_T = \mathcal{L}(y_i).$$

The defect subspace of the adjoint operator T^* can be found similarly; namely,

$$\mathcal{D}_{T^*} = \mathcal{L}(y_{-i}).$$

The next observation is that $Ty_i = 0$ and $T^* y_{-i} = 0$, so that $D_T y_i = y_i$ and $D_{T^*} y_{-i} = y_{-i}$. Now, we use a formula from Remark 1.2.10, namely,

$$D_{T^*}\Theta_T(z) = (zI - T)(I - zT^*)^{-1} D_T,$$

which can be rewritten in terms of $A = \ell_h$ as follows:

$$D_{T^*}\Theta_T(z) = -(A - \zeta I)(A + iI)^{-1}(A^* - iI)(A^* - \zeta I)^{-1} D_T,$$

where $\zeta = i\frac{1+z}{1-z}$. Below, we apply this operator to the vector y_i and use the formula

$$(\ell_h - \lambda I)^{-1} y_\mu = \frac{y_\mu + cy_\lambda}{\mu - \lambda},$$

(it will actually be used twice); here c is chosen to ensure that the above vector belongs to $\mathrm{Dom}(\ell_h)$, that is

$$c = c(\lambda, \mu, h) = -\frac{m(\mu) + h}{m(\lambda) + h}.$$

As a result, we obtain

$$\begin{aligned} D_{T^*}\Theta_T(z)y_i &= -(\ell_h-\zeta I)(\ell_h+iI)^{-1}(\ell_{\bar h}-iI)(\ell_{\bar h}-\zeta I)^{-1}y_i \\ &= -(\ell_h-\zeta I)(\ell_h+iI)^{-1}(\ell_{\bar h}-iI)\frac{y_i+c(\zeta,i,\bar h)y_\zeta}{i-\zeta} \\ &= (\ell_h-\zeta I)(\ell_h+iI)^{-1}c(\zeta,i,\bar h)y_\zeta \\ &= (\ell_h-\zeta I)c(\zeta,i,\bar h)\frac{y_\zeta+c(-i,\zeta,h)y_{-i}}{\zeta+i} \\ &= -c(\zeta,i,\bar h)c(-i,\zeta,h)y_{-i} \\ &= c_0\cdot\frac{m(\zeta)+h}{m(\zeta)+\bar h}y_{-i}, \end{aligned}$$

where $c_0=-\frac{m(i)+\bar h}{m(-i)+h}$ is a unimodular constant. Using $y_{-i}=\overline{y}_i$, and therefore $\|y_i\|=\|y_{-i}\|$, we arrive at the following theorem.

1.4.14. THEOREM. *The characteristic function of the Cayley transform $T=C_+(\ell_h)$ of a dissipative Sturm–Liouville operator ℓ_h is equivalent to*

$$\Theta_{C_+(\ell_h)}(z)=\frac{m(\zeta)+h}{m(\zeta)+\bar h},$$

where $\zeta=i\frac{1+z}{1-z}$ and m is the Weyl function of ℓ_h.

□

1.5. Exercises and Further Results

1.5.1. C^*-algebras and normal operators. The aim of this subsection is to establish the basic spectral theory of normal operators in the classical sense of "diagonalization" of an operator. To this end, we prove the J. von Neumann form of the spectral theorem, describe the commutant of a normal operator N and construct the spectral measure of N. The most natural way to realize this program is to follow von Neumann's approach using (a bit of) C^*-algebras.

Recall that an operator valued set function $\sigma\longmapsto E(\sigma)$ defined on the Borel σ-algebra of $\mathbb{C}$ is called *spectral mesure* of a normal operator

$$N:H\longrightarrow H$$

if (1) $E(\sigma)$ is an orthoprojection on H, $E(\sigma(N))=I$; (2) $E(\sigma\cap\sigma')=E(\sigma)E(\sigma')$ for every σ,σ'; (3) $\sigma\longmapsto(E(\sigma)x,y)$ is a complex Borel measure for every $x,y\in H$ such that

$$(Nx,y)=\int z\,d(Ex,y).$$

A *C^*-algebra* is a Banach algebra A endowed with an *involution* $a\longmapsto a^*$, $a\in A$, that is, a conjugate linear mapping such that

$$(ab)^*=b^*a^*,\quad a^{**}=a,\quad \|a^*\|=\|a\|\text{ and }\|a^*a\|=\|a\|^2.$$

We always suppose that A has a unit e; clearly, $e^*=e$ (since e^* is also a unit) and hence $\|e\|=1$ (because of $\|e\|=\|e^*e\|=\|e\|^2$). The *spectrum* of an element $a\in A$ is denoted by $\sigma(a)=\sigma_A(a)$, the *spectral radius* by $r(a)=lim_n\|a^n\|^{1/n}$, and the *resolvent* by

$$R_\lambda(a)=R_{\lambda,A}(a)=(\lambda e-a)^{-1},\quad \lambda\in\mathbb{C}\backslash\sigma(a).$$

Obvious examples of C^*-algebras are $C(K)$, $L^\infty(\mu)$ (with $f^* = \overline{f}$, the complex conjugate) and $L(H)$, $\mathcal{K} = L(H)/\mathfrak{S}_\infty$ (with the adjoint a^* as an involution).

In the following series, A means a C^*-algebra, if not stated otherwise.

(a) Let $a \in A$ be a *Hermitian element*, that is $a^* = a$. Show that (i) $\sigma(a) \subset \mathbb{R}$ and $\|R_\lambda(a)\| \le |\operatorname{Im}(\lambda)|^{-1}$ for $\lambda \in \mathbb{C}\backslash\mathbb{R}$; (ii) $u = e^{ia}$ is *unitary*, that is, $uu^* = u^*u = e$, whence $\|ub\| = \|bu\| = 1$ for every $b \in A$.

[*Hint:* (i) for $t > 0$ and $\lambda = x + iy \in \mathbb{C}\backslash\mathbb{R}$ with $y > 0$, write $\lambda e - a = iye - b = i(y+t)e - (ite + b)$, where $b^* = b$ and $\|ite + b\| = \|t^2 e + b^2\|^{1/2} \le t(1 + \|b^2\| t^{-2})^{1/2} < t + \frac{1}{2}\|b^2\| t^{-1}$; therefore, $y + t > \|ite + b\|$ for $t > \|b^2\|/2y$ and, hence, $\lambda e - a$ is invertible; moreover,

$$\begin{aligned}\|R_\lambda(a)\| &\le (y+t)^{-1}(1 - (y+t)^{-1}\|ite + b\|)^{-1} = (y + t - \|ite + b\|)^{-1} \\ &\le (y - \frac{1}{2}\|b^2\| t^{-1})^{-1},\end{aligned}$$

and the result follows as $t \longrightarrow \infty$; similarly for $y < 0$; (ii) use $u^* = e^{(ia)^*} = e^{-ia}$; since $\|u\|^2 = \|u^*u\| = 1$, we have $\|b\| = \|u^*ub\| \le \|ub\| \le \|b\|$.]

(b) *Holes in the spectrum.* Let A be a Banach algebra and $B \subset A$ be its closed subalgebra (containing the unit). Show that

$$\sigma_B(x) = \sigma_A(x) \cup (\bigcup_k \omega_k)$$

for every $x \in B$, where ω_k are those bounded connected components of $\mathbb{C}\backslash\sigma_A(x)$ (*holes in* $\sigma_B(x)$) for which $\omega_k \cap \sigma_B(x) \ne \emptyset$; the unbounded component ω_∞ of $\mathbb{C}\backslash\sigma_A(x)$ is always in $\mathbb{C}\backslash\sigma_B(x)$. (Compare with Lemma 2.1.10)

[*Hint:* let $\lambda \in \omega\backslash\sigma_B(x)$, where ω is a connected component of $\mathbb{C}\backslash\sigma_A(x)$, and take $\varphi \in A^*$ such that $\varphi|B = 0$; since $z \longmapsto R_{z,A}(x)$ is holomorphic on ω and

$$R_{z,A}(x) - R_{z',A}(x) = (z' - z)R_{z,A}(x)R_{z',A}(x)$$

for every $z, z' \in \mathbb{C}\backslash\sigma_A(x)$ (the *Hilbert identity*), we get

$$\psi^{(n)}(z) = \varphi((-1)^n n! R_{z,A}(x)^{n+1}),$$

where $\psi(z) = \varphi(R_{z,A}(x))$; since $R_{\lambda,A}(x) \in B$, and hence $R_{\lambda,A}(x)^{n+1} \in B$ for all $n \ge 0$, we get $\psi^{(n)}(\lambda) = 0$ for $n \ge 0$, which implies $\psi|\omega = 0$; the Hahn-Banach theorem shows that $R_{z,A}(x) \in B$ for every $z \in \omega$, which means $\omega \subset \mathbb{C}\backslash\sigma_B(x)$; for $\omega = \omega_\infty$, the latter is clear since $|\lambda| > \|x\|$ implies $\lambda \in \omega\backslash\sigma_B(x)$.]

(c) Let B be a closed C^*-subalgebra of a C^*-algebra A. Show that $\sigma_B(x) = \sigma_A(x)$ for every $x \in B$.

[*Hint:* 1) first, show that $\sigma_B(y) = \sigma_A(y)$ for every Hermitian element $y \in B$ (use (a) and (b)); then, 2) take an invertible in A element $x \in B$ and observe that $(xx^*)((x^{-1})^*x^{-1}) = xx^{-1} = e$; since xx^* is a Hermitian element of B which is invertible in A, the product $(x^{-1})^*x^{-1}$ is its inverse and, by 1), $(x^{-1})^*x^{-1} \in B$; therefore, $x^{-1} = x^*(x^{-1})^*x^{-1} \in B$.]

(d) Let $x \in A$ be a *normal element*, that is $x^*x = xx^*$. Show that $\|x\| = r(x)$.

[*Hint:* first, observe that $\|y\|^2 = \|y^*y\| = \|y^2\|$ for every Hermitian element $y \in A$, and hence $\|y^{2^n}\| = \|y\|^{2^n}$; then, $\|x^{2^n}\|^2 = \|x^{*2^n}x^{2^n}\|^2 = \|(x^*x)^{2^n}\| = \|x^*x\|^{2^n} = \|x\|^{2^{n+1}}$, and the result follows.]

(e) *(I. Gelfand and M. Naimark, 1943)* Let A be a commutative C^*-algebra and $\mathfrak{M}(A)$ its maximal ideal space. Show that the *Gelfand transform*

$$j : x \longmapsto \hat{x}$$

is an isometric homomorphism from A onto $C(\mathfrak{M}(A))$ respecting involutions, i.e. $j(x^*) = \overline{j(x)}$ (complex conjugation) (such a mapping is called $*$-*isomorphism*).

[*Hint:* use (d) to show that j is an isometry; then, taking an element $x \in A$, write $x = a + ib$, where

$$a = \mathrm{Re}(x) = (x + x^*)/2, \quad b = \mathrm{Im}(x) = (x - x^*)/2i,$$

and observe that $a^* = a$, $b^* = b$, so that $ja(t) = \hat{a}(t) \in \sigma(a) \subset \mathbb{R}$ and $jb(t) = \hat{b}(t) \in \sigma(a) \subset \mathbb{R}$ for every $t \in \mathfrak{M}(A)$; this yields $j(x^*) = \overline{j(x)}$; consequently, jA is a closed unital self-adjoint subalgebra of $C(\mathfrak{M}(A))$ separating points of $\mathfrak{M}(A)$ and containing the constant function 1, and hence $jA = C(\mathfrak{M}(A))$ (the Stone–Weierstrass theorem).]

(f) *Function calculus and the spectral mapping theorem.* Let $a \in B$ be a normal element of a C^*-algebra, and let

$$A = \mathrm{alg}(a, a^*) := \mathrm{span}_B(a^n a^{*k} : n, k \geq 0)$$

be the C^*-subalgebra generated by a. Show that

(i) for every polynomial in the variables z and $\overline{z}$, say, $f(z) = p(z, \overline{z})$, the element $f(a) = p(a, a^*)$ is well defined, and $j(f(a)) = f(ja)$, where j is the Gelfand transform;

[*Hint:* use (e) above.]

(ii) the mapping $j_0 : f(a) \longrightarrow f|\sigma(a)$ extends up to a $*$-isomorphism from A onto $C(\sigma(a))$;

[*Hint:* use (e), (i) and $\sigma(a) = ja(\mathfrak{M}(B))$.]

(iii) the inverse of j_0 is a unique $*$-isomorphism $\varphi \longmapsto \varphi(a)$ from $C(\sigma(a))$ onto A (*the function calculus for* a) such that $z \longmapsto a$, $\overline{z} \longmapsto a^*$ and

$$\sigma(\varphi(a)) = \varphi(\sigma(a))$$

for every $\varphi \in C(\sigma(a))$ (the *spectral mapping theorem*)

[*Hint:* use (ii).]

(g) *(B. Fuglede, C. Putnam, 1950/1951)* Let $a, b \in A$ be normal elements of a C^*-algebra A, and let $c \in A$ be such that $ac = cb$. Show that $a^*c = cb^*$.

[*Hint (M. Rosenblum, 1958):* observe, first, that $a^n c = cb^n$ for all $n \geq 0$, and hence $e^{za}c = ce^{zb}$ and $e^{za}ce^{-zb} = c$ for every $z \in \mathbb{C}$; next, show that $e^{\lambda a^*}e^{wa} = e^{\lambda a^* + wa}$ for every $z, w \in \mathbb{C}$ (and similarly for b); finally, consider an entire function

$$f(\lambda) = e^{\lambda a^*}ce^{-\lambda b^*}$$

and write it as

$$\begin{aligned} f(\lambda) &= e^{\lambda a^*} c e^{-\lambda b^*} = e^{2i\,\mathrm{Im}(\lambda a^*)} e^{\overline{\lambda} a} c e^{-\overline{\lambda} b} e^{-2i\,\mathrm{Im}(\lambda b^*)} \\ &= e^{2i\,\mathrm{Im}(\lambda a^*)} c e^{-2i\,\mathrm{Im}(\lambda b^*)}; \end{aligned}$$

since $2\,\mathrm{Im}(\lambda a^*)$ and $2\,\mathrm{Im}(\lambda b^*)$ are Hermitian, it follows from (a) that $\|f(\lambda)\| = \|c\|$ for every $\lambda \in \mathbb{C}$ and, by Liouville's theorem, that $f = \mathrm{const}$; this yields $a^*c - cb^* = 0$ because

$$f(\lambda) = (1 + \lambda a^* + \dots)c(1 - \lambda b^* + \dots) = c + \lambda(a^*c - cb^*) + \dots .]$$

(h) Let $N : H \longrightarrow H$ be a normal operator on a Hilbert space H ($N^*N = NN^*$). Show that there exists a unique $*$-isomorphism

$$\varphi \longmapsto \varphi(N)$$

from $C(\sigma(N))$ onto $\mathrm{alg}(N, N^*)$ (the function calculus for N) such that $z \longmapsto N$, $\overline{z} \longmapsto N^*$ and $\sigma(\varphi(N)) = \varphi(\sigma(N))$ for every $\varphi \in C(\sigma(N))$ (the spectral mapping theorem). In particular, $R_\lambda(N) = (\frac{1}{\lambda - z})(N)$ for $\lambda \in \mathbb{C}\backslash\sigma(N)$.

[*Hint:* use (f)(iii) for $B = L(H)$.]

(i) *Reducing subspaces of a normal operator.* Let $N : H \longrightarrow H$ be a normal operator on a Hilbert space, and let $E \subset H$ be a reducing subspace of N (see A.1.6.1, Volume 1, for the definition).

(i) Show that $\varphi(N)E \subset E$ for every $\varphi \in C(\sigma(N))$.

[*Hint:* φ is a norm limit of polynomials in z and $\overline{z}$.]

(ii) Show that the restriction $N|E$ is a normal operator and $\sigma(N|E) \subset \sigma(N)$.

[*Hint:* use (i) for $\varphi = \frac{1}{\lambda - z}$ and $\lambda \in \mathbb{C}\backslash\sigma(N)$.]

(iii) Let N be a multiplication operator $Nf = zf$ on the space $L^2(H(\cdot), \mu)$ defined in (k) below. Show that there exists a measurable projection valued function $\zeta \longmapsto P'(\zeta)$ such that $E = L^2(H'(\cdot), \mu)$, where $H'(\zeta) = P'(\zeta)H(\zeta)$ a.e. μ.

[*Hint:* this follows from A.1.6.1(c) (Volume 1).]

(j) *Diagonalization of a $*$-cyclic normal operator.* Let $N : H \longrightarrow H$ be a $*$-cyclic normal operator on a Hilbert space H, which means that there exists a $*$-*cyclic vector* $x \in H$, i.e.

$$\mathrm{span}(N^k N^{*l} x : k, l \geq 0) = H.$$

Show that there exists a nonnegative Borel measure μ on $\sigma(N)$ (a *scalar spectral measure* of N) and a unitary mapping $U : L^2(\mu) \longrightarrow H$ such that

$$U^{-1} N U h = zh$$

for every $h \in L^2(\mu)$, where z is the independent variable, $z(\zeta) = \zeta$ for $\zeta \in \sigma(N)$.

[*Hint:* use (h) to show that $\varphi \longmapsto (\varphi(N)x, x)$ is a continuous linear form on $C(\sigma(N))$, then apply the Riesz–Markov theorem to obtain a measure μ such that

$$(\varphi(N)x, x) = \int_{\sigma(N)} \varphi \, d\mu;$$

now define $U\varphi = \varphi(N)x$ for $\varphi \in C(\sigma(N))$ and check that $\|U\varphi\|^2 = \int_{\sigma(N)} |\varphi|^2 \, d\mu$ for every $\varphi \in C(\sigma(N))$; deduce that $\int_{\sigma(N)} \psi \, d\mu \geq 0$ for every $\psi \geq 0$ (and hence that $\mu \geq 0$) and $U(z\varphi) = N(\varphi(N)x)$ (use (h)); the result follows.]

(k) *The von Neumann spectral theorem.* Let $N : H \longrightarrow H$ be a normal operator on a separable Hilbert space H. Show that there exists a nonnegative Borel measure μ on $\sigma(N)$ and a Borel operator valued function $\zeta \longmapsto P(\zeta)$, whose values are orthogonal projections on H, such that

1) $\mu = \chi_\sigma \mu$, where $\sigma = \{\zeta \in \mathbb{C} : P(\zeta) \neq 0\}$, and
2) the operator N is unitarily equivalent to the multiplication operator

$$Z : h \longrightarrow zh$$

on the *direct integral* $L^2(H(\cdot), \mu)$ *of Hilbert spaces* $H(\zeta) = P(\zeta)H$, i.e., on the space

$$L^2(H(\cdot), \mu) = \int_{\sigma(N)} \oplus H(\zeta) \, d\mu(\zeta) := \Big\{ h \in L^2(H, \mu) : h(\zeta) \in H(\zeta)\mu - a.e. \Big\};$$

Z is called the *von Neumann model* of N, and μ is a *scalar spectral measure* of N.

Moreover, the model can be chosen *standard*, that is, satisfying

$$H(\zeta) = \mathrm{span}(e_k : 1 \leq k < d_Z(\zeta) + 1),$$

where d_Z stands for a (multiplicity) function taking values in $\mathbb{Z}_+ \cup \{\infty\}$ and $(e_k)_{k \geq 1}$ is an orthonormal basis in H.

[*Hint:* 1) fix an orthonormal basis $(e_k)_{k \geq 1}$ of H, set $x_1 = e_1$ and $E(x_1) = \mathrm{clos}(Ax_1)$, where $A = \mathrm{alg}(N, N^*)$, and show that $E(x_1)$ is a reducing subspace of N (see A.1.6.1, Volume 1, for the definition);

2) using (j) and (i), show that $N|E(x_1)$ is unitarily equivalent to the multiplication by z in a space $L^2(\mu_1)$ with $\mathrm{supp}(\mu_1) \subset \sigma(N)$;

3) take the least n such that $e_n \notin E(x_1)$, set $x_2 = (I - P_{E(x_1)})e_n$ and repeat steps 1) and 2) for $E(x_2)$ instead of $E(x_1)$ (to obtain a measure μ_2); clearly, $E(x_1) \perp E(x_2)$;

4) proceed by an obvious induction (taking $e_3 \notin E(x_1) \oplus E(x_2)$, $x_3 = (I - P_{E(x_1)} - P_{E(x_1)})e_3$, etc.) and find sequences (eventually finite) of reducing subspaces $E(x_k)$ and measures μ_k such that $\mathrm{supp}(\mu_k) \subset \sigma(N)$, $H = \sum_{k>1} \oplus E(x_k)$ and $N|E(x_k)$ is unitarily equivalent to the multiplication by z on $L^2(\mu_k)$; conclude that N is unitarily equivalent to

$$N'f = (zf_1, zf_2, \dots)$$

acting on the space $\sum_{k \geq 1} \oplus L^2(\mu_k)$ (the space of all functions $f = (f_1, f_2, \dots)$ with $\sum_{k \geq 1} \int |f_k|^2 \, d\mu_k < \infty$);

5) set $\mu = \sum_{k \geq 1} \mu^{(k)}$, where $\mu^{(k)} = k^{-2}\mu_k / \mathrm{Var}(\mu_k)$, and use the Radon–Nikodym theorem to write $\mu^{(k)} = h_k \mu$, where $\sum_{k \geq 1} h_k = 1$ and $0 \leq h_k \leq 1$ μ-a.e.; next, set $\sigma_k = \{\zeta : h_k(\zeta) \neq 0\}$ (σ_k is well defined μ–mod(0)) and show that N' is unitarily eqivalent to $N''f = (zf_1, zf_2, \dots)$ acting on the space

$$\sum_{k \geq 1} \oplus L^2(\chi_{\sigma_k} \mu);$$

and, finally,

6) set $d_Z(\zeta) = \text{card}\{k : \chi_{\sigma_k}(\zeta) \neq 0\}$ and

$$H(\zeta) = \text{span}_H(e_k : 1 \le k < d_Z(\zeta) + 1)$$

(so that $d_Z(\zeta) - \dim H(\zeta)$), and show that N'' (and, hence, N) is unitarily equivalent to the operator Z of the statement.]

(l) *The $L^\infty(\mu)$ function calculus and the spectral measure of a normal operator.* Let $N : H \longrightarrow H$ be a normal operator on a separable Hilbert space H, let Z be its von Neumann model on $\mathcal{H} = L^2(H(\cdot), \mu)$, and let U be a unitary operator $U : \mathcal{H} \longrightarrow H$ such that $U^{-1}NU = Z$.

(i) Let $\varphi \in L^\infty(\mu)$ and $\varphi(N) := U\varphi(Z)U^{-1}$, where $\varphi(Z)f = \varphi f$ is the multiplication operator on $\mathcal{H}$. Show that

$$\varphi \longmapsto \varphi(N)$$

is an isometric $*$-homomorphism from the algebra $L^\infty(\mu)$ to $L(H)$ which coincides on $C(\sigma(N))$ with the calculus of theorem (f) above.

(ii) Let σ be a Borel subset of $\sigma(N)$ and

$$I(\sigma)f = \chi_\sigma f$$

for every $f \in \mathcal{H}$. Show that $\sigma \longmapsto I(\sigma)$ is a spectral measure for Z and $\text{supp}(I(\cdot)) = \sigma(N)$.

(iii) Show that $E_N(\sigma) = U^{-1}I(\sigma)U$ is a spectral measure for N.

(iv) Show that

$$(p(N, N^*)x, y) = \int p d(E_N x, y)$$

for any spectral measure for N, for every $x, y \in H$ and for every polynomial p in z and $\bar{z}$. Consequently, the spectral measure for N is unique and we have

$$(f(N)x, y) = \int_{\sigma(N)} f\, d(E_N x, y)$$

for every $f \in C(\sigma(N))$ and every $x, y \in H$.

[*Hint:* show consecutively that 1) $(N^*x, y) = \overline{(Ny, x)} = \int \bar{z} d(E_N x, y)$;
2) $\nu(\sigma) = (NE_N(\sigma)x, y) = \int_\sigma z\, d(E_N x, y)$ is a measure such that $\int f d\nu = \int zf\, d(E_N x, y)$ for every bounded f;
3) if the formula of (iv) is valid for a polynomial p, then

$$(Np(N, N^*)x, y) = \int p d(NE_N x, y) = \int zp d(E_N x, y);$$

4) adding to 3) its analogue for N^*, complete the proof by induction.]

(m) *Unitary equivalence of two von Neumann models.* Let

$$Z_i : L^2(H_i(\cdot), \mu_i) \longrightarrow L^2(H_i(\cdot), \mu_i) \quad (i = 1, 2)$$

be two von Neumann models, and suppose that the measures are mutually absolutely continuous (in short, $\mu_1 \sim \mu_2$) and $d_{Z_1}(\zeta) = d_{Z_2}(\zeta)$ a.e. μ_i (d_Z is defined in (k) above). Let, further,

$$u(\zeta) : H_1(\zeta) \longrightarrow H_2(\zeta)$$

be a measurable family of unitary operators and $w^2 = d\mu_1/d\mu_2$ (the Radon–Nikodym derivative).

Show that

$$(Uf)(\zeta) = w(\zeta)u(\zeta)f(\zeta)$$

is a unitary operator from $L^2(H_1(\cdot), \mu_1)$ to $L^2(H_2(\cdot), \mu_2)$ such that $UZ_1U^{-1} = Z_2$. In particular, any von Neumann model is unitarily equivalent to a standard one (as defined in (k) above).

(n) *Intertwining operators.* Let $Z_i : L^2(H_i(\cdot), \mu_i) \longrightarrow L^2(H_i(\cdot), \mu_i)$ $(i = 1, 2)$ be two von Neumann models, and let

$$A : L^2(H_1(\cdot), \mu_1) \longrightarrow L^2(H_2(\cdot), \mu_2)$$

be a bounded operator verifying

$$AZ_1 = Z_2A.$$

Next, let $\mu_1 = \mu_1^a + \mu_1^s$ be the Lebesgue decomposition of μ_1 into an absolutely continuous and a singular measure with respect to μ_2.

Show that $A|L^2(H_1(\cdot), \mu_1^s) = 0$ and that there exists an operator valued Borel function $\zeta \longmapsto a(\zeta)$ such that $A = m_a$, where m_a is the multiplication operator

$$(m_af)(\zeta) = a(\zeta)f(\zeta)$$

a.e. μ_1^a for every $f \in L^2(H_1(\cdot), \mu_1^a)$ and $\|A\| = \|m_a\| = \text{ess sup}_{\mu_1^a} \|a(\zeta)\| < \infty$ (in short, $a \in L^\infty(H(\cdot) \longrightarrow H(\cdot), \mu_1^a))$.

Conversely, every such operator A intertwines the operators Z_1 and Z_2.

In particular, every intertwining operator is 0 if and only if $\mu_1^a = 0$, that is $\mu_1 \perp \mu_2$ (R. Douglas, 1969).

[*Hint:* 1) employ (g) to show that $AZ_1^* = Z_2^*A$, next $Af(Z_1) = f(Z_2)A$ for every polynomial $f = p(z, \overline{z})$, and, finally, for every bounded Borel function f on $\mathbb{C}$; taking $f = \chi_S$ where S is the Lebesgue support of the singular part $\mu_1^s = \chi_S\mu_1$, we get $Ag = A\chi_Sg = \chi_SAg = 0$ for every $g \in L^2(H_1(\cdot), \mu_1^s)$;

2) use (m) to show that, without loss of generality, we may suppose that $A|L^2(H_1(\cdot), \mu_1^a)$ is a standard von Neumann model; further, set $\sigma_k = \{\zeta : d_{Z_1}(\zeta) = k\}$, $1 \le k \le \infty$ (d_Z is defined in (k) above), and show that

$$L^2(H_1(\cdot), \mu_1^a) = \sum_{1 \le k \le \infty} \oplus L^2(H_1(\cdot), \chi_{\sigma_k}\mu_1^a);$$

for every restriction $A|L^2(H_1(\cdot), \chi_{\sigma_k}\mu_1^a)$ proceed exactly as in Lemma 1.2.3 (replacing polynomials on $\mathbb{T}$ by polynomials $p(z, \overline{z})$ on $\sigma(Z_1)$), and get a function $a_k(\zeta) : H_1(\zeta) \longrightarrow H_2(\zeta)$ defined a.e. $\chi_{\sigma_k}\mu_1^a$ and such that

$$(Af)(\zeta) = a_k(\zeta)f(\zeta)$$

a.e. $\chi_{\sigma_k}\mu_1^a$ for every $f \in L^2(H_1(\cdot), \chi_{\sigma_k}\mu_1^a)$; finally, set $a(\zeta) = a_k(\zeta)$ for $\zeta \in \sigma_k$, $1 \le k \le \infty$, and $a(\zeta) = 0$ outside of $\bigcup_k \sigma_k$.]

(o) *The commutant and the double commutant.* Let $C \subset L(H)$ and C' be the commutant of C, i.e.

$$C' = \{A : Ac = cA \text{ for all } c \in C\}.$$

Show that (in the notation of Theorems (l)-(n))

$$\begin{aligned}\{N\}' &:= (\{N\})' = (\mathrm{alg}(N, N^*))' = \{m_a : a \in L^\infty(H(\cdot) \longrightarrow H(\cdot), \mu)\},\\ \{N\}'' &= (\mathrm{alg}(N, N^*)')' = \{m_a : a(t) = \varphi(t) I_{H(t)}, \varphi \in L^\infty(\mu)\}\\ &= \{\varphi(N) : \varphi \in L^\infty(\mu)\}.\end{aligned}$$

[*Hint:* the formula for $\{N\}'$ is a special case of (n); the formula for $\{N\}''$ follows because of $(L(H(t) \longrightarrow H(t)))' = \{a(t) : a(t) = \varphi(t) I_{H(t)}$ where $\varphi(t) \in \mathbb{C}\}$.]

(p) *Unitary invariants of normal operators.* Let Z_1, Z_2 be unitarily equivalent von Neumann models. Show that $\mu_1 \sim \mu_2$ and $d_{Z_1}(\zeta) = d_{Z_2}(\zeta)$ a.e. μ_i (d_Z is defined in (k) above).

In particular, if N is a normal operator on a separable Hilbert space H, and Z_1, Z_2 are two associated von Neumann models, then $\mu_1 \sim \mu_2$ and $d_{Z_1}(\zeta) = d_{Z_2}(\zeta)$ a.e. μ_i.

The class of equivalent measures

$$[\mu]_N = \{\mu' : \mu' \sim \mu_1\}$$

is called the *class of the spectral measure* of N, and

$$\zeta \longmapsto d_N(\zeta) = d_{Z_1}(\zeta)$$

is the *(spectral) multiplicity function* of N.

[*Hint:* as Z_1 and Z_2 are unitarily equivalent, they are intertwined by a unitary operator U, i.e., $UZ_1 = Z_2 U$; using the notation of (n), show that $\mu_1^s = 0$, and hence $\mu_1 << \mu_2$, and $Uf(\zeta) = u(\zeta) f(\zeta)$ a.e. for every f; since U is invertible, $\mu_2 << \mu_1$ and $d_{Z_1}(\zeta) = d_{Z_2}(\zeta)$ a.e. μ_i.]

(q) *Unitary equivalence of normal operators.* Let N_1 and N_2 be normal operators. Show that the following assertions are equivalent.

(i) N_2 is a *di-formation* of N_1, that is, there exists an operator A such that $AN_1 = N_2 A$ and $\mathrm{Ker}\, A = \mathrm{Ker}\, A^* = \{0\}$ (see also 1.5.3 below).

(ii) N_1 is unitarily equivalent to N_2.

(iii) $[\mu]_{N_1} = [\mu]_{N_2}$ and $d_{N_1}(\zeta) = d_{N_2}(\zeta)$ a.e. μ.

[*Hint:* for (i) $\Rightarrow$ (ii), use Lemma B.1.5.6 (Volume 1); (ii) $\Rightarrow$ (iii) and (iii) $\Rightarrow$ (i) follow from (p).]

(r) *Compact normal operators (D. Hilbert, E. Schmidt (circa 1906)).* Let N be a compact normal operator on a Hilbert space H, and let χ_ε be the indicator function of $\mathbb{C} \backslash \varepsilon\mathbb{D} = \{\zeta \in \mathbb{C} : |\zeta| \geq \varepsilon\}$ for $\varepsilon > 0$. Show that, in the notation of (k) and (p) above,

1) $\dim(\chi_\varepsilon L^2(H(\cdot), \mu)) < \infty$ for every $\varepsilon > 0$, and therefore,

2) μ is a discrete measure, $\mu = \sum_{k \geq 0} p_k \delta_{\lambda_k}$ with $0 = \lambda_0 = \lim_k \lambda_k$ (and, if $\dim H = \infty$, with $p_k > 0$ for all $k > 0$),

3) $\dim H(\lambda_k) < \infty$ for $k > 0$,

4) $H = \sum_{k \geq 0} \oplus H_k$ where $H_k = \mathrm{Ker}(\lambda_k I - N)$, $k \geq 0$.

[*Hint:* the restriction $Z | \chi_\varepsilon L^2(H(\cdot), \mu)$ is compact and invertible for every $\varepsilon > 0$.]

1.5.2. The residual part of the minimal unitary dilation. Here we consider a c.n.u. contraction $T : H \longrightarrow H$ and its Szőkefalvi-Nagy–Foiaş model $M_\Theta : \mathcal{K}_\Theta \longrightarrow \mathcal{K}_\Theta$, where $M_\Theta = P_\Theta z | \mathcal{K}_\Theta$ and

$$\mathcal{K}_\Theta = \begin{pmatrix} H^2(E_*) \\ \operatorname{clos} \Delta L^2(E) \end{pmatrix} \ominus \begin{pmatrix} \Theta \\ \Delta \end{pmatrix} H^2(E),$$

where $\Theta = \Theta_T$ is the characteristic function of T and $\Delta(\zeta) = (I - \Theta(\zeta)^*\Theta(\zeta))^{1/2}$ for $\zeta \in \mathbb{T}$. The subspace $\mathcal{R}$, defined by

$$\mathcal{R} = \begin{pmatrix} \{0\} \\ \operatorname{clos} \Delta L^2(E) \end{pmatrix} \subset \mathcal{K} = \begin{pmatrix} L^2(E_*) \\ \operatorname{clos} \Delta L^2(E) \end{pmatrix},$$

is called the *residual subspace* of the minimal unitary dilation $Z : \mathcal{K} \longrightarrow \mathcal{K}$ of M_Θ ($Z(F) = zF$ for $F \in \mathcal{K}$), and the restriction $Z|\mathcal{R}$ is the *residual part* of Z. In fact, $\mathcal{R} = (\operatorname{Range} \pi)^\perp$, where π is the corresponding embedding. Let

$$X = P_\Theta | \mathcal{R} = P_{\mathcal{K}_\Theta} | \mathcal{R}.$$

(a) Show that $X : \mathcal{R} \longrightarrow \mathcal{K}_\Theta$ is a contraction such that

$$X(Z|\mathcal{R}) = M_\Theta X.$$

Moreover, $X^* = P_{\mathcal{R}} | \mathcal{K}_\Theta$ and hence

$$X^*(f,g)^{col} = (0,g)^{col} = \lim_n z^n M_\Theta^{*n}(f,g)^{col}$$

for every $(f,g)^{col} \in \mathcal{K}_\Theta$.

[*Hint:* use the inclusion $(I - P_\Theta)\mathcal{R} \subset (\Theta, \Delta)^{col} H^2(E)$ and the fact that the latter subspace is Z-invariant; the equality $(P_\alpha|\beta)^* = P_\beta|\alpha$ is straightforward for any couple of subspaces α, β; the last formula follows from $M_\Theta^{*n}(f,g)^{col} = (P_+\overline{z}^n f, \overline{z}^n g)^{col}$ (see 1.3.7(2)).]

(b) Show that

$$\operatorname{Ker} X^* = \left\{ \begin{pmatrix} f \\ 0 \end{pmatrix} : f \in \mathcal{K}_{\Theta_{inn}} \right\} = \begin{pmatrix} \mathcal{K}_{\Theta_{inn}} \\ 0 \end{pmatrix},$$

where Θ_{inn} comes from the inner-outer factorization $\Theta = \Theta_{inn}\Theta_{out}$ (see 2.1.21, A.1.6.4 (Volume 1) and 1.3.9(1) for the definition and A.1.6.4 for the existence).

In particular, $\operatorname{Ker} X^* = \{0\}$ if and only if Θ is outer.

[*Hint:* following (a), $(f,g)^{col} \in \operatorname{Ker} X^*$ if and only if $g = 0$ and $f \perp \Theta H^2(E)$, i.e. $\Theta_{inn}^* f \perp \Theta_{out} H^2(E)$; using the definitions and notation of A.1.6.4, show that this is equivalent to $\Theta_{inn}^* f \perp H^2(F)$, that is, to $f \in \mathcal{K}_{\Theta_{inn}}$.]

(c) Show that $\operatorname{Ker} X = \{(0,g)^{col} : g \in H^2(E), \Theta g = 0\}$.

[*Hint:* using 1.3.7(1) observe that $X(0,g)^{col} = 0 \Leftrightarrow (\Theta P_+ \Delta g = 0$ and $g - \Delta P_+ \Delta g = 0)$; this gives $\Delta g = \Delta^2 P_+ \Delta g = (I - \Theta^*\Theta) P_+ \Delta g = P_+ \Delta g \in H^2(E)$, and hence $g = \Delta^2 g$, i.e. $\Theta^*\Theta g = 0$; moreover, $g = \Delta^{2n} g$ for every $n \geq 0$, which yields $p(1)g = p(\Delta^2)g$ for every polynomial p and, finally, $g = \Delta g \in H^2(E)$; the converse is straightforward.]

(d) *Holomorphic kernels of holomorphic operator functions.*

(i) Assume that there exists $\lambda \in \mathbb{D}$ such that $\operatorname{Ker} \Theta(\lambda) = \{0\}$ (which is equivalent to $\sigma_p(M_\Theta) \neq \mathbb{D}$ (see 1.3.13)). Show that $\operatorname{Ker} X = \{0\}$.

[*Hint:* use (c) to show that $\Theta(\lambda)g(\lambda) = 0$ for every $(0, g)^{col} \in \operatorname{Ker} X$, and hence $g(\lambda) = 0$; setting $g_1 = (z - \lambda)^{-1} g \in H^2(E)$, obtain $\Theta g_1 = 0$, and then $g'(\lambda) = g_1(\lambda) = 0$, etc., $g^{(n)}(\lambda) = 0$ for every $n \geq 0$.]

(ii) Let $\Theta_*(\zeta) = \Theta(\bar{\zeta})^*$ for $\zeta \in \mathbb{D}$, and suppose that Θ_* is an outer function. Show that $\operatorname{Ker} \Theta(\zeta) = \{0\}$ for every $\zeta \in \mathbb{D}$, and hence $\operatorname{Ker} X = \{0\}$ (and also $\sigma_p(M_\Theta) = \emptyset$).

[*Hint:* by definition, $\Theta_* H^2(E_*)$ is dense in $H^2(E)$, and hence $\Theta_*(\zeta)E_*$ is dense in E.]

(iii) Let $\Theta \in \operatorname{Hol}(\mathbb{D}, E \longrightarrow E_*)$ (the space of all holomorphic $L(E, E_*)$ valued functions in $\mathbb{D}$) and suppose that d is an integer with $\dim \Theta(z)E < d \leq \dim E_*$ for every $z \in \mathbb{D}$; show that there exists a function $g \in \operatorname{Hol}(\mathbb{D}, E)$ with $\Theta(\zeta)g(\zeta) = 0$ and $g(\zeta) \neq 0$ for every $\zeta \in \mathbb{D}$. If $\Theta \in H^\infty(\mathbb{D}, E \longrightarrow E_*)$, then the function g can be chosen in $H^\infty(E)$.

[*Hint:* take any subspace $F \subset E$ with $\dim F = d$ and, using the Kramer rule for systems of linear equations, find a holomorphic non zero solution g with $g(\zeta) \in F$ satisfying the following degenerated system of linear equations $(g(\zeta), P_F \Theta(\zeta)^* E_*) = 0$; multiply by the corresponding principal minor to get $g \in H^\infty(E)$.]

(iv) Exhibit an example of a function $\Theta \in H^\infty(\mathbb{D}, E \longrightarrow E_*)$ with $\operatorname{Ker} \Theta(\zeta) \neq \{0\}$ for every $\zeta \in \mathbb{D}$ and such that for no function g holomorphic at a point $\zeta \in \mathbb{D}$ and with $g(\zeta) \neq 0$ we have $\Theta(\zeta)g(\zeta) = 0$,

[*Hint:* take $E = E_* = W_2^2(\mathbb{D})$ and $\Theta(\lambda) = \lambda I - R^*$, where $R : E \longrightarrow E_*$ is the multiplication operator $Rf(\lambda) = \lambda f(\lambda)$ on the *Sobolev space*

$$W_2^2(\mathbb{D}) = \left\{ f : \int_{\mathbb{D}} \left(|f|^2 + \left| \frac{\partial^2 f}{\partial x^2} \right|^2 + \left| \frac{\partial^2 f}{\partial y^2} \right|^2 \right) dxdy < \infty \right\};$$

due to the Sobolev embedding Lemma (see, for example, W. Rudin [**Ru6**]), the evaluation functionals $f \longmapsto f(\lambda)$ are continuous for $\lambda \in \mathbb{D}$, and the corresponding reproducing kernels $k_\lambda \in W_2^2(\mathbb{D})$ are simple eigenvectors of R^*, i.e. $\Theta(\lambda)k_\lambda = 0$ for every λ; show that there is no nonzero scalar function φ making the product $\lambda \longmapsto \varphi(\lambda)k_\lambda$ holomorphic.]

(v) Exhibit an example of a function $\Theta \in H^\infty(\mathbb{D}, E \longrightarrow E_*)$ with a holomorphic nonzero solution g of the equation $\Theta g = 0$, but with no such solutions in $H^2(E)$.

[*Hint:* take $E = E_* = H^2(\mathbb{D})$ and $\Theta = zI - S^*$ with the shift operator $S : H^2(\mathbb{D}) \longrightarrow H^2(\mathbb{D})$, check that necessarily $g(\zeta) = \varphi(\zeta)k_\zeta$, where $k_\zeta = (1 - z\zeta)^{-1}$, and show that $g \in H^2(E)$ implies that $\varphi = 0$.]

(e) *The residual part and the characteristic function.* Let

$$\mathcal{K}_+ = \begin{pmatrix} H^2(E_*) \\ \operatorname{clos} \Delta L^2(E) \end{pmatrix}$$

be the space of the minimal isometric dilation and let $\mathcal{K}_\Theta^\perp = \mathcal{K}_+ \ominus \mathcal{K}_\Theta$, $\mathcal{R}^\perp = \mathcal{K}_+ \ominus \mathcal{R}$. Show that there exist unitary operators $U : H^2(E) \longrightarrow \mathcal{K}_\Theta^\perp$ and $V : \mathcal{R}^\perp \longrightarrow H^2(E_*)$ such that

$$V(P_{\mathcal{R}^\perp} | \mathcal{K}_\Theta^\perp)U = \Theta_{H^2},$$

where $\Theta_{H^2} : H^2(E) \longrightarrow H^2(E_*)$ stands for the multiplication operator by the function Θ. In fact, $U = \pi$ and $V = \pi_*^*$.

[*Hint:* take $Uh = \binom{\Theta}{\Delta}h$ and $V(f,0)^{col} = f$.]

(f) *The spectrum of the residual part $U|\mathcal{R}$ (B. Sz.-Nagy and C. Foiaş, 1962)*

(i) Let V be a pure isometry on a Hilbert space K, that is, $\bigcap_{n\geq 0} V^n K = \{0\}$. Show that $\lim_n \|P_{V^nK}x\| = 0$ for every $x \in K$, where P_S stands for the orthoprojection onto a subspace S.

[*Hint:* use the Wold–Kolmogorov decompositions

$$K = \sum_{j\geq 0} \oplus V^j(K \ominus VK) \quad \text{and} \quad V^n K = \sum_{j\geq n} \oplus V^j(K \ominus VK),$$

then take a vector $x \in K$, $x = \sum_{j\geq 0} \oplus V^j a_j$ with $a_j \in (K \ominus VK)$, and get $\|x\|^2 = \sum_{j\geq 0} \|a_j\|^2 < \infty$ and $\|P_{V^nK}x\|^2 = \sum_{j\geq n} \|a_j\|^2$.]

(ii) Let $x \in \mathcal{R}$. Show that there exist vectors $x_n \in \mathcal{K}_\Theta$ such that $x = \lim_n U^n x_n$.

[*Hint:* using the notation from (e), observe that $\mathcal{R} \subset \mathcal{K}_+ = \mathcal{K}_\Theta \oplus K$, where $K = \begin{pmatrix}\Theta\\ \Delta\end{pmatrix} H^2(E)$, and hence $\mathcal{R} = U^n\mathcal{R} \subset= U^n\mathcal{K}_\Theta \oplus U^n K$ for every n, $n \geq 0$; therefore, $x = U^n x_n + y_n$ for every $x \in \mathcal{R}$ and $n \geq 0$, where $x_n \in \mathcal{K}_\Theta$ and $y_n = P_{U^nK}x$; now, apply (i) to $V = U|K$ and get $\lim_n \|y_n\| = 0$.]

(iii) Show that $\sigma(U|\mathcal{R}) \subset \sigma(T) \cap \mathbb{T}$.

[*Hint:* let $\lambda \in \mathbb{T}\backslash\sigma(T)$ and $x \in \mathcal{R}$; using (ii) and the definition of the dilation, we have

$$\begin{aligned}
\|(U - \lambda I)x\| &= \lim_n \|(U - \lambda I)U^n x_n\| = \lim_n \|U^n(U - \lambda I)x_n\| \\
&= \lim_n \|(U - \lambda I)x_n\| \geq \lim_n \|P_\Theta(U - \lambda I)x_n\| \\
&= \lim_n \|(T - \lambda I)x_n\| \geq c \cdot \lim_n \|x_n\| \\
&= c\|x\|,
\end{aligned}$$

where $c = \|(T - \lambda I)^{-1}\|^{-1}$; thus, $\lambda \in \mathbb{T}\backslash\sigma(U)$.]

1.5.3. Di-formations, and quasi-similarity to a unitary operator. A bounded linear operator Y is a *d-formation* if $\operatorname{Ker} Y^* = \{0\}$ (where "*d*" is for "dense"), an *i-formation* if $\operatorname{Ker} Y = \{0\}$ ("*i*" is for injective), and a *di-formation* (or, a *quasi-affinity*) if $\operatorname{Ker} Y = \{0\}$ and $\operatorname{Ker} Y^* = \{0\}$.

We say that an operator A is a *d-formation of B* (respectively, is a *i- or di-formation of B*) and write

$$A \preceq_d B$$

(resp., $A \preceq_i B$ or $A \preceq B$), if there exists a d-formation (resp., i- or di-formation) Y such that

$$AY = YB;$$

and finally, A and B are *quasi-similar* if $A \preceq B$ and $B \preceq A$.

(a) Show that $A \preceq_d B$ and $B \preceq_d C$ entail $A \preceq_d C$, and similarly for $\preceq_i$ and $\preceq$. If A and B are normal operators and $A \preceq B$, then A is unitarily equivalent to B.

[*Hint:* for the latter, see B.1.5.6, Volume 1.]

(b) Show that the following assertions are equivalent (in the notation of 1.5.2).

(1) $M_\Theta \preceq_d V$, where V is a unitary operator.
(2) $\lim_n \|M_\Theta^{*n} F\| > 0$ for every $F \in \mathcal{K}_\Theta \backslash \{0\}$.
(3) $\operatorname{Ker} X^* = \{0\}$.
(4) Θ_T is an outer function.
(5) $M_\Theta \preceq_d Z|\mathcal{R}$.

[*Hint:* (1) $\Rightarrow$ (2), $M_\Theta Y = YV$ yields $Y^* M_\Theta^{*n} = V^{*n} Y^*$, whence $\lim_n \|M_\Theta^{*n} F\| \geq \|Y^*\|^{-1} \|Y^* F\|$ for every $F \in \mathcal{K}_\Theta$; (2) $\Rightarrow$ (3) by 1.5.2(a); for (3) $\Rightarrow$ (4) see 1.5.2(b); the implication (4) $\Rightarrow$ (5) follows from 1.5.2(a) and (b); (5) $\Rightarrow$ (1) is obvious.]

(c) Suppose $V \preceq_i M_\Theta$, where V is a unitary operator. Show that $M_\Theta \preceq_i Z|\mathcal{R}$.

[*Hint:* $YM_\Theta = VY$, where $\operatorname{Ker} Y = \{0\}$, implies $\operatorname{Ker} M_\Theta = \{0\}$; therefore, $\operatorname{Ker} X = \{0\}$ (see 1.5.2(d)(i)), and the result follows from 1.5.2(a).]

(d) Suppose that $V' \preceq_i M_\Theta \preceq_d V$, where V and V' are unitary operators. Show that $M_\Theta \preceq Z|\mathcal{R}$.

[*Hint:* note that (b) and (c) entail $\operatorname{Ker} X = \operatorname{Ker} X^* = \{0\}$.]

(e) Suppose that $V' \preceq M_\Theta \preceq V$, where V and V' are normal operators. Show that V and V' are unitarily equivalent, and hence, M_Θ is quasi-similar to a normal operator. If V or V' is unitary, then M_Θ is quasi-similar to $Z|\mathcal{R}$.

[*Hint:* since $V' \preceq V$ (see (a) above), Lemma B.1.5.6 (Volume 1) entails that V and V' are unitarily equivalent; if one of V, V' is unitary, the other is also; next, using (d) we get $V' \preceq M_\Theta \preceq Z|\mathcal{R}$, which, in turn, implies the equivalence of $Z|\mathcal{R}$ and V'.]

(f) Suppose that $\sigma_p(M_\Theta) \neq \mathbb{D}$ and $\lim_n \|M_\Theta^{*n} F\| > 0$ for every $F \in \mathcal{K}_\Theta \backslash \{0\}$ (or any other of equivalent conditions of (b) above). Show that $M_\Theta \preceq Z|\mathcal{R}$, and hence, if M_Θ is quasi-similar to a normal operator, it is quasi-similar to a unitary one, namely to $Z|\mathcal{R}$.

[*Hint:* observe that (b) and 1.5.2(d)(i) yield $\operatorname{Ker} X^* = \{0\}$ and $\operatorname{Ker} X = \{0\}$, respectively, whence $M_\Theta \preceq Z|\mathcal{R}$; the rest follows from (e).]

1.5.4. The classes $C_{\alpha\beta}$ and quasi-similarity to a unitary operator: a unitarily invariant approach. Let $T : H \longrightarrow H$ be a c.n.u. contraction, $U : \mathcal{H} \longrightarrow \mathcal{H}$ its minimal unitary dilation, and let $M_\Theta : \mathcal{K}_\Theta \longrightarrow \mathcal{K}_\Theta$ be the Szőkefalvi-Nagy–Foiaş model of T, where $\Theta = \Theta_T$ stands for the characteristic function of T. Denote by $V : \mathcal{H} \longrightarrow \mathcal{K}$ a unitary operator (see 1.5.2 above for the definition of $\mathcal{K}$ and the other notation), implicitely described in Subsections 1.3.3-1.3.5, which identify H with $\mathcal{K}_\Theta$, and U and T with $Z : \mathcal{K} \longrightarrow \mathcal{K}$ and M_Θ, respectively.

(a) Let $L_* = \operatorname{clos}((I - UT^*)H)$. Show that L_* is a wandering subspace of U (see Section A.1.5, Volume 1, for the definition), and, moreover,

$$V(L_*) = \begin{pmatrix} E_* \\ 0 \end{pmatrix}, \quad V(\mathcal{R}_T) := V\left((\operatorname{span}(U^n L_* : n \in \mathbb{Z}))^\perp\right) = \mathcal{R},$$

where $\mathcal{R}$ is defined in 1.5.3 above.

The space $\mathcal{R}_T$ is called the *residual subspace* of $\mathcal{H}$, and the restriction $U|\mathcal{R}_T$ is the *residual part* of U.

[*Hint:* passing to VUV^{-1} and VTV^{-1}, we find $(I-ZM_{\Theta}^*)(f,g)^{col} = (f(0),0)^{col}$ for every $(f,g)^{col} \in \mathcal{K}_\Theta$; next, by the formula of 1.3.7(1), we have

$$\begin{pmatrix} f \\ g \end{pmatrix} = P_\Theta \begin{pmatrix} e \\ 0 \end{pmatrix} = \begin{pmatrix} (I-\Theta\Theta(0)^*)e \\ -\Delta\Theta(0)^*e \end{pmatrix} \in \mathcal{K}_\Theta$$

for every $e \in E_*$, and hence

$$E_* \supset \operatorname{clos}(f(0) : (f,g)^{col} \in \mathcal{K}_\Theta) \supset \operatorname{clos}(I-\Theta(0)\Theta(0)^*)E_* = E_*$$

(since Θ is a pure contractive function); this shows the first formula, the second one is its obvious consequence.]

(b) *Asymptotic classification of contractions.* Define the *classes* $C_{\alpha\beta}$ of Hilbert space operators by the following equalities

$$\begin{aligned} C_{0\cdot} &= \left\{T : \left\|T\right\| \le 1, \lim_n \left\|T^n x\right\| = 0 \text{ for every } x\right\}, \\ C_{\cdot 0} &= \left\{T : \left\|T\right\| \le 1, \lim_n \left\|T^{*n} x\right\| = 0 \text{ for every } x\right\}, \\ C_{1\cdot} &= \left\{T : \left\|T\right\| \le 1, \lim_n \left\|T^n x\right\| > 0 \text{ for every } x \ne 0\right\}, \\ C_{\cdot 1} &= \left\{T : \left\|T\right\| \le 1, \lim_n \left\|T^{*n} x\right\| > 0 \text{ for every } x \ne 0\right\}, \end{aligned}$$

and

$$C_{\alpha\beta} = C_{\alpha\cdot} \cap C_{\cdot\beta}.$$

(i) Show that

(1) $T \in C_{\cdot 0}$ if and only if Θ_T is an inner function ($\Theta_T(\zeta)$ is isometric a.e. $\zeta \in \mathbb{T}$);

(2) $T \in C_{\cdot 1}$ if and only if Θ_T is an outer function ($\operatorname{clos}\Theta_T H^2(E) = H^2(E_*)$);

(3) $T \in C_{0\cdot}$ if and only if Θ_T is a $*$-inner function ($(\Theta_T)_* = \Theta_T(\bar{\zeta})^*$ is isometric a.e. $\zeta \in \mathbb{T}$);

(4) $T \in C_{1\cdot}$ if and only if Θ_T is a $*$-outer function ($\operatorname{clos}(\Theta_T)_* H^2(E_*) = H^2(E)$).

[*Hint:* the first two assertions follow from 1.5.2(b) and 1.5.3(b) since $T \in C_{\cdot 0} \Leftrightarrow \operatorname{Ker} X^* = \mathcal{K}_\Theta$, and $T \in C_{\cdot 1} \Leftrightarrow \operatorname{Ker} X^* = \{0\}$; the last two assertions now follow from the equality $(\Theta_T)_* = \Theta_{T^*}$, see 2.5.5(a) below.]

(ii) Let $T : H \longrightarrow H$ be a contraction. Show that there exist orthogonal decompositions (a) $H = H_0 \oplus H_1$, and (a*) $H = H_0^* \oplus H_1^*$ that triangularize the operator T in the sense

$$(a)\, T \in \begin{pmatrix} C_{0\cdot} & * \\ 0 & C_{1\cdot} \end{pmatrix}, \quad (a^*)\, T \in \begin{pmatrix} C_{\cdot 1} & * \\ 0 & C_{\cdot 0} \end{pmatrix}.$$

This also yields a triangular representation of the form

$$T \in \begin{pmatrix} C_{01} & * & * & * & * \\ 0 & C_{00} & * & * & * \\ 0 & 0 & C_{11} & * & * \\ 0 & 0 & 0 & C_{00} & * \\ 0 & 0 & 0 & 0 & C_{10} \end{pmatrix}.$$

[*Hint:* without loss of generality, T is a c.n.u. contraction that is unitarily equivalent to the adjoint M_Θ^* of a model operator M_Θ (which means that M_Θ is the model of T^*); in order to prove (a), set

$$H_0 = \{x \in H : \lim_n M_\Theta^{*n} x = 0\} = \operatorname{Ker} X^*,$$

and obtain (obviously) $M_\Theta^*|H_0 \in C_{0\cdot}$; next, observe that

$$\begin{aligned} X^*F &= \lim_n z^{m+n} M_\Theta^{*n} M_\Theta^{*m} F \\ &= z^m \cdot \lim_n z^n M_\Theta^{*n} P_{H_0} M_\Theta^{*m} F + z^m \cdot \lim_n z^n M_\Theta^{*n} P_{H_1} M_\Theta^{*m} F \\ &= z^m \cdot \lim_n z^n M_\Theta^{*n} P_{H_1} M_\Theta^{*m} F \\ &= z^m X^* P_{H_1} M_\Theta^{*m} F, \end{aligned}$$

and therefore,

$$\lim_m \|P_{H_1} M_\Theta^{*m} F\| \geq \lim_m \|X^* P_{H_1} M_\Theta^{*m} F\| = \|X^* F\| > 0$$

for every $F \in H_1 \backslash \{0\}$, which means that $(P_{H_1} M_\Theta^* | H_1) \in C_{1\cdot}$, and the triangulation (a) follows;

(a*) is (a) for T^* with $H_\alpha^* = H_{1-\alpha}(T^*)$;

the triangulation into a 5×5 matrix follows by an iteration of the previous triangulations and taking into account that 1) $T_{ii} \in C_{0\cdot}$ for every triangulation $T = (T_{ij})$ of a $C_{0\cdot}$ operator (with $T_{ij} = 0$ for $i > j$), and 2) $T_{11} \in C_{1\cdot}$ for every triangulation $T = (T_{ij})$ of a $C_{1\cdot}$ operator.]

(c) Suppose that $\sigma_p(T) \neq \mathbb{D}$ and that $T \in C_{\cdot 1}$. Show that $T \preceq U|\mathcal{R}_T$, and hence, if T is quasi-similar to a normal operator, it is quasi-similar to a unitary one, namely to $U|\mathcal{R}_T$.

[*Hint:* this is a unitarily invariant form of 1.5.3(f).]

(d) Suppose that $\sigma_p(T^*) \neq \mathbb{D}$ and that $T \in C_{1\cdot}$. Show that $U|\mathcal{R}_{T^*} \preceq T$, and hence, if T is quasi-similar to a normal operator, it is quasi-similar to a unitary one, namely to $U|\mathcal{R}_{T^*}$.

[*Hint:* this is (c) applied to T^*.]

(e) *(B. Szőkefalvi-Nagy and C. Foiaş, 1963)* Show that the following assertions are equivalent.

(1) T is quasi-similar to a unitary operator.

(2) Θ and Θ_* are outer functions (Θ_* is defined in 1.5.2(d)(ii)).

(3) $T \in C_{11}$.

(4) $U|\mathcal{R}_{T^*} \preceq T \preceq U|\mathcal{R}_T$, and hence, $U|\mathcal{R}_{T^*}$ and $U|\mathcal{R}_T$ are quasi-similar, and T is quasi-similar to the residual part $U|\mathcal{R}_T$ of its minimal unitary dilation U.

[*Hint:* to show that (1) $\Rightarrow$ (2), apply 1.5.3(b) to T and T^* and obtain that $\Theta = \Theta_T$ and Θ_{T^*} are outer, but $\Theta_{T^*} = \Theta_*$ (see 2.5.5(a) below); similarly, (2) $\Rightarrow$ (3); (3) $\Rightarrow$ (4) by (b) and (c); (4) $\Rightarrow$ (1) is obvious.]

(f) *(B. Sz.-Nagy and C. Foiaş, 1962)* Let T be a c.n.u. contraction such that $m(\sigma(T) \cap \mathbb{T}) = 0$. Show that the residual subspaces $\mathcal{R}_T$ and $\mathcal{R}_{T^*}$ (see (a) for the definition) are zero, and hence $T \in C_{00}$.

[*Hint:* since $\sigma(U|\mathcal{R}_T) \subset \sigma(T) \cap \mathbb{T}$ (see 1.5.2(f)) and $m(\sigma(T) \cap \mathbb{T}) = 0$, the restriction $U|\mathcal{R}_T$ has purely singular spectrum; since U has absolutely continuous spectrum (see 1.3.3), we get $\mathcal{R}_T = \{0\}$; now, Propositions 1.5.4(b)(i) and 1.5.2(a) imply that $T \in C_{\cdot 0}$; applying the same reasoning to T^*, we get $T \in C_{0\cdot}$, and the result follows.]

1.5.5. Similarity to a unitary operator: a function model approach. We use the notation as in 1.5.2 to 1.5.4. In particular, given a pure contractive function Θ from $H^\infty(E \longrightarrow E_*)$, M_Θ denotes the corresponding model operator on $\mathcal{K}_\Theta$. Recall that two operators A and B are *similar* if there exists an invertible operator Y such that $YA = BY$.

(a) Show that the following assertions are equivalent.

(1) M_Θ is similar to a co-isometry.

(2) M_Θ^* is similar to an isometry.

(3) There exists a constant $\delta > 0$ such that $\lim_n \|M_\Theta^{*n} F\| \geq \delta \|F\|$ for every $F \in \mathcal{K}_\Theta$.

(4) M_Θ^* is similar to the restiction $Z^*|Range(X^*)$ of the adjoint to the residual part $Z|\mathcal{R}$ of its minimal unitary dilation.

[*Hint:* obviously, (1) $\Leftrightarrow$ (2); to prove (2) $\Rightarrow$ (3), use the same reasoning as in 1.5.3(b) above; to show that (3) $\Rightarrow$ (4), employ 1.5.2(a) and observe that X^* is an isomorphic injection of $\mathcal{K}_\Theta$ into $\mathcal{R}$; (4) $\Rightarrow$ (1) is easy.]

(b) Show that the following assertions are equivalent.

(1) M_Θ is similar to a unitary operator.

(2) $\sigma_p(M_\Theta) \neq \mathbb{D}$ and there exists a constant $\delta > 0$ such that $\lim_n \|M_\Theta^{*n} F\| \geq \delta \|F\|$ for every $F \in \mathcal{K}_\Theta$.

(3) M_Θ is similar to the residual part $Z|\mathcal{R}$ of its minimal unitary dilation.

[*Hint:* as in (a), (1) $\Rightarrow$ (2); to show that (2) $\Rightarrow$ (3), use (a) and notice that now X^* is surjective because of 1.5.2(d)(i); (3) $\Rightarrow$ (1) is obvious.]

(c) *Lemma on neighbouring subspaces.* Let $H = K \oplus K^\perp = L \oplus L^\perp$ be two orthogonal decompositions of a Hilbert space H, and let $P = P_K|L$ and $P_\perp = P_{K^\perp}|L^\perp$.

I. Show that the following are equivalent.

(1) P is an isomorphism onto its range $PL \subset K$ (i.e., P is left invertible).

(2) $\|P_{K^\perp} P_L\| < 1$.

(3) $P_\perp$ is right invertible (i.e., $P_\perp^*$ is an isomorphism onto its range).

[*Hint:* notice that P is an isomorphism on its range if and only if $\|P_K x\|^2 \geq \delta^2 \|x\|^2$ for every $x \in L$ and for an $\delta > 0$; since $\|x\|^2 = \|P_K x\|^2 + \|P_{K^\perp} x\|^2$, this is equivalent to $\|P_{K^\perp}x\|^2 \le (1-\delta^2)\|x\|^2$, and hence to $\|P_{K^\perp}P_L\| < 1$; since $\|P_{K^\perp}P_L\| = \|(P_{K^\perp}P_L)^*\| = \|P_L P_{K^\perp}\|$, the equivalence (3) $\Leftrightarrow$ (2) also follows.]

II. Show that the following are equivalent.
(1') P is a bijection of L onto K.
(2') $\|P_{K^\perp}P_L\| < 1$ and $K \cap L^\perp = \{0\}$.
(3') $P_\perp$ is a bijection of $L^\perp$ onto $K^\perp$.

[*Hint:* use the previous series of equivalences and show that $\mathrm{Ker}(P_K|L)^* = \mathrm{Ker}(P_L|K) = K \cap L^\perp$ and $K \cap L^\perp = (K^\perp)^\perp \cap L^\perp$.]

(d) *(B. Szőkefalvi-Nagy and C. Foiaş, 1965 and 1973)* Let T be a c.n.u. contraction.

I. *Similarity to an isometry (respectively, to a co-isometry).* Show that the following are equivalent.

(1) T is similar to a co-isometry (respectively, to an isometry).
(2) There exists a constant $\delta > 0$ such that $\lim_n \|T^{*n}x\| \ge \delta\|x\|$ (respectively, $\lim_n \|T^n x\| \ge \delta\|x\|$) for every $x \in H$.
(3) Θ_T is right- (respectively, left-) invertible, that is, there exists a function $\Omega \in H^\infty(E_* \longrightarrow E)$ such that

$$\Theta_T(z)\Omega(z) = I_{E_*}$$

(respectively, $\Omega(z)\Theta_T(z) = I_E$) for every $z \in \mathbb{D}$.

[*Hint:* In the case of similarity to a coisometry, the equivalence (1) $\Leftrightarrow$ (2) is a unitarily invariant form of (a) above; the same for similarity to an isometry, but for T^* instead of T (see also 2.5.5(a) below);

in order to join assertion (3), observe that (2) is equivalent to the fact that $X^* = P_{\mathcal{R}_T}|H$ is an isomorphism of H onto the range X^*H, or (in its model form), $P_{\mathcal{R}}|\mathcal{K}_\Theta : \mathcal{K}_\Theta \longrightarrow \mathcal{R}$ is left invertible; by Lemma (c) above, the latter is the same as $P_{\mathcal{R}^\perp}|\mathcal{K}_\Theta^\perp$ is right invertible; now, use 1.5.2(e) getting that the multiplication operator $\Theta_{H^2} : H^2(E) \longrightarrow H^2(E_*)$ is right invertible; Theorem B.9.2.1 (Volume 1) yields the solvability of the mentioned Bezout equation.]

II. *Similarity to a unitary operator.* Show that the following are equivalent.

(1) T is similar to a unitary operator.
(2) $\sigma_p(T) \ne \mathbb{D}$ and there exists a constant $\delta > 0$ such that $\lim_n \|T^{*n}x\| \ge \delta\|x\|$ for every $x \in H$.
(3) $\sigma_p(T^*) \ne \mathbb{D}$ and there exists a constant $\delta > 0$ such that $\lim_n \|T^n x\| \ge \delta\|x\|$ for every $x \in H$.
(4) $\Theta_T(\zeta)$ is an invertible operator for every $\zeta \in \mathbb{D}$ and $\Theta^{-1} \in H^\infty(E_* \longrightarrow E)$.
(5) T is similar to the residual part $U|\mathcal{R}_T$ of its minimal unitary dilation U.

[*Hints:* the equivalences (1) $\Leftrightarrow$ (2) $\Leftrightarrow$ (5) are unitarily invariant forms of (b) above; the same for (1) $\Leftrightarrow$ (3) but for T^* instead of T (see also 2.5.5(a) below); (4) follows from (2), (3) and the part I (since $\Theta_T(\zeta)$ is invertible for every $\zeta \in \mathbb{D}$, being right and left invertible).]

1.5.6. Similarity to a unitary operator: a harmonic analysis approach. The first assertion below is of a general nature and is not related to operators.

(a) *Banach generalized limits* (S. Banach, 1932). Let $x = (x_k)_{k\geq 0} \in l^\infty$,

$$p(x) = \overline{\lim}_n \left| \frac{1}{n+1} \sum_{k=0}^{n} x_k \right|,$$

and let X be the subset of l^∞, where the limit

$$M(x) = \lim_n \frac{1}{n+1} \sum_{k=0}^{n} x_k$$

exists.

(i) Show that p is a continuous seminorm on l^∞, that X is a closed subspace of l^∞, and that $|M(x)| = p(x)$ for $x \in X$.

(ii) Let $c = \{x \in l^\infty : \exists \lim_k x_k\}$ and let $L(x) = \lim_k x_k$ for $x \in c$; show that $c \subset X$ and that $M(x) = L(x)$ for every $x \in c$.

(iii) Denote by GLIM a Hahn-Banach extension of M to the space l^∞ satisfying

$$|\operatorname{GLIM}(x)| \leq p(x)$$

for every $x \in l^\infty$ (first, consider M and GLIM on the space of real sequences $\operatorname{Re}(l^\infty)$, then use the complexification of GLIM).

Show that GLIM satisfies the following properties.

1) GLIM is a linear continuous functional on l^∞ with $\|\operatorname{GLIM}\| = 1$; moreover, $|\operatorname{GLIM}(x)| \leq \overline{\lim}_k \left|x_k\right|$ for every $x \in l^\infty$;

2) if $x_k \in \mathbb{R}_+$ for $k \geq N$, then $\operatorname{GLIM}(x) \geq 0$;

[*Hint:* if $x_k \geq 0$ for all k, then $\frac{1}{2}\|x\|_\infty - \operatorname{GLIM}(x) \leq \|(x_k - \frac{1}{2}\|x\|_\infty)_k\|_\infty \leq \frac{1}{2}\|x\|_\infty$.]

2) $\operatorname{GLIM}(Sx) = \operatorname{GLIM}(x)$ for every $x \in l^\infty$, where S is the shift operator on l^∞;

3) $\operatorname{GLIM}((x_k y_k)_k) = \operatorname{GLIM}(x)L(y)$ for $x \in l^\infty$, $y \in c$.

(b) *(B. Szőkefalvi-Nagy, 1947)* Let T be an invertible operator on a Hilbert space H such that $\sup_{n\in\mathbb{Z}} \|T^n\| =: K < \infty$; let also GLIM be a Banach generalized limit (see (a) above). Show that

(i) $(x, y)_T := \operatorname{GLIM}((T^k x, T^k y)_{k\geq 0})$ is a scalar product on H satisfying

$$K^{-2}(x, x) \leq (x, x)_T \leq K^2(x, x)$$

for every $x \in H$;

(ii) $U : x \longmapsto Tx$ is a unitary transformation on the Hilbert space $H_T = (H, (\cdot, \cdot)_T)$;

(iii) the operator $j : H \longrightarrow H_T$ defined by $jx = x$ is an isomorphism, and $Uj = jT$. Therefore, T is similar to U.

(c) Let T be an invertible operator on a Hilbert space H. Show that the following assertions are equivalent.

(i) T is similar to a unitary operator.

(ii) For every $x, y \in H$, there exists a Borel complex measure $\mu_{x,y}$ on $\mathbb{T}$ such that $(T^n x, y) = \hat{\mu}_{x,y}(n)$ for $n \in \mathbb{Z}$.

(iii) $\sup_{n\in\mathbb{Z}} |(T^n x, y)| < \infty$ for every $x, y \in H$.

(iv) *(B. Szőkefalvi-Nagy, 1947)* $\sup_{n\in\mathbb{Z}} \|T^n\| < \infty$.

[*Hint:* (i) $\Rightarrow$ (ii) since $T = VUV^{-1}$ implies $(T^k x, y) = (U^k V^{-1}x, V^* y) = \int_{\mathbb{T}} \zeta^k (V^{-1}x(\zeta), V^* y(\zeta))\, d\mu(\zeta)$ by the spectral theorem 1.5.1(k); obviously, (ii) $\Rightarrow$ (iii) and (iii) $\Rightarrow$ (iv); (iv) $\Rightarrow$ (i) by (b) above.]

(d) *Resolvent and power growth criteria.* Let T be an invertible operator on a Hilbert space H. Consider the following series of conditions on T.

Sim(T) : T is similar to a unitary operator;

wPMean(T) : $\sup_{0\le r<1} \|(\mathcal{P}_r(T)(\cdot)x, y)\|_1 < \infty$ for every $x, y \in H$, i.e. the weak *Poisson mean boundedness* of T, where

$$\mathcal{P}_r(T)(\zeta) = \sum_{n\in Z} r^{|n|}\zeta^n T^n$$

is supposed to be convergent for all $0 \le r < 1$, $\zeta \in \mathbb{T}$.

PowBd(T) : $\sup_{n\ge 0} \|T^n x\| < \infty$, i.e. the *power boundedness* of T;

AMean^2(T) : $\sup_{n\ge 0}(n+1)^{-1}\sum_{k=0}^{n} \|T^k x\|^2 < \infty$ for every $x \in H$, i.e. the *arithmetic mean square boundedness* for T;

LMean^2 $RG(T)$:

$$\begin{aligned} \left\|R_{r\cdot}(T)x\right\|_{H^2}^2 &= \int_{\mathbb{T}} \left\|R_{r\zeta}(T)x\right\|^2 dm(\zeta) \\ &= \sum_{k\ge 0} r^{-2(k+1)} \left\|T^k x\right\|^2 \\ &\le c^2\|x\|^2(r^2-1)^{-1} \end{aligned}$$

for every $r > 1$ and every $x \in H$, i.e. the *linear mean square resolvent growth* for T;

LRG(T) : $\|R_\lambda x\| \le c\|x\|(|\lambda| - 1)^{-1}$ for $|\lambda| > 1$ and for every $x \in H$, i.e. the $\mathbb{D}_-$ *linear resolvent growth* for T as $|\lambda| \longrightarrow 1$, where R_λ stands for the resolvent of T,

$$R_\lambda = R_\lambda(T) = (\lambda I - T)^{-1};$$

(i) Show that

(1) Sim(T) $\Leftrightarrow$ Sim(T^*) $\Leftrightarrow$ Sim(T^{-1}),

(2) LRG(T) $\Leftrightarrow$ LRG(T^*),

(3) wPMean(T) $\Leftrightarrow$ wPMean(T^*) $\Leftrightarrow$ wPMean(T^{-1}),

(4) Sim(T) $\Rightarrow$ PowBd(T) $\Rightarrow$ AMean^2(T) $\Rightarrow$ LMean^2 $RG(T)$ $\Rightarrow$ LRG(T),

(5) AMean^2(T) $\Rightarrow$ AMean^1(T) $\Rightarrow$ LRG(T), where AMean^1(T) means the condition

$$\sup_{n\ge 0}(n+1)^{-1}\sum_{k=0}^{n} \|T^k x\| < \infty, \qquad \text{for every } x \in H;$$

moreover, none of the converses hold.

[*Hint:* the implications $\mathrm{AMean}^2(T) \Rightarrow \mathrm{LMean}^2\, RG(T)$ and $\mathrm{AMean}^1(T) \Rightarrow \mathrm{LRG}(T)$ follow from the classical *Abel transform formula*:

$$\sum_{n\geq 0} a_n x^n = (1-x)\sum_{n\geq 0}(\sum_{k=0}^{n} a_k)x^n,$$

whereas the implication $\mathrm{LMean}^2\, RG(T) \Rightarrow \mathrm{LRG}(T)$ arises from the standard growth estimate

$$\|f(\zeta)\|_X \leq (1-|\zeta|^2)^{-1/2}\|f\|_{H^2(E)}$$

for E valued H^2-functions (for the scalar case see A.3, and especially A.3.10.1(d), Volume 1); namely, take $\lambda \in \mathbb{D}$, $\rho^2 = \frac{1}{2}(1+|\lambda|^2)$, $f(z) = R_{1/\rho z}x$ for $|z|<1$, $x \in H$, and apply the above inequality for $\zeta = \lambda/\rho$:

$$\Big\|R_{1/\lambda}x\Big\| \leq (1-|\zeta|^2)^{-1/2}\Big\|f\Big\|_{H^2} = (\frac{1+|\lambda|^2}{1-|\lambda|^2})^{1/2}\Big\|f\Big\|_{H^2} \leq c\cdot\frac{1+|\lambda|^2}{1-|\lambda|^2}\Big\|x\Big\|;$$

for counterexamples to the last claim see the method of (e) below.]

(ii) Show that the following are equivalent.

(A) $\mathrm{Sim}(T)$;

(B) $\mathrm{wPMean}(T)$;

(C) *(S. Naboko, 1984)* $\mathrm{LMean}^2\, RG(T)$ and $\mathrm{LMean}^2\, RG(T^{*-1})$;

(D) *(J. van Casteren, 1983)* $\mathrm{LRG}(T)$, $\mathrm{LMean}^2\, RG(T^{*-1})$, and $\mathrm{LMean}^2\, RG(T^{-1})$;

(E) any similar (or stronger, see (i)) combination of the above properties for T^*, or T^{-1}, or T^{*-1} in place of T.

[*Hint:* (A) is equivalent to (B), since $\mathrm{wPMean}(T)$ is equivalent to (c)(ii) above (see A.3.3 and the proof of A.3.4.1, Volume 1);
$\mathrm{Sim}(T) = \mathrm{Sim}(T^{*-1})$ obviously implies (C), (D), and (E);
(C) implies (B) since

$$\mathcal{P}_r(T)(\zeta) = -\frac{1-r^2}{r\zeta}TR_{r\bar{\zeta}}(T)R_{1/r\zeta}(T) = \frac{1-r^2}{r^2}R_{1/r\bar{\zeta}}(T^{-1})R_{1/r\zeta}(T),$$

and hence

$$\|(\mathcal{P}_r(T)(\cdot)x,y)\|_1 \leq \frac{1-r^2}{r^2}\|R_{1/r\cdot}(T)x\|_2\|R_{1/r\cdot}(T^{*-1})y\|_2 \leq \text{const};$$

similarly, (D) implies (B) since

$$\begin{aligned}\Big(\frac{1}{\lambda}-z\Big)^{-1} &= \frac{\lambda}{z}\cdot\frac{1}{z^{-1}-\lambda} = \frac{\lambda}{z}\cdot\frac{1}{\overline{\lambda}^{-1}-z^{-1}}\cdot\frac{\overline{\lambda}^{-1}-z^{-1}}{z^{-1}-\lambda}\\ &= \frac{1}{z\overline{\lambda}}\cdot(\overline{\lambda}^{-1}-z^{-1})^{-1}\{-1+\frac{1-|\lambda|^2}{\lambda}\cdot(\lambda^{-1}-z)^{-1}\},\end{aligned}$$

and therefore

$$\begin{aligned}\mathcal{P}_r(T)(\zeta) &= \frac{1-r^2}{r^2}R_{1/r\bar{\zeta}}(T^{-1})R_{1/r\zeta}(T)\\ &= \frac{1-r^2}{r^4}R_{1/r\bar{\zeta}}(T^{-1})T^{-1}R_{1/r\bar{\zeta}}(T^{-1})\Big\{(1-r^2)R_{1/r\zeta}(T)-r\zeta I\Big\},\end{aligned}$$

which implies, as before,

$$|(\mathcal{P}_r(T)(\zeta)x,y)| \leq c(1-r^2)\|R_{1/r\zeta}(T)x\|\cdot\|R_{1/r\zeta}(T^{*-1})y\|,$$

and the result follows.]

(iii) *Corollaries.* (A) *(S. Naboko, J. van Casteren, 1983/1984).* PowBd(T) and LRG(T^{-1}), that is,

$$\sup_{n\geq 0}\|T^n\| < \infty \text{ and } \|R_\lambda(T)\| \leq c(1-|\lambda|)^{-1} \text{ for } |\lambda| < 1,$$

imply Sim(T) (which improves (b) above claiming that (PowBd(T)& PowBd(T^{-1})) $\Rightarrow$ Sim(T)).

(B) LRG(T) and uLMean^2 $RG(T^{-1})$ yield Sim(T), where we have used uLMean^2 $RG(T^{-1})$ to designate the following *uniform form* of LMean^2 $RG(T^{-1})$:

$$\int_{\mathbb{T}} \left\|R_{r\zeta}(T)\right\|^2 dm(\zeta) \leq c^2(1-r^2)^{-1}, \quad r < 1;$$

(C) the uniform versions uAMean$^2(T)$ and uAMean$^2(T^{-1})$ imply Sim(T) (compare with (e)(i)); here uAMean$^2(T)$ means $\sup_{n\geq 0}(n+1)^{-1}\sum_{k=0}^n \|T^k\|^2 < \infty$.

[*Hint:* uLMean^2 $RG(T^{-1})$ = uLMean^2 $RG(T^{*-1})$ implies LMean^2 $RG(T^{-1})$ and LMean^2 $RG(T^{*-1})$, so (A), (B) follow from (ii)(D); (C) results from (ii)(C).]

(e) *Counterexamples.* (i) Show that there exists an operator T satisfying AMean$^2(T)$ and AMean$^2(T^{-1})$ that is not similar to a unitary operator.

[*Hint:* let $T : l^2(\mathbb{Z}) \longrightarrow l^2(\mathbb{Z})$ be a weighted bilateral shift defined on the elements of the 0-1 standard basis by $Te_k = \lambda_k e_{k+1}$, $k \in \mathbb{Z}$, where $\lambda_0 = 1$, $\lambda_k^{-1} = \lambda_{-k}$ for all $k \in \mathbb{Z}$, and

$$\lambda_{-k} = \begin{cases} \mu_s & \text{for } k = M_s = \sum_{j=1}^s N_j \\ 1 & \text{for } k \neq M_s \ (k > 0) \end{cases},$$

where $\mu_s \searrow 1$ and $N_j \nearrow \infty$ are such that $\prod_{s\geq 1}\mu_s = \infty$, and

$$\Big(\prod_{j=1}^{s-1}\mu_j\Big)^2 M_{s-1} \leq N_s$$

for $s \geq 1$; now, taking $x = \sum_{j\in\mathbb{Z}} x_j e_j \in l^2(\mathbb{Z})$, observe that

$$\sum_{k=0}^n \|T^k x\|^2 = \sum_{k=0}^n \sum_{j\geq 0} |x_j \lambda_j \dots \lambda_{j+k-1}|^2 = \sum_{j\geq 0} |x_j|^2 S_{j,n},$$

where $S_{j,n} = \sum_{k=0}^n (\lambda_j \dots \lambda_{j+k-1})^2$; supposing that $N_s \leq n < N_{s+1}$, check consecutively

1) for $j \leq -M_{s+1}$, each product $\lambda_j \dots \lambda_{j+k-1}$ contains at most one factor μ_r, and hence $S_{j,n} \leq n\mu_1^2$;

2) for $-M_{s+1} < j \leq -M_s$: $\lambda_j \dots \lambda_{j+k-1} = 1$ for $j+k-1 < -M_s$;
$\lambda_j \dots \lambda_{j+k-1} = \mu_s$ for $-M_s \leq j+k-1 < -M_{s-1}$;
$\lambda_j \dots \lambda_{j+k-1} \leq \prod_{l=1}^{s-1}\mu_l$ for $-M_{s-1} \leq j+k-1 < M_{s-1}$;
$\lambda_j \dots \lambda_{j+k-1} \leq \mu_s$ for $j+k-1 \geq M_{s-1}$,
and hence $S_{j,n} \leq n + N_s\mu_s^2 + (\prod_{l=1}^{s-1}\mu_l)^2 2M_{s-1} + n\mu_s^2 \leq n + n\mu_1^2 + 2n + n\mu_1^2$;

3) for $j > -M_s$, similarly, $S_{j,n} \leq n + (\prod_{l=1}^{s-1}\mu_l)^2 2M_{s-1} + n \leq 4n$; and finally, $S_{j,n} \leq cn$ for every j and n, and so we get AMean$^2(T)$; the operator T^{-1}

has exactly the same structure, which gives $\mathrm{AMean}^2(T^{-1})$; on the other hand, T is not similar to a unitary operator since

$$\|T^{M_s}\| \geq \|T^{M_s} e_{M_s}\| = \prod_{l=1}^{s} \mu_l,$$

which tends to infinity.]

(ii) In particular, (A) $\mathrm{LMean}^2\, RG(T)$ and $\mathrm{LMean}^2\, RG(T^{-1})$ do not imply $\mathrm{Sim}(T)$ *(S. Naboko, 1984)*,

(B) *a fortiori*, $\mathrm{LRG}(T)$ and $\mathrm{LRG}(T^{-1})$ do not imply $\mathrm{Sim}(T)$ *(A. Markus, 1966)*, and $\mathrm{LRG}(T)$ and $\mathrm{LMean}^2\, RG(T^{*-1})$ do not imply $\mathrm{Sim}(T)$.

[*Hint:* use (i) and (d)(i).]

(f) *Similarity to an isometry.*

(i) Let $A_n \searrow$, $A_n : H \longrightarrow H$, be a decreasing sequence of positive Hilbert space operators (i.e. $0 \leq (A_{n+1}x, x) \leq (A_n x, x)$ for every $n \geq 1$ and every $x \in H$). Show that $(A_n)_{n\geq 1}$ converges strongly to a positive operator A ($\lim_n \|A_n x - Ax\| = 0$ for every $x \in H$); we write

$$(s) \lim_n A_n = A.$$

[*Hint:* clearly, there exists a weak limit $\lim_n (A_n x, y) = (Ax, y)$, and $B_n = A_n - A \searrow 0$; hence, $\|B_n^{1/2} x\|^2 = (B_n x, x) \longrightarrow 0$ as $n \longrightarrow \infty$, whence $\lim_n \|B_n^{1/2} B_n^{1/2} x\| = 0$ for every $x \in H$.]

(ii) *(R. Douglas, 1968)* Let $T : H \longrightarrow H$ be a Hilbert space contraction. Show that there exists strong limits $(s) \lim_n T^{*n} T^n = A$ and $(s) \lim_n T^n T^{*n} = A_*$. Show that

$$(T \text{ is similar to an isometry }) \Leftrightarrow (A \text{ is invertible }) \Leftrightarrow (\inf_n \|T^n x\| \geq \delta \|x\|)$$

$$(T \text{ is similar to a coisometry }) \Leftrightarrow (A_* \text{ is invertible }) \Leftrightarrow (\inf_n \|T^{*n} x\| \geq \delta \|x\|).$$

[*Hint:* $T^* A T = A$, and so, if A is invertible, then $A^{-1/2} T^* A^{1/2} A^{1/2} T A^{-1/2} = I$; conversely, if $T = BVB^{-1}$, where V is an isometry, then we have $\|T^n x\| = \|BV^n B^{-1} x\| \geq \|B\| \cdot \|B^{-1}\| \cdot \|x\|$.]

1.5.7. More about unitary dilations. (a) *A simple proof of the unitary dilation theorem.* (i) Let $T : H \longrightarrow H$ be a contraction on a Hilbert space H which we identify by $x \longmapsto (x, 0, 0, \dots)$ with the subspace of the space $H' =: l^2(H)$ of sequences $(x_k)_{k\geq 0}$ with $\sum_{k\geq 0} \|x_k\|^2 < \infty$. Setting

$$V(x_0, x_1, \dots) = (Tx_0, D_T x_0, x_1, \dots),$$

show that V is an isometric dilation of T. Choosing an arbitrary unitary operator $V' : H' \longrightarrow (H' \ominus VH') \oplus H'$, show that $U = V \oplus V' : H' \oplus H' \longrightarrow H' \oplus H'$ is a unitary dilation of T and its suitable restriction is a minimal unitary dilation.

[*Hint:* take $U|\mathcal{H}$ for such a restriction, where $\mathcal{H} = \mathrm{span}(U^k(H \oplus 0) : k \in \mathbb{Z})$.]

(ii) Show that, similarly, any operator of the form

$$V_T(x_0, x_1, \dots) = (Tx_0, D_T x_0, W(x_1, \dots))$$

is an isometric dilation of T, whenever $W : l^2(H) \longrightarrow l^2(H)$ is an isometry.

(iii) Let $V : l^2(H) \longrightarrow l^2(H)$ be an isometric dilation of T, and let $\Phi_k : l^2(H) \longrightarrow l^2(H)$ be any isometries with $(\Phi_k x)_0 = x_0$ for every $x \in l^2(H)$. Show that $\Phi_1 V \Phi_2$ is an isometric dilation of T.

(b)* *Unitary A-dilations (H. Langer, 1967).* Let A be a selfadjoint operator $A : H \longrightarrow H$ satisfying $A \geq \delta I$ for a $\delta > 0$, and let

$$(x, y)_{A-I} = ((A - I)x, y)$$

be a new (indefinite) scalar product on H. The following are equivalent.

(i) There exists a unitary operator U on a larger Hilbert space $\mathcal{H} \supset H$ (called a *unitary A-dilation* of T) such that

$$A^{-1/2} T^n A^{-1/2} = P_H U^n | H$$

for every $n \geq 0$, or equivalently,

$$(T^n x, y) = (U^n A^{1/2} x, A^{1/2} y)$$

for every $n \geq 0$ and every $x, y \in H$.

(ii) $\|\zeta T x\|^2 \leq \|x\|^2 + \|x - \zeta T x\|_{A-I}$ for every $x \in H$ and every $\zeta \in \mathbb{D}$.

(c) *Unitary ρ-dilations (B. Szőkefalvi-Nagy and C. Foiaş, 1966).* Let $A = \rho I$ with $\rho > 0$. In this case, an A-dilation from (b)(ii) is called a *unitary ρ-dilation* of T; denote by $\mathcal{C}_\rho$ the *set of operators having a unitary ρ-dilation.* Prove the following statements.

(i) $T \in \mathcal{C}_\rho$ if and only if

$$\frac{2-\rho}{\rho} \|\zeta T x\|^2 + 2 \frac{\rho - 1}{\rho} \operatorname{Re}(\zeta T x, x) \leq \|x\|^2$$

for every $x \in H$ and every $\zeta \in \mathbb{D}$.

(ii) $T \in \mathcal{C}_1 \Leftrightarrow \|T\| \leq 1$.

(iii) *(C. Berger, 1965)* $T \in \mathcal{C}_2 \Leftrightarrow w(T) \leq 1$, where

$$w(T) = \sup\{|(Tx, x)| : x \in H, \|x\| \leq 1\}$$

is the so-called *numerical radius* of T.

(iv) For $\rho \in (0, 2) \setminus \{1\}$,

$$T \in \mathcal{C}_\rho \Leftrightarrow \|\lambda I - T\| \leq \frac{|\lambda|}{|\rho - 1|} \text{ for every } |\lambda| \geq \frac{|\rho - 1|}{2 - \rho},$$

and for $1 < \rho < 2$,

$$T \in \mathcal{C}_\rho \Leftrightarrow \|\lambda I - T\| \leq |\lambda| + 1 \text{ for every } |\lambda| \geq \frac{\rho - 1}{2 - \rho} = \frac{1}{2 - \rho} - 1.$$

(v) For $\rho > 2$,

$$T \in \mathcal{C}_\rho \Leftrightarrow \left(\|(\lambda I - T)x\| \geq \frac{|\lambda|}{\rho - 1} \|x\| \text{ for } |\lambda| \geq \frac{\rho - 1}{\rho - 2} \text{ and } x \in H \right),$$

$$T \in \mathcal{C}_\rho \Leftrightarrow \left(\sigma(T) \subset \overline{\mathbb{D}} \text{ and } \|R_\lambda(T)\| \leq \frac{1}{|\lambda| - 1} \text{ for } 1 < |\lambda| < \frac{\rho - 1}{\rho - 2} \right).$$

[*Hint:* (i) is a straightforward rewriting of (b)(ii); (ii) and (iii) are immediate consequences of (i); next, justify (iv) and the first equivalence of (v), multiply (i) by $\frac{\rho}{(\rho-2)|\zeta|^2}$ and set $\lambda = \frac{\rho-1}{(\rho-2)\zeta}$;

for the second equivalence of (v), deduce from (b)(i) that

$$r(T) = \lim_n \|T^n\|^{1/n} \leq 1$$

for every $T \in \mathcal{C}_\rho$, and hence $\sigma(T) \subset \overline{\mathbb{D}}$; next, use the first equivalence of (v) and set $r = \frac{\rho-1}{\rho-2}$ and $1 < |\lambda| < r$ to obtain

$$\|(\lambda I - T)x\| \geq \|(\frac{\lambda r}{|\lambda|}I - T)x\| - \|(\lambda - \frac{\lambda r}{|\lambda|})x\| \geq (|\lambda| - 1)\|x\|$$

for every $x \in H$; conversely, if the last inequality holds, we get $\|R_\lambda(T)\| \leq \frac{1}{r-1} = \frac{\rho-1}{|\lambda|}$ for $|\lambda| = r$, and hence (by the maximum principle) $\|R_\lambda(T)\| \leq \frac{\rho-1}{|\lambda|}$ for $|\lambda| \geq r$, which implies that $T \in \mathcal{C}_\rho$ by the first equivalence.]

(d) $\mathcal{C}_{\rho'} \subset \mathcal{C}_\rho$ and $\mathcal{C}_\rho \neq \mathcal{C}_{\rho'}$ for $0 < \rho' < \rho$ (and $\dim H > 1$).

[*Hint:* for the inclusion, rewrite (c)(i) as $(\rho - 2)\|(I - \zeta T)x\|^2 + 2\,\mathrm{Re}((I - \zeta T)x, x) \geq 0$; in order to show that the inclusion is strict, it suffices to exhibit an operator $T \in \mathcal{C}_\rho$ with $\|T\| = \rho$ (since $\|T\| \leq \rho$ for every $T \in \mathcal{C}_\rho$); to this end, write H in the form $H = L \oplus H'$ with $\dim L = 2$ and identify it unitarily with a subspace of $\mathcal{H} = l^2(\mathbb{Z}) \oplus l^2(\mathbb{Z}, H')$ setting $L = \mathcal{L}in(e_0, e_1) \subset l^2(\mathbb{Z})$ and $H' = e_0 \cdot H'$, where (e_k) stands for the standard orthonormal basis of $l^2(\mathbb{Z})$; next, define U as the bilateral shift on $\mathcal{H}$, set $T = \rho P_H U|H$ and check that $T^2 = 0$, $P_H U^2|H = 0$ and $\|Te_0\| = \rho$.]

(e)* *(B. Szőkefalvi-Nagy and C. Foiaş, 1967)* Let $T \in \mathcal{C}_\rho$ where $\rho > 0$. Then T is similar to a contraction; moreover, if $\rho \geq 1$, then there exists an invertible operator V such that $\|V^{-1}TV\| \leq 1$ and $\|V^{-1}\| \cdot \|V\| \leq 2\rho - 1$.

(f) Show that $T \in \bigcup_{\rho > 0} \mathcal{C}_\rho$ implies $\sup_{n \geq 0} \|T^n\| < \infty$, but the converse is false.

[*Hint:* compare (e) and B.1.6.6 (Volume 1).]

(g)* *(J. Holbrook, 1968)* For every T with $\sup_{n \geq 0} \|T^n\| < \infty$ there exists a sequence of operators $(T_j)_{j \geq 1}$ such that $T_j \in \bigcup_{\rho > 0} \mathcal{C}_\rho$ and $\lim_j \|T - T_j\| = 0$.

1.5.8. Dilations of several commuting operators. (a) *Noncommuting unitary and isometric dilations.* Let $T_k : H \longrightarrow H$ be contractions, $k = 1, 2, \ldots$. Show that

$$T_1^{n_1} T_2^{n_2} \cdots T_k^{n_k} = P_H U_{T_1}^{n_1} U_{T_2}^{n_2} \cdots U_{T_k}^{n_k}|H$$

for every k and every $n_i \geq 0$, where U_T stands for the unitary dilation constructed in 1.1.18. Similarly,

$$T_1^{n_1} T_2^{n_2} \cdots T_k^{n_k} = P_H V_{T_1}^{n_1} V_{T_2}^{n_2} \cdots V_{T_k}^{n_k}|H$$

for isometric dilations V_{T_i} from 1.5.7(a)(ii) above.

(b) *Commuting unitary dilations (T. Ando, 1963).* Let $T_k : H \longrightarrow H$ $(k = 1, 2)$ be commuting contractions, i.e. contractions satisfying $T_1 T_2 = T_2 T_1$. Show that there exist commuting unitary dilations, that is, unitary operators U_k $(k = 1, 2)$ defined on the same space $\mathcal{H} \supset H$ such that $U_1U_2 = U_2U_1$ and

$$T_1^{n_1} T_2^{n_2} = P_H U_1^{n_1} U_2^{n_2} \Big| H$$

for every $n_i \geq 0$.

[*Hint:* 1) let V_{T_k} $(k = 1, 2)$ be isometric dilations of T exhibited in 1.5.7(a) (ii) with $W = S$ (the shift operator on $l^2(H)$); further, let Ψ be a unitary operator on the space $H \oplus H \oplus H \oplus H$ and

$$\Phi(x_0, x_1, \dots) = (x_0, \Psi(x_1, \dots, x_4), \Psi(x_5, \dots, x_8), \dots)$$

be a unitary operator on $l^2(H)$; show that

$$V_1 = \Phi V_{T_1} \text{ and } V_2 = V_{T_2}\Phi^{-1}$$

are isometric dilations of T satisfying the equalities

$$\begin{aligned} V_1V_2(x_0, x_1, \dots) &= (T_1T_2x_0, \Psi(D_{T_1}T_2x_0, 0, D_{T_2}x_0, 0), x_1, x_2, \dots), \\ V_2V_1(x_0, x_1, \dots) &= (T_2T_1x_0, D_{T_2}T_1x_0, 0, D_{T_1}x_0, 0, x_1, x_2, \dots); \end{aligned}$$

2) show that

$$\|D_{T_1}T_2x_0\|^2 + \|D_{T_2}x_0\|^2 = \|x_0\|^2 - \|T_1T_2x_0\|^2 = \|D_{T_2}T_1x_0\|^2 + \|D_{T_1}x_0\|^2$$

and hence the map $\Psi' : (D_{T_1}T_2x_0, 0, D_{T_2}x_0, 0) \longrightarrow (D_{T_2}T_1x_0, 0, D_{T_1}x_0, 0)$ is a well defined isometry;

3) show that $\dim(\mathrm{Dom}\,\Psi')^\perp = \dim(\Psi'(Dom\Psi'))^\perp$, where $\mathrm{Dom}\,\Psi'$ stands for the subspace where Ψ' is defined, and hence Ψ' can be completed to a unitary operator Ψ on $H \oplus H \oplus H \oplus H$;

4) show that with the above defined Ψ the operators V_k are commuting isometric dilations of T_k;

5) show that any pair of commuting isometric operators V_k can be extended to a commuting pair of unitary operators U_k (consider the Wold–Kolmogorov decompositions of V_k).]

(c) *Nonexistence of commuting unitary dilations (S. Parrott, 1970.)* Show that there exist three pairwise commuting contractions $T_j : H \longrightarrow H$, $j = 1, 2, 3$, without commuting unitary dilations.

More precisely, let $V_j : K \longrightarrow K$ $(j = 1, 2, 3)$ be unitary operators on a Hilbert space K, $\dim K > 1$, with $V_3 = I$, and let $H = K \oplus K$ and

$$T_j(x, y) = (0, V_jx), \quad j = 1, 2, 3.$$

Show that

(i) $\|T_j\| = 1$ and $T_iT_j = 0$ for $i, j = 1, 2, 3$;

(ii) if $\mathcal{H} \supset H$ and $U_j : \mathcal{H} \longrightarrow H$ are unitary operators with

$$P_H U_j | H = T_j$$

then $U_j(x, 0) = T_j(x, 0) = (0, V_jx)$ for every j and $x \in K$;

[*Hint:*$\|U_j(x, 0)\| = \|x\|_K = \|T_j(x, 0)\|$.];

(iii) since $U_3(x, 0) = (0, x)$, we get $U_iU_3^{-1}U_j(x, 0) = (0, V_iV_jx)$ for $x \in K$, and therefore, V_1, V_2 commute if U_1, U_2 commute;

(iv) finally, in order to construct the counterexample, take V_i with $V_1V_2 \neq V_2V_1$.

(d)* *Regular commuting unitary dilations (S. Brehmer, 1961).* Let $(T_i)_{i\in I}$ be a commutative family of contractions $T_i : H \longrightarrow H$, and let $(U_i)_{i\in I}$ be a commutative family of unitary operators on a larger space $\mathcal{H}$; the latter family is called a *regular unitary dilation* of $(T_i)_{i\in I}$ if

$$\prod_{i\in\sigma} T_i^{*k_i} \prod_{j\in\sigma'} T_j^{n_j} = P_H\Big(\prod_{i\in\sigma} U_i^{*k_i} \prod_{j\in\sigma'} U_j^{n_j}\Big)\Big|H$$

for every $k_i \geq 0, n_j \geq 0$ and for every finite subsets $\sigma, \sigma' \subset I$ with $\sigma \cap \sigma' = \emptyset$.

The following are equivalent.

(i) There exists a regular unitary dilation of $(T_i)_{i\in I}$.

(ii) $S(\sigma) := \sum_{\sigma'\subset\sigma}(-1)^{|\sigma'|}(T^{\chi_{\sigma'}})^* T^{\chi_{\sigma'}} \geq 0$ for every finite subset $\sigma \subset I$, where σ' runs over all subsets of σ and $T^{\chi_{\sigma'}}$ means $T_{i_1}\dots T_{i_n}$ for $\sigma' = \{i_1, \dots, i_n\}$; or, equivalently,

$$\begin{aligned}&(S(\sigma)x, x)\\ &\quad = \|x\|^2 - \sum_{1\leq i\leq n} \|T_i x\|^2 + \sum_{1\leq i<j\leq n} \|T_i T_j x\|^2 - \dots + (-1)^n \|T_1 T_2 \cdots T_n x\|^2\\ &\quad \geq 0\end{aligned}$$

for every $x \in H$ and every subfamily $\{T_1, \dots, T_n\} \subset (T_i)_{i\in I}$.

(e) *(S. Brehmer, 1961)* Show that regular commuting unitary dilations exist in each of the following cases:

(i) $(T_i)_{i\in I}$ is a commutative family of isometries;

[*Hint:* use (d) and the fact that $\sum_{\sigma'\subset\sigma}(-1)^{|\sigma'|} = \sum_{k=0}^{n}(-1)^k\binom{n}{k} = (1-1)^n = 0$.]

(ii) $T_i T_j = T_j T_i$ and $T_i^* T_j = T_j T_i^*$ for every i, j with $i \neq j$ (the so-called *doubly commuting contractions*);

[*Hint:* $(S(\sigma)x, x) = (\prod_{j=1}^{n}(I - T_j^* T_j)x, x) \geq 0$ since $I - T_j^* T_j$ commute.]

(iii) $\sum_{i\in I} \|T_i\|^2 \leq 1$.

[*Hint:* in order to apply (d), show that

$$\begin{aligned}(S_k(\sigma)x, x) &:= \sum_{\sigma'\subset\sigma, |\sigma'|=k} \|T^{\chi_{\sigma'}} x\|^2 \leq \sum_{\sigma''\subset\sigma, |\sigma''|=k-1} \|T^{\chi_{\sigma''}} x\|^2 \big(\sum_{i\in I} \|T_i\|^2\big)\\ &\leq (S_{k-1}(\sigma)x, x),\end{aligned}$$

hence $(S(\sigma)x, x) = (S_0(\sigma)x, x) - (S_1(\sigma)x, x) + (S_3(\sigma)x, x) - \dots + (-1)^n(S_n(\sigma)x, x) \geq 0$.]

1.5.9. J. von Neumann type inequalities and spectral sets. (a) *The J. von Neumann inequality, a short proof via dilations.* Let T be a Hilbert space contraction. Deduce from 1.5.7(a) that

$$\|f(T)\| \leq \|f\|_\infty = \max_{\zeta\in\mathbb{T}} |f(\zeta)|$$

for every polynomial $f \in \mathcal{P}ol_+$.

[*Hint:* using 1.5.7(a) and 1.5.1(d) we have

$$\|f(T)\| = \|P_H f(U)\| \leq \|f(U)\| = r(f(U)) = \max_{\lambda\in\sigma(f(U))} |\lambda|;$$

since

$$\sigma(f(U)) \subset f(\sigma(U)),$$

the latter norm does not exceed

$$\max_{\lambda \in f(\sigma(U))} |\lambda| = \max_{\zeta \in \sigma(U)} |f(\zeta)| \leq \|f\|_\infty;$$

the mentioned spectral inclusion is also trivial: take $\mu \in \sigma(f(U))$ and factorize the polynomial $\mu - f$ as $\mu - f = a\prod_k (z - z_k)$, which shows that at least one of the factors in $\mu I - f(U) = a\prod_k (U - z_k I)$ is not invertible.]

(b) *A proof of the von Neumann inequality employing Blaschke approximations. (J. von Neumann, 1951; S. Drury, 1983)* Let T be a Hilbert space contraction.

(i) Show that $\|(\lambda I - T)x\|^2 - \|(I - \overline{\lambda}T)x\|^2 = (\|Tx\|^2 - \|x\|^2)(1 - |\lambda|^2)$ for every $x \in H$ and every $\lambda \in \mathbb{D}$, and hence $\|b_\lambda(T)\| \leq 1$, where b_λ stands for the Blaschke factor $b_\lambda = (\lambda - z)/(1 - \overline{\lambda}z)$.

(ii) Use the Schur theorem B.3.3.3 to show that $\|f(T)\| \leq \|f\|_\infty$ for every polynomial $f \in \mathcal{P}ol_+$.

[*Hint:* suppose $\|f\|_\infty \leq 1$, and let B_n be finite Blaschke products converging to f (see B.3.3.3); using the Riesz–Dunford calculus (2.5.1 below) show that $\lim_n \|f(rT) - B_n(rT)\| = 0$ for $0 \leq r < 1$, which gives $\|f(rT)\| \leq 1$; next, pass to the limit as $r \longrightarrow 1$.]

(c) *A proof of the von Neumann inequality based on the maximum principle (E. Nelson, 1961).*

(i) Let $T : H \longrightarrow H$ be a unitary operator on a finite dimensional Hilbert space. Diagonalizing T, show that $\|f(T)\| \leq \|f\|_\infty$ for every polynomial $f \in \mathcal{P}ol_+$.

(ii) Let $T : H \longrightarrow H$ be a contraction on a finite dimensional Hilbert space, let $T = U|T|$ be the polar decomposition of T (see B.1.5.5). Diagonalizing $|T|$, write

$$T = T(\lambda_1, \ldots, \lambda_n) = Uv^* \begin{pmatrix} \lambda_1 & & 0 \\ & \cdot & \\ & & \cdot \\ 0 & & \lambda_n \end{pmatrix} v,$$

where v is unitary and $0 \leq \lambda_k \leq 1$ for all k. Show that $(z_1, \ldots, z_n) \longmapsto T(z_1, \ldots, z_n)$ is an entire holomorphic function in each variable, and hence $(z_1, \ldots, z_n) \longmapsto f(T(z_1, \ldots, z_n))$ is, where f is a polynomial.

Next, applying the maximum principle in each variable, show that

$$|(f(T(\lambda_1, \ldots, \lambda_n))x, y)| \leq \sup\{|(f(T(z_1, \ldots, z_n))x, y)| : |z_1| = \cdots = |z_n| = 1\}$$

for every $x, y \in H$. Deduce from (i) that $\|f(T)\| \leq \|f\|_\infty$ for every polynomial $f \in \mathcal{P}ol_+$.

[*Hint:* use (i) for operators $T(z_1, \ldots, z_n)$ which are unitary for $|z_1| = \cdots = |z_n| = 1$.]

(iii) Let $T : H \longrightarrow H$ be a contraction on a separable Hilbert space, and $(P_n)_{n \geq 1}$ be an increasing sequence of finite rank orthoprojectors such that $\lim_n P_n x = x$ for every $x \in H$. Setting $T_n = P_n T P_n$, show that $\lim_n T_n x = Tx$

and $\lim_n T_n^k x = T^k x$ for every $k \geq 1$ and every $x \in H$. Deduce from (ii) that $\|f(T)\| \leq \underline{\lim}_n \|f(T_n)\| \leq \|f\|_\infty$.

(d) Let $\rho > 0$ and $T \in \mathcal{C}_\rho$ (see 1.5.7(c)). Show that

(i) $\|f(T)\| \leq \|\rho f + (1-\rho) f(0)\|_\infty$ for every polynomial $f \in \mathcal{P}ol_+$;

[*Hint:* use the definition of $\mathcal{C}_\rho$ and apply (a).]

(ii) *(J. Stampfli, 1965)* $f(T) \in \mathcal{C}_\rho$ for every polynomial $f \in \mathcal{P}ol_+$ satisfying $f(0) = 0$ and $\|f\|_\infty \leq 1$;

[*Hint:* let U_T be a unitary ρ-dilation of T and let U be a unitary dilation of $f(U_T)$; check that $f(T)^n = \rho P_H U^n | H$ for every $n \geq 1$.]

(iii) if $w(T) \leq 1$ (see 1.5.7(c)), then $w(f(T)) \leq 1$ for every polynomial $f \in \mathcal{P}ol_+$ with $f(0) = 0$ and $\|f\|_\infty \leq 1$; in particular, $w(T^n) \leq 1$ for every $n \geq 1$.

[*Hint:* use 1.5.7(c)(iii) and apply (ii) for $\rho = 2$.]

(e) Let $U = (U_1, \dots, U_n)$ be an n-tuple of commuting unitary operators $U_k : \mathcal{H} \longrightarrow \mathcal{H}$ $(k = 1, \dots, n)$. Show that

$$\|f(U)\| \leq \|f\|_{\mathbb{T}^n} = \max_{\zeta \in \mathbb{T}^n} \left| f(\zeta) \right|$$

for every polynomial $f \in \mathcal{P}ol_+(\mathbb{C}^n)$ of n variables $f(z) = f(z_1, \dots, z_n)$.

[*Hint:* let E_{U_1} be the spectral measure of U_1 (see 1.5.1(l)); show that

$$f(U)x = \int_{\mathbb{T}} f(z_1, U_2, \dots, U_n)\, dE_{U_1} x$$

for every $x \in \mathcal{H}$; next, observe that for every Riemann sum Σ we have

$$\begin{aligned} &\left\| \sum_j f(\zeta_j, U_2, \dots, U_n) E_{U_1}(\Delta_j) x \right\|^2 \\ &\quad = \sum_j \|f(\zeta_j, U_2, \dots, U_n) E_{U_1}(\Delta_j) x\|^2 \\ &\quad \leq \sup_{\zeta_1 \in \mathbb{T}} \|f(\zeta_1, U_2, \dots, U_n)\|^2 \sum_j \|E_{U_1}(\Delta_j) x\|^2, \end{aligned}$$

and hence $\|f(U)x\| \leq \|x\| \cdot sup_{\zeta_1 \in \mathbb{T}} \|f(\zeta_1, U_2, \dots, U_n)\|$; complete the proof by induction.]

(f) Let $T = (T_1, \dots, T_n)$ be a n-tuple of commuting contractions on a Hilbert space H having a commutative unitary dilation $U = (U_1, \dots, U_n)$. Show that the analogue of the *von Neumann inequality* holds, i.e.

$$\|f(T_1, \dots, T_n)\| \leq \|f\|_{\mathbb{T}^n}$$

for every polynomial f in $\mathbb{C}^n$.

[*Hint:* use (e).]

(g) Show that the von Neumann inequality (f) holds for any of the following n-tuples $T = (T_1, \dots, T_n)$:

(i) for a commutative family of isometries;
(ii) for a doubly commutative family of contractions (see 1.5.8(e));
(iii) for a commutative family T with $\sum_{i=1}^n \|T_i\|^2 \leq 1$;
(iv) for every pair of commuting contractions ($n = 2$).

[*Hint:* apply (f), 1.5.8(b) and 1.5.8(e).]

(h) Let $T = (T_1, T_2, T_3)$ be the triplet from the Parrott counterexample 1.5.8(c). Show that the von Neumann inequality

$$\|f(T_1, T_2, T_3)\| \leq \|f\|_{\mathbb{T}^3}$$

holds for every polynomial f in $\mathbb{C}^3$.

[*Hint:* let $f(z) = a_0 + \sum_{j=1}^{3} a_j z_j + \dots$, $A = \sum_{j=1}^{3} |a_j|$, and let

$$\Gamma(x \oplus y) = 0 \oplus (A^{-1} \sum_{j=1}^{3} a_j V_j x)$$

be a contraction on $H = K \oplus K$ (see 1.5.8(c) for the notation); choosing ε_j in such a way that $a_j \varepsilon_j = |a_j|$ and setting $g(\zeta) = f(\varepsilon_1 \zeta, \varepsilon_2 \zeta, \varepsilon_3 \zeta)$ for $\zeta \in \mathbb{C}$, we get

$$\|f(T_1, T_2, T_3)(x \oplus y)\| = \|g(\Gamma)(x \oplus y)\| \leq \|g\|_{\mathbb{T}} \|x \oplus y\| \leq \|f\|_{\mathbb{T}^3} \|x \oplus y\|.]$$

(i) *(M. Crabb and A. Davie, 1975)* Show that the von Neumann inequality fails for the following three commuting contractions on $\mathbb{C}^8$. Let $e, f_1, f_2, f_3, g_1, g_2, g_3, h$ be the standard orthonormal basis of $\mathbb{C}^8$ and let $T_k : \mathbb{C}^8 \longrightarrow \mathbb{C}^8$ $(k = 1, 2, 3)$ be defined as follows.

$$T_1 : e \longrightarrow f_1 \longrightarrow (-g_1) \longrightarrow (-h) \longrightarrow 0, \quad f_2 \longrightarrow g_3 \longrightarrow 0 \text{ and } f_3 \longrightarrow g_2 \longrightarrow 0;$$
$$T_2 : e \longrightarrow f_2 \longrightarrow (-g_2) \longrightarrow (-h) \longrightarrow 0, \quad f_1 \longrightarrow g_3 \longrightarrow 0 \text{ and } f_3 \longrightarrow g_1 \longrightarrow 0;$$
$$T_3 : e \longrightarrow f_3 \longrightarrow (-g_3) \longrightarrow (-h) \longrightarrow 0, \quad f_1 \longrightarrow g_2 \longrightarrow 0 \text{ and } f_2 \longrightarrow g_1 \longrightarrow 0.$$

Show that

1) T_k are contractions;
2) $T_i T_j = T_j T_i$ for all i, j;
[*Hint:* only the values on e, f_1, f_2, f_3 of $T_i T_j$ are needed to check];
3) $f(T_1, T_2, T_3)e = 4h$ for $f(z_1, z_2, z_3) = z_1 z_2 z_3 - z_1^3 - z_2^3 - z_3^3$, whereas $\|f\|_{\mathbb{T}^3} < 4$.

[*Hint:* $|\sum_j a_j| = \sum_j |a_j|$ if and only if $a_j = \varepsilon b_j$ for a unimodular $\varepsilon \in \mathbb{C}$ and $b_j \geq 0$.]

(j) *A spherically isometric Hardy space in the ball $\mathbb{B}^n$.* Let $\mathbb{B}^n = \{z \in \mathbb{C}^n : \|z\|^2 = \sum_{k=1}^{n} |z_k|^2 < 1\}$ be the open unit ball in $\mathbb{C}^n$, and let

$$H^2_{\mathbb{B}^n} = H^2_n$$

be the space of power series f such that

$$f = \sum_{\alpha \geq 0} a_\alpha z^\alpha, \quad \|f\|^2_{H^2_n} = \sum_{\alpha \geq 0} |a_\alpha|^2 \frac{\alpha!}{|\alpha|!} < \infty,$$

where $\alpha = (\alpha_1, \dots, \alpha_n) \in \mathbb{Z}^n_+$, $|\alpha| = \sum_{k=1}^{n} \alpha_k$ and $\alpha! = \prod_{k=1}^{n} (\alpha_k!)$.

(i) Show that H^2_n is a RKHS (see B.6.5.2, Volume 1) with the reproducing kernel

$$k_\lambda(z) = \frac{1}{1 - (z, \lambda)},$$

where $z, \lambda \in \mathbb{B}^n$ and $(z, \lambda) = \sum_{k=1}^{n} z_k \overline{\lambda}_k$

[*Hint:* use the Taylor formula $f = \sum_{\alpha\geq 0} \frac{\partial^\alpha f(0)}{\alpha!} z^\alpha$ to show that $k_\lambda(z) = \sum_{\alpha\geq 0} \frac{|\alpha|!}{\alpha!} \overline{\lambda}^\alpha z^\alpha$.]

(ii) Let $(S_j f)(z) = z_j f(z)$, where $z = (z_1, \ldots, z_n)$ and $f \in H^2_n$. Show that

$$P_0 + \sum_{j=1}^{n} S_j S_j^* = I,$$

where $P_0 f = f(0)1$ is the orthoprojection onto the constant functions. (The property $\sum_{j=0}^{n} S_j S_j^* = I$ is called *spherical isometry* of a given family $(S_j)_{0\leq j\leq n}$)

[*Hint:* check the equality $\sum_{j=1}^{n} (S_j^* k_\lambda, S_j^* k_\mu) = (\mu, \lambda)_{\mathbb{C}^n} (k_\lambda, k_\mu)$.]

(iii) *(W. Arveson (1999), S. Drury (1978))* Show that $H^\infty(\mathbb{B}^n) \not\subseteq H^2_{\mathbb{B}^n}$, where $H^\infty_n := H^\infty(\mathbb{B}^n)$ is the space of all bounded holomorphic functions on $\mathbb{B}^n$, and moreover, there exists no constant $c < \infty$ such that $\|f\|_{H^2_n} \leq c\|f\|_{H^\infty_n}$ for every polynomial f.

[*Hint:* the geometric mean inequality shows that $\|z^\alpha\|^2_{H^\infty_n} = n^{-n}$ and hence $\|z^{j\alpha}\|^2_{H^\infty_n} = n^{-jn}$, where $\alpha = (1, 1, \ldots, 1)$ and $j = 1, 2, \ldots$; by the Stirling formula

$$\|z^{j\alpha}\|^2_{H^2_n} = \frac{(j\alpha)!}{|j\alpha|!} \sim c_n \cdot j^{(n-1)/2} n^{-jn},$$

and the result follows as $j \longrightarrow \infty$.]

(k)* *(W. Arveson (1999), G. Popescu (1991))* Let a commutative n-tuple $T = (T_1, \ldots, T_n)$ on a Hilbert space H satisfy $\sum_{j=1}^{n} T_j T_j^* \leq I$, or equivalently,

$$\Big\| \sum_{j=1}^{n} T_j x_j \Big\|^2 \leq \sum_{j=1}^{n} \|x_j\|^2$$

for all $x_j \in H$ (T is called a *n-contraction*). Then

$$\|f(T_1, \ldots, T_n)\| \leq \|f(S_1, \ldots, S_n)\|$$

for every polynomial f, where S_j are the shift operators defined in **(j)**(ii).

[*Hint:* (k) and (l) (below) are dual forms of each other.]

(l)* *(S. Drury (1978))* Let $T = (T_1, \ldots, T_n)$ be a commutative n-tuple of operators on a Hilbert space H satisfying $\sum_{j=1}^{n} T_j^* T_j \leq I$, or equivalently,

$$\sum_{j=1}^{n} \|T_j x\|^2 \leq \|x\|^2$$

for every $x \in H$. Then

$$\|f(T_1, \ldots, T_n)\| \leq \|f(S_1^*, \ldots, S_n^*)\|$$

for every polynomial f, where S_j^* are the backward shift operators on the space $(H^2_{\mathbb{B}^n})^* = (H^2_n)^*$, the Cauchy dual to $H^2_{\mathbb{B}^n}$ defined in (j) above; this means that $f = \sum_{\alpha\geq 0} a_\alpha z^\alpha \in (H^2_n)^*$ if and only if

$$\|f\|^2_{(H^2_n)^*} = \sum_{\alpha\geq 0} |a_\alpha|^2 \frac{|\alpha|!}{\alpha!} < \infty,$$

and $a_\alpha(S_j^* f) = a_{\alpha+\delta_j}(f)$ for all $\alpha \geq 0$, where $\delta_j = (\delta_{jk})_{k=1}^n$.

[*Hint:* (l) and (k) are dual forms of each other.]

(m) *(M. Bozejko, 1989)* Show that

$$\|f(T_1, \dots, T_n)\| \leq \|f(U_{T_1}, \dots, U_{T_n})\|$$

for every (noncommuting) n-tuple $(T_1, \dots, T_n)$ of Hilbert space contractions and every formal (noncommuting) polynomial $f(X_1, \dots, X_n)$ in n variables; here we write $(U_{T_1}, \dots, U_{T_n})$ for the standard unitary dilation from 1.5.8(a).

[*Hint:* straightforward from 1.5.8(a).]

(n)* *(W. Mlak, 1978)* Let $A_k : H \longrightarrow H$ be bounded operators, $k = 0, \dots, n$, and $T : H \longrightarrow H$ be a contraction such that $TA_k = A_k T$, $TA_k^* = A_k^* T$ for every k. Then

$$\Big\| \sum_{j=0}^{n} A_j T^j \Big\| \leq \sup_{|z| \leq 1} \Big\| \sum_{j=0}^{n} A_j z^j \Big\|.$$

(o) *(R. Hirschfeld, 1972)* Let $T = (T_1, \dots, T_n)$ be an n-tuple of commuting operators on a Hilbert space H such that $r(T_j) \leq 1$ for every j, $1 \leq j \leq n$. Define an *operator Poisson kernel* in $T = (T_1, \dots, T_n)$ setting

$$P(\rho T, \zeta) = \frac{1}{n!} \sum_{\sigma} P_{\sigma(1)}(\rho T_{\sigma(1)}, \zeta_{\sigma(1)}) \cdots P_{\sigma(n)}(\rho T_{\sigma(n)}, \zeta_{\sigma(n)}),$$

where σ runs over all permutations of $(1, \dots, n)$ and

$$P_j(\rho T_j, \zeta_j) = \mathrm{Re}\left((\zeta_j I + \rho T_j)(\zeta_j I - \rho T_j)^{-1} \right)$$

is the Poisson kernel in one variable, $0 \leq \rho < 1$, $\zeta_j \in \mathbb{T}$. (Notice that $P_j(\rho T_j, \zeta_j)$ and $P(\rho T, \zeta)$ are all selfadjoint). The *Poisson radius* $\rho(T)$ is defined by

$$\rho(T) = \sup \left\{ \rho \in [0, 1) : P(\rho T, \zeta) \geq 0 \text{ for every } \zeta \in \mathbb{T}^n \right\}.$$

(i) Show that $0 < \rho(T) \leq 1$.

(ii) Show that

$$\|f(\rho(T)T)\| \leq \|f\|_{\mathbb{T}^n}$$

for every complex polynomial f in n variables.

[*Hint (J. Holbrook):* take $0 < \rho < \rho(T)$ and expand P into a multiple Fourier series $P(\rho T, \zeta) = \sum_{k \in \mathbb{Z}^n} T(k) \rho^{|k|} \zeta^k$; show that the series is absolutely convergent and $T(k) = T^k = T^{k_1} \cdots T^{k_n}$ for $k \in \mathbb{Z}_+^n$; next, show that the sesquilinear form

$$(\alpha, \beta)_{\mathcal{H}} = \sum_{k,l \in \mathbb{Z}^n} \rho^{|k-l|} (T(k-l)\alpha_k, \beta_l)_H$$

is positive definite on the vector space of all finitely supported sequences $\alpha = (\alpha_k)_{k \in \mathbb{Z}^n}$;

let $\mathcal{H}$ be the standard Hilbert space associated to the (semi)scalar product $(\cdot, \cdot)_{\mathcal{H}}$ (see B.6.5.2, Volume 1); H is isometrically embedded into $\mathcal{H}$ by $x \longmapsto (\delta_{0k} x)_{k \in \mathbb{Z}^n}$;

denote by S_j the bilateral shift on the j-th component of a multi-index k, show that $(S_j\alpha, S_j\beta)_{\mathcal{H}} = (\alpha, \beta)_{\mathcal{H}}$ for all j, α, β, and

$$(S_1^{k_1} \cdots S_n^{k_n} x, y)_{\mathcal{H}} = \rho^{|k|}(T_1^{k_1} \cdots T_n^{k_n} x, y)_H;$$

finally, apply (f) to ρT.]

1.6. Notes and Remarks

General information. Function models and their applications to perturbation theory and differential operators started in 1946 with M. Livshic' paper [**Liv2**]. Being motivated mostly by the extension theory of nonselfadjoint operators, M. Livshic and other mathematicians of the Soviet operator theory school (M. Krein, M. Brodskii, V. Potapov, Yu. Shmulyan, A. Shtraus, ...) developed the theory having in mind continuous versions of the Schur and Jordan type triangular models mentioned in the Foreword to Part C. However, the final statements of these developments, obtained by 1960, are equivalent to those of the function model theory started 4-5 years later by B. Szőkefalvi-Nagy and C. Foiaş, and also by P. Lax and R. Phillips, and L. de Branges and J. Rovnyak. Here we mean the general definition of the characteristic function (Yu. Shmulyan (1953), A. Shtraus (1959)), and the theorem saying that operators are unitarily equivalent if and only if their characteristic functions coincide (A. Shtraus (1960)), and the construction of a function model coinciding with that of this chapter up to some equivalences: the Cayley transform, the replacement of $\mathbb{T}$ by $\mathbb{R}$, and the Fourier transform (M. Brodskii and M. Livshic (1958)). The reader can consult M. Brodskii and M. Livshic [**BrL**], M. Brodskii [**Bro3**], and I. Gohberg and M. Krein [**GK2**] for this setting of the model theory.

The crucial advantage of the Szőkefalvi-Nagy and Foiaş approach is the systematic use of dilation techniques and the H^∞-calculus arising from it (see Chapter 2 below), as well as some technical points (e.g., the Lebesgue measure of $\mathbb{T}$ is finite). The Szőkefalvi-Nagy and Foiaş model based on the shift operator $Zf = zf$ on $H^2(E)$ and on the previously constructed dilation theory (M. Naimark (1940/1943), B. Szőkefalvi-Nagy (1953)) was formulated in 1964 (see comments below). At approximately the same time, other function models relying on similar ideas appeared, namely, the P. Lax and R. Phillips model for stable contractive semigroups, and the L. de Branges and J. Rovnyak model for completely nonisometric contractions. General sources for these theories are B. Szőkefalvi-Nagy and C. Foiaş [**SzNF4**], F. Riesz and B. Szőkefalvi-Nagy [**RSzN**], B. Szőkefalvi-Nagy [**SzN5**], H. Helson [**Hel**], R. Douglas [**Dou7**], P. Lax and R. Phillips [**LPh1**], L. de Branges and J. Rovnyak [**dBR1**], [**dBR2**], N. Nikolski [**N19**], N. Nikolski and V. Vasyunin [**NVa8**].

Dilations, characteristic functions and function models. The characteristic function and a function model were first introduced by M.S. Livshic for the case of operators with $\operatorname{rank} D_T = \operatorname{rank} D_{T^*} = 1$ as early as in 1946, [**Liv2**], in the form

$$\Theta(z) = (-TJ_T + zD_{T^*}(I - zT^*)^{-1}D_T)|\mathcal{D}_T$$

ready for further generalizations; here $D_T = |I - T^*T|^{1/2}$, $J_T = \operatorname{sign}(I - T^*T)$, $\mathcal{D}_T = \operatorname{clos}(I - T^*T)H$. As mentioned above, the idea comes from the extension theory of symmetric operators, especially extensions outcoming the initial space developed by M. Naimark, M. Krein, and others. In particular, in 1940/1943 M.

Naimark proved [**Nai1**], [**Nai2**] the following theorem which plays a fundamental role in extension and dilation theory.

THEOREM. *Every operator valued measure E, where $E(\sigma) : H \longrightarrow H$ and $0 \le E(\sigma) \le I$ for all σ from a σ-algebra, can be dilated up to an orthoprojection valued measure $\mathcal{E}$ on a larger Hilbert space $\mathcal{H}$, so that $H \subset \mathcal{H}$ and $E(\sigma) = P_H \mathcal{E}(\sigma)|H$ for all σ.*

The existence theorem for the unitary dilation was deduced from Naimark's theorem by B. Szőkefalvi-Nagy [**SzN3**] in 1953. In fact, around that time, there was a series of interesting results by B. Szőkefalvi-Nagy related to the extension-dilation method. For one of them see B.6.5.3 (Volume 1); yet another result is the following theorem [**SzN2**] on operator valued moment problems also related to Chapter B6.

THEOREM. *For every sequence $A_0 = I, \dots, A_n, \dots$ of operators acting on a Hilbert space H and such that $\sum_n c_n A_n \ge 0$ for every polynomial $p(x) = \sum_n c_n x^n$, which is positive for $x \in [-M, M]$, there exists a self-adjoint operator A on a larger Hilbert space $\mathcal{H}$ such that $A_n = P_H A^n|H$ for $n \ge 0$.*

During the next decade (1955-1965), many papers were published about geometric structures hidden behind the minimal unitary dilation; for all these results and original references we refer to the above mentioned sources, and especially to B. Szőkefalvi-Nagy and C. Foiaş [**SzNF4**], F. Riesz and B. Szőkefalvi-Nagy [**RSzN**], and W. Mlak [**Ml**]. However, it is worth mentioning the two following results anticipating dilation theory: in 1939, A. Plessner [**Pl2**] showed that (using the modern language) an isometric operator always has a unitary dilation; and, in 1944, G. Julia [**Ju**] constructed a "one-step dilation" for an arbitrary Hilbert space contraction $T : H \longrightarrow H$ by exhibiting a unitary operator $U : \mathcal{H} \longrightarrow H$ with $\mathcal{H} \supset H$ and $T = P_H U|H$.

Several far reaching generalizations of the unitary dilation theorem were found using two other different techniques, namely the technique of indefinite scalar product spaces and the technique of C^*-algebras. Below we briefly describe both these approaches, but here we shall quote a theorem about normal dilations instead of unitary ones. Namely, the following theorem is proved by C. Foiaş [**Foi2**], C. Berger [**Berge**], and A. Lebow [**Leb**]; for a proof via C^*-algebra techniques and for several variables versions see W. Arveson [**Arv1**], V. Paulsen [**Pau**].

THEOREM. *Let σ be a compact set in $\mathbb{C}$ and let $T : H \longrightarrow H$ be an operator with* spectral set σ, *i.e., such that*

$$\|f(T)\| \le \|f\|_\sigma$$

for every $f \in \mathcal{R}at(\sigma)$ (rational functions with poles in $\mathbb{C} \backslash \sigma$); if $\mathcal{R}at(\sigma) + \overline{\mathcal{R}at(\sigma)}$ is dense in $C(\partial\sigma)$, then there exists a normal $\partial\sigma$-dilation *for T, that is, a normal operator $N : \mathcal{H} \longrightarrow \mathcal{H}$ on a larger Hilbert space $\mathcal{H}$ such that $\sigma(N) \subset \partial\sigma$ and*

$$f(T) = P_H f(N)|H \text{ for every } f \in \mathcal{R}at(\sigma).$$

Yet another (trivial) remark: every Hilbert space operator $T : H \longrightarrow H$ has a normal dilation, for example, $T^n = P_H(\|T\|U)^n|H$ for $n \ge 0$ with a suitable unitary operator U. But this is useless if T is not a "small perturbation" of $\|T\|U$.

When constructing a model for dissipative or accretive operators in place of contractions, one of the possibilities is to use the Cayley transform, as in Section

1.4 above. By this mean, all principal facts of the model theory can be restated for these operators, but such a "translation" is not automatic. Another way is to proceed independently, somehow repeating the model construction for contractions, that is, first find a self-adjoint dilation, and then represent the initial operator as its compression onto a co-invariant subspace. The latter approach is convenient when treating problems of scattering theory, i.e. perturbations of continuous semigroups, wave limits, etc. See P. Lax and R. Phillips [**LPh1**], B. Pavlov [**P1**], S. Naboko [**Nab2**], B. Solomyak [**So2**].

Coordinate free model. The idea of the coordinate free construction of model operators has first appeared in 1977 in V. Vasyunin [**Vas2**]. The final version of this construction, presented in this chapter, is from N. Nikolski and V. Vasyunin [**NVa5**]; see also [**NVa8**].

The transcriprion equations of Subsection 1.3.4 allows one more solution that is responsible for yet another pioneering model (together with the Livshic–Brodskii model and the Sz.-Nagy–Foiaş model), namely, for the *de Branges–Rovnyak model* [**dBR2**]. One of its advantages is that the model space $\mathcal{K}_\Theta$ consists of analytic and co-analytic functions only. Namely, setting

$$\Pi^* = id,$$

we get $\Pi = W_\Theta$ and $W = W_\Theta^{[-1]}$, where W_Θ is as in Subsection 1.3.8 and $W_\Theta^{[-1]}$ means the left inverse to W_Θ on $\operatorname{Range} W_\Theta$ and the zero operator on $\operatorname{Ker} W_\Theta$. Therefore,

$$\mathcal{H} = L^2(E_* \oplus E, W_\Theta^{[-1]}), \quad G = \begin{pmatrix} \Theta \\ I \end{pmatrix} H^2(E), \quad G_* = \begin{pmatrix} I \\ \Theta^* \end{pmatrix} H^2_-(E_*)$$

$$\mathcal{K}_\Theta = \left\{ \begin{pmatrix} f \\ g \end{pmatrix} : f \in H^2(E), \quad g \in H^2_-(E_*), \quad g - \Theta^* f \in \Delta L^2(E) \right\},$$

and the adjoint of the model operator M_Θ acts as

$$M_\Theta^*(f,g)^{col} = (S^*f, \bar{z}g - \bar{z}\Theta^* f(0))^{col},$$

where $Sf = zf$ is the shift operator. For details of this form of the function model we refer to N. Nikolski and V. Vasyunin [**NVa5**], [**NVa4**].

The de Branges–Rovnyak model. The original writing of the de Branges–Rovnyak model is based on the restrictions of the shift operator $S : f \longrightarrow zf$ to the range spaces $\mathcal{E}(\Theta) = \Theta H^2(E)$ endowed with the range norm, see B.6.5.2(n). Denote by $\mathcal{E}(\Theta)^c$ the complementary L. de Branges space of $\mathcal{E}(\Theta)$ (as defined in B.6.5.2(g)), i.e.,

$$\mathcal{E}(\Theta)^c = (I - \Theta P_+ \Theta^*)^{1/2} H^2(E_*),$$

also equipped with the range norm. It is shown in N. Nikolski and V. Vasyunin [**NVa4**] that $S^*\mathcal{E}(\Theta)^c \subset \mathcal{E}(\Theta)^c$ and the adjoint operator $(S_\Theta^*)^*$ of the restriction $S_\Theta^* = S^*|\mathcal{E}(\Theta)^c$ is unitarily equivalent to the completely noncoisometric part $M_\Theta|\mathcal{K}_\Theta^*$ of the model operator M_Θ, where

$$\mathcal{K}_\Theta^* = \begin{pmatrix} H^2(E_*) \\ \operatorname{clos} \Delta H^2(E) \end{pmatrix} \ominus \begin{pmatrix} \Theta \\ \Delta \end{pmatrix} H^2(E).$$

Therefore, a contraction T such that T^* contains no isometric part is unitarily equivalent to the operator $(S_\Theta^*)^*$, which is called *(short) de Branges–Rovnyak model* of T. It is also shown that, in fact,

$$(S_\Theta^*)^* f = Sf - \Theta j^* f$$

for $f \in \mathcal{E}(\Theta)^c$, where $j : E \longrightarrow \mathcal{E}(\Theta)^c$ is defined by $je = S^*\Theta e$ for $e \in E$.

The *complete de Branges–Rovnyak model*, i.e., a model of the above type for an arbitrary c.n.u. contraction, is more complicated and acts on a two component space, as this is the case for its Sz.-Nagy–Foiaş counterpart. See L. de Branges and J. Rovnyak [**dBR2**] and N. Nikolski and V. Vasyunin [**NVa4**]. Another identification of the short de Branges–Rovnyak model and the Szőkefalvi-Nagy–Foiaş model for completely noncoisometric contractions is suggested by R. Douglas and is already given in C. Foiaş [**Foi3**]; see also J. Ball [**Ball**] for such an identification.

It is interesting to note that the problem of characterizing those operator functions $\Theta = \Theta_T$ for which the complete de Branges–Rovnyak model is reduced to the short one, i.e., the question when $\mathcal{K}_\Theta^* = \mathcal{K}_\Theta$, seems to be open. It is clear that the latter equality holds if and only if

$$\operatorname{clos} \Delta H^2(E) = \operatorname{clos} \Delta L^2(E).$$

For many other equivalent conditions and references see A.4.8.8 and A.4.9 (Volume 1). In particular, the question is closely related to the ability of the weight $W = \Delta \geq 0$ to be factored in an Hermitian square

$$W = f^* f$$

of a function $f \in H^\infty(E \longrightarrow F)$. The latter problem was intensively studied in the theory of stationary Gaussian processes. Namely, it is known to be equivalent to the fact that a multivariate process with the spectral density $W dm$ is nondeterministic (regular), that is

$$\bigcap_{n \leq 0} [\operatorname{span}_{L^2(E,W)}(z^k E : k \leq n)] = \{0\};$$

for more see A.2.3, A.2.9 and B.9.1.5, B.9.4, and also A.4.8.8, A.4.9 (Volume 1).

Models for noncontractions. One of the approaches to construct a model for a generic Hilbert space operator $T : H \longrightarrow H$ is to exploit the symplectic structure associated to the operator $J = \operatorname{sign}(I - T^*T)$. The space H endowed with an *indefinite scalar product*

$$[x, y] = (Jx, y)$$

is called *Krein space* (M. Krein (1965), [**Kr5**]), or *Pontryagin space* if, in addition, $\operatorname{rank}(I - J) < \infty$ (L. Pontryagin (1944), [**Pon**]).

Since $J = P^+ - P^-$ for two complementary orthogonal projections $P^+, P^- = I - P^+$, and $J(T^*T) = (T^*T)J$, we get

$$[Tx, Tx] \leq [x, x]$$

for every $x \in H$. Such an operator T is called a *J-contraction*, and one can prove (Ch. Davis [**Davi**]) that there exists a J-unitary dilation of T, say $U : \mathcal{H} \longrightarrow \mathcal{H}$, that is, a dilation of T on a Krein space $(\mathcal{H}, \mathcal{J})$ satisfying $UU^{[*]} = U^{[*]}U = I$, where $U^{[*]}$ is defined by $[Ux, y] = [x, U^{[*]}y]$ for every $x, y \in \mathcal{H}$. Therefore, this U is invertible and, moreover,

1) $\mathcal{J}x = x$ for all $x \in H$,
2) $[U^k x, y] = (T^k x, y)$ for every $x, y \in H$ and every $k \geq 0$,
3) $[U^k x, U^k y] = [x, y]$ for every $x, y \in \mathcal{H}$ and every $k \in \mathbb{Z}$, and

4) $\mathrm{span}(U^k H : k \in \mathbb{Z}) = \mathcal{H}$.

The structure of the operator U is exactly the same as in Theorem 1.1.16, with the only replacement of $-VT^*V_*$ by $-VT^*J_{T^*}V_*$. The operator $\mathcal{J}$ is defined as the identity on H, and as a coefficientwise application of J_{T^*} on G_* and of J_T on G (see the decompositions in Section 1.2). For details, we refer to Ch. Davis [**Davi**], C. Foiaş [**Foi3**], Ch. Davis and C. Foiaş [**DF**], where a function model based on J-unitary dilation is also constructed.

A similar dilation theorem holds for every strongly continuous semigroup $T(s) : H \longrightarrow H$, which in addition is holomorphic in a sector $\{s \in \mathbb{C} : |\arg(s)| < \frac{\pi}{2} - \varepsilon\}$ (B. McEnnis [**McE**]).

Yet another approach for studying general operators by means of function models, especially, "smooth" perturbations $T = U + K$ of unitary operators, was proposed by S. Naboko [**Nab1**] and developed by N. Makarov and V. Vasyunin [**MV1**]. Namely, given an operator T, these authors consider a contraction $T_0 = T\varphi(|T|)$, where $\varphi(t) = 1$ for $t \le 1$ and $\varphi(t) = t^{-2}$ for $t > 1$, and explicitely represent the operator T on the function model of T_0. It is shown that for some problems of spectral analysis this "model" of T is more efficient than the model based on Krein spaces. Some useful modifications of this approach are introduced by V. Kapustin [**Kap2**].

One more approach for modelling nonselfadjoint operators and finite commutative families of such operators is developed by M. Livshic, N. Kravitsky, A. Markus, and V. Vinnikov in [**LKMV**]. For example, let (A_1, A_2) be a pair of commuting operators on a Hilbert space H having finite rank imaginary parts $(A_k - A_k^*)/2i$. The *discriminant polynomial* is defined by

$$\Delta(z_1, z_2) = \det\Big(z_1(A_2 - A_2^*) - z_2(A_1 - A_1^*) + A_1A_2^* - A_2A_1\Big)\Big|G,$$

where $G = (A_1 - A_1^*)H + (A_2 - A_2^*)H$. It is shown that $\Delta(A_1, A_2) = 0$ on the so-called *principal subspace* (the orthogonal complement of the most A_1, A_2 reducing subspace of H, if A_k are selfadjoint); this equation is considered as an analogue of the classical Cayley–Hamilton theorem. It is also proved that the joint spectrum $\sigma(A_1, A_2)$ is included in the so-called *discriminant variety* $D = \Big\{(z_1, z_2) \in \mathbb{C}^2 : \Delta(z_1, z_2) = 0\Big\}$, which is an algebraic surface in $\mathbb{C}^2$. Advanced parts of the spectral analysis for the pair (A_1, A_2) are developed on the discriminant variety: the characteristic function of (A_1, A_2) (of two variables) is defined; being restricted to the discriminant variety, it permits a kind of factorization theory which is responsible for invariant subspaces of (A_1, A_2), the triangular form of (A_1, A_2), solutions of some inverse problems, and so on. For more details we refer to [**LKMV**] and a survey of V. Vinnikov [**Vinn**].

For more information about models and characteristic functions, see the sources listed above, and also the survey article by N. Nikolski [**N8**].

Several commuting operators: dilations and models. For facts about simultaneous unitary dilations of several contractions, presented in Subsections 1.5.8 and 1.5.9, and for many related topics we refer to B. Szőkefalvi-Nagy and C. Foiaş [**SzNF4**]; see also F.-H. Vasilescu [**Vasil**] for other aspects of multivariate operator theory.

Dilations just mentioned rely on harmonic analysis on the torus $\mathbb{T}^n$ in the same way as the Sz.-Nagy–Foiaş theory for one operator is based on the function theory on the circle $\mathbb{T}$. There exist other dilation theories and other models employing function theory in other domains of $\mathbb{C}^n$. In particular, the spherically isometric Hardy space $H^2_{\mathbb{B}^n}$ described in 1.5.9(i) gives rise to a dilation and model theory for commutative n-tuples $T = (T_1, \dots, T_n)$ satisfying the *spherical contraction condition*

$$\sum_{k=1}^{n} T_k T_k^* \leq I.$$

W. Arveson [**Arv5**] proved that such a T is a compression

$$P_H \left(\sum S \oplus U \right) |H$$

on a co-invariant subspace of an orthogonal sum of the shift n-tuple S from 1.5.9(i)-(j) and a commutative spherically isometric n-tuple U satisfying $\sum_{k=1}^{n} U_k U_k^* = I$.

The dilation theory for the so-called *abstract Hardy spaces* (see T. Gamelin [**Gam**] and K. Barbey and H. König [**BKö**]) is developed in a series of papers by T. Nakazi, including applications to several commuting operators, normal dilations, ρ-dilations, the von Neumann type inequalities, etc.; see T. Nakazi [**Nak1**], [**Nak2**] for details and further references.

It is worth mentioning that there exist corresponding dilation and model theories for *noncommuting n-tuples* relying on a noncommutative version of the above space $H^2_{\mathbb{B}^n}$ (the so-called *full Fock space* with the *creation operators* in place of the shifts S_k). Moreover, this noncommutative version, in fact, preceded the commutative one and the latter was extracted from the former as a "maximal commutative part", see W. Arveson [**Arv5**]. We refer to A. Frazho [**Fra**], J. Agler [**Ag1**], J. Bunce [**Bun**], G. Popescu (a series of papers, see [**Pop2**]), and K. Davidson and D. Pitts [**DPi2**].

The noncommutative dilation and extension theories mentioned above are relyed on a C^*-algebra approach, which extends the subject in many aspects; we will briefly describe it in the next paragraphs.

C^*-algebra approach and completely bounded maps. In 1955, W. Stinespring [**Sti**] introduced the notion of completely positive maps between C^*-algebras and found a far reaching generalization of the Bochner–Naimark theory of positive definite functions (see B.6.6 for the latter). Namely, let A and B be C^*-algebras containing units, and let $K \subset A$ be a selfadjoint subspace (i.e. $x \in K \Rightarrow x^* \in K$) containing the unit; a linear mapping $\varphi : K \longrightarrow B$ is called *completely positive* (c.p.) if all of its "breedings"

$$\varphi_n : M_n(K) \longrightarrow M_n(B)$$

are positive, $n = 1, 2, \dots$ (i.e., $X \geq 0 \Rightarrow \varphi_n(X) \geq 0$). Here $M_n(B)$ stands for the C^*-algebra of $n \times n$ matrices (a_{ij}) with entries from B and

$$\varphi_n((a_{ij})) := (\varphi(a_{ij})).$$

In general the expression "completely (P)" means that all φ_n enjoy the property (P); for example, *completely isometric* (c.i.), *completely contractive* (c.c.), or *completely bounded* (c.b.). In the latter case, it is supposed that φ_n are uniformly bounded, that is,

$$\sup_{n\geq 1} \|\varphi_n\| =: \|\varphi\|_{cb} < \infty.$$

In this language, the *Stinespring representation theorem* can be stated as follows.

THEOREM. *A mapping $\varphi : A \longrightarrow L(H)$ is c.p. if and only if there exists a Hilbert space F, a $*$-representation $\pi : A \longrightarrow L(F)$ and a bounded operator $V : H \longrightarrow F$ such that*

$$\varphi(a) = V^*\pi(a)V.$$

The theorem is generalized by V. Paulsen (see [**Pau**]) for c.b. mappings in the form

$$a \longmapsto V_1^*\pi(a)V_2.$$

The Stinespring theorem implies many other representation theorems based on positivity, see V. Paulsen [**Pau**] for an excellent exposition. For example, this is the case for Bochner's and Naimark's theorem mentioned above, as well as for the following general B. Szőkefalvi-Nagy theorem about positive definite operator functions on a group (see B. Szőkefalvi-Nagy and C. Foiaş [**SzNF4**] for original proof and references).

THEOREM. *A function $T : G \longrightarrow L(H)$ on a (multiplicative) group G is positive definite, i.e.,*

$$\sum_{r,s\in G} (T(s^{-1}r)h(r), h(s)) \geq 0$$

for every finitely supported function $h : G \longrightarrow H$, if and only if there exists a unitary representation $U : G \longrightarrow L(\mathcal{H})$ on a larger space $\mathcal{H} \supset H$ with $T(s) = P_H U|\mathcal{H}$ for $s \in G$.

The existence of a unitary dilation for one parameter contractive semigroups $T_+(\cdot)$ follows from this theorem after proving that $T(s) = T_+(s)$ for $s \geq 0$, $T(-s) = T(s)^*$ is a positive definite function.

The basic facts for many concrete dilation theorems are the following *Arveson extension theorem*, [**Arv1**], see also [**Arv4**] and V. Paulsen [**Pau**] for another proof and more references.

THEOREM. *1) Let $K \subset A$ be as above, and $\varphi : K \longrightarrow L(H)$ be a c.p. map. Then there exists a c.p. map $\psi : A \longrightarrow L(H)$ extending φ.*

2) Let B be a unital C^*-algebra and A a unital subalgebra (not necessarily $*$-closed), let $\varphi : A \longrightarrow L(H)$ be a unital homomorphism, and let $\tilde{\varphi} : A + A^* \longrightarrow L(H)$ be a positive extension of φ. Then the following are equivalent.

(a) φ has a B-dilation, that is, there exists a unital $*$-homomorphism $\pi : B \longrightarrow L(\mathcal{H})$, where $\mathcal{H}$ is a Hilbert space containing H, such that

$$\varphi(a) = P_H\pi(a)\Big|\mathcal{H}$$

for all a in A.

(b) φ is a c.c. map.

(c) $\tilde{\varphi}$ is a c.p. map.

In particular, using the above results and his generalisation of the Stinespring theorem, V. Paulsen (see [**Pau**]) proved that every c.b. homomorphism from 2) above is similar to a c.c. homomorphism. Namely, the expression

$$|x| = \inf\left\{\left\|\sum \pi(a_j)V_2 x_j\right\| : \sum \varphi(a_j)x_j = x, a_j \in A, x_j \in H\right\}$$

defines an equivalent Hilbert norm on H and $\varphi : A \longrightarrow L(H, |\cdot|)$ is a c.c. homomorphism. The following characterization of Hilbert space operators similar to a contraction, due to V. Paulsen, is an immediate consequence of this theorem (see [**Pau**]).

THEOREM. *An operator $T : H \longrightarrow H$ is similar to a contraction if and only if the polynomial calculus $f \longmapsto f(T)$ is a c.b. homomorphism from $(\mathcal{P}ol_+, \|\cdot\|_{\mathbb{T}})$ to $L(H)$.*

Exellent systematic expositions of the similarity theme, as well as of the entire theory of c.b. mappings, can be found in G. Pisier [**Pi1**] and V. Paulsen [**Pau**]. In the latter, some attention is also paid to K-spectral sets and the Arveson theory of joint spectral sets of multivariate operators. In particular, it is shown that, given a Hilbert space operator $T : H \longrightarrow H$, the existence of a normal dilation with spectrum on $\partial\sigma$ is equivalent to the fact that σ is a "complete spectral set" for T, i.e., the calculus $f \longmapsto f(T)$ is a c.b. map from $\mathcal{R}at(\sigma)$ to $L(H)$. There is also a multivariate analogue of this characterization.

The von Neumann inequality became from long ago a touchstone of the operator theory related to dilations and functional calculi. The first proof appeared in J. von Neumann [**vN5**] and is close to that of 1.5.9(b) (instead of I. Schur's theorem B.3.3.3, von Neumann employed the Schur theorem about "Schur parameters", see B.3.4.2(b) (Volume 1) and D.4.7.4 below). Essentially the same proof appeared in Drury's survey [**Dru2**], but using the D. Marshall theorem (see [**Mar**]) on uniform approximations by convex combinations of Blaschke products. The proof based on the properties of the Poisson kernel is proposed by E. Heinz [**Hei**], and the proof from 1.5.9(c) appeared in E. Nelson [**Nel**]. A proof relyed on unitary dilations was invented by B. Szőkefalvi-Nagy already in 1953, see [**SzNF4**] for more discussions and references on this approach.

Further developments of these techniques, up to extensions of completely bounded maps, are briefly described above in the comments about dilations of multi-operators, and below in the *Formal credits* of these notes. Inequalities similar to the von Neumann's one are discussed in Chapter 2 below, when considering functional calculi for a given operator.

More detailed information can be found in B. Sz.-Nagy and Foiaş [**SzNF4**], F. Riesz and B. Sz.-Nagy [**RSzN**] (an Appendix by B. Szőkefalvi-Nagy), B. Sz.-Nagy [**SzN5**], S. Drury [**Dru2**], G. Pisier [**Pi1**], V. Paulsen [**Pau**], as well as in the papers quoted below in the *Formal credits*.

Formal credits. As already mentioned, for the construction of unitary dilations and the function model, we closely follow the coordinate free approach described above, especially as presented in N. Nikolski and V. Vasyunin [**NVa5**] and [**NVa8**]. We refer to these papers for more explanations and for a kind of "philosophy" of free models, and to B. Szőkefalvi-Nagy and C. Foiaş [**SzNF4**] and N. Nikolski [**N19**] for history of the subject and more references.

Lemma 1.1.2 comes from D. Sarason [**S2**].

Lemma 1.2.3 in the scalar case $\dim E = \dim E_* = 1$ was already known to N. Wiener (see A.7.2.1, A.7.2.3 and A.7.6); the vector valued form given in the text is usually referred to Y. Fourès and I. Segal [**FoS**], see also P. Lax and R. Phillips [**LPh1**] and B. Sz.-Nagy and C. Foiaş [**SzNF4**]. For yet another proof of the same fact but for the case of Hardy spaces $H^2(E)$ in place of $L^2(E)$ see B.6.5.2(p). The general theorem 1.5.1(o) on the commutant of a normal operator should be well-known, but we cannot specify a reference.

For the original proof of the Langer lemma 1.2.6 and for corresponding references see B. Sz.-Nagy and C. Foiaş [**SzNF4**]. Note that the unitary subspace H_u can be written in the form

$$H_u = \bigcap_{n\in\mathbb{Z}} \operatorname{Ker} D_{T_n}^2 = \operatorname{Ker}(I - A) \cap \operatorname{Ker}(I - A_*),$$

where $T_n = T^n$ for $n \geq 0$ and $T_n = T^{*|n|}$ for $n < 0$, and A and A_* are the limit operators from 1.5.6(f)(ii).

The matrix approach for computing the characteristic function, employed in Theorem 1.2.10, is mostly from R. Douglas [**Dou7**].

The transcription problem occupying Section 1.3 is considered in more details in N. Nikolski and V. Vasyunin [**NVa5**]; in particular, yet another transcription related to invariant subspaces of M_Θ is presented. The principal B. Szőkefalvi-Nagy and C. Foiaş transcription 1.3.5 has been introduced in [**SzNF1**], see [**SzNF4**] for more references and detailed history remarks. The B. Pavlov transcription 1.3.8 appeared in B. Pavlov [1975]; this model is perfectly adapted for the needs of scattering theory, where the *incoming and outgoing subspaces* (G_* and G, respectively) and the corresponding evolutions play central roles. The relationships with the L. de Branges–J.Rovnyak model, described above, are studied in N. Nikolski and V. Vasyunin [**NVa4**]; see also J. Ball [**Ball**].

For the special case described in 1.3.9, where the characteristic function Θ_T is inner (which is equivalent to $T^{*n}x \longrightarrow 0$ for avery $x \in H$ (see 1.5.4(b))), the model was introduced already in G.K. Rota [**Rota**]. Moreover, following Rota, an operator $U : \mathcal{H} \longrightarrow \mathcal{H}$ on a (separable) Hilbert space $\mathcal{H}$ is called *universal*, if for every Hilbert space operator $T : H \longrightarrow H$ there exist $\lambda \neq 0$ and a subspace $\mathcal{H}_0 \in \operatorname{Lat}(U)$ such that λT is similar (or even unitarily equivalent) to $U|\mathcal{H}_0$. Rota's theorem tells that the backward shift $S^* : H^2(E) \longrightarrow H^2(E)$ with $\dim E - \infty$ is universal. S. Caradus [**Cara**] showed that an operator $U : \mathcal{H} \longrightarrow \mathcal{H}$ is universal if and only if $U\mathcal{H} = \mathcal{H}$ and $\dim \operatorname{Ker} U = \infty$.

For the spectral theorem, or J. von Neumann model 1.5.1(k), (p) and (q), used in Section 1.3, see comments below.

In Section 1.4 we mostly follow N. Nikolski and V. Vasyunin [**NVa8**], where the case of integration with respect to an arbitrary measure μ is considered. Namely, it is shown that the operator A, given by

$$(Af)(x) = i\int_{[0,x)} f(t)\,d\mu(t) + \frac{i}{2}\mu(\{x\})f(t), \quad x \in [0,1]$$

is a completely nonselfadjoint dissipative operator on $L^2([0,1],\mu)$, whose Cayley transform T has the characteristic function

$$\Theta_T(z) = \mathcal{S}_A\Big(i\frac{1+z}{1-z}\Big),$$

with

$$\mathcal{S}_A(\zeta) = \left(\prod_{0 \le t \le 1} \frac{\zeta - i\mu(\{t\})/2}{\zeta + i\mu(\{t\})/2} \right) \exp\left(-i \frac{\mu_c([0,1])}{\zeta} \right),$$

$\mu_c = \mu - \sum_{t \in [0,1]} \mu(\{t\})\delta_t$ being the continuous part of μ. A criterion of complete nonselfadjointness of an operator of the form $A + M_a$, where $M_a f(x) = a(x)f(x)$, is found in T. Kriete [**Kri1**].

In the computation of the characteristic function of a dissipative Sturm–Liouville operator in 1.4.13, we follow B. Pavlov [**P1**]; see also N. Nikolski and V. Vasyunin [**NVa8**]. It is interesting to note an important *inverse problem* raised by B. Pavlov [**P1**], asking which contractive (even scalar) functions Θ can be represented in the form $\Theta = \Theta_{C_+(\ell_h)}$? Some progress was made by B. Pavlov himself and some others, but in general the problem is still open; see N. Nikolski and S. Hruschev [**HrN**] for a status report.

It is also worth mentioning that the Cayley transform joins contractive and dissipative (or, accretive) theories, but this link does not mean an "automatic translation" of one theory into another (for example, the domains of ℓ_h and the adjoint ℓ_h^* are different, so the standard formula of Lemma 1.4.3 linking defect operators of a bounded operator and its Cayley transform does not work, etc). An approach to this problem is given in B. Solomyak [**So2**]. For general properties of the Cayley transform see, e.g., N. Akhiezer and I. Glazman [**AG**].

C^*-algebras and normal operators. The field we are touching in Subsection 1.5.1 is very large, and we cannot give here an exhausting account. For standard facts on C^*-algebras we refer to M. Naimark [**Nai4**], R. Douglas [**Dou5**], and G. Murphy [**Mur**]. The Gelfand–Naimark theorem 1.5.1(e) was first proved in [**GN**].

B. Fuglede [**Fug**] proved 1.5.1(g) for $a = b$, C. Putnam [**Put1**] obtained the general form, and M. Rosenblum [**Ros3**] found the beautiful proof presented in the text. Note that the uniqueness theorem for vector valued holomorphic functions $f : \Omega \longrightarrow X$ that we have used in the proof, being referred to a connected open set Ω instead of $\Omega = \mathbb{C}$, is equivalent to the strong maximum modulus theorem. The latter one starts with an always true step saying that if $\|f(z_0)\| = \sup_{z \in \Omega} \|f(z)\|$ for a point $z_0 \in \Omega$, then $\|f(z)\| = \text{const}$ (look at $z \longmapsto \varphi(f(z))$, where $\varphi \in X^*$ such that $\|f(z_0)\| = \varphi(f(z_0))/\|\varphi\|$). The next step of the strong maximum theorem ($\|f(z)\| = \text{const} \Rightarrow f(z) = \text{const}$) is true for Hilbert space valued functions ($f(z) \in H$) and for functions taking values in a strictly normed space X (see Subsection A.3.11.4, Volume 1), but fails for many other Banach spaces X. For a Hilbert space, this is a special case of A.8.5.1 (Volume 1), see also A.8.6 for more. The simplest counterexample for operator valued functions is $f(z) \in X = L(\mathbb{C}^2)$, where

$$f(z) = \begin{pmatrix} z & 0 \\ 0 & 1 \end{pmatrix}$$

with $\|f(z)\| \equiv 1$ for all $z \in \mathbb{D}$. It is curious to note that *for the resolvent*

$$f(z) = R_z(A) = (zI - A)^{-1}$$

of a Hilbert space operator $A : H \longrightarrow H$, $H \neq \{0\}$, the above maximum modulus theorem is still true; namely, A. Daniluk proved that

$$\|R_z(A)\| < \sup_{\zeta \in \Omega} \|R_\zeta(A)\|$$

for every open set $\Omega \subset \mathbb{C}\backslash\sigma(A)$ and every $z \in \Omega$ (published in A. Böttcher and B. Silbermann [**BSi3**], Section 3.4). Compare the above facts with the realization theorem (see D.5.6.1) which states that any $L(H)$ valued function f holomorphic for $|z| > R$ and vanishing at ∞ is of the form $f(z) = C(zI - A)^{-1}B$ for suitable A, B and C. On the other hand, the Liouville theorem (for $\Omega = \mathbb{C}$) is (of course) valid for any Banach space X.

The Fuglede–Putnam theorem 1.5.1(g) was generalized for Dunford spectral operators T on a Banach space in the following form: if $AT = TA$ then $AE_T(\Delta) = E_T(\Delta)A$ for every Borel set Δ, where E_T stands for the projector valued measure corresponding to the scalar part of T; see N. Dunford and J. Schwartz [**DS3**].

The spectral mapping theorem for a normal operator, i.e.,

$$\sigma(p(N, N^*)) = p(\sigma(N))$$

for every polynomial $p = p(z, \overline{z})$ (see 1.5.1(h)), is the key starting point for the spectral theorem. (Note that for selfadjoint or unitary operators, where these polynomials depend on *one* variable only, the spectral mapping theorem is a much simpler fact (see 1.5.9(a)), and so the spectral theorem does). Since 1.5.1(d) is proved, we get $\|p(N, N^*))\| = \|p\|_{\sigma(N)}$, which straightforwardly entails the spectral theorem for cyclic normal operators (see 1.5.1(j)). The general form from 1.5.1(k) differs from 1.5.1(j) by technical details.

There are other (and purely elementary) ways to prove $\sigma(p(N, N^*)) = p(\sigma(N))$, and hence to obtain the spectral theorem; for example, see H. Dowson [**Dow**]. The classical approach, going back to J. von Neumann [**vN1**], regards on a normal operator $N = A + iB$ as a commuting pair of selfadjoint operators A and B, and employs the same commutation property for spectral measures E_A and E_B to get the spectral measure for N. For such a proof, see, for example, N. Akhiezer and I. Glazman [**AG**]. It seems that, technically, the proof passing by the Gelfand–Naimark theorem is simpler and gives more; for other realization of the same path see, for example, W. Rudin [**Ru6**], R. Douglas [**Dou5**].

The double commutant part of 1.5.1(o) can be restated as follows. Let $\mathcal{A} = \text{alg}_{\text{WOT}}(N, N^*)$ be the *von Neumann algebra* (i.e., a WOT-closed C^*-algebra) generated by N; then $\mathcal{A}'' = \mathcal{A}$. This is the special case of the von Neumann double commutant theorem [**vN1**] telling that $\mathcal{A}'' = \mathcal{A}$ for every von Neumann algebra containing I. The commutant part of 1.5.1(o) for selfadjoint operators can be found in M. Birman and M. Solomyak [**BiS**], and for normal operators N of finite multiplicity ($\sup_\zeta d_N(\zeta) < \infty$) in S. Foguel [**Fo1**]. The part of 1.5.1(n) that is referred to R. Douglas is from [**Dou4**].

The techniques of direct integrals of Hilbert spaces from 1.5.1(k) is introduced in J. von Neumann [**vN4**]. It gives the exact meaning to the claim that the spectral theorem diagonalizes a normal operator. With other approaches, different from that of direct integrals, corollaries like 1.5.1(p), 1.5.1(q) require complicated measure theory techniques, see A. Brown [**Brown**], for example. The pioneering results on diagonalization of selfadjoint compact operators (see 1.5.1(r)) were obtained by D. Hilbert [**Hil1**], [**Hil2**] and E. Schmidt [**Schm1**]. For this early stage of the history of the spectral theorem see F. Riesz and B. Szőkefalvi-Nagy [**RSzN**] and W. Ricker [**Ric**].

The unitary dilation technique. In Subsections 1.5.2-1.5.5, and in 1.5.7, we treat some model and unitary dilation techniques, mostly originated in the work of B. Szőkefalvi-Nagy and C. Foiaş during the 1960ies; the general source for more results and original references is [**SzNF4**]. However, several particular points are commented below. The main point is the existence of the "wave limits" from 1.5.2(a),

$$P_{\mathcal{R}}|\mathcal{K}_\Theta = \lim_n U^n M_\Theta^{*n}, \quad P_{\mathcal{R}_*}|\mathcal{K}_\Theta = \lim_n U^{-n} M_\Theta^n,$$

where $\mathcal{R}$ and $\mathcal{R}_*$ stand for the residual and $*$-residual parts of the minimal unitary dilation of T (see also 1.5.4). In scattering theory, similar limits establish unitary equivalence between essential parts of a unitary operator and its "small" unitary perturbation, see, for example, P. Lax and R. Phillips [**LPh1**], D. Yafaev [**Yaf**]. Removing the latter adjective "unitary" we fall into the principal "philosophy" of function models: a contractive perturbation $T = U + K$ of its own unitary dilation U is, perhaps, not equivalent to U, but keeps very close relationships with it. Several initial realizations of this principle are shown in Section 1.5. In particular, we refer to B. Szőkefalvi-Nagy and C. Foiaş [**SzNF5**] for the result of 1.5.5(d)I, and to Chapter B.9 for several results and comments about the Bezout equations arising in this similarity criterion. Many other applications of the same techniques are developed by L. de Branges, J. Rovnyak, L. Shulman, S. Naboko, H. Neidthard, A. Tikhonov, and some others. For a comprehensive survey of this scattering type model spectral theory we refer to B. Solomyak [**So3**], where many of the preceding results are strengthened. In particular, for a trace class perturbation $T = U + K$, $K \in \mathfrak{S}_1$, with $\sigma(T) \not\supseteq \mathbb{D}$, the existence of the *wave operators*

$$\lim_{n\to\pm\infty} U^n T^{-n}$$

is shown on dense subsets of a suitably defined absolutely continuous subspace $H_a(T)$. Moreover, in case $\sup_{n\in\mathbb{Z}} \|T^n\| < \infty$, similarity of absolutely continuous parts $T|H_a(T)$ and $U|H_a(U)$ is established.

For the results presented in Subsections 1.5.2-1.5.5 (excepting 1.5.2(d) and 1.5.5(d)I) see [**SzNF4**] and references given therein. The classes $C_{\alpha\beta}$, giving an asymptotic classification of contractions (see 1.5.4(b)), are introduced in B. Szőkefalvi-Nagy and C. Foiaş [**SzNF2**].

"Di-formations" of Subsection 1.5.3 are called *"quasi-affine transformations"* in [**SzNF4**]. Theorems 1.5.4(d) and 1.5.5(d) are the key points of the similarity and quasi-similarity theory constructed by B. Szőkefalvi-Nagy and C. Foiaş. The notion of quasi-similarity was intensively studied in the framework of the model theory by many people. For more about this, and especially for a quasi-similar classification of C_0 contractions (see chapter 2), we refer to the books [**SzNF4**] and H. Bercovici [**Be**], and to the survey of V. Kapustin and A. Lipin [**KLi1**], [**KLi2**]. The latter papers provide a very careful and systematic revision of the literature on the subject and strengthen many of the previuous results. In particular, the authors introduce a useful notion of *pseudo-similarity*, intermediate between quasi-similarity and similarity. Namely, two operators $T_i : H_i \longrightarrow H_i$ $(i = 1, 2)$ are *pseudo-similar* if there exist operators $X : H_1 \longrightarrow H_2$ and $Y : H_2 \longrightarrow H_1$ intertwining T_i, i.e.

$$XT_1 = T_2X, \quad YT_2 = T_1Y,$$

such that $YX \in \operatorname{alg}(T_1)$, $XY \in \operatorname{alg}(T_2)$ and the ideals $YX \cdot \operatorname{alg}(T_1)$ and $XY \cdot \operatorname{alg}(T_2)$ are WOT dense in the algebras $\operatorname{alg}(T_1), \operatorname{alg}(T_2)$, respectively. Here $\operatorname{alg}(T)$ means the WOT closed algebra generated by T and I. In particular, YX and XY are di-formations, which, for many contractions T_i, are functions of these operators. It is shown that, in many aspects, pseudo-similarity is more natural and useful than quasi-similarity. Also, notice the paper V. Vasyunin and N. Makarov [**MV2**] containing a quasi-similar classification of c.n.u. contractions $T \in C_{10}$ with $\partial_T = \operatorname{rank} D_T < \infty$, $\partial_{T^*} = \operatorname{rank} D_{T^*} < \infty$ and $\operatorname{ind} T = \partial_T - \partial_{T^*} = -1$ in terms of finitely generated ideals of the algebra H^∞.

The existence of unitary ρ-dilations was first proved for $\rho = 2$ (1.5.7(c)) by C. Berger [**Berge**] and P. Halmos [**Hal1**], then was studied by B. Szőkefalvi-Nagy and C. Foiaş, and others. We refer to [**SzNF4**] for more details, in particular for the (elsewhere unpublished) results by H. Langer 1.5.7(b) and J. Stampfli 1.5.9(c)(ii).

Dilations and von Neumann type inequalities. After Ando's theorem 1.5.8(b), [**And1**], there were many attempts to prove the existence theorem for several commuting contractions until S. Parrott published the counterexample 1.5.8(c), [**Par1**]. S. Brehmer's sufficient conditions 1.5.8(d) and 1.5.8(e) appeared in [**Bre**]. Now, after the Arveson extension theorem quoted above, we know that the necessary and sufficient condition for the existence of a multi-unitary dilation is that $T = (T_1, \dots, T_n)$ is a completely contractive multi-operator. For more details, we re-refer the reader to sources mentioned in the above general discussion.

Different proofs of the von Neumann inequality are already commented above. We shall add formal references. Namely, 1.5.9(c) is contained in [**SzNF4**], but 1.5.9(c)(ii) is referred to an unpublished manuscript by J. Stampfli. The inequality 1.5.9(d) for unitary multi-operators is classical; usually, it is deduced from a von Neumann theorem on simultaneous diagonalization of a family of commuting normal operators (see comments above). The claim of 1.5.9(f) is from S. Brehmer [**Bre**], and 1.5.8(h) is from S. Parrott [**Par1**]. The ingenious example 1.5.9(i) of M. Crabb and A. Davie [**CrD**] appeared after a more complicated example by N. Varopoulos [**Var3**] relying on techniques of tensor products. The idea to use the spherically isometric Hardy space H^2_n of 1.5.9(j) is due to D. Clark [**Cl1**] (where it appeared for different purposes) and S. Drury [**Dru1**]. Many facts about this space can be found in W. Arveson [**Arv5**]. The claim of 1.5.9**(j)**(iii) can be found in S. Drury [**Dru1**] and, in a dual form, in W. Arveson [**Arv5**]. For the maximality property of the Hardy multi-shift operator given in 1.5.9(k) and 1.5.9(l) see S. Drury [**Dru1**], G. Popescu [**Pop1**] and W. Arveson [**Arv5**]. Arveson's result is, in fact, a "commutative part" of the Popescu's one (in the latter, the shifts S_k are replaced by "spherically isometric" generators of the so-called *Cuntz algebra* (the creation operators on the full Fock space)). For the results of 1.5.9(m) and 1.5.9(n) see M. Bozejko [**Boz**] and W. Mlak [**Ml**], respectively. The harmonic analysis approach presented in 1.5.9(o) and based on an operator Poisson kernel is developed in R. Hirschfeld [**Hir**], but the proof is due to J. Holbrook (first published in [**Hir**]).

For the functional calculus aspects of von Neumann's inequality see also Sections 2.5 and 2.6 below.

A harmonic analysis approach to similarity to a unitary operator. This approach presented in Subsection 1.5.6 is not related to model operators. This is simply to complete the treatment of the similarity problem, considered in 1.5.3-1.5.5 by means of the model theory. The Banach generalized limits from 1.5.6(a) are classical and should be referred to S. Banach [**Ban2**].

B. Szőkefalvi-Nagy's theorem 1.5.6(b) is proved in [**SzN1**] (in a different way). In fact, its analogue is still valid for any bounded *amenable group* of operators on a Hilbert space, J. Dixmier [**Dix**].

The majority of the results presented in 1.5.6(d) through 1.5.6(e) appeared (independently) in S. Naboko [**Nab3**] and J. van Casteren [**VanC1**]. Our treatment strengthens these papers in several points. For example, the counterexample 1.5.6(e) is stated in [**Nab3**] in a weaker form (as is given in (e)(ii)), and the uniform norm conditions (B) and (C) from 1.5.6(d)(iii) are not presented at all. Note that these uniform conditions are not empty even for normal operators, see D.1.6.1(g) below. In fact, the reader can find some more points to compare how the resolvent and power growth are treated in Subsections 1.5.6 and D.1.6.1. It is also worth mentioning that a simple concrete example of a bilateral shift $Te_k = \lambda_k e_k$, $k \in \mathbb{Z}$, with the same properties as in 1.5.6(e) is given in J. van Casteren [**VanC2**]; namely, $\lambda_k = \alpha_{k+1}/\alpha_k$ where $\alpha_k = (1+|k|)^{-\gamma}$, $0 < 2\gamma < 1$.

The very first counterexample of this series, namely 1.5.6(e)(ii)B, was constructed by A. Markus [**Ma1**] using a completely different idea. Namely, let $(x_n)_{n\geq 0}$ be a conditional basis of a Hilbert space (an example: $x_n(t) = e^{int}$ in $H^2(|t|^\alpha\, dm(e^{it}))$ for $0 < \alpha < 1$, see A.5.5) and

$$Ax_n = \lambda_n x_n, \quad n \geq 1,$$

where $\lambda_n = e^{i\theta_n}$ with a monotone converging sequence of arguments $(\theta_n)_{n\geq 0}$. The Abel transformation shows that

$$\|R_\lambda(A)\| \leq C \cdot \operatorname{Var}\left(\frac{1}{\lambda - \lambda_n}\right) \leq \frac{\text{const}}{\operatorname{dist}(\lambda, \mathbb{T})}$$

for $\lambda \in \mathbb{C}\backslash\mathbb{T}$, but A is not similar to a unitary operator since $(x_n)_{n\geq 1}$ is not a Riesz (unconditional) basis (see A.5.6). It is interesting to note that the last inequality holds for any $(\lambda_n)_{n\geq 1}$ situated on a rectifiable curve γ if and only if γ is an Ahlfors (Carleson) curve, see N. Benamara and N. Nikolski [**BN**].

It is also worth mentioning that 1.5.6(d)(iii), which appeared in the papers by J. van Casteren and S. Naboko quoted above, generalizes an earlier result by I. Gohberg and M. Krein [**GK3**]. This result, in turn, claiming that the conditions $\|T\| \leq 1$ and $\mathrm{LRG}(T^{-1})$ imply similarity to a unitary operator, is a model free statement of 1.5.5(d)II proved by B. Sz.-Nagy and C. Foiaş [**SzNF3**].

The simple similarity criterion from 1.5.6(f) is due to R. Douglas [**Dou3**]; see also P. Fillmore [**Fil**].

We will come back to the similarity problem in the next chapter.

CHAPTER 2

Elements of Spectral Theory in the Language of the Characteristic Function

In this chapter we introduce some initial concepts of spectral theory in the function model language. These techniques will then be developped up to unconditionally convergent spectral decompositions and relevant interpolation problems in the subsequent Chapter 3. In the framework of this book, we are able to realize this program for the case of scalar valued characteristic functions only. This is why, even for quite simple, non-technical questions, we restrict ourselves to scalar valued functions, or to scalar-like functions (such as operator valued functions having a scalar multiple).

In Section 2.1 we describe invariant subspaces of a model operator, as well as the shift invariant subspaces of the spaces $L^p(\mathbb{T})$. The latter, for the case $p = \infty$, is used in Section 2.3 when defining the minimal annihilating function m_T.

In Sections 2.2 and 2.3 we develop the H^∞ functional calculus for a completely nonunitary contraction, including the famous von Neumann inequality, define and study the class C_0 of operators satisfying the Hamilton–Cayley theorem, and describe the spectrum $\sigma(T)$ in terms of the singular points of Θ_T.

Section 2.4 is devoted to the commutant lifting theorem (limited to an "around-scalar" case) based on the Nehari theorem B.1.3.2.

Section 2.5, Exercises and Further Results, extends the material of the chapter in several ways, both of classical and model operator theory. Here we present 1) the Riesz–Dunford functional calculus, 2) continuity and discontinuity theorems on the spectrum mapping $T \longmapsto \sigma(T)$, including Kakutani's counterexample; 3) a treatment of the (in)stability problem for the continuous spectrum (including classical results of Weyl, von Neumann, Kuroda, and Gohberg); 4) von Neumann's theory of spectral sets; 5) the Gohberg–Krein similarity criterion; 6) an extended treatment of the H^∞ functional calculus and the commutant $\{M_\Theta\}'$, including some conditions for $\{M_\Theta\}' \cap \mathfrak{S} \neq \{0\}$ for a given operator ideal $\mathfrak{S}$; and 7) an account of C_0 operators (including an elementary proof of the Foiaş decomposability property). The chapter ends with a survey of the state of the theories developed in it (Section 2.6).

2.1. Invariant Subspaces

We begin with the shift invariant subspaces of vector valued $L^2(E)$ spaces and prove the Lax theorem 2.1.3. Recall that in an abridged form this fact already occured as Exercise A.1.6.3. Now, we need more details, and in particular, we deduce from 2.1.3 the invariant subspace theorem for a contraction with inner characteristic function. Then we apply this result to the integration operator and explain the solution to the Gelfand problem 2.1.9.

2.1.1. Information about vector valued H^2 and H^∞ spaces. Here we gather definitions and list some properties of vector valued H^2 and H^∞-spaces on the unit circle $\mathbb{T}$. In fact, we only need the spaces $L^2(E)$, $H^2(E)$ with a separable Hilbert space E, and the space $H^\infty(L(E, E_*)) = H^\infty(E \longrightarrow E_*)$ of operator valued functions. However, most of the properties listed below hold for more general situations, for instance, for separable reflexive spaces as coefficient spaces E. The proofs are usually straightforward analogues of the proofs of Chapter A.3 (Volume 1); we do not repeat but sometimes replace them by short hints to indicate the necessary changes. For more details we refer to Section A.3.11 (Volume 1).

As before, we denote by L^2 and $L^2(E)$, respectively,the scalar and the vector valued Lebesgue spaces with respect to the normalized Lebesgue measure on $\mathbb{T}$, $m\mathbb{T} = 1$. We use $H^2(E)$ to denote the subspace of $L^2(E)$ of functions f with vanishing negatively indexed Fourier coefficients:

$$\widehat{f}(n) = \int_{\mathbb{T}} f(z)\overline{z}^n \, dm(z) = 0, \quad n < 0.$$

Then f can be extended analytically to the unit disk $\mathbb{D}$ by

$$f(z) = \sum_{n \geq 0} \widehat{f}(n) z^n, \quad |z| < 1,$$

and by Fatou's theorem, the boundary limits

$$\lim_{r \to 1} f(r\zeta) = f(\zeta)$$

exist for almost every $\zeta \in \mathbb{T}$ (even as nontangential limits); see Section A.3.11 (Volume 1). Moreover, if $f_r(\zeta) = f(r\zeta)$ stands for the trace of f on the circle $\{z \in \mathbb{C} : |z| = r\}$, then we have mean convergence in the L^2-norm:

$$\lim_{r \to 1} \int_{\mathbb{T}} \|f(\zeta) - f_r(\zeta)\|^2 \, dm(\zeta) = 0.$$

Conversely, given a holomorphic function $g : \mathbb{D} \longrightarrow E$ for which

$$\sup_{0<r<1} \int_{\mathbb{T}} \|g_r(\zeta)\|^2 \, dm(\zeta) < \infty,$$

the $L^2(E)$-limit $\lim_{r \to 1} g_r = b(g)$ exists and belongs to $H^2(E)$. Moreover, $g(r\zeta) = b(g) * P_r(\zeta)$, where P_r stands for the Poisson kernel. As in the scalar case $E = \mathbb{C}$, we always identify $b(g)$ and g.

An analogous identification is possible for the space $H^\infty(E \to E_*)$ of $L(E, E_*)$-valued bounded analytic functions, where $L(E, E_*)$ is the space of bounded linear operators from E to E_*. Another way of regarding this space is to consider it as $\{F \in L^\infty(E \to E_*) : \widehat{F}(k) = \mathbb{O}, \ k < 0\}$. Finally, it is easy to verify that $F \in L^\infty(E \to E_*)$ is in $H^\infty(E \to E_*)$ if and only if $Fe \in H^2(E_*)$ for every constant $e \in E$, or $Fx \in H^2(E_*)$ for every $x \in H^2(E)$.

The proofs can be based on the obvious fact that the function $z \longmapsto F(z)e$, $z \in \mathbb{D}$, is bounded for every $e \in E$, and hence belongs to $H^2(E_*)$; therefore, one can use the properties of H^2 spaces mentioned above. In particular, the boundary limit $\lim_{r \to 1} F_r(\zeta)e = F(\zeta)e$ exists for every $e \in E$, a.e. on $\mathbb{T}$. However, in general, $\lim_{r \to 1} \|F_r(\zeta) - F(\zeta)\| \neq 0$ (as an example, consider the multiplication operators $F(z)(x_n)_{n \geq 0} = (z^n x_n)_{n \geq 0}$ for $(x_n)_{n \geq 0} \in l^2$, $z \in \mathbb{D}$).

2.1.2. Invariant subspaces of the space $L^2(E)$. We begin with the study of z-invariant subspaces of $L^2(E)$, where E is a separable Hilbert space. Recall that we have already considered this problem in Exercise A.1.6.3 (Volume 1), but now we give more details and deduce some consequences of the following description.

2.1.3. THEOREM (P. Lax, 1959). *Let $L \subset L^2(E)$ with $zL \subset L$. Then there is a unique decomposition $L = L_0 \oplus L_1$ such that*

(1) L_0 is a reducing subspace for z, and as every subspace of this type is of the following form

$$L_0 = \{f \in L^2(E) : f(\zeta) \in P(\zeta)E, \ a.e. \ \zeta \in \mathbb{T}\},$$

where $P(\zeta) : E \to E$ is a measurable function on $\mathbb{T}$ whose values are orthogonal projection.

(2) L_1 is completely nonreducing *(that is, contains no reducing subspaces). Such a subspace is given by*

$$L_1 = \Theta H^2(F),$$

where $\dim F \leq \dim E$, *and $\Theta \in L^\infty(F \to E)$ is an operator valued function whose values are isometric almost everywhere on $\mathbb{T}$. This representation is unique in the sense that if $L_1 = \Theta_1 H^2(F_1)$, then there exists a unitary operator $u : F \longrightarrow F_1$ such that $\Theta = \Theta_1 u$.*

PROOF. Every invariant subspace L of an operator T on a Hilbert space splits into an orthogonal sum $L = L_0 \oplus L_1$ of a reducing and a completely nonreducing subspace. In fact, set $L_0 = \mathrm{span}\{L' : L' \subset L, L' \text{ is reducing for } T\}$. Then L_0 is clearly the maximal reducing subspace, and setting $L_1 = L \ominus L_0 = L \cap L_0^\perp$ we obtain a T-invariant completely nonreducing subspace. In the present case we have $zL_0 \subset L_0$, $z^* L_0 \subset L_0$. Using Lemma 2.1.4 below we get

$$zP_{L_0} = P_{L_0} z,$$

where $P_{L_0} : L^2(E) \to L^2(E)$ is the orthogonal projection onto L_0. By Lemma 1.2.3, there exists a function $P(\cdot) \in L^\infty(E \to E)$ such that $P_{L_0} f = P(\zeta) f(\zeta)$, $\zeta \in \mathbb{T}$. Obviously $P_{L_0}^2 = P_{L_0}$ if and only if $P(\zeta)^2 = P(\zeta)$ for almost every $\zeta \in \mathbb{T}$ and $\|P_{L_0}\| \leq 1$ if and only if $\|P(\zeta)\| \leq 1$ a.e. on $\mathbb{T}$. Thus $P(\zeta)$ is an orthogonal projection a.e. on $\mathbb{T}$ and $L_0 = P_{L_0} L^2(E)$, which completes the description of L_0.

Consider the completely nonreducing subspace L_1. Let $F = L_1 \ominus zL_1$ be the defect space and let

$$L_1 = \sum_{n \geq 0} \oplus z^n F \oplus \left(\bigcap_{n \geq 0} z^n L_1 \right)$$

be the Wold–Kolmogorov decomposition of L_1 from Lemma A.1.5.1 (Volume 1). Then $L_\infty := \bigcap_{n \geq 0} z^n L_1$ satisfies $zL_\infty = L_\infty$ and is thus reducing. Consequently $L_\infty = \{0\}$, and we get $L_1 = \sum_{n \geq 0} \oplus z^n F$. The crucial property of the subspace L is the following: $f \in F$ implies that $\|f(\zeta)\|_E = \text{const} = \|f\|_{L^2(E)}$ a.e. on $\mathbb{T}$. Indeed, if $f \in F = L_1 \ominus zL_1$, we have $f \perp z^n f$, $n \geq 1$, that is

$$0 = \int_{\mathbb{T}} z^n (f(z), f(z))_E \, dm(z), \quad n \geq 1,$$

and by conjugation we get the same condition for $n \leq -1$ (note that $(f(z), f(z))_E = \|f(z)\|_E^2 \in \mathbb{R}$). Hence $h(z) := (f(z), f(z))_E \in L^1(\mathbb{T})$ and $\widehat{h}(n) = 0$, $n \in \mathbb{Z} \setminus \{0\}$.

Consequently $h \equiv$ const. By the polarization formula $(x,y) = \frac{1}{4}\sum_{\varepsilon=\pm 1,\pm i} \varepsilon\|x + \varepsilon y\|^2$, we get the more general result

$$(f(\zeta), g(\zeta))_E = \text{const a.e. on } \mathbb{T} \tag{2.1.1}$$

for $f, g \in F$.

A digression (see also the remark in the beginning of the proof of Lemma 1.2.3).

Suppose F is as above and define a function $\Theta(z) : F \to E$, $|z| = 1$, by

$$\Theta(z)f = f(z), \quad f \in F.$$

Assume that Θ is well defined and linear. Then $\Theta(z)$ is bounded. Moreover, it is actually an isometry a.e. on $\mathbb{T}$:

$$\|\Theta(z)f\| = \|f(z)\|_E = \text{const} = \|f\|_{L^2(E)} = \|f\|_F,$$

and multiplication by Θ allows us to identify

$$H^2(F) \to L_1 = \sum_{n\geq 0} \oplus z^n F,$$

because if $g = \sum_{n\geq 0} z^n f_n$ is a polynomial, then

$$(\Theta g)(\zeta) = \sum_{n\geq 0} \zeta^n f_n(\zeta) = \left(\sum_{n\geq 0} \oplus z^n f_n\right)(\zeta).$$

The only doubt in this reasoning is whether the definition of Θ is meaningful. Namely, $\Theta(z)f = f(z)$ is well defined everywhere on $\mathbb{T}$ except on a subset $\sigma_f \subset \mathbb{T}$ of Lebesgue measure zero, $|\sigma_f| = 0$. As σ_f depends on f and the union of all these, $\sigma = \bigcup_{f\in F} \sigma_f$, may cover the whole circle $\mathbb{T}$, we have to proceed in a more rigorous way.

Namely, choose an orthonormal basis $(f_k)_{k\geq 1}$ in F. Set

$$\sigma_1 = \bigcup_{k\geq 1} \sigma_{f_k},$$

and

$$\Theta(\zeta)f_k = f_k(\zeta), \quad \zeta \in \mathbb{T} \setminus \sigma_1,\ k \geq 1.$$

As $(f_k)_{k\geq 1} \subset F$, we get by (2.1.1) that $\big(f_k(\zeta), f_j(\zeta)\big) =$ const a.e. on $\mathbb{T}$. Let $\sigma_{k,j} \subset \mathbb{T}$, $|\sigma_{k,j}| = 0$ $(k \neq j)$, be such that this equality is valid for *every* $\zeta \in \mathbb{T}\setminus\sigma_{k,j}$. Then the orthogonality of $(f_k)_{k\geq 1}$ implies that of $(f_k(\zeta))_{k\geq 1}$ for every $\zeta \notin \sigma_{k,j}$ $(k, j \geq 1)$. Set $\sigma_2 = \bigcup_{k\neq j} \sigma_{k,j}$ and $\sigma = \sigma_1 \cup \sigma_2$. Further, for finitely supported sequences $(a_k)_{k\geq 1} \subset \mathbb{C}$ set

$$\Theta(\zeta)\sum_k a_k f_k = \sum_k a_k f_k(\zeta), \quad \zeta \in \mathbb{T} \setminus \sigma.$$

Then for $\zeta \in \mathbb{T} \setminus \sigma$ we get

$$\begin{aligned}\left\|\Theta(\zeta)\sum_k a_k f_k\right\|_E^2 &= \sum_k |a_k|^2 \|f_k(\zeta)\|_E^2 = \sum_k |a_k|^2 \|f_k\|_{L^2(E)}^2 \\ &= \left\|\sum_k a_k f_k\right\|_{L^2(E)}^2 = \left\|\sum_k a_k f_k\right\|_F^2.\end{aligned}$$

Hence $\Theta(\zeta)$ is an isometry from F to E a.e. on $\mathbb{T}$, and by the above digression $\Theta H^2(F) = L_1$.

The uniqueness claim is left to the reader. □

2.1.4. LEMMA. *A subspace L_0 is reducing for an operator T if and only if $P_{L_0}T = TP_{L_0}$ where P_{L_0} is the orthogonal projection onto L_0.*

PROOF. By definition L_0 is reducing for T if and only if $TL_0 \subset L_0$ and $TL_0^{\perp} \subset L_0^{\perp}$. The lemma follows. □

2.1.5. Some consequences of the Lax theorem. We already know the first of the corollaries below: this is simply Theorem A.1.3.2, but the proof is slightly different. The principal proposition of this subsection is Corollary 2.1.8, which contains a description of the invariant subspaces of a model operator having an inner characteristic function.

2.1.6. COROLLARY. *Let L be a z-invariant subspace of L^2 and $L = L_0 \oplus L_1$ be its decomposition from Theorem 2.1.3 into a reducing subspace L_0 and a completely nonreducing subspace L_1. Then, either $L_0 = \{0\}$ and hence $L = \Theta H^2$, where Θ is a scalar measurable function such that $|\Theta| = 1$ a.e. on $\mathbb{T}$, or $L_1 = \{0\}$ and hence $L = \chi_\sigma L^2$, where χ_σ is the indicator function of some measurable set $\sigma \subset \mathbb{T}$.*

Indeed, if $L_0 \neq \{0\}$, then we have $L_0 = PL^2$, where $(Pf)(\zeta) = P(\zeta)f(\zeta)$ and $P(\zeta) : \mathbb{C} \to \mathbb{C}$ is a projection for almost every $\zeta \in \mathbb{T}$, that is,

$$P(\zeta) \in \{0, 1\}, \quad \text{a.e. on } \mathbb{T}.$$

Consequently, there exists $\sigma \subset \mathbb{T}$ such that $P = \chi_\sigma$, $|\sigma| > 0$, and $L_0 = \chi_\sigma L^2$.

Next, by construction, $f \perp \chi_\sigma L^2$ for every $f \in L_1$, that is $f|\sigma = 0$ a.e. on $\mathbb{T}$ for every $f \in L_1$. On the other hand, $L_1 = \Theta H^2$, where $|\Theta| = 1$ a.e. on $\mathbb{T}$. By the uniqueness theorem (cf. A.1.4.3) we conclude that $f \equiv 0$ for every $f \in L_1$, so $L_1 = \{0\}$.

If we assume that $L_0 = \{0\}$, then $L = L_1 = \Theta H^2(F)$ with $\dim F \leq \dim E = 1$, and since $L \neq \{0\}$ we have $\dim F = 1$. □

2.1.7. COROLLARY. *Let $L \subset H^2(E)$, $zL \subset L$. Then $L = \Theta H^2(F)$, where $\Theta(\zeta) : F \to E$ is inner. This representation is unique in the same sense as in Theorem 2.1.3*

Suppose there exists $f \in L_0$, $f \not\equiv 0$, where L_0 is the reducing part of the decomposition of Theorem 2.1.3. Then for suitable $n \geq 0$ we have $z^{-n}f \notin H^2(E)$, which is impossible. Hence $L_0 = \{0\}$ and $L = L_1 = \Theta H^2(F)$. Since $\Theta H^2(F) \subset H^2(E)$, Θ is inner. □

2.1.8. Corollary. *Let $\Theta : E \to E_*$ be an inner function and let*

$$\mathcal{K}_\Theta = H^2(E_*) \ominus \Theta H^2(E)$$

be the associated model space. Suppose that $K \subset \mathcal{K}_\Theta$ is an M_Θ-invariant subspace. Then there is a unique (up to unitary equivalence) inner function $\Theta_2(z) : F \to E_*$ such that $\Theta = \Theta_2\Theta_1$, where Θ_1 is another inner function $\Theta_1(z) : E \to F$ ($\dim E_* \leq \dim F \leq \dim E$), and

$$K = \Theta_2 H^2(F) \ominus \Theta H^2(E).$$

Conversely, if we have a nonconstant inner factorization $\Theta = \Theta_2\Theta_1$ of Θ, then $M_\Theta K \subset K$ and the characteristic function of the restriction $\Theta_{M_\Theta|K}$ is the pure part of Θ_1.

Indeed, set $L = K \oplus \Theta H^2(E)$. Then

$$zf = P_\Theta zf + (1 - P_\Theta)zf \in L$$

for every $f \in K$, where P_Θ denotes the orthogonal projection onto $\mathcal{K}_\Theta$, as above. If $f \in \Theta H^2(E)$, then we also have $zf \in \Theta H^2(E) \subset L$, and hence $zL \subset L \subset H^2(E_*)$. By Corollary 2.1.7, there is an inner function $\Theta_2(z) : F \to E_*$ such that $L = \Theta_2 H^2(F)$. By construction we have $L = \Theta_2 H^2(F) \supset \Theta H^2(E)$, and hence $\Theta H^2(E) \perp \Theta_2 H^2_-(F)$ (note that $\Theta_2 H^2(F) \perp \Theta_2 H^2_-(F)$), that is, $\Theta_2^*\Theta H^2(E) \perp H^2_-(F)$. We deduce that

$$\Theta_2^*\Theta H^2(E) \subset H^2(F),$$

which is equivalent to $\Theta_1 := \Theta_2^*\Theta$ being an $H^\infty(E \to F)$-function. Obviously, it is contracting:

$$\|\Theta_1(\zeta)x\| \leq \|\Theta_2(\zeta)^*\Theta(\zeta)x\| \leq \|x\|.$$

Moreover, as $\Theta_2(\zeta)$ is isometric for almost every $\zeta \in \mathbb{T}$, the operator $\Theta_2(\zeta)\Theta_2(\zeta)^*$ is an orthogonal projection onto $\operatorname{Range}\Theta_2(\zeta) = \Theta_2(\zeta)F \supset \Theta(\zeta)E$ (the latter inclusion comes from $\Theta E \subset \Theta_2 H^2(F)$). Hence

$$\Theta_2(\zeta)\Theta_1(\zeta) = \Theta_2(\zeta)\Theta_2^*(\zeta)\Theta(\zeta) = \Theta(\zeta).$$

Now $\Theta(\zeta)$ is an isometry and $\Theta_2(\zeta), \Theta_1(\zeta)$ are contracting, so $\Theta_1(\xi)$ is an isometry a.e. on $\mathbb{T}$, that is, an inner function. Clearly,

$$K = L \ominus \Theta H^2(E) = \Theta_2 H^2(F) \ominus \Theta H^2(E).$$

The verification of the converse part, in particular the assertion that the pure part of Θ_1 is equivalent to $\Theta_{M_\Theta|K}$, is left to the reader.

[*Hint*: $K = \Theta_2(H^2(F) \ominus \Theta_1 H^2(E))$.] □

2.1.9. An application: invariant subspaces of the Volterra integration operator. The *Gelfand problem*, raised in 1938, is to characterize the *cyclic* vectors of the Volterra operator $V : L^2(0,a) \longrightarrow L^2(0,a)$ given by

$$Vf(x) = \int_0^x f\,dt, \quad x \in [0,a],$$

that is, to determine those functions $f \in L^2(0,a)$, for which

$$\operatorname{span}(V^n f : n \geq 0) = L^2(0,a).$$

Certainly, this is the special case of the general problem of describing the *lattice of invariant subspaces of* V. We will use the notation $\operatorname{Lat} V$ for this lattice

$$\operatorname{Lat} V = \{E : E \subset L^2(0,a), VE \subset E\}.$$

In order to describe the lattice of V we will rather consider its Cayley transform $T = C^+(V) = (I - V)(I + V)^{-1}$ and the associated function model of T. We start with the following result.

2.1.10. LEMMA. *Let $T : X \to X$ be an operator acting on a Banach space X, let Ω be a connected component of the complement of the spectrum $\widehat{\mathbb{C}} \setminus \sigma(T)$ and let $E = \overline{E}$ be a closed subspace of X. If $R_\lambda(T)E \subset E$ for a point $\lambda \in \Omega$, where $R_\lambda(T) = (\lambda I - T)^{-1}$ is the resolvent of T, then $R_\mu(T)E \subset E$ for every $\mu \in \Omega$ (by definition we will set $R_\infty(T) = T$).*

Moreover, if $E \in \operatorname{Lat}(T)$, then $R_\lambda(T)E \subset E$ if and only if $\lambda \notin \sigma(T|E)$.

PROOF. For $x \in E$, $y \in E^\perp \subset X^*$ and $z \in \Omega$ set $\varphi(z) = \langle R_z(T)x, y\rangle$. Observe that

$$\left(\frac{d}{dz}\right)^n R_z(T) = \left(\frac{d}{dz}\right)^n (zI - T)^{-1} = (-1)^n n! R_z(T)^{n+1}, \quad n \geq 0.$$

In particular, if $R_\lambda E \subset E$ for $z = \lambda \in \Omega$, then $\varphi^{(n)}(\lambda) = 0$, $n \geq 0$. Now φ is holomorphic on the connected set Ω and hence $\varphi \equiv 0$ on Ω, so we deduce that $R_\mu(T)E \subset E$ for every $\mu \in \Omega$.

Consider now the point $z = \infty$. If $R_\infty(T)E = TE \subset E$, then clearly $R_\lambda(T)E \subset E$ for every $|\lambda| > \|T\|$, and as above we get $R_\lambda E \subset E$ for every λ from the unbounded component Ω_∞ of $\hat{\mathbb{C}} \setminus \sigma(T)$. Conversely, if $R_\lambda E \subset E$ for a $\lambda \in \Omega_\infty$, then the same is true for all $|\lambda| > \|T\|$. Since $\lambda(R_\lambda \lambda - 1) \to T$ as $|\lambda| \to \infty$, we also get $TE \subset E$.

In order to prove the last property, observe that $(\lambda I - T)E \subset E$, and if $(\lambda I - T)|E$ is a bijection then $R_\lambda(T)E = R_\lambda(T)(\lambda I - T)E = E$. Conversely, if $R_\lambda(T)E \subset E$ then, obviously, $R_\lambda(T)|E$ is the inverse of $(\lambda I - T)|E$. □

2.1.11. COROLLARY. *For $T = (I - V)(I + V)^{-1}$ we have*

$$\operatorname{Lat} V = \operatorname{Lat} T.$$

Indeed, as $\sigma(V) = \{0\}$ (cf. 1.4.7), Lemma 2.1.10 implies that $\operatorname{Lat} V \subset \operatorname{Lat} T$. Conversely, $V = (I - T)(I + T)^{-1}$ and by the spectral mapping theorem (see 2.5.1(b)) we get $\sigma(T) = \{1\}$. Now Lemma 2.1.10 gives $\operatorname{Lat} T \subset \operatorname{Lat} V$. □

Recall that the contraction $T = (I - V)(I + V)^{-1}$ is unitarily equivalent to M_{Θ_a}, where

$$\Theta_a = \exp\left(-a\,\frac{1+z}{1-z}\right),$$

and note that the invariant subspace lattices of unitary equivalent operators are isomorphic.

2.1.12. LEMMA. *The lattice* $\operatorname{Lat} M_{\Theta_a}$ *is* totally ordered*: if $E_1, E_2 \in \operatorname{Lat} M_{\Theta_a}$, $E_1 \neq E_2$, then either $E_1 \subset E_2$ or $E_2 \subset E_1$.*

PROOF. We have already seen that a subspace $E \subset \mathcal{K}_{\Theta_a}$ is M_{Θ_a}-invariant, $M_{\Theta_a} E \subset E$, if and only if $E = \theta H^2 \ominus \Theta_a H^2$ where θ is an inner divisor of Θ_a. This is only possible for

$$\theta = \exp\left(-\alpha \frac{1+z}{1-z}\right), \quad 0 \le \alpha \le a.$$

The result follows. □

2.1.13. COROLLARY. *Let $VE \subset E \subset L^2(0,a)$. Then there exists α, $0 \le \alpha \le a$, such that $E = L^2_\alpha$, where*

$$L^2_\alpha = \{f \in L^2(0,a) : f = 0 \text{ a.e. on }]0,\alpha[\}.$$

Indeed, by Corollary 2.1.11 and Lemma 2.1.12, the lattice $\operatorname{Lat} V$ is totally ordered. Obviously, L^2_α is invariant, and thus for every $\alpha \in [0,a]$ we either have $L^2_\alpha \subset E$ or $E \subset L^2_\alpha$. Set $A = \{\alpha \in [0,a] : L^2_\alpha \subset E\}$. Then $E_- := \operatorname{clos} \bigcup_{\alpha \in A} L^2_\alpha \subset E \subset \bigcap_{\beta \notin A} L^2_\beta =: E_+$. Clearly $E_- = E_+ = L^2_\alpha$ with $\alpha = \inf A$. □

2.1.14. COROLLARY (solution to the Gelfand problem). *Let $f \in L^2(0,a)$, $E_f = \operatorname{span}(V^n f : n \ge 0)$. Then $E_f = L^2_\alpha$, where $\alpha = \inf \operatorname{supp} f$ and $\operatorname{supp} f$ denotes the closed support of f (the complement of the union of all open sets where $f = 0$ a.e.). In particular, $E_f = L^2(0,a)$ if and only if $0 \in \operatorname{supp} f$.* □

2.1.15. REMARK. In the general case of the function model with noninner characteristic function Θ

$$\mathcal{K}_\Theta = \begin{pmatrix} H^2(E_*) \\ \operatorname{clos} \Delta L^2 \end{pmatrix} \ominus \begin{pmatrix} \Theta \\ \Delta \end{pmatrix} H^2(E),$$

the invariant subspaces $L \subset \mathcal{K}_\Theta$, $M_\Theta L \subset L$ can be characterized in a similar way. Namely, they are given by certain factorizations of the characteristic functions $\Theta = \Theta_2 \Theta_1$, though the description of these admissible factorizations is now much more intricate, see Section 2.6 for more comments.

2.1.16. Invariant subspaces of $L^p(\mathbb{T})$ spaces. For the functional calculus that will be studied in Section 2.2 below we need to know the z-invariant subspaces in L^1 and L^∞. A description of these is provided by the following theorem.

2.1.17. THEOREM. *Let $zL \subset L$, where L is a closed subspace of $L^p(\mathbb{T})$, $1 \le p \le \infty$ (in the weak*-topology $\sigma(L^\infty, L^1)$ in case $p = \infty$). Then*

(1) either $z^{-1}L = L$ and then there is a measurable set $\sigma \subset \mathbb{T}$, $|\sigma| > 0$, such that $L = \chi_\sigma L^p(\mathbb{T})$,

(2) or $z^{-1}L \ne L$, and then there exists a measurable function Θ, $|\Theta| = 1$ a.e. on $\mathbb{T}$, such that $L = \Theta H^p$.

PROOF. We start with two observations. First, $zL \subset L$ if and only if $\bar{z}L^\perp \subset L^\perp$, where, for $1 \le p < \infty$,

$$L^\perp = \left\{ g \in L^{p'} : \int_{\mathbb{T}} f\bar{g}\, dm = 0 \right\}.$$

Here p' is the conjugated exponent of p: $\frac{1}{p} + \frac{1}{p'} = 1$. Secondly, if we define an operator $\mathcal{J} : L^p \longrightarrow L^p$ by $\mathcal{J}f = \bar{z} f(\bar{z})$, then $\mathcal{J}$ is a unitary operator satisfying $\mathcal{J}z = \bar{z}\mathcal{J}$, and hence, z and $\bar{z}$ are unitarily equivalent. Clearly, $\mathcal{J}L^p = L^p$, $\mathcal{J}H^p = H^p_-$ and $\mathcal{J}H^p_- = H^p$. Moreover, $\mathcal{J}$ is an involution: $\mathcal{J}^2 = I$. These two remarks allow us to recover the case $p > 2$ by standard duality arguments from the case $1 \le p \le 2$.

Thus, we may assume that $1 \le p \le 2$. Consider the following alternative.

I. $f(\zeta)\overline{g(\zeta)} = 0$ a.e. for every $f \in L$ and $g \in L^{\perp}$. Then, clearly, $z^{-1}L \subset L$. Choosing a countable dense subset (f_i) in L, and setting $\mathbb{T} \setminus \sigma = \{\zeta \in \mathbb{T} : f_i(\zeta) = 0,\ i = 1, 2, \dots\}$, the reader can easily verify that

$$L = \{f \in L^p : f = 0 \text{ a.e. on } \mathbb{T} \setminus \sigma\} = \chi_\sigma L^p$$

[*Hint:* use $(f, z^n g) = (f\bar{g})\hat{}(n)$, $n \in \mathbb{Z}$.]

II. There exists $f \in L$ and $g \in L^{\perp}$ such that $f(\zeta)\overline{g(\zeta)} \not\equiv 0$. Then $z^{-1}L \not\subset L$. Now, consider the intersection $L \cap L^2(\mathbb{T})$, which is a closed, z-invariant subspace of L^2 (note that $p \le 2$), and by Lemma 2.1.18 below it is dense in L. This implies that the inequality $z^{-1}L \ne L$ remains true for $L \cap L^2(\mathbb{T})$. Using the description of z-invariant subspaces in L^2 we deduce the existence of a unimodular function Θ such that $L \cap L^2(\mathbb{T}) = \Theta H^2$. Finally, as $|\Theta| = 1$ a.e. on $\mathbb{T}$, we get

$$L = \operatorname{clos}_{L^p}(L \cap L^2(\mathbb{T})) = \operatorname{clos}_{L^p} \Theta H^2 = \Theta H^p.$$

□

2.1.18. LEMMA. *$L \cap L^2(\mathbb{T})$ is dense in L.*

PROOF. For $f \in L$ and $n \ge 1$, let φ_n be the outer function verifying

$$|\varphi_n(\zeta)| = \begin{cases} 1 & \text{if } |f(\zeta)| < n, \\ \frac{1}{|f(\zeta)|} & \text{if } |f(\zeta)| \ge n, \end{cases}$$

for a.e. $\zeta \in \mathbb{T}$ (see A.3.9.1, Volume 1). As $|\varphi_n| \le 1$ and $0 \le \log|\varphi_n|^{-1} \le \max(0, \log|f|)$ we get $\log|\varphi_n|^{-1} \in L^1(\mathbb{T})$. Moreover, we have $\varphi_n \in H^\infty$. Using Lemma 2.1.19(2) below, we get $\varphi_n f \in L$ for every $n \ge 1$. Clearly,

$$|\varphi_n f| = \begin{cases} |f| & \text{if } |f| < n \\ 1 & \text{if } |f| \ge n, \end{cases}$$

and hence $\varphi_n f \in L^\infty \cap L \subset L^2 \cap L$.

It remains to verify the convergence $\varphi_n f \to f$ in L^p. As $m\{|f| \ge n\} \to 0$ for $n \to \infty$ we have

$$\int_{\mathbb{T}} |\varphi_n|^2 \to 1,$$

and $|\varphi_n| \nearrow 1$ for a.e. $\zeta \in \mathbb{T}$. Now, the dominated convergence theorem (note that $0 \le -\log\varphi_n \le \log^+|f|$, and $\log^+|f| \in L^1$) implies that

$$\varphi_n(z) = \exp\left(\int_{\mathbb{T}} \frac{\zeta + z}{\zeta - z} \log|\varphi_n(\zeta)|\, dm(\zeta)\right) \to 1$$

for every $z \in \mathbb{D}$. In particular, $\varphi_n(0) \to 1$ as $n \longrightarrow \infty$, and thus

$$\begin{aligned} \|\varphi_n - 1\|_{H^2}^2 &= \|\varphi_n\|^2 + 1 - 2\operatorname{Re}(\varphi_n, 1) = 2 + o(1) - 2\int_{\mathbb{T}} \varphi_n\, dm \\ &= 2(1 - \varphi_n(0)) + o(1) \to 0 \end{aligned}$$

for $n \longrightarrow \infty$. Therefore, there is a subsequence $(\varphi_{n_k})_k$ converging a.e. on $\mathbb{T}$ to 1. As $|\varphi_{n_k} f| \le |f| \in L^p$ and $\varphi_{n_k} f \to f$ a.e. on $\mathbb{T}$, we can again apply the dominated convergence theorem to get

$$\|\varphi_{n_k} f - f\|_p^p = \int_{\mathbb{T}} |\varphi_{n_k} f - f|^p \, dm \to 0.$$

□

2.1.19. LEMMA. *Let $zL \subset L$, $L \subset L^p$.*
(1) If $g \in L^{p'}$, then $g \in L^\perp$ if and only if $f\overline{g} \in H_0^1$ for every $f \in L$.
(2) If $\varphi \in H^\infty$, then $\varphi L \subset L$.

PROOF. (1) Suppose that $f\overline{g} \in H_0^1$ for every $f \in L$. Then

$$\int f\overline{g}\, dm = \widehat{(f\overline{g})}(0) = (f\overline{g})(0) = 0,$$

and hence $f \perp g$ for every $f \in L$, that is, $g \in L^\perp$.

Conversely, let $f \in L$ and $g \in L^\perp$. As $zL \subset L$, we get by induction $z^n f \perp g$, $n \ge 0$, that is, $\int z^n f\overline{g} = 0$, $n \ge 0$. Hence $\widehat{(f\overline{g})}(-n) = 0$, $n \ge 0$, and so $f\overline{g} \in H_0^1$.

(2) Take now $\varphi \in H^\infty$, $f \in L$ and $g \in L^\perp$. By (1) we have $f\overline{g} \in H_0^1$ and hence $\varphi f\overline{g} \in H_0^1$, that is — again by (1) — $\varphi f \perp g$. As $g \in L^\perp$ is arbitrary, we get $\varphi f \perp L^\perp$ and finally $\varphi f \in L$. □

2.1.20. COROLLARY. *If $L = \overline{L} \subset H^p$, $1 \le p \le \infty$, $zL \subset L$ and $L \ne \{0\}$, then there is an inner function Θ such that $L = \Theta H^p$.* □

2.1.21. The canonical inner-outer factorization. This subsection is simply to summarize several facts and definitions about inner-outer factorization of operator valued functions scattered through the preceding part of the book; we refer to the above Subsections A.1.6.4 (Volume 1), 1.3.9 and 1.5.4. Recall that a function $\Phi \in H^\infty(E \longrightarrow E_*)$ is *inner* if the boundary values $\Phi(\zeta) : E \longrightarrow E_*$ are isometries a.e. on $\mathbb{T}$, and *outer* if $\operatorname{clos}(\Phi H^2(E)) = H^2(E_*)$. It is called $*$*-inner* ($*$*-outer*) if the function $\Phi_*(z) = \Phi(\overline{z})^*$ is inner (outer), and *two-sided inner* (respectively, *two-sided outer*) if it is inner and $*$-inner (respectively, outer and $*$-outer). In particular, Φ is two-sided inner if and only if $\Phi(\zeta) : E \longrightarrow E_*$ are unitary operators a.e. on $\mathbb{T}$. The following proposition contains what is called the *canonical (respectively, $*$-canonical) inner-outer factorization.* It immediately follows already from A.1.6.4 and, on the other hand, is a simple consequence of Corollary 2.1.7.

2.1.22. COROLLARY (B.Sz.-Nagy and C.Foiaş, 1963). *Let $\Phi \in H^\infty(E \longrightarrow E_*)$. There exist an inner function $\Theta \in H^\infty(F \longrightarrow E_*)$ and an outer function $\Psi \in H^\infty(E \longrightarrow F)$ such that*

$$\Phi = \Theta\Psi.$$

This representation is unique in the following sens: if $\Phi = \Theta'\Psi'$ is another inner-outer factorization with $\Theta' \in H^\infty(F' \longrightarrow E_*)$ and $\Psi' \in H^\infty(E \longrightarrow F')$, then there exists a unitary operator $u : F \longrightarrow F'$ such that $\Psi' = u\Psi$ and $\Theta' = \Theta u^{-1}$.

Moreover, there exist a $*$-inner function $\Theta_* \in H^\infty(E \longrightarrow F_*)$ and a $*$-outer function $\Psi_* \in H^\infty(F_* \longrightarrow E_*)$ such that $\Phi = \Psi_*\Theta_*$.

Indeed, the existence of such a factorization is proved already in A.1.6.4 (Volume 1) (recall that Θ is defined as an inner function from Corollary 2.1.7 satisfying $\Theta H^2(F) = L$, where $L = \mathrm{clos}(\Phi H^2(E))$), and the uniqueness property follows from the uniqueness in Corollary 2.1.7.

The $*$-outer- $*$-inner representation follows from the inner-outer representation applied to the function $\Theta_*(z) = \Theta(\overline{z})^*$. □

2.2. The H^∞ Functional Calculus

In this section we extend the standard polynomial calculus $p \longmapsto p(T)$ to the so-called Sz.-Nagy–Foiaş calculus defined on the algebra H^∞, assuming that T is a completely nonunitary Hilbert space contraction. Recall that for any bounded operator $T : X \longrightarrow X$, the polynomial calculus admits an extension to a homomorphism $\mathrm{Hol}(\sigma(T)) \longrightarrow L(X)$, where $\mathrm{Hol}(\sigma(T))$ denotes the algebra of all functions that are holomorphic in a neighbourhood of the spectrum $\sigma(T)$ (the so-called Riesz–Dunford calculus, see 2.5.1 below). The difference is that the latter is defined by the usual Cauchy formula, whereas the H^∞ calculus requires the existence of an (absolutely continuous) unitary dilation and, in general, is defined for functions that are not holomorphic on the spectrum. To construct and to study the H^∞ calculus, we use the function model of T, that is, an operator M_Θ which is unitarily equivalent to T. In fact, these calculi are compatible (Property 2.2.2(2) below); for the reader's convenience we recall in Exercise 2.5.1 the scope of the Riesz–Dunford calculus.

2.2.1. DEFINITION. Let M_Θ be a model operator on the space

$$\mathcal{K}_\Theta = \begin{pmatrix} H^2(E_*) \\ \mathrm{clos}\,\Delta L^2(E) \end{pmatrix} \ominus \begin{pmatrix} \Theta \\ \Delta \end{pmatrix} H^2(E)$$

where $\Theta \in H^\infty(E \to E_*)$ is a pure contractive function,

$$M_\Theta f = P_\Theta z f, \quad f \in \mathcal{K}_\Theta.$$

For a function $\varphi \in H^\infty$ we temporarily set

$$[\varphi] M_\Theta = P_\Theta \varphi | \mathcal{K}_\Theta.$$

2.2.2. First properties. *(1) The mapping $h : \varphi \longmapsto [\varphi] M_\Theta$ is an algebra homomorphism from H^∞ into $L(\mathcal{K}_\Theta)$, that is, a linear multiplicative contractive mapping. It is called the H^∞* calculus *for M_Θ.*

Indeed, the norm of the multiplication operator $[\varphi] : f \longmapsto \varphi f$ in $L^2(E_* \oplus E)$ coincides with the sup-norm

$$\|\varphi\|_\infty = \mathrm{ess\,sup}_{\mathbb{T}} |\varphi| = \|\varphi\|_{H^\infty} = \sup_{\mathbb{D}} |\varphi| .$$

Hence

$$\|[\varphi] M_\Theta\| = \|P_\Theta \varphi | \mathcal{K}_\Theta\| \le \|P_\Theta\| \cdot \|\varphi\| = \|\varphi\|_\infty .$$

Clearly h is linear, and the multiplicativity is justified by the following. First,

$$\begin{pmatrix} \Theta \\ \Delta \end{pmatrix} H^2(E)$$

is invariant with respect to multiplication by $\varphi \in H^\infty$. Indeed, pick

$$x \in \begin{pmatrix} \Theta \\ \Delta \end{pmatrix} H^2(E),$$

that is,

$$x = \begin{pmatrix} \Theta \\ \Delta \end{pmatrix} y$$

for some $y \in H^2(E)$. Then, as φ is scalar valued, we get

$$\varphi x = \begin{pmatrix} \Theta \\ \Delta \end{pmatrix} \varphi y,$$

and $\varphi y \in H^2(E)$, and the result follows.

Choose now $f, g \in H^\infty$. Then for arbitrary $x \in \mathcal{K}_\Theta$ we get by the claim

$$f P_\Theta^\perp g x \in f \begin{pmatrix} \Theta \\ \Delta \end{pmatrix} H^2(E) \subset \begin{pmatrix} \Theta \\ \Delta \end{pmatrix} H^2(E).$$

This implies that for every $x \in \mathcal{K}_\Theta$

$$P_\Theta f P_\Theta g x = P_\Theta f (P_\Theta + P_\Theta^\perp) g x = P_\Theta f g x,$$

and we conclude that

$$([f]M_\Theta)([g]M_\Theta) = [fg]M_\Theta.$$

□

(2) The calculus is hereditary *with respect to the polynomial calculus: if* $p = \sum_{k=0}^n a_k z^k$, *then* $[p]M_\Theta = \sum_{k=0}^n a_k M_\Theta^k$.

This is a direct consequence of the linearity and the multiplicativity. The above formula may be directly extended to functions whose series of Fourier coefficients converges absolutely: $[f]M_\Theta = \sum_{k\geq 0} \widehat{f}(k) M_\Theta^k$ if $\sum_{k\geq 0} |\widehat{f}(k)| < \infty$. □

The calculus is also hereditary with respect to the Riesz–Dunford calculus*: if a function f, $f \in H^\infty$, is holomorphic in a closed neighbourhood V of the spectrum $\sigma(M_\Theta)$ then*

$$[f]M_\Theta = (2\pi i)^{-1} \int_{\partial V} f(z) R_z(M_\Theta)\, dz.$$

Here $R_z(M_\Theta)$ is the resolvent of M_Θ.

Indeed, as $\sigma(M_\Theta) \subset \operatorname{clos} \mathbb{D}$, the neighbourhood V may be chosen starlike. Then, if we set $f_r(z) = f(rz)$, $r < 1, z \in V$, it is clear that f_r tends uniformly to f on ∂V as $r \longrightarrow 1$. By the continuity of the Riesz–Dunford calculus with respect to uniform convergence we get $f_r(M_\theta) \longrightarrow f(M_\Theta)$ as $r \longrightarrow 1$ (we use the notation of Exercice 2.5.1 for the Riesz–Dunford calculus). Clearly,

$$f_r(M_\Theta) = (2\pi i)^{-1} \int_{\partial V} f_r(z) R_z(M_\Theta)\, dz = \sum_{k\geq 0} \widehat{f}(k) r^k M_\Theta^k = [f_r]M_\Theta.$$

Since we also have weak convergence in H^∞ of f_r to f, we get by Property (3) below that $[f_r]M_\theta \longrightarrow [f]M_\Theta$ as $r \longrightarrow 1$. The assertion follows. □

(3) The calculus h is continuous in the weak topologies ($\sigma(H^\infty, L^1)$ for H^∞ and the weak operator topology on $L(\mathcal{K}_\Theta)$).

Indeed, let (f_α) be a net in H^∞ that tends weakly to zero, that is,

$$\int_{\mathbb{T}} f_\alpha g \to 0$$

for every $g \in L^1$. Then for $x, y \in \mathcal{K}_\Theta$ we have

$$\begin{aligned}([f_\alpha]M_\Theta x, y)_{\mathcal{K}_\Theta} &= (P_\Theta f_\alpha x, y)_{\mathcal{K}_\Theta} = (f_\alpha x, y)_{L^2(E_*\oplus E)} \\ &= \int_{\mathbb{T}} f_\alpha(\zeta)(x(\zeta), y(\zeta))_{E_*\oplus E}\, dm(\zeta).\end{aligned}$$

Setting $g(\zeta) = (x(\zeta), y(\zeta))$ we get an L^1-function $g \in L^1(\mathbb{T})$. Hence

$$\lim_\alpha ([f_\alpha]M_\Theta x, y)_{\mathcal{K}_\Theta} = 0.$$

□

NOTATION. In the sequel we use the notation $[f]M_\Theta = f(M_\Theta)$ for $f \in H^\infty$ (and call the latter an H^∞-function of M_Θ).

(4) The H^∞-calculus is well-defined (satisfying the aformentioned properties) for an arbitrary c.n.u. contraction T, and is expressed by

$$f(T) = Uf(M_\Theta)U^{-1},$$

where $\Theta = \Theta_T$ is the characteristic function of T and U stands for the unitary equivalence of T and M_Θ: $T = UM_\Theta U^{-1}$.

Being an obvious consequence of the model theorem 1.3.2, this property, in fact, can be deduced directly from the existence of the minimal unitary dilation. Indeed, as is observed in Subsection 1.3.3, the minimal unitary dilation U of a c.n.u. contraction $T : H \longrightarrow H$ has an absolutely continuous spectrum. This implies that the mapping

$$f \longmapsto [f](T) = P_H f(U)|H$$

is well defined and contractive for $f \in L^\infty(\mathbb{T})$; see 1.5.1(l) for the definition of $f(U)$. Obviously, it coincides with the polynomial calculus for $f \in \mathcal{P}ol_+$ and satisfies the above weak continuity property (3) (with the same proof). It follows that it is multiplicative for $f \in H^\infty$. □

It is also worth mentioning that the above reasoning works for every Hilbert space contraction having an abosolutely continuous unitary dilation. The Langer lemma 1.2.5 shows that such a contraction is of the form $V \oplus T_0$, where V is a unitary operator with absolutely continuous spectrum and T_0 is a c.n.u. contraction.

(5) The kernel of the H^∞ calculus
is $\{0\}$ if the function $\Theta = \Theta_T$ is not inner or $*$-inner (i.e., if $\Delta \neq 0$ or $\Delta_* \neq 0$, where $\Delta_* = I - \Theta\Theta^*$); if Θ is inner then $\varphi(M_\Theta) = 0$ if and only if $\varphi I = \Theta\Omega$ for a function $\Omega \in H^\infty(E_* \longrightarrow E)$.

Indeed, if $\Delta \neq 0$ we obtain from the result of 1.5.2 that $X(Z|\mathcal{R}) = M_\Theta X$ (in the notation of the quoted subsection), and hence $X(\varphi(Z)|\mathcal{R}) = \varphi(M_\Theta)X$. Now, if $\varphi(M_\Theta) = 0$, we get $\varphi(Z)^*X^* = 0$. Since $\mathrm{Ker}(\varphi(Z)^*)$ is a reducing subspace of the unitary operator Z, we can use the result (and the notation) of 1.5.1(i) and get $\overline{\varphi(\zeta)}P'(\zeta) = 0$ a.e. m, where m is the Lebesgue measure on $\mathbb{T}$ (see 1.3.3 above). Since $X^* \neq 0$, and hence $P' \neq 0$, the function $\varphi \in H^\infty$ vanishes on a set of positive Lebesgue mesure, and therefore, $\varphi = 0$.

The same reasoning is applicable to $\varphi(M_\Theta)^* = \varphi_*(M_\Theta^*) = \varphi_*(M_{\Theta_*})$ (see 2.5.5(a) below for the latter equality), where $\Theta_*(\overline{z}) = \Theta(z)^*$. This shows that the kernel of the H^∞ calculus is $\{0\}$ if Θ_T is not a $*$-inner function.

If Θ is inner, then $\varphi(M_\Theta) = P_\Theta \varphi | \mathcal{K}_\Theta = 0$ if and only if $\varphi \mathcal{K}_\Theta \subset \Theta H^2(E)$. Since $\varphi \Theta H^2(E) \subset \Theta H^2(E)$, the latter is equivalent to $\varphi H^2(E_*) \subset \Theta H^2(E)$, or $\Theta^* \varphi H^2(E_*) \subset H^2(E)$, where $\Theta^*(\zeta) = \Theta(\zeta)^*$ for every $\zeta \in \mathbb{T}$. On the other hand, if $\varphi \neq 0$, then $\operatorname{Range}\Theta(\zeta)$ equals to E_* a.e. on $\mathbb{T}$, that is, $\Theta(\zeta)$ is unitary, and we get $\varphi I = \Theta \Omega$ with $\Omega = \Theta^* \varphi$. □

See also Subsection 2.5.6(c) below for a strenghtened form of this property.

2.2.3. The von Neumann inequality. Let T be a contraction on a Hilbert space, $T : H \to H$, and let p be a polynomial. Then

$$\|p(T)\| \leq \|p\|_\infty = \sup_{|z|=1} |p(z)|.$$

Indeed, let $T = T_0 \oplus T_u$ be the canonical decomposition of the contraction T into a unitary operator T_u and a c.n.u. operator T_0. Then

$$\|p(T)\| = \|p(T_0) \oplus p(T_u)\| = \max(\|p(T_0)\|, \|p(T_u)\|).$$

The H^∞-calculus gives $\|p(T_0)\| \leq \|p\|_\infty$ and by the spectral theorem (see 1.5.1(1)) we have $\|p(T_u)\| = \sup_{\sigma(T_u)} |p| \leq \|p\|_\infty$. □

2.3. The Class C_0, Minimal Annihilators, and the Spectrum of M_Θ

The H^∞ calculus allows us to distinguish a class of contractions T with "small" spectrum (the class C_0 below), which permits a natural generalization of the classical *Hamilton–Cayley theorem*. Recall that this theorem claims that $p(T) = 0$ for the characteristic polynomial $p(z) = \det(zI - T)$ of an operator T on a finite-dimensional vector space. The characteristic polynomial, and its refinement, the minimal polynomial, plays an important role in the spectral theory of operators on finite dimensional spaces. Similarly, the minimal annihilating functions, defined in 2.3.2(2) below, are used to describe many properties of C_0-operators.

2.3.1. DEFINITION. The class C_0 consists of those c.n.u. contractions T which have an *annihilating function* $\varphi \in H^\infty$, $\varphi \not\equiv 0$, that is, $\varphi(T) = \mathbb{O}$.

2.3.2. First properties of C_0-operators. *(1) A c.n.u. contraction $T : H \to H$ of finite rank,* $\operatorname{rank} T = \dim\, TH < \infty$, *is in C_0.*

Indeed, H splits into an orthogonal sum $H = \overline{TH} \oplus (\overline{TH})^\perp$. Clearly, $T(\overline{TH}^\perp) \subset \overline{TH}$ and $TH \subset \overline{TH}$. If we find an H^∞-function $\psi \not\equiv 0$ such that $\psi(T|\overline{TH}) = \mathbb{O}$, it is sufficient to set $\varphi = z\psi$, and by the multiplicativity of the H^∞-calculus the result will follow. As $\overline{TH}$ is of finite dimension, we can choose for ψ the characteristic polynomial of $T|\overline{TH}$. □

Tis property means that the class C_0 really generalizes the Hamilton–Cayley theorem that we have used in the previous proof.

(2) Existence of the minimal annihilator. *Let T be a c.n.u. contraction in C_0. Then there is a unique (up to a multiplicative constant) function $m_T \in H^\infty$ such that*

(i) $m_T(T) = \mathbb{O}$;
(ii) m_T is inner;
(iii) if $\varphi \in H^\infty$ annihilates T, $\varphi(T) = \mathbb{O}$, then $\varphi / m_T \in H^\infty$.

The function m_T is called the *minimal annihilator* of T.

Indeed, let $L = \{\varphi \in H^\infty : \varphi(T) = \mathbb{O}\}$. As $T \in C_0$, we have $L \neq \{0\}$. Obviously, L is a subspace of H^∞ which is weakly closed: if $(f_\alpha)_\alpha \subset L$ tends weakly (in $\sigma(H^\infty, L^1)$) to f, then $(f_\alpha(T)x, y) \to (f(T)x, y)$ for arbitrary $x, y \in H$ (cf. 2.2.2(3)). Consequently $f(T) = \mathbb{O}$ and $f \in L$.

The multiplicativity of the calculus shows that L is an ideal: $\varphi \in L$ and $\psi \in H^\infty$ imply that $\varphi\psi \in L$. In particular, L is then z-invariant. By Corollary 2.1.20 there is an inner function, say m_T, such that $L = m_T H^\infty$. Clearly, $m_T(T) = \mathbb{O}$, and the other properties are now obvious.

It remains to verify the uniqueness. Suppose m' is another minimal annihilating function. Then m_T/m', $m'/m_T = \overline{(m_T/m')} \in H^\infty$, and necessarily $m_T/m' =$ const. □

(3) Let T be a c.n.u. contraction having a two-sided inner characteristic function $\Theta = \Theta_T$ (that is, $\Theta_T(\zeta)$ is unitary a.e. on $\mathbb{T}$ and E, E_* may be identified). Suppose that $\varphi \in H^\infty$. The following assertions are equivalent.

(i) $\varphi(T) = \mathbb{O}$.

(ii) There is a function $\Omega \in H^\infty(E \to E)$ such that $\Omega\Theta_T \equiv \Theta_T\Omega \equiv \varphi(z)I$.

For (i) $\Longrightarrow$ (ii) set $\mathcal{K}_\Theta = H^2(E) \ominus \Theta H^2(E)$. Then $\varphi(M_\Theta) = \mathbb{O}$ if and only if $P_\Theta \varphi \mathcal{K}_\Theta = \{0\}$, and thus if and only if $P_\Theta \varphi H^2(E) = P_\Theta \varphi(\mathcal{K}_\Theta + \mathcal{K}_\Theta^\perp) = \{0\}$. The latter is equivalent to $\varphi H^2(E) \subset \Theta H^2(E)$, that is, $\Theta^* \varphi H^2(E) \subset H^2(E)$. Setting now $\Omega = \Theta^*\varphi \in H^\infty(E \to E)$ we get the desired result.

The implication (ii) $\Longleftarrow$ (i) follows from $\varphi(M_\Theta) = P_\Theta \varphi | \mathcal{K}_\Theta = P_\Theta \Theta\Omega | \mathcal{K}_\Theta = 0$. □

(4) Suppose that Θ_T is inner and $*$-inner and that $\partial_T, \partial_{T^*} < \infty$. Then $\partial_T = \partial_{T^*} = n$ and $T \in C_0$. Moreover, m_T is a divisor of the scalar function $d_T = \det \Theta_T(z)$, and d_T in turn divides m_T^n.

Indeed, since $\Theta_T(z) : \mathcal{D}_T \to \mathcal{D}_{T^*}$ is unitary a.e. on $\mathbb{T}$, we have $\partial_T = \dim \mathcal{D}_T = \dim\ \mathcal{D}_{T^*} = \partial_{T^*}$. Let $\Omega(z) = (\Omega_{ij}(z))_{ij}$, where Ω_{ij} is the complementary minor of the matrix entry $(\Theta_T(z))_{ij}$. Then

$$\Theta_T(z)\Omega(z) = \Omega(z)\Theta_T(z) = d_T(z)I, \quad z \in \mathbb{D},$$

and (3) shows that $d_T(T) = \mathbb{O}$. Hence $T \in C_0$ and $d_T/m_T \in H^\infty$. Using again (3) we get a matrix valued function $\Omega_1 \in H^\infty(E \longrightarrow E_*)$ such that $\Omega_1\Theta_T \equiv \Theta_T\Omega_1 \equiv m_T I$. This gives $\det \Omega_1 \cdot \det \Theta_T = m_T^n$, from which we deduce that $m_T^n/d_T \in H^\infty$. □

A more precise analysis of the above proof shows that m_T actually coincides with the LCM of all order-$(n-1)$ minors of Θ_T; see Section 2.6 for references.

2.3.3. Definition. Let Θ be an operator valued holomorphic function $\Theta(z) : E \to E_*$, $|z| < 1$. If $\Theta(z)$ is invertible at $z \in \mathbb{D}$ (and hence in a certain neighbourhood $\{\zeta \in \mathbb{D} : |z - \zeta| < \varepsilon\}$), then we call z *regular.* A point of the boundary $z \in \mathbb{T}$ is called *regular* if Θ is unitary in a neighbourhood $\gamma = \{\zeta \in \mathbb{T} : |\zeta - z| < \varepsilon\} \subset \mathbb{T}$ of z and admits an analytic continuation across γ. The complement of all regular points in $\operatorname{clos} \mathbb{D}$ is called the *spectrum* of Θ and is denoted by $\sigma(\Theta)$.

For the reader's convenience we recall in Subsection 2.3.6 below a description of the spectrum $\sigma(\Theta)$ for a scalar inner function Θ.

2.3.4. THEOREM. *Let T be a c.n.u. contraction. Then*

$$\sigma(T) = \sigma(\Theta_T).$$

Moreover,

$$\sigma_p(T) = \{\lambda \in \mathbb{D} : \operatorname{Ker} \Theta_T(\lambda) \neq \{0\}\}.$$

If $T \in C_0$, we further have

$$\sigma(T) = \sigma(m_T), \quad \sigma(T) \cap \mathbb{D} = \{\lambda : m_T(\lambda) = 0\} = \sigma_p(T).$$

PROOF. In this book, we restrict ourselves to the case $T \in C_0$ and give in Subsection 2.3.7 some indications for the general case.

Pick $\lambda \in \overline{\mathbb{D}} \setminus \sigma(m_T)$. Then m_T is holomorphic at λ and invertible, that is, $m_T(\lambda) \neq 0$. Then

$$u = \frac{1 - m_T/m_T(\lambda)}{\lambda - z} \in H^\infty.$$

As $u(\lambda - z) = (\lambda - z)u = 1 - m_T/m_T(\lambda)$ and $m_T(T) = \mathbb{O}$, the functional calculus gives

$$u(T)(\lambda I - T) = (\lambda I - T)u(T) = I,$$

that is, $\lambda \in \overline{\mathbb{D}} \setminus \sigma(T)$ (and $R_\lambda(T) = u(T)$). We deduce that $\sigma(T) \subset \sigma(m_T)$.

Take now $\lambda \in \sigma(m_T) \cap \mathbb{D}$. Consequently, $m_T(\lambda) = 0$, and m_T factorizes into $m_T = b_\lambda \cdot \varphi$, $\varphi \in H^\infty$, where $b_\lambda = (\lambda - z)/(1 - \overline{\lambda} z)$ is the standard Blaschke factor. Hence $\mathbb{O} = m_T(T) = b_\lambda(T)\varphi(T) = \varphi(T)b_\lambda(T)$. In view of the minimality of m_T, the kernel of $b_\lambda(T)$ cannot be trivial, $\operatorname{Ker} b_\lambda(T) \neq \{\mathbb{O}\}$. Clearly, $\operatorname{Ker} b_\lambda(T) = \operatorname{Ker}(\lambda I - T)$, and hence $\sigma(m_T) \cap \mathbb{D} \subset \sigma_p(T)$. Since T is a c.n.u. contraction, we get $\sigma_p(T) \subset \mathbb{D}$. Thus, $\sigma_p(T) \subset \mathbb{D} \cap \sigma(T) \subset \mathbb{D} \cap \sigma(m_T)$, and therefore $\sigma_p(T) = \mathbb{D} \cap \sigma(m_T)$.

It remains to prove that $\sigma(m_T) \cap \mathbb{T} \subset \sigma(T)$. To this end, we suppose that there exists $\lambda \in \mathbb{T} \setminus \sigma(T)$ with $\lambda \in \sigma(m_T)$. For a neighbourhood V of λ define a singular inner function by

$$m_1(z) = \exp\left(-\int_{\mathbb{T} \cap V} \frac{\zeta + z}{\zeta - z}\, d\mu(\zeta)\right),$$

where μ is the representing measure of the singular inner factor of m_T (cf. A.3.9.3, Volume 1). By A.4.3.9 (Volume 1), $\mu(\mathbb{T} \cap V) > 0$, and hence, $m_1 \not\equiv$ const. If V is such that $V \cap \sigma(T) = \emptyset$, then m_1 is holomorphic in a suitable neighbourhood of $\sigma(T)$ and, moreover, $m_1(z) \neq 0$ in this neighbourhood, which by the Riesz–Dunford calculus implies that $m_1(T)$ is invertible (see 2.5.1 below). As $m_T = m_1 \cdot m_2$ with $m_1, m_2 \in H^\infty$, and $m_2(T) = m_1(T)^{-1} m_T(T) = \mathbb{O}$, we get a contradiction to the minimality of m_T (observe that $m_2/m_T \notin H^\infty$). □

2.3.5. COROLLARY. *Let $\Theta \in L(E \longrightarrow E_*)$ be a pure inner function (in the sense of 1.2.12). Then every function $f \in \mathcal{K}_\Theta = H^2(E_*) \ominus \Theta H^2(E)$ can be analytically continued across $\mathbb{T} \setminus \sigma(\Theta)$.*

Indeed, $\sigma(\Theta) = \sigma(M_\Theta)$ and, hence, the *Fredholm resolvent*

$$z \longmapsto (I - zM_\Theta^*)^{-1}, \quad |z| < 1,$$

can be analytically continued across $(\mathbb{T} \setminus \sigma(M_\Theta^*))^- = \mathbb{T} \setminus \sigma(M_\Theta)$, where $(\cdot)^-$ means the complex conjugate. On the other hand, given a vector $e_* \in E_*$ and a function

$f \in \mathcal{K}_\Theta$, we have

$$\begin{aligned}((I - zM_\Theta^*)^{-1} f, e_*) &= \sum_{n \geq 0} z^n (M_\Theta^{*n} f, e_*) = \sum_{n \geq 0} z^n (f, M_\Theta^n e_*) \\ &= \sum_{n \geq 0} z^n (f, z^n e_*) = \sum_{n \geq 0} z^n (\hat{f}(n), e_*) \\ &= (f(z), e_*)\end{aligned}$$

for all $z \in \mathbb{D}$. Therefore, f is weakly, and hence strongly (see A.3.11, Volume 1), analytically continued across $\mathbb{T} \setminus \sigma(M_\Theta)$. □

2.3.6. The spectrum $\sigma(\Theta)$ and the canonical factorization. For scalar functions Θ, we can easily describe the spectrum in terms of the canonical Riesz–Smirnov factorization (Theorem A.3.9.5, Volume 1). Namely, the following claims are immediate consequences of the theory of Sections A.3 and A.4 of Volume 1 (more precisely, of Theorems A.3.9.1, A.3.9.5, and A.4.3.3, and of Lemmas A.4.3.4 and A.3.7.3); see also A.4.3.9.

I. Given a function $\Theta \in H^\infty$ and a point $\lambda \in \operatorname{clos} \mathbb{D}$, the following assertions are equivalent:

(1) $\lambda \notin \sigma(\Theta)$.

(2) Either $\lambda \in \mathbb{D}$, and then $\Theta(\lambda) \neq 0$, or $\lambda \in \mathbb{T}$ and there exists $\varepsilon > 0$ such that $|\Theta(\zeta)| = 1$ for a.e. $\zeta \in \mathbb{T} \cap D(\lambda, \varepsilon)$ and $\inf\{|\Theta(\zeta)| : \zeta \in \mathbb{D} \cap D(\lambda, \varepsilon)\} > 0$.

II. In particular, for an inner function Θ,

$$\sigma(\Theta) = \{\lambda \in \operatorname{clos} \mathbb{D} : \underline{\lim}_{z \to \lambda} |\Theta(z)| = 0\} = (\operatorname{clos} Z(\Theta)) \cup (\operatorname{supp} \mu_\Theta),$$

where $Z(\Theta) = \{\lambda \in \mathbb{D} : \Theta(\lambda) = 0\}$ and μ_Θ is the measure representing the singular part of Θ.

2.3.7. Hints for the proof of the formula $\sigma(T) = \sigma(\Theta_T)$ in the general case. (1) For $\lambda \in \operatorname{clos} \mathbb{D} \setminus \sigma(\Theta)$, $\Theta = \Theta_T$, set

$$u = \frac{1 - \Theta\Theta(\lambda)^{-1}}{\lambda - z}$$

and $A = P_\Theta u | \mathcal{K}_\Theta$. Again we get $A(\lambda I - T) = (\lambda I - T)A = I$, and $\lambda \in \operatorname{clos} \mathbb{D} \setminus \sigma(T)$. Hence $\sigma(T) \subset \sigma(\Theta)$.

(2) Let $|\lambda| < 1$. We claim that $\Theta(\lambda)$ is invertible if and only if $b_\lambda(T)$ is invertible, that is, if and only if $(\lambda I - T)$ is invertible. Indeed, for $\lambda = 0$ this equivalence is a straightforward consequence of the formula $\Theta_T(0) = -T|\mathcal{D}_T$. The general case is reduced to the former one by the formula $\Theta_{b_\lambda(T)} = \Theta_T \circ b_\lambda$, where $b_\lambda = (\lambda - z)(1 - \overline{\lambda} z)^{-1}$.

(3) The previous arguments show that $\operatorname{Ker} T \neq \{0\}$ if and only if $\operatorname{Ker} \Theta(0) \neq \{0\}$.

Applying again the transformation formula $\Theta_{b_\lambda(T)} = \Theta_T \circ b_\lambda$ we get

(4) $\operatorname{Ker} \Theta(\lambda) \neq \{0\}$ if and only if $\lambda \in \sigma_p(T)$.

(5) It remains to prove $\mathbb{T} \cap \sigma(\Theta) \subset \sigma(T)$. Assume that $\lambda \in \mathbb{T} \setminus \sigma(T)$. Then the resolvent $R_z(T)$ is holomorphic in a neighbourhood of λ, and hence so is $(I - zT^*)^{-1}$.

By definition of Θ_T we also get analyticity of Θ_T in a neighbourhood of λ, and the following identity (which is proved in 2.5.5(c))

$$\|x\|^2 - \|\Theta_T(z)x\|^2 = (1-|z|^2)\cdot\|(I-zT^*)^{-1}D_Tx\|^2, \quad |z|<1,$$

shows that Θ_T is isometric in this neighbourhood. A similar reasoning for $\Theta_T^*(\overline{z}) = \Theta_{T^*}(z)$ shows that Θ_{T^*} is also isometric, and the result follows. □

2.4. The Commutant Lifting Theorem

This important theorem describes the commutant of a model operator as the set of compressions of certain multiplication operators. We have already proved a special case of this theorem in Chapter B.3 (Volume 1). Now, we give the complete statement of the theorem and prove it for the case of inner operator valued characteristic functions. This is sufficient for the applications of this theorem presented in the next Chapter 3, as well as in Chapter B.9 (Volume 1). The most general case is commented on in Section 2.6 below.

2.4.1. The problem. The main purpose of this section is to give a description of the commutant of a c.n.u. contraction in terms of its characteristic function. It is clear that this is equivalent to describing the commutant of a model operator.

Let $\Theta\in H^\infty(E\to E_*)$ be a pure contractive function, and let $M_\Theta:\mathcal{K}_\Theta\to\mathcal{K}_\Theta$ be the associated model operator. The *commutant* of M_Θ is defined by

$$\{M_\Theta\}' = \{A:\mathcal{K}_\Theta\to\mathcal{K}_\Theta : M_\Theta A = AM_\Theta\}.$$

In particular, every function $A=f(M_\Theta)$ with $f\in H^\infty$ commutes with M_Θ. However, usually, there exist operators commuting with M_Θ that are not in the range of the functional calculus.

The following result gives a complete description of the commutant.

2.4.2. Theorem (B. Sz.-Nagy, C. Foiaş). *Let $A\in\{M_\Theta\}'$. Then there exists a measurable function Y,*

$$Y(\zeta): E_*\oplus\operatorname{clos}\Delta(\zeta)E \to E_*\oplus\operatorname{clos}\Delta(\zeta)E,$$

such that

$$Y\begin{pmatrix}\Theta\\ \Delta\end{pmatrix}H^2(E)\subset\begin{pmatrix}\Theta\\ \Delta\end{pmatrix}H^2(E),$$
$$Y\begin{pmatrix}H^2(E_*)\\ \operatorname{clos}(\Delta L^2(E))\end{pmatrix}\subset\begin{pmatrix}H^2(E_*)\\ \operatorname{clos}(\Delta L^2(E))\end{pmatrix} \tag{2.4.1}$$

and $A=P_\Theta Y|\mathcal{K}_\Theta$. Moreover,

$$\|A\| = \inf\left\{\left\|Y+\begin{pmatrix}\Theta g & \mathbb{O}\\ \Delta g & \mathbb{O}\end{pmatrix}\right\|_\infty : g\in H^\infty(E_*\to E)\right\},$$

and the distance is attained.

Conversely, if Y satisfies (2.4.1), *then we have $A=P_\Theta Y|\mathcal{K}_\Theta\in\{M_\Theta\}'$ and $\|A\|\le\|Y\|_\infty$.*

2.4.3. Corollary (a special case). *Let Θ be inner (that is, $\Delta\equiv 0$). Then $A\in\{M_\Theta\}'$ if and only if $A=P_\Theta Y|\mathcal{K}_\Theta$, where $Y\in H^\infty(E_*\to E_*)$ is such that $Y\Theta H^2(E)\subset\Theta H^2(E_*)$ (that is, $\Theta^*Y\Theta\in H^\infty(E\to E)$) and $\|A\| = \operatorname{dist}(Y,\Theta H^\infty)$ (and the distance is attained).* □

2.4.4. Intertwining operators. Before proving Theorem 2.4.2, we point out a useful generalization. Namely, if we consider an operator A acting between two model spaces $A : \mathcal{K}_{\Theta_1} \to \mathcal{K}_{\Theta_2}$ and intertwining the corresponding model operators, $M_{\Theta_2}A = AM_{\Theta_1}$, then there exists a measurable function $Y(\zeta) : E^1_* \oplus \operatorname{clos}\Delta_1(\zeta)E^1 \to E^2_* \oplus \operatorname{clos}\Delta_2(\zeta)E^2$, $|\zeta| = 1$, such that

$$A = P_{\Theta_2}Y|\mathcal{K}_{\Theta_1},$$

and

$$Y\begin{pmatrix}\Theta_1\\ \Delta_1\end{pmatrix}H^2(E^1) \subset \begin{pmatrix}\Theta_2\\ \Delta_2\end{pmatrix}H^2(E^2),$$
$$Y\begin{pmatrix}H^2(E^1_*)\\ \operatorname{clos}\Delta_1 L^2(E^1)\end{pmatrix} \subset \begin{pmatrix}H^2(E^2_*)\\ \operatorname{clos}\Delta_2 L^2(E^2)\end{pmatrix}.$$

The following commutative diagram illustrates the situation.

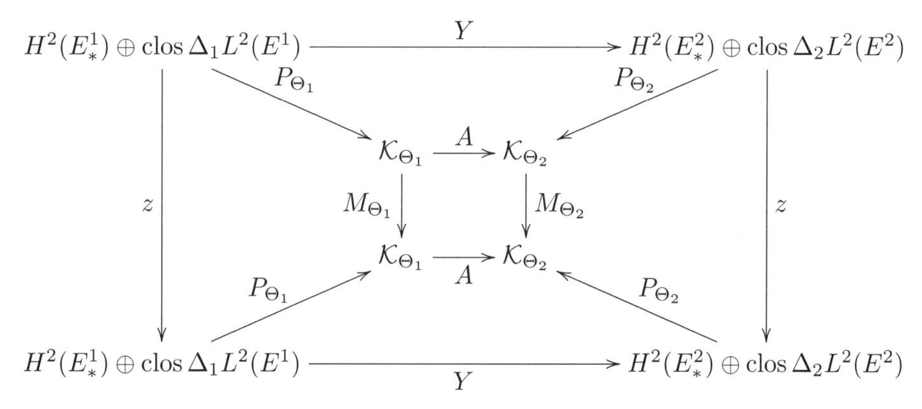

This means that we lift the intertwining relation $AP_{\Theta_1}z = P_{\Theta_2}zA$ up to the equation $Yz = zY$ for the operator Y, where $Y\mathcal{K}^{\perp}_{\Theta_1} \subset \mathcal{K}^{\perp}_{\Theta_2}$, which actually shows that Y is the multiplication operator by $Y(\zeta)$.

2.4.5. Proof of the theorem on intertwining operators in the case where Θ_i are inner and $*$-inner. The proof follows the same lines as for the scalar case, see Theorem B.3.1.11 (Volume 1). We start with the operator valued analogue of Lemma B.3.1.10.

2.4.6. LEMMA. *Let Θ_1, Θ_2 be two inner and $*$-inner functions, $A : \mathcal{K}_{\Theta_1} \to \mathcal{K}_{\Theta_2}$. Then $M_{\Theta_2}A = AM_{\Theta_1}$ if and only if the operator $A_* = \Theta_2^* AP_{\Theta_1} : H^2(E^1_*) \to H^2_-(E^2)$ satisfies the* Hankel equation

$$P_- zA_* = A_* z.$$

PROOF. Clearly, $M_{\Theta_2}A = AM_{\Theta_1}$ if and only if $P_{\Theta_2}zA|\mathcal{K}_{\Theta_1} = AP_{\Theta_1}z|\mathcal{K}_{\Theta_1}$. The latter is equivalent to $P_{\Theta_2}zAP_{\Theta_1} = AP_{\Theta_1}z$ on $H^2(E^1_*)$, and also to

$$P_- z\Theta_2^* AP_{\Theta_1} = \Theta_2^* AP_{\Theta_1}z. \tag{2.4.2}$$

Indeed, note that $P_{\Theta_2} = \Theta_2 P_- \Theta_2^*$ and $\Theta_2^*\Theta_2 \equiv 1$ which gives "$\Longrightarrow$", whereas the converse implication can be obtained multiplying the equation by $\Theta_2 P_-$: we get $\Theta_2 P_- z\Theta_2^* AP_{\Theta_1} = \Theta_2 P_- \Theta_2^* AP_{\Theta_1}z = AP_{\Theta_1}z$ (we have used the fact that Range $A \subset \mathcal{K}_{\Theta_2}$). By definition of A_* we now get the equivalence of (2.4.2) and $P_- zA_* = A_* z$ on $H^2(E^1_*)$. $\square$

Now, we need a vector valued version of Nehari's theorem as it was proved in B.1.6.1 (Volume 1). For the reader's convenience, we reproduce the statement.

2.4.7. THEOREM (L. Page, 1970). *Let* $\Gamma : H^2(E_1) \to H^2(E_2)$ *be a bounded linear operator. The following assertions are equivalent.*

(1) $P_- z\Gamma = \Gamma z$.

(2) There exists a function $f \in L^\infty(E_1 \to E_2)$ *such that* $\Gamma x = P_- f x$, $x \in H^2(E_1)$. *Moreover,*

$$\|\Gamma\| = \inf_{P_- \tilde{f} = \Gamma} \|\tilde{f}\| = \inf_{g \in H^\infty(E_1 \to E_2)} \|f + g\|_{H^\infty} = \operatorname{dist}(f, H^\infty).$$

We resume the proof of Theorem 2.4.2. Setting

$$\Gamma = A_* = \Theta_2^* A P_{\Theta_1} : H^2(E_*^1) \to H_-^2(E^2),$$

we obtain a Hankel operator by Lemma 2.4.6. Now, as in the scalar case, by Page's theorem 2.4.7 there is a function $f \in L^\infty(E_*^1 \to E^2)$ such that

$$A_* x = P_- f x,$$

for every $x \in H^2(E_*^1)$. Hence $\Gamma = \Theta_2^* A P_{\Theta_1} = P_- f$. As $\Theta_2^* A P_{\Theta_1} H^2 \subset H_-^2$, we have

$$\Theta_2 \Gamma = \Theta_2 \Theta_2^* A P_{\Theta_1} = \Theta_2 P_- \Theta_2^* A P_{\Theta_1} = P_{\Theta_2} A P_{\Theta_1} = A P_{\Theta_1}.$$

Hence $\Theta_2 \Gamma H^2 = \Theta_2 P_- f H^2 \subset \mathcal{K}_{\Theta_2} \subset H^2(E_*^2)$. This, together with the obvious inclusion $\Theta_2 P_+ f H^2 \subset H^2$, teams up in $\Theta_2 f H^2(E_*^1) \subset H^2(E_*^2)$. Setting $\varphi := \Theta_2 f \in H^\infty(E_*^1 \longrightarrow E_*^2)$ we get $f = \Theta_2^* \varphi$ with $\varphi \in H^\infty$.

On the other hand, by definition of Γ, we have $\Gamma \Theta_1 H^2 = \{0\}$, and hence $P_- \Theta_2^* \varphi \Theta_1 H^2 = \{0\}$. The latter identity is equivalent to $\Theta_2^* \varphi \Theta_1 H^2 \subset H^2$, which gives $\varphi \Theta_1 H^2 \subset \Theta_2 H^2$.

Now

$$Ax = \Theta_2 \Gamma_{\Theta_2^* \varphi} x = \Theta_2 P_- \Theta_2^* \varphi x = P_{\Theta_2} \varphi x$$

for every $x \in \mathcal{K}_{\Theta_1}$.

It remains to show the norm identity:

$$\begin{aligned} \|A\| &= \|\Gamma | \mathcal{K}_{\Theta_1}\| = \|\Gamma\| = \operatorname{dist}(\Theta_2^* \varphi, H^\infty) = \operatorname{dist}(\Theta_2 \Theta_2^* \varphi, \Theta_2 H^\infty) \\ &= \operatorname{dist}(\varphi, \Theta_2 H^\infty). \end{aligned}$$

Here we have successively used the definition of Γ, the fact that $\Gamma \Theta_1 H^2 = \{0\}$, Page's theorem and that Θ_2 is an isometry. □

2.4.8. REMARK. Using Montel's theorem one can verify that $\inf_h \|\varphi - \Theta_2 h\|_\infty$ is attained. In fact, if $\|f_n\|_\infty \le \text{const}$, then there is a subsequence f_{n_k} converging in WOT to a holomorphic function f and

$$\|f\|_\infty \le \limsup \|f_n\|.$$

2.5. Exercises and Further Results

2.5.1. The Riesz–Dunford functional calculus. Let $T : X \longrightarrow X$ be a bounded operator on a Banach space X, and let $K \subset \mathbb{C}$ be a compact set having piecewise smooth and positively oriented boundary ∂K and such that $\sigma(T) \subset int(K)$ (the interior of K); we call these K *admissible.* Note that for every open set $\mathcal{O} \supset \sigma(T)$ there exists an admissible K such that $\mathcal{O} \supset K \supset \sigma(T)$. Denote by $\operatorname{Hol}(\sigma(T))$ the algebra of functions holomorphic on (a neighbourhood of) $\sigma(T)$ endowed with the usual convergence (a sequence $(f_n)_{n \ge 1}$ converges in $\operatorname{Hol}(\sigma(T))$ if there exists a neighbourhood of $\sigma(T)$, where the functions f_n are well defined and uniformly convergent). For elements of integration of vector valued functions we refer to Section A.3.11 (Volume 1), or, for example, to the textbook of W. Rudin [**Ru6**].

(a) Show that the integral

$$f(T) = \frac{1}{2\pi i} \int_{\partial K} f(z) R_z(T) \, dz,$$

defined for $f \in \mathrm{Hol}(\sigma(T))$ and an admissible K, does not depend on the choice of K, and the mapping

$$f \longmapsto f(T)$$

is a continuous homomorphism $\mathrm{Hol}(\sigma(T)) \longrightarrow L(X)$ (called the *Riesz–Dunford calculus*) which is consistent with the polynomial calculus, that is,

$$z^n(T) = T^n \text{ for } n \geq 1, \quad 1(T) = I.$$

(As usual, $R_z(T) = (zI - T)^{-1}$ stands for the resolvent, $z \in \mathbb{C} \backslash \sigma(T)$).

In particular,

$$(\lambda - z)^{-1}(T) = R_\lambda(T) \text{ and } f(T) = p(T)(q(T))^{-1}$$

for any rational function $f = p/q \in \mathrm{Hol}(\sigma(T))$, and if $AT = TA$ then $Af(T) = f(T)A$ for every $f \in \mathrm{Hol}(\sigma(T))$.

[*Hint:* the mapping is linear and continuous since

$$\|f(T)\| \leq \frac{|\partial K|}{2\pi} \|f\|_{\partial K} \sup_{z \in \partial K} \|R_z(T)\|;$$

$f(T)$ does not depend on the choice of K in the sense that

$$\int_{\partial K} f(z)(R_z(T)x, y) \, dz - \int_{\partial K'} f(z)(R_z(T)x, y) \, dz = 0$$

for every $x \in X$, $y \in X^*$ and every admissible K and K' such that $\sigma(T) \subset \mathrm{int}(K')$ and $K' \subset \mathrm{int}(K)$ (due to the standard Cauchy theorem); using such a compact K' and the *Hilbert identity*

$$R_z(T) R_\lambda(T) = \frac{1}{\lambda - z} \Big(R_z(T) - R_\lambda(T) \Big)$$

(which is obvious), show that $f(T)g(T) = (fg)(T)$, as follows,

$$\begin{aligned}
f(T)g(T) &= \frac{1}{2\pi i} \int_{\partial K} f(z) R_z(T) g(T) \, dz \\
&= \frac{1}{2\pi i} \int_{\partial K} f(z) \left(\frac{1}{2\pi i} \int_{\partial K'} g(\lambda) R_z(T) R_\lambda(T) d\lambda \right) dz \\
&= \frac{1}{2\pi i} \int_{\partial K} f(z) \frac{1}{2\pi i} \int_{\partial K'} g(\lambda) \left\{ \frac{1}{\lambda - z} (R_z(T) - R_\lambda(T)) \right\} d\lambda \, dz \\
&= \frac{1}{2\pi i} \int_{\partial K} f(z) \frac{1}{2\pi i} \int_{\partial K'} g(\lambda) \left\{ -\frac{1}{\lambda - z} R_\lambda(T) \right\} d\lambda \, dz \\
&= \frac{1}{2\pi i} \int_{\partial K'} g(\lambda) R_\lambda(T) \frac{1}{2\pi i} \int_{\partial K} f(z) \frac{1}{z - \lambda} \, dz \, d\lambda = (fg)(T);
\end{aligned}$$

finally, let $r > 0$ such that $\sigma(T) \subset \{z : |z| < r\}$, then

$$z(T) = \frac{1}{2\pi i}\int_{|z|=r} zR_z(T)\,dz = \frac{1}{2\pi i}\int_{|z|=r}\Big(\sum_{k\geq 0} z^{-k}T^k\Big)\,dz = T;$$

the stated commutation property follows since $AR_\lambda(T) = R_\lambda(T)A$ for every $\lambda \in \mathbb{C}\backslash\sigma(T)$ and $A(\int\varphi)B = \int(A\varphi B)$ for every operator valued function φ and every pair of operators A and B (between suitable spaces).]

(b) *The spectral mapping theorem.* Show that $\sigma(f(T)) = f(\sigma(T))$ for every $f \in \mathrm{Hol}(\sigma(T))$. Moreover, if $h \in \mathrm{Hol}(f(\sigma(T)))$, then $h \circ f \in \mathrm{Hol}(\sigma(T))$ and $(h \circ f)(T) = h(f(T))$.

[*Hint:* let $\lambda \in \sigma(T)$ and $g = (f(\lambda) - f)(\lambda - z)^{-1}$; then $g \in \mathrm{Hol}(\sigma(T))$ and $f(\lambda)I - f(T) = (\lambda I - T)g(T)$, which shows that $f(\lambda)I - f(T)$ is not invertible; the inclusion $f(\sigma(T)) \subset \sigma(f(T))$ follows;

conversely, let $\mu \notin f(\sigma(T))$; since $g := (\mu - f)^{-1} \in \mathrm{Hol}(\sigma(T))$, one has $g(T)(\mu I - f(T)) = (\mu I - f(T))g(T) = I$, and hence $\mu \notin \sigma(f(T))$;

for the second claim, let $\mu \notin f(\sigma(T))$ and set $f_\mu = (\mu - f)^{-1}$, then we have $f_\mu(T) = R_\mu(f(T))$; now, by changing the order of integration, we get $h(f(T)) = (2\pi i)^{-1}\int f_\mu(T)h(\mu)\,d\mu = h(f(T))$.]

(c) *The Riesz projections and maximal spectral subspaces.* Let $\sigma(T) = \sigma_1 \cup \sigma_2$, where σ_i are closed and $\sigma_1 \cap \sigma_2 = \emptyset$. Let $\mathcal{O}_i$ be open neighbourhoods of σ_i with $\mathcal{O}_1 \cap \mathcal{O}_2 = \emptyset$, and $\mathcal{O} = \mathcal{O}_1 \cup \mathcal{O}_2$. Show that the operators

$$\mathcal{P}_{\sigma_i} = \chi_{\mathcal{O}_i}(T)$$

satisfy the following properties.

1) $\mathcal{P}_{\sigma_i}^2 = \mathcal{P}_{\sigma_i}$ for $i = 1, 2$, $\mathcal{P}_{\sigma_1} + \mathcal{P}_{\sigma_2} = I$, and $\mathcal{P}_{\sigma_1}X + \mathcal{P}_{\sigma_2}X = X$ (a direct sum);

2) $f(T)\mathcal{P}_{\sigma_i} = \mathcal{P}_{\sigma_i}f(T)$ for $i = 1, 2$ and every $f \in \mathrm{Hol}(\sigma(T))$;

3) $\sigma(T\big|\mathcal{P}_{\sigma_i}X) = \sigma_i$ for $i = 1, 2$, and

4) if $TE \subset E$ such that $\sigma(T\big|E) \subset \sigma_1$, then $E \subset \mathcal{P}_{\sigma_1}X$.

Subspaces of the latter type (i.e. subspaces $X(\sigma) \in \mathrm{Lat}(T)$ such that $E \in \mathrm{Lat}(T)$ and $\sigma(T|E) \subset \sigma(T|X(\sigma)) = \sigma$ imply $E \subset X(\sigma)$) are called *maximal spectral subspaces* (over σ).

[*Hint:* 1) and 2) are obvious from (a); in particular, the subspaces $\mathcal{P}_{\sigma_i}X$ are $f(T)$-invariant for every $f \in \mathrm{Hol}(\sigma(T))$;

in order to prove 3), observe first that if $\lambda \notin \sigma_i$, we can slightly diminish $\mathcal{O}_i$ in such a way that $\lambda \notin \mathcal{O}_i$; then $f := (\lambda - z)^{-1}\chi_{\mathcal{O}_i} \in \mathrm{Hol}(\sigma(T))$ and hence $(\lambda I - T)g(T) = g(T)(\lambda I - T) = \mathcal{P}_{\sigma_i}$, which means that $\lambda \notin \sigma(T\big|\mathcal{P}_{\sigma_i}X)$; if $\lambda \in \sigma_1$ then $\lambda \notin \sigma_2$, and so

$$\lambda I - T = (\lambda I - T)\Big|(\mathcal{P}_{\sigma_1}X) + (\lambda I - T)\Big|(\mathcal{P}_{\sigma_2}X),$$

where the second summand is invertible, whence $\lambda \in \sigma(T\big|\mathcal{P}_{\sigma_1}X)$;

for 4), use Lemma 2.1.10 and get $R_\lambda(T)E \subset E$ and $R_\lambda(T)|E = R_\lambda(T|E)\mathcal{P}_{\sigma_i} = \chi_{\mathcal{O}_i}(T)$ for every $\lambda \in \mathbb{C}\backslash\sigma(T)$; therefore, $I_E = \chi_{\mathcal{O}_1}(T|E) = \chi_{\mathcal{O}_1}(T)|E = \mathcal{P}_{\sigma_1}|E$ which means that $E \subset \mathcal{P}_{\sigma_1}X$.]

(d) Let $S-T \in \mathfrak{S}_\infty$ and $f \in \mathrm{Hol}(\sigma(S)\bigcup\sigma(T))$. Show that $f(S)-f(T) \in \mathfrak{S}_\infty$. The same is true for any other normed ideal of operators $\mathfrak{S}$ instead of $\mathfrak{S}_\infty$.

[*Hint:* show first that $R_\lambda(S) - R_\lambda(T) = R_\lambda(S)(S-T)R_\lambda(T)$ for $\lambda \notin \sigma(S) \cup \sigma(T)$.]

(e) Show that propositions similar to (a) and (b) hold for any unital Banach algebra A instead of $A = L(H)$. Namely, the integral

$$f(x) = \frac{1}{2\pi i}\int_{\partial K} f(z)R_z(x)\,dz$$

defines a continuous homomorphism $\mathrm{Hol}(\sigma(x)) \longrightarrow A$ satisfying $\zeta(x) = x$ and the spectral mapping property $\sigma(f(x)) = f(\sigma(x))$. Here $x \in A$, $f \in \mathrm{Hol}(\sigma(x))$, $\sigma(x)$ is the spectrum of x in A, ζ stands for the independent variable, and $R_z(x) = (ze - x)^{-1}$ is its resolvent.

2.5.2. Spectral convergences. In this subsection we collect several continuity properties of the spectrum mapping $x \longmapsto \sigma(x)$ defined on a Banach algebra A. The principal case is the algebra $A = L(H)$ of bounded operators on a Hilbert space H, but we will pay some attention to the general case too.

Denote by $\mathbb{K} = \mathbb{K}(\mathbb{C})$ the set of all compact subsets of $\mathbb{C}$ endowed with the *Hausdorff metric* d,

$$d(\sigma,\sigma') = \max\Big(\rho(\sigma,\sigma'),\rho(\sigma',\sigma)\Big),$$

where

$$\rho(\sigma,\sigma') = \max_{\lambda\in\sigma}\mathrm{dist}(\lambda,\sigma').$$

Given a unital Banach algebra A, consider the map

$$\sigma(\cdot) : A \longrightarrow \mathbb{K}(\mathbb{C})$$

taking an element $x \in A$ to its spectrum $\sigma(x) = \sigma_A(x)$ in A. As usual, $R_\lambda(x) = (\lambda e - x)^{-1}$ means the resolvent of $x \in A$. By $r(\cdot)$ we denote the *spectral radius function*,

$$r(x) = \max\{|\lambda| : \lambda \in \sigma(x)\} = \lim_n \|x^n\|^{1/n},$$

where the last equality is the famous Gelfand formula. It is clear that $r(x) \le \|x\|$, $r(xy) = \lim_n \|x^n y^n\|^{1/n} \le r(x)r(y)$ for two commuting elements x, y, and

$$|r(x) - r(y)| \le d(\sigma(x),\sigma(y))$$

for every $x, y \in A$.

(a) Show the following properties of ρ and d.

(i) $\rho(\sigma,\sigma') < \varepsilon$ if and only if $\sigma \subset (\sigma')_\varepsilon$, where $(\sigma')_\varepsilon = \bigcup_{\mu\in\sigma'} D(\mu,\varepsilon)$ is the ε-neighbourhood of σ'.

(ii) If $\sigma_j \in \mathbb{K}(\mathbb{C})$ are uniformly bounded (i.e., $\sigma_j \subset D(0,R)$ for all j and some $R > 0$), then

$$\Big(\lim_n \rho(\sigma_n,\sigma') = 0\Big) \quad\Leftrightarrow\quad \Big(\overline{\lim}_n\, \sigma_n \subset \sigma'\Big),$$

where

$$\overline{\lim}_n\, \sigma_n = \Big\{\zeta \in \mathbb{C} : \underline{\lim}_n\, \mathrm{dist}(\zeta,\sigma_n) = 0\Big\}.$$

[*Hint:* the implication "$\Rightarrow$" follows from $\mathrm{dist}(\zeta_n, \sigma') \leq \rho(\sigma_n, \sigma')$ for every $\zeta_n \in \sigma_n$;

to prove "$\Leftarrow$", suppose on the contrary that $\rho(\sigma_{n_i}, \sigma') \geq \varepsilon > 0$ for an infinite sequence n_i; taking suitable $\zeta_i \in \sigma_{n_i}$ we get $\mathrm{dist}(\zeta_i, \sigma') \geq \varepsilon/2$, and any convergent subsequence of (ζ_i) contradicts the inclusion $\overline{\lim}_n \sigma_n \subset \sigma'$.]

(iii) $\left(\lim_n \rho(\sigma', \sigma_n) = 0\right) \Leftrightarrow \left(\sigma' \subset \underline{\lim}_n \sigma_n\right)$, where

$$\underline{\lim}_n \sigma_n = \left\{\zeta \in \mathbb{C} : \lim_n \mathrm{dist}(\zeta, \sigma_n) = 0\right\}.$$

[*Hint:* the implication "$\Rightarrow$" is obvious since $\mathrm{dist}(\zeta, \sigma_n) \leq \rho(\sigma', \sigma_n)$ for every $\zeta \in \sigma'$;

conversely, the inclusion $\sigma' \subset \underline{\lim}_n \sigma_n$ means that

$$\lim_n \mathrm{dist}(\zeta, \sigma_n) = 0$$

for all $\zeta \in \sigma'$; in fact, the convergence is uniform in $\zeta \in \sigma'$ since the functions in question $f_n(\zeta) = \mathrm{dist}(\zeta, \sigma_n)$ are uniformly Lipschitz, and hence $(f_n)_{n \geq 1}$ is a precompact set in $C(\sigma')$ (indeed, $|f(\zeta) - f(\zeta')| \leq |\zeta - \zeta'|$ for $f(\zeta) = \mathrm{dist}(\zeta, \sigma)$ and for every $\sigma \subset \mathbb{C}$).]

(iv) $\lim_n d(\sigma_n, \sigma') = 0$ if and only if $\lim_n \sigma_n = \sigma'$ in the sense of the following uniform convergence: for every $\varepsilon > 0$ there exists N such that

$$\sigma_n \subset (\sigma')_\varepsilon, \quad \sigma' \subset (\sigma_n)_\varepsilon$$

for $n > N$.

[*Hint:* use (i).]

(b) Let A be a unital Banach algebra. Then

$$d(\sigma(x), \sigma(y)) \leq r(x - y)$$

for every pair of commuting elements $x, y \in A$ (i.e., $xy = yx$). In particular, if $x = \lim_n x_n$ and all x_n commute with x ($xx_n = x_n x$), then the spectra $\sigma(x_n)$ converge uniformly to $\sigma(x)$ (in the sense of (a)(iv)).

[*Hint:* if $\mathrm{dist}(\lambda, \sigma(y)) > r(x - y)$, then $\lambda e - x = (\lambda e - y)(e - R_\lambda(y)(x - y))$; therefore, $\lambda \notin \sigma(x)$ since the second parantheses are invertible due to $r(R_\lambda(y)(x - y)) \leq r(R_\lambda(y)) r((x - y)) < 1$; this means that $\mathrm{dist}(\mu, \sigma(y)) \leq r(x - y)$ for every $\mu \in \sigma(x)$, whence $\rho(\sigma(x), \sigma(y)) \leq r(x - y)$; a symmetry argument gives the result.]

(c) *The upper semi-continuity of the spectrum mapping.* Let A be a unital Banach algebra. Show that the spectrum function $\sigma(\cdot)$ is upper semi-continuous in the following sense.

(i) $\mathrm{dist}(\lambda, \sigma(y)) \geq \|R_\lambda(x)\|^{-1} - \|x - y\|$ whatever are $x, y \in A$ and $\lambda \in \mathbb{C} \backslash \sigma(x)$. Moreover, if $\|R_\lambda(x)\|^{-1} - \|x - y\| > 0$ then

$$\|R_\lambda(x) - R_\lambda(y)\| \leq \|R_\lambda(x)\|^2 \|x - y\| (1 - \|R_\lambda(x)\| \cdot \|x - y\|)^{-1}.$$

(ii) $\sigma(y) \subset (\sigma(x))_\varepsilon$ if $\|x-y\| < M(x,\varepsilon)^{-1}$, where $\varepsilon > 0$ and

$$M(x,\varepsilon) = \sup\{\|R_\lambda(x)\| : \operatorname{dist}(\lambda, \sigma(x)) \geq \varepsilon\}.$$

(iii) If $x, x_n \in A$ and $\lim_n x_n = x$ then the following equivalent properties hold

$$\lim_n \rho(\sigma(x_n), \sigma(x)) = 0, \quad \overline{\lim}_n\, \sigma(x_n) \subset \sigma(x).$$

[*Hint:* for (i), write $\lambda e - y = \lambda e - x + x - y = (\lambda e - x)(e - R_\lambda(x)(y-x))$, whence

$$\frac{1}{\operatorname{dist}(\lambda, \sigma(y))} \leq \|R_\lambda(y)\| \leq \|R_\lambda(x)\|(1 - \|R_\lambda(x)\| \cdot \|x-y\|)^{-1},$$

if $1 - \|R_\lambda(x)\| \cdot \|x-y\| > 0$; similarly for $R_\lambda(x) - R_\lambda(y)$;

for (ii), deduce from (i) that $\operatorname{dist}(\lambda, \sigma(y)) > 0$ if $\operatorname{dist}(\lambda, \sigma(x)) \geq \varepsilon$ and $\|x-y\| < M(x,\varepsilon)^{-1}$;

for (iii), employ (ii) and (a) above.]

(d) *Some continuity points of the spectrum mapping.* Let $A = L(H)$ and $T \in L(H)$.

(i) *Continuity of isolated parts of the spectrum.* Let $\sigma \subset \sigma(T)$ be a *clopen subset* of $\sigma(T)$ (i.e., closed and (relatively) open), and let $\varepsilon > 0$ be such that the closures of ε-neighbourhoods $\operatorname{clos}((\sigma)_\varepsilon)$ and $\operatorname{clos}((\sigma(T)\backslash\sigma)_\varepsilon)$ are disjoint. Show that there exists $\delta > 0$ such that $\sigma' = \sigma(T') \cap (\sigma)_\varepsilon$ is a clopen subset of $\sigma(T')$ and

$$\|\mathcal{P}_\sigma(T) - \mathcal{P}_{\sigma'}(T')\| < \varepsilon$$

for every $T' \in L(H)$ with $\|T - T'\| < \delta$; the spectral projections $\mathcal{P}_\sigma$ are defined in 2.5.1(c).

[*Hint:* use (c) to show that $\sigma(T') \subset (\sigma(T))_\varepsilon = (\sigma)_\varepsilon \cup (\sigma(T)\backslash\sigma)_\varepsilon$ for $\delta > 0$ small enough; then, following the definition of $\mathcal{P}_\sigma$ form 2.5.1(c), fix open neighbourhoods $\mathcal{O}_i$, $i = 1, 2$, of $\operatorname{clos}((\sigma)_\varepsilon)$ and $\operatorname{clos}((\sigma(T)\backslash\sigma)_\varepsilon)$, respectively, such that $\overline{\mathcal{O}}_1 \cap \overline{\mathcal{O}}_2 = \emptyset$; it follows from (c)(i) that $\|R_\lambda(T) - R_\lambda(T')\| < \varepsilon$ for $\lambda \in \partial\mathcal{O}_i$ for sufficiently small $\delta > 0$, and hence $\|\mathcal{P}_\sigma(T) - \mathcal{P}_{\sigma'}(T')\| < |\partial\mathcal{O}_1|\varepsilon$.]

(ii) Suppose the spectrum $\sigma(T)$ is *densely disconnected* , which means that for every open $\mathcal{O} \subset \mathbb{C}$ with $\sigma(T) \cap \mathcal{O} \neq \emptyset$ there exists a clopen subset $\sigma \subset \sigma(T)$ such that $\sigma \neq \emptyset$ and $\sigma \subset \mathcal{O}$. Show that the mapping $\sigma(\cdot) : L(H) \longrightarrow (\mathbb{K}(\mathbb{C}), d)$ is continuous at the point T.

[*Hint:* taking $T_n \longrightarrow T$ we have from (c) that $\lim_n \rho(\sigma(T_n), \sigma(T)) = 0$; in order to check the lower semi-continuity property $\lim_n \rho(\sigma(T), \sigma(T_n)) = 0$, take $0 < \varepsilon < 1$ and consider nonempty clopen subsets $\sigma_i \subset \sigma(T)$, $i = 1, \ldots, m$, forming an ε-net for $\sigma(T)$; using (i), show that there exists N such that

$$\|\mathcal{P}_{\sigma_i}(T) - \mathcal{P}_{\sigma_i'}(T_n)\| < \varepsilon$$

for all i and all $n > N$; since $\|\mathcal{P}_{\sigma_i}(T)\| \geq 1$, we get $\sigma_i' = \sigma(T_n) \cap (\sigma_i)_\varepsilon \neq \emptyset$ for every i; therefore, for any $\lambda \in \sigma(T)$ and suitable $\lambda_i \in \sigma_i$ and $\mu_i \in \sigma_i'$, we have

$$\operatorname{dist}(\lambda, \sigma(T_n)) \leq |\lambda - \mu_i| \leq |\lambda - \lambda_i| + |\lambda_i - \mu_i| < \varepsilon + \varepsilon = 2\varepsilon,$$

whence $\rho(\sigma(T), \sigma(T_n)) \leq 2\varepsilon$ for every $n > N$.]

(iii) Let $T \in L(H)$ and suppose that the isolated points of $\sigma(T)$ are dense in $\sigma(T)$. Show that the spectrum mapping $\sigma(\cdot)$ is continuous at T. In particular, every compact operator $T \in L(H)$ is a continuity point of $\sigma(\cdot)$.

[*Hint:* apply (ii).]

(iv) Let $T \in L(H)$ be a compact operator, $r > 0$, and $r' > 0$ be such that $\overline{D}(\lambda, r') \cap \sigma(T) = \{\lambda\}$ for every $\lambda \in \sigma$, where

$$\sigma = \sigma(T) \cap \{\zeta \in \mathbb{C} : |\zeta| > r\}.$$

Show that for every $\varepsilon > 0$ there exists $\delta > 0$ such that any operator T' with $\|T - T'\| < \delta$ satisfies the following properties:

a) $\sigma(T') \cap \{\zeta \in \mathbb{C} : |\zeta| > r\}$ is a finite set contained in $\bigcup_{\lambda\in\sigma} D(\lambda, r')$;

b) $\|\mathcal{P}_\tau(T) - \mathcal{P}_{\tau(T')}(T')\| < \varepsilon$ for every $\tau \subset \sigma$, where $\tau(T') = \mathcal{O} \cap \sigma(T')$ and $\mathcal{O} = \bigcup_{\lambda\in\tau} D(\lambda, r')$;

c) $\operatorname{rank} \mathcal{P}_\tau(T) = \operatorname{rank} \mathcal{P}_{\tau(T')}(T')$ for every $\tau \subset \sigma$, if ε (from b)) is sufficiently small;

d) let $\lambda \in \sigma$ be such that $\operatorname{rank} \mathcal{P}_{\{\lambda\}}(T) = 1$, and so $\mathcal{P}_{\{\lambda\}}(T) = (\cdot, x)y$ for some $x, y \in H$; then $\mathcal{P}_{\{\lambda\}(T')}(T') = (\cdot, x')y'$, and normalizing by $\|y\| = \|y'\| = 1$, one has $\|y - \alpha y'\| < 3\varepsilon$ for some $|\alpha| = 1$ and for any ε as small as in c).

[*Hint:* a) and b) are immediate consequences of (i);

c) follows from lemma (v) below;

in order to prove d), apply $\mathcal{P}_{\{\lambda\}}(T) - \mathcal{P}_{\{\lambda\}(T')}(T')$ to y and y' and use b) getting $\|y - (y, x)y'\| < \varepsilon$, $\|(y', x)y - y'\| < \varepsilon$; since the orthogonal projection onto a subspace is closer to a given vector than any other vector of the subspace, we have $\|y - (y, y')y'\| < \varepsilon$, $\|(y', y)y - y'\| < \varepsilon$; this implies $\|(y', y)y - |(y, y')|^2 y'\| < \varepsilon$, and then $1 - |(y, y')|^2 = \||(y, y')|^2 y' - y'\| < 2\varepsilon$; setting $\alpha = (y, y')/|(y, y')|$, we get $\|y - \alpha y'\| \le \|y - (y, y')y'\| + (1 - |(y, y')|) < 3\varepsilon$.]

(v) *Lemma on neighbouring projections.* Let P and Q be bounded projections on a Banach space X. Show that

a) $\operatorname{rank} Q \le \operatorname{rank} P$ if $QX \cap \operatorname{Ker} P = \{0\}$, and, in particular,

b) $\operatorname{rank} P = \operatorname{rank} Q$ if $\|P - Q\| < 1$.

[*Hint:* the condition $QX \cap \operatorname{Ker} P = \{0\}$ means that $P|(QX)$ is an injection of QX into PX, and a) follows;

for b), take $x \in QX$, $x \ne 0$, and write $\|Px - x\| = \|(P - Q)x\| < \|x\|$, which yields $Px \ne 0$; to complete, apply a) and use the symmetry argument.]

(e) *Some discontinuity points of the spectrum mapping.*

(i) Let $S : H^2 \longrightarrow H^2$ be the shift operator on the Hardy space H^2 and $T_0 = S \oplus S^*$. Show that there exists a sequence $T_n = T_0 + K_n$ of rank one perturbations of T_0 such that $\lim_n \|T_0 - T_n\| = 0$ and $\sigma(T_n) \subset \mathbb{T}$ for every n, whereas $\sigma(T_0) = \overline{\mathbb{D}}$.

[*Hint:* take $K_n(0, 1) = (\varepsilon_n, 0)$, where $0 < \varepsilon_n < 1$, and $K_n = 0$ on the orthogonal complement of the element $(0, 1) \in H^2 \oplus H^2$; it is easy to see that $\|T_n^k\| \le 1$ for $k \ge 0$, and $T_n^{-1} = S^* \oplus S + K_n'$ (check on elements of the standard orthonormal basis $e_i \oplus e_j$), where $K_n'(1, 0) = (0, \varepsilon_n^{-1})$ and $K_n' = 0$ on the orthogonal complement; therefore, $\|T_n^{-k}\| \le \varepsilon_n^{-1}$ for every $k \ge 1$, and hence T_n is similar to a unitary operator by the Sz.-Nagy theorem 1.5.6(b); the result follows.]

(ii) *(S. Kakutani, 1960)* Show that there exists a sequence of nilpotent Hilbert space operators $T_s \in L(H)$, $T_s^{k_s} = 0$, such that $\lim_s \|T_s - T\| = 0$ and $r(T) > 0$.

Namely, take $H = l^2 = l^2(\mathbb{Z}_+)$ and let $T = S\Lambda$ be a *weighted shift operator* defined by

$$Te_k = \lambda_k e_{k+1}$$

for elements of the standard basis $(e_k)_{k\geq 0}$ of l^2, where

$$\lambda_k = e^{-s}$$

for k of the form $k = 2^s(2l+1)$ with some $s \geq 0$, $l \geq 0$. The points a), b) and c) below do not depend on the concrete choice of λ_k.

a) Show that $\|T\| = \sup_k |\lambda_k|$; show that T is unitarily equivalent to the shift operator $Se_k = e_{k+1}$ on the weighted space $l^2(\mathbb{Z}_+, w_k)$, where

$$l^2(\mathbb{Z}_+, w_k^2) = \Big\{x = (x_k)_{k\geq 0} : \|x\|_w^2 = \sum_{k\geq 0} |x_k w_k|^2 < \infty\Big\}$$

and $w_k = \lambda_0 \ldots \lambda_{k-1}$ for $k \geq 0$.

b) Show that the dual space of $l^2(\mathbb{Z}_+, w_k^2)$ with respect to the Cauchy pairing $(x, y) = \sum_{k\geq 0} x_k \overline{y}_k$ is $l^2(\mathbb{Z}_+, w_k^{-2})$, and the adjoint operator S^* is the backward shift on $l^2(\mathbb{Z}_+, w_k^{-2})$, defined as $S^*e_0 = 0$ and $S^*e_k = e_{k-1}$ for $k \geq 1$.

c) Show that any eigenvector x of S^*, $S^*x = \lambda x$, is of the form $x = c(\lambda^k)_{k\geq 0}$, and hence

$$\{0\} \cup \{\lambda \in \mathbb{C} : |\lambda| < r_1(S^*)\} \subset \sigma_p(S^*) \subset \{\lambda \in \mathbb{C} : |\lambda| \leq r_1(S^*)\},$$

where $r_1(S^*) = \underline{\lim}_k (w_k)^{1/k}$.

[*Hint:* look at $\sum_{k\geq 0} |\lambda^k w_k^{-1}|^2 < \infty$.]

d) Show that $r_1(S^*) = \lim_n (w_n)^{1/n} = \exp(-\sum_{s\geq 0} \frac{s}{2^{s+1}}) = e^{-1}$.

[*Hint:*

$$n^{-1} \sum_{k=0}^{n-1} \log \lambda_k = -n^{-1} \sum_{s\geq 0} s \cdot a(s, n),$$

where $a(s, n) = 0$, if $2^s \geq n$, and $a(s, n) = \mathrm{card}\{l \geq 0 : 2^s(2l+1) < n\} = n2^{-s-1} + \Delta(s, n)$ with $-1/2 \leq \Delta(s, n) \leq 1/2$, otherwise; since $n^{-1} \sum_{2^s<n} s = o(1)$ as $n \longrightarrow \infty$, it follows that the limit in question exists and takes the value given above; in order to sum the series, use $f(x) = \sum_{s\geq 1} x^s 2^{-s-1}$.]

e) Show that $\{\lambda \in \mathbb{C} : |\lambda| < r_1(S^*)\} \subset \sigma_p(T^*)$, and hence, $r(T) = r(S^*) \geq r_1(S^*) > 0$, where $r(T)$ stands (as above) for the spectral radius of T.

f) Let $T_s : l^2 \longrightarrow l^2$ be a weighted shift defined by $T_s e_k = 0$ for k of the form $k = 2^s(2l+1)$, and $T_s e_k = Te_k$ otherwise. Show that $T_s^{2^s} = 0$, and therefore, $\sigma(T_s) = \{0\}$ and $r(T_s) = 0$, whereas $\lim_s \|T - T_s\| = 0$.

[*Hint:* $\|T - T_s\| = e^{-s}$.]

2.5.3. (In)stability of the (weakly) continuous spectrum. Following I. Gohberg and M. Krein, we call $\lambda \in \sigma(T)$ a *normal eigenvalue* of T if λ is an isolated point of the spectrum and the corresponding Riesz projection $\mathcal{P}_\lambda$ is of finite rank. We denote by $\sigma_{np}(T)$ the *set of all normal eigenvalues* of T and call

$$\sigma_{wc}(T) = \sigma(T) \backslash \sigma_{np}(T)$$

the *(weakly) continuous spectrum* of T.

(a) Show that $\sigma_{wc}(T)$ is a compact set, and $\sigma(T)\backslash\sigma_{wc}(T)$ consists of an at most countable sequence with cluster points at $\sigma_{wc}(T)$; moreover, $\sigma_F(T) \subset \sigma_{wc}(T)$ (see B.2.5.1, Volume 1, for the definition of the Fredholm spectrum), and hence $\sigma_{wc}(T) \neq \emptyset$ if $\dim X = \infty$.

(b) *(I. Gohberg, 1951)* Let X be a Banach space, $\Omega \subset \mathbb{C}$ a connected open set and let $F : \Omega \longrightarrow L(X)$ be a holomorphic operator valued function such that $F(\zeta)$ is a Fredholm operator for every $\zeta \in \Omega$ and $\operatorname{ind} F(\lambda) = 0$ for a certain point $\lambda \in \Omega$ (see B.2.5.1, Volume 1, about that).

(i) Show that there exists an $\varepsilon > 0$ such that

$$\dim \operatorname{Ker} F(\zeta) = \text{const} \leq \dim \operatorname{Ker} F(\lambda) < \infty$$

for all $\zeta \in D(\lambda, \varepsilon)\backslash\{\lambda\} = \{\zeta \in \mathbb{C} : 0 < |\zeta - \lambda| < \varepsilon\}$;

(ii) Show that $\dim \operatorname{Ker} F(\zeta) = m := \min\{\dim \operatorname{Ker} F(z) : z \in \Omega\}$ for all $\zeta \in \Omega$ but (possibly) a countable sequence $\{\zeta_n\}$ accumulating to $\partial\Omega$, where $m < \dim \operatorname{Ker} F(\zeta_n) < \infty$.

[*Hint:* let $K \in L(X)$ be a finite rank operator such that $G(\lambda) = F(\lambda) + K$ is invertible, and let $r_1 > 0$ be such that $G(\zeta) = F(\zeta) + K$ is still invertible for all $\zeta \in D(\lambda, r_1)$; using

$$\begin{aligned} \dim \operatorname{Ker} F(\zeta) &= \dim \operatorname{Ker} F(\zeta)G(\zeta)^{-1} = \dim \operatorname{Ker}(I - KG(\zeta)^{-1}) \\ &= \dim \operatorname{Ker}((I - KG(\zeta)^{-1})|X_1), \end{aligned}$$

where $X_1 = \operatorname{Range} K = K(X)$, $\dim X_1 < \infty$, define a holomorphic function $\Phi(\zeta) = (I - KG(\zeta)^{-1})|X_1$, $\Phi : D(\lambda, r_1) \longrightarrow L(X_1)$; next, choosing $\zeta_1 \in D(\lambda, r_1)$ with

$$\operatorname{rank} \Phi(\zeta_1) = \max\{\operatorname{rank} \Phi(\zeta) : \zeta \in D(\lambda, r_1)\} =: m$$

and using Kramer's rule from linear algebra, find a projection $P : X_1 \longrightarrow X_1$ such that $\operatorname{rank} P = m$ and $\det(P\Phi(\zeta_1)P) \neq 0$; since Φ is holomorphic, there exists $\varepsilon > 0$ such that $\det(P\Phi(\zeta)P) \neq 0$ for $0 < |\zeta - \lambda| < \varepsilon$, and hence $\operatorname{rank} \Phi(\zeta) = m$ and $\dim \operatorname{Ker} \Phi(\zeta) = \dim X_1 - m$ for $0 < |\zeta - \lambda| < \varepsilon$.]

(c) *(I. Gohberg, 1951)* Let X be a Banach space, $A \in L(X)$, $K \in \mathfrak{S}_\infty$ and $T = A + K$. Next, let Ω be a connected component of the complement $\mathbb{C}\backslash\sigma_{wc}(A)$ such that $\Omega\backslash\sigma(T) \neq \emptyset$. Show that

$$\Omega \subset \mathbb{C}\backslash\sigma_{wc}(T)$$

(in particular, $\Omega \cap \sigma(T)$ is an at most countable sequence of eigenvalues of T tending to $\partial\Omega$).

[*Hint:* apply (b) for $F(\zeta) = \zeta I - T = \zeta I - A - K$, $\zeta \in \Omega$ (which is Fredholm valued and satisfies $\operatorname{ind} F(\zeta) = 0$, see (a) above and B.2.5.1, Volume 1) and obtain that $\operatorname{Ker} F(\zeta) = \{0\}$ in Ω excepting (possibly) a sequence (λ_n) having no limit points in Ω; hence, $\zeta I - T$ is invertible in $\Omega\backslash(\lambda_n)$;

now, let $\lambda = \lambda_n$ and $\varphi_\lambda = \chi_D$, where $D = \overline{D}(\lambda, \varepsilon)$ is a small closed disk containing no other points of $\sigma(T) \cup \sigma(A)$ exept λ; due to 2.5.1(c) and 2.5.1(d),

$$\mathcal{P}_\lambda(T) - \mathcal{P}_\lambda(A) = \varphi_\lambda(T) - \varphi_\lambda(A) \in \mathfrak{S}_\infty,$$

where $\mathcal{P}_\lambda = \mathcal{P}_{\{\lambda\}}$ and since $\operatorname{rank} \mathcal{P}_\lambda(A) < \infty$, we get $\mathcal{P}_\lambda(T) \in \mathfrak{S}_\infty$ and hence $\operatorname{rank} \mathcal{P}_\lambda(T) < \infty$; the latter means that $\Omega \cap \sigma(T) \subset \sigma_{np}(T)$.]

(d) Let X be a Banach space, $A \in L(X)$, $K \in \mathfrak{S}_\infty$ and $T = A + K$. Let further $\Omega_\infty(A)$ be the unbounded connected component of $\mathbb{C}\backslash\sigma_{wc}(A)$. Show that $\Omega_\infty(A) = \Omega_\infty(T)$. In particular,

$$\sigma_{wc}(T) = \sigma_{wc}(A)$$

if the spectrum $\sigma(A)$ is simply connected (i.e., if $\mathbb{C}\backslash\sigma(A) = \Omega_\infty(A)$).

[*Hint:* apply (c) to T and to $A = T - K$.]

(e) Let $T : H \longrightarrow H$ be a Hilbert space operator.

(i) Show that T can be represented in the form

$$T = U + K,$$

where U is unitary and $K \in \mathfrak{S}_\infty$, if and only if $I - T^*T \in \mathfrak{S}_\infty$ and $\lambda I - T$ is a Fredholm operator with $\operatorname{ind}(\lambda I - T) = 0$ for some point (for all points) $\lambda \in \mathbb{D}$.

Moreover, if $I - T^*T \in \mathfrak{S}_\infty$ and $\mathbb{D}\backslash\sigma(T) \neq \emptyset$ then $\sigma_{wc}(T) = \sigma_{wc}(U) \subset \mathbb{T}$.

[*Hint:* the "only if" part is an immediate consequence of B.2.5.1 (Volume 1);

conversely, if $(I - |T|)(I + |T|) = I - T^*T = K_1 \in \mathfrak{S}_\infty$, then the polar representation of T is of the form $T = V|T| = V + VK_2$, where $K_2 \in \mathfrak{S}_\infty$ and V is a partial isometry with $\operatorname{Ker} V = \operatorname{Ker} T$ and $\operatorname{Ker} V^* = \operatorname{Ker} T^*$ (see B.1.5.5, Volume 1); therefore, if T is Fredholm and $\operatorname{ind} T = 0$, there exists a finite rank operator $K_3 : \operatorname{Ker} V \longrightarrow \operatorname{Ker} V^*$ such that $T = (V + K_3) + VK_2 - K_3$ and $V + K_3$ is unitary;

in general, apply (d) to $T = V + VK_2$, where $\|V\| \leq 1$, and obtain that $\sigma(T) \cap \{\zeta : |\zeta| > 1\}$ is an at most countable sequence accumulating to the unit circle $\mathbb{T}$; since $\zeta I - T$ is a Fredholm operator and $\operatorname{ind}(\zeta I - T) = 0$ for all $\zeta \in D(\lambda, \varepsilon)$ if $D(\lambda, \varepsilon) \subset \mathbb{D}$ is small enough (see B.2.5.1, Volume 1), one can find $\zeta \in D(\lambda, \varepsilon)$ with $\frac{1}{\bar\zeta} \notin \sigma(T)$ and employ the previous reasoning for $S = b_\zeta(T)$, where $b_\zeta = (\zeta - z)(1 - \bar\zeta z)^{-1}$ (in order to check the hypothesis, use formula (ii) below); getting $S = U + K_4$ with $K_4 \in \mathfrak{S}_\infty$, use 2.5.1(d) to obtain $T = b_\zeta^{-1}(U) + K$, $K \in \mathfrak{S}_\infty$;

in order to prove $\sigma_{wc}(T) = \sigma_{wc}(U)$, apply (c) and (d) to the components of $\mathbb{C}\backslash\sigma_{wc}(U) \supset \mathbb{C}\backslash\mathbb{T}$.]

(ii) Let $\lambda \in \mathbb{D}$ such that $1/\bar\lambda \notin \sigma(T)$. Show that

$$I - T_\lambda^* T_\lambda = V^*(I - T^*T)V \text{ and } I - T_\lambda T_\lambda^* = V(I - T^*T)V^*,$$

where $V = (1 - |\lambda|^2)(I - \bar\lambda T)^{-1}$; in particular, $I - T_\lambda^* T_\lambda \in \mathfrak{S}_\infty$ if and only if $I - T^*T \in \mathfrak{S}_\infty$, and T_λ is a Fredholm operator if and only if $\lambda I - T$ is, and $\operatorname{ind} T_\lambda = \operatorname{ind}(\lambda I - T)$.

(iii) Let $1 \leq p < \infty$. Show that T can be represented in the form

$$T = U + K,$$

where U is unitary and $K \in \mathfrak{S}_p$, if and only if $I - T^*T \in \mathfrak{S}_p$ and $\lambda I - T$ is a Fredholm operator with $\operatorname{ind}(\lambda I - T) = 0$ for some point (for all points) $\lambda \in \mathbb{D}$.

[*Hint:* the same proof as in (i).]

(f) *(J. von Neumann, 1935)* Let $\Lambda = (\lambda_i)_{i \geq 0}$ and $M = (\mu_j)_{j \geq 0}$ be two sequences of complex numbers such that the set $\operatorname{clos} \Lambda = \operatorname{clos} M$ contains no isolated points. Fix numbers $\varepsilon_n > 0$, $n = 0, 1, \ldots$.

(i) Show that there exist bijections $n \longmapsto i(n)$ and $n \longmapsto j(n)$ such that $|\lambda_{i(n)} - \mu_{j(n)}| < \varepsilon_n$ for all $n \geq 0$.

[*Hint:* let $i'(0) = 0$ and $j'(0)$ with $|\lambda_{i'(0)} - \mu_{j'(0)}| < \varepsilon_0$, and then $j''(0) = min(\mathbb{N}_* \backslash \{j'(0)\})$ and $i''(0) \in \mathbb{N}_* \backslash \{i'(0)\}$ such that $|\lambda_{i''(0)} - \mu_{j''(0)}| < \varepsilon_1$; set $I_0 = \{i'(0), i''(0)\}$ and $J_0 = \{j'(0), j''(0)\}$;

let $i'(1) = \min(\mathbb{N}_* \backslash I_0)$ and $j'(1) \in \mathbb{N}_* \backslash J_0$ such that $|\lambda_{i'(1)} - \mu_{j'(1)}| < \varepsilon_2$, and then $j''(1) = min(\mathbb{N}_* \backslash (J_0 \cup \{j'(1)\}))$ and $i''(1) \in \mathbb{N}_* \backslash (I_0 \cup \{i'(1)\})$ such that $|\lambda_{i''(1)} - \mu_{j''(1)}| < \varepsilon_3$; set $I_1 = I_0 \cup \{i'(1), i''(1)\}$ and $J_1 = J_0 \cup \{j'(1), j''(1)\}$; by induction,

let $i'(p) = \min(\mathbb{N}_* \backslash I_{p-1})$ and $j'(p) \in \mathbb{N}_* \backslash J_{p-1}$ such that

$$|\lambda_{i'(p)} - \mu_{j'(p)}| < \varepsilon_{2p},$$

and then $j''(p) = \min(\mathbb{N}_* \backslash (J_{p-1} \cup \{j'(p)\}))$ and $i''(p) \in \mathbb{N}_* \backslash (I_{p-1} \cup \{i'(p)\})$ such that

$$|\lambda_{i''(p)} - \mu_{j''(p)}| < \varepsilon_{2p+1};$$

set $I_p = I_{p-1} \cup \{i'(p), i''(p)\}$ and $J_p = J_{p-1} \cup \{j'(p), j''(p)\}$;

it is clear that the mappings $2p \longmapsto i'(p)$, $2p+1 \longmapsto i''(p)$ and $2p \longmapsto j'(p)$, $2p+1 \longmapsto j''(p)$ establish the required bijections.]

(ii) Let $\varepsilon_n \searrow 0$, and let N_1 and N_2 be normal operators having *purely discrete spectra* (i.e., eigenvectors of N_i form a complete system in the corresponding Hilbert space), and such that the set $\sigma(N_1) = \sigma(N_2)$ has no isolated points. Show that there exists a unitary operator V such that

$$s_n(N_1 - V^* N_2 V) \leq \varepsilon_n$$

for every $n \geq 0$, where $s_n(\cdot)$ stands for the n-th singular number.

[*Hint:* N_1 and N_2 have orthonormal bases of eigenvectors $N_1 f_i = \lambda_i f_i$, $N_2 g_j = \mu_j g_j$ with $(\lambda_i)_{i \geq 0}$ and $(\mu_j)_{j \geq 0}$ satisfying (i).]

(g) *Instability of the non-discrete spectrum of selfadjoint and unitary operators. (H. Weyl (1909), J. von Neumann (1935), S. Kuroda (1958))*

(i) *(J. von Neumann)* Let $A : H \longrightarrow H$ be a bounded selfadjoint operator, $x \in H$ and $n \geq 1$. Show that there exists a subspace $E_n \subset H$ and an operator $K_n = K_n^* \in \mathcal{F}$ such that

a) $x \in E_n$; $(A + K_n)E_n \subset E_n$,
b) $\dim E_n \leq n$,
c) $\operatorname{rank} K_n \leq 2n$; $\|K_n\| \leq n^{-1}\|A\|$.

[*Hint:* set $\sigma_j = -\|A\| + 2\|A\|n^{-1}[j, j+1)$ for $j = 0, 1, ..., n-1$ and

$$E_n = \operatorname{span}(x_j : 0 \leq j < n),$$

where $x_j = E(\sigma_j)x$ and $E(\cdot)$ stands for the spectral measure of A (see 1.5.1(l) above); write A in matrix form with respect to the decomposition $H = E_n \oplus E_n^\perp$ and define $A + K_n$ as the diagonal part of A, that is,

$$A = \begin{pmatrix} a & b \\ c & d \end{pmatrix}, \quad A + K_n = \begin{pmatrix} a & 0 \\ 0 & d \end{pmatrix};$$

then, $\|K_n\| = max(\|b\|, \|c\|)$, and denoting by λ_j the center of the interval σ_j one obtains

$$\|Ax_j - \lambda_j x_j\|^2 = \int_{\sigma_j} |z - \lambda_j|^2 \, d(E(z)x_j, x_j) \leq (\|A\|n^{-1})^2 \|x_j\|^2$$

for every j, and hence

$$\begin{aligned}\|c(\sum_j \alpha_j x_j)\|^2 &= \|(I - P_{E_n})\sum_j \alpha_j A x_j\|^2 = \|(I - P_{E_n})\sum_j \alpha_j (Ax_j - \lambda_j x_j)\|^2 \\ &\leq \sum_j |\alpha_j|^2 \|Ax_j - \lambda_j x_j\|^2 \leq (\|A\|n^{-1})^2 \sum_j |\alpha_j|^2 \|x_j\|^2 \\ &= (\|A\|n^{-1})^2 \|\sum_j \alpha_j x_j\|^2;\end{aligned}$$

this means that $\|c\| \leq n^{-1}\|A\|$, and similarly for $\|b\|$.]

(ii) Let $A : H \longrightarrow H$ be a selfadjoint operator on a Hilbert space H (maybe, unbounded; see the *Hint* below), and $\mathfrak{S}$ be a symmetrically normed (cross-normed) ideal of compact operators (see B.7.5.2(g), Volume 1) such that $\mathfrak{S} \neq S_1$ (recall that always $\mathfrak{S}_1 \subset \mathfrak{S}$). Let $\varepsilon > 0$. Show that there exists an operator $K = K^* \in \mathfrak{S}$ such that $\|K\|_{\mathfrak{S}} < \varepsilon$ and the operator $A + K$ has a pure discrete spectrum (i.e., the eigenvectors of $A + K$ form a complete system in H).

[*Hint:* without entering into details, by an unbounded selfadjoint operator we mean $A = \sum_m \oplus A_m$ acting on $H = \sum_m \oplus H_m$, where $A_m : H_m \longrightarrow H_m$ are bounded and selfadjoint (see also the spectral theory in C1.5.1 above); it suffices to prove the theorem for every A_m with $\|K_m\|_{\mathfrak{S}} < \varepsilon_m$, $\sum_m \varepsilon_m < \varepsilon$;

let $A : H \longrightarrow H$ be a bounded selfadjoint operator and $(x_s)_{s\geq 1}$ a complete sequence in H; applying (i) with $x = x_1$, one gets a subspace $H_1 = E_n$ containing x_1 and a finite rank operator $K_1' = K_n = K_n^*$ such that $(A + K_1')H_1 \subset H_1$ and

$$\|K_1'\|_{\mathfrak{S}} \leq \varphi_{\mathfrak{S}}(\operatorname{rank} K_1')\|K_1'\| \leq \varphi_{\mathfrak{S}}(2n)n^{-1}\|A\|$$

(see B.7.5.2(g), Volume 1, for the notation); since $\varphi_{\mathfrak{S}}(2n)n^{-1} = o(1)$ as $n \longrightarrow \infty$ (see Kuroda's theorem in B.7.5.2(g), Volume 1) one can choose n such that $\|K_1'\|_{\mathfrak{S}} < \varepsilon 2^{-1}$;

applying (i) again with $A' = (A + K_1')|H_1^{\perp}$ and $x = P_{H_1^{\perp}} x_2$, and continuing by induction, one gets finite dimensional subspaces H_s and selfadjoint operators K_s' such that $\|K_s'\|_{\mathfrak{S}} < \varepsilon 2^{-s}$, $K_s'|H_t = 0$ for $t \neq s$, and

$$(A + K_s')H_s = (A + K)H_s \subset H_s$$

for every $s \geq 1$, where $K = \sum_{s\geq 1} K_s'$; moreover, the inclusions $x_s \in \operatorname{span}(H_t : 1 \leq t \leq s)$ yield $H = \sum_{s\geq 1} \oplus H_s$, and the result follows since $\dim H_s < \infty$ for every $s \geq 1$.]

(iii) Let $U : H \longrightarrow H$ be a unitary operator on a Hilbert space H, and $\mathfrak{S}$ be a symmetrically normed (cross-normed) ideal of compact operators as in (ii) above. Let $\varepsilon > 0$. Show that there exists an operator $K = K^* \in \mathfrak{S}$ such that $\|K\|_{\mathfrak{S}} < \varepsilon$ and the operator $U + K$ is unitary and has a pure discrete spectrum (i.e., eigenvectors of $U + K$ form a complete system in H).

[*Hint:* suppose (without loss of generality) that -1 is not an eigenvalue of U, and set $A = i(I - U)(I + U)^{-1}$, the (selfadjoint) Cayley transform of U (see C1.4 and the remarks in C1.6); applying (ii), obtain $K' = (K')^* \in \mathfrak{S}$ such that $A' = A + K'$ has a pure discrete spectrum; clearly, the Cayley transform $U' = (iI - A')(iI + A')^{-1}$ also has a pure discrete spectrum and

$$U' - U = -2i(iI + A')^{-1}K'(iI + A)^{-1};$$

the result follows.]

(h) *(N. Nikolski, 1969)* Let U be a unitary operator on a Hilbert space H. Show that the following are equivalent.

(i) $\sigma_{wc}(U) = \sigma_{wc}(U + K)$ for every $K \in \mathfrak{S}_\infty$;

(ii) $\sigma_{wc}(U) = \sigma_{wc}(U + K)$ for every $K \in \mathfrak{S}$, where $\mathfrak{S}$ is any symmetrically normed (cross-normed) ideal of compact operators, $\mathfrak{S} \neq \mathfrak{S}_1$;

(iii) $\sigma(U) \neq \mathbb{T}$.

[*Hint:* (iii) $\Rightarrow$ (i) by (d) above; obviously, (i) $\Rightarrow$ (ii);

to prove (ii) $\Rightarrow$ (iii), denote by $U_0 : H \longrightarrow H$ an operator unitarily equivalent to the bilateral shift in $L^2(\mathbb{T})$ (which means, $U_0 e_j = e_{j+1}$ for an orthonormal basis $(e_j)_{j\in\mathbb{Z}}$); supposing that $\sigma(U) = \mathbb{T}$ and using (g) (iii), one gets $K_0, K_1 \in \mathfrak{S}$ such that $U_0 + K_0$, $U + K_1$ are unitary operators with pure discrete spectrum; it is clear that $\sigma(U_0 + K_0) = \sigma(U + K_1) = \mathbb{T}$; employing (f) (ii), find a unitary operator V such that

$$U_0 + K_0 = V^*(U + K_1)V + K_2,$$

where $K_2 \in \mathfrak{S}_1$; now, as in 2.5.2(e) above, set $K_3 = -K_0 + (\cdot, e_{-1})e_0$ and obtain $\sigma_{wc}(V^*(U+K_1)V+K_2+K_3) = \sigma_{wc}(U_0 - (\cdot, e_{-1})e_0) = \sigma_{wc}(S \oplus S^*) = \overline{\mathbb{D}}$; therefore, $\sigma_{wc}(U + K) = \overline{\mathbb{D}} \neq \sigma_{wc}(U)$ for $K = K_1 + V(K_2 + K_3)V^* \in \mathfrak{S}$.]

(i)* *(N. Nikolski, 1969)* Let U be a unitary operator on a Hilbert space and E_U its spectral measure (see 1.5.1 above). The following are equivalent.

(i) $\sigma_{wc}(U) = \sigma_{wc}(U + K)$ for every operator K with $\operatorname{rank} K = 1$;

(ii) $\sigma_{wc}(U) = \sigma_{wc}(U + K)$ for every $K \in \mathfrak{S}_1$;

(iii) $U^{-1} \in \operatorname{alg}^*(U)$ (the latter means the WOT closed algebra genarated by U and I);

(iv) there exists a Borel subset $\sigma \subset \mathbb{T}$ such that $m(\sigma) > 0$ and $E_U(\sigma) = 0$ (the Lebesgue measure m is not absolutely continuous with respect to E_U);

(v) every U-invariant subspace reduces U (i.e., $\operatorname{Lat}(U) = \operatorname{Lat}(U^*)$).

(j)* *(N. Makarov and V. Vasyunin, 1981)* Let T be an "almost unitary" operator on a Hilbert space, that is, $I - T^*T \in \mathfrak{S}_1$ and $\mathbb{D}\backslash\sigma(T) \neq \emptyset$. The following are equivalent.

(i) $\sigma_{wc}(T) = \sigma_{wc}(T + K)$ for every operator K with $\operatorname{rank} K = 1$;

(ii) $\sigma_{wc}(T) = \sigma_{wc}(T + K)$ for every $K \in \mathfrak{S}_1$;

(iii) $(\lambda I - T)^{-1} \in \operatorname{alg}^*(T)$ for some (all) $\lambda \in \mathbb{D}\backslash\sigma(T)$;

(iv) there exists a Borel subset $\sigma \subset \mathbb{T}$ such that $m(\sigma) > 0$ and $E_U(\sigma) = 0$, where U is a unitary operator such that $T = U + K$, $K \in \mathfrak{S}_1$ (see (e)).

(k) (i) Let T be an operator such that $I - T^*T \in \mathfrak{S}_\infty$ and $\mathbb{D} \not\subset \sigma(T)$; show that $\sigma_{wc}(T) \subset \mathbb{T}$, that is, $\sigma(T)\backslash\mathbb{T}$ consists at most of a sequence of normal eigenvalues tending to $\mathbb{T}$.

(ii) Let A be an operator such that $\mathrm{Im}(A) = (A - A^*)/2i \in \mathfrak{S}_\infty$; show that $\sigma_{wc}(A) \subset \mathbb{R}$, that is, $\sigma(A)\backslash\mathbb{R}$ consists at most of a sequence of normal eigenvalues tending to $\mathbb{R}$.

[*Hint:* apply (e) to get (i), and (d) to get (ii).]

2.5.4. Spectral and K-spectral sets according to J. von Neumann. Let $T : H \longrightarrow H$ be an operator on a Hilbert space H; a closed set $\sigma \subset \mathbb{C}$ is called *(von Neumann) spectral set* for T if

$$\|f(T)\| \leq \|f\|_\sigma = \sup_\sigma |f|$$

for every rational function $f \in \mathcal{R}at(\sigma)$ which is regular on σ; given a constant $K \geq 1$, the set σ is K*-spectral set* for T

$$\|f(T)\| \leq K\|f\|_\sigma$$

for every $f \in \mathcal{R}at(\sigma)$.

In fact, the same definitions can be addressed to any *Banach space operator* $T \in L(X)$. Theorems (a), (b), (c) below are due to J. von Neumann, 1951.

(a) Let $T \in L(X)$ and σ be a spectral set of T. Show that $\sigma(T) \subset \sigma$ and $f(\sigma)$ is a spectral set for $f(T)$ for every $f \in \mathcal{R}at(\sigma)$. Moreover, if $f_n \in \mathcal{R}at(\sigma)$, $\lim_n \|f_n - f\|_{C(\sigma)} = 0$ and $\lim_n \|f_n(T) - A\| = 0$, then $f(\sigma)$ is a spectral set for A.

[*Hint:* use 2.5.1(b); for the last assertion use 2.5.2(b).]

(b) Let $T \in L(H)$ be a Hilbert space operator. Show that

1) the disk $\overline{D}(0, r) = \{\zeta \in \mathbb{C} : |\zeta| \leq r\}$ is a spectral set for T if and only if $\|T\| \leq r$;

2) the set $\{\zeta \in \mathbb{C} : |\zeta - \lambda| \geq r\}$ is a spectral set for T if and only if $\lambda \in \mathbb{C}\backslash\sigma(T)$ and $r \leq \|R_\lambda(T)\|^{-1}$;

3) the set $\mathbb{C}^+ = \{\zeta \in \mathbb{C} : \mathrm{Re}(\zeta) \geq 0\}$ is a spectral set for T if and only if $\|(T - I)(T + I)^{-1}\| \leq 1$, and so if and only if $\mathrm{Re}\, T \geq 0$ (for the latter, see Section 1.4 above);

4) $\sigma(T) = \bigcap \sigma$, where σ runs over all spectral sets of T.

[*Hint:* for 1), the inequality is equivalent to the von Neumann inequality 1.5.9(a), and the converse is obvious;

property 2) follows from 1) and (a) applied for $\sigma = \overline{D}(0, \|R_\lambda(T)\|)$, $f = \lambda - z^{-1}$ and $R_\lambda(T)$ instead of T;

similarly, 3) follows from 1) and (a) applied to $\sigma = \overline{D}(0, 1)$, $f = (1 - z)(1 + z)^{-1}$ and $(T - I)(T + I)^{-1}$ instead of T;

4) is an immediate consequence of (a) and 2).]

(c) Let $T \in L(H)$ be a Hilbert space operator. Show that

1) $\sigma = \mathbb{T}$ is a spectral set for T if and only if T is unitary;

2) $\sigma = \mathbb{R}$ is a spectral set for T if and only if T is selfadjoint;

3) if σ is a compact spectral set for T and $\mathrm{clos}\, \mathcal{R}at(\sigma) = C(\sigma)$, then T is a normal operator and $\sigma(T) \subset \sigma$; conversely, the spectrum $\sigma(N)$ of a normal operator N is its spectral set.

[*Hint:* for 1), if $\mathbb{T}$ is a spectral set then $\|T\| \leq 1$ and $\|T^{-1}\| \leq 1$, and hence $\|Tx\| = \|x\|$ for every $x \in H$; the converse is in 1.5.1;

for 2), apply (b) 3) for $\pm iT$ and get $\mathrm{Re}(iT) = 0$ which means that $T = T^*$;

for 3), observe first that the rational calculus for T extends to $C(\sigma)$; next, set $h(z) = \bar{z}$ and use (a) for $f = (z+h)/2$ and $g = (z-h)/2i$; it follows that $f(T)$ and $g(T)$ have $\mathbb{R}$ as a spectral set; now, 2) implies that $f(T)$ and $g(T)$ are commuting selfadjoint operators, and hence $T = f(T)+ig(T)$ is normal; the converse is obvious from 2.5.1.]

(d) *Spectral sets in Banach spaces.*

(i)* *(C. Foiaş, 1957)* Let X be a Banach space such that $\|f(T)\| \le \|f\|_\infty = max_{\mathbb{T}}|f|$ for every polynomial f and every operator $T : X \longrightarrow X$ with $\|T\| = 1$. Then X is a Hilbert space, i.e. the norm $\|\cdot\|_X$ is generated by a scalar product.

(ii) *(H. Bohr, M. Riesz, I. Schur and F. Wiener, 1914)* Let $\rho > 0$, let X be a Banach space, and let $T : X \longrightarrow X$ be such that $\|T\| \le \rho$. Show that

$$\|f(T)\| \le \sum_{k\ge 0} |\hat{f}(k)|\rho^k = \|f(\rho S)\|$$

for every polynomial f, where $S : l^1 \longrightarrow l^1$ is the shift operator on the space l^1 (which we identify with the Wiener algebra

$$W_A = \{f = \sum_{k\ge 0} \hat{f}(k)z^k : \sum_{k\ge 0} |\hat{f}(k)| < \infty\}).$$

Next, show that

$$\|f(\rho S)\| \le \|f\|_\infty = \max_{\mathbb{T}} |f|$$

for every polynomial $f \in \mathcal{P}ol_+$ if and only if $\rho \le 1/3$. An equivalent form is

$$\|f(A)\| \le \|f\|_{3\mathbb{D}} = \max_{\zeta\in 3\mathbb{D}} |f(\zeta)|$$

for every polynomial $f \in \mathcal{P}ol_+$ and every contraction $A : X \longrightarrow X$ on any Banach space X; the constant 3 is sharp.

[*Hint: (V. Katsnel'son and V. Matsaev, 1967)* The first claim is obvious since $\|f(\rho S)\| \ge \|f_\rho\|_{W_A} = \sum_{k\ge 0} |\hat{f}(k)|\rho^k$, where $f_\rho(z) = f(\rho z)$;

in order to prove the necessity part of the second claim, take a Blaschke factor $f = b_\lambda = (\zeta - z)(1 - \bar{\lambda}z)^{-1}$ and show that, for $0 < \rho < 1$,

$$\varphi(|\lambda|) := \|(b_\lambda)_\rho\|_{W_A} = \frac{|\lambda| + \rho - 2\rho|\lambda|^2}{1 - \rho|\lambda|}$$

and $\varphi'(t) = 2 - \frac{1-\rho^2}{(1-\rho t)^2} \ge 0$ for all $0 < t < 1$ if and only if $\rho \le 1/3$; if $\rho > 1/3$, then there is a solution $\varphi'(t) = 0$ with $0 < t < 1$, which gives

$$\sup_{|\lambda|<1} \|(b_\lambda)_\rho\|_{W_A} = \max_{0<t<1} \varphi(t) = \frac{3}{\rho} - \frac{2(2(1-\rho^2))^{1/2}}{\rho} > 1$$

since $(3 - \rho)^2 > 8 - 8\rho^2$;

to prove the sufficiency part, observe that if $\rho \le 1/3$, then $\|(b_\lambda)_\rho\|_{W_A} \le \varphi(1) = 1$ and hence $\|B_\rho\|_{W_A} \le 1$ for every finite Blaschke product B; next, use the Schur theorem B.3.3.3 (Volume 1).]

(iii) *(V. Katsnel'son and V. Matsaev, 1967)* Let $\sigma \subset \mathbb{D}$ be a finite set, and $f \in H^\infty$. Show that there exists a function $g \in W_A$ such that

$$g|\sigma = f|\sigma \text{ and } \|g\|_{W_A} \le (\pi|\sigma| + 1)\|f\|_\infty.$$

Therefore, for every contraction $T : X \longrightarrow X$ acting on a Banach space X with $\dim X = n < \infty$, we have

$$\|f(T)\| \leq (\pi n + 1)\|f\|_\infty$$

for every polynomial $f \in \mathcal{P}ol_+$.

[*Hint:* use B.3.3.2 (Volume 1) to show that for every function f with $\|f\|_\infty < 1$ there exists a Blaschke product B such that $B|\sigma = f|\sigma$ and $\deg B \leq |\sigma|$; now, the result follows from B.8.7.4(c), Volume 1, and the remarks in B.8.8, Volume 1, which gives $\|B\|_{W_A} \leq \pi|\sigma| + 1$;

for the second claim, suppose without loss of generality that the spectrum $\sigma = \sigma(T)$ consists of n distinct points, then use $g(T) = f(T)$ for f, g satisfying $g|\sigma = f|\sigma$.]

2.5.5. The resolvent and the characteristic function. In this subsection we consider the characteristic function Θ_T of a c.n.u. contraction T as defined by the formula

$$\Theta_T(\lambda) = \{-T + \lambda D_{T^*}(I - \lambda T^*)^{-1} D_T\}|\mathcal{D}_T$$

on the set of all $\lambda \in \mathbb{C}$ such that $\frac{1}{\lambda} \notin \sigma(T^*)$ (see Remarks 1.2.10); obviously, this set contains the disk $\mathbb{D}$. (Note that the analytic continuation of $\Theta_T(\lambda)$, $|\lambda| < 1$, if it exists, does not necessarily coincide with $\Theta_T(\lambda)$, $|\lambda| > 1$.)

(a) Let $\frac{1}{\lambda} \notin \sigma(T^*)$; show that $\Theta_T(\lambda)^* = \Theta_{T^*}(\overline{\lambda})$. If, in addition, $\lambda \notin \sigma(T)$, show that $\Theta_T(\lambda)$ is invertible and $\Theta_T(\lambda)^{-1} = \Theta_{T^*}(\frac{1}{\lambda})$. In particular, $\Theta_T(\lambda)$ is unitary for $\lambda \in \mathbb{T} \backslash \sigma(T)$.

[*Hint:* the first formula is straightforward from the above definition; for the second one, write $\Theta_{T^*}(\frac{1}{\lambda})$ in its Möbius form as in Remarks 1.2.10, i.e.

$$\Theta_{T^*}\left(\frac{1}{\lambda}\right) D_{T^*} = D_T \left(I - \frac{1}{\lambda}T\right)^{-1} \left(\frac{1}{\lambda} - T^*\right),$$

multiply it from the left by $\Theta_T(\lambda)$ and use the same formula from 1.2.10 to get $\Theta_T(\lambda)\Theta_{T^*}(\frac{1}{\lambda})D_{T^*} = D_{T^*}$; by symmetry, $\Theta_{T^*}(\frac{1}{\lambda})\Theta_T(\lambda)D_T = D_T$.]

(b) Let $\frac{1}{\lambda} \notin \sigma(T^*)$; show that

$$D_T^2 - D_T\Theta_T(\lambda)^*\Theta_T(\lambda)D_T = (1 - |\lambda|^2)D_T^2(I - \overline{\lambda}T)^{-1}(I - \lambda T^*)^{-1}D_T^2.$$

[*Hint:* 1) employ the formula of 1.2.10 for both $(\Theta_T(\lambda)D_T)^*$ and $\Theta_T(\lambda)D_T$ in order to transform the left hand side into

$$D_T^2 - (\overline{\lambda}I - T^*)(I - \overline{\lambda}T)^{-1}D_{T^*}^2(I - \lambda T^*)^{-1}(\lambda I - T);$$

2) next, use

$$D_{T^*}^2(I - \lambda T^*)^{-1}(\lambda I - T) = (\lambda I - T)(I - \lambda T^*)^{-1}D_T^2.$$

This identity, in turn, can be justified by multiplying $(I - \lambda T^*)^{-1}(\lambda I - T) = -T + \lambda(I - \lambda T^*)^{-1}(I - T^*T)$ by $D_{T^*}^2$ from the left and $(\lambda I - T)(I - \lambda T^*)^{-1} = -T + \lambda(I - T^*T)(I - \lambda T^*)^{-1}$ by D_T^2 from the right; together with 1), this yields

$$\begin{aligned} &D_T^2 - D_T\Theta_T(\lambda)^*\Theta_T(\lambda)D_T \\ &\quad = \{I - (\overline{\lambda}I - T^*)(I - \overline{\lambda}T)^{-1}(\lambda I - T)(I - \lambda T^*)^{-1}\}D_T^2 \\ &\quad = \{(I - \lambda T^*)(I - \overline{\lambda}T) - (\overline{\lambda}I - T^*)(\lambda I - T)\}(I - \overline{\lambda}T)^{-1}(I - \lambda T^*)^{-1}D_T^2; \end{aligned}$$

3) finally, the equality $(I-\lambda T^*)(I-\overline{\lambda}T)-(\overline{\lambda}I-T^*)(\lambda I-T)=(1-|\lambda|^2)D_T^2$ completes the proof.]

(c) Show that $\|x\|^2-\|\Theta_T(\lambda)x\|^2=(1-|\lambda|^2)\|(I-\lambda T^*)^{-1}D_Tx\|^2$ for every $x\in\mathcal{D}_T$ and $\frac{1}{\lambda}\notin\sigma(T^*)$.

[*Hint:* compute the quadratic form for both sides of the identity (b) for an element $y\in H$ and get the required property setting $x=D_Ty$.]

(d) Show that $\|\Theta_T(\lambda)^{-1}\|=\|b_\lambda(T)^{-1}\|$ for every $\lambda\in\mathbb{D}\backslash\sigma(T)$, where $b_\lambda(z)=(\lambda-z)(1-\overline{\lambda}z)^{-1}$, and hence

$$\frac{\|\Theta_T(\lambda)^{-1}\|-|\lambda|}{1-|\lambda|^2}\le\|R_\lambda(T)\|\le\frac{\|\Theta_T(\lambda)^{-1}\|}{1-|\lambda|}.$$

[*Hint:* using (a), rewrite formula (c) in the form $\|\Theta_T(\lambda)^{-1}x\|^2=\|x\|^2+\|Ax\|^2$, where $A=(1-|\lambda|^2)^{1/2}R_\lambda(T)D_{T^*}$; next, observe that $\|\Theta_T(\lambda)^{-1}\|^2=1+\|A\|^2=1+\|A^*\|^2=\|I+AA^*\|$, and, on the other hand, $I+AA^*=b_\lambda(T)^{-1}(b_\lambda(T)^{-1})^*$, which gives the first formula;

to prove the estimates, observe that $b_\lambda(T)^{-1}=\overline{\lambda}I+(1-|\lambda|^2)R_\lambda(T)=R_\lambda(T)-\overline{\lambda}TR_\lambda(T)$, and hence

$$\begin{aligned}\|\Theta_T(\lambda)^{-1}\| &\le |\lambda|+(1-|\lambda|^2)\|R_\lambda(T)\|,\\ \|\Theta_T(\lambda)^{-1}\| &\ge \|R_\lambda(T)\|-|\lambda|\cdot\|R_\lambda(T)\|.\end{aligned}$$
]

(e) *(I. Gohberg and M. Krein, 1966)* Show that a Hilbert space contraction T is similar to a unitary operator if and only if $\sigma(T)\subset\mathbb{T}$ and

$$sup_{\lambda\in\mathbb{D}}(1-|\lambda|)\|R_\lambda(T)\|<\infty.$$

[*Hint:* use (d) and 1.5.5(c).]

2.5.6. The H^∞ calculus and smooth operators in $\{T\}'$ and $H^\infty(T)$. Here T is a c.n.u. contraction on a Hilbert space H, identified (if necessary) with its function model M_Θ, $\Theta=\Theta_T$ being the characteristic function of T. Denote by

$$\{T\}'=\{M_\Theta\}'=\{A:AM_\Theta=M_\Theta A\}$$

the *commutant* of T, and by $H^\infty(T)=\{\varphi(T):\varphi\in H^\infty\}\subset\{T\}'$ the set of all H^∞ functions of T.

(a) Let $\varphi\in H^\infty$. Show that there exists a net of polynomials (p_α) such that WOT-$\lim_\alpha p_\alpha(T)\varphi(T)=\varphi_{inn}(T)$. In particular, $\operatorname{Ker}\varphi(T)=\operatorname{Ker}\varphi_{inn}(T)$, and, therefore, $\varphi(T)$ is a di-formation for every outer function φ (see 1.5.3 for the definition).

[*Hint:* in order to prove the first assertion, apply the same reasoning as in 2.3.2(2) (using 2.2.2(3)); the latter assertion follows since $\operatorname{Ker}\varphi(T)=\{0\}$ and $\operatorname{Ker}\varphi(T)^*=\operatorname{Ker}\varphi_*(T^*)=\{0\}$, where $\varphi_*(\overline{z})=\overline{\varphi(z)}$.]

(b) Let Θ_T be a $*$-inner function (or, equivalently, $T\in C_{0\cdot}$; see 1.5.4(b)). Show that the calculus $f\longmapsto f(T)$ is sequentially continuous from $(H^\infty,\sigma(H^\infty,L^1))$ to

$L(H)$ endowed with the *strong operator topology*, that is, $\sigma\text{-}\lim_n f_n = 0$ entails $\lim_n \|f_n(T)x\| = 0$ for every $x \in H$.

[*Hint:* replacing T by $M_{\Theta_*}^* = P_+\bar{z}|\mathcal{K}_{\Theta_*}$ on the model space $\mathcal{K}_{\Theta_*} = H^2(E) \ominus \Theta_* H^2(E_*)$, where $\Theta_* = \Theta_{T^*}$, one can show more, namely, that $\lim_n \|P_+ f_n(\bar{z})x\| = 0$ for every $x \in H^2(E)$; indeed, $P_+ f_n(\bar{z})k_\lambda e = f_n(\bar{\lambda})k_\lambda e$ for every $\lambda \in \mathbb{D}$ and $e \in E$, where $k_\lambda e = (1 - \bar{\lambda}z)^{-1}e$ is the reproducing kernel of $H^2(E)$; since $f_n \longrightarrow 0$ $*$-weakly, one has $\lim_n \|P_+ f_n(\bar{z})k_\lambda e\| = 0$; the result follows by using $\sup_n \|f_n\|_\infty < \infty$ and $\text{span}(k_\lambda e : \lambda \in \mathbb{D}, e \in E) = H^2(E)$.]

(c) *(P. Vitse, 2001)* Suppose that $H^\infty(T) \cap \mathfrak{S}_\infty \neq \{0\}$. Show that Θ_T is a two-sided inner function.

[*Hint:* clearly, it suffices to prove that Θ_T is inner (since $(\Theta_T)_* = \Theta_{T^*}$ and $\varphi_*(T^*) = \varphi(T)^*$, where $\varphi_*(z) = \varphi(\bar{z})^*$);

using 1.5.2(a) and its notation, get

$$M_\Theta^{*n}(f,g)^{col} = (0, \bar{z}^n g) + o(1)$$

as $n \longrightarrow \infty$ for every $(f,g)^{col} \in \mathcal{K}_\Theta$; for $\varphi \in H^\infty$, this yields (see 1.3.7 for the formula for P_Θ)

$$\begin{aligned} \|\varphi(M_\Theta)^* M_\Theta^{*n}(f,g)^{col}\|^2 &= \|\bar{z}^n\bar{\varphi}g\|^2 - \|P_+\bar{z}^n\Delta\bar{\varphi}g\|^2 + o(1) \\ &= \|\bar{\varphi}g\|^2 + o(1) \end{aligned}$$

as $n \longrightarrow \infty$;

supposing that $\varphi(T) \in \mathfrak{S}_\infty$, $\varphi \neq 0$, and using the fact that $M_\Theta^{*n} \longrightarrow 0$ weakly, the last expression tends to zero, whence $\bar{\varphi}g = 0$, and so, $g = 0$, for every $(f,g)^{col} \in \mathcal{K}_\Theta$; this implies

$$\mathcal{K}_\Theta \subset \begin{pmatrix} H^2(E_*) \\ 0 \end{pmatrix} \text{ and } \begin{pmatrix} \Theta \\ \Delta \end{pmatrix} H^2(E) = \mathcal{K}_\Theta^\perp \supset \begin{pmatrix} 0 \\ \text{clos}(\Delta L^2(E)) \end{pmatrix},$$

and hence $\Delta = 0$, since the latter subspace is $\bar{z}$-invariant and the former contains no non zero $\bar{z}$-invariant parts.]

(d) *(P. Vitse, 2001)* Assume that $I - T^*T \in \mathfrak{S}_\infty$ and $\mathbb{D}\backslash\sigma(T) \neq \emptyset$. Show that

$$\{T\}' \cap \mathcal{F} \neq \{0\}$$

if and only if $\mathbb{D} \cap \sigma(T)(=\sigma_p(T)) \neq \emptyset$; here $\mathcal{F}$ stands for the ideal of finite rank operators. The condition $\mathbb{D}\backslash\sigma(T) \neq \emptyset$ cannot be omitted.

[*Hint:* if $A \in \{T\}' \cap \mathcal{F}$ and $A \neq 0$, then $\text{Range}\, A = AH \neq \{0\}$ is a finite dimensional T-invariant subspace, and hence $T|AH$ has an eigenvalue (which is necessarily in $\mathbb{D}$);

conversely, 2.5.3(e) (i) yields $\mathbb{D} \cap \sigma(T) = \sigma_p(T)$, and taking $\lambda \in \sigma_p(T)$ and using again 2.5.3(e) (i) we get $0 \neq \mathcal{P}_\lambda \in \{T\}' \cap \mathcal{F}$;

a conterexample for $\sigma(T) \supset \mathbb{D}$ may be provided by the shift operator $T = S$ on H^2.]

(e) Let Θ_T be an inner and $*$-inner function and let $A \in \{M_\Theta\}'$ be represented via the commutant lifting theorem $A = P_\Theta Y|\mathcal{K}_\Theta$ with $\Theta^* Y \Theta \in H^\infty(E \longrightarrow E)$ (see 2.4.3 above). Show that

(i) $A \oplus (0|\Theta H^2(E)) = \Theta H_{\Theta^* Y}$, where $H_\varphi = P_-\varphi : H^2(E) \longrightarrow H^2_-(E_*)$ stands for a Hankel operator;

(ii) $A \in \mathfrak{S} \Leftrightarrow H_{\Theta^* Y} \in \mathfrak{S}$ for every set of operators $\mathfrak{S}$ defined in terms of the modulus $|A|$ of an operator (like the ideals $\mathfrak{S}_p$ or $\mathcal{F}$).

The following characterizations can be proved using the techniques of Hankel operators, see Section B.8.7 (Volume 1).

(iii)* *(L. Page, 1970)* $A \in \mathfrak{S}_\infty \Leftrightarrow H_{\Theta^* Y} \in \mathfrak{S}_\infty \Leftrightarrow JP_-\Theta^* Y \in VMO_-(\mathfrak{S}_\infty) = P_+ C(\mathbb{T}, \mathfrak{S}_\infty)$, where $Jf(z) = f(\overline{z})$;

(iv)* *(V. Peller, 1983)* $A \in \mathfrak{S}_p \Leftrightarrow H_{\Theta^* Y} \in \mathfrak{S}_p \Leftrightarrow JP_-\Theta^* Y \in B_p^{1/p} A(\mathfrak{S}_p)$, where

$$B_p^{1/p} A(\mathfrak{S}_p) = \{\varphi \in \mathrm{Hol}(\mathbb{D}, \mathfrak{S}_p) : \int_{\mathbb{D}} \|\varphi''(z)\|_{\mathfrak{S}_p}^p (1 - |z|^2)^{p-1}\, dxdy < \infty\}.$$

2.5.7. The scalar model ($\partial_T = \partial_{T^*} = 1$): commutant, H^∞-calculus, and more. Here we consider c.n.u. contractions $T : H \longrightarrow H$ satisfying $\partial_T = \partial_{T^*} = 1$, and hence that are unitarily equivalent to a scalar model $M_\Theta = P_\Theta z | \mathcal{K}_\Theta$, $\Theta = \Theta_T \in H^\infty$,

$$\mathcal{K}_\Theta = \begin{pmatrix} H^2 \\ \mathrm{clos}(\Delta L^2(\mathbb{T})) \end{pmatrix} \ominus \begin{pmatrix} \Theta \\ \Delta \end{pmatrix} H^2,$$

where $\Delta^2(\zeta) = 1 - |\Theta(\zeta)|^2$ for $\zeta \in \mathbb{T}$. We always suppose that $\Theta \neq 0$.

Properties (a)-(f) below are due to N. Nikolski and S. Hruschev, 1987, and (g)-(j) are due to P. Vitse, 2001.

(a) Show that a matrix function Y,

$$Y = \begin{pmatrix} a & x \\ b & c \end{pmatrix},$$

defines an operator $A = P_\Theta Y | \mathcal{K}_\Theta$ in the commutant $\{M_\Theta\}'$ if and only if

$$a \in H^\infty, \quad x = 0, \quad b = \Delta(a - c)/\Theta \in L^\infty(\Delta\, dm) \text{ and } c \in L^\infty(\Delta\, dm),$$

and if and only if

$$a \in H^\infty, \quad x = 0, \quad (a - c)/\Theta \in L^\infty(\Delta\, dm) \text{ and } c \in L^\infty(\Delta\, dm);$$

moreover,

$$A = 0 \quad \Longleftrightarrow \quad \big(Y = Y_h \text{ with } c = 0, \quad a = \Theta h \text{ and } b = \Delta h\big)$$

for some function $h \in H^\infty$.

[*Hint:* using $Y \begin{pmatrix} H^2 \\ \mathrm{clos}(\Delta L^2) \end{pmatrix} \subset \begin{pmatrix} H^2 \\ \mathrm{clos}(\Delta L^2) \end{pmatrix}$ of Theorem 2.4.2, one obtains $x = 0$, $a \in H^\infty$ and $b \in L^\infty(\Delta\, dm)$; then, use $Y \begin{pmatrix} \Theta \\ \Delta \end{pmatrix} H^2 \subset \begin{pmatrix} \Theta \\ \Delta \end{pmatrix} H^2$, $b\Theta + c\Delta = \Delta a$, to get the first combination of the boundedness conditions;

for the second version of the boundedness conditions, observe that for those points where $\Delta^2 \leq 1/5$, $|(a-c)/\Theta| \leq \frac{\sqrt{5}}{2}(\|a\|_\infty + \|c\|_\Delta) \leq \sqrt{5}\|Y\|$, and $|(a-c)/\Theta| = |b/\Delta| \leq \sqrt{5}|b| \leq \sqrt{5}\|Y\|$ for points where $\Delta^2 > 1/5$;

$A = 0$ if and only if $Y \begin{pmatrix} H^2 \\ \mathrm{clos}(\Delta L^2) \end{pmatrix} \subset \begin{pmatrix} \Theta \\ \Delta \end{pmatrix} H^2$.]

(b) Using the notation of (a), show that

$$\max\Big(\mathrm{dist}\Big(\frac{a - \Delta^2 c}{\Theta}, H^\infty\Big), \|c\|_\Delta\Big) \leq \|A\| \leq \mathrm{dist}\Big(\frac{a - \Delta^2 c}{\Theta}, H^\infty\Big) + 2\|c\|_\Delta,$$

where $\|c\|_\Delta = \|c\|_{L^\infty(\Delta\, dm)}$.

[*Hint*: $\|A\| = \inf_{h\in H^\infty} \|Y + Y_h\|_{L^\infty(\mathbb{C}^2 \longrightarrow \mathbb{C}^2)}$, where Y_h is from (a) above; since

$$\max(\alpha, \beta) \le \|Y(\zeta) + Y_h(\zeta)\|_{\mathbb{C}^2 \longrightarrow \mathbb{C}^2} \le \alpha + \beta,$$

where

$$\begin{aligned} \alpha^2 &= \|(Y(\zeta) + Y_h(\zeta))e_1\|^2 = |a(\zeta) + \Theta h(\zeta)|^2 + \left|\frac{\Delta(a-c)(\zeta)}{\Theta(\zeta)} + \Delta h(\zeta)\right|^2 \\ &= \left|\frac{(a-\Delta^2 c)(\zeta)}{\Theta(\zeta)} + h(\zeta)\right|^2 + \Delta^2(\zeta)|c(\zeta)|^2, \\ \beta &= \|(Y(\zeta) + Y_h(\zeta))e_2\| = |c(\zeta)| \end{aligned}$$

(e_j stand for the basic vectors of $\mathbb{C}^2$), the result follows.]

(c) Let A be an operator as in (a). Show that $A \in H^\infty(M_\Theta)$ if and only if $a = c + \Theta h$ and $b = \Delta h$, where $h \in H^\infty$; in this case, taking $a = c$, $A = a(M_\Theta)$ and

$$\max\Big(\operatorname{dist}\big(a\overline{\Theta}, H^\infty\big), \|a\|_\Delta\Big) \le \|a(M_\Theta)\| \le \operatorname{dist}\big(a\overline{\Theta}, H^\infty\big) + 2\|a\|_\Delta.$$

(d) Let A be an operator as in (a). Show that the following are equivalent.

(i) A is invertible.

(ii) There exist functions $a', h \in H^\infty$ and $c' \in L^\infty(\Delta\, dm)$ such that

$$aa' + \Theta h = 1 \text{ and } cc' = 1 \ a.e.\ \Delta\, dm.$$

(iii) $\inf_{\zeta\in\mathbb{D}} \Big(|a(\zeta)| + |\Theta(\zeta)|\Big) > 0$ and $\operatorname{ess\,inf}_{\Delta\, dm} |c(\zeta)| > 0$.

(For a quantitative version of the criterion (i)$\Longleftrightarrow$(iii) see (m) below).

[*Hint:* if A is invertible, then $A^{-1} \in \{M_\Theta\}'$ and by (a), $A^{-1} = P_\Theta Y'|\mathcal{K}_\Theta$, with a matrix Y' of the above form determined by some functions $a' \in H^\infty$, $c' \in L^\infty(\Delta\, dm)$; since $P_\Theta Y Y'|\mathcal{K}_\Theta = I$, it follows that $aa' + \Theta h = 1$ for some $h \in H^\infty$ and $cc' = 1$ a.e. $\Delta\, dm$; therefore, (i) $\Rightarrow$ (ii);

conversely, a', c' from (ii) determine the operator $P_\Theta Y'|\mathcal{K}_\Theta$ inverse to A (it is easy to see that $(a' - c')/\Theta \in L^\infty(\Delta dm)$);

(iii) is equivalent to (ii) by virtue of the Carleson corona theorem (see B.9.2.4, Volume 1, and 3.2.9 below).]

(e) *An extended spectral mapping theorem.* Let A be an operator from (a) above. Show that

$$\sigma(A) = \mathrm{Range}_\Theta(a) \cup \mathrm{Range}_\Delta(c),$$

where

$$\begin{aligned} \mathrm{Range}_\Theta(a) &= \Big\{\lambda \in \mathbb{C} : \inf_{\zeta\in\mathbb{D}} \Big(|\lambda - a(\zeta)| + |\Theta(\zeta)|\Big) = 0\Big\}, \\ \mathrm{Range}_\Delta(c) &= \Big\{\lambda \in \mathbb{C} : \operatorname{ess\,inf}_{\Delta dm} |\lambda - c(\zeta)| = 0\Big\}. \end{aligned}$$

In particular, $\sigma(\varphi(M_\Theta)) = \mathrm{Range}_\Theta(\varphi) \cup \mathrm{Range}_\Delta(\varphi)$ for $\varphi \in H^\infty$.

[*Hint:* apply (d) to $\lambda I - A$.]

(f)* *A local function calculus.* Let A be an operator from (a) above.

(i) Given an open set $\mathcal{O} \subset \mathbb{C}$ with $\sigma(\Theta) = \sigma(M_\Theta) \subset \mathcal{O}$, there exists $k = k(\mathcal{O})$ (depending only on $\mathcal{O}$) such that

$$\|A\| \le \sup_{\mathcal{O}\cap\mathbb{T}} \left|\frac{a - \Delta^2 c}{\Theta}\right| + k \cdot \sup_{\partial\mathcal{O}\cap\mathbb{D}} |a| + 2\|c\|_\Delta.$$

(ii) Let $A = a(M_\Theta)$ for a function $a \in H^\infty$, and suppose that $\Theta = \Theta_T$ allows a pseudocontinuation across $\mathbb{T}$ as a Nevanlinna function in $\mathbb{C}\backslash\overline{\mathbb{D}}$, namely, let $\overline{\Theta} = \Theta_1/\Theta_2$, where $\Theta_i \in H^\infty$, Θ_2 is inner and $(\Theta_1)_{inn} \wedge \Theta_2 = 1$ (see A.4.8.7, Volume 1, about this; in particular, Θ can be any inner function). Further, let $L(\Theta_2, \varepsilon)$ be the *level set*

$$L(\Theta_2, \varepsilon) = \Big\{\zeta \in \mathbb{D} : |\Theta_2(\zeta)| \le \varepsilon\Big\},$$

where $0 < \varepsilon < 1$. There exists a constant $K = K(\varepsilon)$ (depending on ε, but neither on a nor on Θ) such that

$$\|a(M_\Theta)\| \le K \cdot \sup_{L(\Theta_2,\varepsilon)} |a\Theta_1| + 2\|a\|_\Delta \le K \cdot \sup_{L(\Theta_2,\varepsilon)} |a| + 2\|a\|_\Delta.$$

(iii) Let Θ be an inner function and $0 < \varepsilon < 1$. There exists $p > 1$ and a rectifiable curve $\gamma_\varepsilon \subset \mathbb{D}$ separating the level sets $L(\Theta, \varepsilon^p)$ and $\mathbb{D}\backslash L(\Theta, \varepsilon)$ such that the formula

$$[a](M_\Theta)f = \frac{\Theta(z)}{2\pi i}\int_{\gamma_\varepsilon} \frac{a(\zeta)f(\zeta)}{\Theta(\zeta)(\zeta - z)}\, d\zeta, \quad z \in \mathbb{T},$$

where $f \in \mathcal{K}_\Theta$, defines a bounded operator $[a](M_\Theta) : \mathcal{K}_\Theta \longrightarrow \mathcal{K}_\Theta$ for every $a \in H^\infty(L(\Theta, \varepsilon))$. Moreover, $[a](M_\Theta) \in \{M_\Theta\}'$ for every a, and the mapping

$$a \longmapsto [a](M_\Theta)$$

is a continuous homomorphism from $H^\infty(L(\Theta, \varepsilon))$ to $\{M_\Theta\}'$ such that

- $[a](M_\Theta) = a(M_\Theta)$ for $a \in H^\infty$;
- $[a](M_\Theta) = 0$ iff $a = \Theta h$ for a function $h \in H^\infty(L(\Theta, \varepsilon))$;
- considered on $\bigcup_{\varepsilon>0} H^\infty(L(\Theta, \varepsilon))$, it contains the Riesz–Dunford calculus for M_Θ (for the latter, see 2.5.1).

(g) Let A be an operator as in (a).

(i) Show that $A \in \mathfrak{S}_\infty$ if and only if $c = 0$ and $\frac{a}{\Theta} \in H^\infty + C(\mathbb{T})$. Moreover, if $A \in \mathfrak{S}_\infty$ then

$$A(f, g)^{col} = \begin{pmatrix}\Theta\\ \Delta\end{pmatrix} P_- \Theta^{-1} a f$$

for every $(f, g)^{col} \in \mathcal{K}_\Theta$, and therefore the modulus $|A|$ is unitarily equivalent to $|H_{a/\Theta}|$.

[*Hint:* since $Y \begin{pmatrix}\Theta\\ \Delta\end{pmatrix} H^2 \subset \begin{pmatrix}\Theta\\ \Delta\end{pmatrix} H^2$, A is compact if and only if $P_\Theta Y \Big| \begin{pmatrix}H^2\\ \operatorname{clos}\Delta L^2\end{pmatrix}$ is; therefore, $A \in \mathfrak{S}_\infty$ entails that the mapping

$$(0, g)^{col} \longmapsto P_\Theta Y(0, g)^{col} = P_\Theta(0, cg)^{col}$$

is a compact operator from $\operatorname{clos}(\Delta L^2)$ to $\mathcal{K}_\Theta$; using formulas 1.3.7, we get

$$\begin{aligned}\|P_\Theta(0,cg)^{col}\|^2 &= \|cg\|^2 - \|(I-P_\Theta)(0,cg)^{col}\|^2 \\ &= \|cg\|^2 - \|\Theta P_+\Delta cg\|^2 - \|\Delta P_+\Delta cg\|^2 \\ &= \|cg\|^2 - \|P_+\Delta cg\|^2 = \|\Theta cg\|^2 + \|P_-\Delta cg\|^2 \\ &\geq \|\Theta cg\|^2\end{aligned}$$

which implies that $g \longmapsto \Theta cg$ is compact; whence, $c = 0$; next, using formulas 1.3.7, check that

$$P_\Theta Y(f,g)^{col} = (af - \Theta P_+\Theta^{-1}af, \Delta\Theta^{-1}af - \Delta P_+\Theta^{-1}af)^{col} = \begin{pmatrix}\Theta \\ \Delta\end{pmatrix} P_-\Theta^{-1}af,$$

and hence the Hankel operator $f \longmapsto P_-\Theta^{-1}af$ is compact; now, Hartman's theorem B.2.2.5 (Volume 1) shows that $\frac{a}{\Theta} \in H^\infty + C(\mathbb{T})$;
the converse follows, too.]

(ii) Show that $\{M_\Theta\}' \cap \mathfrak{S}_\infty = \{0\}$ if and only if Θ is outer.

[*Hint:* if $a \in H^\infty$, Θ is outer and $a\Theta^{-1} \in L^\infty$, then $a\Theta^{-1} \in H^\infty$ (see A.2.6.1 or A.4.4.5, Volume 1); therefore, $A \in \mathfrak{S}_\infty \Rightarrow A = 0$ (see (a) above);
conversely, let Θ be not outer and let $\Theta = \Theta_{out}BV$ be its Riesz–Smirnov canonical factorization (see A.3.9, Volume 1);
if $B \neq 1$, then $\sigma_p(M_\Theta) \neq \emptyset$, and hence even $\{M_\Theta\}' \cap \mathcal{F} \neq \{0\}$ (see 2.5.6(d) above);
if $V \neq 1$ (the singular inner part of Θ), consider a factorization $\Theta = \Theta_{out}V_1\Theta_2$, where V_1 is a nonconstant singular inner factor of V such that $m(\sigma(V_1)) = 0$ where m is Lebesgue measure on $\mathbb{T}$ (see 2.3.3 and 2.3.6 for the spectrum of an inner function); now, take an outer $g \in C_A$ such that $g|\sigma(V_1) = 0$ (see A.4.3.8(d), Volume 1) and set

$$a = \Theta_{out}\Theta_2 g;$$

then $a\Theta^{-1} = g\overline{V}_1 \in C(\mathbb{T})$ and hence $A \in \mathfrak{S}_\infty$, but $A \neq 0$ because of $g\overline{V}_1 \notin H^\infty$ and (a) above.]

(h) Show that $H^\infty(M_\Theta) \cap \mathfrak{S}_\infty = \{0\}$ if and only if Θ is not inner.

[*Hint:* for sufficiency see 2.5.6(c); conversely, if Θ is inner then Θ is not outer ($\Theta = \Theta_T$ is not a unimodular constant, since it is purely contractive, see 1.2.12), and also $H^\infty(M_\Theta) = \{M_\Theta\}'$; by (g), $H^\infty(M_\Theta) \cap \mathfrak{S}_\infty \neq \{0\}$]

(i) Show that the following are equivalent.

(i) $\{M_\Theta\}' \cap \mathcal{F}$ is norm dense in $\{M_\Theta\}' \cap \mathfrak{S}_\infty$.
(ii) $H^\infty(M_{\Theta_{inn}}) \cap \mathcal{F}$ is norm dense in $H^\infty(M_{\Theta_{inn}}) \cap \mathfrak{S}_\infty$.
(iii) Θ_{inn} is a Blaschke product.

[*Hint:* (iii) $\Rightarrow$ (i): given $A \in \{M_\Theta\}' \cap \mathfrak{S}_\infty$, use (g)(i) and the notation from there and get $\frac{a}{\Theta} \in H^\infty + C$, and hence $\frac{a}{\Theta_{out}} \in H^\infty + C$, which implies that $a = \Theta_{out}h$ with $h \in H^\infty$ (see (g)(ii)); now, let $\Theta_{inn} = B = \prod_{s\geq 1} b_{\lambda_s}^{k_s}$ and $B_n = \prod_{s\geq n} b_{\lambda_s}^{k_s}$ for $n = 1, 2, ...$; then,

$$\lim_n \|B_n(M_\Theta)F - F\| = \lim_n \|P_\Theta(B_nF - F)\| \leq \lim_n \|(B_n - 1)F\| = 0$$

for every $F \in \mathcal{K}_\Theta$ (the latter follows from $\lim_n \int_{\mathbb{T}} |B_n(\zeta)-1|^2\, dm(\zeta) = 0$ (see A.3.7.3, Volume 1) and $\|(B_n - 1)F\|^2 = \int_{\mathbb{T}} |B_n(\zeta) - 1|^2 \|F(\zeta)\|^2\, dm(\zeta)$), and therefore

$$\lim_n \|B_n(M_\Theta)A - A\| = 0;$$

on the other hand, $B_n(M_\Theta)A$ are finite rank operators, since $|B_n(M_\Theta)A| = |H_{B_n a/\Theta}| = |H_{B_n h/B}|$ and the symbol of the latter Hankel operator is rational, $B_n h/B \in H^\infty + \mathcal{R}$;

similarly, (iii) $\Rightarrow$ (ii), since $A \in H^\infty(M_{\Theta_{inn}})$ clearly implies $B_n(M_{\Theta_{inn}})A \in H^\infty(M_{\Theta_{inn}})$;

in order to show that (i) $\Rightarrow$ (iii), observe that $B(M_\Theta)A = 0$ for every $A \in \{M_\Theta\}' \cap \mathcal{F}$ (see the reasonings in 2.5.6(d)); clearly, the same is true for limits of these (finite rank) operators; on the other hand, if Θ has a nontrivial singular part V, the reasoning in (g)(ii) above exhibits an operator $A \in \{M_\Theta\}' \bigcap \mathfrak{S}_\infty$ with $a\Theta^{-1} = g\overline{V}_1 \in C(\mathbb{T})$ and $g\overline{V}_1 \notin H^\infty$, where g is an outer function; the upper left entry of the corresponding Y matrix for $B(M_\Theta)A$ is $Ba = B\Theta g\overline{V}_1$, and hence $Ba \notin \Theta H^\infty$ (otherwise, $g = V_1 h$ with $h \in H^\infty$ which is impossible), i.e. $B(M_\Theta)A \neq 0$ (see (a) above); the contradiction shows that $V = 1$;

the implication (ii) $\Rightarrow$ (iii) is also proved since $H^\infty(M_{\Theta_{inn}}) = \{M_{\Theta_{inn}}\}'$.]

(j) Let $\Theta = \Theta_T$ satisfy at least one of the following properties 1) $\Theta(\lambda) = 0$ for a point $\lambda \in \mathbb{D}$, 2) there exists a Beurling–Carleson set $\sigma \subset \mathbb{T}$ such that $m(\sigma) = 0$ and $\mu(\sigma) > 0$, where μ is the representing measure of the singular factor of Θ_{inn} (see Section A.3.9, Volume 1).

Show that $\{M_\Theta\}' \cap \mathfrak{S}_p \neq \{0\}$ for every $p > 0$.

A closed subset $\sigma \subset \mathbb{T}$ is called *Beurling–Carleson set* if

$$\int_{\mathbb{T}} \log(\operatorname{dist}(\zeta, \sigma))\, dm(\zeta) > -\infty,$$

or, equivalently, if $\sum_{k\ge1} |I_k| \log |I_k|^{-1} < \infty$, where $(I_k)_{k\ge1}$ is the sequence of the complementary arcs of σ.

[*Hint:* property 1) and 2.5.6(d) yield even $\{M_\Theta\}' \cap \mathcal{F} \neq \{0\}$;

supposing 2) and using the notation of A.4.3.4 (Volume 1), let V_σ be the singular inner function

$$V_\sigma(z) = \exp\Big(-\int_\sigma \frac{\zeta+z}{\zeta-z}\, d\mu(\zeta)\Big),$$

and let g be an outer function such that $g\overline{V}_\sigma \in C^{(n)}(\mathbb{T})$ for a large enough $n \ge 1$ (a theorem by Carleson states that the outer function with $|g(\zeta)| = (\operatorname{dist}(\zeta, \sigma))^N$, $\zeta \in \mathbb{T}$, and N sufficiently large, satisfies this property, see V. Havin and B. Jöricke, [**HJ**], Section II.3.1); now, as in **(i)** above, set $a = g\Theta\overline{V}_1$ and $c = 0$ to obtain an operator $A \in \{M_\Theta\}'$ such that $|A| = |H_{a/\Theta}|$ (see (a) above); as in **(i)**above, we can see that $A \neq 0$; finally, since $C^{(n)}(\mathbb{T}) \subset B_p^{1/p}$ Peller's theorems B.8.7.2(f) and B.8.7.2(g) (Volume 1) imply $A \in \mathfrak{S}_p$.]

(k) *Cyclicity of* M_Θ. Show that M_Θ is cyclic if and only if $\Theta \neq 0$.

[*Hint:* if $\partial_T = \partial_{T^*} = 1$ and $\Theta = 0$, then $M_\Theta = S \oplus S^*$, where S is the shift operator on H^2; this operator has no cyclic vectors since $(-g_*) \oplus f_*$ is orthogonal to $S^n f \oplus S^{*n} g$ for $n \ge 0$, where $f_*(z) = \overline{f}(\overline{z})$;

for $\Theta \neq 0$, show that $x = P_\Theta y$ is a cyclic vector if

$$y = (\Theta_{out}, \Delta u \overline{\Theta}_{inn})^{col}$$

and u is a (bounded) function such that $1-u$ is S-cyclic in $L^2(\mathbb{T})$ (that is $1-u \neq 0$ a.e. and $\int_{\mathbb{T}} \log|1-u|dm = -\infty$, see A.4.1.2, Volume 1); indeed, it suffices to prove that

$$E =: \operatorname{span}(z^n y, (\Theta f, \Delta f) : n \geq 0, f \in H^2) = \binom{H^2}{\operatorname{clos} \Delta L^2};$$

but $\Theta_{inn} f y = (\Theta f, \Delta u f)^{col} \in E$, and hence $(0, \Delta(1-u)f)^{col} \in E$, for every $f \in H^2$, whence $\binom{0}{\operatorname{clos} \Delta L^2} \subset E$; the result follows.]

(l) *A characterization of finite rank model operators with simple spectrum.*

(i) Let $\Theta = \Theta_T$ be a (scalar) inner function and denote by $\deg \Theta$ *the degree of* Θ, that is, $\deg \Theta = \sum d(\lambda)$ if Θ is a finite Blaschke product $B = \prod b_\lambda^{d(\lambda)}$, and $\deg \Theta = \infty$ otherwise. Show that

$$\|B(T)\| = 1$$

for every Blaschke product B with $\deg B < \deg \Theta$.

[*Hint:* if Θ is not a finite Blaschke product, then $\sigma(T) \cap \mathbb{T} = \sigma(\Theta) \cap \mathbb{T} \neq \emptyset$ and hence $\|B(T)\| \geq 1$ since $\sigma(B(T)) \supset B(\sigma(T) \cap \mathbb{T})$; if Θ is a finite Blaschke product, then use $\|B(T)\| = \operatorname{dist}(B\overline{\Theta}, H^\infty)$, and observe that for every $h \in H^\infty$, $\|B - \Theta h\|_\infty \geq 1$ since otherwise B has at most the same number of zeros as Θ.]

(ii) *(B. Cole, K. Lewis and J. Wermer, 1993)* Let $T : H \longrightarrow H$ be a c.n.u. contraction, $n = \dim H < \infty$, and $\operatorname{card}(\sigma(T)) = n$. Show that the following are equivalent.

a) T is unitarily equivalent to M_B, where $B = \prod_{\lambda \in \sigma(T)} b_\lambda$;
b) $\|T^{n-1}\| = 1$.

[*Hint:* a) $\Rightarrow$ b) by (i); conversely, if there exists $x \in H$ such that $\|x\| = \|T^{n-1}x\| = 1$, then $\|x\| = \|Tx\| = \|T^{n-1}x\| = 1$, and hence by 2.5.9(c) below the vectors $T^j x$, $j < n-1$ are linearly independent and are contained in $\operatorname{Ker}(I - T^*T)$; hence $\operatorname{rank}(I - T^*T) = 1$, and similarly $\operatorname{rank}(I - TT^*) = 1$, and T is unitarily equivalent to a scalar model $M_\Theta : \mathcal{K}_\Theta \longrightarrow \mathcal{K}_\Theta$; since $\dim H = \dim \mathcal{K}_\Theta < \infty$, it is clear that $\Theta = B$ (see B.2.4.4, Volume 1).]

(m) *An estimate of inverses in terms of the "corona data".* Let Θ be a scalar inner function and $\varphi \in H^\infty$ such that $\|\varphi\|_\infty \leq 1$ and $\delta = \inf_{z \in \mathbb{D}}(|\varphi(z)| + |\Theta(z)|) > 0$. Show that

$$\max(1, \frac{1}{\delta\sqrt{2}}) \leq \|\varphi(M_\Theta)^{-1}\| \leq \frac{2}{\delta} + \frac{A}{\delta^2} \log \frac{1}{\delta},$$

where A (≤ 100) is a numerical constant.

[*Hint:* the right hand side inequality follows from (d) above (see equivalence (i) $\Leftrightarrow$ (ii)) and an estimate for solutions of the corona equation $g\varphi + h\Theta = 1$ quoted in Section B.9.4 (Volume 1);

for a lower bound for $\|\varphi(M_\Theta)^{-1}\|$, look at $\|\varphi(M_\Theta)^*P_\Theta k_\lambda\|$, where $k_\lambda = (1-\overline{\lambda}z)^{-1}$ and $\lambda \in \mathbb{D}$, as follows,

$$\begin{aligned}\|\varphi(M_\Theta)^*P_\Theta k_\lambda\| &= \|P_+\overline{\varphi}(1-\overline{\Theta(\lambda)}\Theta)k_\lambda\| = \|\overline{\varphi(\lambda)}k_\lambda - \overline{\Theta(\lambda)}P_+(\overline{\varphi}\Theta k_\lambda)\| \\ &\le |\varphi(\lambda)|\cdot\|k_\lambda\| + |\Theta(\lambda)|\cdot\|k_\lambda\| \\ &= \Big(|\varphi(\lambda)|+|\Theta(\lambda)|\Big)\cdot\Big\|P_\Theta k_\lambda\Big\|\cdot\Big(1-\Big|\Theta(\lambda)\Big|^2\Big)^{-1/2};\end{aligned}$$

this gives $\|\varphi(M_\Theta)^{-1}\| \ge \Big(|\varphi(\lambda)|+|\Theta(\lambda)|\Big)^{-1}\Big(1-|\Theta(\lambda)|^2\Big)^{1/2}$ for every $\lambda \in \mathbb{D}$; now, use $\|\varphi(M_\Theta)^{-1}\| \ge 1$ (for the case $\delta \ge 1/\sqrt{2}$) and $(1-|\Theta(\lambda)|^2) \ge 1/2$ (elsewhere).]

(n) *The Foguel–Hankel operators.* Let $\Theta = 0$ (the zero operator from $\mathbb{C}$ to $\mathbb{C}$). Show that the model operator M_Θ is unitarily equivalent to the operator $T = S \oplus S^*$ on the space $\mathcal{K} = H^2 \oplus H^2$ and the commutant $\{T\}'$ consists of all operators A of the form

$$A = \begin{pmatrix} f(S) & 0 \\ \Gamma_\varphi & g(S^*) \end{pmatrix},$$

where $f, g \in H^\infty$ and $\Gamma_\varphi : H^2 \longrightarrow H^2$ is any bounded Hankel operator, $\Gamma_\varphi x = \mathcal{J}H_\varphi x = \mathcal{J}P_-\varphi x$ for every $x \in H^2$, $\varphi \in L^\infty(\mathbb{T})$.

(Recall that the operators of this form are called Foguel–Hankel operators, see B.1.6.6 and B.1.8, Volume 1.)

[*Hint:* take an operator $A \in \{T\}'$ in its matrix form

$$A = \begin{pmatrix} a & b \\ c & d \end{pmatrix}$$

and rewrite the commutation relation $TA = AT$ as $Sa = aS$, $S^*d = dS^*$, $S^*c = cS$ and $Sb = bS^*$; the latter equation implies $S^n b = bS^{*n}$ for every $n \ge 0$, and hence $\|b(x)\| = \|S^n b(x)\| = \|bS^{*n}x\| \longrightarrow 0$ as $n \longrightarrow 0$ for every $x \in H^2$; it remains to use Lemma 1.2.3 (even in its scalar version A.7.2.1, Volume 1) and the Nehari theorem B.1.3.2 (Volume1).]

2.5.8. More about C_0 contractions. Here we list several properties of c.n.u. contractions of the class C_0; see Section 2.3 for the definition, notation and initial properties. Let θ be a scalar inner function and

$$\theta = \Big(\prod_{\lambda\in\mathbb{D}} b_\lambda^{d(\lambda)}\Big)\cdot\exp\Big(-\int_{\mathbb{T}}\frac{\zeta+z}{\zeta-z}\,d\mu(\zeta)\Big)$$

be its canonical factorization (see A.3.9, Volume 1). As in Section A.4.3 (Volume 1), given a Borel subset $\sigma \subset \overline{\mathbb{D}}$, we denote by θ_σ the function

$$\theta_\sigma = \Big(\prod_{\lambda\in\sigma\cap\mathbb{D}} b_\lambda^{d(\lambda)}\Big)\cdot\exp\Big(-\int_{\sigma\cap\mathbb{T}}\frac{\zeta+z}{\zeta-z}\,d\mu(\zeta)\Big)$$

and call it *maximal divisor of θ supported by σ*. Recall that the spectrum of θ is $\sigma(\theta) = \operatorname{supp}(\mu) \cup \operatorname{clos}\{\lambda \in \mathbb{D} : d(\lambda) > 0\}$, see 2.3.6 above. It is clear that if θ' (an inner function) is a divisor of θ, then θ' is a divisor of $\theta_{\sigma(\theta')}$. For the greatest common divisor and the least common multiple of two inner functions we use

$$\begin{aligned}\theta\wedge\theta' &= \mathrm{GCD}(\theta,\theta'),\\ \theta\vee\theta' &= \mathrm{LCM}(\theta,\theta'),\end{aligned}$$

and similarly for a family of inner functions; see A.2.5 and A.3.12.4 (Volume 1) for the definitions and initial properties.

(a) Show that $C_0 \subset C_{00}$.

[*Hint:* this is a straightforward consequence of 2.2.2(5) and 1.5.4(b).]

(b) *Kernels, invariant subspaces and divisors.*

(i) Let T, T' be Hilbert space contractions such that $T \preceq T'$ (see Subsection 1.5.3 for the definition of di-formations), and let $\varphi \in H^\infty$. Show that $\varphi(T) \preceq \varphi(T')$ and, in particular,

$$\dim \operatorname{Ker}(\varphi(T)) = \dim \operatorname{Ker}(\varphi(T')); \quad \varphi(T) = 0 \Leftrightarrow \varphi(T') = 0; \tag{2.5.1}$$

$$T \in C_0 \Leftrightarrow T' \in C_0 \text{ and } m_T = m_{T'}. \tag{2.5.2}$$

(ii) Let $\varphi \in H^\infty$, $T \in C_0$ and $K = \operatorname{Ker} \varphi(T)$. Show that $(T|K) \in C_0$ and $m_{T|K} = \varphi_{inn} \wedge m_T$; in particular, $\operatorname{Ker} \varphi(T) = \{0\} \Leftrightarrow \varphi \wedge m_T = 1$.

[*Hint:* as in 2.3.2, $m_{T|K}$ divides φ_{inn} and m_T, and hence divides $\psi = \varphi_{inn} \wedge m_T$, which yields $\operatorname{Ker} \varphi(T) \supset \operatorname{Ker} \psi(T) \supset m'(T)H$ where $m' = m_T/\psi$; this shows that $m_{T|K}(T)m'(T) = 0$ and hence ψ divides $m_{T|K}$.]

(iii) Let $\varphi \in H^\infty$, $T \in C_0$. Show that $\operatorname{Ker} \varphi(T) = \{0\} \Leftrightarrow \operatorname{Ker} \varphi(T)^* = \{0\}$ ($\Leftrightarrow \operatorname{clos}(Range(\varphi(T))) = H$).

[*Hint:* clear from (ii) and $m_{T^*} = (m_T)_*$.]

(iv) Let T be a c.n.u. contraction and $E_i \in \operatorname{Lat}(T)$ such that $(T|E_i) \in C_0$ for $i \in I$ and $H = \operatorname{span}(E_i : i \in I)$. Show that $T \in C_0$ if and only if $m =: \vee_{i \in I}(m_{T|E_i}) \neq 0$, and in this case $m = m_T$. Similarly, $T \in C_0$ if and only if $T_1 =: (T|E_1) \in C_0$ and $T_2 =: (P_{E_1^\perp} T|E_1^\perp) \in C_0$, and in this case, $m_T = m_{T_1} \vee m_{T_2}$.

[*Hint:* the proof is similar to that of above points.]

(v) Let $T \in C_0$ and $E \in \operatorname{Lat}(T)$. Show that $\sigma(T|E) \subset \sigma(T)$.

[*Hint:* clear from Theorem 2.3.4 and the fact that $m_{T|E}$ divides m_T.]

(c) *Maximal spectral subspaces and decomposability.*

(i) Let $T \in C_0$ and $\sigma \subset \overline{\mathbb{D}}$ be a closed set and $m_\sigma = (m_T)_\sigma$ (see above). Show that $K(\sigma) = \operatorname{Ker}(m_\sigma(T))$ is a maximal spectral subspace over $\sigma \cap \sigma(T)$, as defined in 2.5.1(c) above.

[*Hint:* since $m_\sigma = m_{T|K(\sigma)}$ (see (b)), we have $\sigma(T|K(\sigma)) = \sigma \cap \sigma(T)$ by Theorem 2.3.4; given $E \in \operatorname{Lat}(T)$ such that $\sigma(T|E) \subset \sigma \cap \sigma(T)$, one can use a property from A.3.12.4 (Volume 1): since $m_{T|E}$ divides m_T and $\sigma(m_{T|E}) \subset \sigma$, $m_{T|E}$ divides m_σ, whence $E \subset K(\sigma) = \operatorname{Ker}(m_\sigma(T))$.]

(ii) Every $T \in C_0$ is a *decomposable operator* (in the C. Foiaş sens), that is, for every open covering of the spectrum $\sigma(T) \subset \mathcal{O}_1 \cup \cdots \cup \mathcal{O}_n$ there exist maximal spectral subspaces $E_1, \ldots, E_n$ such that

$$\sigma(T|E_i) \subset \mathcal{O}_i, \quad i = 1, \ldots, n, \text{ and } H = E_1 + \cdots + E_n$$

(the sum is not supposed to be direct).

[*Hint:* let $(\mathcal{O}'_i)$ and $(\mathcal{O}''_i)$ be two other open coverings of $\sigma(T)$ such that $\overline{\mathcal{O}}''_i \subset \mathcal{O}'_i \subset \overline{\mathcal{O}}'_i \subset \mathcal{O}_i$ for every i; let further

$$\sigma_i = \overline{\mathcal{O}}'_i \cap \sigma(T), \tag{2.5.3}$$

$$f_i = m/m_{\sigma_i} = m_{\sigma(T)\backslash\sigma_i}, \tag{2.5.4}$$

where m_σ has the same meaning (for the minimal annihilating function $m = m_T$) as for θ_σ defined in the beginning of this subsection; since $f_i \in \mathrm{Hol}(\mathcal{O}'_i)$ (without loss of generality we suppose a kind of symmetry of the $\mathcal{O}'_i$ such that $\mathcal{O}'_i\backslash\overline{\mathbb{D}} \subset \{1/\bar{\zeta} : \zeta \in \mathcal{O}'_i\}$) and $f_i(z) \neq 0$ for $z \in \mathcal{O}'_i$, one has $\inf_{\mathcal{O}''_i} |f_i| > 0$ for every i, and hence

$$\inf_{\mathcal{O}''\cap\mathbb{D}} \sum_{i=1}^{n} |f_i|^2 > 0,$$

where $\mathcal{O}'' = \cup\mathcal{O}''_i$;

now, the proof can be completed by simply applying the Carleson corona theorem (see B.9.2.4, Volume 1, or 3.2.10 for the statement; in fact, here we need only the elementary version proved in B.9.3.3, Volume 1, see the next paragraph); indeed, let $g_i \in H^\infty$ such that $\sum_{i=1}^n f_i g_i = 1$; then

$$\sum_{i=1}^{n} f_i(T)g_i(T)x = x$$

for every $x \in H$, where $f_i(T)g_i(T)x \in \mathrm{Ker}\, m_{\sigma_i}(T)$ for every i, and the result follows;

finally, let us deduce the existence of g_i from the elementary version of the corona theorem presented in B.9.3.3 (Volume 1); first, without loss of generality, we can suppose that all $\mathcal{O}'_i$ are bounded by a finite number of smooth curves, say γ_i, satisfying $\gamma_i \cap \gamma_j \cap \sigma(T) = \emptyset$ for every $i \neq j$, and such that $\mathcal{O}'_i$ and their complements in $\overline{\mathbb{C}}$ are simply connected (use the fact that $\sigma(T) \cap \mathbb{D}$ is a sequence tending to $\mathbb{T}$); then, set

$$\theta = m_{\sigma_1} \vee m_{\sigma_2} = m_{\sigma_1\cup\sigma_2} \text{ and } \varphi_i = \theta/m_{\sigma_i} \text{ for } i = 1, 2;$$

then, the singularities of φ_1 are in $(\sigma_1 \cup \sigma_2)\backslash\overline{\mathcal{O}}'_1 \subset \sigma(T) \cap (\overline{\mathcal{O}}'_2\backslash\mathcal{O}'_1)$ and those of φ_2 are in $(\sigma_1 \cup \sigma_2)\backslash\overline{\mathcal{O}}'_2 \subset \sigma(T) \cap (\overline{\mathcal{O}}'_1\backslash\mathcal{O}_2)$; the intersection of these two closed sets is contained in $\gamma_1 \cap \gamma_2 \cap \sigma(T)$, and hence is void; therefore, the functions φ_1, φ_2 satisfy the hypotheses of B.9.3.3 (Volume 1), and hence there exist $h_i \in H^\infty$ such that $h_1\varphi_1 + h_2\varphi_2 = 1$; multiplying by m/θ, we get $h_1 f_1 + h_2 f_2 = m/m_{\sigma_1\cup\sigma_2} =: F_2$; now, proceeding in the same way with $F_2, f_3, ..., f_n$, etc., we finally solve the above Bezout equation for the data f_i.]

(d) *Maximal vectors and locally C_0 operators.*

(i) Let T be an operator of class C_0 on a Hilbert space H, and $m_x = m_{T|E(x)}$, where $E(x) = \mathrm{span}(T^k x : k \geq 0)$ for $x \in H$. Show that the set $A = \{x \in H : m_x \neq m_T\}$ is an F_σ set of the first category in H.

The elements of $H\backslash A$ are called *maximal vectors of* T.

[*Hint:* assuming $m_T(0) > 0$ and setting $A_n = \{x \in H : m_x(0) \geq m_T(0) + \frac{1}{n}\}$, observe that $A = \bigcup_{n\geq 1} A_n$ and show that each A_n is closed and contains no interior point;

indeed, if $x_k \in A_n$ and $x = \lim_k x_k$, one can assume that there exists a limit $\lim_k m_{x_k}(\zeta) = \varphi(\zeta)$, $\varphi \in H^\infty$; it follows from 2.5.6(b) that $0 = \lim_k m_{x_k}(T)x_k =$

$\varphi(T)x$, whence $\varphi_{inn}(T)x = 0$, which gives $m_x(0) \geq \varphi_{inn}(0) \geq \varphi(0) \geq m_T(0) + \frac{1}{n}$, and so $x \in A_n$;

to prove $\mathrm{int}(A) = \emptyset$, show first that for given $x, y \in H$ there exists a convex combination $x(t) = tx + (1-t)y$, $0 \leq t \leq 1$, such that $m_{x(t)} = m_x \vee m_y$; to this end, observe that $m_{x(t)}$ divides $m_x \vee m_y$ and $m_{x(t)} \vee m_{x(t')} = m_x \vee m_y$ for $t \neq t'$; property A.3.12.4, Volume 1, applied to the family of ratios $m_x \vee m_y / m_{x(t)}$ shows that all of them, excepting at most a countable set, are equal to 1;

now, assume that there exists a ball $B \subset A_n$ and choose a sequence $x_1, x_2, \ldots$ dense in B, apply the above procedure to obtain $y_k \in B$ such that $m_{y_k} = m_{x_1} \vee \ldots \vee m_{x_k}$; this contradicts to the definition of A_n since clearly $m_T(0) = \lim_k (m_{x_1} \vee \cdots \vee m_{x_k})(0)$.]

(ii) Let T be an operator on a Hilbert space H, which is *locally of class* C_0, that is $(T|E(x)) \in C_0$ for every $x \in H$, where $E(x) = \mathrm{span}(T^k x : k \geq 0)$. Show that $T \in C_0$.

[*Hint:* as in (i), show that the set $A(\lambda) = \{x \in H : |m_x(\lambda)| > \inf_{y \in H} |m_y(\lambda)|\}$ is an F_σ set of first category in H, for every $\lambda \in \mathbb{D}$; let (λ_k) be dense in $\mathbb{D}$ and $x \in H \backslash (\bigcup_k A(\lambda_k))$; then, $|m_x(\lambda_k)| \leq |m_y(\lambda_k)|$ for every $y \in H$ and every λ_k, whence $|m_x(z)| \leq |m_y(z)|$ in $\mathbb{D}$; this means that m_y divides m_x, and hence $m_x(T)y = 0$ for every $y \in H$.]

(e)* *Multiplicities and Jordan models.* An operator J is called *Jordan operator* if

$$J = J(\theta_i) = \sum_{0 \leq i < \mu} \oplus M_{\theta_i},$$

where θ_i are scalar inner functions such that $\frac{\theta_i}{\theta_{i+1}} \in H^\infty$ and $\theta_i \neq \mathrm{const}$ for all $i < \mu \leq \infty$. Below, we list some properties of C_0 operators related to quasi-similarity to a Jordan operator.

(i) *(B. Sz.-Nagy and C. Foiaş, 1970)* Let $T : H \longrightarrow H$ be an operator of the class C_0. There exists a unique Jordan operator $J = J(\theta_i)$ quasi-similar to T (J is called the *Jordan model* of T); moreover, $\theta_0 = m_T$ and $\mu = \mu(T)$, where $\mu(T)$ stands for the *multiplicity of the spectrum (or, multicyclicity)* of T,

$$\mu(T) = \inf\{\mathrm{card}(C) : \mathrm{span}(T^k C : k \geq 0) = H\}.$$

In particular, $\mu(T) = \mu(T^*)$.

(ii) Let J and J' be two Jordan operators and $J \preceq J'$. Then $J = J'$.

(iii) *(B. Sz.-Nagy and C. Foiaş, 1970)* Let $T : H \longrightarrow H$ be an operator of the class C_0. The following are equivalent.

a) $\mu(T) = 1$.
b) $\mu(T^*) = 1$.
c) T is quasimilar to M_θ, $\theta = m_T$.
d) For every $E \in \mathrm{Lat}(T)$ there exists $\varphi \in H^\infty$ such that $E = \mathrm{Ker}\, \varphi(T)$.
e) A vector $x \in H$ is cyclic (i.e., $H = E(x)$ in the notation of (d) above) if and only if $m_x = m_T$.
f) $\{T\}'$ is commutative.

g) $\{T\}' = H^\infty(T)/H^\infty(T)$, where

$$H^\infty(T)/H^\infty(T) = \{A : A = f(T)g(T)^{-1} \in L(H), \text{ where } f, g \in H^\infty \text{ and } g_{inn} \wedge m_T = 1\}.$$

h) $m_T = \det \Theta_T$ (assuming $I - T^*T \in \mathfrak{S}_1$).

(iv) Let $T = \sum_{i \geq 1} \oplus M_{\theta_i}$, where θ_i are scalar inner functions. Then $\mu(T) = \sup\{\operatorname{card}(I) : \bigwedge_{i \in I} \theta_i \neq 1\}$.

(v) If $T \in C_0$ and $E \in \operatorname{Lat}(T)$ then $\mu(T|E) \leq \mu(T)$.

[*Hint:* by (i), $\mu(T|E) = \mu((T|E)^*) \leq \mu(T^*) = \mu(T)$, where the inequality follows from D.2.3.1(4), see below.]

(vi) Let $T, T' \in C_0$. The *multiplicity function* ν_T is defined on the set $\mathcal{I}$ of all inner functions by

$$\nu_T(\theta) = \mu\Big(T|\operatorname{Range}(\theta(T))\Big),$$

where (as always) $\operatorname{Range} A = \operatorname{clos}(AH)$. Then, $T \sim T'$ if and only if $\nu_T = \nu_{T'}$.

(vii) C_0 *weak contractions. (B. Moore and E. Nordgren; H. Bercovici and D. Voiculescu).* A c.n.u. contraction T is called a *weak contraction* if $\mathbb{D} \backslash \sigma(T) \neq \emptyset$ and $I - T^*T \in \mathfrak{S}_1$. Given a weak contraction T, the characteristic function Θ_T satisfies $I - \Theta_T(\zeta) \in \mathfrak{S}_1$ for $\zeta \in \mathbb{D}$, and the determinant

$$\det \Theta_T(\zeta) = \prod_{j \geq 1} \lambda_j(\Theta_T(\zeta))$$

is well defined and is an H^∞ function. If $T \in C_{00}$ ($\Leftrightarrow \Theta_T$ is inner) then $T \in C_0$, and its Jordan model is $J = J(\theta_j)$, where

$$\theta_j = \delta_j / \delta_{j+1}$$

and δ_j is the GCD of all minors of corank j of Θ_T, $j \geq 0$ (that is, of the functions $\det(P_F \Theta_T(\zeta)|F)$ where F is subspace of codimension j). In particular, $\mu(T) = \sup\{j : \theta_j = 1\}$.

(viii) *Two examples.* Let θ_j be inner functions. The Jordan model of $M_{\theta_1} \oplus M_{\theta_2}$ is $M_{\theta_1 \vee \theta_2} \oplus M_{\theta_1 \wedge \theta_2}$. If θ_j are pairwise relatively prime and $\theta = \vee_j \theta_j$ exists ($\neq 0$), then the Jordan model of $\sum_{j \geq 1} \oplus M_{\theta_j}$ is M_θ.

(f)* *The point spectrum and completeness of C_0 operators.* Let T be an operator of the class C_0 on a Hilbert space H, and $d(\lambda) = \operatorname{rank} \mathcal{P}_\lambda$ be the rank of the Riesz projection at a point $\lambda \in \sigma_p(T)$.

(i) The following are equivalent.

a) T is *complete*, that is, $H = \operatorname{span}(\operatorname{Ker}(\lambda I - T)^k : k \geq 1, \lambda \in \sigma_p(T))$.
b) T^* is complete.
c) m_T is a Blaschke product.

Assume moreover that $I - T^*T \in \mathfrak{S}_1$, then the preceding assertions are also equivalent to

d) $\det \Theta_T$ is a Blaschke product.
e) $\det(T^*T) = \det |\Theta_T(0)|^2 = \prod_{\lambda \in \sigma_p} |\lambda|^{2d(\lambda)}$ (assuming $0 \notin \sigma(T)$).

(ii) Assume that $I - T^*T \in \mathfrak{S}_1$. Then

$$\det(T_\zeta^* T_\zeta) = \det |\Theta_T(\zeta)|^2 \leq \prod_{\lambda \in \sigma_p} |b_\lambda(\zeta)|^{2d(\lambda)}$$

for $\zeta \in \mathbb{D}$, where $T_\lambda = b_\lambda(T)$ and $b_\lambda = (\lambda - z)(1 - \overline{\lambda} z)^{-1}$, $\lambda \in \mathbb{D}$; in particular,

$$\sum_{\lambda \in \sigma_p} d(\lambda)(1 - |b_\lambda(\zeta)|^2) \leq -\log(det(T_\zeta^* T_\zeta)) \leq c(\zeta) \cdot \mathrm{Trace}(I - T_\zeta^* T_\zeta),$$

where $c(\zeta) = \max(3, \log \frac{1}{\delta})$ and $\delta = \delta(\zeta) = \min\{|b_\zeta(\lambda)| : \lambda \in \sigma_p(T)\}$. (See also B.7.5.3(b), Volume 1, for a similar inequality).

[*Hint:* for the first equality see 2.5.9(b) (iv) below; then, verify that the Blaschke product $B = \prod_{\lambda \in \sigma_p} b_\lambda^{d(\lambda)}$ divides $\det \Theta_T$, and use $\log \frac{1}{a} = (1 - a) \cdot (1 - a)^{-1} \log \frac{1}{a}$ for $a = \lambda_n(T^*T)$]

(g) *The spectral synthesis property.* Let T be a complete operator of the class C_0 on a Hilbert space H.

(i) Let $B_n = \prod_{|\lambda| > r_n} b_\lambda^{d(\lambda)}$, where $r_n \nearrow 1$ and $d(\lambda)$ is defined in (f) above; show that

$$\mathrm{Range}\, B_n(T) \subset \mathrm{span}(\mathrm{Ker}(\lambda I - T)^k : k \geq 1, \lambda \in \sigma_p(T), |\lambda| \leq r_n)$$

and $\lim_n B_n(T)x = x$ for every $x \in H$.

[*Hint:* see 2.5.6(b).]

(ii) Show that $T|E$ is a complete operator for every $E \in \mathrm{Lat}(T)$.

[*Hint:* $m_{T|E}$ is a divisor of m_T.]

(h)* *About the commutant.* Let $T \in C_0$. (i) *(B. Sz.-Nagy, 1974)* Always, $\{T\}' \cap \mathfrak{S}_\infty \neq \{0\}$, but it may happen that $H^\infty(T) \cap \mathfrak{S}_\infty = \{0\}$.

[*Hint:* for the latter, look at $T = 0$ on a space H with $\dim H = \infty$]

(ii) *(E. Nordgren, 1975)* If $I - T^*T \in \mathfrak{S}_\infty$ then $H^\infty(T) \cap \mathfrak{S}_\infty \neq \{0\}$, and moreover, there exists a sequence $(\varphi_n)_{n \geq 1} \subset H^\infty$ such that $\|\varphi_n\|_\infty \leq 1$, $\varphi_n(T) \in \mathfrak{S}_\infty$ for $n \geq 1$, and $\lim_n \varphi_n(T) = I$ (WOT).

[*Hint:* for the case of a complete operator T, see (g) above.]

(iii) If $\partial_T < \infty$ then $\{T\}' \cap \mathfrak{S}_p \neq \{0\}$ for every $p > 0$; moreover, the operators $\varphi_n(T)$ from the previous theorem (ii) can be chosen in $\mathfrak{S}_p$.

[*Hint:* for the Blaschke factor of m_T, proceed as in (g) above; for the singular factor of m_T proceed as for Θ_T in 2.5.7(j) above - namely, choose a sequence of Beurling–Carleson subsets $\sigma_n \subset \mathbb{T}$ such that $\mu(\mathbb{T} \backslash \bigcup_n \sigma_n) = 0$ and consider H^∞ functions $\varphi_n = a$ (in the notation of 2.5.7(j)) corresponding to σ_n as in 2.5.7(j).]

(iv) (See (e)(iii) above for a criterion of commutativity of $\{T\}'$).

2.5.9. Miscellaneous facts. **(a)** Let T be a Hilbert space contraction. Show that T is a Fredholm operator if and only if $\Theta_T(0)$ is, and $\mathrm{ind}(T) = \mathrm{ind}(\Theta_T(0))$. In particular, if $\partial_T = \mathrm{rank}(I - T^*T) < \infty$ and $\partial_{T^*} < \infty$, then T is Fredholm and $\mathrm{ind}(T) = \partial_T - \partial_{T^*}$.

[*Hint:* $T : H = \mathcal{D}_T \oplus \mathcal{D}_T^\perp \longrightarrow H = \mathcal{D}_{T^*} \oplus \mathcal{D}_{T^*}^\perp$, where as usual $\mathcal{D}_T = \mathrm{clos}(I - T^*T)H$, and $T|\mathcal{D}_T^\perp : \mathcal{D}_T^\perp \longrightarrow \mathcal{D}_{T^*}^\perp$ is a unitary operator; for the rest see B.2.5.1, Volume 1.]

(b) Let $T : H \longrightarrow H$ be a Hilbert space contraction and $T_\lambda = b_\lambda(T)$, where $b_\lambda = (\lambda - z)(1 - \overline{\lambda}z)^{-1}$ and $\lambda \in \mathbb{D}$. Show consecutively that

(i) $\|D_{T_\lambda}x\|^2 = \|D_T Vx\|^2$ and $\|D_{T_\lambda^*}x\|^2 = \|D_{T^*}V^*x\|^2$ for every $x \in H$;

[*Hint:* employ the formulas from 2.5.3(e) (ii).]

(ii) there exist unitary operators $X : \mathcal{D}_{T_\lambda} \longrightarrow \mathcal{D}_T$ and $Y : \mathcal{D}_{T_\lambda^*} \longrightarrow \mathcal{D}_{T^*}$ such that $XD_{T_\lambda} = D_T V$ and $YD_{T_\lambda^*} = D_{T^*}V^*$;

(iii) $Y\Theta_{T_\lambda}(z)X^{-1}D_T = Y\Theta_{T_\lambda}(z)D_{T_\lambda}V^{-1} = D_{T^*}(I - \zeta T^*)^{-1}(\zeta I - T) = \Theta_T(\zeta)D_T$, where $\zeta = b_\lambda^{-1}(z) = (\lambda + z)(1 + \overline{\lambda}z)^{-1}$, and therefore

$$Y\Theta_{T_\lambda}(z)X^{-1} = \Theta_T(\zeta);$$

[*Hint:* use (ii) and formulas of 1.2.11.]

(iv) the operators $I - T_\lambda^*T_\lambda$ and $I - \Theta_T(\lambda)^*\Theta_T(\lambda)$ are unitarily equivalent for every $\lambda \in \mathbb{D}$.

[*Hint:* by (iii), it suffices to check that $I - T^*T$ and $I - \Theta_T(0)^*\Theta_T(0)$ are unitarily equivalent, which is obviuous from the formula for Θ_T.]

(c) Let $T : H \longrightarrow H$ be a c.n.u. contraction and $x \in H$ such that $0 < \|x\| = \|Tx\| = \cdots = \|T^kx\|$. Show that $(T^jx)_{0\le j\le k}$ is linearly independent.

[*Hint:* if not, the subspace $H' = \mathrm{span}(T^jx : 0 \le j < k)$ is T-invariant; since $\|T^jx\|^2 = \|TT^jx\|^2$, one has $T^jx \in \mathrm{Ker}(I - T^*T)$ whence $H' \subset \mathrm{Ker}(I - T^*T)$, and so $T|H'$ is unitary, which is impossible.]

(d) *Invariant subspaces of* $L^p(\mathbb{T}, \mu)$. Let μ be a (finite) Borel measure on $\mathbb{T}$, $1 \le p \le \infty$, and let $E \subset L^p(\mathbb{T}, \mu)$ be a closed z-invariant subspace ($*$-closed for $p = \infty$), that is, $f \in E \Rightarrow zf \in E$.

(i) Assume that $zE = E$ (which is equivalent to $zE \subset E$ *and* $\overline{z}E \subset E$). Show that there exists a Borel subset $\sigma \subset \mathbb{T}$ such that $E = \chi_\sigma L^p(\mathbb{T}, \mu)$.

[*Hint:* as in the proofs of 2.1.17 and A.1.5.3 (Volume 1), observe that if $f \in E$ and $g \in E^\perp = \{y \in L^{p'}(\mathbb{T}, \mu) : \int x\overline{y}\, d\mu = 0 \text{ for every } x \in E\}$ then $\int z^n x\overline{y}\, d\mu = 0$ for every $n \in \mathbb{Z}$, and hence $x\overline{y} = 0$ a.e. μ; then take a countable dense subset $(f_k) \subset E$ ($*$-dense for $p = \infty$) and set $\sigma = \bigcup_k \{\zeta \in \mathbb{T} : f_k(\zeta) \ne 0\}$.]

(ii) Let $L^p(\mathbb{T}, \mu) = L^p(\mathbb{T}, \mu_s) \oplus L^p(\mathbb{T}, \mu_a)$ be the same decomposition in the singular and absolutely continuous subspaces as in A.1.5.2 (Volume 1). Show that $E = E_s \oplus E_a$, which means $E = \{f + g : f \in E_s,\ g \in E_a\}$, where $E_s \subset L^p(\mathbb{T}, \mu_s)$ and $E_a \subset L^p(\mathbb{T}, \mu_a)$ are z-invariant subspaces. Moreover, $zE_s = E_s$.

[*Hint:* taking $f \in E$ and $g \in E^\perp$ and writing $\int z^n f\overline{g}\, d\mu = 0$ for $n \ge 0$, use the Riesz brothers theorem A.2.7.3.]

(iii) Let w be the Radon-Nikodym derivative of μ, that is $d\mu_a = w\,dm$, where m stands for the Lebesgue measure (see again A.1.5.2, Volume 1). Assume that $zE_a \neq E_a$. Show that there exists a function Θ such that $E_a = \Theta H^p$ and $|\Theta|^p w = 1$ a.e. m.

[*Hint:* show that the set $E_p = \{fw^{1/p} : f \in E\}$ is a z-invariant subspace of $L^p(\mathbb{T})$ which is not $\bar{z}$-invariant; then use Theorem 2.1.17 to get a unimodular function θ such that $E_p = \theta H^p$; finally, set $\Theta = \theta / w^{1/p}$.]

2.6. Notes and Remarks

General information. There are several books and surveys exploiting the function model language to treat some problems of nonclassical spectral theory (that is, nonselfadjoint and nonunitary). Let us mention M. Brodskii and M. Livshic [**BrL**], L. de Branges and J. Rovnyak [**dBR1**], [**dBR2**], L. de Branges [**dB2**], B. Szőkefalvi-Nagy and C. Foiaş [**SzNF4**], M. Livshic [**Liv3**], M. Brodskii [**Bro3**], P. Lax and R. Phillips [**LPh1**], [**LPh2**], P. Fuhrmann [**Fuh3**], N. Nikolski [**N19**], N. Nikolski and S. Hruschev [**HrN**], N. Nikolski and V. Vasyunin [**NVa8**], H. Bercovici [**Be**]. Sometimes, these sources work with different realizations (transcriptions) of the model but, essentially, the technique is always the same, as explained in Chapter 1 above. The most developed applications of the function model techniques are in scattering theory, complex interpolation, and control and system theory. But in this chapter, we deal with inner problems of the spectral theory itself.

General spectral theory. The basic facts of general spectral theory that are presented in Section 2.5 are mostly classical. For historical remarks and original references on the holomorphic Riesz–Dunford functional calculus we refer to N. Dunford and J. Schwartz [**DS1**], W. Rudin [**Ru6**], F. Bonsall and J. Duncan [**BoD**], H. Dowson [**Dow**]. In particular, the spectral mapping theorem 2.5.1(b) is due to N. Dunford. For several variables holomorphic calculus for several commuting operators see N. Bourbaki [**Bour**], F.-H. Vasilescu [**Vasil**]. A calculus for several noncommuting selfadjoint operators is constructed and exploited in V. Maslov [**Masl**].

Several generalized *extended calculi* are known for various specific classes of operators. The main goal of a construction of such a calculus is to built a kind of spectral resolution using some "decomposition of unity" permitted by the calculus. As an illustration for this trend we refer to the Foiaş decomposability property proved for C_0 operators in 2.5.8(c) above. For (much) more about decomposable operators see I. Colojoara and C. Foiaş [**CF**]. Other frequently used classes, which allow extended calculi and hence some "decomposability properties", are N. Dunford spectral and scalar operators (N. Dunford and J. Schwartz [**DS3**]) and the so-called well-bounded, Hermitian and other operators (see H. Dowson [**Dow**], F. Bonsall and J. Duncan [**BoD**]).

The most important approach to extended non-holomorphic calculi was proposed by E. Dyn'kin in 1970, [**Dyn1**]. The idea is to define functions $f \longmapsto f(T)$ of a Banach space operator T for f which is initially defined on the spectrum $\sigma(T)$ only, but has an extension $\tilde{f}$ to $\mathbb{C}$ as a differentiable function with prescribed decrease of the $\bar{\partial}$-derivative,

$$\left|\frac{\partial \tilde{f}}{\partial \bar{z}}\right| \leq M(z),$$

where $M(z)$ is often given in the form $M(z) = \varphi(\text{dist}(z, \sigma(T))$. The *Dyn'kin calculus* is then defined by the Cauchy–Green formula

$$f(T) = \frac{1}{2\pi i}\int_{\partial K} \tilde{f}(z) R_z(T)\, dz - \frac{1}{\pi}\int_{K\setminus\sigma(T)} \frac{\partial \tilde{f}}{\partial \bar{z}}(z) R_z(T)\, dx\, dy,$$

where $R_z(T)$ stands for the resolvent of T and $\operatorname{int} K \supset \sigma(T)$. It is proved that functions of several standard classes, like the *Gevrey classes* $C(\{M_n\}, \sigma(T))$ (in the Whitney jet sense), admits extentions with well controlled $\overline{\partial}\tilde{f}$, and this allows to correctly define a homomorphism $f \longmapsto f(T)$, which is independent of the extension $\tilde{f}$ used, satisfies the usual calculus properties, as the spectral mapping theorem, and is hereditary with respect to the Riesz–Dunford calculus.

For reviews on the Dyn'kin calculus and several of its applications we refer to N. Nikolski [**N8**], E. Dyn'kin [**Dyn2**], [**Dyn4**], K. Kellay and M. Zarrabi [**KZ**]. It is also worth mentioning that Dyn'kin's pioneering approach was several times rediscovered for some specific classes of operators (often with no proper references), mostly for partial differential operators, and was successfully applied for spectral analysis of these operators. See E. Davies [**Davies**] for these applications and for further references.

The Szőkefalvi-Nagy–Foiaş H^∞ calculus of Section 2.2 is the main technical tool for this and the next chapters. It is based on the existence of the minimal unitary dilation and the absolute continuity of its spectral measure. Specific points of this calculus will be commented separately. For general discussions and original refernces we refer to B. Szőkefalvi-Nagy and C. Foiaş [**SzNF4**], R. Douglas [**Dou7**], N. Nikolski [**N19**], N. Nikolski and V. Vasyunin [**NVa8**]. The main idea of this approach to calculi is still fruitful for operators posessing a normal dilation, see J. Conway [**Con**].

Perturbations of spectra. This subject is one of the main streams of spectral analysis in the 20th century. Subsections 2.5.2 and 2.5.3 give a first introduction to studies of variations of the spectrum under small perturbations. Continuity of isolated parts of the spectrum as proved in 2.5.2(d) and, in particular, of isolated eigenvalues and the corresponding eigenvectors, is the most applied fact of this analysis. Usually, the applications require more, namely, differentiable or holomorphic dependences of the spectrum on the perturbation parameter. The latter is the classical observation of F. Rellich (1936), see N. Dunford and J. Schwartz [**DS1**], Ch.VII, T. Kato [**Kat**], Ch.VII or H. Baumgärtel [**Bau**]. For differentiability of an isolated eigenvalue see, for example, R. Mennicken and M. Möller [**MeM**]. Remarks on specific results presented in Subsections 2.5.2 and 2.5.3 are given several paragraphs later.

General continuity properties of the spectrum mapping $x \longmapsto \sigma(x)$ are systematically treated in Ch. Rickart [**Rick**], B. Aupetit [**Au**]; see also a survey L. Burlando [**Bur**]. The *stability problem for the (weakly) continuous spectrum* considered in Subsection 2.5.3 (i.e., the question when is

$$\sigma_{wc}(A+K) = \sigma_{wc}(A)$$

for a class of perturbations K) was initiated by H. Weyl in 1909, [**Weyl**] who proved that $\sigma_{wc}(A) = \sigma_{wc}(B)$ for bounded selfadjoint operators A, B with $K = B - A \in \mathfrak{S}_\infty$. I. Gohberg showed (1951; see [**GK1**]) that the result is still valid for any A having connected complement $\mathbb{C}\setminus\sigma(A)$ of the spectrum (and for any

$K \in \mathfrak{S}_\infty$), and N. Nikolski [**N5**] observed (1969) that without this condition the result fails (precisely, for every selfadjoint (respectively, unitary) A with $\sigma(A) = \mathbb{R}$ (respectively, $\sigma(A) = \mathbb{T}$) and for every symmetrically normed (cross-normed) ideal $\mathfrak{S}$ such that $\mathfrak{S} \neq \mathfrak{S}_1$ there exists $K \in \mathfrak{S}$ such that $\sigma_{wc}(A+K) \neq \sigma_{wc}(A)$).

In 1935, J. von Neumann [**vN2**] proved a striking converse to Weyl's theorem showing that selfadjoint operators A and B are unitarily equivalent modulo $\mathfrak{S}_\infty$ if $\sigma_{wc}(A) = \sigma_{wc}(B)$, that is,

$$A - W^*BW \in \mathfrak{S}_\infty$$

for a unitary operator W. In particular, any selfadjoint operator is unitarily equivalent modulo $\mathfrak{S}_\infty$ to an operator with pure discrete spectrum. I. Berg [**B.I**] extended these results to normal operators, and S. Kuroda [**Ku**] replaced the ideal $\mathfrak{S}_\infty$ in the von Neumann theorem by an arbitrary symmetrically normed (cross-normed) ideal $\mathfrak{S}$ such that $\mathfrak{S} \neq S_1$. These results are presented in Subsection 2.5.3 above; for more technical comments see "Formal credits" below. For further generalizations (due to C. Apostol, C. Foiaş, D. Voiculescu, and others) we refer to D. Herrero's book [**Herr2**] (especially to Theorem 3.4.9 and comments to it), and for quantitative versions of these perturbation theorems to J. Holbrook [**Hol**], and E. Gorin and M. Karahaniav [**GorK**].

For the trace ideal $\mathfrak{S}_1$ the situation is different: the Kato–Rosenblum theorem (1957) ensures that the absolutely continuous parts A_a, B_a of two selfadjoint or unitary operators are unitarily equivalent if $A - B \in \mathfrak{S}_1$. (The absolutely continuous part A_a of a selfadjoint (or unitary) operator A realized in its von Neumann model $H(A) = L^2(H(\cdot), \mu)$ (see 1.5.1(k) for the notation) is the restriction $A|H_a(A)$, where $H_a(A) = L^2(H(\cdot), \mu_a)$ and $\mu = \mu_s + \mu_a$ stands for the Lebesgue decomposition of the scalar spectral measure of A). This theorem is the basis of the mathematical scattering theory and can be found in many sources, see N. Akhiezer and I. Glazman [**AG**], T. Kato [**Kat**], B. Simon [**Si1**], D. Yafaev [**Yaf**]. It is important that the mentioned unitary equivalence of A_a and B_a can be realized by explicitely constructed unitary operators $W_\pm$, the so-called *wave operators* (or, *wave limits*)

$$W_\pm(B, A) = \lim_{t \to \pm\infty} e^{-tB} e^{tA} | H_a(A),$$

for the case of selfadjoint operators, and similarly for two unitary operators, by $W_\pm(B, A) = \lim_{n \longrightarrow \pm\infty} B^{-n} A^n | H_a$. Indeed, if these limits exist, then obviously $BW_\pm = W_\pm A$. Moreover, the *invariance principle* (M. Birman) entails that the limits $W_\pm(B, A)$ and $W_\pm(\varphi(B), \varphi(A))$ exist simultaneously and, in fact, are the same; here φ is any increasing real valued function on $\mathbb{R}$ with absolutely continuous derivative φ' satisfying some further technical conditions. Historically, wave limits first appeared in C. Moller [**Mol**] in 1945.

These results, now classical, are presented in all the above mentioned books. Moreover, they can be treated in the framework of the function model spaces and the model operators developed in this book. For these approaches to the scattering theory see L. de Branges and L. Schulman [**dBS**], P. Lax and R. Phillips [**LPh1**], S. Naboko [**Nab2**].

Beyond the framework of normal operators there is no conventional concept for the "continuous" spectrum: various definitions, which are equivalent for normal operators, differ for general ones. As examples of such definitions one can remind 1) the (weakly) continuous spectrum $\sigma_{wc}(A)$, as it is defined in Subsection 2.5.3, 2) the essential (or Fredholm) spectrum $\sigma_{ess}(A)$ (see Section B.2.1, Volume 1, for the definition), 3) the continuous spectrum $\sigma_c(A)$ (see B.4.2.1, Volume 1, for the definition), and 4) the *approximate spectrum*

$$\sigma_{ap}(A) = \{\lambda \in \mathbb{C} : \text{ there exists an orthonormal sequence } (e_k) \text{ such that } \lim_k \|(\lambda I - A)e_k\| = 0\}.$$

We choose the first of these notions because of the particular interest of isolated eigenvalues for applied perturbation theory; see for example T. Kato [**Kat**], especially theorems related to the so-called Weinstein–Aronszajn formulas (chapter 4, §6).

The stability problem $\sigma_{wc}(A) = \sigma_{wc}(A + K)$ for the case of generic trace class perturbations $K \in \mathfrak{S}_1$ and a unitary initial operator A is considered by N. Nikolski (1969). It turns out that such a stability holds for the entire class $\mathfrak{S}_1$ if and only if it holds for all rank one perturbations, and if and only if A has a measurable "spectral gap" of positive Lebesgue measure (see 2.5.3(i) for the exact statement). Later on, N. Makarov and V. Vasyunin obtained a similar result in its natural generality, namely for operators A which, in turn, are trace class perturbations of unitary operators, i.e., such that $A - U \in \mathfrak{S}_1$ for a unitary U (see 2.5.3(j) for the statement and "Formal credits" below for references).

The (in)stability problems for several other parts of the spectrum are also studied; we refer to D. Herrero [**Herr2**] and C. Apostol, L. Fialkow, D. Herrero, and D. Voiculescu [**AFHV**] for a survey and references. For the (in)stability problem for the point spectrum $\sigma_p(A)$ see also L. Nikolskaia [**N.L**], D. Herrero, Th. Taylor, and Z. Wang [**HTW**] and further references in the latter paper.

The commutant lifting theorem (the CLT, in short) of Section 2.4 is the peak point of the approach based on unitary dilations. In an abridged form it states that an operator A commuting with a contraction T, $AT = TA$, can be lifted up to an operator $Y : \mathcal{H} \longrightarrow \mathcal{H}$ acting on the space of the minimal isometric dilation of T, say $V : \mathcal{H} \longrightarrow \mathcal{H}$, in such a way that $YV = VY$ and $A = P_H Y|H$. As mentioned above, there is a form of this theorem symmetric with respect to T and A (the Ando dilation theorem, 1.5.8(b) above), and another "intertwining" form lifting an operator A such that $AT_1 = T_2A$ up to an operator Y intertwining isometric dilations of T_1 and T_2, see Subsection 2.4.4. However, a conceivable symmetric form of the latter result, i.e., existence of commuting dilations for three commuting contractions, is already false (see S. Parrott's example 1.5.8(c)).

The most important fact for applications is the functional structure of liftings Y, see 2.5.7 for a sample of such applications. The links between the above lifting process and the Nehari theorem B.1.3.2, Volume 1, giving a Laurent extension (lifting) of a Hankel type form are also fundamental. We refer to B. Sz.-Nagy and C. Foiaş [**SzNF4**] for more about liftings, to N. Nikolski [**N19**], N. Nikolski and V. Vasyunin [**NVa8**], and V. Pták and P. Vrbová [**PV**] for the interplay between Nehari type theorems and the CLT, and to M. Cotlar and C. Sadosky [**CS3**], [**CS4**], [**CS5**], C. Sadosky [**Sad2**] (a survey with further references) for an extensive theory unifying these two techniques. A concise survey of the CLT technique and some of its applications is contained in C. Foiaş' address [**Foi3**], and a far reaching extension theory, generalizing the CLT, can be found in V. Paulsen [**Pau**] (see also Section 1.6 above).

It is also worth mentioning the fundamental role of the CLT and related techniques for control theory and constrained interpolation of the Carathéodory and

Nevanlinna–Pick type by analytic, rational, and other functions. Some basic examples of such applications are given in chapter B.3 (Volume 1), for some others see Chapter 3 below. For more about these vast fields of applications and related techniques, like the Adamyan–Arov–Krein step-by-step extension process (see Chapter B.1, Volume 1), choice sequences by C. Apostol, C. Foiaş, et al., Schur parameters, etc., we refer to C. Foiaş and A. Frazho [**FF**], J. Ball, I. Gohberg, and L. Rodman [**BGR**], M. Bakonyi and T. Constantinescu [**BaC**], D. Alpay [**Alp**], and N. Nikolski [**N19**]. A version of the CLT in terms of the free function model of Chapter 1 is presented in N. Nikolski and V. Vasyunin [**NVa8**]; free parameters of the liftings are also studied.

Characteristic function as an analysis tool. The characteristic function Θ_T is the central object of the function model theory. The canonical factorization of Θ_T, the Riesz–Smirnov parameters of this factorization (see Chapter A.3, Volume 1), holomorphic extendability properties, - all these function theory subjects (scalar, matrix, or operator valued) obtain adequate interpretations in the language of spectral properties of the given contraction T. This cross fertilization of operator theory and complex analysis is the main advantage of the function model approach to the study of Hilbert space operators. The known achievements of this approach, in scattering theory (see references above), in the spectrum stability problem (see above), invariant subspaces (see below), similarity theory (see 2.5.5), etc., verify the above statements. In particular, our treatment of the free interpolation problem in H^∞ and H^2, Chapter 3 below, is also based on the interplay between spectral subspaces of a model operator and their complex analysis meaning.

Historically, the first appearance of the characteristic function (and the very term) was in M. Livshic's paper of 1946 [**Liv2**]. In particular, the spectrum of an operator was identified with the zero set of the characteristic function, but in its full generality Theorem 2.3.4 was proved much later (1964) by B. Sz.-Nagy and C. Foiaş and H. Helson, see [**N19**], [**SzNF4**] for original references. The characteristic function was extensively studied in the 1950s and 1960s by the Soviet mathematical school (M. Brodskii, M. Livshic, M. Krein, V. Potapov, and many others) in its dissipative (halfplane) framework. In particular, it was identified with the transfer function coming from system and control theory, see M. Brodskii and M. Livshic [**BrL**] and M. Livshic [**Liv3**], as well as with the scattering matrix in the framework of scattering theory, P. Lax and R. Phillips [**LPh1**]. Then, the notion of the characteristic function was included in the B. Szőkefalvi-Nagy and C. Foiaş approach to model operators (in its contractive realization). See the same references as in the *General information.*

Invariant subspaces. Invariant subspaces of Hilbert space operators, their classification and description, play the central role in operator theory, especially when the latter is based on the function model approach. Indeed, all kinds of spectral decompositions, from classical orthogonal resolutions of identity (see 1.5.1(l)) up to Dunford-Schwartz spectral operators, or Foiaş decomposable operators, use invariant subspaces (as maximal spectral subspaces defined in 2.5.1(c) above). Moreover, descriptions of invariant subspaces in terms of factorizations of the characteristic function (whose partial case is presented in Section 2.1) make it possible to use Hardy space techniques in order to treat these spectral decompositions. Some concrete results on invariant subspaces are commented later on. Here we only quote the Sz.Nagy-Foiaş theorem (general version) giving a description of Lat M_Θ. For details and complete references see [**SzNF4**].

Recall first that a representation $\Theta_T = \Theta = \Theta_2\Theta_1$ is called *regular factorization* if the functions $\Theta_1 \in H^\infty(E \longrightarrow F)$, $\Theta_2 \in H^\infty(F \longrightarrow E_*)$ are contraction valued and the operator Z defined below is unitary:

$$Z(\Delta u) = \begin{pmatrix} Z_1 \\ Z_2 \end{pmatrix} \Delta u = \begin{pmatrix} \Delta_1 u \\ \Delta_2 \Theta_1 u \end{pmatrix},$$

where $\Delta_i = (I - \Theta_i^*\Theta_i)^{1/2}$, and all functions are regarded as multiplication operators on the corresponding L^2 spaces. Now, the Sz.Nagy-Foiaş *invariant subspace theorem* can be stated as follows.

THEOREM. *The following formula establishes a bijection between the set of all regular factorizations* $\Theta_T = \Theta = \Theta_2\Theta_1$ *and the lattice* Lat(M_Θ) *of invariant subspaces of a model operator* M_Θ*:*

$$\mathcal{K}_{\Theta_2,\Theta_1} = \begin{pmatrix} \Theta_2 \\ Z_2^*\Delta_2 \end{pmatrix} H^2(F) \oplus \begin{pmatrix} 0 \\ Z_1^* \end{pmatrix} \operatorname{clos} \Delta_1 L^2(E) \ominus \begin{pmatrix} \Theta \\ \Delta \end{pmatrix} H^2(E).$$

The spectral synthesis property 2.5.8(g) is from [**N19**]. For general comments about this property we refer to N. Nikolski [**N11**], [**N8**], [**N19**]. It is worth mentioning that there exists a lot of Hilbert space operators $T : H \longrightarrow H$ with a complete family of eigenvectors

$$\mathcal{X} = (x_\lambda)_{\lambda\in\sigma_p}, \quad \sigma_p = \sigma_p(T),$$

but without the synthesis property (SP, for short). For example, take the backward shift $T = S^* : H^2 \longrightarrow H^2$ on the Hardy space, and choose an invariant subspace $\mathcal{K}_\Theta = H^2 \ominus \Theta H^2$, where Θ is not a Blaschke product; then $T|\mathcal{K}_\Theta$ is not a complete operator (or, equivalently, $\mathcal{K}_\Theta$ is not generated by eigenvectors of S^* contained in it), and therefore, SP fails. If a family $\mathcal{X}$ is minimal and

$$\mathcal{X}' = (x'_\lambda)_{\lambda\in\sigma_p}$$

is its biorthogonal family (x'_λ are eigenvectors of T^*), then the necessary and sufficient condition for SP is that it is hereditarily complete (see A.5.6.4, Volume 1, for the definition), i.e., $H_\sigma = H^\sigma$ for every $\sigma \subset \sigma_p$, where

$$H_\sigma = \operatorname{span}(x_\lambda : \lambda \in \sigma), \quad H^\sigma = \{x \in H : x \perp x'_\lambda, \lambda \in \sigma_p \backslash \sigma\},$$

(A. Markus, [**Ma2**]); see also survey papers N. Nikolski [**N11**], [**N8**]. In particular, $\mathcal{X}'$ should be a complete family if T satisfies the SP. In the beginning of the 1950s H. Hamburger constructed complete compact operators with span $\mathcal{X}' \neq H$, and hence, without SP (see A.5.7.1, Volume 1, and [**N8**] or [**N11**] for further references). The first example of a compact operator T satisfying span $\mathcal{X}$ = *span*$\mathcal{X}' = H$ but not allowing the SP is shown in A. Markus [**Ma2**]; it is also proved that all known sufficient completeness conditions (in terms of the resolvent growth and values of the quadratic form of an operator), in fact, guarantee SP. Later on, many simple examples of (bi)complete minimal families avoiding the SP were provided by L. Dovbysh, N. Nikolski, and V. Sudakov [**DN**], [**DNS**]. Moreover, it is shown that a restriction $T|E$ of a complete compact operator on an invariant subspace $E \in$ Lat(T) can be "arbitrary", which means that for any compact operator $A : E \longrightarrow E$ there exists a *complete* compact operator $T : H \longrightarrow H$ such that $A = T|E$ (it is supposed that $\dim(H \ominus E) = \infty$), see N. Nikolski [**N4**].

For a complete normal operator N, the spectral synthesis problem is transformed into the question whether all invariant subspaces of N are reducing. For this property see A.1.6.1, A.1.6.2, A.5.7.8 (Volume 1) and references therein. Finally, observe that 2.5.8(g), A.5.7.8(c) and A.5.7.8(b)(iii) all together imply the following.

THEOREM. *The following properties of a symmetrically normed ideal $\mathfrak{S}$ of compact operators are equivalent:*

i) every complete contraction T with $I - T^*T \in \mathfrak{S}$ and $\mathbb{D} \not\subseteq \sigma(T)$ satisfies SP;

ii) $\mathfrak{S} = \mathfrak{S}_1$.

To overcome the above mentioned lack of the SP, an *approximate synthesis* concept is proposed in N. Nikolski [**N13**]. It consists of proving, or disproving, the possibility of representing every invariant subspace E of a given complete operator $T : X \longrightarrow X$ as a *lower limit of a sequence of subspaces E_n*,

$$E = \underline{\lim}_n E_n = \{x \in X : \lim_n \operatorname{dist}(x, E_n) = 0\},$$

such that $E_n \in \operatorname{Lat}(T)$ and $\dim E_n < \infty$ for every $n, n \geq 1$. (Note that every E_n is spanned by eigen- and rootvectors of T; SP corresponds to the case where $E_n \subset E$ for every $n \geq 1$). It is shown in N. Nikolski [**N19**], M. Gribov and N. Nikolski [**GrN**], and in S. Shimorin [**Shim2**] that invariant subspaces of the backward shift S^* allow such an approximate synthesis for many spaces of holomorphic functions, including Hardy spaces H^p and Bergman spaces $L^p_a(\mathbb{D})$ (S. Shimorin). A visible obstacle for the approximate synthesis of invariant subspaces of a given operator T is their *index* $i(E) = \dim(E/\overline{TE})$: if $i(E) > 1$ then E can not be approximated by finite dimensional subspaces $E_n \in \operatorname{Lat}(T)$, since $i(E_n) = 1$ for every n, and the index is stable with respect to lower limits; see N. Nikolski [**N23**]. For further references we refer to the latter four sources. For a discussion and references on the index phenomenon, see Section A.1.7 (Volume 1).

Some general remarks on invariant subspaces have to be made. The general problem whether $\operatorname{Lat}(T)$ is nontrivial, i.e. $\operatorname{Lat}(T) \neq \{0, H\}$, for every bounded operator $T : H \longrightarrow H$ on a Hilbert space H with $\dim H > 1$ seems still to be unsolved. Several different techniques are used to prove the existence of invariant subspaces for specific classes of operators. For status reports until 1974 we refer to surveys N. Nikolski [**N8**], M. Naimark, A. Loginov, and V. Shulman [**LNS**], H. Radjavi and P. Rosenthal [**RR**], and C. Pearcy and A. Shields [**PSh**]. For more recent developments see C. Pearcy [**Pea**], H. Bercovici, C. Foiaş, and C. Pearcy [**BFP**], A. Simonic [**Simo**], J. Esterle and A. Volberg [**EV**], K. Kellay and M. Zarrabi [**KZ**], and L. de Branges [**dB7**].

The study of invariant subspaces is a very vast field of operator theory; below, we will briefly mention three of the many interesting questions related to this topic: when $\operatorname{Lat}(T)$ is linearly ordered by inclusion (so, T has "a few invariant subspaces"); when T is *reflexive* (and so, T has "many invariant subspaces"), that is,

$$\operatorname{alg} \operatorname{Lat}(T) = \operatorname{alg}(T),$$

where $\operatorname{alg} \operatorname{Lat}(T) = \{A \in L(H) : \operatorname{Lat}(A) \supset \operatorname{Lat}(T)\}$ and $\operatorname{alg}(T)$ is the WOT closed algebra generated by T; and, when $\operatorname{Lat} \varphi(T) = \operatorname{Lat}(T)$ for a function of T. See also some remarks in the paragraph *"The class C_0"* below in this section.

The first problem (unicellularity) is shortly commented below in the "Formal credits" part.

The reflexivity problem was intensively studied from the beginning of the 1970s, see H. Radjavi and P. Rosenthal [**RR**]. In the function model language, the problem was reduced (mostly by P. Wu) to the case of a scalar inner characteristic function, where it was solved by V. Kapustin as follows (see V. Kapustin and A. Lipin [**KLi2**] or V. Kapustin [**Kap1**]).

THEOREM. *Let Θ be a scalar inner function; then the following are equivalent.*

1) The model operator M_Θ is reflexive.

2) The linear hull of idempotent elements of the algebra $H^\infty/\Theta H^\infty$ is weak dense in $H^\infty/\Theta H^\infty$.*

3) $\Theta = BS_\mu$, where B is a Blaschke product with simple zeros and the representing measure μ of the singular part S_μ vanishes on every Beurling–Carleson subset of $\mathbb{T}$ of Lebesgue measure zero (see 2.5.7(j) for the definition).

For many other details and references related to the reflexivity property we refer to the interesting and complete survey papers by V. Kapustin and A. Lipin [**KLi1**] and [**KLi2**].

The third problem (when $\operatorname{Lat}\varphi(M_\Theta) = \operatorname{Lat}(M_\Theta)$) is related to weak generators of the algebra $H^\infty/\Theta H^\infty$. It is clear that $\operatorname{Lat}\varphi(M_\Theta) \supset \operatorname{Lat}(M_\Theta)$ for every $\varphi \in H^\infty$ and every Θ. On the other hand, if the coset $\varphi + \Theta H^\infty$ of a function $\varphi \in H^\infty$ is a weak*generator of $H^\infty/\Theta H^\infty$ (in short, φ is a *weak generator*), then $\operatorname{Lat}\varphi(M_\Theta) \subset \operatorname{Lat}(M_\Theta)$. For a scalar inner function Θ, the converse is also true, that is, $\operatorname{Lat}\varphi(M_\Theta) = \operatorname{Lat}(M_\Theta)$ if and only if φ is a weak generator of $H^\infty/\Theta H^\infty$, see R. Frankfurt and J. Rovnyak [**FR1**]. The problem of weak generators is to describe φ satisfying this property (still open). As follows from 2.1.9-2.1.14, a special case of this problem (for $\Theta = \exp(-a\frac{1+z}{1-z})$) is to describe the measures μ on an interval $(0, a)$ such that the convolution operator C_μ,

$$C_\mu f(x) = \int_0^x f(x-t)\,d\mu(t),$$

is *unicellular* on $L^2(0,a)$ (or, equivalently, C_μ is a weak generator of the algebra of all convolution operators on $L^2(0,a)$). The problem is still open even for $d\mu(t) = k(t)\,dt$, $k \in L^1(0,a)$. The condition $\inf \operatorname{supp}(\mu) = 0$ is necessary for C_μ to be a weak generator but far from being sufficient (J. Ginsberg and D. Newman [**GiN**]).

Many sufficient conditions to be a generator of $H^\infty/\Theta H^\infty$, open problems of the field and references are listed in R. Frankfurt and J. Rovnyak [**FR2**] and M. Gamal [**Gama1**], [**Gama2**]. In particular, it is shown in [**Gama1**] that for a Blaschke product with simple zeros $\Theta = \prod_{\lambda\in\Lambda} b_\lambda$ a necessary condition for a function $\varphi \in H^\infty$ to be a generator of $H^\infty/\Theta H^\infty$ is that the restriction $\varphi|\Lambda$ is a weak generator of the algebra $l^\infty(\Lambda)$; these are completely described in D. Sarason [**S5**] ($\varphi|\Lambda$ should be an injection and the range sequence $\varphi(\Lambda)$ should not be dominating for any Jordan domain). It is also proved that for Carleson (interpolating, see chapter 3 below) sets Λ this condition is also sufficient; an example of a union of two Carleson sequences is given where this criterion fails.

Our last remark is that, of course, if φ is a weak generator of the entire algebra H^∞ it is *a fortiori* a weak generator of $H^\infty/\Theta H^\infty$; generators of H^∞ are described in D. Sarason [**S3**]. However, as is shown by B. Solomyak, there exists Θ and a generator φ of $H^\infty/\Theta H^\infty$ such that the coset $\varphi + \Theta H^\infty$ contains no generators of H^∞. For more information see sources mentioned above and also N. Nikolski [**N19**], Chapter 4.

Formal credits. Translation invariant subspaces of $L^2(E)$, as treated in Section 2.1, are described by P. Lax [**Lax2**]. For a similar representation of invariant subspaces of an arbitrary unitary operaor U written in its von Neumann spectral form on a space $L^2(H(\cdot),\mu)$ see N. Nikolski [**N1**], R. Douglas [**Dou7**]. About invariant subspaces of a general model space $\mathcal{K}_\Theta$ see comments above. For original references for the canonical inner-outer factorization from 2.1.22, see B. Sz.-Nagy and C. Foiaş [**SzNF4**]; see also comments in A.1.7 (Volume 1).

The Gelfand problem considered in Subsections 2.1.9-2.1.14 is raised in [**Gel1**] and has a long history. It is clear that the problem amounts to prove (or disprove) that the lattice $\mathrm{Lat}(V)$ is linearly ordered by inclusion, that is, for every pair of two subspaces $E, E' \in \mathrm{Lat}(V)$, either $E \subset E'$ or $E' \subset E$. Operators having linearly ordered lattice Lat are called *unicellular*. An operator $V : X \longrightarrow X$ on a finite dimensional space X is unicellular if and only if the canonical Jordan form of V consists of one Jordan block.

The Gelfand problem was first solved by S. Agmon [**Agm**] in 1949 who showed that $\mathrm{Lat}(V) = \mathrm{Lat}(T_\lambda)_{\lambda>0}$ and that this lattice is linearly ordered (V is considered on any space $L^p(0,a)$ or $C([0,a])$); here

$$T_\lambda f(x) = f(x-\lambda) \text{ for } x \geq \lambda \text{ and } T_\lambda f(x) = 0 \text{ for } x < \lambda.$$

Independently, pure operator theory proofs were found by M. Brodskii [**Bro1**] and L. Sakhnovich [**Sakh**]. W. Donoghue [**Don1**] gave a proof based on the *Titchmarch convolution theorem* ($\inf\mathrm{supp}(f*g) = \inf\mathrm{supp}(f) + \inf\mathrm{supp}(g)$ if both the latter quantities are finite). Later on, G. Kalisch [**Kal**] and M. Brodskii [**Bro2**] found elementary proofs and deduced from them the Titchmarch theorem itself. V. Daniel [**Dan**] extended these results (including Agmon's one) to some weighted spaces $L^p(\mathbb{R}_+, w\,dx)$, and R. Leggett [**Leg**] described invariant subspaces of the integration operator

$$Af(x) = \int_0^x f(t)\,d\mu(t)$$

acting on $L^2((0,a),\mu)$. For a function model proof of this result see N. Nikolski and V. Vasyunin [**NVa8**]. The function model proof of Corollaries 2.1.13 and 2.1.14 (and hence a solution of the Gelfand problem) was first given by D. Sarason [**S4**].

In the 1960ies and 1970ies, unicellularity of operators of Volterra type

$$Vf(x) = \int_0^x k(x,t)f(t)\,dt$$

was intensively studied for kernels k satisfying some conditions. Many of the results obtained were based on abstract unicellularity criteria proved by that time, mostly due to M. Brodskii and G. Kisilevskii. For the latter criteria see M. Brodskii [**Bro3**], and for a survey of the above mentioned results (and many others, related to unicellularity or not) and an extensive bibliography, N. Nikolski [**N8**]. A unicellularity criterion for C_0 operators is obtained by B. Sz.-Nagy and C. Foiaş, see [**SzNF4**] or [**N19**].

Theorem 2.1.17 on invariant subspaces is due to H. Helson and T. Srinivasan (see [**Hel**]) but the proof is from N. Nikolski [**N19**], where a similar fact is proved for any function lattice on the circle $\mathbb{T}$ instead of $L^p(\mathbb{T})$.

H^∞ calculus, von Neumann's inequality and spectral sets. The propositions presented in Section 2.2 and Subsections 2.5.4 and 2.5.7 are classical subjects of dilation and model theory. As explained in Chapter 1 (see 1.5.9 and 1.6), the Sz.-Nagy-Foiaş H^∞ calculus depends only on the existence of a unitary dilation with absolutely continuous spectrum. Comments on these matters can be found in Section 1.6. In Subsection 2.5.4 of this chapter we gathered several facts about the polynomial and rational calculi and the von Neumann inequality, which are independent of the dilation techniques. (For those depending on dilations, and in particular for several commuting operators, see 2.5.6 and also Sections 1.5.9 and 1.6 above). It is curious to note that the very first use of the H^∞ calculus in spectral theory is for A. Plessner (1939) (for the case of partially isometric operators), see [**Pl2**] and references therein.

The sharp estimates for H^∞ functions of M_Θ and for elements of the commutant $\{M_\Theta\}'$ given in 2.5.7(a)-(f) are mostly from N. Nikolski and S. Hruschev [**HrN**]. The use of the Carleson corona theorem is the key point of these calculations. The local estimates of 2.5.7(f), initiated in [**HrN**], differ to each other by their nature: (i) is elementary, whereas the estimates of (ii) and (iii) depend on an analysis of the Carleson contours involved in Carleson's proof of the corona theorem. The latter involvement is also important for the proof of S. Treil's *stable rank 1 theorem* (see [**Tre12**]) which strengthens the corona theorem: for given $\varphi, \Theta \in H^\infty$ with $\inf_{\mathbb{D}}(|\varphi| + |\Theta|) > 0$, there exist an *invertible* function f in H^∞ and a function $g \in H^\infty$ such that $f\varphi + g\Theta = 1$. Therefore, an invertible element of $\{M_\Theta\}'$ can be lifted up to an invertible element of H^∞.

An excellent introduction to von Neumann's spectral sets can be found in F. Riesz and B. Sz.-Nagy [**RSzN**]; see also J. Conway [**Con**], and for several variables versions V. Paulsen [**Pau**]. The properties 2.5.4(a), (b), and (c) are proved by J. von Neumann [**vN5**], and 2.5.4(d)(i) by C. Foiaş [**Foi1**]. Still for Hilbert spaces, several approaches were proposed to treat von Neumann spectral sets and von Neumann type inequalities. For approaches based on dilation techniques and harmonic analysis see Subection 1.5.9 and Section 1.6 above. The fact that the disk $|z| \le \|T\|$ is a spectral set for any Hilbert space operator T provoked the conjecture that the annulus

$$a(T) = \{z \in \mathbb{C} : r = \|T^{-1}\|^{-1} \le |z| \le R = \|T\|\}$$

is a spectral set for an invertible operator T. This was disproved by J. Williams (1967) (even for the case $r < R$), but A. Shields proved in 1974 that $a(T)$ is a K-spectral set (as defined in 2.5.4), where $K = 2 + \big((R+r)/(R-r)\big)^{1/2}$. See A. Shields [**Sh1**] for proofs and further references.

Nontrivial calculus estimates for some other classes of Hilbert space operators, such as power bounded and power like operators, were initiated by N. Varopoulos [**Var2**] and continued by V. Peller and others; see V. Peller [**Pe3**] and references therein.

The estimates 2.5.4(d) (ii) and (iii) are Banach space von Neumann type inequalities. Theorem 2.5.4(d)(ii) is proved by H. Bohr [**Boh**] as early as in 1914; about the participation of M. Riesz, I. Schur and F. Wiener, as well as for a new approach to the problem, see L. Aizenberg, A. Aytuna and P. Dyakov [**AizAD**]. This estimate, as well as 2.5.4(d)(iii), were rediscovered in V. Katsnel'son and V. Matsaev [**KM**], and our proof follows these authors. N. Dunford proved that an operator T having a real spectrum and acting on a weakly complete Banach space is a scalar type spectral operator (i.e., is similar to a selfadjoint operator) if and only if its spectrum $\sigma(T)$ is a K-spectral set for T, see N. Dunford and J. Schwartz [**DS3**], Chapter 17. Some particular functions of Banach space operators attract special attention. For example, the norms of associated matrices $(\det A)A^{-1}$ were studied (a problem suggested by B. van der Waerden). It is proved by J. Schäffer and by E. Gluskin, Y. Meyer, and A. Pajor that

$$a\sqrt{n} \le k_n \le b\sqrt{n}$$

for some constants $a, b > 0$, where

$$k_n = \sup\Big[\sup\big\{|\det T|\cdot\|T^{-1}\| : T : X \longrightarrow X, \|T\| \le 1\big\} : \dim X = n\Big]$$

and the exterior supremum is taken over all normed spaces X of dimension n. Clearly, $k_n = \sup\big\{\frac{|\det A|}{\|A\|^{n-1}}\cdot\|A^{-1}\| : A : X \longrightarrow X, \dim X = n\big\}$. It is also shown that $k_n = r_n(W_A)$, where W_A is as in 2.5.4(d) and

$$r_n(Y) = \sup\Big[\inf\big\{\|f\| : f \in Y, f = B_E g, g(0) = 1\big\} : E \subset \mathbb{D}, \#E = n\Big]$$

for a given Banach space $Y \subset \mathrm{Hol}(\mathbb{D})$; $B_E = \prod_{\lambda\in E} b_\lambda$ stands for the Blaschke product with zero set E. The quantities $r_n(Y)$, and some other quantities related to function calculi are studied for various Banach spaces of holomorphic functions. See I. Videnskii and N. Shirokov [**ShV**] for recent results and further references (in particular, for authors mentioned above).

For the case of particular Banach spaces, it is worth mentioning the *Matsaev conjecture* claiming that

$$\|f(T)\| \le \|f(S)\| = \|f\|_{\mathrm{Mult}(l^p)}$$

for every contraction $T : L^p \longrightarrow L^p$ and every polynomial f, where $S : l^p \longrightarrow l^p$ is the shift operator on the space l^p and $\mathrm{Mult}(l^p)$ stands for l^p-(Fourier-) multipliers (see B.4.7.2, Volume 1). The conjecture is still unsolved; it has first appeared as early as in 1971 in N. Nikolski [**N7**], where the problem of L^p estimates is discussed. The conjecture is justified for $p = 1, \infty$ (classical), for $p = 2$ (J. von Neumann), as well as for isometries on any space L^p (A. Kitover, 1975), for absolute contractions T ($\|T\|_{L^1\longrightarrow L^1} \le 1$ and $\|T\|_{L^\infty\longrightarrow L^\infty} \le 1$) (V. Peller (1978), R. Coifman, R. Rochberg and G. Weiss (1978)), and for some other classes of L^p contractions. See V. Peller [**Pe2**] for references and more information. For some other special estimates, including the Kreiss matrix inequality, see D.1.6.3 and D.1.7 below.

The commutant lifting theorem of Section 2.4 is commented above; see also comments in Section 1.6.

The class C_0 and the spectrum of M_Θ are discussed in Section 2.3 and Subsection 2.5.8. A function $\varphi \in H^\infty$ such that $\Omega\Theta_T = \varphi\cdot I_E$ and $\Theta_T\Omega = \varphi\cdot I_{E_*}$, where $\Omega \in H^\infty(E_* \longrightarrow E)$, is called *scalar multiple* of the characteristic function Θ_T. The property 2.3.2(3) shows that in case $T \in C_{00}$, the existence of a scalar multiple is equivalent to the existence of an annihilating function. The latter always implies that $T \in C_0$ (see 2.5.8(a)), but the former does not (if $T \notin C_{00}$). If $T \notin C_0$ but the scalar multiple exists, then it can play the same role as an annihilator: to determine the spectrum, (some) invariant subspaces, etc. In particular, weak contractions (i.e., such that $\mathbb{D} \not\subseteq \sigma(T)$ and $I - T^*T \in \mathfrak{S}_1$) have scalar multiples, see B. Sz.Nagy-C. Foiaş [**SzNF4**], N. Nikolski [**N19**], H. Bercovici [**Be**]. The fact that $(\det\Theta_T)(T) = 0$ for a weak contraction from C_{00} (see 2.5.8(e)(vii)) is an analogue of the classical Hamilton–Cayley theorem for matrices. The very term of weak contractions was introduced by M. Krein in [**Kr6**].

The properties of C_0-contractions collected in Section 2.3 and Subsection 2.5.8 are mostly due to B. Sz.-Nagy and C. Foiaş and H. Bercovici; we again refer to [**SzNF4**], [**SzN5**], [**Be**], and [**N19**] for proofs and original references. The main ingredient of this theory is the theory of Jordan operators presented in 2.5.8(e); see [**SzN5**], [**N19**], and (especially) [**Be**] for proofs and references. In fact, quasi-similarity of a C_0 operator to its Jordan model can be strengthen to *pseudo-similarity* defined and studied by V. Kapustin and V. Lipin [**KLi1**], [**KLi2**] (see Section 1.6); in particular, the lattices Lat T_1, Lat T_2 of two pseudo-similar operators are isomorphic. M. Gamal [**Gama3**] proved that the lattices of two quasi-similar *weak* contractions T_1, T_2 are isomorphic (T_1 and T_2 may be not pseudo-similar).

References and additional information for 2.5.8(d), 2.5.8(f) and 2.5.8(h) can be found in H. Bercovici [**Be**] and N. Nikolski [**N19**].

Theorem 2.3.4 is due to B. Sz.-Nagy and C. Foiaş (1964), see [**SzNF4**] for original references. However, in the framework of the "dissipative" Livshic–Brodskii theory (see above) the fact was known much earlier. For instance, for scalar characteristic functions Θ, it was proved already in 1946, M. Livshic [**Liv2**] (and then rediscovered for the unit disk situation by J. Moeller [**Moe**]).

The spectral synthesis property 2.5.8(g) is commented above.

For an explicit description of *maximal spectral subspaces* of a scalar model operator see Subsection 3.3.6 below. The proof of the Foiaş decomposability property given in 2.5.8(c) is different from the original one (for which we refer to [**SzNF4**] and [**Be**]); in particular, we have used an elementary version of the corona theorem only (see B.9.3.3, Volume 1), and not its complete statement as is the case in the original proof. For a survey on decomposable operators see M. Putinar [**Pu**]. The interesting property 2.5.8(h) (ii) is proved by E. Nordgren [**Nor**]; for this and related results see also H. Bercovici [**Be**].

Subsection 2.5.1 is commented above.

General comments on Subsection 2.5.2 are given above. For properties of the Hausdorff metric we refer to F. Hausdorff [**Hau**] and T. Kato [**Kat**]. The first sytematic analysis of continuity properties of the spectra in a Banach algebra is contained in the work of J. Newburgh [**New**]. Much later, D. Herrero, based on results of C. Apostol, S. Kakutani, B. Morrel, and many others, described the continuity points of the spectrum mapping $T \longmapsto \sigma(T)$ of the algebra $L(H)$. For these results, and for much more, as well as for nice historical accounts of the subject, see the books by C. Apostol, L. Fialkow, D. Herrero, and D. Voiculescu [**AFHV**] (especially, Notes to Chapter 14), B. Aupetit [**Au**], and a survey M. Burlando [**Bur**]. We list some other properties of the mappings $x \longmapsto \sigma_A(x)$, $x \longmapsto r_A(x)$, which can be found in the aforementioned sources:

1) the function $z \longmapsto \log(r_A(ze-x)^{-1})$ is subharmonic on $\mathbb{C}\backslash\sigma(x)$ (*E. Vesentini*; in fact, it is obvious, since $r(\cdot)$ is a seminorm on the commutative subalgebra $A_x \subset A$ generated by the resolvent $(ze-x)^{-1}$, $z \in \mathbb{C}\backslash\sigma(x)$);

2) there exists a Banach algebra A such that $x \longmapsto r_A(x)$ is continuous and $x \longmapsto \sigma_A(x)$ is not *(C. Apostol)*;

3) the mapping $x \longmapsto r_A(x)$ is uniformly continuous if and only if $x \longmapsto \sigma_A(x)$ does, and if and only if $A/\operatorname{Rad}(A)$ is commutative *(B. Aupetit; V. Pták and J. Zemanek)*;

4) if $f : L(H) \longrightarrow Y$ is a continuous mapping into a topological space Y, such that $f(UAU^{-1}) = f(A)$ for every $A \in L(H)$ and every invertible operator U, then $f = \text{const}$ (D. Herrero).

See also the theorem of R. Bhatia, Ch. Davis, and A. McIntosh about the uniform closeness of the spectra of normal operators quoted in Section B.7.6 (Volume 1).

From another series on variations of spectra, it is worth mentioning the so-called *ε-pseudospectrum*, defined by

$$\sigma_\varepsilon(T) = \{\lambda \in \mathbb{C} : \|R_\lambda(T)\| \geq \frac{1}{\varepsilon}\}$$

for an operator $T : X \longrightarrow X$. The ε-pseudospectrum, important in applied mathematics, often exhibits better continuity properties than the usual spectrum. In particular, it is shown that $\sigma_\varepsilon(T) = \bigcup_{\|T'\| \leq \varepsilon} \sigma(T + T')$ and

$$\sigma_\varepsilon(T_\varphi) = \lim_n \sigma_\varepsilon(P_n T_\varphi | \mathcal{P}_n) \text{ for every } \varepsilon > 0,$$

where $P_n T_\varphi | \mathcal{P}_n$, $n \geq 1$, stand for finite section approximations of a Toeplitz operator T_φ with piecewise continuous symbol φ (this continuity property does not hold for the usual spectrum); see A. Böttcher and B. Silbermann [**BSi3**]. See also Notes and Remarks B.4.8 (Volume 1) for the finite section method.

The example 2.5.2(e)(i) belongs to mathematical folklore; the fact that the operator $T_n = S \oplus S^* + K_n$ of this example is similar to a unitary operator can be seen directly (without Sz.-Nagy's theorem): T_n is a weighted bilateral shift $S\Lambda$ on $l^2(\mathbb{Z})$ with $\lambda_k = 1$ for $k \neq -1$, so it is similar to the standard (flat) shift S (since the products $\prod_{k<0} |\lambda_k|, \prod_{k \geq 0} |\lambda_k|$ converge). S. Kakutani's example 2.5.2(e)(ii) was first published in C. Rickart [**Rick**]; our proof exploits some simple facts about weighted shifts $S\Lambda$ (points a), b), c)) which are known from quite long ago, see for example N. Nikolski [**N3**], [**N9**], A. Shields [**Sh1**]. It is worth adding that for a unilateral weighted shift $T = S\Lambda$, the spectrum $\sigma(T)$ is always the disk $\{z \in \mathbb{C} : |z| \leq r(T)\}$, and the approximate point spectrum is the annulus $\{z \in \mathbb{C} : r_1(T) \leq |z| \leq r(T)\}$, see 2.5.2(e) for the definition of $r_1(T)$.

General sources for further references and comments about Subsection 2.5.3 are I. Gohberg and M. Krein [**GK1**], N. Akhiezer and I. Glazman [**AG**], T. Kato [**Kat**]. The properties 2.5.3(b)-(d) are due to I. Gohberg, see references in [**GK1**]. Proposition 2.5.3(e) is a special case of the theory of essentially normal operators by L. Brown, R. Douglas and P. Fillmore, [**BDF**]. An operator $T : H \longrightarrow H$ is *essentially normal* if $[T, T^*] \in \mathfrak{S}_\infty$; let $\mathcal{N}$ and $\mathcal{E}N$ be the set of normal and, respectively, essentially normal operators. It is proved in [**BDF**] that

1) the set $\mathcal{N} + \mathfrak{S}_\infty$ is norm closed;

2) let $T \in \mathcal{E}N$, then $T \in \mathcal{N} + \mathfrak{S}_\infty$ if and only if $\text{ind}(\lambda I - T) = 0$ for all $\lambda \in \mathbb{C} \backslash \sigma_{ess}(T)$;

3) let $T_1, T_2 \in \mathcal{E}N$, then there exists a unitary U such that

$$(UT_1U^{-1} - T_2) \in \mathfrak{S}_\infty$$

if and only if $\sigma_{ess}(T_1) = \sigma_{ess}(T_2)$ and $\text{ind}(\lambda I - T_1) = \text{ind}(\lambda I - T_2)$ for all $\lambda \in \mathbb{C} \backslash \sigma_{ess}(T_1)$.

Theorems 2.5.3(f) and (g) are already commented above. It is worth mentioning however that the presented forms of these results are simplified and strenghten the corresponding classical statements. In particular, the role of the Kuroda characterization B.7.5.2(g) (Volume 1) of the ideal $\mathfrak{S}_1$ ($\mathfrak{S} = \mathfrak{S}_1 \Leftrightarrow \overline{\lim}_n(nc_n(\mathfrak{S})) > 0$) is especially stressed. Theorems 2.5.3(h) and (i) are from N. Nikolski [**N5**], and 2.5.3(j) is due to N. Makarov and V. Vasyunin [**MV1**]. The proof of the latter result heavily relies on a function model for non-contractions briefly mentioned in Section 1.6.

Subsection 2.5.4 is commented above.

The calculations from 2.5.5 are borrowed from B. Sz.-Nagy and C. Foiaş [**SzNF4**]. Theorem 2.5.5(e) is from I. Gohberg and M. Krein [**GK3**]; see also the survey paper by M. Krein [**Kr6**], where many dissipative integral operators are considered as an illustration of general criteria of similarity to a selfadjoint operator. Later on, these results were extended to the resolvent tests for similarity to a *normal* operator. Namely, it is proved that the *linear resolvent growth* condition (for short, (LRG)), i.e.,

$$\|R_\lambda(T)\| \leq \frac{\text{const}}{\text{dist}(\lambda, \sigma(T))}, \quad \lambda \in \mathbb{C}\backslash\sigma(T),$$

implies that T is similar to a normal operator in each of the following two cases:

(i) $\|T\| \leq 1$, $\partial_T = \text{rank}(I - T^*T) < \infty$ *and* $\mathbb{D} \not\subseteq \sigma(T)$ (N. Nikolski and S. Hruschev [**HrN**] for $\partial_T = 1$; N. Benamara and N. Nikolski [**BN**] for $\partial_T > 1$)*;*

(ii) $\|T\| \leq 1$, $\sup_{\lambda\in\mathbb{D}} \text{Trace}(I - \Theta(\lambda)^*\Theta(\lambda)) = \sup_{\lambda\in\mathbb{D}} \text{Trace}(I - b_\lambda(T)^* b_\lambda(T)) < \infty$ *and* $\mathbb{D} \not\subseteq \sigma(T)$ (S. Kupin [**Kup**]).

On the other hand, given an increasing function $t \longmapsto \psi(t)$, $t \geq 0$, the following are equivalent (S. Kupin and S. Treil [**KuT**]):

(1) there exists an operator T satisfying (LRG) which is not similar to a normal operator, and such that $\|T\| \leq 1$, $\mathbb{D} \not\subseteq \sigma(T)$ and $T \in \mathfrak{S}_\psi$, where

$$\mathfrak{S}_\psi = \{T : \|T\| \leq 1, (I - T^*T) \in \mathfrak{S}_\infty, \sum_{j\geq 0} \psi(s_j(I - T^*T)) < \infty\};$$

(2) $\lim_{t\to 0} \psi(t) = 0$.

Observe that $\mathcal{F} = \bigcap_\psi \mathfrak{S}_\psi$, where ψ runs over all functions satisfying (2), and hence (i) cannot be strengthen in terms of "smoothness" of the defect operator $I - T^*T$. It is also known that for every unitary operator U with absolutely continuous spectrum there exists a perturbation K such that $\text{rank}(K) = 1$, $T = U + K$ satisfies (LRG), and $\sigma(T) \subset \mathbb{T}$, but T is not similar to a normal operator (N. Nikolski and S. Treil, [**NTr**]).

Propositions 2.5.6(a) and 2.5.6(b) are standard, see H. Bercovici [**Be**] for different proofs and references. For 2.5.6(c) and 2.5.6(d) we refer to P. Vitse [**Vit2**]. Results similar to 2.5.6(c) can also be found in E. Nordgren [**Nor**]. Property 2.5.6(e) (i) is a straightforward consequence of Lemma 2.4.6, and the smoothness criteria 2.5.6(e) (ii) and (iii) can be deduced from (i) and, respectively, Hartman's theorem B.2.2.5 and Peller's theorems B.8.1.1 and B.8.6.2. The original references are L. Page [**Pag1**] (see P. Muhly [**Muh**], F. Bonsall and S. Power [**BP**] for different proofs) for 2.5.6(e)(iii), and V. Peller [**Pe4**] for 2.5.6(e)(iv).

The facts presented in 2.5.7(a)-2.5.7(f) are from N. Nikolski and S. Hruschev [**HrN**]. Some results similar to 2.5.7(h) can be found in E. Nordgren [**Nor**] and H.

Bercovici [**Be**]. The properties 2.5.7(g)-2.5.7(j) are from P. Vitse [**Vit2**]; the sufficient condition from 2.5.7(j) is likely necessary. It is worth mentioning that these properties of the commutant $\{M_\Theta\}'$ are not automatic consequences of the criteria from 2.5.6(e), since the main point in 2.5.7(g)-(j) is to separate the contributions of the factors Θ^* and Y to the smoothness properties of $P_-\Theta^*Y$ (we use here the notations of 2.5.6(e)).

A proof of the cyclicity property, different from that of 2.5.7(k) and based on an intertwining with a unitary operator, is given in D.2.3.5. The general formula for the *multicyclicity* (multiplicity) of a c.n.u. contraction with finite defects $\partial_T < \infty$, $\partial_{T^*} < \infty$ is found in V. Vasyunin [**Vas5**]. It is curious to note that the exceptional case of 2.5.7(k), where $\Theta = 0$, answers a question of P. Halmos concerning the operator $S \oplus S^*$, see N. Nikolski [**N17**].

The property 2.5.7(l)(ii) is from B. Cole, K. Lewis, and J. Wermer [**CLW**], where it is proved in a different way.

Subsection 2.5.8 is already commented above.

Subsection 2.5.9 collects some technical results used elsewhere. The observation 2.5.9(c) is from B. Cole, K. Lewis and J.Wermer [**CLW**]; see also 2.5.7(l) (ii).

CHAPTER 3

Decompositions into Invariant Subspaces and Free Interpolation

Thanks to the analytic structure of a model space $\mathcal{K}_\Theta$, we can define the so-called "prespectral" invariant subspaces $\mathcal{K}_\Theta(\sigma) \subset \mathcal{K}_\Theta$ of the model operator $M_\Theta : \mathcal{K}_\Theta \longrightarrow \mathcal{K}_\Theta$ and study decompositions of $\mathcal{K}_\Theta$ into series of such subspaces, $\mathcal{K}_\Theta = \sum_i \mathcal{K}_\Theta(\sigma_i)$. Such an intention is naturally motivated for several reasons. First, one can mimic the well-known finite dimensional techniques, which try to decompose an operator into a sum of eigenspaces, or, more generally, into a sum of its Jordan blocks. Secondly, one can have in mind to generalize the main tool of classical spectral theory of normal operators, namely, the orthogonal spectral resolution $N = \int z\, d\mathcal{E}(\cdot)$ of such an operator $N : K \longrightarrow K$.

Considering a general operator T, we have to find, first, a suitable substitute for the spectral measure $\mathcal{E}(\cdot)$ and/or for the spectral subspaces $\mathcal{E}(\sigma)K$, $\sigma \subset \mathbb{C}$. For *closed* subsets $\sigma \subset \mathbb{C}$ there are at least two equivalent definitions. Namely, the spectral subspace $K(\sigma)$ over σ can be defined as

$$K(\sigma) = \{x : \text{the local resolvent } \lambda \longmapsto (\lambda I - T)^{-1}x \text{ admits an analytic extension to } \mathbb{C} \setminus \sigma\}. \tag{3.0.1}$$

The second definition postulates the invariance and maximality properties of a *spectral subspace*: this is a subspace satisfying $TK(\sigma) \subset K(\sigma)$ and the corresponding spectral inclusion $\sigma(T|K(\sigma)) \subset \sigma$, and such that $TE \subset E$ and $\sigma(T|E) \subset \sigma$ imply that $E \subset K(\sigma)$. It can be shown that for normal operators, the subspaces $K(\sigma)$ coincide with the ranges of the spectral measure, $K(\sigma) = \mathcal{E}(\sigma)K$, and therefore they exist for σ taken from the whole Borel σ-algebra $\mathfrak{B}$. However, for general operators, it is not clear how to define the subspaces $K(\sigma)$ for σ from a more or less rich σ-algebra of subsets of the complex plane $\mathbb{C}$. On the other hand, any suitable substitute for the spectral measure should be defined on the entire algebra $\mathfrak{B}$.

For model operators equipped with an additional analytic structure, it is possible to define an analogue of spectral subspaces for every Borel set σ, at least for every contraction whose characteristic function admits a scalar multiple. In this chapter, we restrict ourselves to the simplest case of (inner) scalar characteristic functions $\Theta = \Theta_T$. For convenience, we consider the adjoint operator M_Θ^* and the corresponding M_Θ^*-invariant subspaces; recall that the latter were described in Corollary 2.1.8 as the subspaces $\mathcal{K}_\theta$ where θ is an inner divisor of Θ. Theorem 2.3.4 implies that $\mathcal{K}_\theta$ is a maximal spectral subspace over a closed set $\sigma \subset \operatorname{clos}\mathbb{D}$ (in the above sense) if and only if $\theta = \Theta_\sigma$, where

$$\Theta_\sigma(z) = \Big(\prod_{\lambda\in\sigma} b_\lambda^{k(\lambda)}(z)\Big) \cdot \exp\Big(-\int_{\sigma\cap\mathbb{T}} \frac{\zeta + z}{\zeta - z}\, d\mu(\zeta)\Big), \quad |z| < 1,$$

and

$$\Theta_T(z) = \Big(\prod_{\lambda \subset \mathbb{D}} b_\lambda^{k(\lambda)}(z)\Big) \cdot \exp\Big(-\int_{\mathbb{T}} \frac{\zeta + z}{\zeta - z}\, d\mu(\zeta)\Big)$$

stands for the canonical Riesz–Smirnov representation of Θ_T, see Chapter A.3 (Volume 1).

It is clear that we can now *define* an analogue of the spectral subspaces for M_Θ^*, called *prespectral*, simply by setting

$$\mathcal{K}_\Theta(\sigma) = \mathcal{K}_{\Theta_\sigma}$$

for every Borel subset σ of the disk $\operatorname{clos}\mathbb{D}$. This is a simple but principal step, because it supplies us with a "space valued analogue" of spectral measures.

Now, we are ready to state the main problem of this chapter: given an inner function Θ and a Borel partition $(\sigma_i)_i$ of $\operatorname{clos}\mathbb{D}$, find necessary and sufficient conditions for an *unconditionally convergent decomposition*

$$\mathcal{K}_\Theta = \sum_i \mathcal{K}_\Theta(\sigma_i);$$

this means that every element $f \in \mathcal{K}_\Theta$ allows a unique and unconditionally convergent representation $f = \sum_i f_i$, where $f_i \in \mathcal{K}_\Theta(\sigma_i)$. The solution to this problem depends on (and, in fact, coincides with) the solution to a free interpolation problem.

The *free interpolation problem* of complex analysis, stated for a class X of functions holomorphic on a domain Ω, consists of a description of the subsets $\Lambda \subset \Omega$ for which the restriction space $X|\Lambda$ is free of traces of holomorphy, that is, it is an *ideal space* of functions on Λ, in the sense that $a \in X|\Lambda$ and $|b(z)| \le |a(z)|$, for $z \in \Lambda$, imply that $b \in X|\Lambda$.

We are joining spectral decompositions and complex interpolation in two ways. The first way uses the commutant lifting theorem (CLT) and the H^∞-calculus. Namely, unconditional convergence of the above decomposition is equivalent to the fact that the multiplier $A : \sum_i f_i \longmapsto \sum_i \bar{a}_i f_i$ is a bounded operator for every $(a_i) \in l^\infty$. Since A commutes with M_Θ^*, the CLT implies that there exists a function $f \in H^\infty$ "interpolating" the operator A^* in the sense that $A^* = f(M_\Theta)$. This equality is equivalent to

$$(f - a_i) \in \Theta_i H^\infty$$

for every i. In case $\Theta_i = b_{\lambda_i}$, $\lambda_i \in \Lambda$, are simple Blaschke factors, these inclusions simply mean that $f|\Lambda = a$, and hence $H^\infty|\Lambda = l^\infty$. In more general situations they lead to what we call *generalized interpolation*, namely, to interpolation by germs of H^∞- and H^2-functions (for Blaschke products Θ_i) and to a kind of asymptotic interpolation (for point mass singular inner functions Θ_i). Continuing in this way and using the Carleson corona theorem, we give a necessary and sufficient condition for generalized free interpolation, namely, the so-called *generalized Carleson condition (GC)*:

$$|\Theta(z)| \ge \varepsilon \cdot \inf_k |\Theta_k(z)|, \quad z \in \mathbb{D}$$

where $\Theta = \prod_i \Theta_i$.

Another way to join spectral decompositions and interpolation is to use the classical identification of *unconditional bases* and *Riesz bases*. Namely, a sum of subspaces $\mathcal{K}_\Theta = \sum_i \mathcal{K}_\Theta(\sigma_i)$ represents an unconditionally convergent decomposition of $\mathrm{span}(\mathcal{K}_\Theta(\sigma_i) : i \geq 1)$ if and only if an approximate Parseval identity holds, that is, there exists a constant $c > 0$ such that

$$c \sum_i \|x_i\|^2 \leq \left\| \sum_i x_i \right\|^2 \leq c^{-1} \sum_i \|x_i\|^2$$

for all $x_i \in \mathcal{K}_\Theta(\sigma_i)$, $i \geq 1$. This leads to the following notion of *generalized free interpolation* in H^2: a sequence of functions $(f_i)_{i \geq 1} \subset H^2$ satisfies $\sum_i \|f_i\|^2_{H^2/\Theta_i H^2} < \infty$ if and only if there exists a function $f \in H^2$ interpolating f_i in the sense that $f - f_i \in \Theta_i H^2$ for all $i \geq 1$. In particular, we have (under the condition (GC)) a *generalized embedding theorem*

$$\sum_i \|f\|^2_{H^2/\Theta_i H^2} < \infty$$

for every $f \in H^2$.

The chapter is organized as follows. Section 3.1 contains the prerequisites on unconditional convergence. Section 3.2 is devoted to a solution of the generalized free interpolation problem, which is equivalent to the unconditional convergence of the corresponding spectral decomposition. Several special cases are considered (classical Carleson interpolation, various multiple interpolations, asymptotic interpolation at boundary points).

Section 3.3, Exercises and Further Results, contains the theory of Gram matices with an application to Müntz type completeness theorems, a generalization of the Cotlar-Stein almost orthogonality lemma, and a selfcontained elementary (i.e., without use of the Carleson corona theorem) treatment of classical interpolation problems, including the question on linear interpolation operators (following G. Airapetyan, P. Jones, and S. Vinogradov). There are also several informations about geometric nature of interpolating (Carleson) sequences, as well as on boundary free interpolation, interpolation over the level sets, and Dunford spectral model operators.

3.1. Unconditional Bases

In this section we give some auxiliary facts about unconditional bases that we will use in Section 3.2, together with the commutant lifting theorem, to solve generalized free interpolation problems. Recall that some of the concepts that we discuss below have previously appeared in Part A, Sections 5.1 and 5.6.2 (Volume 1). We start by generalizing the notion of bases introduced in A.5.

3.1.1. DEFINITION. Let X be a Banach space and $\mathcal{X} = (X_n)_{n \geq 1}$ a sequence of subspaces of X (always assumed to be closed). The family $\mathcal{X}$ is called a *basis (of subspaces)* of X if for every $x \in X$ there exists a unique decomposition of x into a convergent series

$$x = \sum_{n \geq 1} x_n, \quad x_n \in X_n.$$

It is called an *unconditional basis (of subspaces)* if this expansion converges unconditionally.

Recall that by definition the above sum converges unconditionally if for every $\varepsilon > 0$ there is a finite subset $\sigma \in \mathbb{N}$ such that

$$\left\| x - \sum_{n \in \sigma'} x_n \right\| < \varepsilon$$

where σ' is any other finite subset of $\mathbb{N}$ with $\sigma \subset \sigma' \subset \mathbb{N}$.

3.1.2. First properties. The following properties are straightforward consequences of classical principles in functional analysis.

(1) Let $\mathcal{X}$ be a basis of subspaces, $x \in X$, and let $x = \sum_{n \geq 1} x_n$ be its decomposition with respect to $\mathcal{X}$. Then the mapping $\mathcal{P}_n : X \longrightarrow X$ defined by $\mathcal{P}_n x = x_n$, $x \in X$, is well-defined, linear and continuous (the classical Banach theorem, see, for example, P. Wojtaszczyk [**Woj**], II.B for the proof).

For a general sequence of subspaces $\mathcal{X}$, we define $\mathcal{P}_n$ on linear combinations $x = \sum x_k \in \mathcal{L}\mathrm{in}(\mathcal{X})$ by setting $\mathcal{P}_n(\sum x_k) = x_n$. The operator $\mathcal{P}_n$ is called a *coordinate projection.* If the coordinate projections of a family $\mathcal{X}$ are well-defined and continuous on $\operatorname{span} \mathcal{X}$, then $\mathcal{X}$ is called *minimal* (cf. the classical case A.5.1.1, Volume 1). Note that for this definition we do not require completeness of $\mathcal{X}$. In particular, if $\mathcal{X}$ is minimal, then $\mathcal{P}_n | X_n = I$ and $\mathcal{P}_n | X_k = \mathbb{O}$, $k \neq n$, and $X_n \cap \operatorname{span}(X_k : k \neq n) = \{0\}$, $n \geq 1$.

(2) If $\mathcal{X}$ is a basis, then

$$\sup_{n \geq 1} \|\mathcal{P}_n\| < \infty, \tag{3.1.1}$$

which is an immediate consequence of the Banach-Steinhaus theorem. A family $\mathcal{X}$ such that the coordinate projections satisfy (3.1.1) is called *uniformly minimal.*

(3) Let $\mathcal{X} = (X_n)_{n \geq 1}$ be minimal, and suppose that it is complete in X, i.e. $X = \operatorname{span} \mathcal{X} = \operatorname{span}(X_n)_{n \geq 1}$. Then $\mathcal{X}$ is a basis (respectively, an unconditional basis) if and only if $\sup_{n \geq 1} \|P_n\| < \infty$, where $P_n = \sum_{k=1}^{n} \mathcal{P}_k$ (respectively, $\sup_{\sigma} \|\mathcal{P}_\sigma\| < \infty$, where $\mathcal{P}_\sigma = \sum_{k \in \sigma} \mathcal{P}_k$ and $\sigma \subset \mathbb{N}$ is finite).

(4) Let $\mathcal{X} = (X_n)_{n \geq 1}$ be minimal and let $\mu_n : X_n \to X_n$, $n \geq 1$, be a sequence of bounded operators such that the operator $\mu : \mathcal{L}\mathrm{in}(\mathcal{X}) \longrightarrow \mathcal{L}\mathrm{in}(\mathcal{X})$ defined by

$$\mu\Big(\sum_n x_n\Big) = \sum_n \mu_n x_n$$

extends continuously to $\operatorname{span} \mathcal{X}$. Then

$$\sup_{n \geq 1} \|\mu_n\| \leq \|\mu\| < \infty. \tag{3.1.2}$$

Such a sequence $\mu = (\mu_n)_{n \geq 1}$ is called a *multiplier* of $\mathcal{X}$. Denote by $\mathrm{Mult}(\mathcal{X})$ the vector space of all multipliers of $\mathcal{X}$ endowed with the *multiplier norm* $\|\mu\|$, which is the norm of the corresponding operator on $\mathcal{L}\mathrm{in}(\mathcal{X})$. Note that the above property (3.1.2) shows that $\mathrm{Mult}(\mathcal{X}) \subset l^\infty(X_n \to X_n)$.

We will also use the notion of *scalar multipliers* of $\mathcal{X}$:

$$\mathrm{mult}(\mathcal{X}) = \{\mu \in \mathrm{Mult}(\mathcal{X}) : \mu_n = \lambda_n I_{X_n},\ \lambda_n \in \mathbb{C}\}.$$

Obviously $\mathrm{mult}(\mathcal{X}) \subset l^\infty$.

(5) Let $\mathcal{X} = (X_n)_{n \geq 1}$ be minimal in a Banach space X and suppose that $X = \operatorname{span} \mathcal{X}$. Then $\mathcal{X}$ is a basis if and only if $bv \subset \mathrm{mult}(\mathcal{X})$ where bv is the space of sequences $\lambda = (\lambda_n)_{n \geq 1}$ of bounded variation: $\sum_{n \geq 1} |\lambda_n - \lambda_{n+1}| < \infty$.

[Hint: use the Abel transformation and the closed graph theorem for the inclusion $bv \subset \text{mult}(\mathcal{X})$, see also the proof of Theorem 3.1.4 below.]

(6) Let $\mathcal{X} = (X_n)_{n\geq 1}$ be minimal in X and suppose that $X = \operatorname{span} \mathcal{X}$. Define the *polar set* of X_n by

$$X_n^{\perp} = \{f \in X^* : f|X_n \equiv 0\}.$$

Then the sequence $\mathcal{X}' = (X_n')_{n\geq 1}$, where $X_n' = \bigcap_{k\neq n} X_k^{\perp}$, $n \geq 1$, is minimal in X^*, and the adjoint projections $\mathcal{P}_n^*$ are the corresponding coordinate projections of $\mathcal{X}'$.

Indeed, for finite sums $f = \sum_k f_k$, $f_k \in X_k'$, and $x = \sum_k x_k$, $x_k \in X_k$, we have

$$\begin{aligned} (\mathcal{P}_n^* f, x) &= (f, \mathcal{P}_n x) = \Big(\sum_k f_k, x_n\Big) = (f_n, x_n) = \Big(f_n, \sum_k x_k\Big) \\ &= (f_n, x). \end{aligned}$$

Consequently, $\mathcal{P}_n^* f = f_n$. As the projections $\mathcal{P}_n^*$ are bounded, the sequence $\mathcal{X}'$ is minimal.

(7) **The scalar case**. If we have $\dim X_n = 1$, $n \geq 1$, we recover the definitions and results of A.5.1, Volume 1 (by choosing a non-zero vector $x_n \in X_n$ for every $n \geq 1$). Note that in this case $\dim X_n' = 1$ and that the coordinate functionals f_n introduced in A.5.1 (Volume 1) are elements of X_n': $f_n \in X_n'$, $n \geq 1$. Moreover, they can be chosen in such a way that $\langle x_k, f_n\rangle = \delta_{kn}$ for all $k, n \geq 1$. We call $(f_n)_{n\geq 1}$ the sequence biorthogonal to $\mathcal{X}$ (with a certain abuse of language). We have

$$\mathcal{P}_n = \langle \cdot, f_n\rangle x_n, \quad n \geq 1,$$

and hence (cf. A.5.1)

$$\|\mathcal{P}_n\|^{-1} = \operatorname{dist}\left(\frac{x_n}{\|x_n\|}, \operatorname{span}(x_k : k \neq n)\right).$$

3.1.3. The principal characterization. Now we are in a position to prove the main theorem of this section characterizing unconditional bases. This will be done for Hilbert spaces, but the first three assertions are still equivalent in the Banach space setting.

We call a family $\mathcal{X} = (X_n)_{n\geq 1}$ in a Hilbert space X a *quasi-orthogonal* (or *Riesz) sequence* if there are two constants $c, C > 0$ such that

$$c^2 \sum_n \|x_n\|^2 \leq \left\| \sum_n x_n \right\|^2 \leq C^2 \sum_n \|x_n\|^2$$

for every finite family $x_n \in X_n$, $n \geq 1$. In case the family is moreover complete, $X = \operatorname{span} \mathcal{X}$, we shall say that it is a *Riesz basis*.

3.1.4. THEOREM. *Let $\mathcal{X} = (X_n)_{n\geq 1}$ be a minimal sequence of subspaces of a Hilbert space X and suppose that $X = \operatorname{span} \mathcal{X}$. The following assertions are equivalent.*

(1) $\mathcal{X}$ is an unconditional basis in X.

(2) $\mathcal{X}'$ is an unconditional basis in X (in the case of Banach spaces it is an unconditional basis for X^* in the weak-topology).*

(3) $\text{mult}(\mathcal{X}) = l^{\infty}$.

(4) $\text{Mult}(\mathcal{X}) = l^{\infty}(X_n \to X_n)$.

(5) $\mathcal{X}$ is a Riesz basis *in X.*

(6) There is an orthogonal basis of subspaces $(H_n)_{n\geq 1}$ of a Hilbert space H and an isomorphism V from X onto H such that $VX_n = H_n$, $n \geq 1$ (and hence $V^{-1}H_n = X_n$).

PROOF. The scheme of the proof is as follows: (1)⇔(2); (1)⇒(3)⇒(5)⇒(6)⇒(1); (5)⇒(4)⇒(3).

(1)⇔(2). By Property 3.1.2(3) and the Banach-Steinhaus theorem, it only remains to verify the weak* completeness of the dual sequence $\mathcal{X}'$. To this end suppose that $x \in X$ vanishes on $\mathcal{L}\text{in}\,\mathcal{X}'$, that is, $x \perp X_n' = \mathcal{P}_n^* X^*$. Then $0 = \langle x, \mathcal{P}_n^* x^* \rangle = \langle \mathcal{P}_n x, x^* \rangle$ for every $x^* \in X^*$ and $n \geq 1$, which gives $\mathcal{P}_n x = 0$, $n \geq 1$, and hence $x = 0$.

(1)⇒(3). Let $(\mu_n)_{n\geq 1} \in l^\infty$ and $x \in \mathcal{L}\text{in}(\mathcal{X})$, $x = \sum_{k\geq 1} x_k$, $x_k \in X_k$. Obviously,

$$\mu x = \sum_{k\geq 1} \mu_k x_k = \lim_n \sum_{k=1}^{n} \mu_k x_k .$$

It suffices to show that $\|\mu\| \leq \text{const}\, \|(\mu_n)_{n\geq 1}\|_\infty$ for every finitely supported sequence on $\mathbb{N}$ with values in $\mathbb{R}$ (we will then derive the complex case in a standard way). Let $(\mu_n)_{n\geq 1}$ be such a sequence. We can represent it in the following way:

$$\mu = \sum_{i=1}^{n} \mu_i \chi_{\sigma_i},$$

where $\sigma_i \cap \sigma_j = \emptyset$, $i \neq j$, and $\mu_1 < \mu_2 < \ldots < \mu_n$ (here σ_i are finite subsets of $\mathbb{N}$). Setting $\tau_k = \bigcup_{i=k}^{n} \sigma_i$, $1 \leq k \leq n$, and $\mu_0 = 0$, we obtain

$$\mu = \sum_{k=1}^{n} (\mu_k - \mu_{k-1}) \chi_{\tau_k}.$$

Defining $C = \sup_\sigma \|\mathcal{P}_\sigma\| = \sup_\sigma \|\chi_\sigma\|$, which is finite by the assumption, we get

$$\begin{aligned} \|\mu\| &\leq \sum_{k=1}^{n} |\mu_k - \mu_{k-1}| \cdot \|\chi_{\tau_k}\| \leq C(|\mu_1| + \mu_n - \mu_1) \\ &\leq 3C\|\mu\|_\infty \end{aligned}$$

(here $\|\mu\|$ stands for the multiplier norm). For complex valued sequences, we obtain

$$\|\mu\| \leq 6C\|\mu\|_\infty.$$

(3)⇒(5). By the assumption and the closed graph theorem, there exists a constant C such that $\|\mu\| \leq C\|\mu\|_\infty$ for every $\mu \in l^\infty$. Here $\|\cdot\|$ is the multiplier norm $\|\mu\| = \|\mu\|_{\text{Mult}}$. Let $\varepsilon = (\varepsilon_n)_{n\geq 1}$ be an arbitrary unimodular sequence, $|\varepsilon_n| = 1$, $n \geq 1$. Then

$$\begin{aligned} \Big\| \sum_{n\geq 1} \varepsilon_n x_n \Big\| &= \Big\| \varepsilon \sum_{n\geq 1} x_n \Big\| \leq C \Big\| \sum_{n\geq 1} x_n \Big\| = C \Big\| \sum_{n\geq 1} \bar{\varepsilon}_n \varepsilon_n x_n \Big\| = C \Big\| \bar{\varepsilon} \sum_{n\geq 1} \varepsilon_n x_n \Big\| \\ &\leq C^2 \Big\| \sum_{n\geq 1} \varepsilon_n x_n \Big\| \end{aligned}$$

for every finite family of vectors $(x_n)_{n\geq 1}$, $x_n \in X_n$, $n \geq 1$. Now, we apply Orlicz' Lemma 3.1.5 (see below):

$$\begin{aligned}\sum_{n\geq 1} \|x_n\|^2 &\leq \max\left\{\Big\|\sum_{n\geq 1} \varepsilon_n x_n\Big\|^2 : |\varepsilon_n| = 1, n \geq 1\right\} \leq C^2 \Big\|\sum_{n\geq 1} x_n\Big\|^2 \\ &\leq C^4 \cdot \min\left\{\Big\|\sum_{n\geq 1} \varepsilon_n x_n\Big\|^2 : |\varepsilon_n| = 1, n \geq 1\right\} \leq C^4 \sum_{n\geq 1} \|x_n\|^2,\end{aligned}$$

which shows (5).

(5)⇒(6). Let $l^2(X_n)$ be the space of those X valued square-summable sequences $x = (x_n)_{n\geq 1} \in l^2(X)$ for which $x_n \in X_n$, $n \geq 1$. Define an operator V from the linear hull $\mathcal{L}\mathrm{in}(X_n : n \geq 1)$ to $l^2(X_n)$ by

$$V\left(\sum_{n\geq 1} x_n\right) = (x_n)_{n\geq 1}.$$

By assumption, the operator V extends to an isomorphism from X onto $l^2(X_n)$. Clearly, $VX_n = H_n := \{x = (x_i)_{i\geq 1} : x_i = 0 \text{ for } i \neq n\}$, and with this definition we obviously have $H_n \perp H_k$, $n \neq k$. This shows (6).

(6)⇒(1) is obvious, and so is (4)⇒(3).

Let us prove the remaining implication (5)⇒(4). Let $\mu_n : X_n \to X_n$ and $A = \sup_{n\geq 1} \|\mu_n\| < \infty$. We need to show that the operator $\mu(\sum_{n\geq 1} x_n) = \sum_{n\geq 1} \mu_n x_n$ extends continuously to X. Indeed

$$\begin{aligned}\left\|\mu\left(\sum_{n\geq 1} x_n\right)\right\|^2 &= \Big\|\sum_{n\geq 1} \mu_n x_n\Big\|^2 \leq C^2 \sum_{n\geq 1} \|\mu_n x_n\|^2 \leq C^2 A^2 \sum_{n\geq 1} \|x_n\|^2 \\ &\leq (C^2 A^2 / c^2) \|\sum_{n\geq 1} x_n\|^2\end{aligned}$$

for every finite family of vectors $x_n \in X_n$, $n \geq 1$. □

3.1.5. Lemma (W. Orlicz, 1933). *Let $(x_n)_{n\geq 1}$ be a finite family of vectors in a Hilbert space X. Then*

$$\begin{aligned}&\min\left\{\Big\|\sum_{n\geq 1} \varepsilon_n x_n\Big\|^2 : |\varepsilon_n| = 1, n \geq 1\right\} \\ &\leq \textstyle\sum_{n\geq 1} \|x_n\|^2 \leq \max\left\{\Big\|\sum_{n\geq 1} \varepsilon_n x_n\Big\|^2 : |\varepsilon_n| = 1, n \geq 1\right\}.\end{aligned}$$

Proof. We have

$$\begin{aligned}\|x\|^2 + \|y\|^2 \leq \|x + \varepsilon y\|^2 \quad &\text{if } \varepsilon = (x, y)/|(x, y)|, \\ \|x + \varepsilon y\|^2 \leq \|x\|^2 + \|y\|^2 \quad &\text{if } \varepsilon = -(x, y)/|(x, y)|\end{aligned}$$

for every $x, y \in X$. The result follows by induction. □

Using 3.1.2(7), we get the following corollary.

3.1.6. Corollary. *Let H be a Hilbert space, $(x_n)_{n\geq 1}$ an unconditional basis of vectors, and let $\mathcal{F}x = (a_n)_{n\geq 1}$ where $x = \sum_{n\geq 1} a_n x_n \in H$. Then*

$$\mathcal{F}H = l^2\big(\|x_n\|^2\big) = \Big\{(a_n)_{n\geq 1} : \sum_{n\geq 1} |a_n|^2 \|x_n\|^2 < \infty\Big\}.$$

□

3.2. Generalized Free Interpolation

Now, we realize the ideas outlined in the introduction to this Chapter, applying, first, the commutant lifting theorem, and then the Carleson corona theorem (Theorem 3.2.10 below). Some other technical tools, such as the elementary Lemma 3.2.12, are also quite substantial.

3.2.1. Divisibility in the algebra H^∞ as an operator equation. In this Section, we consider the model operator M_Θ for scalar inner functions $\Theta \in H^\infty$, $|\Theta| = 1$ a.e. on $\mathbb{T}$. Then

$$\begin{aligned} \mathcal{K}_\Theta &= H^2 \ominus \Theta H^2, \\ M_\Theta &= P_\Theta z | \mathcal{K}_\Theta. \end{aligned}$$

An obvious but useful observation is the following. If $Sf = zf$, $f \in H^2$, is the shift operator and $S^* f = P_+ \bar{z} f$ its adjoint, then, as ΘH^2 is invariant for S, the space $\mathcal{K}_\Theta$ is invariant for S^*:

$$S^* \mathcal{K}_\Theta \subset \mathcal{K}_\Theta.$$

Moreover, this property extends to arbitrary H^∞-functions: $P_+ \overline{\varphi} \mathcal{K}_\Theta \subset \mathcal{K}_\Theta$, $\varphi \in H^\infty$.

3.2.2. Lemma. *Let Θ be inner and Θ_1 an inner divisor of Θ, that is, $\Theta/\Theta_1 \in H^\infty$. Then*

(1) $M_\Theta^* = S^* | \mathcal{K}_\Theta$,

(2) $\mathcal{K}_{\Theta_1} \subset \mathcal{K}_\Theta$, $M_\Theta^* \mathcal{K}_{\Theta_1} \subset \mathcal{K}_{\Theta_1}$ and $M_\Theta^* | \mathcal{K}_{\Theta_1} = M_{\Theta_1^*}$. Moreover, $\varphi(M_\Theta)^* | \mathcal{K}_{\Theta_1} = \varphi(M_{\Theta_1})^*$ if $\varphi \in H^\infty$.

Proof. Since $S^* f \in \mathcal{K}_\Theta$ for every $f \in \mathcal{K}_\Theta$, we have

$$S^* f = P_+ \bar{z} f = P_\Theta P_+ \bar{z} f = P_\Theta \bar{z} f = M_\Theta^* f,$$

which proves (1).

Using now the above mentioned inclusion $P_+ \overline{\varphi} \mathcal{K}_{\Theta_1} \subset \mathcal{K}_{\Theta_1}$, we get

$$\varphi(M_\Theta^*) | \mathcal{K}_{\Theta_1} = P_\Theta \overline{\varphi} | \mathcal{K}_{\Theta_1} = P_+ \overline{\varphi} | \mathcal{K}_{\Theta_1} = \varphi(M_{\Theta_1})^*.$$

□

3.2.3. Lemma. *Let Θ be inner, Θ_1 an inner divisor of Θ and $\varphi \in H^\infty$. Then*

(1) $\varphi(M_\Theta)^* = \mathbb{O}$ if and only if $\varphi/\Theta \in H^\infty$;

(2) $\varphi(M_\Theta)^* | \mathcal{K}_{\Theta_1} = \mathbb{O}$ if and only if $\varphi/\Theta_1 \in H^\infty$.

Proof. This is immediate from 2.2.2(5) and 3.2.2. □

3.2.4. Unconditional bases of model subspaces. We have seen above that Θ_1 is an inner divisor of Θ if and only if $\mathcal{K}_{\Theta_1} \subset \mathcal{K}_\Theta$. For a family of inner divisors $\{\Theta_\alpha\}_\alpha$ of Θ, we thus obtain a family $\{\mathcal{K}_{\Theta_\alpha}\}_{\alpha\in A}$ of subspaces of $\mathcal{K}_\Theta$, which is complete, $\operatorname{span}(\mathcal{K}_{\Theta_\alpha} : \alpha \in A) = \mathcal{K}_\Theta$, if and only if $\Theta = \operatorname{LCM}\{\Theta_\alpha\}_{\alpha\in A}$ (this is a consequence of A.2.5, Volume 1, applied to the orthogonal complements).

The commutant lifting theorem enables us now to relate bases of model subspaces and generalized interpolation (cf. also A.3.3 Volume 1).

3.2.5. THEOREM. *Let Θ and Θ_n, $n \geq 1$, be inner functions such that $\Theta = \operatorname{LCM}\{\Theta_n\}_{n\geq 1}$. The following assertions are equivalent.*

(1) The family $(\mathcal{K}_{\Theta_n})_{n\geq 1}$ is an unconditional basis in $\mathcal{K}_\Theta$.

(2) For every $\mu = (\mu_n)_{n\geq 1} \in l^\infty$, there exists a function $\varphi \in H^\infty$ such that $\varphi - \mu_n \in \Theta_n H^\infty$, $n \geq 1$.

(3) For every bounded sequence $(f_n)_{n\geq 1}$ of H^∞ functions, $\sup_{n\geq 1} \|f_n\|_\infty < \infty$, there exists $\varphi \in H^\infty$ such that $\varphi - f_n \in \Theta_n H^\infty$, $n \geq 1$.

(4) For every sequence $(f_n)_{n\geq 1}$ of H^∞ functions with $\sup_{n\geq 1} \operatorname{dist}(f_n\overline{\Theta_n}, H^\infty) < \infty$, there exists a function $\varphi \in H^\infty$ such that $\varphi - f_n \in \Theta_n H^\infty$, $n \geq 1$.

Note that the condition on the sequence $(f_n)_{n\geq 1}$ introduced in (4) is apparently weaker than that of (3). It actually states that it is sufficient to have uniform boundedness of the quotient norms of $(f_n)_{n\geq 1}$: $\sup_{n\geq 1} \|f_n\|_{H^\infty/\Theta_n H^\infty} < \infty$.

PROOF. Obviously, we have (4)$\Longrightarrow$(3)$\Longrightarrow$(2). Let $T_n = f_n(M_\Theta)^*$, $n \geq 1$. Clearly, if the operator T,

$$T\Big(\sum x_n\Big) = \sum T_n x_n,$$

where $x_n \in \mathcal{K}_{\Theta_n}$ (finite sum), is well-defined and bounded, then it extends to a mapping on $\mathcal{K}_\Theta$ which commutes with M_Θ^* (in order to see this, one can observe that $M_\Theta^* \mathcal{K}_{\Theta_n} \subset \mathcal{K}_{\Theta_n}$, $M_\Theta^* | \mathcal{K}_{\Theta_n} = M_{\Theta_n}^*$, $n \geq 1$). Applying now the commutant lifting theorem 2.4.2, we obtain a function $\varphi \in H^\infty$ such that $T = \varphi(M_\Theta)^*$, i.e. such that $f_n(M_\Theta)^* = \varphi(M_\Theta)^* | \mathcal{K}_{\Theta_n}$, $n \geq 1$. By Lemma 3.2.3, this means that $\varphi - f_n \in \Theta_n H^\infty$, $n \geq 1$.

The proof can now be achieved by using the equivalences (1)$\Leftrightarrow$(3) and (1)$\Leftrightarrow$(4) of Theorem 3.1.4, the formula

$$\|f_n(M_{\Theta_n})\| = \operatorname{dist}(f_n\overline{\Theta}_n, H^\infty)$$

(see Theorem 2.4.7) and the obvious estimate $\|f_n(M_{\Theta_n})\| \leq \|f_n\|$. $\square$

3.2.6. REMARK. (1) The sequences $(\mu_n)_{n\geq 1}$ and $(f_n)_{n\geq 1}$ appearing in the preceding theorem are called *interpolation data* with respect to the sequence $(\Theta_n)_{n\geq 1}$. The major difference between the assertions (2), (3) and (4) is thus the amount of a priori admissible interpolation data.

(2) If one (and hence all) of the assertions is fulfilled, we clearly have $\mathcal{K}_{\Theta_n} \cap \mathcal{K}_{\Theta_m} = \{0\}$, $n \neq m$. This means that the functions Θ_n and Θ_m are relatively prime, $\operatorname{GCD}(\Theta_n, \Theta_m) = 1$, and then $\operatorname{LCM}\{\Theta_n\}_{n\geq 1} = \prod_{k\geq 1} \Theta_k$.

3.2.7. Unconditional bases and Bezout equations. Our next objective is to reduce the condition on unconditional bases to Bezout type equations in H^∞ with norm estimates of the solutions.

We use the same notation and assumptions as in Theorem 3.2.5.

3.2.8. LEMMA. *The following assertions are equivalent.*

(1) $(\mathcal{K}_{\Theta_n})_{n\geq 1}$ *is an unconditional basis in* $\mathcal{K}_\Theta$.

(2) There exists a constant $C < \infty$ *such that for every finite set* $\sigma \subset \mathbb{N}$ *there are functions* $f_\sigma, g_\sigma \in H^\infty$ *satisfying*

$$f_\sigma \Theta_\sigma + g_\sigma \Theta_{\sigma'} = 1, \quad \|f_\sigma\|_\infty \leq C, \tag{3.2.1}$$

where $\Theta_\sigma = \prod_{n\in\sigma} \Theta_n$, $\sigma' = \mathbb{N} \setminus \sigma$ *(hence* $\Theta_{\sigma'} = \prod_{n\in\mathbb{N}\setminus\sigma} \Theta_n = \Theta/\Theta_\sigma$*).*

PROOF. Suppose that $\mathcal{X}$ is an unconditional basis. Then by Remark 3.2.6(2) we have $\operatorname{span}(\mathcal{K}_{\Theta_n} : n \in \sigma) = \mathcal{K}_{\Theta_\sigma}$, and clearly the projection $\mathcal{P}_\sigma$ satisfies

$$\mathcal{P}_\sigma | \mathcal{K}_{\Theta_\sigma} = I, \quad \mathcal{P}_\sigma | \mathcal{K}_{\Theta_{\sigma'}} = \mathbb{O}.$$

Then $\mathcal{P}_\sigma$ commutes with M_Θ^*, and the commutant lifting theorem implies the existence of a function $\varphi_\sigma \in H^\infty$ such that $\mathcal{P}_\sigma = \varphi_\sigma(M_\Theta)^*$ and $\|\mathcal{P}_\sigma\| = \|\varphi_\sigma\|_\infty$. Using Lemma 3.2.3 we get $f_\sigma, g_\sigma \in H^\infty$ such that $\varphi_\sigma = \Theta_\sigma g_\sigma$ and $1 - \varphi_\sigma = \Theta_{\sigma'} f_\sigma$. Moreover, $\|f_\sigma\| = \|\mathcal{P}_\sigma\|$. Obviously, $1 = \Theta_\sigma f_\sigma + \Theta_{\sigma'} g_\sigma$ and, in view of 3.1.2(3), we also have $\sup_\sigma \|f_\sigma\| = \sup_\sigma \|\mathcal{P}_\sigma\| = C < \infty$.

The converse also follows from 3.1.2(3) by setting $\varphi_\sigma = \Theta_{\sigma'} g_\sigma$ and $\mathcal{P}_\sigma = \varphi_\sigma(M_\Theta)^*$. □

3.2.9. Prerequisites for the generalized Carleson condition (GC). We are now looking for a more customary formulation of condition (2) of the preceding lemma involving only the inner functions Θ_σ, $\sigma \in \mathbb{N}$. This will be achieved through Carleson's corona theorem (see Chapter B.9 (Volume1) and also Section 3.4 for comments and references).

3.2.10. THEOREM (L. Carleson, 1962). *Let* $f_i \in H^\infty$, $1 \leq i \leq n$, *and assume that for some* $\delta > 0$ *we have*

$$\delta \leq \left(\sum_{i=1}^n |f_i(z)|^2 \right)^{1/2} \leq 1, \quad z \in \mathbb{D}.$$

Then there exists a family of functions $g_i \in H^\infty$, $1 \leq i \leq n$, *such that*

$$\sum_{i=1}^n f_i(z) g_i(z) \equiv 1; \quad \left(\sum_{i=1}^n |g_i(z)|^2 \right)^{1/2} \leq A\delta^{-2} \log \frac{1}{\delta}, \quad z \in \mathbb{D},$$

where A *is an absolute constant.*

In fact, we now immediately get the following fact.

3.2.11. COROLLARY. *Under the assumptions of Theorem 3.2.5 and the notation of Lemma 3.2.8 we have that* $\mathcal{X} = \{\mathcal{K}_{\Theta_n}\}_{n\geq 1}$ *is an unconditional basis of* $\mathcal{K}_\Theta$ *if and only if*

$$\inf_{\sigma\subset\mathbb{N}} \inf_{z\in\mathbb{D}} (|\Theta_\sigma(z)| + |\Theta_{\sigma'}(z)|) > 0.$$

□

This actually gives a condition involving only the family $(\Theta_\sigma)_\sigma$. Now, we shall seek another form of this condition that is closer to the Carleson condition for free interpolation in H^∞ (see 3.2.17 and 3.3.3 below).

3.2.12. LEMMA. *Let $\alpha_n \geq 0$, $\alpha = \sum_{n\geq 1} \alpha_n$, $\alpha_0 = \max_{n\geq 1} \alpha_n$. Then*

$$\frac{\alpha - \alpha_0}{2} \leq \sup_{\sigma \subset \mathbb{N}} \min \left(\sum_{n\in\sigma} \alpha_n, \sum_{n\in\sigma'} \alpha_n \right) \leq \alpha - \alpha_0,$$

where $\sigma \neq \emptyset$, $\sigma \neq \mathbb{N}$, $\sigma' = \mathbb{N} \setminus \sigma$.

PROOF. Without loss of generality we can assume that the sequence $(\alpha_n)_{n\geq 1}$ is finitely supported, say, $(\alpha_n)_{n\geq 1} = (\alpha_1, \dots, \alpha_N)$ for some $N \geq 1$. Choose $\sigma \subset \mathbb{N}$ with $\sigma, \sigma' \neq \emptyset$, and $m \in \mathbb{N}$ such that $\alpha_0 = \alpha_m$ for some $m \geq 1$. Then either $m \in \sigma$ or $m \in \sigma'$, and as $\sum_{n\in\sigma'} \alpha_n = \alpha - \sum_{n\in\sigma} \alpha_n$, we get

$$\min \left(\sum_{n\in\sigma} \alpha_n, \sum_{n\in\sigma'} \alpha_n \right) \leq \alpha - \alpha_0.$$

This gives the right-hand side inequality.

Consider the left-hand side inequality. As $(\alpha_n)_{n\geq 1}$ is finitely supported, there is σ_0 such that

$$\sum_{n\in\sigma_0} \alpha_n = \sup_\sigma \min \left(\sum_\sigma \alpha_n, \sum_{\sigma'} \alpha_n \right) = M.$$

Assume that $M < \frac{\alpha - \alpha_0}{2}$. For $k \in \sigma_0'$ $(= \mathbb{N} \setminus \sigma_0)$ with $\alpha_k \neq 0$ define a partition

$$\sigma_1 = \sigma_0 \cup \{k\}, \quad \sigma_1' = \sigma_0' \setminus \{k\}.$$

As $\sum_{n\in\sigma_1} \alpha_n = \sum_{n\in\sigma_0} \alpha_n + \alpha_k > \sum_{n\in\sigma_0} \alpha_n = M$, we get by definition of M that $\sum_{n\in\sigma_1'} \alpha_n \leq M$. Now, using the assumption, we get

$$\begin{aligned} \alpha &= \sum_{n\in\sigma_1} \alpha_n + \sum_{n\in\sigma_1'} \alpha_n = \sum_{n\in\sigma_0} \alpha_n + \alpha_k + \sum_{n\in\sigma_1'} \alpha_n \leq M + \alpha_k + M \\ &= 2M + \alpha_k < \alpha - \alpha_0 + \alpha_k \leq \alpha - \alpha_0 + \alpha_0 = \alpha, \end{aligned}$$

which is impossible. □

3.2.13. COROLLARY. *Let Θ_n, $n \geq 1$, be H^∞ functions such that $\|\Theta_n\| \leq 1$ for every n, and assume that the product $\Theta = \prod_n \Theta_n$ converges. Then*

$$\frac{|\Theta(z)|}{\inf_n |\Theta_n(z)|} \leq \inf_\sigma (|\Theta_\sigma(z)| + |\Theta_{\sigma'}(z)|) \leq 2 \left(\frac{|\Theta(z)|}{\inf_n |\Theta_n(z)|} \right)^{1/2}$$

for every $z \in \mathbb{D}$.

Indeed, it is clear that it suffices to consider points $z \in \mathbb{D}$ where $\Theta(z) \neq 0$. Set $\alpha_n = \log \frac{1}{|\Theta_n(z)|}$. Then the previous lemma gives

$$\begin{aligned} \frac{|\Theta(z)|}{\inf_n |\Theta_n(z)|} &\leq \inf_\sigma \max (|\Theta_\sigma(z)|, |\Theta_{\sigma'}(z)|) \leq \inf_\sigma (|\Theta_\sigma(z)| + |\Theta_{\sigma'}(z)|) \\ &\leq 2 \inf_\sigma \max (|\Theta_\sigma(z)|, |\Theta_{\sigma'}(z)|) \leq 2 \left(\frac{|\Theta(z)|}{\inf_n |\Theta_n(z)|} \right)^{1/2}. \end{aligned}$$

□

The following theorem recapitulates the preceding results.

3.2.14. THEOREM. *Let Θ, $\Theta_n \in H^\infty$, $n \geq 1$, be inner functions and let $\Theta = \prod_{n\geq 1} \Theta_n$. The following assertions are equivalent.*

(1) The family $(\mathcal{K}_{\Theta_n})_{n\geq 1}$ is an unconditional basis in $\mathcal{K}_\Theta$.

(2) For every $\mu = (\mu_n)_{n\geq 1} \in l^\infty$, there exists a function $\varphi \in H^\infty$ such that

$$\varphi - \mu_n \in \Theta_n H^\infty, \quad n \geq 1.$$

(3) For every sequence of H^∞ functions $(f_n)_{n\geq 1}$ with $\sup_{n\geq 1} \|f_n\|_\infty < \infty$, there is a function $\varphi \in H^\infty$ satisfying

$$\varphi - f_n \in \Theta_n H^\infty, \quad n \geq 1.$$

(4) For every sequence of H^∞ functions $(f_n)_{n\geq 1}$ with $\sup_{n\geq 1} \operatorname{dist}(f_n \overline{\Theta}_n, H^\infty) < \infty$ (or equivalently, $\sup_{n\geq 1} \|f_n\|_{H^\infty/\Theta_n H^\infty} < \infty$), there is a function $\varphi \in H^\infty$ such that

$$\varphi - f_n \in \Theta_n H^\infty, \quad n \geq 1.$$

(5) There exists $\delta > 0$ such that

$$\text{(GC)} \qquad |\Theta(z)| \geq \delta \cdot \inf_n |\Theta_n(z)|, \quad z \in \mathbb{D}.$$

The condition (GC) is called *generalized Carleson condition.* For a sequence of inner functions $(\Theta_n)_{n\geq 1}$ satisfying (GC) we will write $(\Theta_n)_{n\geq 1} \in$ (GC).

3.2.15. Remarks on the local meaning of generalized interpolation. Here, we briefly comment on the meaning of the inclusions $f - f_n \in \Theta_n H^\infty$, $n \geq 1$, for some special cases (cf. also B.3.3).

(1) **Classical interpolation.**

Let $\Theta = B = \prod_{n\geq 1} b_{\lambda_n}$ be a Blaschke product with simple zeros $\lambda_n \in \mathbb{D}$, $\sum_{n\geq 1}(1 - |\lambda_n|) < \infty$. Set $\Theta_n = b_{\lambda_n}$. Then the interpolation condition

$$f - \mu_n \in \Theta_n H^\infty = b_{\lambda_n} H^\infty, \quad n \geq 1,$$

becomes

$$f(\lambda_n) = \mu_n, \quad n \geq 1.$$

Hence condition (2) of Theorem 3.2.14 (and consequently all assertions of this theorem) is equivalent to

$$H^\infty |\{\lambda_n\} = l^\infty$$

(note that the inclusion $H^\infty|\{\lambda_n\} \subset l^\infty$ is obvious).

In this case, interpolation of a sequence of complex numbers $(\mu_n)_n$ or H^∞ functions $(f_n)_n$ makes no difference (cf. conditions (2) and (3) of Theorem 3.2.5).

(2) **Multiple interpolation.**

Let $\Theta = B = \prod_{n\geq 1} b_{\lambda_n}^{d_n}$ be a Blaschke product with multiple zeros $\lambda_n \in \mathbb{D}$, $\lambda_n \neq \lambda_m$, $n \neq m$, $k_n \geq 1$, $\sum_{n\geq 1} d_n(1 - |\lambda_n|) < \infty$ (the Blaschke condition). Let further

$$\Theta_n = b_{\lambda_n}^{d_n}, \quad n \geq 1.$$

Then

$$f - f_n \in \Theta_n H^\infty = b_{\lambda_n}^{d_n} H^\infty, \quad n \geq 1$$

if and only if

$$f^{(j)}(\lambda_n) = f_n^{(j)}(\lambda_n), \quad 0 \leq j < d_n, \; n \geq 1.$$

In this case generalized interpolation means that we interpolate the values of f_n and of their derivatives. In other words, the germs of f and f_n of order d_n at λ_n coincide: their Taylor expansions $T_{d_n}(\cdot, \lambda_n)$ at λ_n of order d_n are the same:

$$T_{d_n}(f, \lambda_n) = T_{d_n}(f_n, \lambda_n), \quad n \geq 1.$$

Provided the sequence $(\Theta_n)_{n\geq 1}$ verifies the generalized Carleson condition, a necessary and sufficient condition for $(f_n)_{n\geq 1} \subset H^\infty$ to be interpolated by a function $f \in H^\infty$ is then that

$$\inf_{g\in H^\infty} \|f_n + \Theta_n g\|_\infty = \|f_n\|_{H^\infty/\Theta_n H^\infty} \leq \text{const}\,.$$

The space $H^\infty/b_{\lambda_n}^{d_n} H^\infty$ is sometimes called the *space of germs of height* d_n.

(3) **Asymptotic interpolation.**
Let

$$\Theta = \exp\left(-\int_{\mathbb{T}} \frac{\zeta + z}{\zeta - z}\, d\mu(\zeta)\right)$$

be a singular inner function associated with the discrete measure

$$\mu = \sum_{k\geq 1} a_k \delta_{\zeta_k},$$

where $\zeta_k \in \mathbb{T}$, $a_k > 0$, $\sum_{k\geq 1} a_k < \infty$ (cf. A.3.9.3, Volume 1). Define the inner function Θ_n associated with the singly supported measure $a_n\delta_{\zeta_n}$ by

$$\Theta_n = \Theta_{\zeta_n} = \exp\left(-\frac{\zeta_n + z}{\zeta_n - z}\cdot a_n\right), \quad n \geq 1.$$

Then

$$\Theta = \prod_{n\geq 1} \Theta_{\zeta_n}.$$

We claim that the interpolation condition $f - f_n \in \Theta_{\zeta_n} H^\infty$, $n \geq 1$, is now equivalent to

$$f(r\zeta_n) - f_n(r\zeta_n) = O\left(e^{-\frac{2a_n}{1-r}}\right), \quad r \to 1.$$

The necessity of this condition is obvious. The proof of the sufficiency is based on the Phragmen–Lindelöf principle (see Chapter A.4, Volume 1). In fact, using A.4.6, we get that Θ_{ζ_n} is outer on the half disks

$$\begin{aligned}\mathbb{D}_+ &= \{z \in \mathbb{D} : \operatorname{Im}(\overline{\zeta}_n z) > 0\},\\ \mathbb{D}_- &= \{z \in \mathbb{D} : \operatorname{Im}(\overline{\zeta}_n z) < 0\}.\end{aligned}$$

By assumption, the quotient $(f - f_n)/\Theta_{\zeta_n}^{1-\varepsilon}$ is bounded on the radius $[0, \zeta_n[$ for every $\varepsilon > 0$, and it is bounded on $\mathbb{T}$ by $\|f - f_n\|_\infty$. By the generalized maximum principle cited above (A.4.6), we get that $|(f - f_n)/\Theta_{\zeta_n}^{1-\varepsilon}| \leq C$ (with C not depending on ε) on $\mathbb{D}_\pm$ and hence on $\mathbb{D}$. Finally, $(f - f_n)/\Theta_{\zeta_n} \in H^\infty$ for every n. □

3.2.16. Corollary. *Suppose that $(\Theta_n)_{n\geq 1} \in (GC)$. Then for every sequence $(f_n)_{n\geq 1} \subset H^\infty$ with* $\sup_{n\geq 1}\|f_n\|_\infty < \infty$, *there exists $f \in H^\infty$ such that*

$$f(r\zeta_n) = f_n(r\zeta_n) + O(e^{-2a_n/1-r}), \quad n \geq 1.$$

□

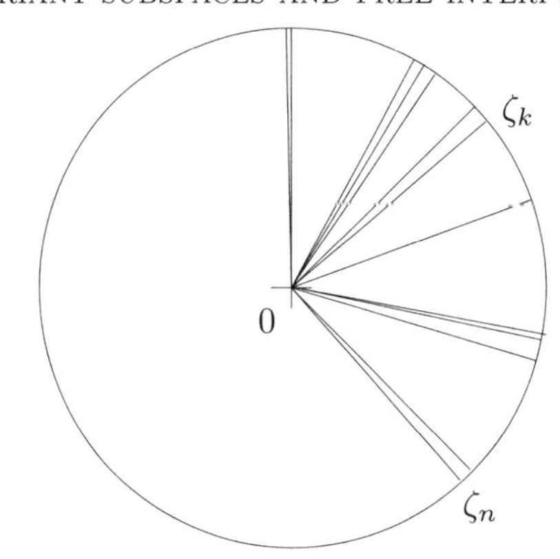

FIGURE 3.2.1

In other words, f behaves asymptotically as f_n on the radii $[0, \zeta_n[$ (which can be everywhere dense in $\mathbb{D}$) up to a given exponential remainder term (see Figure 3.2.1).

3.2.17. The (GC) condition and uniform minimality. Let $\Theta = \prod_n \Theta_n \in H^\infty$. We have already proved that the generalized Carleson condition

$$|\Theta(z)| \geq \delta \cdot \inf_{n\geq 1} |\Theta_n(z)|, \quad z \in \mathbb{D},$$

is equivalent to

$$\inf_{\sigma\subset\mathbb{N}} \inf_{z\in\mathbb{D}} \left(|\Theta_\sigma(z)| + |\Theta_{\sigma'}(z)|\right) > 0,$$

where $\Theta_\sigma = \prod_{n\in\sigma} \Theta_n$ for $\sigma \subset \mathbb{N}$, $\sigma' = \mathbb{N} \setminus \sigma$. By the relation between Θ_σ and $\mathcal{P}_\sigma$ (cf. 3.2.8, 3.2.13), this means that $\sup_\sigma \|\mathcal{P}_\sigma\| < \infty$. As in A.5.1 (Volume 1), we call the family $(\mathcal{K}_{\Theta_n})_{n\geq 1}$ *uniformly minimal* if the spectral projections $\mathcal{P}_n = \mathcal{P}_{\{n\}}$, $n \geq 1$, satisfy $\sup_{n\geq 1} \|\mathcal{P}_n\| < \infty$.

Here, we consider the special case $\Theta_n = b_{\lambda_n}$ and

$$\Theta = B = \prod_{n\geq 1} b_{\lambda_n},$$

where $(\lambda_n)_{n\geq 1}$ is a sequence of distinct points satisfying the Blaschke condition. The sequence $\Lambda = (\lambda_n)_{n\geq 1} \subset \mathbb{D}$ (or the sequence $(b_{\lambda_n})_{n\geq 1}$) is said to satisfy the *Carleson condition* if

$$\text{(C)} \qquad \delta = \inf_{n\geq 1} \prod_{k\neq n} |b_{\lambda_k}(\lambda_n)| > 0.$$

Setting $B_{\lambda_n} = B/b_{\lambda_n}$ this can be reformulated in the following way

$$\delta = \inf_{n\geq 1} |B_{\lambda_n}(\lambda_n)| > 0.$$

Using the same arguments as in Lemma 3.2.8, it is easily seen that this condition is equivalent to the uniform minimality of $(\mathcal{K}_{b_{\lambda_n}})_{n\geq 1}$, that is, $\sup_{n\geq 1} \|\mathcal{P}_n\| < \infty$. The following result shows that the a priori weaker uniform minimality condition already implies the unconditional basis property for the particular family $(\mathcal{K}_{b_{\lambda_n}})_{n\geq 1}$.

3.2.18. LEMMA. *Let $\Lambda = (\lambda_n)_{n\geq 1} \subset \mathbb{D}$ and $\Theta_n = b_{\lambda_n}$. Then*

$$(b_{\lambda_n})_{n\geq 1} \in (GC) \iff \Lambda \in (C).$$

PROOF. Consider first (GC)$\Longrightarrow$(C). Suppose that $z \longrightarrow \lambda_n$, then it is clear that

$$\inf_k |b_{\lambda_k}(z)| = |b_{\lambda_n}(z)|$$

(for z sufficiently close to λ_n). Hence, (GC) implies that $|B(z)| \geq \delta \cdot |b_{\lambda_n}(z)|$ for $|z - \lambda_n| < \eta$. We get

$$|\frac{B}{b_{\lambda_n}}(z)| \geq \delta$$

for $z - \lambda_n < \eta$, which yields $|B_{\lambda_n}(\lambda_n)| \geq \delta$.

Conversely, suppose that Λ satisfies (C). It suffices to prove (GC) for finite products $B = \prod_{k=1}^N b_{\lambda_k}$ (with δ independent of N). Recall first the following elementary facts:

(1) if $|b_\lambda(z)| < \varepsilon < 1$, then $|\lambda - z| < \frac{\varepsilon}{1-\varepsilon} \cdot \min\left((1 - |\lambda|^2), (1 - |z|^2)\right)$ (see 3.3.4 below);

(2) $|\varphi'(\xi)| \leq \|\varphi\|_\infty/(1 - |\xi|^2)$ for $|\xi| < 1$, $\varphi \in H^\infty$.

Define disks D_{λ_n} by

$$D_{\lambda_n} = \{z : |b_{\lambda_n}(z)| \leq \delta/3\}.$$

For $z \in D_{\lambda_n}$ there is $\xi \in D_{\lambda_n}$ such that

$$\begin{aligned} |B_{\lambda_n}(z)| &\geq |B_{\lambda_n}(\lambda_n)| - |B_{\lambda_n}(z) - B_{\lambda_n}(\lambda_n)| \geq \delta - |B'_{\lambda_n}(\xi)| \cdot |\xi - \lambda_n| \\ &\geq \delta - \frac{|\xi - \lambda_n|}{1 - |\xi|^2} \geq \delta - \delta/2 \\ &= \delta/2, \end{aligned}$$

where we have used $|\xi - \lambda_n| < \frac{\delta/3}{1-\delta/3} \cdot \min\left(1 - |\xi|^2, 1 - |\lambda_n|^2\right)$. Hence

$$|B(z)| = |B_{\lambda_n}(z)| \cdot |b_{\lambda_n}(z)| \geq \frac{\delta}{2} \cdot |b_{\lambda_n}(z)|$$

for $z \in \bigcup_{n=1}^N D_{\lambda_n}$. It remains to check the case $z \in \mathbb{D} \setminus \bigcup_{n=1}^N D_{\lambda_n}$. Suppose first that

$$z \in \partial\left(\bigcup_{n=1}^N D_{\lambda_n}\right) \subset \bigcup_{n=1}^N \partial D_{\lambda_n}.$$

Then there exists $n \in \{1, \dots, N\}$ such that $z \in \partial D_{\lambda_n}$, i.e. $|b_{\lambda_n}(z)| = \delta/3$. This yields

$$|B(z)| = |B_{\lambda_n}(z) b_{\lambda_n}(z)| = \frac{\delta}{3}|B_{\lambda_n}(z)| \geq \delta^2/6.$$

As $|B| \geq \delta^2/6$ on the boundary $\partial(\mathbb{D} \setminus \bigcup_1^N D_{\lambda_n})$ (and $|B| = 1$ on $\mathbb{T}$), and B is free of zeros in $\mathbb{D} \setminus \bigcup_1^N D_{\lambda_n}$, the classical maximum modulus principle now shows that $|B| \geq \delta^2/6$ on $\mathbb{D} \setminus \bigcup_1^N D_{\lambda_n}$. □

3.2.19. REMARK. In fact, it can be shown that the equivalence between the uniform minimality and the unconditional basis property remains true for a general family $(\mathcal{K}_{\Theta_n})_{n\geq 1}$ corresponding to a given factorization $\Theta = \prod \Theta_n$ of an inner function Θ into the product of inner functions Θ_n. Namely, the following assertions are equivalent.

(a) $(\Theta_n)_{n\geq 1} \in$ (GC).

(b) $\inf_n \inf_{\mathbb{D}}(|\Theta_n(z)| + |\Theta_n^c(z)|) > 0$, where $\Theta_n^c = \prod_{k\neq n} \Theta_k$.
(c) $\sup_n \|\mathcal{P}_n\| < \infty$.
(d) The sequence $(\mathcal{K}_{\Theta_n})_{n\geq 1}$ is uniformly minimal.
(e) The sequence $(\mathcal{K}_{\Theta_n})_{n\geq 1}$ is an unconditional basis in $\mathcal{K}_\Theta$.

See 3.3.5(a) below for the proof of (a)$\Longleftrightarrow$(b) and hence all other equivalences. See also the comments and references in Section 3.4.

It is worth mentioning once more that, from a general point of view, the properties of a sequence of subspaces to be minimal and to be an unconditional (Riesz) basis are quite far from each other. Why they coincide for the particular families $(\mathcal{K}_{\Theta_n})_{n\geq 1}$ is still not completely understood.

3.2.20. Free interpolation in H^2. We have seen that free interpolation in H^∞ is equivalent to $(\mathcal{K}_{\Theta_n})_{n\geq 1}$ being an unconditional basis of $\mathcal{K}_\Theta$, $\Theta = \prod_{n\geq 1} \Theta_n$. (As above, Θ and Θ_n are scalar inner functions). Now, we translate this condition into a kind of interpolation in H^2, free in a naturel sense. To this end, set

$$\mathcal{F}x = (\mathcal{P}_n x)_{n\geq 1},$$

where $x \in \mathcal{K}_\Theta$ and $\mathcal{P}_n$ are the coordinate (spectral) projections associated with $(\mathcal{K}_{\Theta_n})_{n\geq 1}$. Then, if $(\mathcal{K}_{\Theta_n})_{n\geq 1}$ is an unconditional basis in $\mathcal{K}_\Theta$, Theorems 3.1.4 and 3.2.5 assert that $\mathcal{F}\mathcal{K}_\Theta = l^2(\mathcal{K}_{\Theta_n})$ (which can be viewed as a kind of Riesz–Fischer theorem on orthogonal decompositions). It is clear that we have the same equality for the dual basis $(\mathcal{P}_n^*\mathcal{K}_\Theta)_{n\geq 1} = (\mathcal{K}'_{\Theta_n})_{n\geq 1}$, that is, $\mathcal{F}_*\mathcal{K}_\Theta = l^2(\mathcal{K}'_{\Theta_n})$, where $\mathcal{F}_*x = (\mathcal{P}_n^*x)_{n\geq 1}$. In other words, for every family $(f_n)_{n\geq 1}$, $f_n \in \mathcal{K}'_{\Theta_n}$, satisfying $\sum_{n\geq 1} \|f_n\|^2 < \infty$ there is a function $f \in H^2$ such that $\mathcal{F}_*f = (f_n)_{n\geq 1}$. Conversely, if $f \in H^2$, then $\sum_{n\geq 1} \|\mathcal{P}_n^*f\|^2 < \infty$, where the projections $\mathcal{P}_n^*$ have been trivially extended to the entire space H^2 by setting $\mathcal{P}_n^*|\Theta_n H^2 = \mathbb{O}$.

3.2.21. LEMMA. *With the above extension of $\mathcal{P}_n$ we have* $\operatorname{Ker} \mathcal{P}_n^* = \Theta_n H^2$.

PROOF. $\operatorname{Ker} \mathcal{P}_n^* = (\mathcal{P}_n H^2)^\perp = (\mathcal{K}_{\Theta_n})^\perp = \Theta_n H^2$. □

3.2.22. COROLLARY. *For $f, f_n \in H^2$, the following assertions are equivalent.*
(1) $\mathcal{P}_n^*f = \mathcal{P}_n^*f_n$.
(2) $f - f_n \in \Theta_n H^2$. □

3.2.23. THEOREM. *The following assertions are equivalent.*
(1) $(\Theta_n)_{n\geq 1} \in (GC)$.
(2) For every family $(f_n)_{n\geq 1}$ with $\sum_{n\geq 1} \|f_n\|^2 < \infty$ there is a function $f \in H^2$ such that $f - f_n \in \Theta_n H^2$.
Moreover, if $(\Theta_n)_{n\geq 1} \in (GC)$, and if $f \in H^2$, then

$$\sum_{n\geq 1} \|f\|^2_{H^2/\Theta_n H^2} < \infty.$$

PROOF. In view of the preceding observations, it is sufficient to remark that $\|f\|_{H^2/\Theta_n H^2} \leq \|\mathcal{P}_n^*f\|$ for every $f \in H^2$ (since $\mathcal{P}_n^*f - f \in \Theta_n H^2$; cf. Lemma 3.2.21). □

The assertion (2) gives the definition of *free interpolation in H^2*.

This theorem again expresses the meaning of *freedom* in this context. The only restriction that we have to impose on the interpolation data f_n is that concerning the norm. For instance, let us consider multiple interpolation: $\Theta = B = \prod_{n\geq 1} b_{\lambda_n}^{d_n}$, $\Theta_n = b_{\lambda_n}^{d_n}$ (cf. 3.2.15(2)), and suppose that $(\Theta_n)_{n\geq 1} \in (GC)$. If $(f_n)_{n\geq 1} \subset H^2$ and if there is $f \in H^2$ such that

$$\|f_n\|_{H^2/b_{\lambda_n}^{d_n}H^2} \leq C\|f\|_{H^2/b_{\lambda_n}^{d_n}H^2}$$

(which, in particular implies that $\sum_{n\geq 1} \|P_{b_{\lambda_n}^{d_n}} f\|_{H^2}^2 = \sum_{n\geq 1} \|f_n\|^2_{H^2/b_{\lambda_n}^{d_n}H^2} < \infty$), then there is $g \in H^2$ such that

$$g^{(j)}(\lambda_n) = f_n^{(j)}(\lambda_n), \quad 0 \leq j < d_n, n \geq 1.$$

3.2.24. Final remarks. (1) **Rearrangement of terms.**

What can be said about the interpolation data if we do not have free interpolation? For example, consider the classical case of a sequence of reproducing kernels $(k_{\lambda_n})_{n\geq 1}$ for which the corresponding sequence $\Lambda = (\lambda_n)_{n\geq 1}$ does not satisfy the Carleson condition. Then the generalized Fourier series

$$\sum_{n\geq 1} \mathcal{P}_n f = \sum_{n\geq 1} a_n k_{\lambda_n},$$

will not converge in general. Now one could think of rearranging the terms of this series in a suitable way

$$\sum_{n\geq 1} a_n k_{\lambda_n} = \sum_{s} \left(\sum_{n\in\sigma_s} a_n k_{\lambda_n} \right),$$

where $\{\sigma_s\}_{s\geq 1}$ is a partition of $\mathbb{N}$, to recover convergence. This fact can be formulated in a more geometric language. Consider

$$\mathcal{K}_{B_s} = \operatorname{span}\left(\mathcal{K}_{b_{\lambda_n}} : n \in \sigma_s\right),$$

where

$$B_s = \prod_{i\in\sigma_s} b_{\lambda_i}.$$

Then it seems more likely that the sequence $(B_s)_{s\geq 1}$ satisfies (GC) than the initial one $(b_{\lambda_n})_{n\geq 1}$.

(2) **Automatic rearrangements.**

Consider the level set

$$L_\varepsilon = \{z : |\Theta(z)| < \varepsilon\},$$

$0 < \varepsilon < 1$, and its partition into connected components $\{\Omega_n\}_{n\geq 1}$:

$$L_\varepsilon = \bigcup_n \Omega_n.$$

The inner function Θ decomposes into

$$\Theta = \prod_{k\geq 1} b_{\lambda_k} \cdot \exp\left(-\int_{\mathbb{T}} \frac{\zeta + z}{\zeta - z}\, d\mu(\zeta) \right).$$

Let now

$$\Theta_n = \Theta_{\Omega_n} = \prod_{\lambda_k\in\Omega_n} b_{\lambda_k} \cdot \exp\left(-\int_{\Omega_n\cap\mathbb{T}} \frac{\zeta + z}{\zeta - z}\, d\mu(\zeta) \right)$$

be the inner factor of Θ supported by the component Ω_n (see 3.3.5(c) below for the meaning of the set $\Omega_n \cap \mathbb{T}$). In the special case of a Blaschke product $\Theta = B = \prod_{n\geq 1} b_{\lambda_n}$ the factors $\Theta_n = B_n$ correspond to a kind of the above mentioned arrangement of the series (see (1)). It can be shown that the sequence $(\Theta_n)_{n\geq 1}$ satisfies the condition (GC); see 3.3.5(c) and comments in Section 3.4 below.

(3) **Local estimates of interpolation data**

The following estimate can be proved (see Subsection 2.5.7 above):

$$\|f\|_{H^\infty/\Theta_n H^\infty} = \mathrm{dist}(f\overline{\Theta_n}, H^\infty) \leq C(\varepsilon) \cdot \sup_{z\in L_\varepsilon(\Theta_n)} |f(z)|.$$

It enables us to consider more general interpolation data. In fact, if the family $(f_n)_{n\geq 1}$ is locally uniformly bounded, that is, $f_n \in H^\infty(L_\varepsilon(\Theta_n))$, $\|f_n\|_\infty \leq C$, then there is a function $f \in H^\infty(\mathbb{D})$ such that $f - f_n \in \Theta_n H^\infty(L_\varepsilon(\Theta_n))$, $n \geq 1$.

For comments, see Section 3.4 below.

3.3. Exercises and Further Results

3.3.1. Gram matrix and ortogonalizers. In this subsection, H is a Hilbert space and $X = (x_k)_{k\geq 1}$ a sequence in H. Define the *Gram matrix* of X by

$$G(X) = ((x_k, x_j))_{k,j\geq 1}$$

(a) (i) Show that $G = G(X)$ is a positive definite kernel on $\mathbb{N}$ (see B.6.5.2, Volume 1); in fact,

$$(Ga, a)_{l^2} = \|\sum_{k\geq 1} a_k x_k\|_H^2 = \|W(a)\|_H^2$$

for every $a \in l_0(\mathbb{N})$ (finitely supported sequences on $\mathbb{N}$); here

$$W = W_X : l_0(\mathbb{N}) \longrightarrow H$$

is a linear operator defined on basic vectors $e_k = (\delta_{kj})_{j\geq 1}$ of the space l^2 by $We_k = x_k$.

(ii) Let $Y = (y_k)_{k\geq 1} \subset H$; show that X is *unitarily equivalent* to Y, i.e., there exists a unitary operator $U : \mathrm{span}(X) \longrightarrow \mathrm{span}(Y)$ such that $Ux_k = y_k$ for every $k \geq 1$, if and only if $G(X) = G(Y)$.

(iii) Show that X is *isomorphic* to Y, i.e., there exists an isomorphism $U : \mathrm{span}(X) \longrightarrow \mathrm{span}(Y)$ such that $Ux_k = y_k$ for every $k \geq 1$, if and only if

$$c^{-1}G(Y) \leq G(X) \leq bG(Y)$$

for some $b, c > 0$.

(iv) In particular, X is isomorphic to an orthonormal sequence if and only if $G(X)$ is bounded and boundedly invertible.

(In this case X is called a *Riesz sequence* that means a Riesz basis in $\mathrm{span}(X)$; for Riesz bases, see the introduction to this chapter and Theorem 3.1.4).

(b) (i) Show that G extends to a bounded operator on $l^2(\mathbb{N})$, i.e.,

$$(Ga, a)_{l^2} = \|\sum_{k\geq 1} a_k x_k\|_H^2 \leq c^2\|a\|_{l^2}$$

for every $a \in l_0(\mathbb{N})$, if and only if X is the image of an orthonormal basis by a bounded operator, and if and only if W_X is bounded.

(In this case, X is called *"Hilbert system of vectors"*, and $\|G\| = c^2 = \|W\|^2$, where W is defined in (a)).

(ii) Show that X is linearly independent if and only if $(Ga, a)_{l^2} > 0$ for every $a \in l_0(\mathbb{N})$, $a \neq 0$. The operator G is bounded from below, i.e.

$$(Ga, a)_{l^2} = \| \sum_{k \geq 1} a_k x_k \|_H^2 \geq c^{-2} \|a\|_{l^2}$$

for every $a \in l_0(\mathbb{N})$, where $c > 0$, if and only if the *orthogonalizer*

$$V_X : x_k \longrightarrow e_k, \quad k \geq 1,$$

extends to a bounded operator from $\text{span}(X)$ to l^2; moreover, in this case, $(x_k)_{k\geq 1}$ is minimal and

$$V_X y = ((y, f_k))_{k \geq 1}$$

for every $y \in \text{span}(X)$, where $(f_k)_{k\geq 1}$ is a biorthogonal sequence. If $\sup_k \|x_k\| < \infty$ and G is bounded from below (i.e., V_X is bounded) then X is uniformly minimal (see A.5.1, Volume 1).

(If V_X is bounded, X is called a *"Bessel system of vectors"*, and $\|G^{-1}\| = c^2 = \|V_X\|^2$).

(iii) Show that X is uniformly minimal (see A.5.1, Volume 1) if and only if

$$c := \sup_{k\geq 1} \sup \Big| \sum_{j, j\neq k} a_j (x_j, \frac{x_k}{\|x_k\|}) \Big| < 1,$$

where the interior sup is taken over all finite families (a_j), $a_j \in \mathbb{C}$ with $\| \sum_{j,j\neq k} a_j x_j \| \leq 1$. In particular, if X is uniformly minimal and normalized, $\|x_k\| = 1$ for $k = 1, 2, ...$, then

$$\sup_{j \neq k} |(x_j, x_k)| < 1.$$

[*Hint:* by A.5.1.11 and A.5.1.12 (Volume 1), $c^2 = \sup_{k\geq 1}(1 - \|\mathcal{P}_k\|^{-2})$, where $\mathcal{P}_k$ is defined in 3.1.2 (and in A.5.1.9); since, by Lemma A.5.1.12 (Volume 1),

$$\|\mathcal{P}_k\|^{-2} \geq \inf_{k\geq 1} \|\mathcal{P}_k\|^{-2} = \inf_{k\geq 1} \text{dist} \left(\frac{x_k}{\|x_k\|}, \text{span}(x_j : j \neq k) \right)^2 > 0;$$

the result follows.]

(c) Let X be a minimal sequence and $X' = (f_k)_{k\geq 1}$ its biorthogonal such that $\text{span}(X') \subset \text{span}(X)$. Show that

1)

$$(W_X a, W_{X'} b)_H = (a, b)_{l^2}$$

for every $a, b \in l_0(\mathbb{N})$.

2) $W_X V_X = I$ on $\mathcal{L}\text{in}(X)$, $V_X W_X = I$ on $l_0(\mathbb{N})$, $W_{X'} V_{X'} = I$ on $\mathcal{L}\text{in}(X')$, $V_{X'} W_{X'} = I$ on $l_0(\mathbb{N})$, where

$$V_{X'} y = ((y, x_k))_{k\geq 1};$$

3) $(u, v)_H = (V_X u, V_{X'} v)_{l^2}$ for every $u \in \mathcal{L}\text{in}(X)$ and $v \in \mathcal{L}\text{in}(X')$.

(d) Let X be a minimal sequence and let $X' = (f_k)_{k\geq 1}$ be its biorthogonal such that span(X') = span(X). Show that the following are equivalent.

1) X is a Riesz sequence;
2) $G(X)$ extends to a bounded operator on $l^2(\mathbb{N})$ having bounded inverse;
3) W_X and $W_{X'}$ are bounded;
4) V_X and $V_{X'}$ are bounded;
5) W_X and V_X are bounded;
6) $W_{X'}$and $V_{X'}$ are bounded;
7) X' is a Riesz sequence.

[*Hint:* for 1) $\Leftrightarrow$ 2) see (a) (iv) above; 2) $\Leftrightarrow$ 3) follows from (a) and (c), 1); 1) $\Leftrightarrow$ 4) follows from (c), 3); 5) $\Leftrightarrow$ 1) (or 6) $\Leftrightarrow$ 1)) follows from (c), 2); 6) $\Leftrightarrow$ 7) by the previous equivalences.]

(e)* *Vector valued Gram matrices (N. Nikolski, 1978).* Let $\mathcal{X} = (X_n)_{n\geq 1}$ be a minimal sequence of subspaces of a Hilbert space H, and let $\mathcal{P}_n$ be the corresponding skew projection onto X_n (see 3.1.2 above for the definition), $\mathcal{J}_{\mathcal{X}}x = (\mathcal{P}_n x)_{n\geq 1}$ and $J_{\mathcal{X}}x = (P_n x)_{n\geq 1}$, where P_n stands for the orthogonal projection onto X_n. Let further $G_{\mathcal{X}} = J_{\mathcal{X}}J_{\mathcal{X}}^*$ be the *Gram operator* of $\mathcal{X}$ defined on finitely supported sequences of the space

$$l^2(X_n) = \{(x_n)_{n\geq 1} : x_n \in X_n, \sum_{n\geq 1} \|x_n\|^2 < \infty\}.$$

The following are equivalent.

(i) $\mathcal{X}$ is an unconditional basis of $H(\mathcal{X}) = span(\mathcal{X})$.
(ii) $\mathcal{X}$ is a Riesz basis of $H(\mathcal{X})$.
(iii) $\mathcal{J}_{\mathcal{X}}H \subset l^2(H)$ and $\mathcal{J}_{\mathcal{X}'} \subset l^2(H)$, where $\mathcal{X}' = (X'_n)_{n\geq 1}$ is the biorthogonal sequence (see 3.1.2 for the definition).
(iv) $G_{\mathcal{X}}$ extends to a bounded and invertible transformation from $l^2(X_n)$ onto itself.
(v) $J_{\mathcal{X}}H = l^2(X_n)$.
(vi) $J_{\mathcal{X}}H \supset l^2(X_n)$.

(f) Let E, H be Hilbert spaces, $A \in L(E \longrightarrow H)$ and $x_k = Ae_k$, where $(e_k)_{k\geq 1}$ is an orthonormal basis in E (so that, $X = (x_k)_{k\geq 1}$ is a Hilbert system). Show that

(i) X is minimal if and only if $e_k \in A^*H$ for every $k \geq 1$;
(ii) X is uniformly minimal if and only if $\sup_{k\geq 1} \|e_k\|_{A^*} < \infty$, where $\|e_k\|_{A^*} = \inf\{\|h\| : A^*h = e_k\}$ is the range norm on A^*H ($= \|A^{*-1}e_k\|$ if A^* is injective);
(iii) X is a Bessel system (and, hence, a Riesz sequence) if and only if $A^*H = E$, and if and only if the norms $\|\cdot\|_E$ and $\|\cdot\|_{A^*}$ are equivalent (and, assuming that A^* is further injective, if and only if A^{*-1}, and/or A^{-1}, is bounded).

(g) *Best approximation and orthogonalization.* Let X be a linearly independent sequence, $L_n = \mathcal{L}\text{in}(X_n)$ where $X_n = (x_k)_{k=1}^n$, and $G(X_n)$, $G_n = \det G(X_n)$ be the n-th finite section of $G(X)$ and its determinant, respectively.

(i) Let $x \in H$. Show that

$$(x', y) = -\frac{1}{G_n} \cdot \det \begin{pmatrix} G(X_n) & (X_n^{col}, y) \\ (x, X_n) & 0 \end{pmatrix}$$

for every $y \in H$, where $x' = P_{L_n} x$ is the orthogonal projection of x on L_n and (x, X_n) means the line $(x, x_1), ..., (x, x_n)$ (and (X_N^{col}, y) is defined in a similar way); moreover,

$$(x - x', y) = \frac{1}{G_n} \cdot \det \begin{pmatrix} G(X_n) & (X_n^{col}, y) \\ (x, X_n) & (x, y) \end{pmatrix}$$

for every $y \in H$.

[*Hint:* let $x' = P_{L_n} x = \sum_{k=1}^n \lambda_k x_k$ be the orthogonal projection of x on L_n; the equations $\sum_{k=1}^n \lambda_k (x_k, x_j) + (-1)(x, x_j) = 0$ $(j = 1, ..., n)$ and $\sum_{k=1}^n \lambda_k (x_k, y) + (-1)(x', y) = 0$ mean that

$$\det \begin{pmatrix} G(X_n) & (X_n^{col}, y) \\ (x, X_n) & (x', y) \end{pmatrix} = 0,$$

which implies the first formula; the second follows similarly.]

(ii) Let $d = \operatorname{dist}(x, L_n)$. Show that

$$d^2 = \frac{\det G[X_n, x]}{\det G(X_n)},$$

where $[X_n, x] = (x_1, ..., x_n, x)$.

[*Hint:* as in (i), the equations $(x', x_j) + (-1)(x, x_j) = 0$ $(j = 1, ..., n)$, $(x', x) + (-1)((x, x) - d^2) = 0$ imply that

$$\det \begin{pmatrix} G(X_n) & (X_n^{col}, x) \\ (x, X_n) & (x, x) - d^2 \end{pmatrix} = 0,$$

and hence the result.]

(iii) *The Gram–Schmidt orthogonalization process.* Show that the *orthogonalization* of X_n (i.e., the orthonormal sequence $(e_k)_{k=1}^n$ defined by $\mathcal{L}in(x_1, ...x_k) = \mathcal{L}in(e_1, ..., e_k)$ and $(x_k, e_k) > 0$ for $k = 1, ..., n$) is given by the symbolic determinants (see (i) for their linearized form)

$$e_k = \frac{1}{(G_{k-1} G_k)^{1/2}} \cdot \det \begin{pmatrix} G(X_{k-1}) & X_{k-1}^{col} \\ (x_k, X_{k-1}) & x_k \end{pmatrix}$$

for $k = 1, ..., n$.

[*Hint:* write $e_k = (x_k - P_{L_{k-1}} x_k)/\|x_k - P_{L_{k-1}} x_k\|$ and apply (i).]

(h) *Hadamard's inequality.*

(i) Show that in the notation of (g)

$$\det G(X_n) \le \prod_{k=1}^n \|x_k\|^2,$$

and that equality holds if and only if vectors x_k, $1 \le k \le n$, are pairwise orthogonal.

[*Hint:* by definition of $d = \mathrm{dist}(x_n, L_{n-1})$ and by (g) (ii) we get $\det G(X_n) \le \|x_n\|^2 \det G(X_{n-1})$, and the equality holds if and only if $x_n \perp L_{n-1}$; the result follows by induction.]

(ii*) $\det G(X_{1,n}) \le \det G(X_{1,k}) \cdot \det G(X_{k+1,n})$, where $X_{j,k} = (x_i)_{j \le i \le k}$ and $1 \le k < n$.

(i)* *(C. Müntz (1914), O. Szasz (1916))* Let $H = L^2(0,1)$; $\alpha, \alpha_k \in \mathbb{C}$ with $\mathrm{Re}(\alpha) > -1/2$, $\mathrm{Re}(\alpha_k) > -1/2$, $k = 1, 2, ...$, and let $t \longmapsto t^\alpha$ be a function in $L^2(0,1)$. Then,

$$\det G(t^{\alpha_1}, ..., t^{\alpha_n}) = \frac{\prod_{1 \le k < j \le n} |\alpha_j - \alpha_k|^2}{\prod_{1 \le j,k \le n} (1 + \alpha_j + \overline{\alpha}_k)}$$

and hence

$$\mathrm{dist}(t^\alpha, \mathcal{L}\mathrm{in}(t^{\alpha_1} ..., t^{\alpha_n})) = \frac{1}{1 + 2\,\mathrm{Re}(\alpha)} \cdot \prod_{k=1}^{n} \left| \frac{\alpha - \alpha_k}{1 + \alpha + \overline{\alpha}_k} \right|^2.$$

In particular, $\mathrm{span}(t^{\alpha_k} : k \ge 1) = L^2(0,1)$ if and only if

$$\sum_{k \ge 1} \frac{\mathrm{Re}(2\alpha_k + 1)}{1 + |2\alpha_k + 1|^2} = \infty.$$

[*Hint:* in order to get the completeness theorem, apply the second formula to $\alpha = 0, 1, 2, ...$, and use the classical Weierstrass theorem; for another proof see D.4.7.2(a) below.]

3.3.2. The Cotlar–Stein almost orthogonality lemma. (a) *(M. Cotlar (1955); E. Stein (1967))* Let $(T_j)_{1 \le j \le n}$ be a family of (bounded) Hilbert space operators, and let

$$a(i,j) = \|T_i^* T_j\|^{1/2}, \quad b(i,j) = \|T_i T_j^*\|^{1/2}.$$

Let A and B be the operators with matrices a and b, respectively, acting on $E = (\mathbb{C}^n, \|\cdot\|_E)$ endowed with a Banach space norm $\|\cdot\|_E$ (for instance, it may be the usual Euclidean norm). Show that

$$\Big\| \sum_j T_j \Big\| \le \|AB\|^{1/2} \le \Big(\|A\| \cdot \|B\| \Big)^{1/2}.$$

[*Hint:* let $T = \sum_j T_j$; since $\|T^*T\| = \|T\|^2$, and hence

$$\|(T^*T)^m\| = \|T\|^{2m},$$

it suffices to estimate this last norm; we have

$$(T^*T)^m = \sum T_{j_1}^* T_{j_2} T_{j_3}^* ... T_{j_{2m}},$$

where the summation is over all j_k running over $1, ..., n$; next,

$$\begin{aligned} \|T_{j_1}^* T_{j_2} T_{j_3}^* ... T_{j_{2m}}\| &\le \|T_{j_1}^* T_{j_2}\| \cdot \|T_{j_3}^* T_{j_4}\| ... \|T_{j_{2m-1}}^* T_{j_{2m}}\| \\ \|T_{j_1}^* T_{j_2} T_{j_3}^* ... T_{j_{2m}}\| &\le \|T_{j_1}^*\| \cdot \|T_{j_2} T_{j_3}^*\| \cdot \|T_{j_4} T_{j_5}^*\| ... \|T_{j_{2m-2}} T_{j_{2m-1}}^*\| \cdot \|T_{j_{2m}}\|, \end{aligned}$$

or, taking the geometric mean of these estimates and denoting by $x(j) = \|T_j\|^{1/2} = a(j,j)^{1/2} = b(j,j)^{1/2}$,

$\|T^*_{j_1} T_{j_2} T^*_{j_3} ... T_{j_{2m}}\| \le x(j_1)a(j_1,j_2)b(j_2,j_3)...b(j_{2m-2},j_{2m-1})a(j_{2m-1},j_{2m})x(j_{2m})$;

summing up these estimates, we get

$\|(T^*T)^m\| \le \sum \|T^*_{j_1} T_{j_2} T^*_{j_3} ... T_{j_{2m}}\| \le (x, AB...BAx) \le \|AB\|^{m-1}\|x\|_{E^*}\|Ax\|_E$,

where $(\cdot,\cdot)$ stands for the standard $\mathbb{C}^n$ scalar product and E^* means the dual space of E with respect to $(\cdot,\cdot)$; finally, let $m \longrightarrow \infty$ in $\|T\| = \|(T^*T)^m\|^{1/2m}$, to get the the result.]

(b) In the notation of (a), show that

(i) the least norms $\|A\|_{E\longrightarrow E}$ and $\|B\|_{E\longrightarrow E}$ are attained for $E = \mathbb{C}^n$ endowed with the standard Euclidean norm;

(ii) for $E = \mathbb{C}^n$, $\|A\| \le \sup_i \sum_j a(i,j)$ and $\|B\| \le \sup_i \sum_j b(i,j)$; in fact,

$$\|A\| \le \sup_i (x_i^{-1} \sum_j a(i,j)x_j)$$

for every positive vector $x \in \mathbb{C}^n$ ($x_i > 0$ for every i), and there exists x for which the equality holds (and similarly for B);

(iii) the norm $\|C\|$, where $C = AB = (c((i,j))$, can be estimated similarly, $\|C\| \le \sqrt{\alpha\beta}$, where $\alpha = \sup_i(y_i^{-1}\sum_j c(i,j)x_j)$, $\beta = \sup_j(x_j^{-1}\sum_i c(i,j)y_j)$ for some positive vectors x, y.

[*Hint:* for (i), observe that A and B are selfadjoint on $\mathbb{C}^n$, and hence $\|A\| = r_{\mathbb{C}^n}(A) = r_E(A) \le \|A\|_{E\longrightarrow E}$, where $r_E(A)$ means the spectral radius of A;

for (ii) and (iii), apply the Schur test (see Section B.5.3 or B.5.6.7, Volume 1) and the Frobenius theorem (see B.5.6.7, Volume 1).]

(c) Let $x_k \in H$ for $k \ge 1$, where H is a Hilbert space, and suppose that

$$\sup \|\sum \varepsilon_k x_k\| < \infty,$$

where the supremum is taken over all families of complex numbers ε_k, with $|\varepsilon_k| \le 1$ and with $\varepsilon_k \ne 0$ for only finitely many k; show that 1) $\sum_{k\ge1}\|x_k\|^2 < \infty$, and 2) the series $\sum_{k\ge1} x_k$ unconditionally converges.

[*Hint:* property 1) follows from Orlicz's lemma 3.1.5; in order to prove 2), observe first that the hypothesis is unordered, and so, it suffices to check simple convergence; assuming that the series diverges, there exist infinitely many disjoint finite sums $y_j = \sum_{k\in I(j)} x_k$ satisfying $\|y_j\| > \delta > 0$; this contradicts to 1) applied to $\sum y_j$.]

(d) *A series version of (a).* Let $(T_j)_{j\ge1}$ be a family of (bounded) Hilbert space operators, and suppose that every finite sequence $(T_j)_{1\le j\le n}$ satisfies the assumptions of (a), with $\sup_n \|A_nB_n\|^{1/2} < \infty$ (in an obvious adaptation of the notation of (a)). Show that the series $\sum_j T_j x$ converges unconditionally for every $x \in H$, and

$$\Big\|\sum_j T_j\Big\| \le \sup_n \|A_nB_n\|^{1/2} \le (\sup_n \|A_n\| \cdot \|B_n\|)^{1/2}.$$

[*Hint:* this is an immediate consequence of (a) and (c).]

3.3.3. Classical free interpolation and bases of reproducing kernels. In this subsection, we present several other approaches to the free interpolation problem and the reproducing bases problem which are independent of the Carleson corona theorem as it was the case in Section 2 of this chapter. Below, Λ means an at most countable subset of the unit disk $\mathbb{D} = \{\zeta \in \mathbb{C} : |\zeta| < 1\}$; sometimes, we enumerate Λ as a sequence $\Lambda = (\lambda_n)_{n\geq 1}$ so that $\lambda_j \neq \lambda_k$ for $j \neq k$.

(a) *Minimal families of reproducing kernels.*
Let $k_\lambda(z) = (1 - \overline{\lambda}z)^{-1}$ for $\lambda \in \mathbb{D}$, and $1 \leq p \leq \infty$.

(i) Show that the following are equivalent.

(1) The family $(k_\lambda)_{\lambda\in\Lambda}$ is minimal.
(2) The family $(k_\lambda)_{\lambda\in\Lambda}$ is not complete in H^p (is not weak* complete for $p = \infty$).
(3) Λ is a Blaschke sequence (see A.3.7, Volume 1).

[*Hint:* for (2) $\Leftrightarrow$ (3), see A.5.7.3(n), Volume 1, for $1 < p < \infty$ and A.5.8, Volume 1, for $p = 1, \infty$; (1) is equivalent to (3) by the same reason since minimality means existence of functions f_λ in $H^{p'}$, $p^{-1} + {p'}^{-1} = 1$ (or $BMOA$, or $P_+L^1(\mathbb{T})$, see A.5.8, Volume 1, for comments) such that $(f_\lambda, k_\mu) = f_\lambda(\mu) = \delta_{\lambda,\mu}$ for $\lambda, \mu \in \Lambda$.]

(ii) Let Λ be a Blaschke sequence and let $B = \prod_{n\geq 1} b_{\lambda_n}$, where (as always) $b_\lambda(z) = \varepsilon_\lambda(\lambda - z)(1 - \overline{\lambda}z)^{-1}$ and $\varepsilon_\lambda = |\lambda|/\lambda$. Let further

$$\mathcal{K}_B^p = \operatorname{span}_{H^p}(k_\lambda : \lambda \in \Lambda).$$

Show that $\mathcal{K}_B^p = H^p \cap (B\overline{H_-^p})$ and that there exists a unique biorthogonal system (h_λ) contained in $\mathcal{K}_B^{p'}$, namely,

$$h_\lambda = (1 - |\lambda|^2)\frac{B_\lambda}{B_\lambda(\lambda)} \cdot k_\lambda,$$

where $B_\lambda = B/b_\lambda$.

[*Hint:* clearly, $h_\lambda \in \mathcal{K}_B^{p'}$ and $(k_\mu, h_\lambda) = \delta_{\lambda,\mu}$; any other biorthogonal g_λ from $\mathcal{K}_B^{p'}$ satisfies $h_\lambda - g_\lambda \in BH^{p'} \cap \overline{H_-^{p'}}$, and hence $g_\lambda = h_\lambda$.]

(iii) Let $p = 2$, and let (h_λ) be the family from point (ii) (above) written in the form

$$h_\lambda = c_\lambda g_\lambda, \quad g_\lambda = \frac{B}{z - \lambda},$$

where $c_\lambda = -\overline{\varepsilon}_\lambda(1 - |\lambda|^2)/B_\lambda(\lambda)$. Show that $Vk_\lambda = g_\lambda$ for every $\lambda \in \Lambda$, where V is the conjugate linear isometric mapping from $\mathcal{K}_B^2$ onto $\mathcal{K}_B^2$ given by

$$Vf = \overline{z}B\overline{f}, \quad f \in \mathcal{K}_B^2.$$

In particular, $\operatorname{span}(h_\lambda : \lambda \in \Lambda) = \mathcal{K}_B^2$.

[*Hint:* since $\mathcal{K}_B^p = H^p \cap (B\overline{H_-^p})$, we have $V\mathcal{K}_B^p \subset \mathcal{K}_B^p$; in fact, V is onto because it is an involution, $V^2 = I$.]

(b) Let $x_\lambda = k_\lambda/\|k_\lambda\|_2 = (1-|\lambda|^2)^{1/2}k_\lambda$ be normalized reproducing kernels in H^2. Show that the Gram matrix of $\mathcal{X} = (x_\lambda)_{\lambda\in\Lambda}$ is given by

$$(x_\lambda, x_\mu) = \frac{(1-|\lambda|^2)^{1/2}(1-|\mu|^2)^{1/2}}{1-\overline{\lambda}\mu};$$

if $\mathcal{X}$ is minimal then the biorthogonal system $\mathcal{X}' = (f_\lambda)$ from $\mathcal{K}_B = \mathcal{K}_B^2$ is

$$f_\lambda = (1-|\lambda|^2)^{1/2}\frac{B_\lambda}{B_\lambda(\lambda)}\cdot k_\lambda$$

(see (a) (ii)), and hence

$$\|x_\lambda\|\cdot\|f_\lambda\| = |B_\lambda(\lambda)|^{-1},$$

and the Gram matrix of $\mathcal{X}'$ is given by

$$(f_\lambda, f_\mu) = \frac{(1-|\lambda|^2)^{1/2}(1-|\mu|^2)^{1/2}}{1-\overline{\mu}\lambda}\cdot\frac{\overline{\varepsilon}_\lambda\varepsilon_\mu}{B_\lambda(\lambda)\overline{B}_\mu(\mu)}.$$

(c) *Interpolation, bases, and the Carleson embeddings.* Let $\Lambda = (\lambda_n)_{n\geq 1}$ be a Blaschke sequence and $\mathcal{X} = (x_\lambda)_{\lambda\in\Lambda} = (k_\lambda/\|k_\lambda\|_2)_{\lambda\in\Lambda}$ be a family of normalized reproducing kernels of H^2. Show that the following are equivalent.

1) $\mathcal{X}$ is a Riesz basis in $\mathcal{K}_B$.
2) $\mathcal{X}$ is a uniformly minimal sequence.
3) Λ satisfies the *Carleson condition (C)*

(C) $$\delta = \delta(\Lambda) = \inf_{\lambda\in\Lambda}|B_\lambda(\lambda)| > 0$$

4) Λ satisfies (C) and the restriction operator $j(f) = f|\Lambda$ is bounded on normalized reproducing kernels x_λ, $\lambda\in\Lambda$, as an operator from H^2 to $l^2(w)$, where $w(n) = 1-|\lambda_n|^2$, $n\geq 1$; this means that

$$\sup_{\lambda\in\Lambda}\|j(x_\lambda)\|_{l^2(w)} < \infty.$$

5) Λ satisfies (C) and the restriction operator $j : H^2 \longrightarrow l^2(w)$ is bounded.
6) $H^2|\Lambda = l^2(w)$.
7) $H^2|\Lambda \supset l^2(w)$.
8) The space $H^2|\Lambda$ is a lattice, i.e. $a\in H^2|\Lambda$ and $|b(\lambda)|\leq|a(\lambda)|$ for every $\lambda\in\Lambda$ imply that $b\in H^2|\Lambda$.
9) The series

$$Pf = \sum_{n\geq 1} f(\lambda_n)\frac{B}{B'(\lambda_n)(z-\lambda_n)}$$

converges unconditionally for every $f\in H^2$ (and represent the orthogonal projection onto $\mathcal{K}_B$).
10) The *interpolation operator* In,

$$(In)a = \sum_{n\geq 1} a_{\lambda_n}(1-|\lambda_n|^2)\cdot k_{\lambda_n}\cdot\frac{B_{\lambda_n}}{B_{\lambda_n}(\lambda_n)},$$

is bounded from $l^2(w)$ (see 4) above) into $\mathcal{K}_B$ and

$$(In)a|\Lambda = a, \quad a\in l^2(w).$$

11) $H^\infty|\Lambda = l^\infty$.

[*Hint:* clearly, 1) $\Rightarrow$ 2); 2) $\Leftrightarrow$ 3) since $\|x_\lambda\| \cdot \|f_\lambda\| = |B_\lambda(\lambda)|^{-1}$ (see (b)); 3) $\Rightarrow$ 4) because

$$\begin{aligned} -\infty &< \log(\delta(\Lambda)^2) \leq \log|B_{\lambda_n}(\lambda_n)|^2 = \sum_{k\neq n} \log|\frac{\lambda_k - \lambda_n}{1-\overline{\lambda}_k\lambda_n}|^2 \\ &= \sum_{k\neq n} \log(1 - \frac{(1-|\lambda_k|^2)(1-|\lambda_n|^2)}{|1-\overline{\lambda}_k\lambda_n|^2}) \\ &\leq -\sum_{k\neq n} \frac{(1-|\lambda_k|^2)(1-|\lambda_n|^2)}{|1-\overline{\lambda}_k\lambda_n|^2}; \end{aligned}$$

4) $\Rightarrow$ 5) because (C) $\Rightarrow$ (GC) (Lemma 3.2.18), whence $|B(z)| \geq \delta \cdot \inf_n |b_{\lambda_n}(z)|$, which implies (as above) that

$$\begin{aligned} -\infty &< \log(\delta^2) \leq \sup_n \sum_{k\neq n} \log|b_{\lambda_k}(z)|^2 \leq \sup_n \left(-\sum_{k\neq n} \frac{(1-|\lambda_k|^2)(1-|z|^2)}{|1-\overline{\lambda}_k z|^2} \right) \\ &\leq 1 - \sum_{k\geq 1} \frac{(1-|\lambda_k|^2)(1-|z|^2)}{|1-\overline{\lambda}_k z|^2} \end{aligned}$$

for every $z \in \mathbb{D}$; the latter inequality means that $\sup_{z\in\mathbb{D}} \|j(x_z)\|_2 < \infty$, and hence (by Theorem A.5.7.2(b), Volume 1) the embedding operator j is bounded;

5) $\Rightarrow$ 1) by 3.3.1(d), since j is equivalent to $V_{\mathcal{X}'}$ and $V_{\mathcal{X}}$ is equivalent to $V_{\mathcal{X}'}$ in view of (C) and property (a) (iii);

next, 1) $\Rightarrow$ 6), since by 3.3.1(a) (iv), $c\|f\|_2 \leq \|V_{\mathcal{X}'}f\|_{l^2} \leq b\|f\|_2$ for some constants $b, c > 0$ and for every finite sum $f = \sum a_k f_{\lambda_k}$ (since $V_{\mathcal{X}'}f = a$ is an arbitrary finitely supported sequence, the result follows);

obviously, 6) $\Rightarrow$ 7), whereas 7) implies 2), because by the closed graph theorem the emebdding $l^2(w) \subset H^2|\Lambda$ is continuous (the latter space is endowed with the norm of the quotient space H^2/BH^2), and hence there exists functions $f_n \in H^2$ whose norm is comparable with $\|e_n\|_{l^2(w)} = (1-|\lambda_n|^2)^{1/2}$ and such that $f_n(\lambda_k) = \delta_{n,k}$; therefore, 7) $\Rightarrow$ 2) $\Rightarrow$ 1);

to join 8), observe that obviously 6) $\Rightarrow$ 8), and, conversely, the Orlicz lemma 3.1.5 entails that a Hilbert space lattice on $\mathbb{N}$ is necessarily a weighted l^2 space, i.e. $l^2(w)$; in order to check that the weight w is equivalent to $(1-|\lambda_n|^2)$, use the same reasoning as in the previous paragraph;

the equivalences 1) $\Leftrightarrow$ 9) $\Leftrightarrow$ 11) are special cases of Theorem 3.2.5;

and finally, 1) $\Rightarrow$ 10) by 3.3.1(a) (more precisely, by 3.3.1(a)(i), 3.3.1(a)(iv)) since $In = W_{\mathcal{X}'}$, and (obviously) 10) $\Rightarrow$ 3) $\Rightarrow$ 1).]

(d) *Reproducing kernel bases via the Cotlar–Stein lemma.* In the notation of (b) and (c), let $\mathcal{P}_\lambda = (\cdot, f_\lambda)k_\lambda$ be the skew projections corresponding to the biorthogonal system $(k_\lambda)_{\lambda\in\Lambda}$, $(f_\lambda)_{\lambda\in\Lambda}$ (see Section 3.2). Show that the entries of the Cotlar matrices $(\|\mathcal{P}_\lambda^*\mathcal{P}_\mu\|)_{\lambda,\mu}$ and $(\|\mathcal{P}_\lambda\mathcal{P}_\mu^*\|)_{\lambda,\mu}$ coincide with the modulus of the entries of the Gram marix given in (b) above; namely,

$$\|\mathcal{P}_\lambda^*\mathcal{P}_\mu\| = \|\mathcal{P}_\lambda\mathcal{P}_\mu^*\| = \frac{(1-|\lambda|^2)^{1/2}(1-|\mu|^2)^{1/2}}{|1-\overline{\mu}\lambda| \cdot |B_\lambda(\lambda)| \cdot |\overline{B}_\mu(\mu)|}.$$

Therefore, $(k_\lambda)_{\lambda\in\Lambda}$ is an unconditional (Riesz) basis in $\mathcal{K}_B$ if the Cotlar matrix $C = (\|\mathcal{P}_\lambda^*\mathcal{P}_\mu\|)_{\lambda,\mu}$ is bounded on $l^2(\Lambda)$.

[*Hint:* the formula for $\|\mathcal{P}_\lambda^*\mathcal{P}_\mu\|$ is straightforward; the sufficiency of $\|C\| < \infty$ follows from the Cotlar–Stein lemma 3.3.2(a).]

(e) *The interpolation operator of Jones–Vinogradov, the spaces $\mathcal{K}_B^p$ and the Frostman condition.*

Given an inner function Θ, let $\mathcal{K}_\Theta^p = H^p \cap \Theta H_-^p$ for $1 \le p \le \infty$.

(i) *(P. Jones; S. Vinogradov (1983))* Let $\Lambda = (\lambda_n)_{n\ge1}$ be a Blaschke sequence ordered in such a way that $|\lambda_n| \le |\lambda_{n+1}|$ for $n \ge 1$ and satisfying the Carleson condition (C); denote by $B = B_\Lambda$ the corresponding Blaschke product. Let

$$JVa = \sum_{n\ge1} a_{\lambda_n} v_n,$$

where $a = (a_{\lambda_n})_{n\ge1}$ and

$$v_n = (1-|\lambda_n|^2)\cdot k_{\lambda_n}\cdot \frac{B_{\lambda_n}}{B_{\lambda_n}(\lambda_n)}\cdot\frac{C_{\lambda_n}}{C_{\lambda_n}(\lambda_n)}$$

with the "correcting factor" C_{λ_n} defined by

$$C_{\lambda_n} = (1-|\lambda_n|^2)\cdot k_{\lambda_n}\cdot D_{\lambda_n};\quad D_{\lambda_n} = \prod_{k>n} d_{\lambda_k};\quad d_{\lambda_k} = |\lambda_k|^{1/2}\frac{1-(\overline{\lambda}_k z/|\lambda_k|)}{1-\overline{\lambda}_k z}.$$

Show that JV is a bounded linear mapping

$$JV : l^p(\Lambda, w) \longrightarrow \mathcal{K}_{B^2}^p$$

such that $(JVa)|\Lambda = a$ for every $a \in l^p(\Lambda, w)$, where

$$l^p(\Lambda, w) := l^p(\Lambda, 1-|\lambda_n|^2) = \{a : \sum_{n\ge1} |a_{\lambda_n}|^p(1-|\lambda_n|^2) < \infty\}$$

and $1 \le p \le \infty$. In particular, $\mathcal{K}_{B^2}^p|\Lambda = l^p(\Lambda, w)$.

Moreover, $\|JV\| \le 8\delta(\Lambda)^{-2}$, where $\delta(\Lambda)$ is the Carleson constant of Λ defined in point 3) of (c) (see also Subsection 3.2.17).

[*Hint:* in order to see that the products D_{λ_n} converge, observe that $d_\lambda(z) = b_{\lambda/\sqrt{|\lambda|}}(z\sqrt{|\lambda|})$, and use the general fact that $\prod_{k\ge1} b_{z_k}(zw_k)$ converges for any Blaschke sequences (z_k), (w_k) (for this, see the formula for $|b_{z_k}(zw_k) - 1|$ in the proof of A.3.7.3, Volume 1); it is also easy to see that $v_n \in \mathcal{K}_{B^2}^\infty = H^\infty \cap B^2 H_-^\infty$, since $v_n = c_n(k_{\lambda_n}B_{\lambda_n})(k_{\lambda_n}D_{\lambda_n})$ and $k_{\lambda_n}\overline{b}_{\lambda_n} \in H_-^\infty$ and $(d_{\lambda_k}\overline{b}_{\lambda_k} - |\lambda_k|^{1/2}) \in H_-^\infty$ for every n and k;

in order to estimate $|v_n(z)|$, observe that $|\lambda_n| \le |\lambda_k|$ for $n < k$, and hence

$$\begin{aligned}
&|\lambda_k - \lambda_n|\lambda_k||^2 - |\lambda_k||\lambda_n - \lambda_k|^2\\
&\quad= |\lambda_k|^2 + |\lambda_n|^2|\lambda_k|^2 - 2\,\mathrm{Re}(\lambda_k\overline{\lambda}_n|\lambda_k|) - |\lambda_k||\lambda_n|^2 - |\lambda_k|^3 + 2\,\mathrm{Re}(\lambda_k\overline{\lambda}_n|\lambda_k|)\\
&\quad= |\lambda_k|(1-|\lambda_k|)(|\lambda_k| - |\lambda_n|^2) \ge 0;
\end{aligned}$$

this yields

$$|d_{\lambda_k}(\lambda_n)| = |\lambda_k|^{-1/2}\Big|\frac{\lambda_k - |\lambda_k|\lambda_n}{1-\overline{\lambda}_k\lambda_n}\Big| \ge \Big|\frac{\lambda_k - \lambda_n}{1-\overline{\lambda}_k\lambda_n}\Big| = |b_{\lambda_k}(\lambda_n)|,$$

whence $|D_{\lambda_n}(\lambda_n)| = \prod_{k>n} |d_{\lambda_k}(\lambda_n)| \geq |B_{\lambda_n}(\lambda_n)| \geq \delta = \delta(\Lambda)$; now,

$$\begin{aligned}\sum_{n\geq 1} |v_n(z)| &\leq \delta^{-2} \sum_{n\geq 1} \frac{(1-|\lambda_n|^2)^2}{|1 \quad \overline{\lambda}_n z|^2} \cdot |D_{\lambda_n}(z)| \\ &\leq 4\delta^{-2} \sum_{n\geq 1} \frac{(1-|\lambda_n|)(1-|\lambda_n||z|^2)}{|1-\overline{\lambda}_n z|^2} \cdot |D_{\lambda_n}(z)|;\end{aligned}$$

here we can use the formula

$$1 \geq |d_\lambda(z)|^2 = |b_{\lambda/\sqrt{|\lambda|}}(z\sqrt{|\lambda|})|^2 = 1 - \frac{(1-|\lambda|)(1-|\lambda||z|^2)}{|1-\overline{\lambda}z|^2} \geq 0$$

to get

$$\begin{aligned}\sum_{n\geq 1} \frac{(1-|\lambda_n|)(1-|\lambda_n||z|^2)}{|1-\overline{\lambda}_n z|^2} \cdot |D_{\lambda_n}(z)| &= \sum_{n\geq 1} (1-|d_{\lambda_n}(z)|^2) \prod_{k>n} |d_{\lambda_k}(z)| \\ &\leq 2\sum_{n\geq 1} (1-|d_{\lambda_n}(z)|) \prod_{k>n} |d_{\lambda_k}(z)| \\ &= 2\sum_{n\geq 1} \Big(\prod_{k>n} |d_{\lambda_k}(z)| - \prod_{k\geq n} |d_{\lambda_k}(z)| \Big) \\ &= 2\Big(1 - \prod_{k\geq 1} |d_{\lambda_k}(z)|\Big) \\ &\leq 2,\end{aligned}$$

and finally,

$$\sum_{n\geq 1} \Big| v_n(z) \Big| \leq 8\delta^{-2}$$

for every $z \in \mathbb{D}$; this implies that $|JVa(z)| \leq 8\delta^{-2} \|a\|_{l^\infty}$ for every $a \in l^\infty(\Lambda)$;
on the other hand, the estimate for $p = 1$ is immediate

$$\|JVa\|_{H^1} \leq \delta^{-2} \sum_{n\geq 1} |a_{\lambda_n}|(1-|\lambda_n|^2) \Big\| \frac{1-|\lambda_n|^2}{(1-\overline{\lambda}_n z)^2} \Big\|_{H^1} = \delta^{-2} \|a\|_{l^1(w)},$$

and, therefore, the operator $JV : l^p(\Lambda, w) \longrightarrow H^p$ is bounded for $p = 1$ and $p = \infty$; the result now follows from the Riesz–Thorin interpolation theorem (see A.5.7.4 and comments in A.5.8, Volume 1).]

(ii) *Interpolation in* $\mathcal{K}_B^p$ *for* $1 < p < \infty$.

1) Show that, for an arbitrary interpolation datum $a \in l^p(\Lambda; w)$, the interpolating function f in $\mathcal{K}_B^p$ (i.e., such that $f|\Lambda = a$) is unique (if there exists any). Hence, at least for finitely supported data, it coincides with the interpolation operator $(In)a$ from (c) (10) above;

2) Show that if $l^p(\Lambda; w) \subset \mathcal{K}_B^p|\Lambda$ then the series $(In)a$ from (c)(10) converges unconditionally in H^p for every $a \in l^p(\Lambda; w)$, and hence $\Lambda \in (C)$;

3) Conversely, show that if $\Lambda \in (C)$ then the series $(In)a$ from (c)(10) converges unconditionally in H^p for every $a \in l^p(\Lambda; w)$ and satisfies $(In)a|\Lambda = a$.

[*Hint:* for 1), use $\mathcal{K}_B^p \cap BH^p = \{0\}$; for 2), employ the same reasoning as in (c) above, when proving (7) $\Rightarrow$ (2); for 3), use (i), which gives $H^p|\Lambda = l^p(\Lambda; w)$, then use 1), and deduce $(In)a = P_B(JVa)$ observing that $f|\Lambda = (P_B f)|\Lambda$ for every $f \in H^p$, where

$$f \longmapsto P_B f = BP_-(\overline{B}f)$$

is the projection onto $\mathcal{K}_B^p$ parallel to BH^p (P_- stands for the Riesz projection from $L^p(\mathbb{T})$ onto H_-^p; see Riesz' theorem in A.5.7.3, Volume 1).]

(iii) *(G. Airapetyan, 1977)* Show that $\mathcal{K}_B^\infty|\Lambda = l^\infty(\Lambda)$ if and only if Λ satisfies the following *uniform Frostman condition*

$$\text{(F)} \qquad F(\Lambda) := \sup_{\zeta\in\mathbb{T}} \sum_{n\geq 1} \frac{1-|\lambda_n|^2}{|\zeta-\lambda_n|} < \infty$$

and, of course, the Carleson condition (C).

[*Hint:* clearly, $(C)\&(F)$ imply that the operator In (see (c) (10) above) is bounded from $l^\infty(\Lambda)$ to $\mathcal{K}_B^\infty$;

conversely, if $\mathcal{K}_B^\infty|\Lambda = l^\infty(\Lambda)$ then, as in (ii), 1), there exists a constant $c > 0$ such that

$$|(In)a(z)| = \Big|\sum_{n\geq 1} a_{\lambda_n}(1-|\lambda_n|^2)k_{\lambda_n}(z)\frac{B_{\lambda_n}(z)}{B_{\lambda_n}(\lambda_n)}\Big| \leq c\|a\|_{l^\infty(\Lambda)}$$

for every $z \in \mathbb{D}$ and every $a \in l^\infty(\Lambda)$, and hence

$$\sum_{n\geq 1} \frac{1-|\lambda_n|^2}{|1-\overline{\lambda}_n z|}\frac{|B_{\lambda_n}(z)|}{|B_{\lambda_n}(\lambda_n)|} \leq c;$$

now, using the last inequality from the proof of Lemma 3.2.18, we have $|B_{\lambda_n}(z)| \geq |B(z)| \geq \delta^2/6$, and hence

$$\sum_{n\geq 1} \frac{1-|\lambda_n|^2}{|1-\overline{\lambda}_n z|}\frac{1}{|B_{\lambda_n}(\lambda_n)|} \leq 6c/\delta^2,$$

for $z \in \Omega = \mathbb{D}\backslash(\bigcup_n D_{\lambda_n})$ where $D_{\lambda_n} = \{\zeta : |b_{\lambda_n}(\zeta)| \leq \delta/3\}$; since these disks are pairwise disjoint for small δ (see 3.3.3(b) below), we can pass to the radial limit along a sequence from Ω at any point $\zeta \in \mathbb{T}$; the result follows.]

(f) *A summation method for biorthogonal decompositions.* Let Λ be a Blaschke sequence in $\mathbb{D}$ and $1 < p < \infty$. In the notation of (b) and (c), show that the generalized Fourier series

$$\sum_{n\geq 1}(f, x_{\lambda_n})f_{\lambda_n} = \sum_{n\geq 1} f(\lambda_n)(1-|\lambda_n|^2)\cdot k_{\lambda_n}\cdot\frac{B_{\lambda_n}}{B_{\lambda_n}(\lambda_n)}$$

is summable with sum $P_B f$ (see (e) above) by the following summation method, $\lim_m \|P_B f - S_m(f)\|_p = 0$ for every $f \in H^p$, where

$$S_m(f) = \sum_{n=1}^{m} B_m(\lambda_n)(f, x_{\lambda_n})f_{\lambda_n}, \quad B_m = \prod_{n>m} b_{\lambda_n}.$$

[*Hint:* since $S_m(f) \in \mathcal{K}_B^p$ and $B_m f - S_m(f)$ vanishes at every point λ_n, $n \geq 1$, we have $P_B(B_m f) = S_m(f)$ and hence $\|S_m(f)\|_p \leq C_p\|f\|_p$ for every $f \in H^p$; on the other hand, on the dense subset $BH^p + \mathcal{L}\mathrm{in}(f_{\lambda_n} : n \geq 1)$, the required convergence is obvious.]

(g) *Sublattices in* $H^\infty|\Lambda$. Let $\Lambda = (\lambda_n)_{n\geq 1}$ be a Blaschke sequence in $\mathbb{D}$ and let $\delta_n = |B_{\lambda_n}(\lambda_n)| = \prod_{k\neq n}|b_{\lambda_k}(\lambda_n)|$ for $n = 1, 2, \ldots$

(i)* *(J. Garnett, 1977)* Let h be a positive decreasing function on $[1,\infty)$. The following are equivalent.

1) $H^\infty|\Lambda$ contains every sequence $(a_{\lambda_n})_{n\geq 1}$ satisfying $|a_{\lambda_n}| \leq \delta_n h(\log\frac{e}{\delta_n})$, $n \geq 1$.

2) $\int_1^\infty h(t)\,dt = \infty$.

(ii) Without loss of generality we can assume that $|\lambda_n| \leq |\lambda_k|$ for $n \leq k$. Show that

1) $(a_{\lambda_n})_{n\geq 1} \in H^\infty|\Lambda$ if $|a_{\lambda_n}| \leq C\delta_n\varepsilon_n \cdot \exp(-4\varepsilon_n \log\frac{e}{\delta_n})$ for a positive decreasing sequence $(\varepsilon_n)_{n\geq 1}$;

2) in particular, $(a_{\lambda_n})_{n\geq 1} \in H^\infty|\Lambda$ if $|a_{\lambda_n}| \leq C\delta_n^{1+\varepsilon}$ for a positive $\varepsilon > 0$;

3) if, moreover, $(\delta_n)_{n\geq 1}$ is (possibly) decreasing and $|a_{\lambda_n}| \leq C\delta_n/\log\frac{e}{\delta_n}$ then $(a_{\lambda_n})_{n\geq 1} \in H^\infty|\Lambda$.

[*Hint (S. Vinogradov, E. Gorin, S. Hruschev, 1981):* for 1), let

$$f = \sum_{n\geq 1} a_{\lambda_n} v_n,$$

where

$$v_n = \left(\frac{1-|\lambda_n|^2}{1-\overline{\lambda}_n z}\right)^2 \cdot \frac{B_{\lambda_n}(z)}{B_{\lambda_n}(\lambda_n)} \cdot \exp\Big[\varepsilon_n(\alpha_n(\lambda_n) - \alpha_n(z))\Big],$$

with $\alpha_n(z) = \sum_{k\geq n}\frac{1+\overline{\lambda}_k z}{1-\overline{\lambda}_k z}(1-|\lambda_k|^2)$; note that

$$\beta_n := \mathrm{Re}(\alpha_n(z)) = \sum_{k\geq n}\gamma_k(z),$$

where

$$\gamma_k(z) = \frac{1-|\overline{\lambda}_k z|^2}{|1-\overline{\lambda}_k z|^2}(1-|\lambda_k|^2),$$

and $1-|\lambda_n|^2 \leq 1-|\overline{\lambda}_n z|^2$, which imply that $|f(z)| \leq S_1$ with

$$\begin{aligned} S_1 &= \sum_{n\geq 1}\frac{\left|a_{\lambda_n}\right|}{\delta_n}\cdot\frac{1-|\overline{\lambda}_n z|^2}{|1-\overline{\lambda}_n z|^2}(1-|\lambda_n|^2)\exp\Big[\varepsilon_n(\beta_n(\lambda_n)-\beta_n(z))\Big] \\ &= \sum_{n\geq 1}\frac{\left|a_{\lambda_n}\right|}{\delta_n}\cdot(\beta_n(z)-\beta_{n+1}(z))\exp\Big[\varepsilon_n(\beta_n(\lambda_n)-\beta_n(z))\Big]; \end{aligned}$$

now use $t \leq e^t - 1$ for $t = \varepsilon_n(\beta_n(z) - \beta_{n+1}(z))$ to obtain

$$S_1 \leq S_2 = \sum_{n\geq 1}\frac{|a_{\lambda_n}|}{\delta_n\varepsilon_n}\cdot\left(e^{-\varepsilon_n\beta_{n+1}} - e^{-\varepsilon_n\beta_n}\right)\exp[\varepsilon_n\beta_n(\lambda_n)],$$

and use the same estimate as for 4) $\Rightarrow$ 5) of 3.3.3(c) to get

$$\begin{aligned}\beta_n(\lambda_n) &= \sum_{k\geq n}\frac{1-|\overline{\lambda}_k\lambda_n|^2}{|1-\overline{\lambda}_k\lambda_n|^2}(1-|\lambda_k|^2)\leq\sum_{k\geq n}\frac{1-|\lambda_n|^4}{|1-\overline{\lambda}_k\lambda_n|^2}(1-|\lambda_k|^2)\\ &\leq 2\sum_{k\geq n}\frac{(1-|\lambda_n|^2)(1-|\lambda_k|^2)}{|1-\overline{\lambda}_k\lambda_n|^2}\leq 2+4\log|B_{\lambda_n}(\lambda_n)|^{-1}\\ &\leq 4\log\frac{e}{\delta_n};\end{aligned}$$

therefore,

$$S_2\leq C\cdot\lim_N\sum_{n=1}^{N}\left(e^{-\varepsilon_n\beta_{n+1}}-e^{-\varepsilon_n\beta_n}\right)\leq C\cdot\lim_N\left(e^{-\varepsilon_N\beta_{N+1}}-e^{-\varepsilon_1\beta_1}\right)\leq C;$$

finally, $|f(z)|\leq C$ for every $z\in\mathbb{D}$, and the result follows;
assertion 2) is a special case of 1) with $\varepsilon_n=\varepsilon/4$;
assertion 3) is a special case of 1) with $\varepsilon_n=1/\log\frac{e}{\delta_n}$.]

(h) *Yet another linear interpolation operator (A. Bernard; N. Varopoulos, 1971).* Let Ω be an open set in $\mathbb{C}$, and let $\Lambda=(\lambda_n)_{n\geq1}\subset\Omega$ be such that

$$C:=\sup_{\|a\|_\infty\leq1}\inf\left\{\|g\|:g\in H^\infty(\Omega),g(\lambda_j)=a_j,j\geq1\right\}<\infty.$$

Show that there exist functions $v_n\in H^\infty(\Omega)$ such that $\sum_{n\geq1}|v_n(z)|\leq C^2$ for $z\in\Omega$ and $v_n(\lambda_j)=\delta_{nj}$. In particular, $T:a\longrightarrow Ta=\sum_{n\geq1}a_nv_n$ is an interpolation operator, $T:l^\infty\longrightarrow H^\infty(\Omega)$, with $\|T\|\leq C^2$.

[*Hint:* given $\varepsilon>0$ and $N\geq1$, it suffices to construct $u_n\in H^\infty(\Omega)$ such that $\sum_{n=1}^N|u_n(z)|\leq C^2+\varepsilon$ for $z\in\Omega$ and $u_n(\lambda_j)=\delta_{nj}$ for $1\leq n,j\leq N$, and then let $\varepsilon\longrightarrow0$, $N\longrightarrow\infty$ (using Montel's theorem);

now, let $\zeta=e^{2\pi i/N}$ and $\delta>0$; there exist $h_k\in H^\infty(\Omega)$ such that $|h_k(z)|\leq C+\delta$ for $z\in\Omega$, $1\leq k\leq N$, and $h_k(\lambda_j)=\zeta^{jk}$ for $1\leq k,j\leq N$; set

$$u_n=\left(\frac{1}{N}\sum_{k=1}^{N}\zeta^{-kn}h_k\right)^2$$

to get $u_n(\lambda_j)=\left(\frac{1}{N}\sum_{k=1}^N\zeta^{k(j-n)}\right)^2=\delta_{nj}$ and

$$\begin{aligned}\sum_{n=1}^{N}|u_n| &= \sum_{n=1}^{N}\left|\frac{1}{N}\sum_{k=1}^{N}\zeta^{-kn}h_k\right|^2=\frac{1}{N^2}\sum_{n=1}^{N}\sum_{k,l=1}^{N}\zeta^{-kn}h_k\zeta^{ln}\overline{h}_l\\ &= \frac{1}{N^2}\sum_{k,l=1}^{N}h_k\overline{h}_l\sum_{n=1}^{N}\zeta^{(l-k)n}=\frac{1}{N}\sum_{k=1}^{N}|h_k|^2\leq(C+\delta)^2,\end{aligned}$$

and the result follows.]

3.3.4. Geometric characteristics of interpolating sequences. Let $\Lambda=(\lambda_n)_{n\geq1}$ be a sequence of pairwise distinct points in $\mathbb{D}$.

(a) Show that the following are equivalent.

(i) Λ is an interpolating sequence (i.e., $\Lambda\in(C)$).

(ii) Λ is a (hyperbolically) *sparse (separated) sequence*, i.e.

$$\text{(R)} \qquad \Delta(\Lambda) = \Delta = \inf_{k \neq j} |b_{\lambda_k}(\lambda_j)| > 0,$$

and $\mu_\Lambda = \sum_{n \geq 1}(1 - |\lambda_n|^2)\delta_{\lambda_n}$ is a *Carleson measure* (see A.5.7.2(b), Volume 1).

(iii) $\Lambda \in (R)$ and

$$\sup_{z \in \mathbb{D}} \sum_{n \geq 1} (1 - |b_z(\lambda_n)|^2) < \infty.$$

(the *Möbius invariant Blaschke condition*).

(iv) $\Lambda \in (R)$ and

$$C_\Lambda^2 := \sup_{j \geq 1} \sum_{n \geq 1} \frac{(1 - |\lambda_n|^2)(1 - |\lambda_j|^2)}{|1 - \overline{\lambda}_n \lambda_j|^2} < \infty.$$

(v) There exists an ε, $0 < \varepsilon < 1$, such that every connected component of the level set $L(B, \varepsilon) = \{z \in \mathbb{D} : |B(z)| < \varepsilon\}$, $B = \prod_n b_{\lambda_n}$, contains at most one point of Λ (cf. also 3.3.5(c) below).

[*Hint:* for (i) $\Rightarrow$ (ii), use $\Delta \geq \delta(\Lambda)$ and the implication 4) $\Rightarrow$ 5) of 3.3.3(c); (ii) $\Leftrightarrow$ (iii) by Theorem A.5.7.2(b), Volume 1 (see also the formula for $1 - |b_z(\lambda_n)|^2$ in 3.3.3(c) above); (iii) $\Rightarrow$(iv) is obvious; finally, for (iv) $\Rightarrow$ (i), use the series $\log |B_{\lambda_k}(\lambda_k)|^2 = \sum_{n \neq k} \log |b_{\lambda_n}(\lambda_k)|^2$ (as in 3.3.3(c) above) and the inequality $\log(1 - x) \geq -c(\Delta)x$ for $0 \leq x \leq 1 - \Delta^2 < 1$ (with $x = 1 - |b_{\lambda_n}(\lambda_k)|^2$);

for (i) $\Rightarrow$ (v), employ the condition (GC) (see Lemma 3.2.18), i.e.

$$\inf_{n \geq 1} |b_{\lambda_n}(z)| \geq |B(z)| \geq \delta \cdot \inf_{n \geq 1} |b_{\lambda_n}(z)|,$$

which implies that

$$\bigcup_n L(b_{\lambda_n}, \varepsilon) \subset L(B, \varepsilon) \subset \bigcup_n L(b_{\lambda_n}, \varepsilon/\delta),$$

and hence (v), since (R) means that $L(b_{\lambda_n}, \varepsilon/\delta) \cap \Lambda = \{\lambda_n\}$; conversely, let $B' = \prod_{n=1}^N b_{\lambda_n}$; (v) implies that every component Ω'_k of $L(B', \varepsilon)$ contains at most one point of Λ', and hence, for every k, there exists b_{λ_n} such that b_{λ_n}/B' is holomorphic on Ω'_k and $|b_{\lambda_n}/B'| \leq 1/\varepsilon$ on $\partial\Omega'_k$; the maximum principle shows that $\varepsilon \cdot \inf_{n \geq 1} |b_{\lambda_n}(z)| \leq |B'(z)|$ on $\mathbb{D}$; passing to the limit we get (GC).]

(b) Show that if $\Lambda \in (R)$ then the disks $D(\lambda_n, \frac{\Delta(\Lambda)}{2}(1 - |\lambda_n|))$ are pairwise disjoint. Conversely, if $\delta^2 < 3/4$, and

$$|\lambda_n - \lambda_k| \geq 4\delta \cdot \max(1 - |\lambda_n|, 1 - |\lambda_k|), \quad n \neq k,$$

(and, all the more, if the disks $D(\lambda_n, 4\delta(1 - |\lambda_n|))$ are pairwise disjoint) then $\Lambda \in (R)$ and $\Delta(\Lambda) \geq \delta$.

[*Hint:* assume that $D(\lambda_n, \frac{\Delta}{2}(1 - |\lambda_n|)) \cap D(\lambda_k, \frac{\Delta}{2}(1 - |\lambda_k|)) \neq \emptyset$ then $|\lambda_n - \lambda_k| < \Delta \cdot \max(1 - |\lambda_n|, 1 - |\lambda_k|) = \Delta(1 - |\lambda|)$, and hence $|b_{\lambda_n}(\lambda_k)| \leq |\lambda_n - \lambda_k|/(1 - |\lambda|) < \Delta$, which is impossible;

conversely, if

$$|b_{\lambda_n}(\lambda_k)|^{-2} = 1 + \frac{(1 - |\lambda_n|^2)(1 - |\lambda_k|^2)}{|\lambda_n - \lambda_k|^2} > \delta^{-2},$$

then $|\lambda_n - \lambda_k|^2 < \frac{\delta^2}{1 - \delta^2}(1 - |\lambda_n|^2)(1 - |\lambda_k|^2) < 16\delta^2 \max((1 - |\lambda_n|)^2, (1 - |\lambda_k|)^2)$; contradiction.]

(c) *(V. Kabaila (1958), D. Newman (1959))* Show that $\Lambda \in (C)$ if

$$\overline{\lim}_n \frac{1-|\lambda_{n+1}|}{1-|\lambda_n|} < 1.$$

The converse is also true if Λ is an increasingly ordered sequence in the interval $(0,1)$.

[*Hint:* if $0 < q < 1$ and $1-|\lambda_{n+1}| \le q(1-|\lambda_n|)$ for all $n \ge 1$, then $1-|\lambda_{n+k}| \le q^k(1-|\lambda_n|)$ and $|\lambda_{n+k}|-|\lambda_n| \ge (1-q^k)(1-|\lambda_n|)$, $1-|\lambda_{n+k}\lambda_n| = 1-|\lambda_n|+|\lambda_n|(1-|\lambda_{n+k}|) \le (1+q^k)(1-|\lambda_n|)$; since

$$|b_\lambda(z)|^2 = 1 - \frac{(1-|\lambda|^2)(1-|z|^2)}{|1-\overline{\lambda}z|^2} \ge |b_{|\lambda|}(|z|)|,$$

we get

$$\prod_{k\ge1} |b_{\lambda_{n+k}}(\lambda_n)|^2 \ge \prod_{k\ge1} b_{|\lambda_{n+k}|}(|\lambda_n|) \ge \prod_{k\ge1} \frac{1-q^k}{1+q^k} = P,$$

and $\prod_{1\le k<n} |b_{\lambda_{n-k}}(\lambda_n)|^2 \ge \prod_{1\le k<n} \frac{1-q^k}{1+q^k}$, and finally, $|B_{\lambda_n}(\lambda_n)| \ge P^2$ for every n, $n \ge 1$;

conversely,

$$1 - \frac{1-\lambda_{n+1}}{1-\lambda_n} = \frac{\lambda_{n+1}-\lambda_n}{1-\lambda_n} \ge \frac{\lambda_{n+1}-\lambda_n}{1-\lambda_{n+1}\lambda_n} \ge |B_{\lambda_n}(\lambda_n)| \ge \delta(\Lambda).]$$

(d) *(A. Naftalevich, 1956)* Let $0 \le r_n \le r_{n+1} < 1$ and $\sum_{n\ge1}(1-r_n) < \infty$. Show that there exist θ_n such that $\Lambda = (\lambda_n = r_n e^{i\theta_n})_{n\ge1}$ satisfies the condition (C).

[*Hint (V. Vasyunin, 1980)*: let

$$\theta_n = \sum_{k\ge n}(1-r_k);$$

then, for $k > n > N$ and N sufficiently large, $|\lambda_n - \lambda_k| \ge 2^{-1}|\theta_n - \theta_k| = 2^{-1}\sum_{n\le j<k}(1-r_j) \ge 2^{-1}(1-|\lambda_n|)$ and, by (b) above, $(\lambda_n)_{n>N} \in (R)$ with $\Delta(\Lambda) \ge 1/8$;

in order to check that $\mu = \sum_{n>N}(1-|\lambda_n|^2)\delta_{\lambda_n}$ is a Carleson measure, use A.5.7.2(b) (Volume 1) and consider a "square" of the form

$$Q(\zeta, l) = \{z : 1-l \le |z| \le 1, |\arg(z) - \arg(\zeta)| \le l/2\}$$

(instead of "half-disks" as in A.5.7.2(b)(iii), Volume 1, which is obviously equivalent); then

$$\mu(Q(\zeta,l)) = \sum_{m\le n<k}(1-r_n^2) \le 2\sum_{m\le n<k}(1-r_n),$$

where $\{n : m \le n < k\} = \{n : \lambda_n \in Q(\zeta,l)\}$, and therefore, $\mu(Q(\zeta,l)) \le 2(\theta_m - \theta_k) \le 2l$ (by definition, $\theta_\infty = 0$).]

(e) *(S. Vinogradov and M. Goluzina, 1974)* If $F(\Lambda) < \infty$ (see 3.3.3(e) (iii) above) then μ_Λ is a Carleson measure.

[*Hint:* the maximum modulus principle implies that

$$\begin{aligned}\sum_{k\geq 1}\frac{1-|\lambda_k|^2}{|1-\overline{\lambda}_k z|} &= \sup_{|\varepsilon_k|=1}\Big|\sum_{k\geq 1}\varepsilon_k\frac{1-|\lambda_k|^2}{1-\overline{\lambda}_k z}\Big| \\ &\leq \sup_{|\varepsilon_k|=1}\sup_{\zeta\in\mathbb{T}}\Big|\sum_{k\geq 1}\varepsilon_k\frac{1-|\lambda_k|^2}{1-\overline{\lambda}_k\zeta}\Big| = F(\Lambda)\end{aligned}$$

for every $z\in\mathbb{D}$, and hence

$$\sum_{k\geq 1}\frac{(1-|\lambda_k|^2)(1-|z|^2)}{|1-\overline{\lambda}_k z|^2}\leq 2\sum_{k\geq 1}\frac{1-|\lambda_k|^2}{|1-\overline{\lambda}_k z|}\cdot\frac{1-|z|^2}{1-|\overline{\lambda}_k z|}\leq 2F(\Lambda);$$

by A.5.7.2(b), Volume 1, μ_Λ is Carleson.]

(f) If μ_Λ is a Carleson measure (see (a) (iv) for μ_Λ) then Λ is a union of at most $40[I(\mu)]$ sequences satisfying (C) (the "intensity" $I(\mu)$ is defined in A.5.7.2(b), Volume 1).

[*Hint:* let

$$Q(n,k)=\{z:1-2^{-n}\leq|z|\leq 1-2^{-n-1},2\pi k2^{-n}\leq\arg(z)\leq 2\pi(k+1)2^{-n}\},$$

where $0\leq k<2^n$ and $n\geq 0$; then by A.5.7.2(b), Volume 1, $Q(n,k)$ contains at most $10[I(\mu)]$ points of Λ, and we split $\Lambda=\bigcup_{j=1}^{4m}\Lambda_j$ into sequences Λ_j in the following way: Λ_1 takes one point from each non-empty intersection $\Lambda\cap Q(n,k)$ with even n and k, Λ_2 receives by one of the remaining points, etc. up to Λ_m; the sequences $\Lambda_{m+1},\dots,\Lambda_{2m}$ are obtained in the same way by taking squares $Q(n,k)$ with n even and k odd; $\Lambda_{2m+1},\dots,\Lambda_{3m}$ with n odd, k even, and $\Lambda_{3m+1},\dots,\Lambda_{4m}$ with n odd, k odd; clearly, (b) implies that $\Lambda_j\in(R)$ for every j, and by (a), $\Lambda_j\in(C)$.]

3.3.5. More about generalized free interpolation. The following results represent the scope of generalized free interpolation in the unit disk, mostly obtained between 1978 and 1990 by N. Nikolski, V. Vasyunin, and A. Volberg; see Section 3.4 below for comments and references.

(a)* *Free interpolating sequences.* Let $\Theta_n\in H^\infty$, $0\neq|\Theta_n|\leq 1$ for $n=1,\dots$ The following are equivalent.

(i) For every $(c_n)_{n\geq 1}\in l^\infty$ there exists a function $f\in H^\infty$ such that

$$f-c_n\in\Theta_n H^\infty\ (n\geq 1),\ \text{and}\ \sup_{n\geq 1}\Big\|\frac{f-c_n}{\Theta_n}\Big\|_\infty<\infty.$$

(ii) For every bounded sequence $(f_n)_{n\geq 1}\subset H^\infty$ there exists a function $f\in H^\infty$ such that

$$f-f_n\in\Theta_n H^\infty\ (n\geq 1),\ \text{and}\ \sup_{n\geq 1}\Big\|\frac{f-f_n}{\Theta_n}\Big\|_\infty<\infty.$$

(iii) For every sequence $(f_n)_{n\geq 0}\subset H^\infty$ satisfying

$$\sup_{n\geq 1}\operatorname{dist}(\overline{\Theta}_n f_n,H^\infty)<\infty\ \text{and}\ \sup_{n\geq 1}\|f_n\|_{L^\infty(\Delta_n\,dm)}<\infty,$$

where $\Delta_n^2=1-|\Theta_n|^2$, there exists a function $f\in H^\infty$ such that

$$f-f_n\in\Theta_n H^\infty(n\geq 1),\ \text{and}\ \sup_{n\geq 1}\Big\|\frac{f-f_n}{\Theta_n}\Big\|_\infty<\infty.$$

(iv) There exists a function $\Theta \in H^\infty$, $0 \neq |\Theta| \leq 1$, and a Borel partition $\tau = (\sigma_n)_{n\geq 1}$ of the closed unit disk, $\overline{\mathbb{D}} = \bigcup_{n\geq 1} \sigma_n$, $\sigma_n \cap \sigma_k = \emptyset$ for $n \neq k$, such that $\varepsilon|\Theta_{\sigma_n}(z)| \leq |\Theta_n(z)| \leq |\Theta_{\sigma_n}(z)|$ for every $n \geq 1$ and $z \in \mathbb{D}$, where $\varepsilon > 0$ and (see 2.5.8 above for the notation)

$$\Theta_{\sigma_n} = \Big(\prod_{\lambda \in \sigma_n} b_\lambda^{k(\lambda)} \Big) \cdot \exp\Big(- \int_{\sigma_n \cap \mathbb{T}} \frac{\zeta + z}{\zeta - z}\, d\mu \Big),$$

so that $\Theta = \prod_{n\geq 1} \Theta_{\sigma_n}$, and $(\Theta_{\sigma_n})_{n\geq 1}$ satisfies the generalized Carleson condition (GC), i.e.

$$|\Theta(z)| \geq d_\tau(z)$$

for every $z \in \mathbb{D}$, where $d_\tau(z) = \inf_{n\geq 1} |\Theta_{\sigma_n}(z)|$.

(v) The subspaces $\mathcal{K}_{\Theta_n}$, $n \geq 1$, form an unconditional (Riesz) basis of $\mathcal{K}_\Theta$ (for a non-inner Θ, see Chapters 1, 2 for the definition of $\mathcal{K}_\Theta$).

(vi) The *uniform minimality condition* holds (see 3.2.19 above):

$$\varepsilon(\tau) := \inf_{n\geq 1} \inf_{z\in\mathbb{D}} \Big(|\Theta_{\sigma_n}(z)| + |\Theta(z)/\Theta_{\sigma_n}(z)| \Big) > 0.$$

[*Hint for (vi) $\Rightarrow$ (iv) (V. Vasyunin, 1978):* it suffices to show that

$$|\Theta_k(z)| \geq \frac{\varepsilon(\tau(k))^2}{4} \cdot d_{\tau(k)}(z),$$

where $k = 1, 2, \ldots$ and $\tau(k) = (\sigma_n)_{n=1}^k$, $\Theta_k = \prod_{n=1}^k \Theta_{\sigma_n}$, and then let $k \longrightarrow \infty$; indeed, $\varepsilon(\tau(k)) \geq \varepsilon(\tau)$ and $d_{\tau(k)}(z) \geq d_\tau(z)$;

for a finite product Θ_k, observe that if $|\Theta_{\sigma_n}(z)| \leq \varepsilon/2 = \varepsilon(\tau(k))/2$ for some n, then $|\Theta_k(z)/\Theta_{\sigma_n}(z)| \geq \varepsilon - \varepsilon/2 = \varepsilon/2$, and hence

$$|\Theta_k(z)| \geq \frac{\varepsilon}{2} |\Theta_{\sigma_n}(z)| \geq \frac{\varepsilon}{2} \cdot d_{\tau(k)}(z);$$

on the complementary (open) set

$$\Omega = \bigcap_{n=1}^k \{ z \in \mathbb{D} : |\Theta_{\sigma_n}(z)| > \varepsilon/2 \};$$

we have $\inf_\Omega |\Theta_k(z)| > (\varepsilon/2)^k > 0$; if $z \in \mathbb{D} \cap \partial\Omega$ then there exists n such that $|\Theta_{\sigma_n}(z)| = \varepsilon/2$, whence $|\Theta_k(z)| = \frac{\varepsilon}{2} |\Theta_k(z)/\Theta_{\sigma_n}(z)| \geq \varepsilon^2/4$; moreover, for almost all $\zeta \in \mathbb{T}$ the modulus $|\Theta_k(\zeta)|$ coincides with one of the values $|\Theta_{\sigma_n}(\zeta)|$, and hence $|\Theta_k(\zeta)| \geq \varepsilon/2 \geq \varepsilon^2/4$ for almost all points ζ that are nontangential limits of points in Ω; the maximum modulus principle applied to $\Theta_k^{-1}|\Omega$ completes the proof.]

(b)* *Approximate free interpolation.* Let $w_n \in L^\infty(\mathbb{T})$, $0 \leq w_n \leq 1$ for $n = 1, \ldots$

(i) The following are equivalent. (A sequence $(w_n)_{n\geq 1}$ satisfying these conditions is called *approximate free interpolating (AFI)*).

1) For every $(c_n)_{n\geq 1} \in l^\infty$ there exists a function $f \in H^\infty$ and $c > 0$ such that $|f - c_n| \leq c w_n$ a.e. on $\mathbb{T}$ for every $n \geq 1$, $c = c(\delta) \sup_n |c_n|$ and $c(\delta)$ depends only on δ of 3).

2) For every bounded sequence $(f_n)_{n\geq 1} \subset H^\infty$ there exists a function $f \in H^\infty$ and $c > 0$ such that $|f - f_n| \leq c w_n$ a.e. on $\mathbb{T}$ for every $n \geq 1$, $c = c(\delta) \sup_n \|f_n\|_\infty$ and $c(\delta)$ depends only on δ of 3).

3) The outer functions $[w_n]$ satisfy the following *generalized boundary Carleson condition (GBC)*

$$\delta = \inf_{z\in\mathbb{D}} \sup_{n\geq 1} \left|[W_n](z)\right| > 0,$$

where $W_n(\zeta) = \inf\{w_k(\zeta) : k \geq 1, k \neq n\}$ for $\zeta \in \mathbb{T}$ (in particular, $\log(w_n), \log(W_n) \in L^1(\mathbb{T})$ for $n \geq 1$).

4) The outer functions $[w_n]$ satisfy the boundary uniform minimality condition

$$\inf_n \inf_{z\in\mathbb{D}} \left(|[w_n](z)| + |[W_n](z)|\right) > 0.$$

(ii) Let σ_n be arcs on $\mathbb{T}$ and $w_n = \varepsilon_n\chi_{\sigma_n} + \chi_{\mathbb{T}\setminus\sigma_n}$; denote $\{\sigma_{n,}\varepsilon_n\} := (w_n)_{n\geq 1}$. The following assertions are equivalent.

1) $\{\sigma_{n,}\varepsilon_n\}$ is an (AFI) sequence.

2) For every bounded sequence $(f_n)_{n\geq 1} \subset H^\infty$ there exists a function $f \in H^\infty$ and $c > 0$ such that $|f - f_n| \leq c\varepsilon_n$ a.e. on σ_n for every $n \geq 1$, $c = c(K)\sup_n \|f_n\|_\infty$ and $c(K)$ depends only on K of 3).

3)

$$K = \sup_{z\in\mathbb{D}} \inf_n \Big(\sum_{k\geq 1, k\neq n} \omega(z,\sigma_k) \cdot \log\frac{1}{\varepsilon_k}\Big) < \infty,$$

where $\omega(z,\sigma) = \int_\sigma \frac{1-|z|^2}{|\zeta - z|^2}\, dm(\zeta)$ is the harmonic measure of $\sigma \subset \mathbb{T}$ at a point $z \in \mathbb{D}$.

4) The sequence $\{\sigma_{n,}\varepsilon_n\}$ is sparse (see definition in (iii), 1) below) and the Blaschke factors $b_{\lambda_n}^{m_n}$ satisfy (GC), where $m_n = [\log\frac{1}{\varepsilon_n}]$, $\lambda_n = (1 - |\sigma_n|)\zeta_n$ and ζ_n is the center of σ_n.

5) The sequence $\{\sigma_{n,}\varepsilon_n\}$ is sparse (see definition in (iii), 1) below) and the singular inner functions $\theta_n = \exp(-|\sigma_n|\log\frac{1}{\varepsilon_n} \cdot (\zeta_n + z)(\zeta_n - z)^{-1})$ satisfy (GC), where ζ_n is the center of σ_n.

(iii) In the notation of (ii),

1) if $\{\sigma_{n,}\varepsilon_n\}$ is an (AFI) sequence then $\sum_{k\geq 1}|\sigma_k|\log\frac{1}{\varepsilon_k} < \infty$, and the disks $D(a \cdot \log\frac{1}{\varepsilon_n}, \sigma_n)$ are pairwise disjoint for a positive a, where

$$D(R,\sigma)$$

stands for a disk with radius R and center in $\mathbb{D}$, and which meets the circle $\mathbb{T}$ at the endpoints of σ;

(A sequence $\{\sigma_{n,}\varepsilon_n\}$ having the last property is called *sparse*);

2) if there exist $r_n > 0$ such that $\sum_n r_n^{-1} < \infty$ and if the disks $D(r_n \cdot \log\frac{1}{\varepsilon_n}, \sigma_n)$ are pairwise disjoint, then $\{\sigma_{n,}\varepsilon_n\}$ is an (AFI) sequence;

3) if the $\log\frac{1}{\varepsilon_n}$ times enlarged arcs $\sigma_n \cdot \log\frac{1}{\varepsilon_n}$ (with respect to their midpoints) are pairwise disjoint, then $\{\sigma_{n,}\varepsilon_n\}$ is an (AFI) sequence.

(iv) In the notation of (ii), the following are equivalent.

1) There exists $\varepsilon_n > 0$ such that $\lim_n \varepsilon_n = 0$ and $\{\sigma_{n,}\varepsilon_n\}$ is an (AFI) sequence.

2) There exist $R_n > 0$ such that $\lim_n R_n = \infty$, the disks $D(R_n, \sigma_n)$ are pairwise disjoint and the Blaschke product $B = \prod_{n\geq 1} b_{\lambda_n}$ satisfies the *vanishing*

Carleson condition, i.e. $\lim_n |B_{\lambda_n}(\lambda_n)| = 1$, where λ_n has the same meaning as in (ii), 4) above.

(c) *Free interpolation over level sets.* Let $\Theta \in H^\infty$ with $0 \neq |\Theta| \le 1$, $0 < \varepsilon < 1$, and let

$$L(\Theta, \varepsilon) = \{z \in \mathbb{D} : |\Theta(z)| < \varepsilon\} = \bigcup_{n\ge1} \Omega_n$$

be an ε-level set of Θ with the (open) connected components Ω_n; next, let Ω_n^* be the set of $\zeta \in \mathbb{T}$ such that the Stolz angle $S(\zeta, \alpha_\zeta, d_\zeta) = \{z \in \mathbb{D} : |\arg(1 - \overline{\zeta}z)| < \alpha_\zeta, 1 - |z| < d_\zeta\}$ is contained in Ω_n, and finally

$$\Omega_0 = \mathbb{D}\backslash\Big(\bigcup_{n\ge1}\Omega_n\Big),\ \Omega_0^* = \mathbb{T}\backslash\Big(\bigcup_{n\ge1}\Omega_n^*\Big),\ \text{and } \sigma_n = \Omega_n \cup \Omega_n^* \text{ for } n = 0, 1, \dots .$$

Set

$$\Theta_{\sigma_n} = \Big(\prod_{\lambda\in\sigma_n} b_\lambda^{k(\lambda)}\Big)\cdot\exp\Big(-\int_{\sigma_n\cap\mathbb{T}}\frac{\zeta+z}{\zeta-z}\,d\mu\Big)$$

(see 2.5.8 above for the notation).

(i) Show that $\Theta = \prod_{n\ge0}\Theta_{\sigma_n}$ (the *level set factorization* of Θ); the function Θ_{σ_0} is outer, $\varepsilon \le |\Theta_{\sigma_0}(z)| \le 1$ for $z \in \mathbb{D}$, and $|\Theta_{\sigma_n}| \le \varepsilon$ a.e. on $\sigma_n\cap\mathbb{T}$ and $|\Theta_{\sigma_n}| = 1$ a.e. on $\mathbb{T}\backslash\sigma_n$.

(ii)* For every $0 < \varepsilon < 1$, the level set factorization satisfies the generalized Carleson condition (GC), $|\Theta(z)| \ge \varepsilon \cdot \inf_{n\ge0}|\Theta_{\sigma_n}(z)|$ for every $z \in \mathbb{D}$ (and, therefore, the subspaces $\mathcal{K}_{\Theta_n}$, $n \ge 0$, form an unconditional (Riesz) basis of $\mathcal{K}_\Theta$, and generalized free interpolation with respect to $(\Theta_{\sigma_n})_{n\ge1}$ holds (see (a) above)).

[*Hint (for a Blaschke product Θ):* let $\Theta = B = \prod_{\lambda\in\Lambda} b_\lambda^{k(\lambda)}$ be a Blaschke product and σ_n be one of the components above; set $\varphi = B_{\sigma_n}B^{-1}|\Omega_n$ and show that $|\varphi| \le \varepsilon^{-1}$ on $\partial\sigma_n\cap\mathbb{D} = \partial\Omega_n\cap\mathbb{D}$ and $|\varphi| = 1$ a.e. on $\sigma_n\cap\mathbb{T}$; next, observe that φ is an outer Nevanlinna function on Ω_n (which is conformally equivalent to the unit disk) since φ is a limit of rational (and hence outer) functions φ_m, $\varphi_m = \prod_\lambda b_\lambda^{-k(\lambda)}$, where λ runs over the finite set $\{z : |z| \le 1 - m^{-1}\} \cap (\Lambda\backslash\Omega_n)$, and since it satisfies $|\varphi_m(z)| \nearrow |\varphi(z)|$ for $z \in \Omega_n$ (see A.4.5, Volume 1, about that); applying the maximum principle to φ, we get $\inf_{k\ge0}|B_{\sigma_k}(z)| \le |B_{\sigma_n}(z)| \le \varepsilon^{-1}|B(z)|$ for $z \in \Omega_n$; on the complement $\mathbb{D}\backslash L(B, \varepsilon)$ the inequality $|B(z)| \ge \varepsilon\cdot\inf_{n\ge0}|B_{\sigma_n}(z)|$ is obvious.]

(iii) Let $\Theta = \prod_{k\ge1}\Theta_k$ and assume that the level sets $L(\Theta_k, \varepsilon)$ are connected for every ε, $0 < \varepsilon < 1$, and every k, $k \ge 1$ (Examples: $\Theta_k = b_\lambda^{k(\lambda)}$; $\Theta_k = \exp(-a(\zeta + z)(\zeta - z)^{-1})$); show that every level set factorization $\Theta = \prod_{n\ge0}\Theta_{\sigma_n}$ is a grouping of Θ_k, i.e., $\Theta_{\sigma_n} = \prod_{k\in A(n)}\Theta_k$, where $(A(n))_{n\ge1}$ is a partition of positive integers.

[*Hint:* every $L(\Theta_k, \varepsilon)$ is contained exactly in one component Ω_n of $L(\Theta, \varepsilon)$.]

(iv) Let N be an integer, $N \ge 1$, and let $\Lambda \subset \mathbb{D}$ be a Blaschke sequence, $\Lambda = (\lambda_k)_{k\ge1}$ be its renumbering, $B = \prod_{\lambda\in\Lambda} b_\lambda$ the corresponding Blaschke product. Show that the following are equivalent.

1) Λ is the union of at most N interpolating (Carleson) sequences.

2) There exists ε, $0 < \varepsilon < 1$, such that every component Ω_n of the level set $L(B, \varepsilon)$ contains at most N points of Λ.

[*Hint:* use the equivalence (i) $\Leftrightarrow$ (v) of 3.3.4(a).]

(d) *Generalized free interpolation in* H^p, $1 < p < \infty$. Let Θ_n be inner functions and let $\Theta = \prod_{n\geq 1} \Theta_n$, $\Theta \neq 0$. The following are equivalent (*N. Nikolski (1978, $p = 2$), A. Hartmann (1997, $p \neq 2$)*).

(i) For every $f_n \in H^p$ satisfying $\sum_{n\geq 1} \|f_n\|_p^p < \infty$ there exists a function $f \in H^p$ such that $f - f_n \in \Theta_n H^p$ for all $n \geq 1$.

(ii) Given functions $f_n \in H^p$, $n \geq 1$, there exists a function $f \in H^p$ satisfying $f - f_n \in \Theta_n H^p$ for all $n \geq 1$ if and only if $\sum_{n\geq 1} \|f_n\|^p_{H^p/\Theta_n H^p} < \infty$.

(iii) The factorization $\Theta = \prod_{n\geq 1} \Theta_n$ satisfies the (GC) condition.

(iv) $\mathcal{K}_{\Theta_n}$, $n \geq 1$, form an unconditional (Riesz) basis of $\mathcal{K}_\Theta$.

[*Hint (for $p = 2$):* in the notation of 3.3.1(e) with $X_n = \mathcal{K}_{\Theta_n}$, (i) implies that $J_\mathcal{X} H \supset l^2(X_n)$; using 3.3.1(e), this entails that $J_\mathcal{X} H = l^2(X_n)$, and further, that (i) is equivalent to (iv); now, Theorem 3.2.14 implies the other equivalences.]

(e)* *(V. Vasyunin (1984, $p = \infty$), A. Hartmann (1997, $1 < p < \infty$))*

(i) Let $\Lambda \subset \mathbb{D}$ be a finite set, $\Lambda = (\lambda_k)_{k=1}^N$, $B = B_\Lambda$. There exist constants $a > 0$, $b > 0$ depending only on N and p, such that

$$a\|\Delta(f)\|_{l^p(w_n)} \leq \|f\|_{H^p/BH^p} \leq b\|\Delta(f)\|_{l^p(w_n)}$$

for every $f \in H^p$, where $w_n = 1 - |\lambda_n|^2$ and $\Delta(f) = \Delta^k_\Lambda(f|\Lambda) = (\Delta^k(f))_{k=0}^{N-1}$ with

$$\Delta^0(f) = f(\lambda_1); \quad \Delta^1(f) = \Delta^1 f(\lambda_2), \quad \Delta^1 f(z) = \frac{f(z) - f(\lambda_1)}{b_{\lambda_1}(z)}$$

and

$$\Delta^k(f) = \frac{\Delta^{k-1} f(\lambda_{k+1}) - \Delta^{k-1} f(\lambda_k)}{b_{\lambda_k}(\lambda_{k+1})}.$$

(ii) Let $\Lambda \subset \mathbb{D}$ be a finite union of interpolating sequences and $\sigma_n = \Lambda \cap \Omega_n$, where Ω_n are as defined in (c) (iv) above and suppose that $\text{card}(\sigma_n) \leq N$. Then the set of restrictions $H^p|\Lambda = \{f|\Lambda : f \in H^p\}$ consists of all functions $\lambda \longmapsto a(\lambda)$, $\lambda \in \Lambda$, such that

$$\Big(\sum_{n\geq 1} \sum_{\lambda_k \in \sigma_n} (1 - |\lambda_k|^2)|\Delta^k_{\sigma_n}(a)|^p\Big)^{1/p} < \infty,$$

$\Delta^k_{\sigma_n}(a)$ being as defined in (i).

For $p = \infty$, this means that $H^\infty|\Lambda$ consists of those functions a on Λ which are *N-hyperbolically smooth* in the sense that the Nth hyperbolically divided differences $\Delta^{N-1}_\sigma(a)$, are uniformly bounded over all $\sigma \subset \Lambda$, $\text{card}(\sigma) \leq N$.

3.3.6. Spectral subspaces and Dunford spectral model operators. Here Θ is a scalar function, $\Theta \in H^\infty$, $0 \neq |\Theta| \leq 1$. Recall that an operator T on a Hilbert space H is called *spectral in the sense of Dunford* if there exists a bounded Borel projection valued measure $\mathcal{E}$ on the plane $\mathbb{C}$ commuting with T ($\mathcal{E}(A)T = T\mathcal{E}(A)$ for every Borel set A) such that $\mathcal{E}(\mathbb{C}) = I$ and

$$\sigma(T\big|\mathcal{E}(A)H) \subset \text{clos}(A),$$

for every Borel set A. If such a measure exists, then it is unique and its values $\mathcal{E}(A)H$ for closed sets A are maximal spectral subspaces of T, see C2.5.1(c) and C2.5.8(c) for the definitions and some properties.

(a) Show that E is a maximal spectral subspace of M_Θ^* if and only if $E = \mathcal{K}_{\Theta_\sigma}$ for a closed subset σ, $\sigma \subset \overline{\mathbb{D}}$.

[*Hint:* see the introduction to this chapter for the definitions and the proof.]

(b)* *(N. Nikolski and S. Hruschev, 1987)* The following are equivalent.

(i) M_Θ is a Dunford spectral operator.

(ii) M_Θ^* is a Dunford spectral operator.

(iii) $\inf_{z\in\mathbb{D}} |\Theta_{out}(z)| > 0$, the singular measure μ determining the singular part Θ_{inn} is purely discrete, and the factorization $\Theta_{inn} = \prod_{n\geq 1} \Theta_{\sigma_n}$ satisfies the (GC) condition:

$$|\Theta_{inn}(z)| \geq \delta \cdot \inf |\Theta_{\sigma_n}(z)|$$

for $z \in \mathbb{D}$, where σ_n runs over all singleton subsets of $Z = \Lambda \cup \{\zeta \in \mathbb{T} : \mu(\{\zeta\}) > 0\}$, Λ being the zero set of Θ (so that, finally, Θ_{σ_n} is either $b_\lambda^{d(\lambda)}$ with $\lambda \in \Lambda$, or $\Theta_{\sigma_n} = \exp(-\mu(\{\zeta\})\frac{\zeta+z}{\zeta-z})$ with $\zeta \in \mathbb{T}$).

3.4. Notes and Remarks

Unconditional bases and free interpolation. Unconditional bases of subspaces $(X_n)_n$ are defined in 3.1.1 as bases whose series decompositions are all unconditionally (unorderly) convergent; it can be shown that this is equivalent to say that the decompositions $\sum_n x_n$, where $x_n \in X_n$, are permutationally convergent, i.e., are still convergent after any permutation (see A.5.1.5, Volume 1, for the case $\dim X_n = 1$ for every $n \geq 1$). For general properties of bases and unconditional bases, as well as for their use in geometry of Banach spaces and in theory of linear operators, we refer to J. Lindenstrauss and L. Tzafriri [**LT**], I. Singer [**Sin**] and I. Gohberg and M. Krein [**GK1**]. For this chapter, the principal general property is containd in Theorem 3.1.4 which claims, in particular, that any unconditional basis of a Hilbert space is isomorphic to an orthogonal basis. As for many classical results, it is not so easy to attribute this fact to a sole author since various versions of it were rediscovered several times; let us mention G. Köthe and O. Toeplitz [**KT**], E. Lorch [**Lor**] and I. Gelfand [**Gel2**], and refer to comments in N. Nikolski [**N19**], Chapter 6. N. Bari called *Riesz basis* (respectively, *Riesz basis sequence*) any sequence that is linearly isomorphic to an orthogonal basis (respectively, to an orthogonal sequence (eventually incomplete in the space)), [**Bar1**], [**Bar2**]. When speaking about vector sequences $(x_n)_n$, some people replace "orthogonal" by "orthonormal". However, this seems to be less convenient, especially if this special case is placed in the general framework of bases of subspaces. For a stability result for orthonormal bases, and hence for normalized Riesz bases, with respect to mean square perturbations see A.5.7.1(f) (Volume 1).

In an infinite dimensional Banach space X with unconditional basis, there are usually many nonisomorphic bases; more precisely, two normalized unconditional bases in a Banach space X are isomorphic to each other if and only if X is isomorphic to one of the following three spaces c_0, l^1 or l^2 (J. Lindenstrauss and M. Zippin, 1969; see J. Lindenstrauss and L. Tzafriri [**LT**]).

As explained in the introduction to this chapter, free interpolation is related to unconditional bases at least in two ways, both are based on Theorem 3.1.4, namely, via the "lattice property" of the coefficient space (a spacial interpolation), and through the l^∞ property of the multiplier space (an operator interpolation). In the framework of Banach spaces of holomorphic functions, the very first papers using this duality between bases and interpolation were, probably, H. Shapiro and A. Shields [**ShaS**], N. Nikolski and B. Pavlov [**NiP1**], [**NiP2**], and E. Amar [**Am1**]. However, in different contexts the same duality was successfully exploited already by T. Carleman [**Car3**], S. Banach [**Ban1**], and R. Paley and N. Wiener [**PW**] (see, for instance, Theorem 39 in [**PW**]). For a history of these approaches and their realizations for various function spaces (including Frechet and other spaces) see N. Nikolski [**N19**], Chapters 6 and 7, and N. Nikolski and S. Hruschev [**HrN**]. The very name *"free interpolation"* was introduced by V. Havin in the beginning of the 1970ies (probably, V. Havin and S. Vinogradov [**HVin**] is the first written mention) in order to make clear the following general problem:

Given a vector space X, $X \subset \mathrm{Hol}(\Omega)$, describe the subsets $\Lambda \subset \Omega$ such that the trace space $X|\Lambda$ has the lattice property ($a \in X|\Lambda$, $|b| \le |a| \Rightarrow b \in X|\Lambda$), i.e. describe those subsets Λ where the values of f from X are "free from the traces of holomorphy".

Later on, the same wording was adapted for interpolation by derivatives or, generally, germs of anlytic functions. Some other terminology of interpolation theory is also introduced in V. Havin and S. Vinogradov [**HVin**] ("sparse sequence", "Carleson (or, Newman–Carleson) condition", "Carleson intensity of a measure" ($C(\mu)$ of A.5.7.2(b), Volume 1, or C_Λ of 3.3.3), etc.).

It is clear from the Gram matrix approach (Subsection 3.3.1) that the Riesz basis property of a sequence of subspaces $\mathcal{X} = (X_n)_{n\ge1}$ is equivalent to the two embeddings $\mathcal{J}_{\mathcal{X}} H \subset l^2(H)$ and $\mathcal{J}_{\mathcal{X}'} \subset l^2(H)$ (see 3.3.1(d) and 3.3.1(e)). The case of a sequence of eigenspaces of a model operator M_Θ^* is of particular interest. Let

$$M_\Theta : \mathcal{K}_\Theta \longrightarrow \mathcal{K}_\Theta = H^2(E) \ominus \Theta H^2(E)$$

be a model operator having an inner characteristic function Θ, and let $X_n = (1 - \overline{\lambda}_n z)^{-1} \cdot \pi_n E$ be eigenspaces of M_Θ^*, where

$$\Theta = (b_{\lambda_n}\pi_n + (I - \pi_n))\Theta_n$$

stands for a regular factorization corresponding to the invariant subspace $\mathrm{Ker}(M_\Theta^* - \overline{\lambda}_n I)$; here $\pi_n : E \longrightarrow E$ is the orthogonal projection onto $\mathrm{Ker}\,\Theta(\lambda_n)^*$. Assume that the sequence $\mathcal{X} = (X_n)_{n\ge1}$ is minimal and complete in $\mathcal{K}_\Theta$. In this case it is shown in N. Nikolski and B. Pavlov [**NiP1**], [**NiP2**] that the embeddings split into two parts, namely, $\mathcal{X}$ is a Riesz basis in $\mathcal{K}_\Theta$ if and only if

$$\sum_{n\ge1}(1 - |\lambda_n|^2)\|\pi_n f(\lambda_n)\|_E^2 < \infty, \quad f \in H^2(E),$$

$$\sum_{n\ge1}(1 - |\lambda_n|^2)\|\pi_n' f(\overline{\lambda}_n)\|_E^2 < \infty, \quad f \in H^2(E),$$

$$\sup_{n\ge1}\left\|\pi_n'(\Theta_n'(\lambda_n))^{-1}\pi_n\right\| < \infty,$$

where $\Theta = \Theta_n'(b_{\lambda_n}\pi_n' + (I - \pi_n'))$ is a dual factorization corresponding to $\mathrm{Ker}(M_\Theta - \lambda_n I)$. The third of these conditions is equivalent to the uniform minimality of $\mathcal{X}$ and represents a vector valued analog of the Carleson condition (C). In the scalar valued case, $\dim E = 1$, the embeddings coincide (compare with 3.3.3(a) (iii) above), the third condition is nothing but (C), and moreover (C) entails these embeddings (see 3.3.3(c)). In general, this is not the case: there exists an operator $T = M_\Theta$ satisfying the above hypotheses and such that $\mathcal{X}$ is uniformly minimal but not a Riesz basis; moreover, in this example, $I - T^*T \in \mathfrak{S}_1$, $I - TT^* \in \mathfrak{S}_1$, see [**NiP1**]. S. Treil showed [**Tre4**] that the uniform minimaly, again, implies the Riesz basis property for "spacially compact" sequences of eigenvectors, see below. It is worth mentioning that for complete operators of the above type the Riesz basis property of eigenspaces is equivalent to similarity to a normal operator; see N. Nikolski and S. Hruschev [**HrN**] for an approach exploiting this equivalence. Thus, the example mentioned above should be compared with S. Kupin's results [**Kup**] and S. Kupin and S. Treil's results [**KuT**], see Section C2.6.

Generalized free interpolation and spectral decompositions. When passing to bases of invariant subspaces of a model operator M_Θ (or M_Θ^*), a new powerful technique appears, namely, the commutant lifting theorem for operator interpolation mentioned above. As is shown in Subsection 3.2.7, the Carleson corona theorem 3.2.10 is an adequate tool to treat these matters. Simplified proofs of the corona theorem can be found in P. Koosis [**Ko4**] and N. Nikolski [**N19**], Appendix 3, and the original one (based on dyadic techniques) in J. Garnett [**Gar**]. Generalized interpolation in the form of the inclusions $f - f_n \in \Theta_n H^\infty$, $n \geq 1$, first appeared in V. Vasyunin [**Vas3**] (for $f_n = c_n \in \mathbb{C}$) and N. Nikolski [**N14**], [**N15**], and in its final form, including the outer spectrum and a complete treatment of the size of the interpolation data, in N. Nikolski [**N20**], N. Nikolski and S. Hruschev [**HrN**], and N. Nikolski and A. Volberg [**NV**]. In particular, the papers [**HrN**] and [**NV**] contain more information about the local meaning of generalized interpolation (see Subsections 3.2.15 through 3.2.19). In [**N19**] and [**HrN**], one can find a comparison of this approach with preceding results on multiple interpolation and complete lists of references.

Geometrically, the principal meaning of the results presented in this chapter is explained in Remark 3.2.19 and Subsection 3.3.5(a). Namely, the uniform minimality of a family of invariant subspaces $(\mathcal{K}_\Theta(\sigma_n))_{n\geq 1}$ entails (is equivalent to) that, in fact, it is an unconditional (Riesz) sequence. This fundamental fact of the geometry of S^*-invariant subspaces of H^2 is yet not completely understood. Perhaps, it has the same nature as the Reproducing Kernel Thesis (RKT) discussed in Section A.5.8 (Volume 1). For example, it is worth mentioning a specific coincidence occuring in the case of eigenspaces $X_n = \mathcal{K}_\Theta(\sigma_n) = \mathbb{C} \cdot (1 - \overline{\lambda}_n z)^{-1}$, where the boundedness of the Gram operator $W_{\mathcal{X}'} = In$ (see 3.3.1 and 3.3.3(c), 10)) *implies* boundedness of $W_{\mathcal{X}}^* = (W_{\mathcal{X}'})^{-1}$ (and hence, the Riesz basis property). Yet another feature emphasizing the above mentioned resemblance is that the uniform minimality condition (C), in fact, contains the embedding for kernels k_{λ_n} as a straightforward consequence (see 3.3.3(c) above). For more about embedding thereoms of this kind see A.5.7.2 (Volume 1), D.4.7.5 and D.4.8. The following embedding theorem by Treil ([**Tre10**]) generalizes 3.2.23 and shows that the RKT holds also for embeddings associated with an arbitrary measurable family of invariant subspaces over the unit disk.

THEOREM. *Given a measurable family of inner functions $(\Theta_\lambda)_{\lambda\in\mathbb{D}}$ and a positive Borel measure μ in $\mathbb{D}$, the following are equivalent.*

1) $\int_{\mathbb{D}} \|f\|^2_{H^2/\Theta_\lambda H^2}\, d\mu(\lambda) < \infty$ *for every* $f \in H^2$.

2) $\sup_{z\in\mathbb{D}} \int_{\mathbb{D}} (1 - |\Theta_\lambda(z)|^2)\, d\mu(\lambda) < \infty$.

Moreover, S. Treil proved the same implication (uniform minimality $\Rightarrow$ unconditional basis) for families of S^*-invariant subspaces of the vector valued $H^2(E)$ with $\dim E < \infty$ ([**Tre10**]), and for families of eigenspaces $(E_n \cdot (1 - \overline{\lambda}_n z)^{-1})_{n\geq 1}$, $E_n \subset E$, for $\dim E = \infty$ satisfying a "spacial compactness" condition (any normalized sequence $x_n \in E_n$ must be relatively compact), [**Tre4**] and [**Tre9**]. Yet another "abstract" Hilbert space approach to bases-interpolation going up to a kind of Carleson condition is proposed in B. Chalmers [**Cha**]. For free interpolation theorems for some other Banach spaces (Bergman, weighted Bergman, Bargman, and other spaces) see K. Seip [**Sei1**], [**Sei2**] and H. Hedenmalm, B. Korenblum, and K. Zhu [**HKZ**] and references therein. Free interpolating sets for the space $BMOA$ are studied by R. Sundberg, A. Kotochigov, A. Aleksandrov and others; see a survey in A. Aleksandrov [**Aleks8**]. For some other free interpolation theorems, see Chapter D.4 below.

In operator theory, unconditionally convergent decompositions such as $\mathcal{K}_\Theta = \sum_{n\geq 1} \mathcal{K}_\Theta(\sigma_n)$ are interesting because they are responsible for the Dunford spectrality property of the model operator M_Θ, see the introduction to this chapter and Subsection 3.3.6 for details. The first step to the spectrality criterion of 3.3.6 is made by C. Foiaş [**Foi4**] (see also [**N19**] for a slightly refined form), with the subsequent progress made by R. Teodorescu (1975) and P. Wu (1979), but the final form of the result appeared in N. Nikolski and S. Hruschev [**HrN**], where original references can also be found. For a general model operator, such an explicit criterion seems to be unknown.

Formal credits. Theorem 3.1.4 is commented above. Lemma 3.1.5 is from W. Orlicz [**Orl**]. The equivalence (1) $\Leftrightarrow$ (2) of Theorem 3.2.5 (and of Theorem 3.2.14) is from V. Vasyunin [**Vas3**], and (1) $\Leftrightarrow$ (3) $\Leftrightarrow$ (4) are from N. Nikolski [**N14**], [**N15**], [**N20**] and N. Nikolski and S. Hruschev [**HrN**].

The interrelations between spectral theory of model operators and the Carleson corona theorem, as well as its operator valued generalizations (see Chapter B.9, Volume 1), are known from long ago (see P. Fuhrmann, [**Fuh1**], and comments to the spectral mapping theorem in Section 2.6 above). For decompositions into invariant subspaces, the corona type theorems and the Bezout equations are also employed from quite long ago, see V. Vasyunin [**Vas3**], N. Nikolski [**N14**], [**N15**], [**N19**]. On the contrary, the simple technical Lemma 3.2.12 which links directly the condition (GC) to Corollary 3.2.11 was remained unnoticed until [**HrN**].

Theorem 3.2.14 is already commented above. Condition (GC) first appeared in V. Vasyunin [**Vas1**], [**Vas3**] (where it is called generalized Carleson condition; in [**N19**] it is also called Carleson–Vasyunin condition). The important special cases considered in 3.2.15 through 3.2.18 represent an abridged version of a detailed analysis of (GC) contained in [**Vas3**], [**N19**], [**HrN**]. The advanced version of 3.2.19 given in 3.3.5(a) can also be found in these sources.

For the Gram matrices of Subsection 3.3.1 see N. Akhiezer [**Akh3**], F. Gantmacher [**Gan**], S. Kaczmarz and H. Steinhaus [**KSt**] (and, especially, a survey paper by R. Guter and P. Ulyanov added to the Russian translation of this book (GIFML, 1958, Moscow)), and S. Mikhlin [**Mi4**]. Many of the properties of Gram matrices and their applications for best approximations and finite section methods (see B.4.8, Volume 1) go back to the 1930s, see S. Lewin [**Lew**], S. Mikhlin [**Mi4**]. The subspace valued generalization 3.3.3(e) is from N. Nikolski [**N14**]. Concerning the Müntz-Szasz theorem 3.3.1(i) see [**Akh3**], [**KSt**], R. DeVore and G. Lorentz [**DeVL**], and original references therein.

The Cotlar–Stein lemma 3.3.2(a) was originally proved for selfadjoint and mutually commuting operators T_j satisfying $\max(\|T_i^*T_j\|, \|T_iT_j^*\|) \le c(i-j)^2$, where $\sum_n c(n) < \infty$ (M. Cotlar [**Cot1**]), and generalized by E. Stein for (noncommutative) sequences with $\max(\|T_i^*T_j\|, \|T_iT_j^*\|) \le c(i,j)^2$, $\sup_i \sum_j c(i,j) < \infty$; see E. Stein [**St2**], Chapter 7 for more details and references. In particular, [**St2**] contains many important applications of the Cotlar–Stein lemma to the boundedness of singular integrals. For such an integral operator, say J, one typically chooses $T_j = \Delta_j J S_j$, where $S_j f = f * \Phi_{2^j}$ are certain "Fejer type means", and $\Delta_j = S_j - S_{j-1}$ form a kind of Littlewood–Paley decomposition. In 3.3.3(d) we show that the Cotlar matrix associated with a free interpolation problem coincides with the Gram matrix of the corresponding reproducing kernels.

Subsection 3.3.3 contains a treatment of free interpolation problems in H^p spaces independent of the corona theorem. The principal point 3.3.3(c) is based on the embedding theorem A.5.7.2(b), Volume 1. It is worth mentioning that, under the (necessary and sufficient) condition (C), there always exists a *linear interpolating operator.* Several constructions are given. First, in 3.3.3(e)(ii), it is shown that, for $1 < p < \infty$, such an operator can be chosen with values in the "least possible" subspace $\mathcal{K}_B^p$. For $p = \infty$, the Jones–Vinogradov operator 3.3.3(e)(i) gives a "twice larger" subspace $\mathcal{K}_{B^2}^\infty$. The construction of 3.3.3(e)(i) is taken from S. Vinogradov [**Vin5**]; G. Airapetyan's result 3.3.3(e)(iii), [**Air**], shows that this subspace can be diminished, in fact, only for uniform Frostman sequences Λ. The original construction by P. Jones, [**J2**], was simplified by S. Vinogradov, E. Gorin, and S. Hruschev [**GHVin**]. Their modification is used in 3.3.3(g)(ii) in order to describe a "lattice type" vector space of data which can be interpolated even if the condition (C) fails. For Garnett's result 3.3.3(g)(i) see [**Gar**] and references therein; it is curious to notice that under the double monotonicity conditions $|\lambda_n| \nearrow 1$ and $\delta_n \searrow$, the elementary result 3.3.3(g)(ii) is even better than 3.3.3(g)(i) (but in general, this is not the case). The ingenious construction 3.3.3(h) is due to A. Bernard [**BerA**] and N. Varopoulos [**Var1**]; for a finite sequence Λ it is still working for any uniform Banach algebra instead of $H^\infty(\Omega)$. It is also worth mentioning that any interpolation operator $T : l^\infty \longrightarrow H^\infty$ maps l^∞ isomorphically ($\|a\|_{l^\infty} \le \|Ta\|_{H^\infty} \le c\|a\|_{l^\infty}$) onto a complementable subspace of H^∞ ($TR = P$ is a bounded projection onto $T(l^\infty)$, where $Rf = f|\Lambda$). For several other proofs of the Carleson interpolation theorem see J. Garnett [**Gar**]. P. Koosis [**Ko7**] found a direct deduction of this theorem from the Nevanlinna–Pick criterion (Section B.3.2, Volume 1).

Finally, note that 3.3.3(f) is from N. Nikolski [**N19**], and, of course, the above theory of interpolation can be transferred into the Hardy spaces $H^p(\mathbb{C}_+)$ on the halfplane; for some new features in $\mathbb{C}_+$ see D.4.7 below and [**N19**], Chapter 11.

Subsection 3.3.4 is borrowed from N. Nikolski [**N19**], where we refer to for original references.

For 3.3.5(a) see N. Nikolski [**N20**], N. Nikolski and S. Hruschev [**HrN**]; for 3.3.5(b) see the same papers and also [**N21**] and N. Nikolski and A. Volberg [**NV**]. For a generalized version of the maximum modulus principle employed in the implication (vi) $\Rightarrow$ (iv) of 3.3.5(a) and in 3.3.5(c)(ii) see [**HrN**]. The results from 3.3.5(c) are also from [**NV**], [**N21**] and [**HrN**]. Moreover, it is shown in [**HrN**] that the following free interpolation of locally defined data is possible: for every $f \in H^\infty(L(\Theta, \varepsilon))$ there exists a function $g \in H^\infty$ such that $f - g \in \Theta H^\infty(L(\Theta, \varepsilon))$. A similar result for inner functions Θ, in fact, was already known to L. Carleson [**C3**]. S. Vinogradov and S. Rukshin [**RuVin**] found an H^p analog of these results, and also proved that $L = L(\Theta, \varepsilon)$ cannot be replaced by an essentially smaller open set L. All these comments and results concern the problem on how to express the (nonlocal) norm $\text{dist}(\overline{\Theta} f, H^\infty)$ of interpolation data in "local terms"; for this, see also C2.5.7(f) above.

For 3.3.5(d), which coincides with 3.2.23 for $p = 2$, see N. Nikolski [**N14**] and A. Hartmann [**Hart1**], and for 3.3.5(e), see V. Vasyunin [**Vas4**] and A. Hartmann [**Hart1**]. It is worth mentioning that similar results are known also for more general Hardy–Orlicz spaces, A. Hartmann [**Hart2**].

For more about Dunford spectral operators and for the proof of 3.3.6(b) see N. Nikolski and S. Hruschev [**HrN**].

Part D

Analytic Problems in Linear System Theory

Part D

Analytic Problems in Linear System Theory

Foreword to Part D

Part D of the book is devoted to some problems of control in linear dynamic systems acting in Banach and Hilbert spaces. Rather than giving practical applications of control theory to Signal Processing, Electronic Engineering or Partial Differential Equations (PDE), we present here a pure analyst vision of the theory, more precisely, a glance at the system theory from the operator theory and complex analysis point of view. The key points are the links between control theory and linear operators and especially the influence of control theory on operators, complex and harmonic analysis. This point of view essentially determined the choice of topics treated here, in particular of those that suggest new problems and insights to analysts, and that allow us to employ techniques from the aforementioned areas in analysis.

More precisely, we give an easily accessible introduction to problems in linear system control based on standard facts of functional analysis and linear operators. As a source of examples, we consider the Cauchy problem $x' = Ax + Bu$, $x(0) = x_0$, in a suitable vector space with a generator A either of *hyperbolic type* (that is having its spectrum in a strip $|\operatorname{Re} z| < c$) or *parabolic type* (that is having its spectrum in a strip $|\operatorname{Im} z| < c$).

The results turn out to be much more advanced in case the eigenvectors of the generator A form a Riesz basis (which is quite realistic since most of the elliptic generators in PDE's have this property). The configuration of the frequencies (of the point spectrum of A) will be of major importance for the quality of the control admitted by a given dynamical system. Although there is a great variety of results in control theory and of their dependence on the spectrum of the generator A and on the control operator, one can observe a common principle in all of our examples: exact controllability, even in the case of a suitable renormalization of the state space (see Chapter 3), is more often possible for systems of hyperbolic (oscillating) type than for those of parabolic (damped) type, whereas parabolic systems are more easily null controllable.

Let us begin with the terminology of control theory of linear dynamical systems. Let X, Y and $\mathcal{U}$ be Banach spaces and $A : X \longrightarrow X$, $B : \mathcal{U} \longrightarrow X$, $C : X \longrightarrow Y$ and $D : \mathcal{U} \longrightarrow Y$ linear mappings. A *linear dynamical system* is a system of the following form:

$$(0.0.1) \qquad \begin{cases} x'(t) = Ax(t) + Bu(t) \\ x(0) = x_0 \\ y(t) = Cx(t) + Du(t) \end{cases} , \quad t \geq 0,$$

where $x(t) \in X$ and $u(t) \in \mathcal{U}$ for $t \geq 0$. We call

- $x(t)$ the *state* of the system at time t;
- X the *state space*;
- A the *generator operator*;
- B the *control operator* (always assumed to be bounded);
- C, D the *observation operators* (likewise assumed to be bounded);
- u the *input (control)* function;
- y the *output* function;
- x_0 the *initial state.*

The system is called *stationary* if A, B, C and D do not depend on t; in this book, we consider the stationary case only.

Controlling the system (0.0.1) means to act on the system by means of a suitable control function u in such a way that starting from a given initial state x_0 the system attains at a time $\tau > 0$ the in advance given final state $x_1 = x(\tau)$. There are various notions of controllability: exact, null, approximate, optimal (in different senses), feedback control, etc.

We will make precise these notions in due course after having introduced the appropriate language. Now, let us just mention that control in linear dynamical systems, has become, in the last 20 years, the main consumer of operator theory and a new field of applications for Hardy space techniques (earlier, this role was mostly played by perturbation and scattering theory).

In what follows we suppose that

- The function $t \longmapsto x(t)$, $[0, \tau[\longrightarrow X$, is an absolutely continuous function with square-summable derivative $x' \in L^2(0, \tau; X)$;
- The operator $A : X \longrightarrow X$ is the *generator* of a strongly continuous *semigroup* $(S(t))_{t \geq 0}$, that is, $S(0) = I$ and $t \longmapsto S(t)x_0$ is continuous for every x_0, and $A = \lim_{t \to 0} \frac{1}{t}(S(t) - I)$ on $\mathcal{D}(A)$, the domain of definition of A. We write $S(t) = e^{tA}$;
- $u \in L^2(0, \tau; \mathcal{U})$ (finite energy control).

0.0.1. Definition. The system (0.0.1) is called

- *exactly controllable* (at time $\tau \geq 0$) if for every $x_0, x_1 \in X$ there is a function $u \in L^2(0, \tau; \mathcal{U})$ such that
$$\begin{cases} x(\tau) & = x_1 \\ x(0) & = x_0; \end{cases}$$
- *null controllable* (at time $\tau \geq 0$) if for every $x_0 \in X$ there is $u \in L^2(0, \tau; \mathcal{U})$ such that
$$\begin{cases} x(\tau) & = 0 \\ x(0) & = x_0; \end{cases}$$
- *appoximately controllable* (at time moment $\tau \geq 0$) if for every $x_0, x_1 \in X$ and every $\varepsilon > 0$ there is $u \in L^2(0, \tau; \mathcal{U})$ such that
$$\begin{cases} \|x(\tau) - x_1\| < \varepsilon \\ x(0) = x_0; \end{cases}$$

- *appoximately controllable on* $[0, \infty)$ if for every $x_0, x_1 \in X$ and arbitrary $\varepsilon > 0$ there exists $\tau \in [0, +\infty[$ and $u \in L^2(0, \tau; \mathcal{U})$ such that

$$\begin{cases} \|x(\tau) - x_1\| < \varepsilon \\ x(0) = x_0. \end{cases}$$

Later on, we will also introduce controllability at time $\tau = \infty$, renormalized exact controllability, generalized exact and generalized null controllability as well as exact null controllability.

Certain optimizations of control will likewise be discussed, for instance the following.

1. Determine $\tilde{u}$ such that

$$\begin{cases} x(\tau) & = x_1 \\ x(0) & = x_0 \end{cases}$$

and

$$\|\tilde{u}\|_{L^2(0,\tau;\mathcal{U})} = \min\{\|u\|_{L^2(0,\tau;\mathcal{U})} \quad : \quad u \text{ takes the system (0.0.1) from } x_0 \text{ to } x_1 \text{ at time } \tau\ \}.$$

2. For a given constant c, find a control u such that

$$\|u\|_{L^2(0,\tau;\mathcal{U})} \leq c \quad \text{and} \quad \begin{cases} x(\tau) = x_1 \\ \tau \text{ is minimal.} \end{cases}$$

3. Minimization of the output noise by a linear compensator (feedback H^∞-control).
4. Minimization of the number of controls: $\dim \mathcal{U} = \min$ or rather $\dim B\mathcal{U} = \min$.

Let us finally mention that the same problems can be stated also for discrete dynamic systems

$$\begin{cases} x_{n+1} & = Ax_n + Bu_n \\ y_n & = Cx_n + Du_n \end{cases}, \quad n = 0, 1, \dots$$

These can be treated in a very similar way as the continuous ones.

For short, we will use the following abbreviations.

ECO	Exact COntrollability (or Exactly COntrollable)
NCO	Null COntrollability (or Null COntrollable)
ACO	Approximate COntrollability (or Approximately COntrollable)
RECO	Renormalized Exact COntrollability (or Renormalized Exactly COntrollable)
GECO	Generalized Exact COntrollability (or Generalized Exactly COntrollable)
ENCO	Exact Null COntrollability (or Exact Null COntrollable)
GNCO	Generalized Null COntrollability (or Generalized Null COntrollable)

To finish this introduction we recall some basic definitions and notions.

If X and Y are two Banach spaces, let $L(X, Y)$ be the space of linear continuous operators from X to Y equipped with the standard norm. In case $X = Y$, we simply write $L(X) = L(X, X)$. Let $X^* = L(X, \mathbb{C})$ be the standard dual space of X.

If $\Omega \subset \mathbb{R}^n$ is an open subset and E a Banach space, we write $L^2(\Omega, E)$ for the space of E valued square-summable functions on Ω (with respect to Lebesgue measure), equipped with the usual norm $\|f\| = \left(\int_\Omega \|f(x)\|_E^2\,dx\right)^{1/2}$.

The spaces $L^2(\mathbb{T}) = L^2(\mathbb{R}/2\pi\mathbb{Z})$, H^2, $L^2(\mathbb{T}, E)$ and $H^2(E)$ have the same meaning as before (cf., e.g., A.1.1, Volume 1, C.2.2.1 or B.1.6.1). We will also consider the Hardy space H^2_+ in the upper half plane $\mathbb{C}_+ = \{z \in \mathbb{C} : \operatorname{Im} z > 0\}$ (cf. A.6.3, Volume 1) and its vector valued analogue $H^2_+(E)$:

$$H^2_+ = \Big\{f : f \text{ holomorphic in } \mathbb{C}_+, \|f\|^2 = \sup_{y>0} \int_{\mathbb{R}} |f(x+iy)|^2\,dx < \infty\Big\},$$

and

$$H^2_+(E) = \Big\{f \quad : \quad f \text{ holomorphic in } \mathbb{C}_+, E \text{ valued },$$
$$\|f\|^2 = \sup_{y>0} \int_{\mathbb{R}} \|f(x+iy)\|_E^2\,dx < \infty\Big\}.$$

See Section A.3.11 (Volume 1) for more about vector valued functions.

CHAPTER 1

Basic Theory

In this chapter we obtain the basic criteria for exact controllability (ECO), approximate controllability (ACO) and null controllability (NCO) in terms of the semigroup $S(\cdot)$ and the control operator B.

We also consider a refinement of null controllability, namely exact null controllability (cf. Subsection 1.3.24). For a given system, the main attention will be paid to distinguish the roles played by the semigroup $S(\cdot)$ (and its generator A) and the control operator B in different types of control (cf. Section 1.3). This will be illustrated by the example of the heating of a metal bar (cf. 1.5), though the main examples of applications are postponed to Chapter 3.

1.1. The Main Formula

Let

$$x' = Ax + Bu, \quad t \geq 0, \tag{1.1.1}$$

be the system to be controlled. In the sequel we refer to this as the *system* (A, B). Let $S(\cdot)$ be the semigroup associated to A. Then if A is independent of t we have

$$\frac{d}{dt} S(t)x_0 = AS(t)x_0$$

(on $\mathcal{D}(A)$ which is invariant with respect to $S(t)$). Otherwise, if $A = A(t)$ depends on t, then we get the fundamental two parameter solution

$$\begin{cases} S = S(t, r) \\ \dfrac{\partial S(t, r)}{\partial r} = A(r)S(t, r), \quad t < r, \end{cases}$$

and we recover the stationary case with $S(t, r) = S(r - t)$.

Consider the stationary case. Then we obtain the general solution to the problem by

$$x(t) = S(t)x_0 + \int_0^t S(t - r)Bu(r)\, dr =: S(t)x_0 + c(t)u. \tag{1.1.2}$$

The operator

$$c(t) : L^2(0, \tau; \mathcal{U}) \longrightarrow X$$

is called *controllability operator* (or *controllability map*).

1.2. Initial Observations about Controllability

Here we introduce some additional terminology and list a few basic conditions for systems to be controllable.

1.2.1. Reachable states. The collection

$$A(\tau)x_0 = \{x(\tau) : \text{ there is } u \in L^2(0,\tau;\mathcal{U}) \text{ such that } x(0) = x_0\}$$

is called the *set of reachable states from* x_0, and

$$A(\tau) = \{x_1 \in X \quad : \quad \text{there are } u \in L^2(0,\tau;\mathcal{U}), x_0 \in X, \text{ such that } x(0) = x_0 \text{ and } x(\tau) = x_1\}$$

is the set of all reachable states at time τ.

1.2.2. ECO and ACO. Clearly

$$\begin{aligned} A(\tau)x_0 &= S(\tau)x_0 + c(\tau)L^2, \\ A(\tau) &= S(\tau)X + c(\tau)L^2, \end{aligned}$$

and thus a system is exactly controllable at time τ if and only if $A(\tau)x_0 = X$ for every $x_0 \in X$, and approximately controllable if and only if $\operatorname{clos}(A(\tau)x_0) = X$ for every $x_0 \in X$. Note that $c(\tau)L^2$ is a vector subspace of X (not necessarily closed), and hence, for arbitrary $x_0, x_0' \in X$ we get

$$\begin{aligned} A(\tau)x_0 = X &\iff A(\tau)x_0' = X, \\ \operatorname{clos}(A(\tau)x_0) = X &\iff \operatorname{clos}(A(\tau)x_0') = X. \end{aligned}$$

In particular, setting $x_0 = 0$, we obtain the following.

1.2.3. COROLLARY. *The system (A, B) is exactly controllable (respectively approximately controllable) at time moment τ if and only if $c(\tau)L^2 = X$ (respectively $\operatorname{clos} c(\tau)L^2) = X$).*

□

1.2.4. Monotonicity in time. The set of reachable states from $x_0 = \mathbf{0}$ at time τ is monotone in τ: for $\varepsilon > 0$ we have

$$A(\tau + \varepsilon)\mathbf{0} \supset A(\tau)\mathbf{0}.$$

In fact, by 1.2.2 this is equivalent to $c(\tau + \varepsilon)L^2 \supset c(\tau)L^2$, and we verify the last inclusion. Pick an arbitrary control $u \in L^2(0,\tau;\mathcal{U})$ and set

$$u_1(t) = \begin{cases} 0 & \text{if } 0 \le t \le \varepsilon, \\ u(t - \varepsilon) & \text{if } \varepsilon \le t < \tau + \varepsilon, \end{cases}$$

(which means that we do not control before the time ε). Obviously $u_1 \in L^2(0,\tau;\mathcal{U})$ and

$$\begin{aligned} c(\tau)u &= \int_0^\tau S(\tau - s)Bu(s)\,ds = \int_\varepsilon^{\tau+\varepsilon} S(\tau + \varepsilon - r)Bu_1(r)\,dr \\ &= c(\tau + \varepsilon)u_1. \end{aligned}$$

□

1.2.5. Null controllability (NCO). If the system is null controllable at time τ, then there are $x_0 \in X$ and $u \in L^2(0,\tau;\mathcal{U})$ such that $x(0) = x_0$ and $x(\tau) = \mathbf{0}$. The main formula (1.1.2) (with $x(\tau) = 0$) shows the following observation.

OBSERVATION. The system (A, B) is null controllable if and only if

$$S(\tau)X \subset c(\tau)L^2.$$

A priori, null controllability is a weaker property than exact controllability. They coincide if and only if $S(\tau)X = X$. This happens to be true, for instance, if the generator A is bounded, for in this case we have $S(\tau) = e^{\tau A}$ and hence $S(\tau)^{-1} = e^{-\tau A}$ which in particular gives surjectivity of $S(\tau)$.

More generally speaking, to have $NCO = ECO$, it suffices to have invertibility of $S(t)$ for all $t > 0$. In fact, this is the case if and only if the semigroup $S(\cdot)$ extends to a continuous group on $\mathbb{R}$ by $S(-t) = S(t)^{-1}$, $t > 0$. An example of such a group is given by the translation operators $S(t)f = f(x-t)$ on $L^2(\mathbb{R})$.

1.3. Basic Criteria for ACO, ECO, and NCO

We now study the main types of controllability, that is ACO, ECO and NCO, and their dependence on metric properties of the semigroup $S(\cdot)$ and the operator B as well as their adjoints.

We start with a classical result by R. Kalman, one of the founding fathers of the theory of linear dynamical systems.

1.3.1. THEOREM (R. Kalman, 1963). *A system (A, B) is approximately controllable at time τ if and only if*

$$\operatorname{span}(S(t)B\mathcal{U} : 0 \le t \le \tau) = X. \tag{1.3.1}$$

The system (A, B) is approximately controllable on $[0, +\infty[$ if and only if $B\mathcal{U}$ is a cyclic *subspace of $S(\cdot)$:*

$$\operatorname{span}(S(t)B\mathcal{U} : t > 0) = X. \tag{1.3.2}$$

If the generator A is bounded, then (1.3.2) holds if and only if (1.3.1) holds for every (some) $\tau > 0$, and if and only if

$$\operatorname{span}(A^n B\mathcal{U} : n \ge 0) = X. \tag{1.3.3}$$

PROOF. It suffices to show that

$$\operatorname{clos} c(\tau)L^2(0,\tau;\mathcal{U}) = \operatorname{span}(S(t)B\mathcal{U} : 0 \le t \le \tau).$$

The inclusion "$\subset$" is a simple consequence of the fact that the integral $\int_0^\tau S(\tau - r)Bu(r)\,dr$ is the limit of Darboux sums, and so it belongs to span $\big(S(t)B\mathcal{U} : 0 \le t \le \tau\big)$.

For the converse inclusion we use the Hahn-Banach theorem. Suppose that $f \in X^*$ is zero on $c(\tau)L^2$. We will show that it also vanishes on $S(t)B\mathcal{U}$ for arbitrary $0 \le t \le \tau$. To this end, let $u(r) = u_0 \cdot \varphi(r)$, where $u_0 \in \mathcal{U}$ and φ is a scalar valued continuous function. Clearly, $u \in L^2(0,\tau;\mathcal{U})$ and by assumption:

$$\begin{aligned} 0 &= \langle f, c(\tau)u\rangle = \langle f, \int_0^\tau \varphi(r)S(\tau - r)Bu_0\,dr\rangle \\ &= \int_0^\tau \varphi(r)\langle f, S(\tau - r)Bu_0\rangle\,dr. \end{aligned}$$

As φ is arbitrary we deduce that

$$\langle f, S(\tau - r)Bu_0 \rangle = 0, \quad 0 \le r \le \tau,$$

and hence

$$\langle f, S(t)Bu_0 \rangle = 0, \quad 0 \le t \le \tau. \tag{1.3.4}$$

This proves the general case of the theorem (the case $\tau = \infty$ is a consequence of this and the definition of ACO on $[0, \infty)$).

Consider now the case where $S(t) = e^{tA}$ and A are bounded. We differentiate successively (1.3.4) at $t = 0$ to get

$$\langle f, A^n Bu_0 \rangle = 0, \quad n = 0, 1, \dots .$$

Since

$$S(t)B\mathcal{U} = e^{tA}B\mathcal{U} \subset \operatorname{span}(A^n B\mathcal{U} : n \ge 0),$$

we obtain by a reasoning similar to the above that

$$\operatorname{clos} c(\tau)L^2 = \operatorname{span}(A^n B\mathcal{U} : n \ge 0).$$

□

1.3.2. Remark. The characterization (1.3.3) means that, in case A is bounded, the approximate controllability at time τ is actually independent of τ (and the initial state). This is not true in general for unbounded generators A. To have an example, consider the translation group $\big(S(t)f\big)(x) = f(x - t)$ on $L^2(\mathbb{R})$ and $\mathcal{U} = L^2(0,1) = \chi_{]0,1[}L^2(\mathbb{R})$.

Another criterion for approximate controllability is established by the following result.

1.3.3. Theorem. *Let (A, B) be a dynamical system and let*

$$c(\tau)u = \int_0^\tau S(\tau - r)Bu(r)\,dr, \quad u \in L^2(0, \tau; \mathcal{U}),$$

be the associated controllability map at time $\tau > 0$. Then the adjoint operator

$$c(\tau)^* : X^* \longrightarrow L^2(0, \tau; \mathcal{U}^*)$$

is given by

$$\big(c(\tau)^* y\big)(r) = B^* S(\tau - r)^* y, \quad 0 \le r \le \tau;\ y \in X^*,$$

and the system is approximately controllable at time τ if and only if

$$\operatorname{Ker} c(\tau)^* = \{\mathbf{0}\},$$

that is, if and only if

$$B^* S(t)^* y = \mathbf{0} \text{ for } 0 \le t \le \tau \quad \Longrightarrow \quad y = \mathbf{0}$$

for every $y \in X^*$.

Proof. Clearly, $\operatorname{clos}(c(\tau)L^2) = X$ if and only if $\operatorname{Ker} c(\tau)^* = \{\mathbf{0}\}$. In order to find the adjoint of $c(\tau)$, assume that $0 \le t \le \tau$, $y \in X^*$ and $u \in L^2(0, \tau; \mathcal{U})$. Then

$$\begin{aligned} \langle u, c(\tau)^* y \rangle &= \langle c(\tau)u, y \rangle = \int_0^\tau \langle S(\tau - r)Bu(r), y \rangle\,dr \\ &= \int_0^\tau \langle u(r), B^* S(\tau - r)^* y \rangle\,dr, \end{aligned}$$

and hence $c(\tau)^*y$ is given by $r \longmapsto (B^*S(\tau - r)^*y)$ $(\in L^2(0,\tau;\mathcal{U}))$. □

Now we pass to exact controllability. In contrast to approximate controllability, the ECO depends much more on τ and the control operator, even if A is bounded. The basic criterion is the following.

1.3.4. THEOREM. *A system (A, B) is exactly controllable at time τ if and only if there is $\gamma > 0$ such that*

$$\|c(\tau)^*y\| \geq \gamma\|y\| \;, \quad y \in X^*, \tag{1.3.5}$$

that is, if and only if

$$\int_0^\tau \|B^*S(t)^*y\|^2_{\mathcal{U}^*} dt \geq \gamma^2\|y\|^2_{X^*}, \quad y \in X^*.$$

PROOF. By Banach's theorem (cf. also Lemma 1.3.12 below), $c(\tau)L^2(0,\tau;\mathcal{U}) = X$ if and only if $c(\tau)^*$ satisfies (1.3.5). □

The estimate (1.3.5) of Theorem 1.3.4 can be improved by means of the following result.

1.3.5. LEMMA. *Let $S(\cdot)$ be a strongly continuous semigroup on a Banach space X. The following assertions are equivalent.*

(1) There are $\alpha > 0$ and $\varepsilon > 0$ such that

$$\|S(t)^*y\| \geq \varepsilon e^{-\alpha t}\|y\|$$

for every $t > 0$ and $y \in X^*$.

(2) There is $t_0 > 0$ such that

$$\inf_{\|y\|=1} \|S(t_0)^*y\| > 0.$$

PROOF. Obviously (1) implies (2). In order to prove the converse implication, let $t > 0$ and

$$\varepsilon(t) = \inf\Big\{\|S(t)^*y\| : y \in X^*, \; \|y\| = 1\Big\}. \tag{1.3.6}$$

For $y \in X^*$ and $r, t \in \mathbb{R}_+$ the semigroup property gives

$$\|S(t+r)^*y\| = \|S(t)^*S(r)^*y\| \geq \varepsilon(t) \cdot \|S(r)^*y\| \geq \varepsilon(t)\varepsilon(r) \cdot \|y\|,$$

from which we deduce that

$$\varepsilon(t+r) \geq \varepsilon(t)\varepsilon(r). \tag{1.3.7}$$

On the other hand,

$$\|S(t+r)^*y\| \leq \|S(r)^*\| \cdot \|S(t)^*y\|,$$

and hence

$$\varepsilon(t+r) \leq \|S(r)\| \cdot \varepsilon(t).$$

This inequality shows that if there is $t_0 > 0$ such that $\varepsilon(t_0) > 0$, then $\varepsilon(t) > 0$ for all t, $0 < t \leq t_0$, and

$$0 < \varepsilon(t_0) \leq \|S(t_0 - t)\| \cdot \varepsilon(t). \tag{1.3.8}$$

For arbitrary $t > 0$ there exists $n \in \mathbb{Z}_+$ and r, $0 \leq r < t_0$, such that $t = t_0 n + r$. Using now (1.3.7) we get

$$\varepsilon(t) = \varepsilon(nt_0 + r) \geq \varepsilon(nt_0)\varepsilon(r) \geq \varepsilon(t_0)^n\varepsilon(r) > 0.$$

Moreover, (1.3.7) shows that that the function a,

$$a(t) := \frac{1}{\varepsilon(t)}, \quad t \geq 0.$$

has the submultiplicativity property $a(t+r) \leq a(t)a(r)$ for $t, r \geq 0$, and (1.3.8) shows that a is locally bounded: $\sup_{t \leq t_0} a(t) < \infty$, $t_0 > 0$. It is well-known that this implies an estimate of the following type: $a(t) \leq Ce^{\alpha t}$. Indeed, set $M = \sup_{t \leq 1} a(t)$ and $t = [t] + r$, $0 \leq r < 1$. Then

$$a(t) \leq a(r)a([t]) \leq Ma(1)^{[t]} \leq Ce^{\alpha t}$$

for suitable constants $C > 0$, $\alpha > 0$ (cf. also Subsection 2.2.1). □

1.3.6. COROLLARY. *Let $\varepsilon(\cdot)$ be the function defined in (1.3.6). Then either $\varepsilon(t) > 0$ for every $t > 0$, or $\varepsilon(t) \equiv 0$, $t > 0$.* □

We now obtain the following.

1.3.7. THEOREM. *Let (A, B) be a dynamical system (where, as usual, A is the generator of a strongly continuous semigroup).*

(1) If (A, B) is ECO at a time $\tau > 0$, then there are constants $\alpha > 0$, $\varepsilon > 0$ such that

$$\|S(t)^* y\| \geq \varepsilon e^{-\alpha t} \|y\| \tag{1.3.9}$$

for all $t > 0$ and $y \in X^*$.

(2) Conversely, if there exists $t_0 > 0$ such that

$$\inf_{\|y\|=1} \|S(t_0)^* y\| > 0, \tag{1.3.10}$$

then the system (A, B) is ECO at every time $\tau > 0$ provided B satisfies Range $B = X$.

PROOF. (1) Assume that the conclusion is false. Then by Lemma 1.3.5 (and Corollary 1.3.6) the function $\varepsilon(t)$ from (1.3.6) vanishes identically. We will show that this is impossible. Using the semigroup property it is easy to see that

$$\|S(t)\| \leq Me^{ct}, \quad t > 0,$$

for suitable c, M (cf. Subsection 2.2.1). Clearly, for arbitrary $\eta > 0$ there is a $\delta > 0$ such that

$$\delta M^2 e^{2c\delta} < \eta/2.$$

Fix $\tau > 0$. By assumption $\varepsilon(\tau) = 0$, so there exists $y \in X^*$, $\|y\| = 1$, such that

$$\tau M^2 e^{2c\tau} \|S(\delta)^* y\|^2 < \eta/2.$$

Now

$$\begin{aligned} J(y) &:= \int_0^\tau \|B^*S(t)^*y\|^2\,dt \\ &\le \|B^*\|^2\left(\int_0^\delta \|S(t)^*y\|^2\,dt + \int_\delta^\tau \|S(t)^*y\|^2\,dt\right) \\ &\le \|B\|^2\delta M^2e^{2c\delta} + \|B\|^2\int_\delta^\tau \|S(t-\delta)^*\|^2\|S(\delta)^*y\|^2\,dt \\ &\le \|B\|^2\cdot\frac{\eta}{2} + \|B\|^2\tau M^2e^{2c\tau}\|S(\delta)^*y\|^2 \\ &< \|B\|^2\cdot\eta. \end{aligned}$$

As $\|y\| = 1$ and $J(y) < \eta\cdot\|B\|^2$ for arbitrary $\eta > 0$, Theorem 1.3.4 shows that (A, B) cannot be exactly controllable. Contradiction.

(2) By the condition $\operatorname{Range} B = X$ and Banach's theorem (cf. Lemma 1.3.12), there exists $\beta > 0$ such that $\|B^*y\| \ge \beta\|y\|$, $y \in X^*$. Hence

$$J(y) \ge \beta^2\int_0^\tau \|S(t)^*y\|^2\,dt \ge \beta^2\int_0^\tau \varepsilon(t)^2\|y\|^2\,dt,$$

and since $\varepsilon(t_0) > 0$ we get

$$J(y) \ge \gamma(\tau)^2\|y\|^2$$

with some constant $\gamma(\tau) > 0$ independent of $y \in X^*$. Theorem 1.3.4 completes the proof. □

1.3.8. Corollary. *If the system (A, B) is exactly controllable at time $\tau > 0$, then this is likewise the case for a system (A, B') provided that* $\operatorname{Range} B' = X$. □

1.3.9. Corollary. *Let (A, B) be a dynamical system with surjective control operator,* $\operatorname{Range} B = X$. *Then the following assertions are equivalent.*

(1) (A, B) is exactly controllable at a time $\tau > 0$.

(2) (A, B) is exactly controllable at every time $\tau > 0$.

(3) $\inf_{\|y\|=1}\|S(t)^*y\| > 0$ *for every $t > 0$.*

(4) There are constants $\varepsilon, \alpha > 0$ such that $\|S(t)^*y\| \ge \varepsilon e^{-\alpha t}\|y\|$ for every $t > 0$ and $y \in X^*$.

Under the additional condition that $\operatorname{Ker} S(t) = \{\mathbf{0}\}$ *for every $t > 0$, these assertions are equivalent to the following.*

(5) The semigroup $S(\cdot)$ is the restriction to $\mathbb{R}_+$ of a strongly continuous group on $\mathbb{R}$.

In fact, the equivalence of the first four assertions is already proved. Let us show that (5) $\iff$ (4). Actually (5) implies that $\|S(t)^{*-1}\| = \|S(-t)^*\| \le \text{const}\cdot e^{\alpha t}$, $t > 0$ (cf. Subsection 2.2.1 below). This gives (4). Conversely, (4) implies injectivity of $S^*(t)$ which teams up with $\operatorname{Ker} S(t) = \{\mathbf{0}\}$, $t > 0$, in the invertibility of $S(t)^*$. It is easy to see that S can now be extended by $S(-t) = S(t)^{-1}$, $t > 0$, to a continuous group on the whole line $\mathbb{R}$. □

1.3.10. Remark. We can summarize the preceding results by saying that for a "big" control operator (in particular, for a surjective one) ECO does not depend on τ. On the other hand, surjectivity of B is not necessary for exact controllability, and if B fails to be surjective the ECO may depend on τ.

1.3.11. The role of the operator B. Kalman's theorem 1.3.1 states that the approximate control of (A, B) only depends on whether the range $B\mathcal{U}$ is cyclic for the semigroup $S(\cdot)$. Likewise we will see that in the case of exact controllability it is the range of B rather than the operator itself that determines the controllability.

To this end, we need the following theorem which generalizes the classical Banach theorem on surjective operators. For the Hilbert space case, it is much easier to prove and is known as R. Douglas' Lemma. This special case of the lemma is also contained in B.1.4.5.

1.3.12. LEMMA. *Let Q, R and X be Banach spaces. Suppose that Q is reflexive and let $\alpha \in L(Q, X)$, $\beta \in L(R, X)$. Then the following assertions are equivalent.*

(1) $\beta R \subset \alpha Q$.

(2) There exists $\delta_0 \in L(R, Q/\operatorname{Ker}\alpha)$ such that $\beta = \alpha_0\delta_0$ where $\alpha_0 : Q/\operatorname{Ker}\alpha \to X$ is the quotient mapping $\alpha_0(q + \operatorname{Ker}\alpha) = \alpha q$, $q \in Q$.

(3) There is a constant $C > 0$ such that

$$\|\beta^* y\| \leq C\|\alpha^* y\|, \quad y \in X^*.$$

If the subspace $\operatorname{Ker}\alpha$ is complemented, then the assertions (1)-(3) are equivalent to

(4) $\beta = \alpha\delta$ for a suitable $\delta \in L(R, Q)$.

PROOF. (1) $\Longrightarrow$ (2). Since α_0 is injective, the application $\delta_0 : r \longmapsto \alpha_0^{-1}\beta r$ is well-defined and linear on R. Moreover, the operator δ_0 is closed. Indeed, if $\lim_n r_n = \mathbf{0}$ and $q_0 = \lim_n \delta_0 r_n$ then $\alpha_0 q_0 = \lim_n \beta r_n = \mathbf{0}$. Hence $q_0 = \mathbf{0}$. The closed graph theorem now implies that $\delta_0 \in L(R, Q/\operatorname{Ker}\alpha)$. Clearly $\beta = \alpha_0\delta_0$.

(2) $\Longrightarrow$ (3). By (2) we have $\beta = \alpha_0\delta_0$ and hence $\beta^* = \delta_0^*\alpha_0^*$. Let $y \in X^*$. As $(Q/\operatorname{Ker}\alpha)^*$ can be isometrically embedded into Q^* we can identify $\alpha_0^* y$ and $\alpha^* y$. Consequently

$$\|\beta^* y\| \leq \|\delta_0^*\| \cdot \|\alpha_0^* y\|_{(Q/\operatorname{Ker}\alpha)^*} = \|\delta_0^*\| \cdot \|\alpha^* y\|_{Q^*}.$$

(3) $\Longrightarrow$ (1). By (3), the operator $T : \alpha_0^* X^* \longrightarrow R^*$ given by $\alpha_0^* y \longmapsto \beta^* y$, $y \in X^*$, is well-defined and continuous. As $\alpha_0^* X^*$ is weak* dense in $(Q/\operatorname{Ker}\alpha)^*$ and $Q/\operatorname{Ker}\alpha$ is reflexive, the image $\alpha_0^* X^*$ is norm-dense in $(Q/\operatorname{Ker}\alpha)^*$. Hence, we can extend T to an operator $T \in L((Q/\operatorname{Ker}\alpha)^*, R^*)$. Obviously we have $\beta^* = T\alpha_0^*$ and hence $\beta^{**} = \alpha_0^{**}T^* = \alpha_0 T^*$ (using the reflexivity of $Q/\operatorname{Ker}\alpha$). Next, setting $\delta_0 = T^*|R \in L(R, Q/\operatorname{Ker}\alpha)$, we get $\beta = \beta^{**}|R = \alpha_0(T^*|R) = \alpha_0\delta_0$ which implies that $\beta R = \alpha_0\delta_0 R \subset \alpha_0(Q/\operatorname{Ker}\alpha) = \alpha Q$.

For the equivalence (4) $\Longleftrightarrow$ (2) assume that $\operatorname{Ker}\alpha$ is complemented, that is, there is a continuous projection $P \in L(Q)$ having $\operatorname{Ker}\alpha$ as range: $P(Q) = \operatorname{Ker}\alpha$.

First, show that (2) $\Longrightarrow$ (4). Let $\pi : Q \longrightarrow Q/\operatorname{Ker}\alpha$ be the canonical quotient mapping $q \longmapsto q + \operatorname{Ker}\alpha$ which is now right invertible. Indeed, the application $A : Q/\operatorname{Ker}\alpha \longrightarrow Q$, given by $A(q + \operatorname{Ker}\alpha) = (I - P)q$, is well-defined, continuous, and satisfies $\pi \circ A = I_{Q/\operatorname{Ker}\alpha}$. By (2), there is $\delta_0 \in L(R, Q/\operatorname{Ker}\alpha)$ such that $\beta = \alpha_0\delta_0$. Setting $\delta = A\delta_0$ we get $\delta \in L(R, Q)$ and $\pi\delta = \delta_0$. Hence $\beta = \alpha_0\delta_0 = \alpha_0\pi\delta = \alpha\delta$.

The converse implication is obvious (put $\delta_0 = \pi\delta$). $\square$

1.3.13. REMARKS. (1) It is clear from the preceding proof that the implications (1) $\Longleftrightarrow$ (2) $\Longrightarrow$ (3) are still true without the assumption on reflexivity of Q. On the other hand, in case $R = X$, $\beta = I$ (the case of Banach's theorem) one can always conclude that (3) $\Longrightarrow$ (1)&(2). Indeed, in this case $\alpha_0^* X^*$ is weak* closed (note that by (3) $\alpha_0^* X^*$ is weakly sequentially closed), and hence $\alpha_0^* X^* = (Q/\operatorname{Ker}\alpha)^*$, which allows us to finish the proof as before.

The necessity of a topological complement of $\operatorname{Ker}\alpha$ for property (4) is, at least for $\beta = I$, clear: if $\alpha\delta = I$ with $\delta \in L(X, Q)$, then $P = \delta\alpha : Q \longrightarrow Q$ is a continuous projection onto $\operatorname{Ker}\alpha$. Indeed, as $I = \beta = \alpha\delta$, the operator δ is injective, and hence $\operatorname{Ker} P = \operatorname{Ker}\alpha$. Moreover $P^2 = \delta\alpha\delta\alpha = \delta I \alpha = P$.

(2) The lemma enables us to reformulate the (equivalent, and for exact controllability necessary) conditions (1.3.9) and (1.3.10) of Theorem 1.3.7 in the following way

(1.3.9') $$S(t)X = X \text{ for every } t > 0$$

and

(1.3.10') $$\text{there is a } t_0 > 0 \text{ such that } S(t_0)X = X.$$

We now give some more necessary and/or sufficient conditions for exact controllability, where the role of the control operator is explicitely involved. These conditions will be commented on and completed in Subsection 1.3.20 for the case of a group $S(\cdot)$ on $\mathbb{R}$ (cf. also Corollary 1.3.9). See also Subsection 1.3.18 for compact control operators and Subsection 3.4.6 for necessary and sufficient conditions on B in the case where the generator A possesses a basis of eigenvectors.

1.3.14. COROLLARY. *Let $B_i \in L(\mathcal{U}_i, X)$, $i = 1, 2$, be two control operators such that*

$$\operatorname{Range} B_1 \subset \operatorname{Range} B_2.$$

Then the system (A, B_2) is exactly controllable if (A, B_1) is. Hence, if the systems (A, B_i), $i = 1, 2$, have the same control subspaces

$$\operatorname{Range} B_1 = \operatorname{Range} B_2,$$

then they are simultaneously exactly controllable (or not).

Indeed, Lemma 1.3.12 implies that

$$\|B_1^* y\| \leq C \|B_2^* y\|, \quad y \in X^*,$$

(and also $c\|B_2^* y\| \leq \|B_1^* y\|$, $y \in X^*$, if the control subspaces coincide). Theorem 1.3.4 now applies to give the assertion. □

Now we are in a position to sharpen the sufficiency part of Theorem 1.3.7 (i.e., the assertion (2) of that theorem).

1.3.15. THEOREM. *Let $\mathcal{U}$, X be reflexive Banach spaces and (A, B) a dynamical system. Assume that there exist $t_0 > 0$ and $t_1, ..., t_n \geq 0$ such that*

$$S(t_0)X = X,$$

(that is $\|S(t_0)^* y\| \geq \varepsilon \|y\|$, $y \in X^*$, for a suitable $\varepsilon > 0$), and

$$\sum_{i=1}^{n} S(t_i) B\mathcal{U} = X$$

(which is equivalent to

$$\gamma^2 \|y\|^2 \leq \sum_{i=1}^{n} \|B^* S(t_i)^* y\|^2, \quad y \in X^*,$$

for some $\gamma > 0$). Then (A, B) is exactly controllable at every time τ with

$$\tau > \max(t_i : 1 \le i \le n).$$

PROOF. Lemma 1.3.12 shows the equivalences mentioned in parentheses. In particular, for the second one, one has to consider the operator $\alpha = S(t_1)B \oplus \cdots \oplus S(t_n)B : Q := \mathcal{U} \times \cdots \times \mathcal{U} \longrightarrow X$ defined by $(u_1, \dots, u_n) \longmapsto \sum_{i=1}^{n} S(t_i)Bu_i$. Then $\|\alpha^* y\|^2 = \sum_{i=1}^{n} \|B^* S(t_i)^* y\|^2$.

Our next aim is to exactly control the system (A, B) for $\tau > \max(t_i : 1 \le i \le n)$, that is, given $x \in X$, to find a control $u \in L^2(0, \tau; \mathcal{U})$ such that

$$x = c(\tau)u = \int_0^\tau S(t)Bu(t)\, dt.$$

As $S(t_0)X = X$ for some $t_0 > 0$, Theorem 1.3.7 guarantees that $S(t)X = X$ for every $t > 0$ and hence

$$X = S(t)X = \sum_{i=1}^{n} S(t)S(t_i)B\mathcal{U} = \sum_{i=1}^{n} S(t_i + t)B\mathcal{U}.$$

This implies that for every $x \in X$ and $t > 0$, there exists $u_i(t) \in \mathcal{U}$ such that

$$x = \sum_{i=1}^{n} S(t_i + t)Bu_i(t), \quad t \ge 0.$$

Moreover the vector valued function $t \longmapsto (u_i(t))_{i=1}^{n}$ can be chosen measurable and bounded. To see this, we approximate the operator valued function $b(t) = \sum_{i=1}^{n} S(t_i + t)B$, $0 \le t \le T$, by a locally constant function. Namely, let $(\Delta_j)_{j=1}^{m}$ be a sufficiently fine partition of the interval $[0, T]$ and $b_m = \sum_{j=1}^{m} \sum_{i=1}^{n} S(t_i + r_j)\chi_{\Delta_j} B$, where $r_j \in \Delta_j$. Now, we can solve the equation $b_m(t)v_m(t) \equiv x$ for $0 \le t \le T$ with a locally constant solution v_m, and them pass to the weak limit in $L^2(0, T; \mathcal{U})$ as $m \longrightarrow \infty$. This gives a measurable and bounded function u_i.

Finally, we suppose that $t_1 < t_2 < \cdots < t_n$, choose $\varepsilon > 0$ sufficiently small, and set

$$u(t) = \begin{cases} \frac{1}{\varepsilon} u_i(t - t_i), & t \in [t_i, t_i + \varepsilon],\ 1 \le i \le n, \\ 0 & \text{otherwise}. \end{cases}$$

For $\tau \ge t_n + \varepsilon$ we obtain

$$c(\tau)u = \int_0^\tau S(t)Bu(t)\, dt = x,$$

and as $u \in L^2(0, \tau; \mathcal{U})$ the result follows. □

1.3.16. REMARK. The preceding result shows that exact control is compatible with the condition $\operatorname{Ker} B^* \ne \{\mathbf{0}\}$, $\operatorname{Range} B = B\mathcal{U} \ne X$ (even in the case $n = 1$). The following example illustrates this fact. For further examples we refer the reader to 3.4.6.

1.3.17. EXAMPLE. Let $X = \mathcal{U} = L^2(\mathbb{R})$, $S(t)f = f(x + t)$, $x, t \ge 0$ (the group of left translations). Further, let $E \subset L^2(\mathbb{R})$ be the subspace given by

$$E = \Big\{ f \in L^2(\mathbb{R}) : f \equiv 0 \text{ outside } \bigcup_{k \in \mathbb{Z}} [kn, kn + 1] \Big\},$$

where $n \geq 2$ is fixed, and let $B = P_E$ be the orthogonal projection onto E. Clearly $S(t)X = X$ for every $t \geq 0$ and

$$\sum_{j=0}^{n-1} S(j)B\mathcal{U} = X.$$

By the preceding theorem, the system (A, B), where A is the generator of $S(\cdot)$, is exactly controllable at every time $\tau > n-1$. Let us consider the time moment $\tau = n-1$. It is easy to see that

$$\int_0^{n-1} \|B^*S(t)^*y\|^2_{L^2(\mathbb{R})}\,dt = \int_0^{n-1}\int_{\mathbb{R}} \chi(r)\cdot|y(r-t)|^2\,drdt = \int_{\mathbb{R}} |y(s)|^2\varphi(s)\,ds,$$

where χ is the indicator function of the set $\bigcup_{k\in\mathbb{Z}}[kn, kn+1]$ and $\varphi(s) = \inf\{1, |s - kn - 1| : k \in \mathbb{Z}\}$. Hence, $\inf\{\varphi(s) : s \in \mathbb{R}\} = 0$ and Theorem 1.3.4 shows that the system is not exactly controllable. On the other hand, it can be verified that (A, B) is approximately controllable for $\tau = n-1$, but this is no longer true for $\tau < n-1$.

1.3.18. Compact controllability is impossible. The following fact again expresses the principle that in order to have exact controllability, not only the semigroup $S(\cdot)$ has to be "big" (in the sense of 1.3.7) but also the control operator B.

1.3.19. THEOREM. *Let (A, B) be a dynamical system with infinite dimensional state space, $\dim X = \infty$. If B is compact, then there is no ECO at any time moment.*

PROOF. As $\dim X = \infty$, it is enough to show that the controllability operator $c(\tau) : L(0, \tau, \mathcal{U}) \longrightarrow X$ is compact, or equivalently that $c(\tau)^*$ is compact. First, we claim that the operator valued function $t \longmapsto B^*S(t)^*$, $t \geq 0$, with values in $L(X^*, \mathcal{U}^*)$ is continuous in the operator norm. Indeed,

$$\|B^*S(t)^* - B^*S(r)^*\| = \|\left(S(t) - S(r)\right)B\|.$$

By assumption, we have $\lim_{t\to r}\left(S(t) - S(r)\right)x = \mathbf{0}$ for every $x \in X$, and this convergence is uniform on compact sets. As B is a compact operator, and hence $\operatorname{clos}\{x = Bu \ : \ \|u\| \leq 1\}$ is compact, the claim follows.

We now show that for every $\tau > 0$ the image under $c(\tau)^*$ of the unit ball, that is,

$$J = \{c(\tau)^*y : y \in X^*, \quad \|y\| \leq 1\}$$

is precompact in $L^2(0, \tau; \mathcal{U}^*)$, which means that it has a finite ε-net for every $\varepsilon > 0$. The continuity of $t \longmapsto B^*S(t)^*$ implies the existence of a partition $\Delta_1, ..., \Delta_n$ of $[0, \tau]$ such that

$$\|B^*S(t)^* - B^*S(r)^*\| < \frac{\varepsilon}{2\sqrt{\tau}} \tag{1.3.11}$$

for every $t, r \in \Delta_i$, $i = 1, ..., n$. Now observe that the set

$$J_\varepsilon = \left\{j(y) := \sum_{i=1}^{n} B^*S(\tau - t_i)^*y\chi_{\Delta_i} : \|y\| \leq 1, y \in X^*\right\}$$

is an $\varepsilon/2$-net for J. Here $t_i \in \Delta_i$ are arbitrary fixed points. It remains to find a *finite* ε-net in J_ε. Note that the mapping

$$y \longmapsto \alpha(y) := (B^*S(\tau - t_1)^*y, ..., B^*S(\tau - t_n)^*y)$$

from X^* to $\mathcal{U}^* \times \mathcal{U}^* \times \cdots \times \mathcal{U}^*$ (equipped, for instance, with the sup-norm) is compact. Therefore, there exists an $\varepsilon/2\sqrt{\tau}$-net for the range $\{\alpha(y) : \|y\| \le 1\}$, say $\{\alpha(y_1), ..., \alpha(y_m)\}$. Then for every $y \in X^*$, $\|y\| \le 1$, there is k, $1 \le k \le m$, such that $\|\alpha(y) - \alpha(y_k)\| < \frac{\varepsilon}{2\sqrt{\tau}}$, which yields

$$\begin{aligned}\|j(y) - j(y_k)\|_{L^2(0,\tau;\mathcal{U}^*)} &= \left(\sum_{i=1}^{n} \int_{\Delta_i} \|B^*S(\tau - t_i)^* y - B^*S(\tau - t_i)^* y_k\|^2 \, dt\right)^{1/2} \\ &< \frac{\varepsilon}{2\sqrt{\tau}} \cdot \sqrt{\tau} = \frac{\varepsilon}{2}.\end{aligned}$$

Thus $\{j(y_1), ..., j(y_m)\}$ is an $\varepsilon/2$-net for J_ε, and hence, an ε-net for J. □

1.3.20. Remarks on the case where $S(\cdot)$ is a group. We have already encountered the special case when $S(t)$, $t \ge 0$, is the restriction to $\mathbb{R}_+$ of a strongly continuous group on $\mathbb{R}$ in Corollary 1.3.9, as well as in Example 1.3.17. In particular the following facts have been proved.

(1) The exact controllability of a semigroup $S(\cdot)$ together with the injectivity of $S(t)$, $t > 0$, imply that $S(\cdot)$ extends to a strongly continuous group on $\mathbb{R}$.

(2) A group $S(\cdot)$ is always exactly controllable at every time moment $\tau > 0$ if the control operator B is surjective.

(3) There exist groups $S(\cdot)$ that are exactly controllable with non-surjective control operator.

Let us add the following points to this list. First, observe that if $S(\cdot)$ is strongly continuous group on $\mathbb{R}$ such that $(S(t))_{t \ge 0}$ is generated by A, then the operator $-A$ is the generator of $(S(-t))_{t \ge 0} = (S(t)^{-1})_{t \ge 0}$. This implies the following property.

(4) If $S(\cdot)$ is a group with generator A, then the systems (A, B) and $(-A, B)$ are simultaneously exactly controllable (or not).

Indeed,

$$c(\tau)u = c(S, \tau)u = \int_0^\tau S(\tau - t)Bu(t)\,dt = S(\tau)\int_0^\tau S(-t)Bu(t)\,dt$$

and $S(\tau)$ is invertible. The result follows. □

It is also clear (cf. Subsection 2.2.2 below) that the spectrum $\sigma(A)$ of the generator A of a group is in a strip

$$\sigma(A) \subset \{\lambda \in \mathbb{C} : -c_1 \le \operatorname{Re}\lambda \le c_2\}, \tag{1.3.12}$$

where, using the notations of Subsection 2.2.2, $c_2 = \alpha(S)$, $c_1 = \alpha(S^{-1})$. This and Point (1) team up in the following.

(5) If the injective semigroup $S(\cdot)$ is exactly controllable at time $\tau > 0$, then the spectrum of its generator is in a strip of type (1.3.12).

(6) In case $S(\cdot)$ is a group, we refer the reader to Subsection 3.4.6 below for more comments concerning renormalized exact controllability and to the remark after Theorem 1.3.25 for null controllability.

1.3.21. Remarks on discrete-time systems. Here, we briefly comment on the discrete dynamical systems

$$\begin{aligned} x_{n+1} &= Ax_n + Bu_n, \\ y_n &= Cx_n + Du_n, \quad n \geq 0, \end{aligned}$$

where $A \in L(X)$, $B \in L(\mathcal{U}, X)$, $C \in L(X, Y)$ and $D \in L(\mathcal{U}, Y)$. The controllability operator $c(\cdot)$ is given by

$$c(n+1)u = \sum_{k=0}^{n} A^{n-k} B u_k, \quad n \geq 0,$$

where $u = (u_k)_{k \geq 0} \in l^2 = l^2(\mathbb{Z}_+, \mathcal{U})$. The general solution to the system is given by

$$x_{n+1} = A^{n+1} x_0 + c(n+1)u, \quad n \geq 0,$$

and the adjoint $c(n+1)^* : X^* \longrightarrow l^2(\mathbb{Z}_+, \mathcal{U}^*)$ by

$$c(n+1)^* y : k \longmapsto B^* A^{*(n-k)} y, \quad 0 \leq k \leq n, \ y \in X^*.$$

Also observe that $c(n+1)u = Bu_n + AT_n(u_k)_{k=0}^{n-1}$ for a suitable bounded operator T_n. This together with the preceding remarks immediately give the following properties.

(1) The system (A, B) is exactly controllable (respectively approximately controllable) at time $n+1$ if and only if

$$\mathcal{L}\text{in}(A^k B\mathcal{U} : 0 \leq k \leq n) = X$$

(respectively $\text{span}(A^k B\mathcal{U} : 0 \leq k \leq n) = X$*), and also if and only if there is a constant $\gamma > 0$ such that*

$$\gamma^2 \|y\|^2 \leq \sum_{k=0}^{n} \|B^* A^{*k} y\|^2, \quad y \in X^*.$$

(2) The inequality

$$\varepsilon^2 \|y\|^2 \leq \|B^* y\|^2 + \|A^* y\|^2, \quad y \in X^*,$$

with $\varepsilon > 0$, or equivalently the identity $X = B\mathcal{U} + AX$, is necessary for exact controllability.

(3) If B is surjective, then (A, B) is exactly controllable, and if $\text{clos}\, B\mathcal{U} = X$*, then it is approximately controllable.*

(4) In the special case $\mathcal{U} = X$, $B = A$, the system is exactly controllable if and only if A is surjective ($AX = X$), and approximately controllable if and only if $\text{Ker}\, A^* = \{\mathbf{0}\}$ *(that is,* $\text{clos}\, AX = X$*). It is always null controllable.*

1.3.22. Criteria for null and exact null controllability. Here we introduce a refined form of null controllability (called ENCO) and give general criteria for both NCO and ENCO.

1.3.23. THEOREM. *Let $\mathcal{U}$ be a reflexive space. Then the system (A, B) is null controllable at time τ if and only if there exists $\gamma > 0$ such that for every $y \in X^*$ we have*

$$\|c(\tau)^* y\| \geq \gamma \|S(\tau)^* y\|.$$

Proof. Since null controllability is equivalent to $c(\tau)L^2 \supset S(\tau)X$ (cf. Subsection 1.2.5), the theorem follows from Lemma 1.3.12 (with $\alpha = c(\tau)$ and $\beta = S(\tau)$). □

A special situation occurs if the inclusion $c(\tau)L^2 \supset S(\tau)X$ turns out to be an identity.

1.3.24. Definition. A system (A, B) is called *exactly null controllable (ENCO)* at time $\tau > 0$ if

$$S(\tau)X = c(\tau)L^2(0, \tau; \mathcal{U}).$$

1.3.25. Theorem. *The system (A, B) is exactly null controllable (ENCO) at time $\tau > 0$ if and only if there are two positive constants $c, C > 0$ such that*

$$c^2\|S(\tau)^*y\|^2 \leq \int_0^\tau \|B^*S(t)^*y\|^2 dt \leq C^2\|S(\tau)^*y\|^2, \quad y \in X^*$$

Proof. The proof is the same as before (using Lemma 1.3.12 twice). □

We finish the section with some links between NCO, ECO and ACO.

(1) If the semigroup $S(\cdot)$ has dense range (that is $\operatorname{Ker} S(t)^* = \{\mathbf{0}\}$ *for some and hence all $t > 0$) then NCO implies ACO.*

(2) If the identity $S(t)X = X$ holds for some (and hence all) $t > 0$, then the system (A, B) is exactly controllable if and only if it is null controllable. Clearly this occurs if $S(\cdot)$ extends to a continuous group on $\mathbb{R}$.

(3) For null controllability we have the following analogue of Theorem 1.3.15.

Theorem. *If there are points τ, $t_1, ..., t_n$ such that*

$$S(\tau)X \subset \sum_{i=1}^{n} S(t_i)B\mathcal{U},$$

then the system (A, B) is null controllable at every time $\tau' > \max(\tau, t_1, ..., t_n)$.

The proof is the same as that of Theorem 1.3.15. The case of null controllability for a generator with a basis of eigenvectors is treated in 3.5.

1.4. Stable Systems

In this section we briefly discuss exponential stability and some of the relevant properties important for control. However, these are rather far from the mainstream of this book.

A semigroup is called *exponentially stable* if there are constants $\alpha > 0$ and $M > 0$ such that

$$\|S(t)\| \leq Me^{-\alpha t}, \quad t \geq 0.$$

In this case, α is called the *decay rate* of the semigroup $(S(t))$ and the corresponding system (A, B) is also called *stable*.

For such a semigroup we can define the controllability operator at infinity.

1.4.1. Definition. Let S be an exponentially stable semigroup. We set

$$c(\infty)u = \int_0^\infty S(t)Bu(t)\, dt, \quad u \in L^2(0, \infty; \mathcal{U}),$$

and call $c(\infty)$ the controllability map at time $\tau = \infty$.

1.4.2. LEMMA. *If the semigroup $S(\cdot)$ is exponentially stable, then the controllability map $c(\infty) : L^2(0,\infty;\mathcal{U}) \longrightarrow X$ is well-defined and bounded.*

PROOF. By the Cauchy–Schwarz inequality we have

$$\|c(\infty)u\| \le \left(\int_0^\infty \|S(t)\|^2 dt\right)^{1/2} \|B\| \cdot \|u\|_{L^2}.$$

□

There exist several results of Tauberian type that simplify the verification of exponential stability. We give such a result in both versions, continuous and discrete.

1.4.3. THEOREM. *Let $S(\cdot)$ be a continuous semigroup on a Banach space X. If*

$$\int_0^\infty \|S(t)x\|^2 dt < \infty \text{ for all } x \in X,$$

then $S(\cdot)$ is exponentially stable.

If $T \in L(X)$ such that $\sum_{n\ge 0} \|T^n x\|^2 < \infty$ for every $x \in X$, then the semigroup $(T^n)_{n\ge 0}$ is exponentially stable (which here means that there are $\alpha > 0$, $M > 0$ such that $\|T^n\| \le Me^{-\alpha n}$, $n \ge 0$).

PROOF. Here, we give the proof of the discrete case; for the continuous one, see Subsection 1.6.2 below.

A standard application of the closed graph theorem together with the assumption yield

$$\sum_{n\ge 0} \|T^n x\|^2 \le C^2 \|x\|^2, \quad x \in X, \tag{1.4.1}$$

for a suitable constant $C < \infty$. Then the Cauchy–Schwarz inequality gives

$$\Big\|\sum_{n\ge 0} \lambda^n T^n x\Big\| \le C\|x\| \Big(\sum_{n\ge 0} |\lambda|^{2n}\Big)^{1/2} = C\|x\|(1-|\lambda|^2)^{-1/2},$$

for $|\lambda| < 1$. Hence $I - \lambda T$ is invertible and

$$\|(I-\lambda T)^{-1}\| \le C(1-|\lambda|^2)^{-1/2}. \tag{1.4.2}$$

In particular we deduce that the spectrum of T, $\sigma(T)$, is contained in the closed unit disk. Suppose that some $\zeta \in \mathbb{T}$ is a point of the spectrum. Then, using an obvious lower estimate of the resolvent, we get

$$\|(I - r\bar{\zeta}T)^{-1}\| \ge \frac{1}{r} \frac{1}{\operatorname{dist}\left(\frac{\zeta}{r}, \sigma(T)\right)} = \frac{1}{1-r} \tag{1.4.3}$$

for $0 < r < 1$. Letting $r \longrightarrow 1$ we get a contradiction to (1.4.2). Hence $\sigma(T) \subset \mathbb{D}$, and so the spectral radius $r(T) := \lim_n \|T^n\|^{1/n}$ is less than 1: $r(T) < 1$. Now choose $0 < \alpha < \log(1/r(T))$ to get the result. □

1.4.4. REMARK. **(1) The exponential decay rate**. The elementary reasoning from the above proof allows us to estimate the decay rate α, and in particular, its dependence on the boundedness constant C of the functional calculus (1.4.1). Namely, combining (1.4.2) and (1.4.3), we get

$$(1 - r(T)|\lambda|)^{-1}\| \le C(1-|\lambda|^2)^{-1/2},$$

where $r(T)$ is the spectral radius of T, and hence

$$r(T) \leq \left(1 - \frac{1}{C^2}\right)^{1/2}.$$

In particular, we may choose $0 < \alpha < \log(1 - 1/C^2)^{-1/2}$ in Theorem 1.4.3.

(2) Controllability Gramian. For applications, it is important to know $c(\infty)$ without explicitely referring to $S(t)$. To this end, the following *controllability Gramian* of the system is useful

$$L_c = c(\infty)c(\infty)^* : X^* \longrightarrow X^*,$$

where X is now supposed to be a Hilbert space. The following theorem holds.

THEOREM. *Let $\mathcal{U}, X$ be Hilbert spaces and (A, B) an exponentially stable system. Then L_c is the unique self-adjoint solution of the following Lyapunov equations.*

$$\begin{cases} L_c\mathcal{D}(A^*) & \subset \mathcal{D}(A) \\ AL_c + L_cA^* & = -BB^* \text{ on } \mathcal{D}(A^*). \end{cases}$$

Recall that exact controllability at time moment $\tau = \infty$, that is the equality $c(\infty)L^2(0, \infty; \mathcal{U}) = X$, is equivalent to the estimate $\langle L_c x, x\rangle \geq \gamma\|x\|^2$ for every $x \in X$, whereas approximate controllability is characterized by $\operatorname{Ker} L_c = \{\mathbf{0}\}$ (see Section 1.3 above).

(3) NCO and feedback stabilization. In applications, it is important to "stabilize" the given system, that is, to pass from a possibly unstable system to a stable one by modifying some of its elements. Usually, a *feedback stabilization* is used by introducing the so-called *feedback operator* $F : X \longrightarrow \mathcal{U}$ and considering the modified control function $u+Fx$ instead of u (cf. also Section 5.2). This means that we replace the system $x' = Ax+Bu$ by the modified system $x' = Ax+B(u+Fx) = (A+BF)x+Bu$ which can happen to be exponentially stable. There is the following relation between controllability and the existence of a feedback stabilization.

THEOREM. *Null controllability implies the existence of exponential stabilization by feedback.*

1.5. An Example: Heating of a Metal Bar

The preceding results will now be illustrated through a concrete control problem: to control the temperature distribution of a metal bar. It is known that the heating process of a metal bar (of normalized length 1) is described by the following system.

$$\begin{cases} \dfrac{\partial x}{\partial t}(s,t) & = \dfrac{\partial^2 x}{\partial s^2}(s,t) + u(s,t) \\ x(s,0) & = x_0(s), \quad 0 \leq s \leq 1, \\ \dfrac{\partial x}{\partial s}(0,t) & = \dfrac{\partial x}{\partial s}(1,t) \equiv 0, \end{cases}$$

where $x(s, t)$ is the temperature of the bar at point s and at time t, $x_0(s)$ is the initial temperature distribution, and $u(s, t)$ is the intensity of the heating. The boundary conditions $\dfrac{\partial x}{\partial s} \equiv 0$ at $(0, t)$ and $(1, t)$, $t \geq 0$, signify that there is no heat exchange at the endpoints of the bar. We can rewrite these equations in the standard notation for dynamical systems

$$x' = Ax(t) + Bu(t)$$

with $X = L^2(0,1)$ ($x(t) \in L^2(0,1)$), $\mathcal{U} = L^2(0,1)$, where

- $Ah = \dfrac{d^2h}{ds^2}$, $h \in \mathcal{D}(A)$ with

$$\begin{aligned}\mathcal{D}(A) = \{h \in L^2(0,1) \quad : \quad & h, \frac{dh}{ds} \text{ are absolutely continuous }, \frac{d^2h}{ds^2} \in L^2(0,1) \\ & \text{and } \frac{dh}{ds}(0) = \frac{dh}{ds}(1) = 0\},\end{aligned}$$

- $B = I$,
- $x_0 \in L^2(0,1)$ is the initial state.

For $u = 0$, the usual method of separation of variables yields $x(t) = S(t)x_0$, where

$$S(t)x_0 = \int_0^1 G(t,s,r)x_0(r)\,dr$$

and

$$G(t,s,r) = 1 + \sum_{n\geq 1} 2e^{-n^2\pi^2 t}\cos(n\pi s)\cos(n\pi r)$$

is the *Green function* of the equation. For a non-trivial control function $u \not\equiv 0$, $u \in L^2(0,1;\mathcal{U})$, $\mathcal{U} = L^2(0,1)$, formula (1.1.2) yields

$$x(t) = S(t)x_0 + \int_0^t S(t-s)u(s)\,ds.$$

Clearly, the application

$$S(t) : L^2(0,1) \longrightarrow L^2(0,1)$$

is a continuous semigroup, and as

$$S(t)x_0 = \int_0^1 x_0(s)ds + \sum_{n\geq 1} 2e^{-n^2\pi^2 t}\cos(n\pi s)\int_0^1 \cos(n\pi r)x_0(r)\,dr,$$

it is diagonal with respect to the orthonormal basis $\mathcal{B} = \{1\} \cup \{\sqrt{2}\cos(n\pi s)\}_{n\geq 1}$. Moreover, $S(t)$ and B are self-adjoint: $S(t) = S(t)^*$ and $B = B^* = I$. In view of Theorem 1.3.4, the system (A,B) is exactly controllable at time $\tau > 0$ if and only if there is $\gamma > 0$ such that

$$\int_0^\tau \|B^*S(t)^*y\|^2_{\mathcal{U}^*}\,dt \geq \gamma^2\|y\|^2_{X^*}$$

for all $y \in X^* = X = L^2(0,1)$. Using Parseval's identity for the basis $\mathcal{B}$ in $L^2(0,1)$, we rewrite this in an equivalent form

$$\int_0^\tau \left| \int_0^1 y(s)ds \right|^2 dt + 2\sum_{n\geq 1} \left| \int_0^1 \cos(n\pi r)y(r)\,dr \right|^2 \frac{1-e^{-2n^2\pi^2\tau}}{n^2\pi^2}$$
$$\geq \gamma^2 \|y\|^2_{L^2(0,1)}$$
$$= \gamma^2 \left\{ \left| \int_0^1 y(s)\,ds \right|^2 + 2\sum_{n\geq 1} \left| \int_0^1 \cos(n\pi r)y(r)\,dr \right|^2 \right\}, \quad y \in L^2(0,1),$$

which, of course, is impossible whatever is $\gamma > 0$.

1.5.1. COROLLARY. *The heating of a metal bar is* never *exactly controllable (at any time $\tau > 0$ and for any control operator B).* □

On the contrary, the equation is null controllable at every time $\tau > 0$. Indeed, by Theorem 1.3.23 it is sufficient to have

$$\int_0^\tau \|B^* S(t)^* y\|^2_{\mathcal{U}}\, dt \geq \gamma^2 \|S(\tau)^* y\|^2_X$$
$$= \gamma^2 \left\{ \left| \int_0^1 y(s)\,ds \right|^2 + 2\sum_{n\geq 1} e^{-2n^2\pi^2\tau} \left| \int_0^1 \cos(n\pi r)y(r)\,dr \right|^2 \right\}.$$

Substituting $x = 2n^2\pi^2\tau$ in the well-known inequality $\frac{1-e^{-x}}{x} \geq e^{-x}$, $x > 0$, we get the estimate $\frac{1-e^{-2n^2\pi^2\tau}}{2n^2\pi^2} \geq \tau e^{-2n^2\pi^2\tau}$, so $\gamma^2 = \tau$ does the job.

On the other hand, by Theorem 1.3.25, we can never have *exact null controllability* at any time $\tau > 0$.

Finally let us consider approximate controllability. If for $\tau > 0$ and $y \in X^* = L^2(0,1)$ we have $c(\tau)^* y = S(\tau - \cdot)^* y = \mathbf{0}_{L^2(0,1;\mathcal{U})}$, that is, $S(t)^* y = \mathbf{0}$, $0 \leq t \leq \tau$, then

$$\left| \int_0^1 y(s)\,ds \right|^2 + 2\sum_{n\geq 1} e^{-2n^2\pi^2\tau} \left| \int_0^1 \cos(n\pi r)y(r)\,dr \right|^2 = 0.$$

The completeness of $\mathcal{B}$ in $L^2(0,1)$ implies that $y \equiv 0$. This means that $\operatorname{Ker} c(\tau)^* = \{\mathbf{0}\}$, and Theorem 1.3.3 shows that (A, B) is *approximately controllable* at the moment τ (which was chosen arbitrarily). □

1.6. Exercises and Further Results

1.6.1. Weak and strong type Tauberian theorems on the spectrum. Let $T : X \longrightarrow X$ be a bounded operator on a Banach space X and

$$R(\lambda) = \lambda^{-1} R_{1/\lambda}(T) = (I - \lambda T)^{-1}$$

be its *Fredholm resolvent*, where $R_\lambda(T) = (\lambda I - T)^{-1}$ is the (standard) resolvent, $\lambda \in \mathbb{C} \backslash \sigma(T)$. In particular, $R(\lambda)$ is defined for $|\lambda| < 1/r(T)$, where $r(T)$ stands for the spectral radius of T. The aim of this subsection is to show that some ("Tauberian") conditions on the weak resolvent $\langle R(\cdot)x, x^*\rangle_X$ imply already that $r(T) < 1$, and hence $\|T^n\| \leq Ce^{-\omega n}$ for $n \geq 0$ and a positive ω.

In what follows, we assume that $r(T) \le 1$, and we will use the following *weak resolvent condition* $(w\mathcal{R})$:

$$\langle R(\cdot)x, x^*\rangle_X \in \mathcal{R}$$

for every $x \in X$, $x^* \in X^*$; here and below, $\mathcal{R}$ is a Banach space satisfying the following conditions:

1) $\mathcal{R} \subset Hol(\mathbb{D})$ (continuous embedding);
2) $\mathcal{R}$ contains the polynomials $\mathcal{P}ol_+$ as a dense subset and satisfies

$$\overline{\lim}_n \|z^n\|_{\mathcal{R}}^{1/n} \le 1$$

An element of the dual space $\mathcal{R}^*$ is identified with its Cauchy transform, $\varphi = \sum_{n\ge 0} \hat{\varphi}(n) z^n$, where $\hat{\varphi}(n) = \langle z^n, \varphi\rangle_{\mathcal{R}}$ for $n \ge 0$, and the duality is realized via the Cauchy pairing,

$$\langle f, \varphi\rangle_{\mathcal{R}} = \sum_{n\ge 0} \hat{f}(n)\hat{\varphi}(n),$$

at least for $f \in \mathcal{R}$ having a norm convergent Taylor series $f = \sum_{n\ge 0} \hat{f}(n) z^n$. Notice that 1) implies the inclusion $\mathcal{P}ol_+ \subset \mathcal{R}^*$.

(a) *(N. Nikolski, 1977)* Let T be a Banach space operator satisfying $(w\mathcal{R})$ and let $\mathcal{R}$ be as above. Suppose that

3) the functional $V_\zeta : \varphi \longmapsto \varphi(\zeta)$ is *not* bounded on $\mathcal{R}^* \cap \mathcal{P}ol_+$ whenever $\zeta \in \mathbb{T}$.

Show that $r(T) < 1$.

[*Hint:* by the closed graph theorem there is a constant C such that

$$\|\langle R(\cdot)x, x^*\rangle_X\|_{\mathcal{R}} \le C\|x\| \cdot \|x^*\|$$

for every $x \in X$, $x^* \in X^*$, and hence

$$\begin{aligned} \|\varphi(T)\| &= \sup\left\{|\langle \varphi(T)x, x^*\rangle_X| : \|x\|_X \le 1, \|x^*\|_{X^*} \le 1\right\} \\ &= \sup\left\{|\langle\langle R(\cdot)x, x^*\rangle_X, \varphi\rangle_{\mathcal{R}}| : \|x\|_X \le 1, \|x^*\|_{X^*} \le 1\right\} \\ &\le C\|\varphi\|_{\mathcal{R}^*}; \end{aligned}$$

assume that $\zeta \in \sigma(T)$ and use the spectral mapping theorem (C.2.5.1(b)) to get $|V_\zeta(\varphi)| = |\varphi(\zeta)| \le \|\varphi(T)\| \le C\|\varphi\|_{\mathcal{R}^*}$ for any polynomial φ; by 3), $\sigma(T) \cap \mathbb{T} = \emptyset$, and hence $r(T) < 1$.]

(b) Let be a Banach space operator satisfying $T \in (w\mathcal{R})$, $\mathcal{R}$ be as in (a) above, and let

$$M_{\mathcal{R}}(\zeta) = \|(1-\zeta z)^{-1}\|_{\mathcal{R}} = \|V_\zeta\|_{\mathcal{R}^*} = \sup\left\{|\varphi(\zeta)| : \varphi \in \mathcal{R}^* \cap \mathcal{P}ol_+, \|\varphi\|_{\mathcal{R}^*} \le 1\right\}$$

for $|\zeta| < 1$, with the similar meaning for $M_{\mathcal{R}^*}$. Show that

(i) condition 3) of (a) implies that

$$\lim_{|\zeta|\to 1} M_{\mathcal{R}}(\zeta) = \infty;$$

if, moreover, $\mathcal{R}$ is *rotation invariant* ($f \in \mathcal{R} \Rightarrow f(\zeta\cdot) \in \mathcal{R}$ and $\|f(\zeta\cdot)\|_{\mathcal{R}} = \|f\|_{\mathcal{R}}$ for every ζ, $|\zeta| = 1$), then the latter property is equivalent to 3) of (a), and it is also equivalent to

$$\mathcal{R}^* \not\subset H^\infty;$$

(ii) $r(T) \leq m_{\mathcal{R}}^{-1}(C)$, where C is the constant from the proof of (a) and $m_{\mathcal{R}}^{-1}$ is the inverse of the function

$$m_{\mathcal{R}}(r) = \inf_{r \leq |\zeta| < 1} M_{\mathcal{R}}(\zeta);$$

(iii) $(1 - |\zeta|^2)^{-1} \leq M_{\mathcal{R}}(\overline{\zeta}) M_{\mathcal{R}^*}(\zeta)$, and hence, if $\lim_{r \longrightarrow 1} q(r) = 0$,

$$r(T) \leq q^{-1}(1/C)$$

(with the same C as in (ii)), where

$$q(r) = q_{\mathcal{R}}(r) = \sup \left\{ (1 - |\zeta|^2) M_{\mathcal{R}^*}(\zeta) : r \leq |\zeta| < 1 \right\}$$

for $0 \leq r < 1$, and q^{-1} stands for the inverse mapping of q (we assume here that q is a decreasing function);

[*Hint:* for (i), observe that, due to 2), $(1 - \zeta z)^{-1} \in \mathcal{R}$ and $\langle (1 - \zeta z)^{-1}, \varphi \rangle_{\mathcal{R}} = \varphi(\zeta)$ for $|\zeta| < 1$ and $\varphi \in \mathcal{R}^*$ (which justifies the equalities from the definition of $M_{\mathcal{R}}$); the direct implication is easy; the converse follows (for rotation invariant spaces) from the maximum principle: if $|\varphi(\zeta)|$ is big for ζ, $|\zeta| < 1$ and $\varphi \in \mathcal{P}ol_+$, then $|\varphi(\zeta')|$ is big as well for some ζ', $|\zeta'| = 1$, but since $\mathcal{R}$ is rotation invariant the same is still true for every ζ', $|\zeta'| = 1$; in the case of the inclusion $\mathcal{R}^* \subset H^\infty$, apply the closed graph theorem to obtain the rest of this assertion;

for (ii), it suffices to notice that $|\varphi(\zeta)| \leq \|\varphi(T)\| \leq C \|\varphi\|_{\mathcal{R}^*}$ and hence $M_{\mathcal{R}}(\zeta) \leq C$ for every $\zeta \in \sigma(T)$; then $m_{\mathcal{R}}(r(T)) \leq C$, and the result follows;

for (iii), $|q_\zeta(s)| \leq \|q_\zeta(T)\| \leq C \|q_\zeta\|_{\mathcal{R}^*} \leq C q(|\zeta|)$, where $q_\zeta = (1 - |\zeta|^2)/(1 - \zeta z)$ and $s \in \sigma(T)$; take $\zeta = \overline{s}$ with $|s| = r(T)$ to obtain $1 \leq C q(r(T))$.]

(c) Let $L^1(\mathbb{T})/H^1_-$ be the *space of Cauchy integrals* $f(z) = \int_{\mathbb{T}} (\zeta - z)^{-1} h(\zeta)\, dm(\zeta)$, $h \in L^1(\mathbb{T})$, let $T \in (w\mathcal{R})$ and $\mathcal{R}$ be as in (a) above.

(i) Assume that $\mathcal{R}$ is rotation invariant and $\mathcal{R} \subset L^1(\mathbb{T})/H^1_-$. Show that $\mathcal{R} \neq L^1(\mathbb{T})/H^1_-$ if and only if $\mathcal{R}^* \not\subseteq H^\infty$; in this case, $r(T) < 1$.

(ii) Let $\mathcal{R} = H^p$, $p \geq 1$. Show that

$$r(T) \leq (1 - C^{-p'})^{1/2}$$

for $p > 1$, where $\frac{1}{p'} + \frac{1}{p} = 1$, and

$$r(T) \leq 1 - e^{-C/a}$$

for $p = 1$ $(a > 0)$.

(iii) Show that $(R(\cdot)x, x^*) \in L^1(\mathbb{T})/H^1_-$ for every $x, x^* \in H$, where $T : H \longrightarrow H$ is a unitary operator with absolutely continuous spectrum acting on a Hilbert space H (and, of course, $r(T) = 1$).

[*Hint:* for (i), the last claim is contained in (a) and (b)(i); since the polynomials are dense in $\mathcal{R}$, $\mathcal{R} \neq L^1(\mathbb{T})/H^1_-$ if and only if the norm of $\mathcal{R}$ is strictly stronger than that of $L^1(\mathbb{T})/H^1_-$, that is, if and only if there exists a linear form continuous in $\mathcal{R}$ but discontinuous in $L^1(\mathbb{T})/H^1_-$;

for (ii), apply (b)(iii) with $q(r) = q_{H^p}(r) = (1 - r^2) M_{(H^p)^*}(r) = (1 - r^2)^{1/p'}$ for $p > 1$ (see A.3.12.1(d), Volume 1) and, hence, $q^{-1}(t) = (1 - t^{p'})^{1/2}$; for $p = 1$, apply (b)(ii) with $m_{H^1}(r) = M_{H^1}(r) \geq a \cdot \log(1-r)^{-1}$ (see A.5.7.3(m), Volume 1) and, hence, $m^{-1}(t) \leq 1 - e^{t/a}$;

for (iii), apply C.1.5.1, (k) and (n).]

(d) Let $\mathfrak{M}(\mathbb{T})/H^1_-$ be the *space of Cauchy type integrals* in $\mathbb{D}$, i.e., of functions $f(z) = \int_{\mathbb{T}}(\zeta - z)^{-1}\,d\mu(\zeta)$, $\mu \in \mathfrak{M}(\mathbb{T})$ (the space of all complexe Borel measures on $\mathbb{T}$), T be a bounded operator on a Banach space X. Show that

$$(R(\cdot)x, x^*) \in \mathfrak{M}(\mathbb{T})/H^1_-$$

for every $x, x^* \in H$ if and only if T is *polynomially bounded* (see B.1.6.6 and B.1.8, Volume 1, about this latter property).

[*Hint:* use $(C_A(\mathbb{T}))^* = \mathfrak{M}(\mathbb{T})/H^1_-$, where $C_A(\mathbb{T}) = C(\mathbb{T})_+ = \{f \in C(\mathbb{T}) : \hat{f}(n) = 0 \text{ for } n < 0\}$ (the same proof as for A.5.7.8(a)(iv), Volume 1), and the same computations as in the proof of (a).]

(e) (i) Let $X \neq \{0\}$ be a Banach space, $1 \leq p < \infty$ and $w_n > 0$ $(n = 0, 1, ...)$ such that $\overline{\lim}_n w_n^{1/n} \leq 1$. Show that the following are equivalent.

1) If T is a bounded operator on X with $\sum_{n\geq 0} |(T^n x, x^*)|^p w_n < \infty$ for every $x \in X$, $x^* \in X^*$, then $r(T) < 1$.

2) $\sum_{n\geq 0} w_n = \infty$.

[*Hint:* the implication 2) $\Rightarrow$ 1) is a straightforward consequence of (a) for

$$\mathcal{R} = l^p_A(w_n) = \{f = \sum_{n\geq 0} \hat{f}(n) z^n : \sum_{n\geq 0} |\hat{f}(n)|^p w_n < \infty\};$$

to check the converse, assume that $\sum_{n\geq 0} w_n < \infty$ and take $T = I$.]

(ii) Let $w_n = (n+1)^{-\alpha}$, $\alpha \leq 1$, and let T be an operator satisfying condition 1) of (e). Show that

$$r(T) \leq 1 - (A/C)^{p/(1-\alpha)}$$

for $\alpha < 1$, where $C = C(T)$ has the same meaning as in (b) and $A = A(p, \alpha)$ is a constant, and

$$r(T) \leq 1 - \exp\{-(A/C)^p\}$$

for $\alpha = 1$.

[*Hint:* let $\mathcal{R} = l^p_A(w_n)$; show that $m_{\mathcal{R}}(r) \geq A(1-r)^{(\alpha-1)/p}$ for $\alpha > 1$, and that $m_{\mathcal{R}}(r) \geq A(\log\frac{1}{1-r})^{1/p}$ for $\alpha = 1$; then apply (b) (notice that the function q of (b) gives the same result for $\alpha > 1$ but less for $\alpha = 1$).]

(iii) Let $X \neq \{0\}$ be a Banach space, $1 \leq p, q < \infty$ and let $w(r) > 0$ for $0 \leq r < 1$, $\int_0^1 w(r)\,dr < \infty$. Show that the following are equivalent.

1) If T is a bounded operator on X and $\int_0^1 \|(R(r\cdot)x, x^*)\|^q_{H^p} w(r)\,dr < \infty$ for every $x \in X$, $x^* \in X^*$, then $r(T) < 1$.

2) $\int_0^1 w(r)(1-r)^{-q/p'}\,dr = \infty$ if $p > 1$ and $1/p' + 1/p = 1$, or $\int_0^1 w(r)\log^q(1-r)^{-1}\,dr = \infty$ if $p = 1$.

[*Hint:* for 2) $\Rightarrow$ 1), use (b)(i) for the corresponding space $\mathcal{R}$ to obtain that, for $\rho \longrightarrow 1$, $M_{\mathcal{R}}(\rho)^q$ is comparable with $\int_0^1 w(r)(1-\rho r)^{-q/p'}\,dr$ for $1 < p < \infty$, and with $\int_0^1 w(r)\log^q(1-\rho r)^{-1}\,dr$ for $p = 1$;

for 1) $\Rightarrow$ 2), assume the contrary and take $T = I$.]

(f) Let $1 \le p < \infty$ and $(w_n)_{n\ge 0}$ be a sequence with $w_n > 0$ and $\sum_{n\ge 0} w_n = \infty$. Show that there exists a function $f \in L^1(\mathbb{T})$ such that

$$\sum_{n\ge 0} |\hat{f}(n)|^p w_n = \infty.$$

(Recall that for $f \in H^1$ this property fails because of Hardy's inequality, see B.1.6.7(b), B.1.8, Volume 1).

[*Hint:* apply (e) (i) to $T : L^2(\mathbb{T}) \longrightarrow L^2(\mathbb{T})$, $Tf(\zeta) = \zeta f(\zeta)$.]

(g) *The critical growth of L^2-means.* Let φ be a positive continuous function on $[0, 1)$ and

$$\mathcal{R}_\varphi = \{f \in \mathrm{Hol}(\mathbb{D}) : \|f_r\|_{H^2} = o(\varphi(r)) \text{ as } r \longrightarrow 1\}$$

endowed with the norm $\|f\| = \sup_{0\le r<1}(\|f_r\|_{H^2}/\varphi(r))$ (here $f_r(z) = f(rz)$).

(i) Let $\inf_{0\le r<1}((1-r)^{1/2}\varphi(r)) = 0$. Show that $r(T) < 1$ for every Banach space operator satisfying $(w\mathcal{R}_\varphi)$.

[*Hint:* since $\mathcal{R} = \mathcal{R}_\varphi$ is rotation invariant, it suffices to show that $M_\mathcal{R}(\zeta) = \|(1-\zeta z)^{-1}\|_\mathcal{R}$ is not bounded as $|\zeta| \longrightarrow 1$ and to apply (b); clearly $\|(1-\zeta rz)^{-1}\|_{H^2} = (1 - |r\zeta|^2)^{-1/2}$ and hence

$$M_\mathcal{R}(\zeta) = \sup_{0\le r<1} \left((1 - |r\zeta|^2)^{-1/2}/\varphi(r)\right) = \infty$$

and $\sup_{|\zeta|<1} M_\mathcal{R}(\zeta) = \sup_{0\le r<1} \left((1 - r^2)^{-1/2}/\varphi(r)\right) = \infty$.]

(ii) Let $\inf_{0\le r<1}((1 - r)^{1/2}\varphi(r)) > 0$. Show that $T \in (w\mathcal{R}_\varphi)$ for every (Hilbert space) unitary operator with absolutely continuous spectrum (for which $r(T) = 1$).

[*Hint:* by (c), it suffices to show that $L^1(\mathbb{T})/H^1_- \subset \mathcal{R}_\varphi$; for this, show that $\|f_r\|_{H^2} = o((1 - r^2)^{-1/2})$ as $r \longrightarrow 1$ for every $f \in L^1(\mathbb{T})/H^1_-$; indeed, assume first that f and h ($h \in H^1_-$) are continuous in $\overline{\mathbb{D}}$ and write

$$f_r(z) = \sum_{n\ge 0} \hat{f}(n)r^n z^n = \int_\mathbb{T} (f(\overline{\zeta}) + h(\overline{\zeta}))(1 - rz\zeta)^{-1}\, dm(\zeta),$$

which can be viewed as a $L^1(\mathbb{T})/H^1_-$ valued Riemann integral; then

$$\|f_r\|_{H^2} \le \int_\mathbb{T} |f(\overline{\zeta}) + h(\overline{\zeta})| \cdot \|(1 - r\zeta\cdot)^{-1}\|_{H^2}\, dm(\zeta) = \|f + h\|_1(1 - r^2)^{-1/2},$$

or $\|f_r\|_{H^2} \le \|f\|_{L^1/H^1_-}(1-r^2)^{-1/2}$; since on the dense subset of polynomials $\|f_r\|_{H^2}$ is bounded, we get the result.]

(iii) Let T be a Banach space operator with $r(T) \le 1$ and let

$$\varphi_T(r)^2 = \int_\mathbb{T} \|R(r\zeta)\|^2\, dm(\zeta), \quad 0 < r < 1;$$

show that if $\varphi_T(r) = o((1 - r^2)^{-1/2})$ as $r \longrightarrow 1$ then $r(T) < 1$.

[*Hint:* immediate from (i).]

(iv) *Normal Hilbert space operators satisfying* $\varphi_T(r) = O((1-r^2)^{-1/2})$ as $r \longrightarrow 1$ (φ_T is defined in (iii)). Let $N : H \longrightarrow H$ be a normal operator on a Hilbert space H with $\sigma(N) \subset \overline{\mathbb{D}}$. Show that $\varphi_N(r) = O((1-r^2)^{-1/2})$ as $r \longrightarrow 1$ if and only if

$$\operatorname{card}(\sigma(N) \cap \mathbb{T}) < \infty.$$

The "only if" part is still valid for every Banach space operator N.

[*Hint:* by the spectral theorem

$$\|R(\lambda)\| = \sup_{z\in\sigma(N)} |(1-\lambda z)^{-1}| = 1/\operatorname{dist}(\lambda, \sigma_*(N)),$$

where $\sigma_*(N) = \{1/z : z \in \sigma(N)\}$; since

$$\varphi_T(r)^2 = \int_{\mathbb{T}} \|R(r\zeta)\|^2 \, dm(\zeta) \geq \int_{\sigma_*(N)\cap\mathbb{T}} (1-r)^{-2} \, dm(\zeta),$$

it follows that $m(\sigma(N) \cap \mathbb{T}) = m(\sigma_*(N) \cap \mathbb{T}) = 0$ is necessary;

now, assume that $m(\sigma_*(N)\cap\mathbb{T}) = 0$ and observe that $\operatorname{dist}(\lambda, \sigma_*(N))^2 = |re^{it} - e^{i\theta}|^2 = (1-r)^2 + 4r\sin^2\frac{t-\theta}{2}$ for $\lambda = re^{it}$ and some $e^{i\theta} \in \sigma_*(N) \cap \mathbb{T}$; this means that the required estimate is equivalent to the following one

$$\int_0^{2\pi} \frac{dt}{\operatorname{dist}(t+i\varepsilon, S)^2} \leq \frac{C}{\varepsilon},$$

where $S = \{\theta \in [0, 2\pi] : e^{i\theta} \in \sigma_*(N) \cap \mathbb{T}\}$ and $\varepsilon = 1-r$; writing $[0, 2\pi]\backslash S = \cup_{k\geq 1}(a_k, b_k)$, where $\{(a_k, b_k)\}$ are the complementary intervals, and using $|S| = 0$, we get

$$\int_0^{2\pi} \frac{dt}{\operatorname{dist}(t+i\varepsilon, S)^2} = \sum_{k\geq 1} \int_{a_k}^{b_k} \frac{dt}{\operatorname{dist}(t+i\varepsilon, S)^2},$$

and next

$$\int_{a_k}^{b_k} \frac{dt}{\operatorname{dist}(t+i\varepsilon, S)^2} = \int_{a_k}^{c_k} \frac{dt}{(t-a_k)^2+\varepsilon^2} + \int_{c_k}^{b_k} \frac{dt}{(t-b_k)^2+\varepsilon^2},$$

where $c_k = (a_k + b_k)/2$; therefore,

$$\varepsilon \int_0^{2\pi} \frac{dt}{\operatorname{dist}(t+i\varepsilon, S)^2} = \sum_{k\geq 1} \left(\arctan\frac{c_k - a_k}{\varepsilon} + \arctan\frac{b_k - c_k}{\varepsilon} \right);$$

the last expression is bounded as $\varepsilon \longrightarrow 0$ if and only if the set of intervals is finite.]

1.6.2. Tauberian theorems for exponential stability of semigroups. Let $t \longmapsto S(t)$, $t \in \mathbb{R}_+$ be a (strongly) continuous semigroup on a Banach space X, and let

$$\alpha_0(S) = \lim_{t\longrightarrow\infty} \frac{1}{t} \log\|S(t)\| = \inf\left\{\alpha : \alpha \in \mathbb{R}, \sup_t e^{\alpha t}\|S(t)\| < \infty\right\}$$

be the *growth exponent (bound)* of $S(\cdot)$ and

$$s(A) = \sup\left\{ \operatorname{Re}(\lambda) : \lambda \in \sigma(A)\right\}$$

be the *spectral bound* of A, the generator of $S(\cdot)$; (see 2.2.1 and 2.2.2 below for more details; in particular, $s(A) \leq \alpha_0(S)$). Recall that $\alpha_0(S) < 0$ means that S is exponentially stable, see Section 1.4.

(a) *Norm Tauberian conditions.* Let $\mathcal{L}$ be a Banach lattice on $\mathbb{R}_+$ such that the space

$$\mathcal{R} := \Big\{ \sum_{n\geq 0} a_n z^n : \sum_{n\geq 0} a_n \chi_{\Delta_n} \in \mathcal{L} \Big\},$$

where $a_n \in \mathbb{C}$ and $\Delta_n = [n, n+1]$ for $n \geq 0$, is shift invariant ($f \in \mathcal{R} \Rightarrow Wf := zf \in \mathcal{R}$) and satisfies conditions 1), 2) of Subsection 1.6.1, and let $S(\cdot)$ be a continuous semigroup on a Banach space X such that

$$\|S(\cdot)x\| \in \mathcal{L}$$

for every $x \in X$.

(i) Show that if the related space $\mathcal{R}$ satisfies condition 3) of 1.6.1 (or the equivalent conditions from 1.6.1(b) (i)) then $\alpha_0(S) < 0$.

[*Hint:* let $K = \sup_{0\leq t\leq 1} \|S(t)\|$, then

$$\|S(n+1)x\| = \|S(n+1-t)S(t)x\| \leq K\|S(t)x\|$$

for every $x \in X$ and $t \in \Delta_n$, $n \geq 0$, and therefore, $\sum_{n\geq 0} \|S(n)x\| z^n \in \mathcal{R}$; set $T = S(1)$ to get

$$\sum_{n\geq 0} \langle T^n x, x^* \rangle_X z^n = \langle R(\cdot)x, x^* \rangle \in \mathcal{R}$$

for every $x \in X$, $x^* \in X^*$, and hence by 1.6.1(a) $\alpha_0(S) = \lim_{n\longrightarrow\infty} \frac{1}{n} \log \|S(n)\| = \log(r(T)) < 0$.]

(ii) Let S be as in (i) and let c be a constant such that $\|S(\cdot)x\|_{\mathcal{L}} \leq c\|x\|$ for every $x \in X$. Show that

$$\alpha_0(S) \leq \min(\log m_{\mathcal{R}}^{-1}(C), \log q_{\mathcal{R}}^{-1}(1/C)),$$

where $C = cK\|W\|$, K and W are defined above and $m_{\mathcal{R}}^{-1}$, $\log q_{\mathcal{R}}^{-1}$ are from 1.6.1(b).

[*Hint:* since $|\langle T^n x, x^* \rangle| \leq \|S(n)x\| \cdot \|x^*\|$ and $\mathcal{L}$ is a lattice, it follows that the embedding norm from 1.6.1(a), applied to $T = S(1)$, is at most $C = cK\|W\|$; then apply the estimates from 1.6.1(b).]

(b) Let $X \neq \{0\}$ be a Banach space, $1 \leq p < \infty$ and w be a positive function on $\mathbb{R}_+$ such that $\sup_{t>0} w(t+1)/w(t) < \infty$ and $\overline{\lim}_{t\to\infty} w(t)^{1/t} \leq 1$.

(i) Show that the following are equivalent.

1) If S is a continuous semigroup on X such that $\int_{\mathbb{R}_+} \|S(t)x\|^p w(t)\, dt < \infty$ for every $x \in X$, then $\alpha_0(S) < 0$.

2) $\int_{\mathbb{R}_+} w(t)\, dt = \infty$.

[*Hint:* for 2) $\Rightarrow$ 1), set $\mathcal{L} = L^p(\mathbb{R}_+, w) = \{f : \int_{\mathbb{R}_+} |f(t)|^p w(t)\, dt < \infty\}$ and obtain that $\mathcal{R} = l^p_A(w(n))$ (in the notation of (a)); then apply (a) and 1.6.1(e); for 1) $\Rightarrow$ 2) use the same reasoning as in 1.6.1(e) with $S(t) = I$, $t \geq 0$.]

(ii) Let $w(t) = (t+1)^{-\alpha}$, $\alpha \leq 1$, and let S be a semigroup satisfying condition 1) of (i). Show that

$$\alpha_0(S) \leq -(A/cK)^{p/(1-\alpha)}$$

for $\alpha < 1$, where $c = c(S)$ is a constant satisfying $\int_{\mathbb{R}_+} \|S(t)x\|^p w(t)\, dt \le c^p \|x\|^p$ for every $x \in X$, $K = \sup_{0\le t\le 1} \|S(t)\|$, and $A = A(p, \alpha)$ is a constant, and

$$\alpha_0(S) \le -\exp\{-(A/cK)^p\}$$

for $\alpha = 1$.

[*Hint:* since $\alpha_0(S) = \log(r(T)) \le r(T) - 1$ for $T = S(1)$, one can apply an estimate for $r(T)$ from 1.6.1(e) (ii) (because $\mathcal{R} = l^p_A(w_n)$ by (i) above).]

(c)* *Weak Tauberian conditions. (P. Clément (p = 1, 1987), G. Weiss (1988))* Let X be a Banach space, $1 \le p < \infty$ and let S be a continuous semigroup on X with generator A defined on a domain $\mathrm{Dom}(A) \subset X$. If $\int_{\mathbb{R}_+} |\langle S(t)x, x^*\rangle|^p dt < \infty$ for every $x \in X$ and $x^* \in X^*$, then $s(A) < 0$.

(d) *Spectral mapping theorems for semigroups.* Let X be a Banach space and let S a continuous semigroup on X with generator A (symbolically, $S(t) = e^{tA}$, $t > 0$).

(i) Show that $\alpha_0(S) = s(A)$ if there exists $t > 0$ such that $\sigma(S(t)) = \exp(t\sigma(A))$.

[*Hint:* indeed, $s(A) = \sup(\mathrm{Re}(\sigma(A))) = \frac{1}{t}\log(r(S(t)) = \lim_n \frac{1}{nt}\log\|S(nt)\| = \alpha_0(S)$.]

(ii)* *(L. Gearhart, 1978)* Let X be a Hilbert space, $t > 0$ and $\lambda \in \mathbb{C}$. The following are equivalent.

1) $e^{t\lambda} \in \mathbb{C}\backslash\sigma(S(t))$.
2) $\lambda + \frac{2\pi i}{t}\mathbb{Z} \subset \mathbb{C}\backslash\sigma(A)$ and $\sup\left\{\left\|((\lambda + \frac{2\pi k i}{t})I - A)^{-1}\right\| < \infty : k \in \mathbb{Z}\right\}$.

(iii)* The spectral mapping theorem holds for the following classes of (strongly) continuous semigroups:
1) eventually norm continuous semigroups;
2) eventually compact semigroups;
3) eventually strongly differentiable semigroups;
4) analytic semigroups (that is, $S(\cdot)$ is extendable to a holomorphic mapping $z \longmapsto S(z)$ in a sector $\{z \in \mathbb{C} : z = re^{i\theta}, |\theta| < \varepsilon\}$ such that $S(z + z') = S(z)S(z')$ and $\lim_{z\longrightarrow 0} S(z)x = x$ for every $x \in X$).

(iv) *Counterexamples.* 1) *(C. Chicone and Yu. Latushkin, 1999)* Let $X = c_0$ be the space of complexe sequences that converge to zero and

$$S(t)x = S(t)(x_n)_{n\ge1} = (e^{int}x_n)_{n\ge1}$$

for $t \ge 0$. Show that S is a strongly continuous semigroup on X, $\sigma(S(t)) = \mathrm{clos}(\exp(t\sigma(A)))$ for every t, but $\sigma(S(t)) \ne \exp(t\sigma(A))$ for $t > 0$ with $2\pi/t$ irrational.

[*Hint:* $A(x_n)_{n\ge1} = (inx_n)_{n\ge1}$ for x with $(inx_n)_{n\ge1} \in X$, and hence $\sigma(A) = \{in : n \ge 1\}$.]

2) *(J. Zabczyk, 1975)* Show that there exists a continuous Hilbert space semigroup S having a complete set of root vectors and such that $\sigma(A) = 2\pi i\mathbb{N}$ and $\|S(t)\| = e^t$ for every $t > 0$ (and therefore, $s(A) = 0$ and $\alpha_0(S) = 1$).

Namely, let $X = \sum_{n\geq 1} \oplus \mathbb{C}^n = l^2(\mathbb{N}, (\mathbb{C}^n))$ be a Hilbert space and

$$S(t)(x_n)_{n\geq 0} = (e^{2\pi int} e^{tJ_n} x_n)_{n\geq 1}$$

for $t \geq 0$, where J_n is the standard Jordan block in $\mathbb{C}^n$; show that $S(\cdot)$ satisfies the claimed properties.

[*Hint:* first, $A(x_n)_{n\geq 1} = ((2\pi in + J_n)x_n)_{n\geq 1}$ for those $x = (x_n)_{n\geq 1} \in X$ for which $A(x_n)_{n\geq 1} \in X$; hence, $\lambda \in \mathbb{C}\backslash\sigma(A)$ if and only if

$$\sup_{n\geq 1} \|(\lambda - 2\pi in - J_n)^{-1}\| < \infty;$$

it is clear that, whatever is λ, $|\lambda - 2\pi in| > 1$ for all n except at most one, say, for $n = n_1(\lambda)$; since $\|J_n\| = 1$, it follows from von Neumann's inequality (C.2.2.3) that

$$\|(\lambda - 2\pi in - J_n)^{-1}\| \leq \Big(|\lambda - 2\pi in| - 1\Big)^{-1}$$

for $n \neq n_1(\lambda)$, and hence $\sigma(A) = 2\pi i\mathbb{N}$; in particular, $s(A) = 0$;

secondly, it is clear that $\|S(t)\| \leq e^t$; in fact, the equality holds, since there are $x^n \in X$ such that $\|x^n\| = 1$ and $\sup_n \|S(t)x^n\| = e^t$; indeed, let $(x^n)_k = 0$ for $n \neq k$ and $(x^n)_n = n^{-1/2}(1, 1, ..., 1)$, then $\|S(t)x^n\| = \|e^{tJ_n}(x^n)_n\|$ and, since

$$e^{tJ_n} = \sum_{k=0}^{n-1} \frac{t^k}{k!} J_n^k$$

and $J_n^k(x^n)_n = n^{-1/2}(1, ..., 0, 0)$, we have

$$\|e^{tJ_n}(x^n)_n\|^2 = \frac{1}{n} \sum_{j=0}^{n-1} \Big| \sum_{k=0}^{j} \frac{t^k}{k!} \Big|^2;$$

the latter arithmetic mean tends to e^{2t}; therefore, $\|S(t)\| = e^t$ for every t, $t > 0$, and the result follows.]

1.6.3. Power boundedness and bounds for powers. Let T be a bounded operator on a Banach space X. The following mean growth conditions are defined in C.1.5.6(d) (for the case of a Hilbert space X):

$\text{AMean}^2(T)$	:	$\sup_{n\geq 0}(n+1)^{-1}\sum_{k=0}^{n} \|T^k x\|^2 \leq M(T)^2\|x\|^2$ for every $x \in X$, where $M(T) < \infty$;
$\text{uAMean}^2(T)$	:	$\sup_{n\geq 0}(n+1)^{-1}\sum_{k=0}^{n} \|T^k\|^2 < \infty$.
$\text{PowBd}(T)$	:	$PB(T) := \sup_{n\geq 0} \|T^n\| < \infty$.
$\text{LMean}^2\, RG(T)$	:	$\int_{\mathbb{T}} \|R_{r\zeta}(T)x\|^2\, dm(\zeta) \leq c^2\|x\|^2(r^2-1)^{-1}$ for every $r > 1$ and every $x \in X$ (for a Hilbert space X, it is equivalent to $\sum_{k\geq 0} r^{-2(k+1)}\|T^k x\|^2 \leq c^2\|x\|^2(r^2-1)^{-1}$)
$\text{uLMean}^2\, RG(T)$	:	$\int_{\mathbb{T}} \|R_{r\zeta}(T)\|^2\, dm(\zeta) \leq C(T)^2(r^2-1)^{-1}$ for every $r > 1$,
$LRG(T)$	:	$\|R_\lambda\| \leq c(T)(\lvert\lambda\rvert - 1)^{-1}$ for $\lvert\lambda\rvert > 1$, where $R_\lambda = (\lambda I - T)^{-1}$;

see C.1.5.6(d) (i) for some relations between these conditions (for the case of a Hilbert space X). In particular, for a Hilbert space operator T,

$$\begin{aligned} \mathrm{uAMean}^2(T)\&\,\mathrm{uAMean}^2(T^{-1}) &\Rightarrow \mathrm{AMean}^2(T)\&\,\mathrm{AMean}^2(T^{*-1}) \\ &\Rightarrow \mathrm{LMean}^2\, RG(T)\&\,\mathrm{LMean}^2\, RG(T^{*-1}) \\ &\Rightarrow \mathrm{PowBd}(T)\&\,\mathrm{PowBd}(T^{-1}), \end{aligned}$$

whereas $\mathrm{AMean}^2(T)\&\,\mathrm{AMean}^2(T^{-1})$ do not imply $\mathrm{PowBd}(T)$.

(a)* *(J. van Casteren, 1997)* Let T be a bounded operator on a Hilbert space H. Then $\mathrm{AMean}^2(T)\&\,\mathrm{AMean}^2(T^*)$ imply $\mathrm{PowBd}(T)$, and moreover,

$$PB(T) \le e \cdot M(T)M(T^*), \text{ and } \overline{\lim}_n \|T^n\| \le \frac{e}{2} M(T)M(T^*).$$

(b) Let X be a Banach space.

(i) Show that $\mathrm{LRG}(T)$ implies $\|T^n\| \le e \cdot c(T)(n+1)$ for $n \ge 0$. The estimate is sharp up to a constant factor *(A. Shields, 1978)*.

[*Hint:* $\|T^n\| = \|(2\pi i)^{-1} \int_{|z|=r} R_z z^n\, dz\| \le c(T)(r-1)^{-1} r^{n+1} \le e \cdot c(T)(n+1)$ for $r = 1 + (n+1)^{-1}$;

for Shields' example, take

$$X = H_1^1 = \{f : f' \in H^1\}$$

equipped with the norm $\|f\|_X = \|f\|_\infty + \|f'\|_{H^1}$, and show that X is a unital Banach algebra with the standard multiplication; next, set $Tf = zf$ and get $\|T^n\| \ge \|z^n\|_X = n+1$, and

$$\begin{aligned} \|R_\lambda\| &= \|(\lambda - z)^{-1}\|_X = (|\lambda| - 1)^{-1} + \int_{\mathbb{T}} |\lambda - z|^{-2}\, dm \\ &= (|\lambda| - 1)^{-1} + (|\lambda|^2 - 1)^{-1} \\ &\le 2(|\lambda| - 1)^{-1} \end{aligned}$$

for every $|\lambda| > 1$.]

(ii)* *(H.-O. Kreiss, 1962)* Let X be a Banach space, $\dim X < \infty$. Then

$$PB(T) \le e \cdot c(T) \dim X.$$

The estimate is sharp *(M. Spijker, 1991)*.

(c)* *(O. El-Fallah and T. Ransford, 2000)* Let X be a Banach space, $T \in L(X)$.

(i) If $\sigma(T) \subset \mathbb{D} \cup \{1\}$ and

$$\|R_\lambda\| \le c_{\{1\}}(T)|\lambda - 1|^{-1}$$

for $|\lambda| > 1$ then $PB(T) \le c_{\{1\}}(T)^2$.

(ii) Let $\sigma \subset \mathbb{T}$ be a closed set, $\sigma(T) \subset \mathbb{D} \cup \sigma$ and $\|R_\lambda\| \le c_\sigma(T)\, \mathrm{dist}(\lambda, \sigma)^{-1}$ for $|\lambda| > 1$, and let

$$\Phi_\sigma(n) = 2n \cdot m(\sigma_{\pi/2n}),$$

where $\sigma_\varepsilon = \{\zeta \in \mathbb{T} : d(\zeta, \sigma) \le \varepsilon\}$, $d(\cdot,\cdot)$ is the arclength on $\mathbb{T}$, and m is the normalized Lebesgue measure on $\mathbb{T}$. Then

$$\max_{0 \le k < n} \|T^k\| \le 2^{-1} e \cdot c_\sigma(T)^2 \Phi_\sigma(n)$$

for $n \geq 1$. The estimate is sharp up to a constant factor. In particular,

$$PB(T) \leq 2^{-1}e \cdot c_\sigma(T)^2 \operatorname{card}(\sigma)$$

(for a finite set σ), and $\|T^n\| = o(n)$ as $n \longrightarrow \infty$ for σ with $m(\sigma) = 0$.

(d) Let X be a Banach space, $T \in L(X)$.

(i) Show that if $PB(T) < \infty$ then the following $\mathrm{LRG}^n(T)$ condition holds:

$$pc(T) := \sup\{(|\lambda| - 1)^n \|R_\lambda^n\| : n \geq 0, |\lambda| > 1\} < \infty.$$

In fact, $pc(T) \leq PB(T)$.

[*Hint:* use $(-1)^n n! R_\lambda^{n+1} = \frac{d^n R_\lambda}{d\lambda^n}$.]

(ii)* *(Ch. Lubich and O. Nevanlinna, 1991)* Let X be a Banach space, $T \in L(X)$. Then $\mathrm{LRG}^n(T)$ (see (i) above for the definition) is equivalent to the estimate

$$\|e^{zT}\| \leq pc(T) \cdot e^{|z|}$$

for $z \in \mathbb{C}$; both these conditions imply that

$$\|T^n\| \leq pc(T) \cdot \sqrt{2\pi(n+1)}$$

for $n \geq 0$. The estimate is sharp up to a constant factor; for example, if $S : l^\infty \longrightarrow l^\infty$ is the shift operator and $T = b_\lambda(S)$, where $b_\lambda(z) = (\lambda - z)(1 - \overline{\lambda}z)^{-1}$ with $0 < |\lambda| < 1$, then $pc(T) < \infty$ and $\|T^n\| \geq c \cdot \sqrt{n+1}$ for a positive c.

(e) (i) Let X be a Banach space, $T \in L(X)$. Show that the sequence $(w_n) = (\|T^n\|)$ ($n \in \mathbb{Z}$ or $n \in \mathbb{Z}_+$) is *submultiplicative*, that is, $w_0 = 1$ and $w_{n+k} \leq w_n w_k$ for every n, k.

(ii) Conversely, given a submultiplicative sequence (w_n) (on $\mathbb{Z}$ or on $\mathbb{Z}_+$) there exists a Hilbert space operator T (in the case of $\mathbb{Z}$, an invertible operator) such that $w_n = \|T^n\|$ for all n.

[*Hint:* take $X = l^2(w_n^2) = \{x = (x_n) : \|x\|^2 = \sum_n |x_n|^2 w_n^2 < \infty\}$ and $T = S_w$, the (right) shift operator on X.]

(iii) Let $p > 0$ and let T be an invertible Banach space operator satisfying the conditions $\mathrm{uAMean}^p(T)$& $\mathrm{uAMean}^p(T^{-1})$, where

$$\mathrm{uAMean}^p(T) : \sup_{n \geq 0}(n+1)^{-1} \sum_{k=0}^{n} \|T^k\|^p < \infty.$$

Show that $PB(T) < \infty$, $PB(T^{-1}) < \infty$, i.e., $\sup_{n \in \mathbb{Z}} \|T^n\| < \infty$.

[*Hint:* set $w_n = \|T^n\|^{p/2}$ for $n \in \mathbb{Z}$, apply (ii), then use C.1.5.6(d), Corollary (iii), (C), and get $\sup_{n \in \mathbb{Z}} w_n < \infty$.]

1.7. Notes and Remarks

General remarks. The mathematical theory of control has its origins in technical problems (such as "automatic regulators" in mecanical, electrical and electronical engineering) as well as in mathematics itself (such as differential equations, both ordinary and partial). The litterature as well as the subject itself are immense, and their more or less complete description is obviously out of the scope of this book. For example, the case of unbounded operators B, C D, frequently encoutered in the applications, will not be treated. Also, neither non-linear control problems $x' = f(x, \dots)$ nor the case of non-linear optimization functionals are considered. For these reasons we will restrict ourselves only to references related to the topics presented in the text.

Concerning control theory mostly oriented to applications we refer the reader to A. Butkovskii [**Butk**], J. Doyl, B. Francis and A. Tennenbaum [**DFT**], J. Helton [**Helt2**], J. Helton and O. Merino [**HM**] and T. Kailath [**Kai**]. For control problems motivated by partial differential equations the reader should consult J.L. Lions [**Lio**], A. Butkovskii [**Butk**], D. Russel [**Rus**] and the vast litterature cited in these sources.

The operator theory point of view is presented in P. Fuhrmann [**Fuh3**], W. Wonham [**Wo**], S. Dolecki and D. Russel [**DolR**], C. Foiaş and A. Frazho [**FF**], R. Curtain and A. Pritchard [**CPr**], R. Curtain and H. Zwart [**CZ**], A. Feintuch [**Fe**], and the links between control theory and harmonic analysis in S. Avdonin and S. Ivanov [**AI2**], A. Butkovskii [**Butk**], R. Curtain and H. Zwart [**CZ**].

The basic references for operator semigroups (which represent the irreplaceable tool of the control theory) are E. Hille and R. Phillips [**HPh**], A. Pazy [**Paz**].

Stability and exponential stability. Strongly stable semigroups are those satisfying

$$\lim_{t\to\infty} \|S(t)x\| = 0$$

for every $x \in X$, whereas exponentially stable semigroups are characterized by the estimate

$$\|S(t)\| \le Ce^{-\varepsilon t}$$

for an $\varepsilon > 0$. Since both properties are crucially important for the applications of the theory of semigroups, there exists an abundance of publications about them. The following criterion by I. Miyadera (1952), W. Feller, and R. Phillips (1953) describes generators of semigroups satisfying an exponential estimate (often refered to as *"Hille–Yosida's theorem"*, see A. Pazy [**Paz**] for original references and comments).

THEOREM. *The following are equivalent.*

1) A is a generator of a strongly continuous semigroup $S(\cdot)$ on X satisfying $\|S(t)\| \le Ce^{\alpha t}$ for an $\alpha \in \mathbb{R}$ and every $t > 0$;

2) A is a closed operator with dense domain $\mathrm{Dom}(A)$ *and such that $\|R_\lambda(A)^n\| \le C(\lambda - \alpha)^{-n}$ for every $n \ge 1$ and for every $\lambda \in (\alpha, \infty)$.*

A discrete version of this criterion (for $\alpha = 0$) is presented in 1.6.3(d) above. Some other results of Section 1.6 are also related to the same problem, namely, how to recognize exponentially stable semigroups $S(\cdot)$ using as less information on the behaviour of $|\langle S(t)x, x^*\rangle|$, $\|S(t)x\|$ or $\|R_\lambda x\|$ as possible. These results are commented below.

The strong stability property can be deduced from the boundedness of $S(\cdot)$ and a kind of smallness of the imaginary spectrum $\sigma(A) \cap i\mathbb{R}$ of the generator. For example, the *Arendt–Batty–Lyubich–Vũ* theorem states the following, W. Arendt and C. Batty [**ArBa**], Yu. Lyubich and Q. Vũ [**LyuV**].

THEOREM. *Let A be the generator of a strongly continuous semigroup $S(\cdot)$ on a Banach space X satisfying* $\sup_{t>0} \|S(t)\| < \infty$ *and such that $\sigma(A) \cap i\mathbb{R}$ is at most countable and $\sigma_p(A^*) \cap i\mathbb{R} = \emptyset$ (equivalently,* $\mathrm{Range}(itI - A)$ *is dense in X for every $t \in \mathbb{R}$). Then $S(\cdot)$ is strongly stable.*

A discrete semigroup analog is also true (replacing $S(\cdot)$ be a power bounded operator T such that $\sigma(T) \cap \mathbb{T}$ is at most countable and $\sigma_p(T^*) \cap \mathbb{T} = \emptyset$). For another but similar combination of conditions implying stability for discrete time semigroups see C.1.5.4(f) above. It is also clear that the condition on $\sigma(A) \cap i\mathbb{R}$ is sharp: given an uncountable closed set $\sigma \subset \mathbb{R}$, there exists a continuous (atom free) measure μ having bounded support in σ, and hence, there is an *isometric* norm continuous group $S(t) = e^{itx} : L^2(\mu) \longrightarrow L^2(\mu)$ satisfying $\sigma(A) \cap i\mathbb{R} \subset \sigma$ and $\sigma_p(A^*) \cap i\mathbb{R} = \emptyset$. On the other hand, the preceding result can be deduced from the following theorem, which is quoted both in discrete and continuous forms.

THEOREM. 1) Discrete semigroup version *(Y. Katznelson and L. Tzafriri, see* [**KaTz**]*). Let X be a Banach space, $T \in L(X)$ such that* $\sup_{n\geq 0} \|T^n\| < \infty$ *and f be an analytic function from the Wiener class, i.e.,*

$$f \in W_+ = \mathcal{F}l^1(\mathbb{Z}_+) = \{\sum_{n\geq 0} a_n z^n : \sum_{n\geq 0} |a_n| < \infty\},$$

satisfying the spectral synthesis property on the set $\sigma(T) \cap \mathbb{T}$. Then

$$lim_n \|T^n f(T)\| = 0,$$

where $f(T) = \sum_{n\geq 0} a_n T^n$.

2) Continuous semigroup version *(J. Esterle, E. Strouse, and F. Zouakia,* [**ESZ**]*; Q.Ph. Vũ* [**Vu**]*). Let X be a Banach space, $S(\cdot)$ a continuous semigroup satisfying* $\sup_{t>0} \|S(t)\| < \infty$ *and f be an analytic function form the Wiener class, i.e.,*

$$f \in W_+ = \mathcal{F}L^1(\mathbb{R}_+) = \{\int_0^\infty \hat{f}(y) e^{ixy}\, dy : \int_0^\infty |\hat{f}(y)| dy < \infty\},$$

satisfying the spectral synthesis property on the set $-i\sigma(A) \cap \mathbb{R}$. Then

$$lim_n \|T^n f(A)\| = 0,$$

where $f(A) = \int_0^\infty \hat{f}(y) S(y)\, dy$.

Recall that the *spectral synthesis property* on a set σ means that $f|\sigma = 0$ and there exist functions f_n from the (entire) algebra W ($\mathcal{F}l^1(\mathbb{Z}_+)$, or $\mathcal{F}L^1(\mathbb{R}_+)$) such that $\lim_n \|f - f_n\|_W = 0$ and every f_n vanishs on an open neighbourhood of σ (depending on n).

For these and many other results on stability and exponential stability problems, as well as for further references, see A. Pazy [**Paz**], J. Van Neerven [**VanN**], R. Curtain and H. Zwart [**CZ**], C. Chicone and Yu. Latushkin [**ChL**], W. Arendt [**Are1**], C. Batty [**Bat**], C. Batty and S. Yeates [**BY**], Yu. Tomilov [**Tom**]. In particular, various "local" versions of the above results are known (that is, the conclusion $\lim_{t\to\infty} \|S(t)x\| = 0$ is made on the basis of some (limited) information on the individual orbit ($(S(t)x)_{t>0}$ only), as well as stability results relying on assumptions on the resolvent $(\lambda I - A)^{-1}$, etc.

Tauberian theorems on the spectrum and spectral mapping theorems. This circle of problems is closely related to the previous one and can be described as follows: imposing some restrictions on certain means of a semigroup $S(\cdot)$, to deduce stronger properties for (different) means of $S(\cdot)$ and for the spectra $\sigma(S(t))$. Theorem 1.4.3 and the results presented in Section 1.6 are propositions of this type. Incidentially, the expression $(1 - |\lambda|)R(\lambda)$, where $R(\lambda)$ is the Fredholm resolvent of an operator T (see 1.6.1 for the definition), is also a kind of mean of $(T^n)_{n\geq 0}$ (namely, the *Abel–Poisson mean*).

In particular, the results contained in Subsection 1.6.3 are aimed to give sharp estimates of the norms $\|T^n\|$ deduced from properties of the resolvent. These results have mostly emerged from the so-called *Kreiss matrix theorem* given in 1.6.3(b)(ii), H.-O. Kreiss [**Kre**]. For a survey of this field see J. Strikwerda and B. Wade [**StWa**], as well as comments and references below. There are several other sharp estimates of functions of an operator in C.1.5.9, C.2.5.4 and C.2.6 above. For estimates of functions of power bounded operators see V. Peller [**Pe3**], [**Pe6**] and O. Nevanlinna [**Nev.O2**], [**Nev.O1**].

Formal credits. Theorem 1.3.1 is due to R. Kalman, see R. Kalman, P. Falb and M. Arbib [**KFA**] and other monographs quoted above. For a condition close to (5) of Corollary 1.3.9 see S. Dolecki and D. Russel [**DolR**]; for a necessary condition for ECO expressed in terms of the resolvent $R_\lambda(A)$ see D. Russel and G. Weiss [**RW**].

Subsection 1.3.11 and all the points 1.3.14-1.3.21 are aimed to separate the roles of the semigroup $S(\cdot)$ and the control operator B as explicit as possible. Lemma 1.3.12 for the case of Hilbert spaces Q, R, S is for R. Douglas [**Dou2**].

Usually, Theorem 1.3.19 is stated for the case $\operatorname{rank}(B) < \infty$ only, and the known proofs are different from ours (for example, cf. S. Dolecki and D. Russel [**DolR**], R. Curtain and H. Zwart [**CZ**]).

For the discrete-time systems of Subsection 1.3.21 see W. Wonham [**Wo**]. The notion of ENCO as defined in 1.3.24 is probably introduced by S. Avdonin and S. Ivanov [**AI2**].

Theorem 1.4.3 is proved in R. Datko [**Dat**] (for a Hilbert space X) and independetly in N. Nikolski [**N12**]. In fact, [**N12**] contains an axiomatic approach to tauberian theorems similar to 1.4.3 and, in a sense, a necessary and sufficient condition for $r(T) < 1$ expressed in terms of the resolvent (compare points (i) and (iii) of 1.6.1(c)). An extended version of these results is presented in 1.6.1(a)-(f) and 1.6.2(a)-(b). Moreover, using this method we get quantitative versions of 1.4.3 contained in Remark 1.4.4(1) and 1.6.1(a),(ii)-(iii), 1.6.1(c), 1.6.1(e)(ii), 1.6.1(g) and 1.6.2(a)(ii), 1.6.2(b)(ii). For continuous analogs see comments below.

For Example 1.5 and other examples emerged from PDE's see D. Russel [**Rus**], A. Butkovskii [**Butk**], R. Curtain and A. Pritchard [**CPr**], S. Avdonin and S. Ivanov [**AI2**], J.-L. Lions [**Lio**].

Subsection 1.6.1 is taken from N. Nikolski [**N12**]. The results of 1.6.1 mean that the least "natural" space $\mathcal{R}$ containing all weak resolvents $\langle R(\cdot)x, x^*\rangle_X$ and still compatible with $r(T) = 1$ is $\mathcal{R} = L^1(\mathbb{T})/H^1_-$. All other statements of 1.6.1 and 1.6.2(a), (b) are simple consequences of this observation. As was already mentioned, in 1.6.1(f), the space $L^1(\mathbb{T})$ cannot be replaced by H^1 (since for $w_n = (n+1)^{-1}$ and

$p = 1$ it would contradict the Hardy inequality B.1.6.7(b)). Clearly, proposition 1.6.1(f) can be stated as a property of Hadamard multipliers between the spaces $L^1(\mathbb{T})$ and $\mathcal{F}l^p$, see R. Edwards and G. Gaudry [**EG**], R. Larsen [**Lar**].

Properties 1.6.2(a)(i) and 1.6.2(b)(i) are also claimed in N. Nikolski [**N12**]; for the special case $w = 1$ of 1.6.2(b)(i) see also A. Pazy [**Paz**]. Some other exponential stability criteria can be found in R. Curtain and H. Zwart [**CZ**]; for example, the following one.

THEOREM. *A continuous semigroup $S(\cdot)$ on a Hilbert space H is exponentially stable if and only if $R.(A) \in H^\infty(\mathbb{C}^+, L(H))$, where $\mathbb{C}^+ = \{z \in \mathbb{C} : \mathrm{Re}(z) > 0\}$, and if and only if there exists a positive operator $P \in L(H)$ solving the following* Lyapunov equation

$$(Ax, Px) + (Px, Ax) = -(x, x) \textit{ for all } x \in \mathrm{Dom}(A).$$

The weak Tauberian theorem 1.6.2(c) is proved in P. Clément [**Cle**] and G. Weiss [**Wei**].

The spectral mapping theorem of 1.6.2(d)(ii) is proved in L. Gearhart [**Gea**] for contraction semigroups and in J. Howland [**How2**] for the general case. For generalizations to Banach spaces as treated in 1.6.2(d)(ii) see C. Chicone and Yu. Latushkin [**ChL**], Theorem 2.31. Theorem 1.6.2(d)(iii) can be found in R. Nagel [**Nag**]. The elementary example 1) from 1.6.2(d), (iv) is for [**ChL**], and the example 2) is for J. Zabczyk [**Zab**]. In fact, the very first (but somewhat implicit) example of a continuous semigroup $S(\cdot)$, for which the spectrum of the generator A does not determine the growth bound of $S(\cdot)$, that is $s(A) < \alpha_0(S)$, is contained already in E. Hille and R. Phillips [**HPh**]. Similar counterexamples are possible even if the resolvent $R_\lambda(A)$ is compact, see R. Curtain and H. Zwart [**CZ**], A. Pazy [**Paz**], W. Arendt [**Are1**], C. Chicone and Yu. Latushkin [**ChL**].

The Kreiss matrix theorem 1.6.3(b)(ii) is from H.-O. Kreiss [**Kre**] (see also comments above). For 1.6.3(a) see J. van Casteren [**VanC2**]. The latter property together with those of C.1.5.6 and C.1.6 show that in order to obtain power boundedness of an operator T we need some information on both trajectories $T^n x$ and $T^{*n}x$, whereas $T^n x$ and $T^{-n}x$ do not help so much; the situation changes (of course) when we pass to $\|T^n\|$, compare with 1.6.3(e). The theorems of 1.6.3(c) are from O. El-Fallah and T. Ransford [**ElR**]. A. Shields' example of 1.6.3(b)(i) is contained in [**Sh2**]. Several other examples of this type are known; for this, as well as for the propositions 1.6.3(d), (i) and (ii), and for references for Spijker's example from 1.6.3(b)(ii), see O. Nevanlinna [**Nev.O1**].

CHAPTER 2

First Optimizations: Multiplicity of the Spectrum and the DISC

This chapter is devoted to an optimization problem for approximate controllability (ACO). More precisely, we look for the minimal rank of the control operator, $\dim B\mathcal{U}$, ensuring ACO. Two cases will be distinguished: that of *free minimization* (that is, given A, find $\min \dim B\mathcal{U}$) which reduces to the study of the classical multiplicity of the spectrum $\mu(S(\cdot))$ of the semigroup associated to the generator A, and that of *constrained minimization* (that is, given A and a control subspace $B_0\mathcal{U}_0$, find $\min \dim B\mathcal{U}$ ensuring ACO and such that $B\mathcal{U} \subset B_0\mathcal{U}_0$). The latter gives rise to a new numerical characteristic of a semigroup called $\mathrm{DISC}(S(\cdot))$.

2.1. The Least Dimension of Controlling Subspaces

Let

$$\begin{cases} x' &= Ax + Bu \\ y &= Cx + Du \end{cases}$$

be a dynamical system, where $A : X \longrightarrow X$ generates a strongly continuous semigroup $S(\cdot)$ and $B : \mathcal{U} \longrightarrow X$ is a bounded control operator. Recall that by Kalman's theorem 1.3.1 the system (A, B) is approximately controllable on $]0, \infty[$ if and only if the subspace $B\mathcal{U}$ is *cyclic* for $S(\cdot)$:

$$\operatorname{span}(S(t)B\mathcal{U} : t \geq 0) = X.$$

For bounded A, this is equivalent to

$$\operatorname{span}(A^n B\mathcal{U} : n \geq 0) = X.$$

First, we consider the following problem.

2.1.1. PROBLEM. Given A, find the least possible dimension of the control subspace $\dim B\mathcal{U}$ such that (A, B) is ACO:

$$\min\{\dim B\mathcal{U} : (A, B) \text{ is approximately controllable }\}.$$

Recall also that by Theorem 1.3.19, a finitely controlled system (that is, having $\operatorname{rank} B < \infty$), is never exactly controllable (in case $\dim X = \infty$). So, a problem of the above type has no meaning for ECO instead of ACO. Yet, it can occur for null controllability, see Subsection 3.5.6 below.

In this chapter, we deal with approximate controllability only.

2.2. Reduction to Bounded Operators

In this Section we define the multiplicity of the spectrum (or, the multicyclicity) of the generator A of a continuous semigroup. Some bounded functions of A are shown to have the same multiplicity as A; in particular, this is the case for the resolvent and the Cayley transform of A.

2.2.1. The growth bound of a semigroup. Observe first that for a strongly continuous semigroup $S(t)$, $t \geq 0$, there exists $\alpha \in \mathbb{R}$ such that

$$\|S(t)\| \leq ce^{\alpha t}, \quad t \geq 0.$$

Indeed, as $S(\cdot)x$ is continuous for every $x \in X$, $\|S(\theta)\|$ is locally bounded on $[0, \infty)$ (use the Banach–Steinhaus Theorem) and thus bounded for $\theta \in [0, 1]$. We get, setting $M = \sup_{[0,1]} \|S(\theta)\|$,

$$\|S(t)\| = \|S([t] + \theta)\| \leq M\|S(1)\|^{[t]} \leq ce^{\alpha t},$$

where $\alpha = \log \|S(1)\|$ and $[t]$ is the integer part of t. Moreover if we set

$$\alpha_0 = \overline{\lim}_{t \to \infty} \frac{\log \|S(t)\|}{t} < +\infty, \tag{2.2.1}$$

then

$$\|S(t)\| \leq c_\alpha e^{\alpha t}$$

for every $\alpha > \alpha_0$. The value α_0 is the *growth bound* of the semigroup.

In fact, the limit $\lim_{t \to \infty} \frac{\log w(t)}{t}$ exists for any locally bounded *submultiplicative function* w (i.e., such that $w(t + r) \leq w(t)w(r)$). To see this, consider the function $\varphi(t) = \log w(t)$. The submultiplicativity of w implies subadditivity of φ: $\varphi(t+r) \leq \varphi(t) + \varphi(r)$. Consequently

$$\frac{\varphi(s)}{s} = \frac{\varphi(nt + r)}{nt + r} \leq \frac{n\varphi(t)}{nt + r} + \frac{\varphi(r)}{nt + r},$$

where $t > 0$, $s > 0$, and $s = nt + r$, $0 \leq r < t$. Now choose $\alpha > \alpha_0 := \inf_{t>0} \varphi(t)/t$ and fix $t > 0$ such that $\varphi(t)/t < \alpha$. If s tends to infinity (that is $n \longrightarrow \infty$) we get

$$\overline{\lim}_{s \to \infty} \frac{\varphi(s)}{s} \leq \frac{\varphi(t)}{t} < \alpha.$$

As $\alpha > \alpha_0$ was arbitrary we now get

$$\overline{\lim}_{s \to \infty} \frac{\varphi(s)}{s} \leq \alpha_0 \leq \underline{\lim}_{t \longrightarrow \infty} \frac{\varphi(t)}{t},$$

as required. □

Hence, for the *growth bound* α_0, we have

$$\alpha_0 = \lim_{t \to \infty} \frac{\log \|S(t)\|}{t}.$$

Now, we return to the problem of the minimal dimension of control subspaces and make few remarks concerning its relations to the growth bound of a semigroup.

First, making a linear transformation in time $t \longmapsto \gamma t$, for some $\gamma > 0$, we get

$$\alpha_0(S(\gamma \cdot)) = \lim_{t \to \infty} \frac{\log \|S(\gamma t)\|}{t} = \gamma \alpha_0(S).$$

Obviously, this re-parameterization conserves the cyclicity of the control space $B\mathcal{U}$, that is, this space is cyclic for $S(t)$, $t \geq 0$, if and only if it is cyclic for $S(\gamma t)$, $t \geq 0$. Consequently, we can restrict ourselves to the case $\alpha_0 < 1$.

Secondly, we can also modify the semigroup $S(t)$ by $S_\gamma(t) = e^{-\gamma t}S(t)$, $t \geq 0$, without changing the cyclic subspaces (however, note that this modification considerably changes the control system). In this case $\alpha_0(S_\gamma) = \alpha_0(S) - \gamma$.

Thus, dealing with our minimization problem, we may assume without loss of generality that α_0 is as small as we want.

2.2.2. The resolvent. It is well-known (and easy to verify) that for $\lambda \in \mathbb{C}$ with $\operatorname{Re}\lambda > \alpha_0$, the *resolvent*

$$R_\lambda(A) = (\lambda I - A)^{-1}$$

is a well-defined bounded operator and that

$$(\lambda I - A)^{-1}x = \int_0^\infty e^{-\lambda t}S(t)x\,dt.$$

This implies that if $x \in \mathcal{D}(A)$, then $R_\lambda x \in \mathcal{D}(A)$ and moreover

$$(\lambda I - A)R_\lambda x = R_\lambda(\lambda I - A)x = x.$$

Define the spectrum $\sigma(A)$ in the usual way as the complement set $\mathbb{C} \setminus \{\lambda \in \mathbb{C} : (\lambda I - A)^{-1}$ exists$\}$. Then the preceding observation yield

$$\sup \operatorname{Re}\sigma(A) \leq \alpha_0(S), \tag{2.2.2}$$

where $\sup \operatorname{Re}\sigma(A) = \sup\{\operatorname{Re}\lambda : \lambda \in \sigma(A)\}$ is the *spectral bound* of A. The above supremum is a counterpart of the logarithm $\log r(T)$ of the spectral radius $r(T) = \sup\{|\lambda| : \lambda \in \sigma(T)\}$ of a discrete semigroup $(T^n)_{n\geq 0}$, for which $\log r(T) = \sup_{\lambda\in\sigma(T)} \log|\lambda| = \lim_n \frac{\log\|T^n\|}{n}$. In the case of a continuous time parameter the inequality (2.2.2) may be strict (see 1.6.2(d)(iv) for an example), but for many "regular" cases it turns out to be an equality (for instance if A is bounded or if A has a Riesz basis of eigenvectors, see Section 3 below).

Finally, if $S(t)$, $t \geq 0$, is invertible, which means that $S(\cdot)$ is actually a group and $S(-t) = S(t)^{-1}$ for $t \geq 0$, then we have $-\alpha_0(S^{-1}) \leq \alpha_0(S)$ and

$$-\alpha_0(S^{-1}) \leq \inf \operatorname{Re}\sigma(A) \leq \sup \operatorname{Re}\sigma(A) \leq \alpha_0(S).$$

For more information about growth and spectral bounds see Subsections 1.6.1, 1.6.2 and comments in Section 1.7 above.

2.2.3. Invariant subspaces of a semigroup and its resolvent.

THEOREM. *Let $\lambda \in \mathbb{C}$ with $\alpha_0(S) < \operatorname{Re}\lambda$. Then $R_\lambda(A)$ is a bounded operator, and for any closed subspace $E \subset X$ we have*

$$S(t)E \subset E,\ t \geq 0 \iff R_\lambda(A)E \subset E.$$

PROOF. As

$$R_\lambda = \int_0^\infty e^{-\lambda t}S(t)\,dt$$

is norm convergent (considered as limit of Riemann sums) we get the implication "$\Longrightarrow$". For the converse implication we first differentiate the resolvent $R_\mu(A)$ n times with respect to μ, $\operatorname{Re}\mu > \alpha_0$. Then for $\mu = \lambda$ we get

$$(-1)^n n!(\lambda I - A)^{-n-1} = \int_0^\infty (-t)^n e^{-\lambda t} S(t)\, dt,$$

and hence

$$R_\lambda(A)^{n+1} = \frac{1}{n!}\int_0^\infty t^n e^{-\lambda t} S(t)\, dt.$$

Pick now $x \in E$ and $y \in E^\perp \subset X^*$. By assumption, we have $R_\lambda(A)^{n+1}x \in E$ for every $n \geq 1$. Consequently

$$\begin{aligned} 0 &= \; < R_\lambda(A)^{n+1}x, y >= \frac{1}{n!}\int_0^\infty t^n e^{-\lambda t} < S(t)x, y > dt \\ &= \frac{1}{n!}\int_0^\infty t^n e^{-\frac{(\lambda-\alpha_0)}{2}t} e^{-\frac{(\lambda+\alpha_0)}{2}t} < S(t)x, y > dt. \end{aligned}$$

Clearly $e^{-\frac{(\lambda+\alpha_0)}{2}t} < S(t)x, y >$ is a continuous function on $\mathbb{R}_+$ tending to zero at ∞. On the other hand, $\mathcal{L}in(t^n e^{-\gamma t})_{n\geq 0}$ is dense in $C_0(\mathbb{R}_+)$ for every $\gamma > 0$. Hence, the above identity is equivalent to

$$\int_0^\infty \varphi(t) e^{-\frac{(\lambda+\alpha_0)}{2}t} < S(t)x, y > dt = 0$$

for every $\varphi \in C_0(\mathbb{R}_+)$. Then $< S(t)x, y >$ has to be identically equal to zero, which means that $S(t)x \in E$ for every $t \geq 0$ and for every $x \in E$. □

2.2.4. COROLLARY. *The range $B\mathcal{U}$ is a subspace of approximate control for $S(\cdot)$ on $]0, \infty[$ if and only if it is cyclic for R_λ:* $\operatorname{span}(R_\lambda^n B\mathcal{U} : n \geq 0) = X$. □

2.2.5. Multiplicity of the spectrum: the defintion. The quantity

$$\mu(S(\cdot)) := \min(\dim C : \operatorname{span}(S(t)C : t \geq 0) = X)$$

is called the *multiplicity of the spectrum (or the multicyclicity)* of $S(\cdot)$.

For an arbitrary bounded operator $T : X \longrightarrow X$ the *multiplicity of the spectrum (or the multicyclicity)* is defined by

$$\mu(T) := \min\{\dim C : \operatorname{span}(T^n C : n \geq 0) = X\}.$$

We use the notation $\operatorname{Cyc}(T)$ (respectively $\operatorname{Cyc}(S(\cdot))$) for the family of all cyclic subspaces of T (respectively of $S(\cdot)$).

2.2.6. COROLLARY. *$\mu(S(\cdot)) = \mu(R_\lambda(A))$ for every $\lambda > \alpha_0(S)$.* □

Sometimes it is useful to consider the *Cayley transform* of A instead of R_λ :

$$C^-(A) = (I + A)(I - A)^{-1} = 2(I - A)^{-1} - I.$$

(In Section C.1.4, the Cayley transform C^+ was introduced for accretive operators. Here we consider $C^-(A)$ as $-A$ is likely accretive, at least for stable semigroups on a Hilbert space). This operator is bounded if $\alpha_0(S) < 1$ (see the comments at the end of Subsection 2.2.1).

In the Hilbert space case, Theorem C.1.4.2 shows that $\operatorname{Re} A \leq 0$ if and only if $\|C^-(A)\| \leq 1$; it can also be shown that $S(\cdot)$ is a *contraction semigroup* (i.e., $\|S(t)\| \leq 1$ for all $t \geq 0$) if $\operatorname{Re}(Ax, x) \leq 0$ and $\operatorname{Re}(A^*x, x) \leq 0$ for every $x \in \operatorname{Dom}(A)$, see references in Section 2.6 below; for a necessary and sufficient condition see the comments in Section 1.7 above.

Definition 2.2.5 and the theorem of 2.2.3 (or Corollary 2.2.6) immediately give the following result (note that we may assume that $\alpha_0(S(\cdot)) < 1$).

2.2.7. COROLLARY. *If $C^-(A)$ is the Cayley transform of A, then*

$$\begin{aligned}\mu(S(\cdot)) &= \min\{\dim \mathcal{BU} : \mathcal{BU} \text{ is an approximate control subspace }\}\\ &= \mu(C^-(A)).\end{aligned}$$

□

2.3. Multiplicity of the Spectrum

Let $T : X \longrightarrow X$ be a bounded operator on a Banach space X and let

$$\mu(T) = \min\{\dim C : \ C \in \mathrm{Cyc}(T)\}$$

be its multicyclicity (multiplicity), where, as before,

$$\mathrm{Cyc}(T) = \{C : \mathrm{span}(T^n C : n \geq 0) = X\}$$

denotes the collection of cyclic subspaces of T.

In this Section, we start with some general rules for the multicyclicity $\mu(T)$ valid for all operators and explain the behaviour of $\mu(T)$ with respect to customary operations on operators. Then we point out how to compute the multicyclicity for normal operators and for model operators with defect indices one. Recall that some other multicyclicity calculations, mostly related to the shift operators, are contained in A.2.8.2 and A.4.8.1-A.4.8.6 (Volume 1).

2.3.1. General properties of the multiplicity of the spectrum. *(1) The multicyclicity $\mu(T)$ of an operator depends exclusively on its subspace lattice*

$$\mathrm{Lat}\, T = \{E \subset X : TE \subset E\}$$

and not on T itself. In particular, $\mu(T_1) = \mu(T_2)$ whenever $\mathrm{Lat}\, T_1 = \mathrm{Lat}\, T_2$.

Indeed, $C \in \mathrm{Cyc}(T)$ if and only if there is no invariant subspace $E \in \mathrm{Lat}\, T$, $E \neq X$, such that $C \subset E$. □

(2) If T_1 is similar to T_2, that is $T_1 = V^{-1}T_2V$ for some isomorphism V, then $\mu(T_1) = \mu(T_2)$.

Indeed, $C \in \mathrm{Cyc}(T_1) \Longleftrightarrow VC \in \mathrm{Cyc}(T_2)$ and $\dim VC = \dim C$. □

This property can be strengthened as follows.

(3) If T_1 is quasi-similar to T_2 then $\mu(T_1) = \mu(T_2)$. (Recall that quasi-similarity means the existence of di-formations *U and V, that is, injective operators with dense ranges, such that $UT_1 = T_2U$ and $T_1V = VT_2$, see C.1.5.3 for more details).*

Indeed, if $UT_1 = T_2U$, then by induction $UT_1{}^n = T_2{}^nU$, $n \geq 0$. This shows that if $C \in \mathrm{Cyc}(T_1)$, then $UC \in \mathrm{Cyc}(T_2)$, and therefore $\mu(T_2) \leq \mu(T_1)$. By symmetry, $\mu(T_1) \leq \mu(T_2)$, and hence $\mu(T_1) = \mu(T_2)$.

□

EXAMPLE. Let $(f_n)_{n\geq 1}$ be a minimal family in a Hilbert space X with $\mathrm{span}(f_n : n \geq 1) = X$ such that the corresponding coordinate functionals $(g_n)_{n\geq 1}$ are also complete: $\mathrm{span}(g_k : k \geq 1) = X$. By definition we have $\langle f_n, g_k \rangle = \delta_{nk}$. Let $Tf_n = \lambda_n f_n$, $\lambda_i \neq \lambda_j$. Let $(e_n)_{n\geq 1}$ be an orthonormal basis of X. If $\mu_n \neq 0$ tends sufficiently fast to 0, then the operator given by

$$U : \sum_{n\geq 1} a_n f_n \longrightarrow \sum_{n\geq 1} a_n \mu_n e_n,$$

is a continuous di-formation (see C.1.5.3). Define another operator $N : X \longrightarrow X$ by $Ne_n = \lambda_n e_n$, $n \geq 1$, which is a diagonal normal operator. Then $UT = NU$. Let $(\nu_n)_{n\geq 1}$, $\nu_n \neq 0$, ν_n, be another sequence tending sufficiently fast to 0. Then

$$V : \sum_{n\geq 1} a_n e_n \longrightarrow \sum_{n\geq 1} a_n \nu_n f_n$$

is also a continuous di-formation and $TV = VN$. We conclude that T is quasi-similar to the normal diagonal operator N with simple spectrum, and hence $\mu(T) = \mu(N) = 1$ (see Section 2.3.2 below). □

(4) Let $TE \subset E$ and define the quotient operator $T/E : X/E \longrightarrow X/E$ by

$$x + E \longmapsto Tx + E$$

(note that if X is a Hilbert space, then $X/E = E^\perp$ and $T/E = P_{E^\perp} T|E^\perp$ is the compression *of T to $E^\perp$). Then*

$$\mu(T/E) \leq \mu(T) \leq \mu(T/E) + \mu(T|E).$$

Indeed, let $\pi : X \longrightarrow X/E$ be the canonical quotient mapping. If C is cyclic for T, then πC is cyclic for T/E. As $\dim \pi C \leq \dim C$ this gives the first inequality.

For the second one, let $C_1 \in \mathrm{Cyc}(T|E)$ with $\dim C_1 = \mu(T|E)$ and let $C'_2 \in \mathrm{Cyc}(T/E)$ with $\dim C'_2 = \mu(T/E)$. Take now a subspace C_2 in X such that $\pi C_2 = C'_2$, $\dim C_2 = \dim C'_2$. Then it is easy to see that $C = C_1 + C_2$ is cyclic for T. Moreover $\dim C \leq \mu(T/E) + \mu(T|E)$, and the result follows. □

(5) Always,

$$\begin{aligned} \mu(T) &\geq \dim \mathrm{Ker}\, T^* \\ \mu(T) &= \mu(T - \lambda I), \quad \lambda \in \mathbb{C}, \end{aligned}$$

and hence

$$\mu(T) \geq \sup_{\lambda\in\mathbb{C}} \dim \mathrm{Ker}(T^* - \lambda I).$$

Indeed, let $E = \overline{TX}$. Then $T/E = \mathbb{O}$, and hence $\mu(T/E) = \dim(X/E)$. On the other hand, as $\overline{TX} = (\mathrm{Ker}\, T^*)^\perp$ we get $(X/\overline{TX})^* = (TX)^\perp = \mathrm{Ker}\, T^*$. Now we can apply (4) to obtain

$$\mu(T) \geq \mu(T/E) = \dim(X/\overline{TX}) = \dim \mathrm{Ker}\, T^*.$$

□

(6) Let $T : X \longrightarrow X$ and X be of finite dimension. Then the inequality of (5) becomes an equality:

$$\mu(T) = \sup_{\lambda\in\mathbb{C}} \dim \mathrm{Ker}(T - \lambda I).$$

The following points (a)-(f) represent a sketch of the proof of this equality; the details are left to the reader.

(a) Let $X = \sum_{\lambda\in\sigma(T)} \dot{+} X_\lambda$ be the decomposition of X into a direct sum of invariant subspaces (the *rootspaces*, or *generalized eigenspaces*) such that $\sigma(T|X_\lambda) = \{\lambda\}$, $\lambda \in \sigma(T)$.

(b) The polynomial calculus shows that every invariant subspace of E, $TE \subset E$, splits into a direct sum $E = \sum_{\lambda\in\sigma(T)} \dot{+} E_\lambda$ with $E_\lambda \subset X_\lambda$ and $TE_\lambda \subset E_\lambda$. See also A.2.8.2 (Volume 1) about the splitting property of $\operatorname{Lat} T$.

COROLLARY. $\mu(T) = \sup_{\lambda\in\sigma(T)} \mu(T|X_\lambda)$.

Indeed, setting $E_\lambda = \sum_{\mu\in\sigma(T)\setminus\{\lambda\}} \dot{+} X_\mu$, we get by (4) the estimate $\mu(T) \geq \sup_\lambda \mu(T/E_\lambda) = \sup_\lambda \mu(T|X_\lambda)$. On the other hand, (b) and the polynomial calculus yield $\mu(T) \leq \sup_\lambda \mu(T|X_\lambda)$. □

(c) We have $T|X_\lambda = \sum_{k=1}^{d} \dot{+} J_{\lambda,k}$ where $d = \dim \operatorname{Ker}(T - \lambda I)$ and $J_{\lambda,k}$ is a Jordan block of a certain dimension.

(d) $\mu(J_{\lambda,k}) = 1$ (cyclic operator).

(e) Using (7) below we then obtain $\mu(T|X_\lambda) \leq d$ and, by (5), $\mu(T|X_\lambda) = d$.

(f) We conclude that $\mu(T) = \sup_\lambda \mu(T|X_\lambda) = \sup_\lambda \dim \operatorname{Ker}(T - \lambda I)$. □

(7) If $T = T_1 \dot{+} T_2$ is a direct sum of operators defined on a direct sum of spaces $X = X_1 \dot{+} X_2$, then

$$\max(\mu(T_1), \mu(T_2)) \leq \mu(T) \leq \mu(T_1) + \mu(T_2).$$

Indeed, this is an immediate consequence of (4) and the fact that T/X_1 is similar to the restriction $T|X_2$ (respectively T/X_2 is similar to $T|X_1$). □

We have the following generalization of the corollary from (6)(b).

(8) If the lattice of $T = T_1 \dot{+} T_2$ splits into $\operatorname{Lat} T = \operatorname{Lat} T_1 \dot{+} \operatorname{Lat} T_2$, *then*

$$\mu(T) = \max(\mu(T_1), \mu(T_2)).$$

□

2.3.2. The multiplicity of the spectrum of a normal operator. A *normal* operator $T : H \longrightarrow H$ on a Hilbert space H is defined by the property $TT^* = T^*T$. In particular, self-adjoint ($T = T^*$) and unitary ($TT^* = T^*T = I$) operators are normal. The *von Neumann spectral theorem* (see C.1.5.1(k)) states that a normal operator T is unitarily equivalent to the multiplication operator $M : f \longmapsto tf(t)$ on a vector valued L^2 space $\mathcal{H}$ given by

$$\mathcal{H} := \int \oplus H(t)\, d\nu(t) = \{g \in L^2(\sigma(T), \nu; H) : g(t) \in H(t) \text{ a.e. } \nu\};$$

here ν is a Borel measure on $\sigma(T)$ and $H(t) \subset H$ is a measurable family of subspaces (that is, $t \longmapsto P_{H(t)}x$ is measurable for every $x \in H$, $P_{H(t)}$ being the orthogonal projection onto $H(t)$). The dimension $\dim H(t) = d_T(t)$ is called the *local spectral multiplicity* of T. See C.1.5.1 for more about the spectral theorem.

2.3.3. THEOREM (J. Bram, 1955). *If T is a normal operator, then*

$$\mu(T) = \text{ess sup}_\nu d_T.$$

PROOF. First, we show that $\mu(T) \geq \text{ess sup}_\nu d_T$. Let $k \geq 1$ be such that $\nu(\sigma) > 0$, where $\sigma = \{t : d_T(t) = k\}$. Since $t \longrightarrow P_{H(t)}$ is measurable, we can choose an orthonormal basis $\{\delta_1(t), ..., \delta_k(t)\}$ of $H(t)$, $t \in \sigma$, such that $t \longmapsto \delta_i(t)$, $1 \leq i \leq k$, are measurable, and we shall extend them trivially outside σ: $\delta_i(t) = 0$, $t \notin \sigma$. Suppose further that $C \in \text{Cyc}(T) \subset \mathcal{H}$ is of dimension m with a basis $\{f_1, \dots, f_m\}$. Then, there exist polynomials $p_{i,n,l}$ such that for $1 \leq l \leq k$ we have

$$F_{n,l} := \sum_{i=1}^{m} p_{i,n,l}(T) f_i \xrightarrow{L^2} \delta_l, \quad \text{as } n \to \infty,$$

where $p(T)f = p(t)f(t)$, $t \in \sigma$.

By definition $f_i(t) \in H(t)$, $t \in \sigma$, and the element $(F_{n,l}(t))_{l=1}^k$ can be considered as a $k \times k$ matrix in the basis $(\delta_j(t))_{j=1}^k$. Replacing eventually $((F_{n,l}(t))_{l=1}^k)_{n \geq 1}$ by a subsequence, we may assume that

$$(F_{n,l}(t))_{l=1}^k \longrightarrow (\delta_l(t))_{l=1}^k$$

a.e. on σ when $n \to \infty$. Then

$$\lim_n \det(F_{n,l}(t))_{l=1}^k = \det(\delta_l(t))_{l=1}^k \equiv 1 \text{ a.e. on } \sigma.$$

This means that $\det(F_{n,l}(t)) \neq 0$ provided n is sufficiently large. As $F_{n,l}(t) \in \mathcal{L}in(f_1(t), ..., f_m(t))$, we get $m \geq k$. Hence $\mu(T) \geq \text{ess sup}_\nu d_T$.

For the converse estimate $\mu(T) \leq \text{ess sup}_\nu d_T$, suppose that $d = \text{ess sup}_\nu d_T < \infty$ (otherwise there is nothing to prove). Then we may decompose T into the sum

$$T = \sum_{i=1}^{d} \oplus T_i,$$

where $T_i f = t f(t)$, $f \in L^2(\tau_i, \nu)$ and τ_i are suitable Borel subsets of the spectrum $\sigma(T)$. Note that $L^2(\tau_i, \nu)$ is now a space of scalar valued functions. It remains to prove that for i, $1 \leq i \leq d$, we have $\mu(T_i) = 1$. Actually, this has already been shown in A.4.8.3 (Volume 1). Now, point (7) of Subsection 2.3.1 gives $\mu(T) \leq \sum_{i=1}^d \mu(T_i) = d$. □

2.3.4. The multiplicity of the spectrum of a scalar model operator. Here we compute the multicyclicity of a completely nonunitary contraction having a scalar characteristic function. To this end, we exploit the function model theory developed in Part C. Namely, we use the di-formation techniques of C.1.5.3 connecting a model operator and the residual part of its minimal unitary dilation (see C.1.5.2 above). Another proof of the cyclicity theorem 2.3.5 is already contained in C.2.5.7(k) above.

Now, let us recall that a c.n.u. contraction T with one dimensional defects $\partial_T = \partial_{T^*} = 1$ and the characteristic function $\Theta = \Theta_T$ is unitarily equivalent to its function model $M_\Theta : \mathcal{K}_\Theta \longrightarrow \mathcal{K}_\Theta$, where

$$\mathcal{K}_\Theta = \begin{pmatrix} H^2 \\ \text{clos}\, \Delta L^2 \end{pmatrix} \ominus \begin{pmatrix} \Theta \\ \Delta \end{pmatrix} H^2,$$

and where $L^2 = L^2(\mathbb{T})$, H^2 is the Hardy space and

$$M_\Theta f = P_\Theta z f, \quad f \in \mathcal{K}_\Theta.$$

The general case $0 \le \partial_T, \partial_{T^*} < \infty$ is commented on in Section 2.6.

2.3.5. THEOREM. *Let M_Θ be a scalar model operator.*
(1) If $\Theta \not\equiv 0$, then $\mu(M_\Theta) = 1$.
(2) If $\Theta \equiv 0$ ($\Delta \equiv 1$), then $\mathcal{K}_\Theta = H^2 \oplus H^2_-$, where $H^2_- = L^2(\mathbb{T}) \ominus H^2$, and

$$M_\Theta f = P_\Theta z f = (zf_1, P_- zf_2) = Sf_1 \oplus S^* f_2, \quad f_1 \in H^2, f_2 \in H^2_-.$$

Here P_- is the orthogonal projection onto H^2_- and $Sf = zf$, $f \in H^2$, is the shift operator on H^2. In this case, $\mu(M_\Theta) = 2$.

PROOF. (1) As was shown in 2.3.1(3), $\mu(M_\Theta) \le \mu(T)$ for every operator T whose d-formation is M_Θ; the latter property means that there exists an operator X satisfying $M_\Theta X = XT$ and $\operatorname{Ker} X^* = \{0\}$. Therefore, to prove the theorem, it suffices to exhibit such an operator T with simple spectrum (i.e., with $\mu(T) = 1$) and an operator X with the mentioned properties.

To this end, we consider the inner-outer factorization $\Theta = \Theta_{inn}\Theta_{out}$ of Θ and the model space $\mathcal{K}_{\Theta_{inn}} = H^2 \ominus \Theta_{inn} H^2$ corresponding to Θ_{inn}. Further, let $Y : \mathcal{H} \longrightarrow \mathcal{H}$ be an operator of multiplication $F \longmapsto YF$ on the space $\mathcal{H}$ of the minimal isometric dilation of M_Θ,

$$\mathcal{H} = \begin{pmatrix} H^2 \\ \operatorname{clos} \Delta L^2 \end{pmatrix} = H^2 \oplus (\operatorname{clos} \Delta L^2),$$

defined by the following matrix valued function (which we also denote by Y)

$$Y = \begin{pmatrix} \Theta_{out} & \mathbb{O} \\ \Delta \Theta^*_{inn} & I \end{pmatrix}.$$

Now, let $\mathcal{K} \subset \mathcal{H}$ be the subspace

$$\mathcal{K} = \begin{pmatrix} \mathcal{K}_{\Theta_{inn}} \\ \operatorname{clos} \Delta L^2 \end{pmatrix} = \mathcal{K}_{\Theta_{inn}} \oplus (\operatorname{clos} \Delta L^2)$$

and let $T : \mathcal{K} \longrightarrow \mathcal{K}$ be defined as

$$T = M_{\Theta_{inn}} \oplus z = P_{\mathcal{K}} z | \mathcal{K},$$

where z stands for the operator of multiplication by the independent variable on the corresponding function space, $F \longmapsto zF$. We will show that the operator $M_\Theta = P_\Theta z | \mathcal{K}_\Theta$ is a d-formation of T and that $\mu(T) = 1$.

To this end, we set

$$X = P_\Theta Y | \mathcal{K} : \mathcal{K} \longrightarrow \mathcal{K}_\Theta,$$

and check that $XT = M_\Theta X$, and that the range $X\mathcal{K}$ is dense in $\mathcal{K}_\Theta$. Indeed, since

$$Y\mathcal{K}^\perp = Y \begin{pmatrix} \Theta_{inn} H^2 \\ \{0\} \end{pmatrix} = \begin{pmatrix} \Theta \\ \Delta \end{pmatrix} H^2 = (\mathcal{K}_\Theta)^\perp,$$

we have

$$\begin{aligned} XTF &= P_\Theta Y P_{\mathcal{K}} zF = P_\Theta Y (P_{\mathcal{K}} + P_{\mathcal{K}}^\perp) zF = P_\Theta Y zF = P_\Theta z Y F \\ &= P_\Theta z (P_\Theta + P_\Theta^\perp) Y F = P_\Theta z P_\Theta Y F = M_\Theta X F \end{aligned}$$

for every $F \in \mathcal{K}$.

For the density of the range of X, consider $X^* = P_{\mathcal{K}} Y^* | \mathcal{K}_\Theta$ with

$$Y^* = \begin{pmatrix} \Theta^*_{out} & \Theta_{inn}\Delta \\ 0 & I \end{pmatrix}.$$

If $(f,g)^{col} \in \mathcal{K}_\Theta$, where $(x,y)^{col}$ denotes the column vector with entries x, y, and $X^*(f,g)^{col} = 0$, then $\Theta^* f + \Delta g \in H^2_-$ and

$$Y^*(f,g)^{col} = (\Theta^*_{out} f + \Theta_{inn}\Delta g, g)^{col} \in \mathcal{K}^\perp = \Theta_{inn} H^2 \oplus \{0\}.$$

Hence, $g = 0$ and thus $\Theta^* f \in H^2_-$ and $\Theta^*_{out} f \in \Theta_{inn} H^2$. This implies that $\Theta^* f = 0$, and $f = 0$. Therefore, $\operatorname{Ker} X^* = \{0\}$, which means that $\operatorname{clos}(X\mathcal{K}) = \mathcal{K}_\Theta$.

Now, let $F = f \oplus g \in \mathcal{K}$ where $f = P_{\Theta_{inn}} 1 = 1 - \overline{\Theta}_{inn}(0)\Theta_{inn}$ and $g \in \operatorname{clos}(\Delta L^2)$ is a z-cyclic function, that is, $\operatorname{span}(z^n g : n \geq 0) = \operatorname{clos}(\Delta L^2)$ (see A.4.1.2, A.4.1.3, Volume 1, for the existence). Since $\Theta_{inn}(M_{\Theta_{inn}}) = 0$, we obtain

$$\Theta_{inn}(T)(f \oplus g) = 0 \oplus (\Theta_{inn} g) \in \mathcal{K},$$

and hence

$$\operatorname{span}(T^n F : n \geq 0) \supset \{0\} \oplus \operatorname{span}(z^n \Theta_{inn} g : n \geq 0) = \{0\} \oplus \operatorname{clos}(\Delta L^2).$$

This implies that $(f \oplus 0) = (f \oplus g) + (0 \oplus -g) \in \operatorname{span}(T^n F : n \geq 0)$. Since $M^n_{\Theta_{inn}} P_{\Theta_{inn}} 1 = P_{\Theta_{inn}} z^n$ for all $n \geq 0$, the function f is cyclic for $M_{\Theta_{inn}}$. Thus, $\operatorname{span}(T^n F : n \geq 0) \supset \mathcal{K}_{\Theta_{inn}} \oplus \{0\}$, and, finally,

$$\operatorname{span}(T^n F : n \geq 0) = \mathcal{K}_{\Theta_{inn}} \oplus \operatorname{clos}(\Delta L^2) = \mathcal{K},$$

which means that F is a cyclic vector for T, and $\mu(T) = 1$.

(2) The second assertion is a simple exercise: for every $f \oplus g \in H^2 \oplus H^2_-$ the vector $g_* \oplus (-f_*)$ where $g_*(z) = \bar{g}(\bar{z})$, $|z| = 1$, is orthogonal to $S^n f \oplus S^{*n} g$, $n \geq 0$. Therefore, $1 < \mu(M_\Theta) \leq 2$, which means that $\mu(M_\Theta) = 2$. □

2.3.6. Special cases. (1) Inner and outer characteristic functions. These special cases are much easier to prove than the general one. Indeed, for an inner function Θ, the obvious cyclic function $f = P_\Theta 1 = 1 - \overline{\Theta(0)}\Theta$ is exhibited in the proof above. For an outer function Θ, the space $\mathcal{K}$ is reduced to the residual part of the minimal unitary dilation, $\mathcal{K} = \{0\} \oplus \operatorname{clos}(\Delta L^2)$ (see C.1.5.2 for this), and it is easy to check that the restricted projection $XP_\Theta|\mathcal{K}$ is a di-formation intertwining the unitary operator $T = z : \mathcal{K} \longrightarrow \mathcal{K}$ and M_Θ; since $\mu(T) = 1$, the result follows.

(2) Sturm–Liouville operators revisited.
Let

$$Ay = -y'' + Vy \text{ on } [0, \infty[, \quad y \in W^2_2,$$

where W^2_2 is the usual *Sobolev space* $W^2_2 = \{f \in L^2([0,\infty)) : f, f'$ are absolutely continuous and $f'' \in L^2([0,\infty[)\}$. If the potential V is real, then the operator A is symmetric on $\overset{0}{W}{}^2_2$ and has the indices (0,0), (1,1) or (2,2) depending on the behaviour of V at the limit points 0 and ∞. If, for instance, V is continuous and $V(x) \geq -M(x)$ for a sufficiently "regular" function $M > 0$ satisfying $\int_0^\infty M(x)^{-1/2}\,dx = \infty$, then the indices are (1,1) (the case of *Weyl's limit point*).

Assum now that we have the initial condition

$$y'(0) = hy(0),$$

where $h \in \mathbb{C}$, $\operatorname{Im} h > 0$. Then the associated operator A extends to a *dissipative* operator A_h, that is, $\operatorname{Im}(A_h f, f) \geq 0$ for $f \in \mathcal{D}(A_h))$. Its Cayley transform

$$(A_h - iI)(A_h + iI)^{-1} = T_h$$

is a contraction with scalar valued characteristic function Θ_{T_h}, which can be expressed in terms of the Weyl-function of A_h, $\lambda \longmapsto m(\lambda)$ (see C.1.4.13 and C.1.4.14).

Since $\Theta_{T_h} \neq 0$, we get the following result.

2.3.7. COROLLARY. *Let $S(\cdot)$ be the contracting semigroup associated with iA_h ($\mathrm{Re}(iA_h) \leq \mathbb{O}$). Then*

$$\mu(S(\cdot)) = \mu(T_h) = 1.$$

□

2.4. The Minimal Dimension of Constrained (Realizable) Control

In this Section, we consider yet another problem of dimensional optimization of control subspaces. Roughly speaking, we try to find the best correction of a possibly unfortunate first choice of the control subspace of a given system. Mathematically, this leads to a new characteristic $\mathrm{DISC}(T)$, $\mathrm{DISC}(T) \geq \mu(T)$, that is much more subtle than the multicyclicity number $\mu(T)$ studied in the above Section. We present here a brief introduction to the theory of DISC's; see Section 2.6 for further information and references.

2.4.1. A motivation for the DISC. Let

$$x' = Ax + Bu \tag{2.4.1}$$

be a dynamical system in X, $A : X \longrightarrow X$, $B : \mathcal{U} \longrightarrow X$. We are still concerned with the optimization problem 2.1.1

$$\min\{\dim B\mathcal{U} : \text{ the system (2.4.1) is approximately controllable }\},$$

which we now consider for a more realistic situation. We assume that (A, B) is ACO for a certain fixed control subspace $C = B\mathcal{U}$, $\dim C < \infty$, which was chosen for some technical requirements, or just randomly. This space can be very big, $\dim C = N >> \mu(S(\cdot))$, $\mu(S(\cdot))$ being the minimal control dimension. We are now interested in the problem

$$\min\{\dim C' : C' \text{ is an approximate control subspace and } C' \subset C\}. \tag{2.4.2}$$

For the worst initial choice of C, we arrive to the following concept.

2.4.2. DEFINITION. For the semigroup $(S(\cdot))$ associated with a system (A, B) we define the *Dimension of Input Subspace of Control*, abbreviated by DISC, by

$$\mathrm{DISC}(S(\cdot)) = \sup_{\substack{C \in \mathrm{Cyc}(S(\cdot)) \\ \dim C < \infty}} \min\{\dim C' : C' \subset C, C' \in \mathrm{Cyc}(S(\cdot))\}.$$

Given an operator $A : X \longrightarrow X$, the $\mathrm{DISC}(A)$ is defined similarly.

2.4.3. Basic properties of the DISC. *(1) As for the multiplicity μ we get*

$$\mathrm{DISC}(S(\cdot)) = \mathrm{DISC}(R_\lambda(A))$$

for every $\lambda \in \mathbb{C}$ with $\mathrm{Re}\,\lambda > \alpha_0(S)$ ($\alpha_0(S)$ is the growth bound *of the semigroup). Likewise* $\mathrm{DISC}(S(\cdot)) = \mathrm{DISC}(C^-(A))$, *where $C^-(A)$ is the Cayley transform of A introduced after Corollary 2.2.6.*

(2) For a bounded operator $T \in L(X)$, the DISC *is uniquely determined by* $\mathrm{Lat}\, T$.

Indeed, the $\mathrm{DISC}(T)$ is defined only in terms of the set $\mathrm{Cyc}(T)$. □

In particular, if $\operatorname{alg} T$ denotes the algebra generated by T in the weak operator topology, and if $T_1, T_2 \in L(X)$ satisfy $\operatorname{alg} T_1 = \operatorname{alg} T_2$, then we first get $\operatorname{Lat} T_1 = \operatorname{Lat} T_2$ and then $\operatorname{Cyc}(T_1) = \operatorname{Cyc}(T_2)$. Consequently $\mu(T_1) = \mu(T_2)$ and $\operatorname{DISC}(T_1) = \operatorname{DISC}(T_2)$.

(3) Clearly, $\operatorname{DISC}(T) \geq \mu(T)$.

Later on, we shall give some examples where $\operatorname{DISC}(T) > \mu(T)$.

(4) If T_1 and T_2 are similar, then $\operatorname{DISC}(T_1) = \operatorname{DISC}(T_2)$.

The question whether this remains true for quasi-similar operators seems to be open.

(5) $\operatorname{DISC}(T_1 \dot{+} T_2) \geq \max(\operatorname{DISC}(T_1), \operatorname{DISC}(T_2))$, *where $T_1 \dot{+} T_2$ is the direct sum of T_1 and T_2.*

We leave this (simple) exercise to the reader.

(6) If $E \in \operatorname{Lat} T$ and $\mu(T|E) < \infty$, then

$$\operatorname{DISC}(T/E) \leq \operatorname{DISC}(T).$$

The proof of the corresponding property for $\mu(T)$ carries over to this situation.

(7) $\operatorname{DISC}(T_1 \dot{+} T_2) \leq \operatorname{DISC}(T_1) + \operatorname{DISC}(T_2)$ *if the set* $\operatorname{Cyc}(T_1 \dot{+} T_2)$ *splits. The last property means that a subspace $E \subset X_1 \dot{+} X_2$ is cyclic for $T_1 \dot{+} T_2$ if and only if the projections $P_1 E$ and $P_2 E$ are cyclic for T_1 and T_2, respectively.*

See A.2.8.2 and A.4.9 (Volume 1) for more comments about the splitting properties.

(8) $\operatorname{DISC}(T_1 \dot{+} T_2) = \max(\operatorname{DISC}(T_1), \operatorname{DISC}(T_2))$ *if* $\operatorname{Cyc}(T_1 \dot{+} T_2)$ *splits and* T_1, T_2 have many cyclic subspaces. *The last property means that for every finite dimensional cyclic subspace $C \in \operatorname{Cyc}(T)$, verifying $\operatorname{DISC}(T) = d \leq \dim C$, almost every $C' \subset C$ with $\dim C' = d$ (in the sense of Lebesgue measure on the Grassman manifold $G_d(C)$ of all d-dimensional subspaces) is cyclic.*

2.4.4. Examples of DISC**'s, or a DISCO'theca.** Below, we sketch some ideas that are useful for computations of DISC's. We believe that some details of these computations reveal the nature of this characteristic better than a formally written list of DISC's properties does. Avoiding technical complications, we often give a complete treatment for a leading special case and refer the reader to the references cited in Section 2.6 below for an extensive exposition .

2.4.5. Operators on finite dimensional spaces. It is not hard to see that if $\dim X < \infty$ then

$$\operatorname{DISC}(T) = \mu(T) = \sup_\lambda \dim \operatorname{Ker}(T - \lambda I).$$

2.4.6. Complete compact operators. Let $T : X \longrightarrow X$ be a compact operator having a complete family of eigen- and rootspaces:

$$X = \operatorname{span}(\operatorname{Ker}(T - \lambda I)^n : n \geq 1, \lambda \in \mathbb{C} \setminus \{0\}).$$

Then

$$\operatorname{DISC}(T) = \mu(T) = \sup_\lambda \dim \operatorname{Ker}(T - \lambda I).$$

We refer the reader to the comments and references in Section 2.6.

2.4.7. THEOREM (DISC of normal operators). *Let $T \in L(H)$ where H is a Hilbert space.*

(a) If $T = T^*$, then $\mathrm{DISC}(T) = \mu(T)$ ($=$ ess $\sup_\nu d_T(\cdot)$, *where $d_T(\cdot)$ is the above mentioned local spectral multiplicity of T).*

(b) Suppose that $T = U$ is a unitary operator, $U = U_s \oplus U_a$ its decomposition into U_s, the singular, and U_a, the absolutely continuous part with respect to the Lebesgue measure m on $\mathbb{T}$. Let d_{U_a} be the local spectral multiplicity of U_a and let $\underline{\mu}(U_a) = \operatorname{ess\,inf}_m d_{U_a}(\cdot)$ *be its* lower multiplicity. *Then*

$$\mathrm{DISC}(U) = \max\{\mu(U_s), \mu(U_a) + \underline{\mu}(U_a)\}. \tag{2.4.3}$$

(c) Suppose that $T = N$ is normal.

(i) If N is completely reducible *(that is* $\operatorname{Lat} N = \operatorname{Lat} N^*$*), then*

$$\mathrm{DISC}(N) = \mu(N).$$

Moreover if $N' = N|E$ for some $E \in \operatorname{Lat} N$, then

$$\mathrm{DISC}(N') = \mu(N').$$

(ii) Conversely, if for every $N' = N|E$, $E \in \operatorname{Lat} N$ we have $\mathrm{DISC}(N') = \mu(N')$, *then* $\operatorname{Lat} N = \operatorname{Lat} N^*$.

To outline some of the main ingredients of this result, we present here a proof of the assertions (a) and (b) in a special case. For the general case and for the proof of (c) we refer the reader to the comments and references in Section 2.6.

We begin with two general lemmas.

2.4.8. LEMMA. *Let $T : H \longrightarrow H$ be an operator on a Hilbert space H.*

(1) If C is a T-cyclic subspace and P a non-trivial projection commuting with T, $PT = TP$, $P \neq \mathbb{O}$, then $PC \neq \{\mathbf{0}\}$.

(2) Conversely, if T is completely reducible and $PC \neq \{\mathbf{0}\}$ for every non-zero orthogonal projection P commuting with T, then C is cyclic for T.

PROOF. (1) Since $T^n P = PT^n$ for $n \geq 1$, we have $PH = P\operatorname{span}(T^n C : n \geq 0) = \operatorname{span}(T^n PC : n \geq 0)$, and hence $PC \neq \{0\}$.

(2) Let

$$E = H \ominus \operatorname{span}(T^n C : n \geq 0),$$

and let $P = P_E$ be the orthogonal projection onto E. By assumption, E is reducing. Consequently $TP = PT$. But $PC = \{\mathbf{0}\}$ by construction, and hence $P = \mathbb{O}$. Therefore, $C \in \operatorname{Cyc}(T)$. □

2.4.9. LEMMA. *Let ν be a σ-finite measure, and let $C \subset L^1(\nu)$ with $\dim C < \infty$. Then, for almost every function $f \in C$ (with respect to Lebesgue measure on C), the following implication holds:*

$$\int_\sigma |f|\, d\mu = 0 \Longrightarrow \left\{\int_\sigma |g|\, d\mu = 0 \text{ for every } g \in C\right\}$$

for every measurable set σ.

PROOF. Let $f_i \in C$, $i = 1, ..., n$, be such that $C = \mathcal{L}\mathrm{in}(f_1, ..., f_n)$. We can assume that the functions f_i are defined everywhere (by choosing a suitable representant of the equivalence class $[f_i] \in L^2(\nu)$). Let

$$e_i := \operatorname{supp} f_i = \{t : f_i(t) \neq 0\}.$$

It suffices to show that

$$\operatorname{supp}\left(\sum_{i=1}^{s}\alpha_i f_i\right)=\left(\bigcup_{i=1}^{s} e_i\right)\setminus W(\overline{\alpha}) \tag{2.4.4}$$

for every s, $1\le s\le n$ and for almost every $\overline{\alpha}=(\alpha_1,...,\alpha_s)\in\mathbb{C}^s$, where $W(\overline{\alpha})$ is an appropriate set with $\nu W(\overline{\alpha})=0$.

The claim is obviously correct for $s=1$.

Assume now that the claim is true for $s\le n-1$, and consider a point $\overline{\alpha}\in\mathbb{C}^s$ satisfying (2.4.4). Set

$$Y(\alpha)=\left\{t:\sum_{i=1}^{s}\alpha_i f_i(t)+\alpha f_{s+1}(t)=0\right\}$$

for $\alpha\in\mathbb{C}$. Then the sets $Z(\alpha)=Y(\alpha)\cap\left(\bigcup_{i=1}^{s} e_i\setminus W(\overline{\alpha})\right)$, $\alpha\in\mathbb{C}$, are pairwise disjoint: $Z(\alpha)\cap Z(\beta)=\emptyset$, $\alpha\ne\beta$. As the measure ν is σ-finite, the set $A_{\overline{\alpha}}:=\{\alpha\in\mathbb{C}:\nu Z(\alpha)>0\}$ is at most countable. Hence, for every $\alpha\ne 0$, we have

$$\begin{aligned}\left(\bigcup_{i=1}^{s+1} e_i\right)\setminus\operatorname{supp}\left(\sum_{i=1}^{s}\alpha_i f_i+\alpha f_{s+1}\right)&=\left(\bigcup_{i=1}^{s} e_i\right)\cap Y(\alpha)\\&=Z(\alpha)\cup(W(\overline{\alpha})\cap Y(\alpha))=:W(\overline{\alpha},\alpha).\end{aligned}$$

Then

$$\operatorname{supp}\left(\sum_{i=1}^{s}\alpha_i f_i+\alpha f_{s+1}\right)=\left(\bigcup_{i=1}^{s+1} e_i\right)\setminus W(\overline{\alpha},\alpha),$$

and $\nu W(\overline{\alpha},\alpha)=0$ for $\alpha\in\mathbb{C}\setminus(A_{\overline{\alpha}}\cup\{0\})$ and for almost every $\overline{\alpha}\in\mathbb{C}^s$. □

2.4.10. Proof of 2.4.7(a) for the case $\mu(T)=1$. In this case, by the spectral theorem, the operator T is unitarily equivalent to the multiplication operator $f\longmapsto zf$ in $L^2(\nu)$, where ν is a compactly supported Borel measure on $\mathbb{R}$.

Let us prove that $\mathrm{DISC}(T)=1$. To this end, let C be a cyclic subspace of finite dimension. Lemma 2.4.8 states that $PC\ne\{\mathbf{0}\}$ provided that P is a non-zero orthogonal projection commuting with T: $PT=TP$. In particular, if $Pf=\chi_\sigma f$, $f\in L^2(\nu)$, for a set $\sigma\subset\mathbb{R}$ such that $\chi_\sigma\ne 0$ in $L^\infty(\nu)$, then $\chi_\sigma C\ne\{\mathbf{0}\}$. Now, Lemma 2.4.9 implies that there exists a function $f\in C$ such that $\chi_\sigma f\ne\mathbf{0}$ whenever $\sigma\subset\mathbb{R}$ satisfies $\chi_\sigma\ne 0$ in $L^\infty(\nu)$.

Next, observe that T is completely reducible (since $T=T^*$), and every projection commuting with T is of the form $f\longmapsto\chi_\sigma f$ for a suitable σ (see C.1.5.1(o)). Applying Lemma 2.4.8(2), we obtain that f is a cyclic vector of T. Therefore $\mathrm{DISC}(T)=1$. □

2.4.11. Outline of the proof of 2.4.7(b) for the case $\mu(T)=1$. The proof of this case depends on the subsequent five lemmas (2.4.12-2.4.16). Among these, Lemma 2.4.15 is the key point, and we enter into the details of its proof, whereas for Lemmas 2.4.12-2.4.14 we only give brief indications. As Lemma 2.4.16 is used in Lemma 2.4.15 and is of interest in its own right, we also provide a proof for this result. However, we will not assemble these results into a formal proof and refer the reader to comments and references in Section 2.6 for more details.

2.4.12. LEMMA. *With the notation of Theorem 2.4.7 we have*

$$\operatorname{Lat}(U) = \operatorname{Lat}(U_s) \oplus \operatorname{Lat}(U_a),$$

which means that Lat(U) splits; *moreover*

$$\operatorname{DISC}(U) = \max\left(\operatorname{DISC}(U_s), \operatorname{DISC}(U_a)\right).$$

[*Hint*: see A.2.1.1, Volume 1, for details.]

2.4.13. LEMMA. *The operator U_s is completely reducible, and hence* DISC$(U_s) =$ $\mu(U_s)$ $(\leq 1$ *if* $\mu(U) = 1)$.

[*Hint*: use Lemmas 2.4.8 and 2.4.9 as in 2.4.10.]

2.4.14. LEMMA. *Let $\mathcal{S} : L^2(\mathbb{T}) \longrightarrow L^2(\mathbb{T})$, $\mathcal{S}f = zf$, be the two-sided shift. Then the operator U_a is unitarily equivalent to the direct sum $(\sum_{k=1}^{m} \oplus \mathcal{S}) \oplus U_0$ of $m := \underline{\mu}(U_a)$ copies of $\mathcal{S}$, and a unitary completely reducible operator U_0 with absolutely continuous spectrum. If $\mu(U) = 1$, then $\mu(U_a) \leq 1$ and hence either $m = 0$ or $U_0 = \mathbb{O}$.*

[*Hint*: use von Neumann's spectral theorem 1.5.1(k).]

2.4.15. LEMMA. *Let $\mathcal{S}$ be the two-sided shift. Then* DISC$(\mathcal{S}) = 2$. *(whereas $\mu(\mathcal{S}) = 1$).*

Let us explain the roles of Lemmas 2.4.14 and 2.4.15 in the proof of Theorem 2.4.7. To this end, we rewrite the formula (2.4.3) for U_a, $\operatorname{DISC}(U_a) = \mu(U_a) + \underline{\mu}(U_a)$, in the following way: $\operatorname{DISC}(U_a) = 2\underline{\mu} + (\mu - \underline{\mu})$, where we have set $\mu = \mu(U_a)$ and $\underline{\mu} = \underline{\mu}(U_a)$. Lemma 2.4.14 states that U_a contains $\underline{\mu}$ copies of the shift $\mathcal{S}$, and Lemma 2.4.15 says that each one has $\operatorname{DISC}(\mathcal{S}) = 2$ which yields the term $2\underline{\mu}$. Moreover U_0 is completely reducible and hence its contribution to $\operatorname{DISC}(U_a)$ is $\operatorname{DISC}(U_0) = \mu(U_0) = \mu - \underline{\mu}$.

PROOF OF LEMMA 2.4.15. First, we prove that $\operatorname{DISC}(\mathcal{S}) \geq 2$. Set

$$C = \mathcal{L}\mathrm{in}(\chi_{\mathbb{T}_+}, \chi_{\mathbb{T}_-}),$$

where $\mathbb{T}_\pm = \{e^{it} : 0 < \pm t < \pi\}$. Clearly, C is cyclic for $\mathcal{S}$ and two-dimensional. On the other hand, C does not contain any cyclic function, see A.4.1.2 (Volume 1). Thus, $\operatorname{DISC}(\mathcal{S}) \geq 2$.

For the converse estimate $\operatorname{DISC}(\mathcal{S}) \leq 2$, let $C \in \operatorname{Cyc}(\mathcal{S})$, $\dim C < \infty$. Lemma 2.4.8 shows that $\chi_\sigma C \neq \{\mathbf{0}\}$ whenever $\sigma \subset \mathbb{T}$ with $m\sigma > 0$, and Lemma 2.4.9 then gives a function $f \in C$ such that $\chi_\sigma f \neq \mathbf{0}$ for every $\sigma \subset \mathbb{T}$, $m\sigma > 0$. Set

$$E = \operatorname{span}_{L^2(\mathbb{T})}(z^n f : n \geq 0).$$

If $E = L^2(\mathbb{T})$ the proof is finished. If not, $\{\mathbf{0}\} \neq E \neq L^2(\mathbb{T})$, and we can apply the Beurling–Helson theorem A.1.3.2 (Volume 1). Since $\chi_\sigma f \neq \mathbf{0}$ for every set with non-vanishing measure it follows that E is 1-invariant, and hence there is a function $\Theta \in L^\infty(\mathbb{T})$, $|\Theta| = 1$ a.e. on $\mathbb{T}$, such that $E = \Theta H^2$.

Let us now see what happens on the remaining part $L^2/\Theta H^2$. As $E \in \operatorname{Lat}\mathcal{S}$, the quotient operator $\mathcal{S}/E : L^2/E \longrightarrow L^2/E$, $(\mathcal{S}/E)(g + E) = \mathcal{S}g + E$, $g \in L^2$, is well-defined. As before, we denote by $\pi : L^2 \longrightarrow L^2/E$ the canonical quotient mapping. Then the range πC is cyclic for $\mathcal{S}/E$ and $\dim \pi C \leq \dim C < \infty$. Since $E = \Theta H^2$ with $|\Theta| = 1$ a.e. on $\mathbb{T}$, the operator $\mathcal{S}/E$ is unitarily equivalent to $\mathcal{S}/H^2$ defined on L^2/H^2, and the latter is unitarily equivalent to the adjoint operator

$S^* : H^2 \longrightarrow H^2$ where $S = \mathcal{S}|H^2$. Lemma 2.4.16 below states that $\mathrm{DISC}(S^*) = 1$. This means that there is a function $g \in C$ such that πg is cyclic for $\mathcal{S}/E$. Now the subspace $\mathcal{L}\mathrm{in}(f, g) \subset C$ is cyclic for $\mathcal{S}$, and so $\mathrm{DISC}(\mathcal{S}) \leq 2$. □

2.4.16. LEMMA. *The adjoint shift operator $S^* : H^2 \longrightarrow H^2$ satisfies*

$$\mathrm{DISC}(S^*) = 1.$$

PROOF. Let $C \in \mathrm{Cyc}(S^*)$ be an n-dimensional subspace, and let $(f_i)_{i=1}^n$ be a basis of C. Set

$$E_{f_i} = \mathrm{span}_{H^2}(S^{*k} f_i : k \geq 0), \quad i = 1, ..., n,$$

and assume that $E_{f_i} \neq H^2$, $1 \leq i \leq n$ (otherwise there is nothing to prove). We apply the Beurling–Helson theorem A.1.3.2 (Volume 1) to get non-trivial inner functions Θ_i such that $E_{f_i} = H^2 \ominus \Theta_i H^2$, $1 \leq i \leq n$. Hence

$$H^2 = \mathrm{span}(E_{f_i} : \ 1 \leq i \leq n) \subset H^2 \ominus \Theta H^2$$

with $\Theta = \Theta_1 \Theta_2 \cdots \Theta_n$ (cf. Properties A.2.5.2 and A.2.5.5, Volume 1). The contradiction shows that there is an index i, $1 \leq i \leq n$, such that $E_{f_i} = H^2$, that is f_i is cyclic and $\mathrm{DISC}(S^*) = 1$. □

2.4.17. The shift operator on the Hardy space. Let

$$\begin{aligned} S : H^2 &\longrightarrow H^2, \\ f &\longmapsto zf. \end{aligned}$$

Then $\mathrm{DISC}(S) = 2$ (whereas obviously $\mu(S) = 1$).

Let us show here that $\mathrm{DISC}(S) \geq 2$. For some $n \geq 1$ set $C = \mathcal{L}\mathrm{in}(z^n, b_\lambda)$ where $b_\lambda(z) = \frac{\lambda - z}{1 - \bar{\lambda} z}$, $\lambda \in \mathbb{D}$. Moreover choose $\lambda \neq 0$ such that

$$\left(\frac{1 + |\lambda|}{2}\right)^n < \min\left\{|b_\lambda(z)| : |z| = \frac{1 + |\lambda|}{2}\right\}.$$

As $\mathrm{GCD}(z^n, b_\lambda) = 1$ the space C is cyclic (cf. A.2.5.3, Volume 1). We verify that C does not contain any cyclic function. Indeed, let $\alpha, \beta \in \mathbb{C}$ and set $f = \alpha z^n + \beta b_\lambda$. Two cases can be distinguished. Either $|\alpha| \neq |\beta|$, and an application of Rouché's theorem on $\mathbb{T}$ shows that f has a zero in $\mathbb{D}$, or $|\alpha| = |\beta|$ in which case we apply Rouché's theorem on $|z| = \frac{1}{2}(1 + |\lambda|)$ to obtain a zero of f in $|z| < \frac{1}{2}(1 + |\lambda|)$. In both cases f is not cyclic. This means that $\mathrm{DISC}(S) \geq 2$.

The inequality $\mathrm{DISC}(S) \leq 2$ is more delicate and involves arguments similar to those used in the proofs of Lemmas 2.4.14, 2.4.15 (see comments and references in Section 2.6). □

2.4.18. Multiple shifts. Let

$$S_n = S \oplus ... \oplus S : H_n^2 \longrightarrow H_n^2$$

where $H_n^2 = H^2 \oplus \cdots \oplus H^2$ $(= H^2(\mathbb{C}^n))$. Using a reasoning analogous to that of the proof of Bram's theorem 2.3.3 the reader can verify that $\mu(S_n) = n$. However $\mathrm{DISC}(S_n) = n + 1$ (and not $2n$), cf. comments and references in Section 2.6.

2.4.19. Multiple backward shifts. Let $S^* : H^2 \longrightarrow H^2$ be the backward shift given by $S^*f = \frac{f-f(0)}{z}$. As is shown in 2.4.16, $\mathrm{DISC}(S^*) = \mu(S^*) = 1$. We now consider the multiple backward shift

$$S_n^* = S^* \oplus ... \oplus S^* : H_n^2 \longrightarrow H_n^2.$$

To begin with we show that, a bit surprisingly, $\mu(S_n^*) = 1$. Moreover, the operator S_n^* gives an example of so-called hypercyclic operators in the following sense. An operator $T : X \longrightarrow X$ is said to be *hypercyclic* if there exists a vector $x \in X$ such that $\mathrm{clos}(T^n x : n \geq 0) = X$.

2.4.20. LEMMA. *The operator λS_n^* is hypercyclic for every $\lambda \in \mathbb{C}$, $|\lambda| > 1$. In particular, $\mu(S_n^*) = 1$.*

PROOF. Let $(p_k)_{k\geq 1}$ be a sequence of polynomials with coefficients in $\mathbb{C}^n$ such that $\mathrm{clos}\{p_k\}_{k\geq 1} = H_n^2$. Pick $\lambda \in \mathbb{C}$, $|\lambda| > 1$, and define

$$f = \sum_{k\geq 1} \frac{1}{\lambda^{n_k}} z^{n_k} p_k,$$

where the growth of the integers n_k will be determined later. First, we require a growth that guarantees that

$$\sum_{k\geq 1} \frac{1}{|\lambda|^{n_k}} \|p_k\|_{H_n^2} < \infty,$$

$$n_k + \deg p_k < n_{k+1} \quad k \geq 1.$$

Note that the second condition implies that the polynomials $z^{n_k} p_k$ are pairwise orthogonal. Clearly

$$(\lambda S_n^*)^{n_j} f = p_j + \sum_{k>j} \frac{1}{\lambda^{n_k - n_j}} z^{n_k - n_j} p_k,$$

and consequently

$$\|p_j - (\lambda S_n^*)^{n_j} f\|_2^2 = \sum_{k>j} \frac{1}{|\lambda|^{2(n_k - n_j)}} \|p_k\|_2^2.$$

It remains to choose n_k growing sufficiently fast to have

$$\lim_{j\to\infty} \sum_{k>j} \frac{1}{|\lambda|^{2(n_k - n_j)}} \|p_k\|_2^2 = 0.$$

□

2.4.21. REMARK. Another example of a "chaotic behaviour" is shown by the semigroup analogue of the preceding lemma. Namely, for $\alpha > 0$ define the weighted translation semigroup $S(t)f = e^{\alpha t} f(x+t)$, $t \geq 0$, on $L^2(0,\infty)$. Then there is a function $f \in L^2(0,\infty)$ such that $\mathrm{clos}(S(t)f : t \geq 0) = L^2(0,\infty)$.

2.4.22. THEOREM. $\mathrm{DISC}(S_n^*) = n$.

PROOF. Here, we show that $\mathrm{DISC}(S_n^*) \geq n$. Set

$$C = f \cdot \mathbb{C}^n = \{cf : c \in \mathbb{C}^n\},$$

where f is a scalar valued cyclic function for S^*. Then $C \in \mathrm{Cyc}(S_n^*)$. Assume that $C' \subset C$ with $\dim C' < n$. Obviously, $C' = f \cdot E$ with a subspace $E \subset \mathbb{C}^n$,

$\dim E < n$. Then $C' \subset H^2(E)$ and $S_n^{*k}C' \subset H^2(E)$. Consequently, C' is not cyclic, and hence $\mathrm{DISC}(S_n^*) \geq n$.

For the inequality $\mathrm{DISC}(S_n^*) \leq n$ we refer to the comments and references in Section 2.6 below. □

2.4.23. Model operators. As we know, every c.n.u. Hilbert space contraction is unitarily equivalent to a model operator M_Θ (cf. C.1.3.2), and it is more than natural to seek the DISC's for the class of model operators. However, very little is known about that. Among known results we have

$$\mathrm{DISC}(S_n^* \oplus S_m \oplus M_\Theta) = m + \max(1, \mu(M_\Theta), n),$$

where Θ is a matrix valued two-sided inner function. For this result see the references in Section 2.6.

Even the simplest case of a scalar characteristic function Θ seems to be open (except for the aforementioned case of an inner function, or an outer function satisfying $\inf_{\xi \in \mathbb{T}} |\Theta(\xi)| > 0$). See also comments in Section 2.6 below.

2.5. Exercises and Further Results

2.5.1. Multiplicity of some Hilbert space contractions. (a) *(A. Atzmon, 1987)* Let H be a Hilbert space, let $U : H \longrightarrow H$ be a unitary operator whose spectrum is not purely singular, and let

$$Zf = zf, \quad f \in L^2(H(\cdot), \chi_\sigma m) = \int_\sigma \oplus H(\zeta)\, dm(\zeta),$$

where $\sigma \subset \mathbb{T}$, be the von Neumann standard model of its absolutely continuous part, see C.1.5.1(k) above for the notations (in particular, $\dim H(\zeta) \geq 1$ a.e. on σ). Show that

$$\mu(U^n) \geq [n \cdot m(\sigma)]$$

for $n \geq 0$ (here $[\cdot]$ stands for the entire part of $\cdot$).

[*Hint (B. Sz.-Nagy and C. Foiaş, 1983)*: first $U^n = Z^n \oplus U_s^n$, where U_s is the singular part of U (see C.1.5.1), and hence $\mu(U^n) \geq \mu(Z^n)$; by Theorem 2.3.3, it suffices to show that $d_{Z^n}(\zeta) \geq k$ on a set of positive Lebesgue measure whenever $n \cdot m(\sigma) > k$ (here d_N stands for the local spectral multiplicity of N, see 2.3.2 and C.1.5.1(k));

to this end, consider the intersections

$$\sigma(j, n) = \sigma \cap \Delta(j, n)$$

of σ with the arcs $\Delta(j, n) = [\zeta_j, \zeta_{j+1})$, $\zeta_j = e^{2\pi i j/n}$, $j = 0, ..., n-1$, and show that, for $n \cdot m(\sigma) > k$, there exist at least k sets

$$\sigma'(j, n) = \overline{\zeta}_j \sigma(j, n), \quad j \in J \ (\mathrm{card}(J) \geq k),$$

such that

$$\sigma' = \sigma'_J := \bigcap_{j \in J} \sigma'(j, n)$$

has positive Lebesgue measure; in this case, we obtain that the following restriction of Z^n,

$$\sum_{j \in J} \oplus (Z^n | L^2(H(\cdot), \chi_{\zeta_j \sigma'} m)),$$

is unitarily equivalent to $U' = \sum_{j \in J} \oplus (Z^n | L^2(H(\cdot), \chi_{\sigma'} m))$ (indeed, $(\zeta_j \zeta)^n = \zeta^n$ for every $\zeta \in \sigma'$), whereas the latter operator obviously has the local multiplicity $d_{U'}(\zeta) = \text{card}(J) \cdot \dim H(\zeta) \geq \text{card}(J)$ for $\zeta \in \sigma'$;

finally, given an integer n, $n \cdot m(\sigma) > k$, we show that $m(\sigma'_J) > 0$ for an intersection $\sigma' = \sigma'_J$ with $\text{card}(J) \geq k$; to this end, assume that, to the contrary, $\sum_{j=0}^{n-1} \chi_{\sigma'(j,n)}(\zeta) < k$ a.e. on $\Delta(0,n)$; then

$$\frac{k}{n} > \sum_{j=0}^{n-1} \int_{\Delta(0,n)} \chi_{\sigma'(j,n)}(\zeta)\, dm(\zeta) = \sum_{j=0}^{n-1} \int_{\mathbb{T}} \chi_{\sigma(j,n)}(\zeta)\, dm(\zeta) = m(\sigma);$$

the result follows.]

(b) *(B. Sz.-Nagy and C. Foiaş, 1983; A. Atzmon, 1987)* Let T be a c.n.u. contraction such that $T \notin C_{0\cdot}$, i.e. T^n does not tend to zero (in the strong operator sense). Show that there exists a constant $\alpha_T > 0$ such that $\mu(T^n) \geq [n\alpha_T]$ for $n \geq 0$.

[*Hint:* let $V : \mathcal{K} \longrightarrow \mathcal{K}$ be the minimal isometric dilation of T^* and $\mathcal{R}$ be the space of its residual part (see C.1.5.2), σ corresponds to $U = V|\mathcal{R}$ in the same way as in (a) above; observe that, by C.1.5.2, $\mathcal{R} \neq \{0\}$ and hence $m(\sigma) > 0$; adapting the notation of C.1.5.2 for T^* instead of T, we get

$$X^*T = (Z|\mathcal{R})^* X^*,$$

where $X^* = P_{\mathcal{R}}|\mathcal{K}_\Theta : \mathcal{K}_\Theta \longrightarrow \mathcal{R}$; since V is the minimal dilation, we obtain

$$\mathcal{R} = P_{\mathcal{R}}\mathcal{K} = P_{\mathcal{R}}(\text{span}(Z^k \mathcal{K}_\Theta : k \geq 0)) = \text{span}(Z^k P_{\mathcal{R}} \mathcal{K}_\Theta : k \geq 0),$$

and moreover, since $Z^* P_{\mathcal{R}} \mathcal{K}_\Theta = Z^* X^* \mathcal{K}_\Theta = X^* T \mathcal{K}_\Theta \subset P_{\mathcal{R}} \mathcal{K}_\Theta$ we get $Z^k P_{\mathcal{R}} \mathcal{K}_\Theta \subset Z^{k+1} P_{\mathcal{R}} \mathcal{K}_\Theta$ for every k, and hence

$$\mathcal{R} = \text{span}(Z^{k_j} X^* \mathcal{K}_\Theta : j \geq 0)$$

for any sequence of integers $k_j \longrightarrow \infty$ (in what follows, we choose $k_j = nj$); now, if $C \subset \mathcal{K}_\Theta$ is a cyclic subspace for T^n, then

$$X^* \mathcal{K}_\Theta \subset \text{clos}(X^* \mathcal{K}_\Theta) = \text{span}(Z^{*nk} X^* C : k \geq 0),$$

and hence $\text{span}(Z^{nk} X^* C : k \in \mathbb{Z}) = \mathcal{R}$; by (a), $\mu(T^n) = \dim C \geq \mu(Z^n) \geq [n \cdot m(\sigma)]$.]

(c) *(R. Douglas, 1974)* Let T be a contraction on a Hilbert space H such that $\lim_n T^{*n} x = 0$ for every $x \in H$ (i.e., T is a c.n.u. contraction in the class $C_{\cdot 0}$). Show that $\mu(T) \leq \partial_{T^*}$.

[*Hint:* show that $\sum_{k=0}^{n} T^k D_{T^*}^2 T^{*k} = I - T^n T^{*n}$, and then $\sum_{k \geq 0} T^k D_{T^*}^2 T^{*k} x = x$ for every $x \in H$ (norm convergent series).]

(d)* *(V. Vasyunin, 1989)* Let T be a Hilbert space contraction such that $\partial_{T^*} < \infty$, $\partial_T < \infty$, and $T = U \oplus T_{cnu}$, where U is unitary and T_{cnu} is a c.n.u. contraction. Then

$$\mu(T) = \max\Big\{\mu(U_s), \partial_{T^*} - r + \max(\mu_0, \mu_{1a}, \varepsilon, \varepsilon_*)\Big\},$$

where U_s is the singular part of U, and hence,

$$\mu(U_s) = \text{ess sup}(\text{rank}\, E_{U_s}(\zeta)),$$

where $E_{U_s}(\zeta)$ stands for the spectral resolution of U_s,

$$r = \mathrm{rank}(\Theta(\zeta))$$

(which is constant a.e. on $\mathbb{T}$),

$$\mu_0 = \mu(T_0).$$

Here T_0 is a C_0 operator with characteristic function Θ_{00} arising from the following "cascade" factorizations $\Theta = \Theta_{inn}\Theta_{out}$, $\Theta_{inn} = \Theta_{10}\Theta_{00}$ the latter being the $*$-outer-inner factorization (see C.2.1.21). Further,

$$\mu_{1a} = \mu(U_a \oplus T_{11}) = \sup\{\mathrm{rank}\, E_{U_a}(\zeta) + \mathrm{rank}(I - \Theta_{11}(\zeta)^*\Theta_{11}(\zeta)) : \zeta \in \mathbb{T}\},$$

$\Theta_{out} = \Theta_{11}\Theta_{01}$ being the $*$-outer-inner factorization, and

$$\varepsilon_* = \chi_{(0,\infty)}(\partial_T - r).$$

Finally, $\varepsilon = 1$ if the set $\{\sum_\kappa f_\kappa (\Theta_{10})_\kappa : f_\kappa \in H^\infty\}$ contains no outer function and $\varepsilon = 0$ otherwise, $(\Theta_{10})_\kappa$ are the minors of order r of the matrix Θ_{10}.

In the special case

$$T = U \oplus S_n \oplus T_0 \oplus S_m^*,$$

where U is a unitary operator, T_0 is a C_0 operator, and S_n stands for the shift on the space $H^2(\mathbb{C}^n)$, the formula becomes

$$\mu(T) = \max\left\{\mu(U_s), n + \max(\mu_0, \mu_{1a}, \varepsilon_*)\right\}$$

with $\varepsilon_* = 0$ if $m = 0$ and $\varepsilon_* = 1$ otherwise.

2.5.2. Multiplicity of some Toeplitz operators. **(a)** *(N. Nikolski, 1983)* Let $F \in H^\infty(\mathbb{C}^n \longrightarrow \mathbb{C}^n)$ and let $T_{F^*}x = P_+(F^*x)$ be an antianalytic Toeplitz operator, $x \in H^2(\mathbb{C}^n)$.

(i) Show that $\mu(T_{F^*}) = 1$ if $z \longmapsto \det(F(z) - \lambda I)$ is a nonzero function for every $\lambda \in \mathbb{C}$, and $\mu(T_{F^*}) = \infty$ otherwise.

In particular, for a diagonal symbol $F = \mathrm{diag}(f_1, ..., f_n)$, where $f_k \in H^\infty$, $T_{F^*} = \sum_{k=1}^n \oplus T_{\overline{f}_k}$, one has $\mu(T_{F^*}) = 1$ if and only if $f_k \neq$ const for every k.

[*Hint:* assume that the condition is satisfied, and let p be the characteristic polynomial of $F(0)^*$ (take $p = z$ if $F(0) = 0$), and $A = p(T_{F^*}) = T_{p_*(F)^*}$, where $p_*(z) = \overline{p}(\overline{z})$; since $p_*(F)(0)^* = p(F(0)^*) = 0$, it follows that $\mathrm{span}(\mathrm{Ker}\, A^n : n \geq 1) = H^2(\mathbb{C}^n)$; applying 2.5.4(d) (below), we get $\mu(A) = \max(1, \dim \mathrm{Ker}\, A^*)$, where $A^* = T_{p_*(F)}$ is a multiplication operator by a matrix function $p_*(F)$; since $\det(p_*(F(z)) \neq 0$ for all $z \in \mathbb{D}$ excepting a sequence tending to $\partial\mathbb{D}$, $\mathrm{Ker}\, A^* = \{0\}$ and hence $\mu(T_{F^*}) \leq \mu(A) = 1$;

to prove the second assertion, assume (without loss of generality) that $\det(F(z)) \equiv 0$ in $\mathbb{D}$; then, by C.1.5.2(d)(iii), $\mathrm{Ker}(T_{F^*})^* = \mathrm{Ker}\, T_F \neq \{0\}$ and, being z-invariant, this kernel is infinite dimensional; hence $\mu(T_{F^*}) \geq \dim \mathrm{Ker}\, T_F = \infty$.]

(ii) Let $F = \mathrm{diag}(f_1, ..., f_n)$, $f_k \in H^\infty$, $f_k \neq$ const for every k, and take a Banach space operator $A \in L(X)$ such that

$$f_k(\mathbb{D})^* \setminus \mathrm{pch}(\sigma(A)) \neq \emptyset,$$

where $\sigma^* = \{\overline{\lambda} : \lambda \in \sigma \subset \mathbb{C}\}$ and $\mathrm{pch}(\sigma) = \{\lambda \in \mathbb{C} : |p(\lambda)| \leq \sup_\sigma |p|$ for every $p \in \mathcal{P}ol_+\}$ is the polynomially convex hull of σ. Show that

$$\mu(T_{F^*} \oplus A) = \max(\mu(T_{F^*}), \mu(A)) = \mu(A).$$

[*Hint:* find a sequence $(y_j)_{j\geq 1}$ of eigenvectors of T_{F^*} corresponding to eigenvalues from $\mathbb{C}\setminus \mathrm{pch}(\sigma(A))$ whose span coincide with $H^2(\mathbb{C}^n)$; then, using the definition of $\mathrm{pch}(\sigma(A))$ and the Riesz–Dunford calculus (C.2.5.1), show that there exists sufficently rapidly convergent series $y = \sum_{k\geq 1} a_k y_k$ and polynomials p_n such that $\|p_n(A)\| < n^{-1}$, $\|p_n(T_{F^*})y - y_n\| < n^{-1}$; this implies that $\mathrm{span}(p_n(T_{F^*} \oplus A)(x \oplus y) : n \geq 1) \supset \mathbb{O} \oplus H^2(\mathbb{C}^n)$ for every $x \in X$, and the result follows.]

(iii) Let $F = \mathrm{diag}(f_1, ..., f_n)$, $G = \mathrm{diag}(g_1, ..., g_m)$, where $f_k, g_j \in H^\infty$, $f_k \neq \mathrm{const}$ for every k, such that

$$f_k(\mathbb{D})^* \setminus \mathrm{pch}(g_j(\mathbb{D}) : 1 \leq j \leq m) \neq \emptyset, \quad k = 1, ..., n.$$

Show that

$$\mu(T_{F^*} \oplus T_G) = \mu(T_G).$$

[*Hint:* use (ii).]

(iv)* Let $F = \mathrm{diag}(f_1, ..., f_n)$, $f_k \in H^\infty$, and $g \in H^\infty$ be a *univalent* (one-to one) mapping of the disk $\mathbb{D}$ onto a Jordan domain $g(\mathbb{D})$ such that

$$f_k(\mathbb{D})^* \subset g(\mathbb{D}), \quad k = 1, ..., n.$$

Then $\mu(T_{F^*} \oplus T_g) = \mu(T_{F^*}) + \mu(T_g)$.

(v)* Let $\varphi \in H^\infty$ and let

$$v(\varphi) = \sup\{\mathrm{card}(\varphi^{-1}(\lambda)) : \lambda \in \mathbb{C}\}, \quad ev(\varphi) = \sup\{\mathrm{card}(\varphi_{\overline{\mathbb{D}}}^{-1}(\lambda)) : \lambda \in \mathbb{C}\}$$

be the *maximal valency* and the *essential maximal valency* of φ, respectively, where $\varphi_{\overline{\mathbb{D}}} : \overline{\mathbb{D}} \longrightarrow \mathbb{C}$ is the extended map with nontangential boundary values added (essential values of $\varphi_{\overline{\mathbb{D}}}|\mathbb{T}$ are counted only, that is, those $\lambda \in \mathbb{C}$ for which the set $\{\zeta \in \mathbb{T} : |\varphi_{\overline{\mathbb{D}}}(\zeta) - \lambda| < \varepsilon\}$ is of positive Lebesgue measure for every $\varepsilon > 0$). Then, $\mu(T_\varphi) \geq ev(\varphi) \geq v(\varphi)$.

(vi)* Let γ_0, γ_1 be piecewise smooth Jordan curves in $\mathbb{C}$, which are assumed to be simple loops with exactly one common point and such that $\Omega_0 = \mathrm{Int}(\gamma_0) \subset \Omega_1 = \mathrm{Int}(\gamma_1)$. Further, let φ be a conformal mapping of $\mathbb{D}$ onto the crescent domain $\Omega = \Omega_1 \setminus \mathrm{clos}\,\Omega_0$.

Let ω, ω_0 be harmonic measures of Ω and Ω_0, respectively (at any points); then, $\mu(T_\varphi) = 2$ ($> ev(\varphi) = 1$) if $\omega_0 \leq c \cdot \omega|\partial\Omega_0$ for a constant $c > 0$, and $\mu(T_\varphi) = 1$, otherwise.

(b) *(B. Solomyak, 1985)* Let $\varphi \in H^\infty$ be such that $\varphi' \in C_A(\mathbb{D})$ and such that φ is analytic in neighbourhoods of points $\zeta \in \overline{\mathbb{D}}$ where $\varphi'(\zeta) = 0$. Assume further that $\varphi(\mathbb{T})$ is a "generic curve" (which means that it has a finite number of transversal self-intersections). Then $\mu(T_\varphi) = ev(\varphi) = v(\varphi)$.

(c) *(B. Solomyak and A. Volberg, 1989)* Let Ω be a bounded domain in $\mathbb{C}$ such that $\partial\Omega$ is a finite union of analytic Jordan arcs and $\Omega = \mathrm{Int}(\mathrm{clos}\,\Omega)$ (this excludes sliced and punctured domains); next, let φ be analytic on $\mathrm{clos}\,\Omega$. Then

$$ev(\varphi) \leq \mu(T_\varphi) \leq ev(\varphi) + 1,$$

and $\mu(T_\varphi) = ev(\varphi)$ if and only if every "belt domain" B contained in the set $V(\varphi)$ is not "wide" (wide means that $\omega_{cB} \leq C \cdot \omega_B|\partial(cB)$ for a constant C, where cB is any bounded component of the complement $\mathbb{C}\backslash B$ and ω_G stands for the harmonic measure of G (at a point)); here $V(\varphi)$ is the set of maximal valency of φ, $V(\varphi) = \{\lambda \in \mathbb{C} : \mathrm{card}(\varphi_{\overline{\Omega}}^{-1}(\lambda)) = ev(\varphi)\}$ (see (a)(v) for the definition of $\varphi_{\overline{\Omega}}$).

(d)* *(W. Wogen, 1978)* Let $f \in H^2$, $f = \sum_{k\geq 1} a_k z^k$, be such that

$$\lim_n \frac{R^n}{a_n} (\sum_{k>n} |a_k|^2)^{1/2} = 0$$

for every $R > 0$. Then f is a cyclic vector for every antianalytic Toeplitz operator $T_{\overline{\varphi}} : H^2 \longrightarrow H^2$ with $\varphi \in H^\infty$, $\varphi \neq$ const. (Example: $a_{k+1}/a_k = k^{-k}$, $a_0 = 1$).

2.5.3. Multiplicity of some composition operators. Let $\omega \in H^\infty$ with $\|\omega\|_\infty \leq 1$, $\omega \neq$ const, and $C_\omega f = f \circ \omega$, $f \in \mathrm{Hol}(\mathbb{D})$, be a *composition operator.* Recall that

$$C_\omega : H^p \longrightarrow H^p$$

is a bounded operator (see A.3.12.5, Volume 1). The results presented below are mostly due to P. Bourdon and J. Shapiro (1994).

(a) Assume that $\mu(C_\omega) < \infty$. Show that ω is *univalent* (i.e., injective) in $\mathbb{D}$ and *essentially univalent* on $\mathbb{T}$, i.e. $\omega|(\mathbb{T}\backslash\sigma)$ is injective for a subset $\sigma \subset \mathbb{T}$, $m(\sigma) = 0$ ($\omega(\zeta)$ means the nontangential limit at ζ which exists for almost all $\zeta \in \mathbb{T}$).

[*Hint:* if $\omega(z_0) = \omega(\zeta(z_0))$ for $z_0 \neq \zeta(z_0)$, $z_0, \zeta(z_0) \in \mathbb{D}$, then by the open mapping theorem there exist an infinity of $z \in \mathbb{D}$ such that $\omega(z) = \omega(\zeta(z))$ for some $\zeta(z) \neq z$; denoting by φ_z the evaluation functional $\varphi_z : f \longmapsto f(z)$, $f \in H^p$, we get

$$C_\omega^* \varphi_z = \varphi_{\omega(z)};$$

now, the assumption entails that $\dim \mathrm{Ker}\, C_\omega^* = \infty$, and hence $\mu(C_\omega) = \infty$;

assume that ω is not essentially univalent, let $f_1, ..., f_n$ be functions in H^p, and assume that there exist finite sums

$$F_s = \sum_{j,k} f_j \circ \omega^{(k)}, \quad \omega^{(k)} = \omega \circ ... \circ \omega,$$

approximating the functions z^s, $0 \leq s < N$, where $N > n$; then there exists $\sigma \subset \mathbb{T}$, $m(\sigma) = 0$, such that these sequences are well defined and converge to z^s everywhere on $\mathbb{T}\backslash\sigma$; since ω is not essenitially univalent on $\mathbb{T}$, there exist N pairs $\tau = \{\zeta_1, \zeta_1', ..., \zeta_N, \zeta_N'\}$ such that $\omega(\zeta_k) = \omega(\zeta_k')$ for $1 \leq k \leq N$; this implies that $\dim \mathcal{L}\mathrm{in}(F_s|\tau, 0 \leq s < N) \leq N + n$, whereas $\dim \mathcal{L}\mathrm{in}(z^s|\tau, 0 \leq s < 2N) = 2N$; contradiction completes the proof.]

(b)* If C_ω is cyclic (i.e. $\mu(C_\omega) = 1$) then the polynomials in ω ($p \circ \omega$, $p \in \mathcal{P}ol_+$) are dense in H^p. If $\|\omega\|_\infty < 1$, then the converse is also true.

(c) *Cyclicity properties of linear-fractional maps.* Let $\omega(z) = \frac{az+b}{cz+d}$ be a *linear-fractional map* (l-f.m.), $ad - bc \neq 0$, such that

$$\omega(\mathbb{D}) \subset \mathbb{D}.$$

(i) Let ω be an *elliptic l-f.m.*, i.e., there exists a linear-fractional transformation α such that

$$\alpha \circ \omega \circ \alpha^{-1}(z) = \kappa z, \quad |\kappa| = 1.$$

Show that C_ω is cyclic ($\mu(C_\omega) = 1$) if and only if $\kappa = e^{i\theta}$, where θ/π is irrational; otherwise, $\mu(C_\omega) = \infty$.

In other words, if ω has a fixed point $p \in \mathbb{D}$ and $|\omega'(p)| = 1$, then C_ω is cyclic if and only if $(\arg \omega'(p))/\pi$ is irrational.

[*Hint:* clearly, C_ω is similar to $C_{\kappa z} = C_\alpha C_\omega C_\alpha^{-1}$, hence $\mu(C_\omega) = \mu(C_{\kappa z})$; for $C_{\kappa z}$ with an irrational θ/π every $f = \sum_{n\geq 0} a_n z^n$ with $a_n \neq 0$, $n \geq 0$, is cyclic; if θ/π is rational then $C_{\kappa z}^N = I$ for an integer N.]

(ii)* A. Let ω be a non-elliptic l-f.m. Then $\mu(C_\omega) = \infty$ if ω has a fixed point in $\mathbb{D}$ and a fixed point on $\mathbb{T}$; otherwise, $\mu(C_\omega) = 1$.

B. In particular, if ω has no fixed point in $\mathbb{D}$, then $\mu(C_\omega) = 1$.

C. If ω is a *parabolic l-f.m.*, i.e., it has a unique fixed point in $\hat{\mathbb{C}}$, then $\mu(C_\omega) = 1$.

D. If ω has an attractive fixed point in $\mathbb{D}$ and a repulsive fixed point in $\mathbb{C}\backslash\overline{\mathbb{D}}$ then $\mu(C_\omega) = 1$.

(iii) There exist l-f. maps ω_1, ω_2 such that $\omega_1(\mathbb{D}) = \omega_2(\mathbb{D})$ and $\mu(C_{\omega_1}) = \infty$, $\mu(C_{\omega_2}) = 1$ (take $\omega_1 = z/(2-z)$ (which has fixed points 0 and 1) and $\omega_2 = (1+2z)/3$ having fixed points 1 and ∞ (ω_2 has no fixed points in $\mathbb{D}$ and is not parabolic)).

(d)* *Hypercyclicity properties for linear-fractional maps.* Let ω be a non-elliptic l-f.m. (Recall that hypercyclicity means the existence of an orbit $(C_\omega^n f)_{n\geq 0}$ which is dense in H^p).

(i) If ω is an automorphism of $\mathbb{D}$ (i.e., $\omega(\mathbb{D}) = \mathbb{D}$) then C_ω is hypercyclic.

(ii) If ω is not an automorphism of $\mathbb{D}$ and has no fixed point in $\mathbb{D}$, then C_ω is hypercyclic if and only if ω is not a parabolic l-f.m.

(e) Assume that ω fixes a point ζ of $\mathbb{D}$ ($\omega \in H^\infty$, $\|\omega\|_\infty \leq 1$). Show that C_ω is not hypercyclic.

[*Hint:* if $\lim_k \|C_\omega^{n_k} f - g\|_p = 0$ then $g(\zeta) = f(\zeta)$, and hence g cannot be an arbitrary function from H^p.]

2.5.4. Miscellaneous results. (a) Let T be a Banach space operator, and let q be a nonconstant polynomial. Show that $\mu(q(T)) \leq \mu(T)\deg(q)$.

[*Hint:* let $C \in \mathrm{Cyc}(T)$ and $L = \mathcal{L}\mathrm{in}(T^k C : 0 \leq k < \deg(q))$; since every polynomial $p \in \mathcal{P}ol_+$ can be written in the form $p = \sum_{m=0}^N r_m q^m$, where $r_m \in \mathcal{P}ol_+$, $\deg(r_m) < \deg(q)$ (*Gauss' algorithm*, $p = qp_1 + r_0$, $p_1 = qp_2 + r_1$, etc.), it is clear that $p(T)x \in \mathrm{span}(q(T)^m L : m \geq 0)$ for every $x \in C$, and hence $L \in \mathrm{Cyc}(q(T))$.]

(b) Let $T : X \longrightarrow X$ be a Banach space operator having a complete family $(x_k)_{k\geq 1}$ of eigenvectors, $Tx_k = \lambda_k x_k$, $\lambda_k \neq \lambda_j$ ($k \neq j$), $X = \mathrm{span}(x_k : k \geq 1)$.

Show that $\mu(T) = 1$, and moreover, that there exist $\varepsilon_k > 0$ such that each vector of the form

$$x = \sum_{k \geq 1} a_k \frac{x_k}{\|x_k\|}, \quad 0 < |a_k| \leq \varepsilon_k \ (k \geq 1)$$

is cyclic for T. In fact, it suffices that ε_k satisfy $\sum_{k\geq 1} k^{-2} \log \frac{1}{\varepsilon_k} = \infty$.

[*Hint:* choose ε_k as in A.5.7.8(c)(vii), Volume 1; now, if $\varphi \in X^*$ and

$$0 = \langle T^n x, \varphi \rangle = \sum_{k \geq 1} \lambda_k^n a_k \frac{\langle x_k, \varphi \rangle}{\|x_k\|}$$

for every $n \geq 0$, then $\langle x_k, \varphi \rangle = 0$ for every $k \geq 1$, and the result follows; the last assertion follows from R. Sibilev's result quoted in A.5.7.8(c)(viii), Volume 1.]

(c) *(D. Herrero, 1978)* Let $T : X \longrightarrow X$ be a Banach space operator and $\mu_R(T)$ its *rational multiplicity*, that is, $\min\{\dim C\}$ over all *rationally cyclic subspaces* $C \subset X$, i.e., subspaces satisfying

$$X = \operatorname{span}(R_\lambda(T)C : \lambda \in \mathbb{C}\backslash\sigma(T)).$$

Show that

$$\mu_R(T) \leq \mu(T) \leq \mu_R(T) + 1;$$

moreover, for every rationally cyclic subspace C there exists $C' \in \operatorname{Cyc}(T)$ such that $C' \supset C$, $\dim C' \leq \dim C + 1$.

[*Hint (N. Nikolski, 1989):* let $E = \operatorname{span}(T^n C : n \geq 0)$, and T/E be the quotient operator defined in 2.3.1(4); the cosets $x_\lambda = R_\lambda(T)x + E$ are eigenvectors of T/E (if non-zero) for every $x \in C$; moreover, $X/E = \operatorname{span}(x_\lambda : \lambda \in \mathbb{C}\backslash\sigma(T))$ (by the rational cyclicity of C); by (b) above, $\mu(T/E) = 1$, and the inequality $\mu(T) \leq \mu_R(T) + 1$ follows from 2.3.1(4).]

(d)* *(V. Vasyunin and N. Nikolski, 1983)* Let $T : X \longrightarrow X$ be a Banach space operator such that $X = \operatorname{span}(\operatorname{Ker} T^n : n \geq 1)$. Then

$$\mu(T) = \max(1, \dim \operatorname{Ker} T^*).$$

In particular, if $T = (t_{ij})_{i,j\geq 1}$ is a block upper triangular operator on $l^2(E)$ with $t_{ij} = 0$, $j \leq i$, and $\operatorname{Ker}(t_{i,i+1})^* = \{0\}$ for every $i \geq 1$, then $\mu(T) = 1$ (*W. Wogen, 1978*).

(e)* *(N. Nikolski, 1989)* Let $T_i : X_i \longrightarrow X_i$, $1 \leq i \leq n$, be Banach space operators such that $\sigma(T_i) \cap \sigma(T_j) = \emptyset$ for $i \neq j$ and $\mu(T_i) = 1$ for $1 \leq i \leq n$. Then

$$\mu(T_1 \times ... \times T_n) = 1,$$

where the direct product $T_1 \times ... \times T_n$ acts on the space $X = X_1 \times ... \times X_n$.

(f)* *(W. Wogen, 1982)* Given n, $1 \leq n \leq \infty$, there exists a Hilbert space operator $T : H \longrightarrow H$ such that

$$\mu(T) \leq n, \ \mu(T^*) = 1 \quad \text{and} \quad \mu(\{T\}') = n, \ \mu(\{T\}'') = n,$$

where $\{T\}' = \{A : AT = TA\}$ and $\{T\}'' = \{A : AB = BA \text{ for every } B \in \{T\}'\}$ stand for the *commutant* and the *bicommutant* (double commutant) of T, and the multiplicity $\mu(M)$ of a set of operators M is defined by

$$\mu(M) = \min\{\dim C : E_C(M) = H\}$$

$E_C(M)$ is the least (closed) subspace of H containing C and invariant with respect to each operator from M.

(g)* *(D. Herrero, 1978)* Let H be a Hilbert space, $\dim H = \infty$. The set of all operators $T \in L(H)$, for which at least one of the multiplicities $\mu(T)$, $\mu(T^*)$ is finite, is nowhere dense in $L(H)$.

If X is a Banach space, $\dim X = \infty$, then the set of all operators T with $\mu(T) = \infty$ is dense in $L(X)$.

(h)* *(C. Berger and B. Shaw (1973), D. Voiculescu (1980))* Let T be an operator on a Hilbert space whose selfcommutator $[T^*, T] = T^*T - TT^*$ has a trace class negative part, i.e. $[T^*, T] = [T^*, T]_- + [T^*, T]_+$ (here $A = A_- + A_+$ corresponds to the restrictions of the von Neumann spectral integral for a selfadjoint operator A to $\mathbb{R}_- = (-\infty, 0)$ and $\mathbb{R}_+ = [0, \infty)$, respectively) and

$$[T^*, T]_- \in \mathfrak{S}_1, \text{ and } \mu_R(T) < \infty$$

(μ_R is the rational multiplicity, as defined in (c) above). Then $[T^*, T] \in \mathfrak{S}_1$ and

$$\operatorname{Trace}[T^*, T] = \sum_{n \geq 1} \left(\|Te_n\|^2 - \|T^*e_n\|^2 \right) \leq \frac{\mu_R(T)}{\pi} \cdot m_2(\sigma(T)),$$

where m_2 stands for the planar Lebesgue measure and $(e_n)_{n \geq 1}$ is an orthonormal basis of H.

(i) *(D. Voiculescu, 1980)* Let $\varphi \in C(\mathbb{T})$, $m_2(\varphi(\mathbb{T})) = 0$ and let $T_\varphi : H^2 \longrightarrow H^2$ be the corresponding Toeplitz operator. Show that $[T_\varphi^*, T_\varphi] = H_{\overline{\varphi}}^* H_{\overline{\varphi}} - H_\varphi^* H_\varphi$, where H_φ stands for the corresponding Hankel operator (see Chapter B.1, Volume 1). Moreover, if $\sum_{k>0} k|\hat{\varphi}(-k)|^2 < \infty$ and

$$\operatorname{Trace}[T_\varphi^*, T_\varphi] = \sum_{k \geq 0} k|\hat{\varphi}(k)|^2 - \sum_{k>0} k|\hat{\varphi}(-k)|^2 > 0,$$

then the operator T_φ has a nontrivial invariant subspace.

[*Hint:* for the commutator formula see Lemma B.4.4.3 (Volume 1); next, if $\sigma(T_\varphi) \neq \sigma_{ess}(T_\varphi)$, then invariant subspaces obviously exist (see Subsection B.2.5.1, Volume 1); the case $\sigma(T_\varphi) = \sigma_{ess}(T_\varphi)$ $(= \varphi(\mathbb{T}))$ is impossible due to (h).]

(j)* *Multicyclicity of multiple shifts (N. Nikolski, 1970)* Let $1 \leq p \leq \infty$, E be a separable Banach space, let $w_n > 0$ for $n \in \mathbb{Z}$, and

$$l^p(\mathbb{Z}, w_n; E) = \left\{ x = (x_n)_{n \in \mathbb{Z}} : x_n \in E, \sum_{n \in \mathbb{Z}} \|x_n\|_E^p w_n < \infty \right\}$$

(with the ususal modification for $p = \infty$). Let S_E be the right shift operator on $l^p(\mathbb{Z}, w_n; E)$,

$$S_E(x_n)_{n \in \mathbb{Z}} = (x_{n-1})_{n \in \mathbb{Z}}.$$

Assume that $L^\infty(\mathbb{T}) \subset \mathcal{F}l^p(\mathbb{Z}, w_n; E)$ and $\sum_{n \in \mathbb{Z}} (n^2+1)^{-1} |\log \|S_E^n\|| < \infty$, where

$$\mathcal{F}l^p(\mathbb{Z}, w_n; E) = \left\{ \sum_{n \in \mathbb{Z}} a_n e^{int} : a \in l^p(\mathbb{Z}, w_n) \right\}$$

and $l^p(\mathbb{Z}, w_n) = l^p(\mathbb{Z}, w_n; \mathbb{C})$.

Then $\mu(S_E) = \dim E$ (the weak $*$ multicyclicity for $p = \infty$) for every (some) E with $\dim E > 1$ if and only if $l^p(\mathbb{Z}, w_n) \subset l^2(\mathbb{Z})$. If the latter inclusion does not hold then $\mu(S_E) = 1$ for every E.

In particular, for $w_n = 1$ and $\dim E > 1$, $\mu(S_E) = 1$ if and only if $p > 2$.

2.6. Notes and Remarks

General information. For comments and references about comparative properties of semigroups and their generators see Section 1.7 above. The interrelations between the minimal rank of the control operator and the multiplicity $\mu(A)$ are based on Kalman's theorem 1.3.1; they are described in any book treating control theory, see for example P. Fuhrmann [**Fuh3**], W. Wonham [**Wo**].

For a sytematic study of the multiplicity function $T \longmapsto \mu(T)$ (also called multicyclicity) we refer to D. Herrero [**Herr2**], C. Apostol, L. Fialkow, D. Herrero and D. Voiculescu [**AFHV**], H. Radjavi and P. Rosenthal [**RR**], and also N. Nikolski [**N22**], N. Nikolski and V. Vasyunin [**NVa1**], V. Kapustin and A. Lipin [**KLi1**], [**KLi2**], P. Bourdon and J. Shapiro [**BouSh**], and K.-G. Grosse-Erdmann [**Gr-E**]. Several comments about cyclicity of concrete operators (mostly the multiplication operator $f \longmapsto zf$ by the independent complex variable) are scattered in Sections A.2.8, A.4.8, A.4.9, A.7.6 (Volume 1).

Basic facts about the DISC are contained in [**NVa1**], [**NVa3**], [**NVa2**].

Formal credits. Subsections 2.2.1 and 2.2.2: for the critical growth bound and spectral mapping theorems for semigroups we refer to the results and comments in Subsections 1.6.1, 1.6.2 and 1.7 above.

The invariant subspace theorem 2.2.3 is related to a general property of this kind contained in C.2.1.10.

Section 2.3 is an excerpt from N. Nikolski and V. Vasyunin [**NVa1**] and N. Nikolski [**N22**]. In [**N22**], the attention is in particular focused on the problem of computing the multiplicity $\mu(T_1 \oplus T_2)$ of the direct sum of given operators, see 2.3.1, points (7) and (8), and (for more examples) 2.5.1(d), 2.5.2(a) and 2.5.4(e). Theorem 2.3.3 is from J. Bram [**Br**]. Theorem 2.3.5 is, of course, the special case of 2.5.1(d); the idea to use intertwining properties of the residual part of the unitary dilation comes from B. Sz.-Nagy and C. Foiaş, see [**SzNF4**] and especially [**SzNF7**] (it is also used in 2.5.1(b)). For another proof see C.2.5.7(k) (and C.2.5.8 for comments).

Section 2.4 is written as a short survey of the DISC properties, for the omitted details we refer to N. Nikolski and V. Vasyunin [**NVa3**], [**NVa1**], [**NVa2**]. It is worth mentioning that for DISC computations the splitting techniques, which allow to express $\mathrm{DISC}(T_1 + T_2)$ in terms of $\mathrm{DISC}(T_1)$, $\mathrm{DISC}(T_2)$, are even more crucial than they are for the quantity $\mu(T)$. To apply these techniques, the following property is of great importance.

PROPERTY. Given an operator T with $d = \mathrm{DISC}(T) < \infty$ and given a cyclic subspace F, $d \leq \dim F < \infty$, almost all subspaces $E \subset F$ of dimension $d = \dim E$ are still cyclic (almost all means with respect to the invariant measure on the *Grassman manifold* of all d dimensional subspaces of F).

In [**NVa1**], an operator satisfying this property is refered to as *"having many cyclic subspaces"*. All operators listed in Section 2.4, as well as all those treated in [**NVa3**]-[**NVa2**], are proved to have many cyclic subspaces. The problem whether an arbitrary operator with $\text{DISC}(T) < \infty$ satisfies this property seems still to be open. See also D. Herrero and J. McDonald [**HerrMcD**].

Lemma 2.4.20 is due to S. Rolewicz [**Rol**]. An operator $T : X \longrightarrow X$ for which there exists an element $x \in X$ with

$$\text{clos}(T^n x : n \geq 0) = X$$

is called *chaotic*, or *hypercyclic* (terminology by B. Beauzamy, see [**Gr-E**]). The hypercyclicity phenomenon was discovered by G. Birkhoff (1929), [**Bir**], who showed that the translation operator $T_a f(z) = f(z + a)$, where $a \in \mathbb{C}\backslash\{0\}$, is hypercyclic on the (Frechet) space of entire functions $\text{Hol}(\mathbb{C})$. Similarly to λS^*, the operator λT_a, $a > 0$ is hypercyclic on $L^2(0, \infty; E)$ for every separable Hilbert spase E and every $\lambda \in \mathbb{C}$, $|\lambda| > 1$. Many other operators, including many composition operators (see 2.5.3(d)), are hypercyclic. For surveys on the hypercyclicity property and for more references see P. Bourdon and J. Shapiro [**BouSh**], K.-G. Grosse-Erdmann [**Gr-E**].

The formula from 2.4.23 for the DISC of a special model operator (proved in N. Nikolski and V. Vasyunin [**NVa2**]) can be compared with the multiplicity formula quoted in 2.5.1(d). For a general model operator M_Θ, the quantity $\text{DISC}(M_\Theta)$ seems to be unknown, even if Θ is a scalar (contractive) function. Nevertheless, in view of the results given above and the techniques developped in [**NVa1**]–[**NVa3**], the following formula seems to be plausible

$$\text{DISC}(M_\Theta) = \begin{cases} 2 & \text{if } |\Theta(\xi)| < 1 \text{ for almost all } \xi \in \mathbb{T} \\ 1 & \text{otherwise,} \end{cases}$$

where $\Theta \in H^\infty$, $|\Theta| \leq 1$, and M_Θ is the associated model operator. In order to illustrate this formula, consider the Sturm–Liouville operator A from 2.3.6(2) and the corresponding semigroup $S(\cdot)$. It is known from N. Nikolski and S. Hruschev [**HrN**] that for the majority of potentials V the second case of the conjecture occurs, i.e. $|\Theta| = 1$ on a subset of positive measure. If the above formula were true then we would have $\text{DISC}\, S(\cdot) = 1$. This means that we could always control an infinite oscillating string with dissipative boundary conditions by a single control. Moreover, by definition of the DISC, this single control could be found in any, arbitrarily chosen, finite dimensional control subspace.

In a recent talk [**Gama3**], M. Gamal has shown that the lattices $\text{Lat}(T_1)$ and $\text{Lat}(T_2)$ arc isomorphic for every two quasisimilar weak contractions T_1 and T_2. More precisely, there exists a di-formation X such that $XT_1 = T_2 X$ and the mapping $j_X : E \longmapsto \text{clos}(XE)$ is a lattice isomorphism $j_X : \text{Lat}(T_1) \longrightarrow \text{Lat}(T_2)$. In particular, $\text{DISC}(T_1) = \text{DISC}(T_2)$. This implies an explicit formula for $\text{DISC}(T)$ for any weak contraction T since T is quasisimilar to an operator of the form $U \oplus T_0$, where U is unitary and $T_0 \in C_0$. Among other things, it seems that this confirms the above conjecture on $\text{DISC}(M_\Theta)$.

The lower estimates of 2.5.1(a) and 2.5.2(b) are from A. Atzmon [**Atz**]. In fact, they are implicitly contained in B. Sz.-Nagy and C. Foiaş [**SzNF7**], and our proofs follow [**SzNF7**]. In the opposite direction, the obvious upper estimate $\mu(T^n) \leq n\mu(T)$ can be mentioned as a special case of 2.5.4(a). For more general results including a multiplicity formula for normal multiplication operators $Tf = \varphi f$, $f \in L^2(\nu)$, see M. Abrahams and T. Kriete [**AbrKr**], E. Azoff and K. Clancey [**AzCl**]. Other "spreading" procedures for overlapping spectral multiplicities, such as direct sums $T \oplus T \oplus ...$ or tensoring $T \otimes I_E$ (E is a Banach space), are also of interest. For instance,

$$\mu(T \oplus T ... \oplus T) = n\mu(T)$$

for any C_0 Hilbert space contraction T (see the Sz.-Nagy–Foiaş formula for the multiplicity of a C_0 operator, C.2.5.8(e)). A necessary and sufficient condition for

$$\mu(S \otimes I_E) = 1$$

is given in N. Nikolski [**N6**], [**N9**], where S is the shift operator on a Banach lattice $l = l(\mathbb{Z})$ on $\mathbb{Z}$. For example, it is proved that, for $l = l^p(\mathbb{Z})$,

$$\mu(S \oplus S) = 1 \Leftrightarrow \Big(\mu(S \otimes I_E) = 1, \forall E\Big) \Leftrightarrow p > 2.$$

Some consequences for completeness of integer translates in $L^p(\mathbb{R}, w)$ and in other function spaces are contained in [**N27**]. In particular, they include the formula $\mu(T_a) = 1$, where $T_a f(x) = f(x + a)$, $f \in L^p(\mathbb{R}, w)$, which holds under some hypotheses on p and w; see Section A.7.6 (Volume 1) for more details and comments.

The upper estimates from 2.5.1(c) are from R. Douglas [**Dou7**], and the formula from 2.5.1(d) is proved by V. Vasyunin [**Vas5**] (some results from V. Vasyunin and N. Makarov [**MV2**] are used).

The results from 2.5.2(a) are from N. Nikolski [**N18**], [**N17**]. Some of them are strengthened and extended in 2.5.2(b) (B. Solomyak, [**So1**]) and 2.5.2(c) (B. Solomyak and A. Volberg, [**SoV1**]). In fact, these papers contain many other results about multiplicities of Toeplitz operators, in particular, a description of cyclic subspaces of T_φ (under some assumptions on φ). Moreover, in [**SoV2**] a matrix valued analog of 2.5.2(c) is obtained. Another approach to Toeplitz multiplicities, based on Riemann surface similarity models, is developed by D. Yakubovich, see [**Yak1**], [**Yak2**].

The insighting result of 2.5.2(d) is due to W. Wogen [**Wog1**]; it can also be deduced from 2.5.4(b).

As is mentioned above, Subsection 2.5.3 is mostly based on P. Bourdon and J. Shapiro [**BouSh**] and J. Shapiro [**Shap**], where the reader can find references for further and preceding results. The behaviour of the orbits $(\omega^{(k)}(z))_{k \geq 1}$ in the complex plane has a dominant influence on cyclicity properties of C_ω, in particular, for linear-fractional self-maps of $\mathbb{D}$. These can be classified, first, by the number of fixed points in $\hat{\mathbb{C}}$ (one or two), and then, identifying these points with ∞ and 0, by their conjugacy class $\tilde{\omega} = \alpha \circ \omega \circ \alpha^{-1}$ (with a suitable linear-fractional map α) in the following way. Those having one fixed point in $\hat{\mathbb{C}}$: 1) *parabolic* (conjugate to $\tilde{\omega}(z) = z + a$); and those having two fixed points: 2) *elliptic* (conjugate to $\tilde{\omega}(z) = \kappa z$, $|\kappa| = 1$, $\kappa \neq 1$), 3) *hyperbolic* (conjugate to $\tilde{\omega}(z) = \kappa z$, $\kappa > 0$, $\kappa \neq 0$), and 4) *loxodromic* (all others). A variety of results depending on this classification is presented in 2.5.3(c); for further results (not only for l-f.m.) and open problems see the two references mentioned above.

The results of 2.5.3 (d) and (e) concern the hypercycility phenomenon that is briefly commented above. See the survey K.-G. Grosse-Erdmann [**Gr-E**] for an exhaustive reference list (1999) containing, in particular, the results by C. Read (there exists an operator on l^1 for which any nonzero vector is hypercyclic), G. Godefroy and J. Shapiro (any convolution operator on $\text{Hol}(\mathbb{C}^n)$ is hypercyclic), and H. Salas (a criterion of hypercyclicity of weighted shifts on l^p).

The results of 2.5.4(b) and (e) are from N. Nikolski [**N22**], where the formula $\mu(T_1 \times T_2) = \max(\mu(T_1), \mu(T_2))$ is studied for many classes of operators and operator algebras. The estimate from 2.5.4(c) is due to D. Herrero [**Herr1**], the proof presented above is from [**N22**]. The formula from 2.5.4(d) is proved in N. Nikolski and V. Vasyunin [**NVa3**]; the special case of upper triangular block matrices is due to W. Wogen [**Wog1**] (with a different proof). The examples described in 2.5.4(f) were the first ones giving commutants of arbitrary multiplicity, W. Wogen [**Wog2**], see also C. Apostol, L. Fialkow, D. Herrero and D. Voiculescu [**AFHV**]. For 2.5.4(g) (solving a longstanding problem of P. Halmos) see D. Herrero [**Herr1**] and [**AFHV**].

The interesting results from 2.5.4(h) and (i) represent a very short excerpt from the theory of hyponormal, seminormal (... and other "around-normal") operators. *Hyponormal operators* on a Hilbert space are defined as $T \in L(H)$ having non-negative self-commutator, $[T^*, T] \geq 0$ (i.e. $\|Tx\|^2 - \|T^*x\|^2 \geq 0$ for every $x \in H$; in particular, *subnormal operators* (restrictions of normal operators to invariant subspaces $T = N|E$) are hyponormal). C. Putnam discovered in 1971 ([**Put3**]) that

$$\pi\|[T^*, T]\| \leq m_2(\sigma(T))$$

for every hyponormal operator T, and then C. Berger and B. Shaw [**BergeSh**] proved that any hyponormal operator with a cyclic vector has a trace class self-commutator and satisfies a special case of the inequality from 2.5.4(h). In full generality, the results of 2.5.4(h) and (i) are proved by D. Voiculescu [**Voi**]. Now, there is an exciting theory of hypo- and seminormal operators, including concrete integral operator models and many beautiful results; see C. Putnam [**Put2**], K. Clancey [**Clan**], D. Herrero [**Herr2**], D. Xia [**Xia**], D. Yakubovich [**Yak3**].

For 2.5.4(j) and related results on the multiplicity of weighted and vector valued shifts we refer to N. Nikolski [**N6**], [**N9**], [**N10**] and [**N27**], where further references can be found. For the multiplicity of the shifts $f(x) \longmapsto f(x+h)$ on function spaces on the real line $\mathbb{R}$ and on $\mathbb{R}^n$ see Section A.7.6 (Volume 1) and references therein.

CHAPTER 3

Eigenvector Decompositions, Vector Valued Exponentials, and Squared Optimization

In this chapter, we treat several problems of control theory such as ECO, RECO, NCO etc., by using geometric properties of eigenvalues and eigenvectors of the generator A of the equation

$$x' = Ax + Bu.$$

After a brief analysis of the general situation in Section 3.2, we consider the case of a Hilbert space operator A having a Riesz basis of eigenvectors $A\varphi_n = -\lambda_n\varphi_n$, $A^*\psi_n = -\bar{\lambda}_n\psi_n$, $n \geq 1$. For this important special case the theory is quite complete. The corresponding results depend heavily on the behaviour of the vector valued exponentials

$$e^{-\bar{\lambda}_n t}B^*\psi_n,$$

and hence on the distribution of the frequencies $(-\lambda_n)_{n\geq 1}$ in the complex plane.

The main problem considered in this chapter is as follows. Given a system (A, B),

$$x' = Ax + Bu,$$

with prescribed spectrum $\sigma_p(A) = \{-\lambda_n : n \geq 1\}$, find "the smallest" control operator B ensuring one or another type of control. An "ideal" controller uses a sole control channel, which means that B should be a rank one operator. This is possible for some weak form of control only (as null controllability) and under strong additional restrictions on the spectrum (Section 3.5 below). In order to approach to the situation with "small" controllers, we allow ourselves to renormalize the state space and introduce several forms of generalized control. Such a scheme enables us to obtain a series of results guaranteeing the control with "small" B's taken from the Schatten–von Neumann classes $\mathfrak{S}_p$ (Section 3.4).

The classical squared optimization is considered in Section 3.7. Finally, in Section 3.8, we use the theory of exponential bases that will be developed in Chapter 4.

We begin this chapter with classical examples of systems having parabolic or hyperbolic generators (Section 3.1). Such generators have a typical band distribution of the spectrum along the real, respectively the imaginary, axis and possess orthogonal bases of eigenvectors. This situation will serve as a sample model for generalizations in Sections 3.2-3.4.

In this chapter, to avoid some formal (language) complications, *we always assume* that $(A, B) \in ACO$, that is,

$$X = \text{span}(S(t)B\mathcal{U} : t \geq 0).$$

The general case is obviously reduced to this one by restricting the semigroup to the latter span.

3.1. Examples of Parabolic and Hyperbolic Systems

We distinguish two particularly important cases for the control of the Cauchy problem: the case when $-A$ is positive definite and the case when A is left-symmetric ($A^* = -A$). Let us consider some examples.

3.1.1. Parabolic systems. Let $\frac{\partial y}{\partial t} = Ay + Bu$ be a Cauchy problem that we wish to control. Suppose that $-A$ is *elliptic*, that is $< Ay, y > \leq -\gamma \|y\|^2$ for every $y \in \mathcal{D}(A)$. Then the eigenvalues $-\lambda_n$, given by $A\varphi_n = -\lambda_n \varphi_n$, are, if they exist, negative numbers that usually tend to $-\infty$. We then obtain a family of exponentials $(e^{-\lambda_n t} B^* \psi_n)_{n \geq 1}$ with real frequencies $\lambda_n \longrightarrow +\infty$. We refer to the corresponding problem as *parabolic control problem.*

To be more concrete, let $\Omega \subset \mathbb{R}^N$ be an open set with smooth boundary (say, $\partial\Omega \in C^1$). Define a differential expression A by

$$Af = \sum_{i,j=1}^{N} \frac{\partial}{\partial x_i} \left(a_{ij}(x) \frac{\partial f}{\partial x_j}(x) \right) + a_0(x) f(x),$$

and assume that the matrix of the coefficients is real and symmetric, $a_{ij} = a_{ji}$, and a_0, $a_{ij} \in L^\infty(\Omega)$. The expression A defines on the *Sobolev space*

$$\overset{0}{W}{}_2^1 (\Omega) = \mathrm{clos}_{W_2^1(\Omega)} \left\{ f : f \text{ is compactly supported in } \Omega \ , \ \frac{\partial f}{\partial x_j} \in L^2(\Omega) \right\}$$

a continuous sesqui-linear form

$$a[f, g] = \sum_{i,j=1}^{N} \int_\Omega a_{ij}(x) \frac{\partial f}{\partial x_i} \cdot \overline{\frac{\partial g}{\partial x_j}} \, dx + \int_\Omega a_0(x) f \bar{g} \, dx.$$

This form, in turn, gives rise to a self-adjoint operator, also denoted by A,

$$(Af, g)_{L^2(\Omega)} = a[f, g], \quad f \in \mathcal{D}(A) \subset \overset{0}{W}{}_2^1 (\Omega), \ g \in \overset{0}{W}{}_2^1 (\Omega).$$

If we require $-A$ to be elliptic, that is, if there exists $\sigma > 0$ such that

$$\sum_{i,j=1}^{N} a_{ij}(x) \xi_i \xi_j \geq \sigma \|\xi\|^2, \quad \xi \in \mathbb{R}^n,$$

and if a_{ij}, $i, j = 1, \ldots, n$, are sufficiently smooth, then it is known that the point spectrum

$$\sigma_p(-A) = \{\lambda_n : A\varphi_n = -\lambda_n \varphi_n, \ \varphi_n | \partial\Omega = 0\}$$

for the homogeneous Dirichlet problem has the asymptotic behaviour $\lambda_n \asymp n^{2/N}$ as $n \to \infty$. Moreover the eigenfunctions φ_n are orthogonal and complete in $L^2(\Omega)$ (see Section 3.9 for references).

3.1.2. Hyperbolic systems. Let $\frac{\partial^2 y}{\partial t^2} = Ay + Bu$ be a control system of second order (in time). Again we suppose that $-A$ is elliptic. The standard reduction

$$z = \begin{pmatrix} y \\ \dfrac{\partial y}{\partial t} \end{pmatrix}, \quad \mathcal{L} = \begin{pmatrix} 0 & I \\ A & 0 \end{pmatrix}$$

transforms the initial system into the canonical form for control problems

$$\frac{\partial z}{\partial t} = \mathcal{L}z + \begin{pmatrix} 0 \\ Bu \end{pmatrix}.$$

Denote by $-\lambda_n$ the eigenvalues (always negative) of A. Then, setting $\mu_n = \sqrt{\lambda_n}$, we obtain the purely imaginary eigenvalues of $\mathcal{L}$: $\mathcal{L}z_n^{\pm} = \pm i\mu_n z_n^{\pm}$, $n \geq 0$, and hence the exponentials $e^{\pm i\mu_n t}B^*\psi_n{}^{\pm}$ with imaginary frequencies (and thus, oscillating). In this case we have a *hyperbolic control system* (examples of such systems are wave equations, pendula and Schrödinger equations, etc.).

To be more concrete, consider a system of independent pendula (or simply oscillating particles) controlled by a scalar function u:

$$\begin{cases} \ddot{y}_n & = -\lambda_n^2 y_n + u(t), \quad t \geq 0 \\ \dot{y}_n(0) & = b_n \\ y_n(0) & = a_n \end{cases},$$

where $(y_n)_{n\geq 1} = (y_n(t))_{n\geq 1} \in l^2$ (finite energy system), $(a_n)_{n\geq 1}, (b_n)_{n\geq 1} \in l^2$, $\lambda_n > 0$, $\lambda_n \neq \lambda_k$ $(n \neq k)$ and $\sum_{n\geq 1} \frac{1}{\lambda_n{}^2} < \infty$.

Setting $L = \mathrm{diag}(-\lambda_n{}^2)$, $y = (y_n)_{n\geq 1}$ and $\bar{u} = (u, u, ...)$, we get

$$\frac{\partial^2 y}{\partial t^2} = Ly + \bar{u},$$

which, as explained above, turns into the standard form

$$\frac{\partial z}{\partial t} = \mathcal{L}z + \begin{pmatrix} 0 \\ \bar{u} \end{pmatrix},$$

with

$$z = \begin{pmatrix} y \\ \dfrac{\partial y}{\partial t} \end{pmatrix}, \quad \mathcal{L} = \begin{pmatrix} 0 & I \\ L & 0 \end{pmatrix}.$$

The latter differential equation is considered in the space $l^2 \oplus l^2(\frac{1}{\lambda_n^2})$ of vectors $z = (y_0, y_1)$ with $y_0 \in l^2$, $y_1 \in l^2(\frac{1}{\lambda_n^2})$ $(\sum_{n\geq 1} |(y_1)_n|^2 \frac{1}{\lambda_n{}^2} < \infty)$. In particular, $\bar{u} \in l^2(\frac{1}{\lambda_n^2})$. An explicit calculation gives (modulo an obvious rearrangement)

$$S(t) = e^{\mathcal{L}t} = \mathrm{diag}\left\{\begin{pmatrix} \cos \lambda_n t & \dfrac{1}{\lambda_n} \sin \lambda_n t \\ -\lambda_n \sin \lambda_n t & \cos \lambda_n t \end{pmatrix}\right\}_{n\geq 1}.$$

The eigenvector equation is

$$\mathcal{L}\begin{pmatrix} y_0 \\ y_1 \end{pmatrix} = \gamma \begin{pmatrix} y_0 \\ y_1 \end{pmatrix},$$

that is

$$\begin{cases} y_1 & = \gamma y_0 \\ L y_0 & = \gamma y_1 = \gamma^2 y_0. \end{cases}$$

The second equation means that

$$-{\lambda_n}^2 (y_0)_n = \gamma^2 (y_0)_n, \quad n \geq 1.$$

This implies the existence of an integer m such that

$$\begin{aligned} y_0 &= (\delta_{mn})_{n\geq 1} =: \delta_m, \\ \gamma &= \pm i\lambda_m. \end{aligned}$$

Here δ_{mn} is the usual Kronecker symbol. We obtain an orthogonal basis (normalized by 2) of eigenvectors in $l^2 \oplus l^2(\frac{1}{\lambda_n^2})$:

$$\begin{cases} \varphi_m^+ & = \begin{pmatrix} \delta_m \\ i\lambda_m \delta_m \end{pmatrix} \text{ for } \gamma = i\lambda_m \\ \varphi_m^- & = \begin{pmatrix} \delta_m \\ -i\lambda_m \delta_m \end{pmatrix} \text{ for } \gamma = -i\lambda_m. \end{cases}$$

The control operator is of rank 1: $Bc = \begin{pmatrix} 0 \\ c \cdot \mathbf{1} \end{pmatrix}$, where $c \in \mathbb{C}$ and $\mathbf{1} = (1, 1, ...)$. Now, we compute the vector valued system of exponentials mentioned in the introduction to this chapter. First, note that the family $(\varphi_m^\pm)$ is orthogonal, and hence, it is equal — up to a multiplicative constant — to its biorthogonal sequence $(\psi_m^\pm)$. Applying the adjoint controllability map (taken in $\tau - t$ instead of t, cf. Theorem 1.3.3) to this family we obtain

$$\mathcal{E}_m^\pm(t) = B^* S(t)^* \psi_m^\pm = e^{\pm i\lambda_m t} B^* \psi_m^\pm = b_m^\pm e^{\pm i\lambda_m t}, \quad 0 \leq t \leq \tau.$$

It is convenient to rearrange this sequence in the following way

$$\mathcal{E}_n = e^{i\mu_n t}, \quad n \in \mathbb{Z},$$

where

$$\begin{cases} \mu_n & = \lambda_n, \quad n \geq 0, \\ \mu_n & = -\lambda_n, \quad n < 0. \end{cases}$$

We will see in Chapter 4 that the geometric properties of such a family $(\mathcal{E}_n)_{n\in\mathbb{Z}}$ in $L^2(0, \tau)$ depend on the distribution of the frequencies λ_n and the time of control τ. In particular, if the frequencies are well-distributed, then there is a critical time $\tau_0 = 2\pi \cdot \underline{\lim}_{n\to\infty} \frac{\lambda_n}{n}$ called the *optical length* of the system, and which is characterized by the following property. For $\tau < \tau_0$ there is no control even in a generalized sense (cf. Section 3.4), whereas if $\tau > \tau_0$ the system is generalized exactly controllable (that is, there is an efficient description of the reachable states, cf. Section 3.4 for details) and even renormalized exactly controllable (see Sections 3.4, 3.5 and 3.8).

3.1.3. Alternation of axes: Fourier versus Laplace. When studying exponentials, and, more generally, in Fourier analysis, one preferably uses the Fourier transform rather than the Laplace transform. This means that we replace $e^{-\bar{\lambda}_n t}$ by

$$e^{-\bar{\lambda}_n t} = e^{i(i\bar{\lambda}_n)t} = e^{i\mu_n t}.$$

Hence, if $\{-\lambda_n\}_{n\geq 1} = \sigma_p(A)$ is the point spectrum of a continuous semigroup, then

$$\operatorname{Im} \mu_n \geq -C, \quad C \in \mathbb{R},$$

where the last property comes from the assumption $\operatorname{Re}(-\lambda_n) \leq C$. Using the Fourier transform we thus obtain imaginary values μ_n in the parabolic case and real ones in the hyperbolic case.

3.1.4. Another source of examples: convolution operators. Differential operators with constant coefficients are in fact nothing else than convolution operators with suitable distributions. For instance, if we denote by δ_{t_k} the Dirac measure supported by a point $t_k \in \mathbb{R}$ and set $A = \sum_{k,j} a_{k,j}\delta_{t_k}^{(j)}$, then we get

$$(Af)(x) = \sum_{k,j} a_{k,j} f^{(j)}(x - t_k), \quad x \in \mathbb{R},\ f \in C^\infty(\mathbb{R}).$$

This means that the action of the system on the signal f is concentrated in certain isolated fixed points, a rather particular situation. In general, this action may be arbitrarily distributed on the control interval, which gives rise to convolution operators: $Af = f * S$, where S is a function, a measure or even a distribution verifying $f \in X \Longrightarrow S * f \in X$ (a so-called *convolutor* of X). The characteristic equation is then given by

$$S * f = \lambda f$$

for some $f \in X$. In case the Fourier transform makes sense, this is equivalent to the existence of $\mu \in \mathbb{C}$ such that $\hat{S}(\mu) = \lambda$, $f(x) = e^{i\mu x}$.

Now, suppose that we have the so-called *distributed boundary conditions*, which means that we consider the domain

$$\mathcal{D}(A) = \{f \in X : S * f \in X \text{ and } < f, \phi >= 0\},$$

where ϕ is some distribution defined and continuous on X. Then the eigenvalues of A are given by

$$\{\lambda = \hat{S}(\mu) : \mu \in \mathbb{C},\ 0 =< e^{i\mu x}, \phi >= \hat{\phi}(\mu)\}.$$

This approach provides a great number of important examples and a great variety of eigenvalue distributions (see Section 3.9 for references).

3.2. Complete Generators

Let $S(t) : X \longrightarrow X$, $t \geq 0$, be a continuous semigroup, and assume that its generator A has a *minimal* system of eigenvectors complete in X:

$$A\varphi_n = -\lambda_n \varphi_n, \quad n \geq 1,$$
$$\operatorname{span}(\varphi_n : n \geq 1) = X$$

(we refer to Subsection A.5.1.1 (Volume 1) for the definition and properties of minimal families). Let $(\psi_n)_{n\geq 1} \subset X^*$ be the associated *biorthogonal* system

$$\langle \varphi_n, \psi_k \rangle = \delta_{nk},$$

which we also assume to be complete (total) in X^*: $\langle x, \psi_k \rangle = 0$, $k \geq 1$, for arbitrary $x \in X$ implies that $x = \mathbf{0}$. It is easy to see that

$$A^* \psi_n = -\bar{\lambda}_n \psi_n, \quad n \geq 1,$$

(or $A^* \psi_n = -\lambda_n \psi_n$ if the duality of X and X^* is given by a bilinear form). Then

$$(3.2.1) \qquad \begin{aligned} S(t)\varphi_n &= e^{-\lambda_n t} \varphi_n, \quad n \geq 1, \ t \geq 0 \\ S(t)^* \psi_n &= e^{-\bar{\lambda}_n t} \psi_n, \quad n \geq 1, \ t \geq 0. \end{aligned}$$

With every x in X we can now associate its *generalized Fourier series* with respect to $((\varphi_n)_{n \geq 1}, (\psi_k)_{k \geq 1})$:

$$x \sim \sum_{n \geq 1} \langle x, \psi_n \rangle \varphi_n.$$

We get

$$S(t)x \sim \sum_{n \geq 1} e^{-\lambda_n t} \langle x, \psi_n \rangle \varphi_n, \quad t \geq 0.$$

In case the series converges, we have $S(t)x = \sum_{n \geq 1} e^{-\lambda_n t} \langle x, \psi_n \rangle \varphi_n$, and if moreover $X = L^2(\Omega)$, then

$$S(t)x = (x, G(t, \cdot, \cdot))_{L^2(\Omega)},$$

where the *Green function* $G(t, r, s)$ of the system is given by

$$G(t, r, s) = \sum_{n \geq 1} e^{-\lambda_n t} \overline{\psi_n}(r) \varphi_n(s).$$

In this section we assume neither convergence nor summability of these series. The only property that will be used is the uniqueness property: a vector x having vanishing coefficients is equal to zero.

Recall that a minimal family is called *uniformly minimal* (see A.1.5.1, Volume 1) if

$$\inf_{n \geq 1} \operatorname{dist}\left(\varphi_n / \|\varphi_n\|, \operatorname{span}(\varphi_k : k \neq n)\right) > 0,$$

which is equivalent to say that the coordinate projections $\mathcal{P}_n$, defined by the equation $\mathcal{P}_n(\sum a_k \varphi_k) = \langle \sum a_k \varphi_k, \psi_n \rangle \varphi_n = a_n \varphi_n$, are uniformly bounded: $1 \leq \|\mathcal{P}_n\| = \|\varphi_n\| \cdot \|\psi_n\| \leq \text{const}$, $n \geq 1$. In the sequel we will work with some stronger properties than minimality and uniform minimality.

Now, using Fourier series if eigenfunctions, we consider the exact control equation

$$S(\tau)x_0 + c(\tau)u = x_1,$$

or

$$S(\tau)x_0 + \int_0^\tau S(\tau - t)Bu(t)\, dt = x_1.$$

Here $x_0 \in X$ is the initial state, x_1 the final state of the system and $u \in L^2(0, \tau; \mathcal{U})$ is a finite energy control (cf. Chapter 1).

3.2.1. LEMMA. *Let $\mathcal{E}_n \in L^2(0,\tau;\mathcal{U}^*)$ be given by*

$$\mathcal{E}_n(t) = e^{-\bar{\lambda}_n t} B^* \psi_n, \quad n \geq 1, t \geq 0.$$

Then

$$\langle c(\tau)u, \psi_n \rangle_X = \langle u_1, \mathcal{E}_n \rangle_{L^2(0,\tau;\mathcal{U})}, \quad n \geq 1,$$

for every $u \in L^2(0,\tau;\mathcal{U})$, where $u_1(t) = u(\tau - t)$, $0 \leq t \leq \tau$.

PROOF. We have

$$\begin{aligned} \langle c(\tau)u, \psi_n \rangle_X &= \int_0^\tau \langle S(\tau - t)Bu(t), \psi_n \rangle_X \, dt \\ &= \int_0^\tau \langle u(t), B^* S(\tau - t)^* \psi_n \rangle_{\mathcal{U}} \, dt \\ &= \int_0^\tau \langle u(t), e^{-\bar{\lambda}_n (\tau - t)} B^* \psi_n \rangle_{\mathcal{U}} \, dt \\ &= \int_0^\tau \langle u_1(t), e^{-\bar{\lambda}_n s} B^* \psi_n \rangle_{\mathcal{U}} \, ds \\ &= \langle u_1, e^{-\bar{\lambda}_n t} B^* \psi_n \rangle_{L^2(0,\tau;\mathcal{U})} \\ &= \langle u_1, \mathcal{E}_n \rangle. \end{aligned}$$

□

3.2.2. COROLLARY. *Let $x_0 \in X$ be an initial state. The state $x_1 \in X$ is reachable at time $\tau > 0$ starting from x_0 (see 1.2.1 for the definition) if and only if*

$$\langle x_1 - S(\tau)x_0, \psi_n \rangle_X = \langle u_1, e^{-\bar{\lambda}_n t} B^* \psi_n \rangle_{L^2(0,\tau;\mathcal{U})}, \quad n \geq 1.$$

□

This corollary, which works under the completeness condition on $(\psi_n)_{n\geq 1}$, reduces the control problem to the following *moment problem*.

Given the numbers

$$a_n = \langle x_1 - S(\tau)x_0, \psi_n \rangle, \quad n \geq 1,$$

find a function $u_1 \in L^2(0,\tau;\mathcal{U})$ such that

$$\langle u_1, \mathcal{E}_n \rangle = a_n, \quad n \geq 1.$$

For this reason, the approach to control problems by means of Fourier series $\sum_{n\geq 1} \langle x, \psi_n \rangle \varphi_n$ and families of exponentials $(\mathcal{E}_n)_{n\geq 1}$, is often called the *method of moments*. We will enter in some details of this method. Certainly, geometric properties of the family $(\mathcal{E}_n)_{n\geq 1}$ will play an important role. Recall that we always assume that $(A, B) \in ACO$ (see the introduction to this chapter), and this implies that $B^*\psi_n \neq 0$ for every $n \geq 1$, and hence $\mathcal{E}_n \neq 0$, $n \geq 1$.

3.2.3. THEOREM. *Let $S(\cdot)$ be a strongly continuous semigroup with a complete minimal system of eigenvectors $(\varphi_n)_{n\geq 1}$. Suppose that the associated biorthogonal family $(\psi_n)_{n\geq 1}$ is total. If the corresponding system (A, B) is exactly controllable, then*

$$\|\psi_n\| \asymp \|e^{-\bar{\lambda}_n t} B^* \psi_n\|_{L^2(0,\tau;\mathcal{U}^*)}, \quad n \geq 1,$$

that is, there are constants $c, C > 0$ such that

$$c\|\psi_n\| \leq \|e^{-\bar{\lambda}_n t} B^* \psi_n\|_{L^2(0,\tau;\mathcal{U}^*)} \leq C\|\psi_n\|, \quad n \geq 1.$$

PROOF. The left hand estimate $c\|\psi_n\| \leq \|e^{-\bar{\lambda}_n t} B^* \psi_n\|_{L^2(0,\tau;\mathcal{U}^*)}$ is a direct consequence of Theorem 1.3.4 with $y = \psi_n$.

Consider the right hand estimate. As $S(\cdot)$ is continuous, the growth bound $\alpha = \alpha_0(S(\cdot))$ is finite, so $s := \sup_{n\geq 1} \operatorname{Re}(-\lambda_n) \leq \alpha < +\infty$. Hence

$$\begin{aligned} \|\mathcal{E}_n\|_{L^2}^2 &= \int_0^\tau e^{-2\operatorname{Re}\lambda_n t} \|B^*\psi_n\|_{\mathcal{U}^*}^2 \, dt = \|B^*\psi_n\|_{\mathcal{U}^*}^2 \left(\frac{e^{-2\operatorname{Re}\lambda_n \tau} - 1}{-2\operatorname{Re}\lambda_n \tau} \right) \cdot \tau \\ &\leq \text{const} \cdot \|\psi_n\|^2. \end{aligned}$$

(We have used that $\frac{e^x - 1}{x} \leq$ const for $-\infty < x \leq$ const.) □

In what follows, the following lemma will be useful.

3.2.4. LEMMA. *In the the above notation, assume that $(\varphi_n)_{n\geq 1}$ is minimal and that*

$$\varphi_n \in B\mathcal{U}, \quad n \geq 1.$$

Pick $u_n \in \mathcal{U}$ with $\varphi_n = Bu_n$, $n \geq 1$. Then $(\mathcal{E}_n)_{n\geq 1}$ is likewise minimal, and every family $(f_n \cdot u_n)_{n\geq 1}$, where $f_n \in L^2(0,\tau)$ normalized by

$$\int_0^\tau f_n(t) e^{-\lambda_n t} dt = 1, \quad n \geq 1,$$

is a biorthogonal system for $(\mathcal{E}_n)_{n\geq 1}$.

PROOF.

$$\begin{aligned} \langle f_k \cdot u_k, \mathcal{E}_n \rangle_{L^2(0,\tau;\mathcal{U})} &= \int_0^\tau f_k(t) e^{-\lambda_n t} \langle u_k, B^*\psi_n \rangle_{\mathcal{U}} \, dt \\ &= \delta_{nk} \int_0^\tau f_n(t) e^{-\lambda_n t} dt = \delta_{nk}. \end{aligned}$$

□

3.3. Riesz Bases and Exact Controllability

Let $X = H$ and $\mathcal{U}$ be Hilbert spaces and let $(\varphi_n)_{n\geq 1}$ be a sequence of nonzero vectors in H. Recall (see C.3.1.3) that $(\varphi_n)_{n\geq 1}$ is a Riesz sequence (or basis if it spans the entire space H) if there exist constants $c, C > 0$ such that

$$c^2 \sum_{n\geq 1} |a_n|^2 \|\varphi_n\|^2 \leq \left\| \sum_{n\geq 1} a_n \varphi_n \right\|^2 \leq C^2 \sum_{n\geq 1} |a_n|^2 \|\varphi_n\|^2,$$

for every finite family $(a_n)_{n\geq 1}$ of complex (or real if H is a real space) numbers. Recall also that Riesz bases are exactly the same as unconditional bases of a Hilbert space. Moreover, if $(\varphi_n)_{n\geq 1}$ is a Riesz basis in H, then

$$H = \left\{ \sum_{n\geq 1} a_n \varphi_n : \sum_{n\geq 1} |a_n|^2 \|\varphi_n\|^2 < \infty \right\}. \tag{3.3.1}$$

Observe that the associated biorthogonal family $(\psi_n)_{n\geq 1}$ itself is a Riesz basis and satisfies

$$1 \leq \|\varphi_n\| \cdot \|\psi_n\| \leq C/c, \quad n = 1, 2, \ldots.$$

(These estimates mean that $(\varphi_n)_{n\geq 1}$ is uniformly minimal.) Moreover,

$$H = \left\{ \sum_{n\geq 1} a_n \varphi_n : \sum_{n\geq 1} |a_n|^2 \frac{1}{\|\psi_n\|^2} \leq \infty \right\}. \tag{3.3.2}$$

3.3.1. THEOREM. *Let $(\varphi_n)_{n\geq 1}$ be the eigenvectors of the generator A of a semigroup $S(\cdot)$,*

$$A\varphi_n = -\lambda_n \varphi_n, \quad n \geq 1,$$

and suppose that they constitute a Riesz basis in H. Let further $(\psi_n)_{n\geq 1}$ be its biorthogonal system. If the family $(\mathcal{E}_n)_{n\geq 1}$, given by

$$\mathcal{E}_n(t) = e^{-\bar{\lambda}_n t} B^* \psi_n,$$

is a Riesz sequence in $L^2(0,\tau;\mathcal{U})$, then the space Range $c(\tau)$ *of reachable states from the zero initial state $x_0 = 0$ has the following form:*

$$\begin{aligned}
\operatorname{Range} c(\tau) &= \left\{ \sum_{n\geq 1} b_n \varphi_n : b_n = < u_1, \mathcal{E}_n >, \ u_1 \in L^2(0,\tau;\mathcal{U}) \right\} \\
&= \left\{ \sum_{n\geq 1} b_n \varphi_n : \sum_{n\geq 1} |b_n|^2 \frac{1}{\|\mathcal{E}_n\|^2_{L^2}} < \infty \right\}.
\end{aligned}$$

PROOF. This is an immediate consequence of the preceding remarks and Corollary 3.2.2 (see also the remark just before Theorem 3.2.3). □

This theorem gives a half of the following criterion for ECO. Recall that we always assume that $(A, B) \in ACO$.

3.3.2. THEOREM. *In the notation and assumptions of Theorem 3.3.1, consider the associated dynamical system (A, B). The following assertions are equivalent.*

(1) (A, B) is exactly controllable at time $\tau > 0$.

(2) $(\mathcal{E}_n)_{n\geq 1}$ is a Riesz sequence in $L^2(0,\tau;\mathcal{U})$. Moreover,

$$\begin{aligned}
&\sup_{n\geq 1} |\operatorname{Re} \lambda_n| < \infty, \\
&\inf_{n\geq 1} \frac{\|B^*\psi_n\|}{\|\psi_n\|} > 0.
\end{aligned} \tag{3.3.3}$$

PROOF. (2) $\Longrightarrow$ (1). Clearly

$$\|\mathcal{E}_n\|^2_{L^2(0,\tau;\mathcal{U})} = \int_0^\tau e^{-2\operatorname{Re}\lambda_n t} \|B^*\psi_n\|^2_{\mathcal{U}}\, dt = \frac{e^{-2\operatorname{Re}\lambda_n \tau} - 1}{-2\operatorname{Re}\lambda_n} \|B^*\psi_n\|^2_{\mathcal{U}}. \tag{3.3.4}$$

As B is bounded and $\sup(-\operatorname{Re}\lambda_n) < \infty$, we have $\|\mathcal{E}_n\|^2 \asymp \|\psi_n\|^2$, $n \geq 1$. By Theorem 3.3.1 and (3.3.2) we obtain $\operatorname{Im} c(\tau) = H$.

(1) $\Longrightarrow$ (2). The fact that ECO forces $(\mathcal{E}_n)_{n\geq 1}$ to be a Riesz sequence is a consequence of a more general result, see Theorem 3.4.3 below. The reader should verify that there is no loop in the reasoning: Theorem 3.4.3 is independent of the present one. Now, if $(\mathcal{E}_n)_{n\geq 1}$ is a Riesz sequence, then Theorem 3.3.1 and exact controllability imply that $\|\mathcal{E}_n\|^2 \asymp \|\psi_n\|^2$, $n \geq 1$. This together with the identity (3.3.4) imply (3.3.3). □

3.3.3. REMARK. It is worth mentioning that under the assumption that $(\varphi_n)_{n\geq 1}$ and $(\psi_n)_{n\geq 1}$ are Riesz bases, the necessity of conditions (3.3.3) for exact controllability is already clear from the results of Section 1.3, namely from Theorem 1.3.7 and Corollary 1.3.9.

The spectral condition $|\operatorname{Re}\lambda_n| \leq \text{const}$ tells us that the spectrum of A (or $S(\cdot)$) has to be in a strip, in other words, *a spectrum of hyperbolic type is necessary for exact controllability.* Anticipating the results of Chapter 4 we can add that if the scalar exponentials $(e^{-\bar{\lambda}_n t})_{n\geq 1}$ form a Riesz sequence in $L^2(0,\tau)$, then this holds also for $(\mathcal{E}_n)_{n\geq 1}$. Therefore, using the criteria of Chapter 4, we can claim that if the frequencies of hyperbolic type are well-distributed (see Section 4.5) in the mentioned strip, and if moreover $\|B^*\psi_n\| \geq \varepsilon\|\psi_n\|$, $n \geq 1$ for some $\varepsilon > 0$, then the system (A, B) is exactly controllable.

Returning to Theorem 3.3.1, we can observe that since $\operatorname{Re}(-\lambda_n) \leq \text{const}$, we have $\|\mathcal{E}_n\| = \|e^{-\bar{\lambda}_n t}B^*\psi_n\| \leq \|\psi_n\|_X$ which once more shows that $c(\tau)L^2 \subset X$ (cf. Lemma 1.3.12). As a comment to the possible difference between $c(\tau)L^2$ and X, we would like to mention that it can be caused by a difference in the normalization of the bases (φ_n, ψ_n) and $(\mathcal{E}_n, \mathcal{E}_n^*)$. Sometimes, this can explain an eventual lack of control. This is illustrated by the following example and will be studied in details in Section 3.4.

3.3.4. EXAMPLE. Recall the *heating process of a metal bar* treated in Example 1.5 and described by the equation

$$\begin{cases} Ax &= \frac{d^2x}{ds^2} \\ x'(0) &= x'(1) = 0. \end{cases}$$

For the corresponding eigenvalues and eigenfunctions, $A\varphi_n = -\lambda_n\varphi_n$, we obtain

$$\begin{cases} \lambda_n = n^2\pi^2 \\ \varphi_0 = \psi_0 = 1, \quad \varphi_n(s) = \psi_n(s) = \cos\pi ns, \quad \|\varphi_n\| = 1/\sqrt{2}, \quad n = 1, \dots \\ \mathcal{E}_n = e^{-\lambda_n t}\cos\pi ns, \quad \|\mathcal{E}_n\|^2_{L^2(0,\tau;L^2(0,1))} = \dfrac{1}{2}\dfrac{1 - e^{-2\tau\lambda_n}}{2\lambda_n} \asymp \dfrac{1}{\lambda_n}, \quad n \to \infty. \end{cases}$$

Because of the orthogonality, the completeness and the normalization of $(\varphi_n)_{n\geq 1}$ we get

$$H = L^2(0,1) = \left\{ \sum_{n\geq 0} a_n\varphi_n : \sum_{n\geq 0} |a_n|^2 < \infty \right\}.$$

The orthogonality of $(\varphi_n)_{n\geq 0}$ carries over to that of $(\mathcal{E}_n)_{n\geq 0}$. Using Theorem 3.3.1 we obtain

$$\begin{aligned} \operatorname{Range} c(\tau) &= \left\{ \sum_{n\geq 0} a_n\varphi_n : \sum_{n\geq 0} |a_n|^2\lambda_n < \infty \right\} \\ &= \left\{ \sum_{n\geq 0} a_n\varphi_n : \sum_{n\geq 0} |a_n|^2 n^2 < \infty \right\} \\ &= W_2^1(0,1), \end{aligned}$$

where $W_2^1(0,1)$ is the first order *Sobolev space* on $(0,1)$.

3.4. Generalized Controllability and Renormalizations

This section is devoted to a deeper investigation of the controllability of those semigroups $S(\cdot)$ that have a basis of eigenfunctions. We introduce the notion of generalized controllability as well as that of renormalized exact controllability. It turns out that many dynamical systems (hyperbolic rather than parabolic) are exactly controllable after a suitable adjustment of the space the semigroup is acting on. Recall that we always assume that $(A, B) \in ACO$.

3.4.1. The problem of generalized controllability, and renormalizations. For applications it is often sufficient to have an explicit description of the space of reachable states

$$\operatorname{Range} c(\tau) = c(\tau)L^2(0, \tau; \mathcal{U})$$

instead of the much more restrictive requirement of exact controllability $\operatorname{Range} c(\tau) = X$. Moreover, for the control of a dynamical system, the principal objects are the operators A, B and the corresponding semigroup $S(\cdot)$ whereas the underlying space X can be, if necessary, adapted to these objects. For this reason, as in a kind of generalized controllability problem, we will mainly be interested in a description of $\operatorname{Range} c(\tau)$.

In case the generator A admits a complete minimal family $(\varphi_n)_{n\geq 1}$ of eigenvectors, $A\varphi_n = -\lambda_n \varphi_n$, $n \geq 1$, having a complete biorthogonal $(\psi_n)_{n\geq 1}$, every $x \in X$ is defined by its formal Fourier series

$$x \sim \sum_{n\geq 1} \langle x, \psi_n \rangle \varphi_n,$$

and hence by its Fourier coefficients $\langle x, \psi_n \rangle$, $n \geq 1$. It thus seems natural to seek a description of the control space $\operatorname{Range} c(\tau)$ in terms of some weighted Fourier coefficients as is made in the following definition.

3.4.2. Definition. For a sequence $(w_n)_{n\geq 1}$ of positive numbers we set, using the preceding notation,

$$X(w_n) = \left\{ x \in X : \text{ there is } y \in X \text{ such that } \langle x, \psi_n \rangle = \frac{1}{w_n} \langle y, \psi_n \rangle, n \geq 1 \right\}.$$

The system (A, B) is called *generalized exactly controllable* (GECO) at time $\tau > 0$ if there are $w_n > 0$, $n \geq 1$, such that

$$\operatorname{Range} c(\tau) = X(w_n),$$

and *renormalized exactly controllable* (RECO) if there are $w_n > 0$, $n \geq 1$, such that

$$\begin{aligned} S(t)X(w_n) &\subset X(w_n), \quad t > 0, \\ B\mathcal{U} &\subset X(w_n), \end{aligned}$$

and $(A|X(w_n), B)$ is exactly controllable.

In view of the preceding discussion (Sections 3.2 and 3.3), the most natural renormalization corresponds to the multipliers

$$w_n = \frac{\|\psi_n\|}{\|\mathcal{E}_n\|}, \quad n \geq 1.$$

Actually, the following Theorem 3.4.3 justifies this claim, at least for the case of a generator having a Riesz basis of eigenvectors. Then the renormalization not only exists, but is even unique. We wish to point out the difference between GECO and RECO. In fact, GECO means nothing but a natural description of the space of reachable states from $x_0 = 0$, whereas in RECO we achieve the true exact control (and by the same control operator B) simply by replacing the state space X by another one which is better adapted to the semigroup $S(\cdot)$. Before studying GECO and RECO, we mention that null controllability can be generalized in a similar way. The *generalized null controllability problem* (GNCO) at time $\tau > 0$ consists in describing the space

$$\{x \in X : S(\tau)x \in c(\tau)L^2(0,\tau;\mathcal{U})\}.$$

It can be treated in the same way as GECO, RECO and ENCO, but we are rather interested in the latter three control problems. We start with GECO.

3.4.3. THEOREM. *Let $\mathcal{U}$, H be Hilbert spaces, $B \in L(\mathcal{U}, H)$, and let A be the generator of a continuous semigroup in H having a Riesz basis of eigenvectors, $A\varphi_n = -\lambda_n\varphi_n$, $n \geq 1$. Let further $(\psi_n)_{n\geq 1}$ be the biorthogonal family and $\mathcal{E}_n = e^{-\bar{\lambda}_n t}B^*\psi_n$, $n \geq 1$. The following assertions are equivalent.*

(1) The system (A, B) is generalized exactly controllable (GECO) at time $\tau > 0$.

(2) $(\mathcal{E}_n)_{n\geq 1}$ is a Riesz sequence in $L^2(0,\tau;\mathcal{U})$.

PROOF. The implication (2) $\Longrightarrow$ (1) follows from Theorem 3.3.1, (3.3.2) and the definition of GECO by setting $w_n = \|\psi_n\|/\|\mathcal{E}_n\|$.

(1) $\Longrightarrow$ (2). Let w_n, $n \geq 1$, be such that

$$\operatorname{Range} c(\tau) = H(w_n) = \left\{ x = \sum_{n\geq 0} a_n\varphi_n \in H : \sum_{n\geq 0} |a_n|^2 {w_n}^2 \|\varphi_n\|^2 = \|x\|_w^2 < \infty \right\}.$$

Observe first that by definition $H(w_n) \subset H$ and hence $H(w_n) = H(w_n + 1)$. Then, without loss of generality, we may assume that $\inf_{n\geq 1} w_n > 0$. Now $H(w_n)$ equipped with the norm $\|\cdot\|_w$ is a Hilbert space. Let $c_1(\tau) : L^2(0,\tau;\mathcal{U}) \longrightarrow H(w_n)$ be defined by $u \longmapsto c(\tau)u$. Then $c_1(\tau)$ is continuous and surjective. By Banach's theorem (see Lemma 1.3.12), its adjoint admits a lower estimate $\|c_1(\tau)^*y\| \geq \varepsilon\|y\|$, $y \in H(w_n)^*$. Realizing the duality between $H(w_n)$ and $H(w_n)^*$ via the scalar product of H:

$$\langle \sum_{n\geq 1} a_n\varphi_n, \sum_{k\geq 1} b_k\psi_k\rangle_w = \sum_{n\geq 1} a_n\bar{b}_n$$

(at least for finite sums), we get

$$\Big\| \sum_{k\geq 1} b_k\psi_k \Big\|^2_{H(w)^*} \asymp \sum_{k\geq 1} |b_k|^2 \frac{1}{{w_k}^2}\|\psi_k\|^2 \quad \Big(\text{or} \asymp \sum_{k\geq 1} |b_k|^2 \frac{1}{{w_k}^2\|\varphi_k\|^2} \Big).$$

Moreover, with this convention, we have $c_1(\tau)^*|H = c(\tau)^*$ and hence $c_1(\tau)^*\psi_n = B^*S(\tau - t)^*\psi_n = e^{-\bar{\lambda}_n(\tau-t)}B^*\psi_n$, $0 \leq t \leq \tau$, $n \geq 1$. Now, $c_1(\tau)^*$ is an isomorphism onto its range, and as $(\psi_n)_{n\geq 1}$ is a Riesz basis in $H(w_n)^*$, the sequence $(e^{-\bar{\lambda}_n(\tau-t)}B^*\psi_n)_{n\geq 1}$ is also a Riesz sequence in $L^2(0,\tau;\mathcal{U})$. Consequently, this is also true for $(\mathcal{E}_n)_{n\geq 1}$ since

$$f(t) \stackrel{J}{\longmapsto} f(\tau - t)$$

is a unitary operator in $L^2(0,\tau;\mathcal{U})$. □

In order to guarantee renormalized exact controllability we have to add another strong requirement, namely, condition (3.4.1) below. The analysis of this condition will be the purpose of Subsection 3.4.6

3.4.4. THEOREM. *In the notation and assumptions of Theorem 3.4.3. the following assertions are equivalent.*

(1) The system (A,B) is renormalized exactly controllable (RECO) at time $\tau > 0$, and the renormalization $H(w_n)$ is given by

$$c(\tau)L^2(0,\tau;\mathcal{U}) = H(w_n) = \left\{ \sum_{n\geq 1} a_n \varphi_n : \sum_{n\geq 1} |a_n|^2 \frac{1}{\|\mathcal{E}_n\|^2} < \infty \right\}.$$

(2) The family $(\mathcal{E}_n)_{n\geq 1}$ is a Riesz sequence in $L^2(0,\tau;\mathcal{U})$ and

$$B\mathcal{U} \subset H(w_n). \tag{3.4.1}$$

PROOF. (1) $\Longrightarrow$ (2) is a consequence of Theorem 3.4.3 and the definition of RECO.

(2) $\Longrightarrow$ (1). Set $w_n = \|\psi_n\|/\|\mathcal{E}_n\|$. Then Theorems 3.4.3 and 3.3.1 give $\operatorname{Range} c(\tau) = H(w_n)$. We have already seen that $S(t)\varphi_n = e^{-\lambda_n t}\varphi_n$, where $\operatorname{Re}(-\lambda_n) \leq$ const (note that $S(\cdot)$ was supposed to be strongly continuous, and hence the growth bound is finite), and $S(t)^*\psi_n = e^{-\bar{\lambda}_n t}\psi_n$, $t \geq 0$ (cf. (3.2.1)). Consequently, for arbitrary $t \geq 0$ and $n \geq 1$, we get

$$|\langle S(t)y, \psi_n\rangle| = e^{-Re\lambda_n t}|\langle y, \psi_n\rangle| \leq c(t)\cdot|\langle y,\psi_n\rangle|.$$

This implies that every renormalization based on the lattice property of Fourier coefficients $\langle y, \psi_n\rangle$ is invariant under $S(t)$. Moreover if $(\varphi_n)_{n\geq 1}$ is a Riesz basis, then $S(\cdot)$ will be continuous on such a space. The explicit formula for the controllability map

$$c(\tau)u = \int_0^\tau S(\tau - t)Bu(t)\,dt, \quad u \in L^2(0,\tau;\mathcal{U}),$$

shows that it only depends on the control subspace $B\mathcal{U}$ and the action of the semigroup on this subspace. Hence the restriction $S(\cdot)|H(w_n)$ has the same set of reachable states, $\operatorname{Range} c(\tau)$, which, by Theorem 3.3.1, coincides with $H(w_n)$. Hence (A,B) is generalized exactly controllable, and as $B\mathcal{U} \subset H(\omega_n)$ we deduce that $(A|H(w_n), B)$ is exactly controllable. □

3.4.5. EXAMPLE (Heating of a metal bar (3.3.4) revisited). We had

$$\begin{cases} \dfrac{\partial x}{\partial t} &= Ax + Bu, \quad A = \dfrac{d^2}{ds^2}, \quad B = I \\ x'(0) &= x'(1) = 0, \end{cases}$$

with $H = \mathcal{U} = L^2(0,1)$, $w_n \asymp 1/n^2$. Then

$$\begin{aligned} \operatorname{Range} c(\tau) = H(w_n) &= \left\{ \sum_{n\geq 0} a_n \cos \pi n s : \sum_{n\geq 1} |a_n|^2 n^2 < \infty \right\} \\ &= W_2^1(0,1) = \{f : f' \in L^2(0,1)\}, \end{aligned}$$

for every $\tau > 0$. This means that the system is generalized exactly controllable. On the other hand, since $B\mathcal{U} = L^2(0,1) \not\subset H(w_n) = W_2^1$, RECO is impossible.

In what follows we will see that this is a rather typical situation: in general, parabolic systems are neither exactly nor renormalized exactly controllable. In return, they are often null or generalized exact controllable.

3.4.6. Renormalized Exact Controllability (RECO). Here, we analyze the situation described in Theorem 3.4.4 and, above all, condition (3.4.1). This means that we consider Hilbert space semigroups $S(\cdot)$ having a Riesz basis of eigenvectors $(\varphi_n)_{n\geq 1} \subset H$ with corresponding biorthogonal $(\psi_n)_{n\geq 1}$ and control operators $B : \mathcal{U} \longrightarrow H$ such that the vector valued exponentials

$$\mathcal{E}_n(t) = e^{-\bar{\lambda}_n t} B^* \psi_n, \quad n \geq 1,$$

form a Riesz sequence in $L^2(0, \tau; \mathcal{U})$. We shortly call this situation the *Riesz bases hypothesis*, that is, when $(\varphi_n)_{n\geq 1}$ (and hence $(\psi_n)_{n\geq 1}$) is a Riesz basis and when $(\mathcal{E}_n)_{n\geq 1}$ a Riesz sequence.

Notice that $(\mathcal{E}_n)_{n\geq 1}$ is a Riesz sequence if B is surjective and $(\varphi_n)_{n\geq 1}$ is a Riesz basis. Moreover, the sequence $(\mathcal{E}_n)_{n\geq 1}$ is a Riesz sequence for any choice of $B \neq 0$ if and only if $(e^{-\bar{\lambda}_n t})_{n\geq 1}$ is a Riesz sequence in $L^2(0, \tau)$ (see point (9) below in this subsection and Chapter 4 for more details).

Theorem 3.4.3 states that under the Riesz bases hypothesis we already have GECO. If we are now interested in the stronger property RECO, we have to examine the step from GECO to RECO, that is, property (3.4.1). This (necessary!) condition means that the control operator respects the renormalization (at time τ), that is,

$$\operatorname{Range} c(\tau) = H(w_n),$$

$$B\mathcal{U} \subset \operatorname{Range} c(\tau). \tag{3.4.1}$$

The remaining part of this subsection is devoted to an analysis of condition (3.4.1). We always assume the Riesz bases hypothesis.

(1) A necessary and sufficient condition for (3.4.1). *There exists a constant $C_\tau > 0$ such that*

$$\|B^* y\|^2_{\mathcal{U}} \leq C^2_\tau \|c(\tau)^* y\|^2_{L^2(0,\tau;\mathcal{U})} = C^2_\tau \int_0^\tau \|B^* S(\tau - t)^* y\|^2 \, dt$$

for every $y \in H$.

This is clear from Lemma 1.3.12. □

(2) A necessary spectral condition for (3.4.1). *The spectrum of the generator A is in a strip*

$$-c \leq -\operatorname{Re} \lambda_n \leq C, \quad n \geq 1. \tag{3.4.2}$$

Indeed, since the eigenvectors of A, $A\varphi_n = -\lambda_n \varphi_n$, $n \geq 1$, form a Riesz basis, the spectrum $\sigma(A)$ coincides with the closure $\operatorname{clos}\{-\lambda_n : n \geq 1\}$. On the other hand, the semigroup $S(\cdot)$ is injective: indeed, $S(t)x = \sum_{n\geq 1} < x, \psi_n > e^{-\lambda_n t} \varphi_n = \mathbf{0}$ if and only if $x = \mathbf{0}$. Moreover, by Theorem 3.4.4, it is exactly controllable on $H(w_n)$, and hence by 1.3.20(1) it extends to a group. Then, using 1.3.20(5), we see that $\sigma(A)$ (for H or $H(w_n)$) is in a strip. □

In fact, we can specify more explicitly these bounds for the frequencies λ_n. The upper bound $-\operatorname{Re}\lambda_n \leq \alpha_0(S)$ for all $n \geq 1$ is clear. For the lower bound we can use point (1) above. Setting $y = \psi_n$ we get

$$\|B^*\psi_n\|^2 \leq C_\tau^2 \|B^*\psi_n\|^2 \frac{e^{-2\operatorname{Re}\lambda_n \tau} - 1}{-2\operatorname{Re}\lambda_n},$$

and hence

$$\frac{1}{C_\tau^2} \leq \frac{e^{-2\operatorname{Re}\lambda_n \tau} - 1}{-2\operatorname{Re}\lambda_n \tau} \cdot \tau = \tau f(-2\operatorname{Re}\lambda_n \tau),$$

where $f(x) = \frac{e^x - 1}{x}$, $x \in \mathbb{R}$. If g is the inverse mapping of f defined on $\mathbb{R}_+$, then

$$\frac{1}{2\tau} g\left(\frac{1}{\tau C_\tau^2}\right) \leq -\operatorname{Re}\lambda_n, \quad n \geq 1.$$

For a more explicit estimate one can observe that $-s^{-1} < g(s)$ on $(0, 0.5)$, $-1 \leq g(s) \leq 1$ on $[0.5, 1.7]$ and $\log s + \log\log s < g(s)$ for $s > 1,7$. □

(3) Norm estimates for $\mathcal{E}_n$. *Under the assumption* (3.4.2) *(and hence, by (2) above, under the assumption* (3.4.1)*), we have*

$$\|\mathcal{E}_n\| \asymp \|B^*\psi_n\|, \quad n \geq 1.$$

Indeed, $\|\mathcal{E}_n\|^2_{L^2(0,\tau;\mathcal{U})} = \|B^*\psi_n\|^2 \frac{e^{-2\operatorname{Re}\lambda_n \tau} - 1}{-2\operatorname{Re}\lambda_n}$, so the result follows from (2). □

(4) Necessary and sufficient conditions for (3.4.1) *in terms of the frequencies λ_n, $n \geq 1$, and the operator B (we still assume the Riesz bases hypothesis) are:*

- *condition* (3.4.2), *and*
- $$\sum_{n\geq 1} \frac{|(u, B^*\psi_n)|^2}{\|B^*\psi_n\|^2} < \infty$$

  for every $u \in \mathcal{U}$.

Indeed, the necessity of the first condition is proved in (2). Under that condition, (3) gives $\|\mathcal{E}_n\| \asymp \|B^*\psi_n\|$, $n \geq 1$. Hence, $H(w_n) = H(\|\psi_n\|/\|B^*\psi_n\|)$. Then

$$Bu \in H(w_n) \Longleftrightarrow \sum_{n\geq 1} |(Bu, \psi_n)|^2 \frac{1}{\|\psi_n\|^2} \cdot \frac{\|\psi_n\|^2}{\|B^*\psi_n\|^2} < \infty,$$

which shows the necessity.

Clearly this reasoning can be reversed. □

(5) Sufficient conditions for (3.4.1). *Suppose that* (3.4.2) *holds. Then each of the following two conditions is sufficient for* (3.4.1).

(A) There is $\varepsilon > 0$ such that $\|B^*\psi_n\| \geq \varepsilon \|\psi_n\|$, $n \geq 1$.

(B) $\mathcal{U} = H$ and B^* is a diagonal operator in the basis $(\psi_n)_{n\geq 1}$:

$$B^*\psi_n = \gamma_n \psi_n, \quad \gamma_n \neq 0, \ n \geq 1.$$

This in an immediate consequence of (4) (in particular for (B), set $w_n = |\gamma_n| \cdot \|\psi_n\|/\|\mathcal{E}_n\|$). □

As a comment, we recall that in (A) we assume that $(\mathcal{E}_n)_{n\geq 1}$ is a Riesz sequence, whereas in (B) this follows from the fact that $(\psi_n)_{n\geq 1}$ is a Riesz basis (and (3.4.2)). It is also clear, that (A) and (3.4.2) mean that $\|\mathcal{E}_n\| \asymp \|\psi_n\|$, $n \geq 1$, and thus $H = H(w_n)$. This means that we actually have exact controllability in H without any renormalization.

The following example shows that the condition (A) is weaker than the surjectivity requirement $B\mathcal{U} = H$ (which is equivalent to $\|B^*x\| \geq \varepsilon\|x\|$ for all $x \in H$).

(6) Example. Let $\mathcal{U} = H = L^2(0,1)$, $\varphi_n(x) = \cos \pi nx$, $n \geq 0$ (cf. Example 1.5, 3.3.4, 3.4.5). Let

$$Bf = b \cdot f, \quad f \in L^2(0,1),$$

where $b \in L^\infty(0,1)$ is an arbitrary function, $b \neq 0$ in L^∞. The biorthogonal system $(\psi_n)_{n\geq 0}$ is $\psi_0 = 1$, $\psi_n = 2\cos \pi nx$, $n \geq 1$, and hence $\inf_{n\geq 0} \|B^*\psi_n\| = \inf_{n\geq 0} \|\bar{b}\psi_n\| > 0$, which implies condition (A) of (5). Since moreover the sequence $(\mathcal{E}_n)_{n\geq 0}$ is orthogonal, the system (A, B) is renormalized exactly controllable whenever A is a diagonal operator $A\varphi_n = -\lambda_n\varphi_n$, $n \geq 0$, with $\sup_{n\geq 1} |\operatorname{Re} \lambda_n| < \infty$.

On the other hand, if ess inf $|b| = 0$, then B is not surjective.

(7) Renormalized controllability by means of compact operators. In order to measure the "size" of a compact operator, we use below symmetrically quasi-normed operator ideals $\mathfrak{S}$. General properties of these ideals and references can be found in Sections B.7.5, B.7.6 (Volume 1) and 3.9 below. Now we recall that most of the symmetrically quasi-normerd ideals $\mathfrak{S}$ are defined through a function depending on an infinite number of variables, the so-called *fundamental function* Φ, in such a way that $B \in \mathfrak{S}$ if and only if $\lim_n \Phi(0, ..., 0, s_n(B), s_{n+1}(B), ...) = 0$, where $s_n(B)$ are the singular numbers of B (see B.2.1.1, B.7.1.3, Volume 1). The most famous examples are the Schatten–von Neumann classes $\mathfrak{S}_p$ defined by $\Phi(\lambda) = (\sum_{n\geq 0} |\lambda_n|^p)^{1/p}$, $0 < p \leq \infty$ (with the standard modification for $p = \infty$); for the definition of the classes $\mathfrak{S}_p$ see also B.2.1.1 (Volume 1). Using these notions we can prove that RECO is compatible with arbitrary "smallness" of the control operator B in the following sense.

Assume that A satisfies condition (3.4.2)*, and let $\mathfrak{S}$ be a symmetrically quasi-normed ideal. Then there exists an operator $B \in \mathfrak{S}$ such that (A, B) is renormalized exactly controllable (RECO).*

Indeed, using (5) (B), it is enough to take B such that $B^*\psi_n = \gamma_n\psi_n$, $n \geq 1$, with $\gamma_n > 0$ tending sufficiently fast to zero (for instance $\sum_{n\geq 0} |\gamma_n|^p < \infty$ in case $\mathfrak{S} = \mathfrak{S}_p$). □

It is well-known that the intersection of symmetrically quasi-normed ideals is the ideal of finite rank operators $\mathcal{F}$, but the previous property does not hold for $\mathcal{F}$ as is shown in the next point. We recall again that we always suppose that $(A, B) \in ACO$.

(8) Finite rank renormalized exact controllability is impossible. *Namely, property* (3.4.1) *never holds for any finite rank operator B,* $0 < \operatorname{rank} B < \infty$.

Indeed, if $\dim B^*H < \infty$, choose an orthonormal basis $u_1, ..., u_n$ in B^*H. Then

$$\sum_{j=1}^{n}\sum_{k\geq 1} \frac{|(u_j, B^*\psi_k)|^2}{\|B^*\psi_k\|^2} = \sum_{k\geq 1}\sum_{j=1}^{n} \frac{|(u_j, B^*\psi_k)|^2}{\|B^*\psi_k\|^2} = \sum_{k\geq 1} \frac{\|B^*\psi_k\|^2}{\|B^*\psi_k\|^2} = \infty,$$

which contradicts property (4) above. □

It is worth mentioning that this property is also an immediate consequence of the preceding theory without using criterion (4). Indeed, the Riesz bases hypothesis and (3.4.1) imply by Theorem 3.4.4 that the system (A_0, B_0) with $A_0 = A|H(w_n)$, $B_0 : \mathcal{U} \longrightarrow H(w_n)$, $B_0 u = Bu$, $u \in \mathcal{U}$, is exactly controllable. By Theorem 1.3.19 this is impossible since $\dim H(w_n) = \infty$ and $\operatorname{rank} B_0 < \infty$. □

(9) Dependence of RECO on the time moment τ. In this chapter, our principal assumption about the system (A, B) is the Riesz bases hypothesis, which means first that $(\varphi_n)_{n\geq 1}$ is a Riesz basis in H, and secondly, that the exponentials $\mathcal{E}_n = e^{-\bar{\lambda}_n t} B^*\psi_n$, $n \geq 1$, form a Riesz sequence in $L^2(0, \tau; \mathcal{U})$. Then, by (4), Property (3.4.1) is independent of the time. We conclude that RECO does not depend on the time.

Now, anticipating the results of Chapter 4, we will make some comments on the influence of the time τ upon the fact that $(\mathcal{E}_n)_{n\geq 1}$ is a Riesz sequence in $L^2(0, \tau; \mathcal{U})$. We still assume that $(\varphi_n)_{n\geq 1}$ and $(\psi_n)_{n\geq 1}$ are Riesz bases. First we mention that if $B\mathcal{U} = H$, then the family $(\mathcal{E}_n)_{n\geq 1}$ is always a Riesz sequence, regardless of the frequency distribution and the time. The same can be said for diagonal operators $B^*\psi_n = \gamma_n \psi_n$, $n \geq 1$.

For other operators B, this property, in principle, depends on the time. A general result states that if $(\mathcal{E}_n)_{n\geq 1}$ is a Riesz sequence on $[0, \tau]$, then this is also the case on every interval $[0, \tau']$ with $\tau' > \tau$, whereas in general it is not true for $\tau' < \tau$.

Suppose now a frequency distribution of $(\lambda_n)_{n\geq 1}$ such that $(\mathcal{E}_n)_{n\geq 1}$ is a Riesz sequence regardless of B. This is equivalent to $(e^{-\bar{\lambda}_n t})_{n\geq 1}$ being a Riesz sequence in $L^2(0, \tau)$. For this scalar case there exists a critical length of the interval $[0, \tau]$, where $(\mathcal{E}_n)_{n\geq 1}$ could be a Riesz sequence. If we restrict ourselves to a frequency distribution of hyperbolic type, that is $\sup_{n\geq 1} |\operatorname{Re} \lambda_n| < \infty$ (which by (2) is necessary for RECO), then we have an explicit formula for this critical length τ_0 (called the *optical length* of the system):

$$\tau_0 = \lim_{t\to\infty} \frac{\pi}{t} \left(\sum_{|\operatorname{Im} \lambda_n| < t} 1 \right).$$

More precisely, if the exponentials $(e^{-\bar{\lambda}_n t})_{n\geq 1}$ form a Riesz sequence on $[0, \tau]$, then the above limit exists and $\tau \geq \tau_0$; for more comments and references see Section 3.9.

(10) Conclusion. As a conclusion, we can say that if we assume that $S(\cdot)$ has a Riesz basis of eigenvectors, $S(t)\varphi_n = e^{-\lambda_n t}\varphi_n$, $n \geq 1$, then we can guarantee RECO with a suitable control operator if and only if $S(\cdot)$ is of hyperbolic type,

$\sup_{n\geq 1} |\operatorname{Re}\lambda_n| < \infty$. This actually means that $S(\cdot)$ extends to a group, and in this case the control is possible at all times and with arbitrarily small control operators (contained in a symmetrically quasi-normed operator ideal but without being of finite rank).

We will see in Section 3.5 that even for hyperbolic ("oscillating") systems even null controllability is not possible if B is of finite rank.

3.5. Null Controllability (NCO)

We continue the study of various types of controllability always assuming ACO and the *Riesz bases hypothesis*. We will also see that it is possible to work under weaker assumptions. The following characterization for NCO was given in 1.2.5

$$S(\tau)H \subset c(\tau)L^2(0,\tau;\mathcal{U}),$$

and exact null controllability (ENCO) was defined by

$$S(\tau)H = c(\tau)L^2(0,\tau;\mathcal{U})$$

(see 1.3.24). Now, we first establish some criteria for NCO and ENCO in terms of the frequencies λ_n and the control operator B, and then consider some examples, both in the hyperbolic and the parabolic cases.

3.5.1. Basic criteria. Here we transform the above definitions of NCO and ENCO into the language of geometry of vector valued exponentials followed by some spectral asymptotics. As mentioned above, all but the first Lemma 3.5.2, giving a necessary spectral condition, is valid under the Riesz bases hypothesis.

3.5.2. LEMMA. *Let $\sigma_p(A^*)$ be the point spectrum of A^*: $A^*\psi_\lambda = -\bar{\lambda}\psi_\lambda$, $-\bar{\lambda} \in \sigma_p(A^*)$. If (A,B) is NCO at time $\tau > 0$, then there is a constant $C = C_\tau$ such that*

$$\frac{2\operatorname{Re}\lambda}{e^{2\operatorname{Re}\lambda\tau}-1} \leq C_\tau^2 \frac{\|B^*\psi_\lambda\|^2}{\|\psi_\lambda\|^2}, \quad -\bar{\lambda} \in \sigma_p(A^*).$$

PROOF. By Theorem 1.3.23 we have

$$\|S(\tau)^*y\|^2 \leq C_\tau^2 \|c(\tau)^*y\|^2 = C_\tau^2 \int_0^\tau \|B^*S(\tau-t)^*y\|^2\,dt$$

for every $y \in X^*$. Setting $y = \psi_\lambda$, we obtain

$$e^{-2\operatorname{Re}\lambda\tau}\|\psi_\lambda\|^2 \leq C_\tau^2 \|B^*\psi_\lambda\|^2 \frac{e^{-2\operatorname{Re}\lambda\tau}-1}{-2\operatorname{Re}\lambda}, \quad -\bar{\lambda} \in \sigma_p(A^*),$$

as claimed. □

3.5.3. THEOREM. *Let (A,B) be a dynamical system in a Hilbert space H and let $\tau > 0$. Assume that the Riesz bases hypothesis is satisfied. Then the following assertions are equivalent.*

(1) (A,B) is null controllable at time $\tau > 0$ (NCO).

(2) There exists a constant $C_\tau > 0$ such that

$$\|S(\tau)^*y\|^2 \leq C_\tau^2 \int_0^\tau \|B^*S(t)^*y\|^2\,dt, \quad y \in H.$$

(3) There is a constant $C > 0$ such that

$$\frac{2\operatorname{Re}\lambda_n}{e^{2\operatorname{Re}\lambda_n\tau}-1} \leq C^2 \frac{\|B^*\psi_n\|^2}{\|\psi_n\|^2}, \quad n \geq 1,$$

where $A^*\psi_n = -\bar{\lambda}_n\psi_n$ *and* $(\psi_n)_{n\geq 1}$ *is a Riesz basis of eigenvectors of* A^*.

PROOF. The equivalence (1) $\iff$ (2) is contained in Theorem 1.3.23.
That (1) or (2) imply (3) is a consequence of Lemma 3.5.2.
The converse implication (3) $\implies$ (2) is a consequence of the following:

$$\begin{aligned}
\Big\| S(\tau)^* \sum_{n\geq 1} a_n\psi_n \Big\|^2 &\leq c_1 \sum_{n\geq 1} |a_n|^2 e^{-2Re\lambda_n\tau} \|\psi_n\|^2 \\
&\leq c_1 C^2 \sum_{n\geq 1} |a_n|^2 \int_0^\tau \|e^{-\bar{\lambda}_n t} B^*\psi_n\|_{\mathcal{U}}^2\, dt \\
&\leq c_1 C^2 c_2 \Big\| \sum_{n\geq 1} a_n e^{-\bar{\lambda}_n t} B^*\psi_n \Big\|^2_{L^2(0,\tau;\mathcal{U})} \\
&= c_1 C^2 c_2 \Big\| c(\tau)^* \Big(\sum_{n\geq 1} a_n\psi_n \Big) \Big\|^2_{L^2(0,\tau;\mathcal{U})}
\end{aligned}$$

for every finite linear combination $\sum_{n\geq 1} a_n\psi_n$. □

3.5.4. COROLLARY. *Let* (A, B) *be a dynamical system satisfying the Riesz bases hypothesis. Then*

(1) NCO is monotone in τ*;*

(2) NCO holds provided the control operator satisfies

$$\inf_{n\geq 1} \frac{\|B^*\psi_n\|}{\|\psi_n\|} > 0; \tag{3.5.1}$$

(3) if the point spectrum $\sigma_p(A)$ *is of hyperbolic type*

$$\sup_{n\geq 1} |\operatorname{Re}\lambda_n| < \infty, \tag{3.5.2}$$

then (3.5.1) *is equivalent to NCO, and also to ECO.*

Indeed, assertion (1) is a consequence of 1.2.4, (2) follows from Theorem 3.5.3 and the spectral condition $\inf_{n\geq 1} \operatorname{Re}\lambda_n > -\infty$. Finally, assertion (3) is contained in Theorem 3.3.2 and the comment (2) after Theorem 1.3.25. □

We add that, by this theorem, under the hypothesis (3.5.2) we never have compact null control (see Theorem 1.3.19; cf. also Subsection 3.5.6, Theorem 3.5.7 for small controls).

3.5.5. THEOREM. *Let* A *be the generator of a strongly continuous group on* $\mathbb{R}$ *having a Riesz basis of eigenvectors,* $A\varphi_n = -\lambda_n\varphi_n$, $n \geq 1$. *Let further* $(\psi_n)_{n\geq 1}$ *be its biorthogonal and* $B \in L(\mathcal{U}, H)$. *The following assertions are equivalent.*

(1) (A, B) *is exactly null controllable (ENCO) at time* $\tau > 0$.

(2) The system $(\mathcal{E}_n)_{n\geq 1} = (e^{-\bar{\lambda}_n t} B^*\psi_n)_{n\geq 1}$ *is a Riesz sequence and*

$$\frac{2\operatorname{Re}\lambda_n}{e^{2\operatorname{Re}\lambda_n\tau} - 1} \asymp \frac{\|B^*\psi_n\|^2}{\|\psi_n\|^2}, \quad n \geq 1.$$

PROOF. By Theorem 1.3.25, ENCO is equivalent to the following estimate

$$c\|S(\tau)^*y\|^2 \leq \int_0^\tau \|B^*S(t)^*y\|^2\, dt \leq C\|S(\tau)^*y\|^2, \quad y \in H,$$

where $c, C > 0$ are some fixed constants. Taking into account the estimates

$$\|S(\tau)^*y\|^2 = \Big\|S(\tau)^*\Big(\sum_{n\geq 1}(y,\varphi_n)\psi_n\Big)\Big\|^2 \asymp \sum_{n\geq 1}|(y,\varphi_n)|^2 e^{-2Re\lambda_n\tau}\|\psi_n\|^2$$

and

$$\begin{aligned}\int_0^\tau \Big\|\sum_{n\geq 1}(y,\varphi_n)\mathcal{E}_n(t)\Big\|^2 dt &= \int_0^\tau \|B^*S(t)^*y\|^2\,dt \asymp \|S(\tau)^*y\|^2 \\ &\asymp \sum_{n\geq 1}|(y,\varphi_n)|^2 e^{-2\operatorname{Re}\lambda_n\tau}\|\psi_n\|^2,\end{aligned}$$

we get first (setting $y = \psi_n$)

$$\frac{1-e^{-2\operatorname{Re}\lambda_n\tau}}{2\operatorname{Re}\lambda_n}\|B^*\psi_n\|^2 = \|\mathcal{E}_n\|^2_{L^2(0,\tau;\mathcal{U})} \asymp e^{-2\operatorname{Re}\lambda_n\tau}\|\psi_n\|^2$$

and then

$$\Big\|\sum_{n\geq 1}(y,\varphi_n)\mathcal{E}_n\Big\|^2_{L^2(0,\tau;\mathcal{U})} \asymp \sum_{n\geq 1}|(y,\varphi_n)|^2\|\mathcal{E}_n\|^2_{L^2(0,\tau;\mathcal{U})}, \quad y \in H.$$

The latter relation means that $(\mathcal{E}_n)_{n\geq 1}$ is a Riesz sequence. We have proved that (1) $\Longrightarrow$ (2).

The converse follows by reversing the above reasoning. □

3.5.6. Small null controls. It turns out that compact null control is possible if the frequency distribution is "sufficiently parabolic".

3.5.7. THEOREM. *Let $\tau > 0$ and let A be the generator of a strongly continuous group having a Riesz basis of eigenvectors, $A\varphi_n = -\lambda_n\varphi_n$, $n \geq 1$. Suppose that $\mathfrak{S}$ is a symmetrically quasi-normed operator ideal such that*

$$\text{(3.5.3)} \qquad \operatorname{diag}(\gamma_n)_{n\geq 1} := \operatorname{diag}\left(\left(\frac{\operatorname{Re}\lambda_n}{e^{2\operatorname{Re}\lambda_n\tau}-1}\right)^{1/2}\right)_{n\geq 1} \in \mathfrak{S}$$

(i.e. $\operatorname{diag}(\gamma_n)_{n\geq 1} : l^2 \longrightarrow l^2$ is in $\mathfrak{S}$). Then there is a control operator B such that

(1) $B \in \mathfrak{S}$,

(2) (A, B) is exactly null controllable at time τ,

(3) the exponentials $(e^{-\bar{\lambda}_n t}B^*\psi_n)_{n\geq 1}$ form a Riesz sequence.

PROOF. Choose $B^*\psi_n = \gamma_n\psi_n$, $n \geq 1$, and apply Theorem 3.5.5 (note that with this choice of B^* the Riesz bases hypothesis is automatically satisfied; see also the comments after 3.4.6(5)). □

It is worth mentioning that $\lim_n \gamma_n = 0$ for every symmetrically quasi-normed operator ideal $\mathfrak{S}$ (see comments in Section 3.9 below). This means that $\lim_n \operatorname{Re}\lambda_n = +\infty$, so we may rewrite condition (3.5.3) in the following way

$$\text{(3.5.4)} \qquad \operatorname{diag}\left((|\operatorname{Re}\lambda_n| + 1)^{1/2}e^{-\operatorname{Re}\lambda_n\tau}\right)_{n\geq 1} \in \mathfrak{S}.$$

The necessity of condition (3.5.3) or (3.5.4) is more delicate. We begin with the following result.

3.5.8. THEOREM. *Let $2 \le p \le \infty$, and let A be an operator with a Riesz basis of eigenvectors: $A\varphi_n = -\lambda_n \varphi_n$, $n \ge 1$. The following assertions are equivalent.*

(1) There is an operator $B \in \mathfrak{S}_p$ such that (A, B) is null controllable.

(2) The spectrum $(\lambda_n)_{n\ge1}$ verifies (3.5.4) with $\mathfrak{S} = \mathfrak{S}_p$:

$$\left(\sum_{n\ge1} (|\operatorname{Re}\lambda_n| + 1)^{p/2} e^{-p\operatorname{Re}\lambda_n \tau}\right)^{1/p} < \infty \tag{3.5.5}$$

(and $\lim_n (\operatorname{Re}\lambda_n) = +\infty$ for $p = \infty$).

PROOF. The implication (2) $\Longrightarrow$ (1) is a special case of Theorem 3.5.7. Conversely, by Lemma 3.5.2, we get

$$\left(2\operatorname{Re}\lambda_n (e^{2\operatorname{Re}\lambda_n \tau} - 1)^{-1}\right)^{1/2} \le C \left\| B^* \left(\frac{\psi_n}{\|\psi_n\|}\right)\right\|, \qquad n \ge 1.$$

As $(\psi_n/\|\psi_n\|)_{n\ge1}$ is a normalized Riesz basis there exists an isomorphism V and an orthonormal basis $(e_n)_{n\ge1}$ such that $Ve_n = \psi_n/\|\psi_n\|$, $n \ge 1$. As $B^*V \in \mathfrak{S}_p$ and $p \ge 2$, and using the result of B.7.5.1(q), Volume 1, we get $\sum_{n\ge1} \|B^*Ve_n\|^p < \infty$. This completes the proof. $\square$

3.5.9. COROLLARY. *Under the assumptions and the notation of Theorem 3.5.8, set*

$$l = \underline{\lim}_n \frac{\operatorname{Re}\lambda_n}{\log n}, \qquad L = \overline{\lim}_n \frac{\operatorname{Re}\lambda_n}{\log n}.$$

Then, for the existence of an operator $B \in \mathfrak{S}_p$, $2 \le p < \infty$, ensuring null control at time $\tau > 0$ it is

(1) necessary to have $\tau \ge \dfrac{1}{pL}$,

(2) sufficient to have $\tau > \dfrac{1}{pl}$.

This is a simple consequence of a comparison of the series (3.5.5), and the standard series $\sum_{n\ge1} \frac{1}{n^{1\pm\varepsilon}}$, $\varepsilon > 0$. $\square$

3.5.10. Small and large ideals for NCO and ENCO. We can summarize the preceding results in the following way. For "large" symmetrically quasi-normed ideals $\mathfrak{S}$, that is, ideals containing $\mathfrak{S}_2$ (and contained in the ideal of all compact operators $\mathfrak{S}_\infty$), the criterion for existence of a null control operator is expressed in terms of the spectrum of A (cf. (3.5.5)). Namely, it depends on the growth of $\operatorname{Re}\lambda_n$ towards infinity. See also Corollary 3.5.9 which shows that the critical growth turns out to be logarithmic.

On the contrary, for "small" ideals, contained in $\mathfrak{S}_2$, the situation is much more intricate, both for necessary and sufficient conditions. For instance, the *necessity* (for the existence of a control operator $B \in \mathfrak{S}$ ensuring NCO) always involves Lemma 3.5.2 and thus depends on the behaviour of the norms $\|B^*e_n\|$, $n \ge 1$, where $(e_n)_{n\ge1}$ is an orthonormal basis (or a basis equivalent to an orthonormal one). But for the membership $B \in \mathfrak{S}$ in an ideal $\mathfrak{S} \subset \mathfrak{S}_2$ there is no necessary condition expressed only in $\|B^*e_n\|$, $n \ge 1$, excepting for $\mathfrak{S} = \mathfrak{S}_2$, that is, $\sum_{n\ge1} \|B^*e_n\|^2 < \infty$. Indeed, for every $a_n \ge 0$ with $\sum_{n\ge1} {a_n}^2 < \infty$, there exists a rank-one operator B such that $\|B^*e_n\| \ge a_n$, $n \ge 1$.

On the other hand, the *sufficiency part* (see Theorem 3.5.7) is based on the Riesz bases hypothesis, that is, in order to have NCO we have to find a normalized sequence $\{u_n\}_{n\geq 1} \in \mathcal{U}$, $\|u_n\| = 1$, such that

- the family $(e^{-\bar{\lambda}_n t} u_n)_{n\geq 1}$ is a Riesz sequence,
- the operator $B^* e_n = \gamma_n u_n$, $n \geq 1$, is in $\mathfrak{S}$, and $(\gamma_n)_{n\geq 1}$ is given in (3.5.3).

At the actual state of the art, both problems (to find approaches to the above necessary and sufficient conditions) seem to be open. Nevertheless, we are able to manage two special cases which are interesting for applications: rank-one exact null control (see Subsection 3.5.11 below) and finite rank ENCO at time $\tau = \infty$, see Section 3.8.

3.5.11. ENCO by means of rank one operators. Here we prove the existence of smallest, i.e. rank one, null controllers for systems satisfying some spectral sparseness conditions. Namely, in this case, the general criteria 3.5.3 and 3.5.5 can be expressed in terms of the Carleson interpolation condition studied in Chapter C.3. Later on, in Chapter 4, we will describe in detail the links between exponential bases and free interpolation in Hardy spaces developed in Chapter C.3. In particular, for sets of frequencies $(\lambda_n)_{n\geq 1}$ occuring in control theory, namely, in the case where $\inf_{n\geq 1} \operatorname{Re} \lambda_n > -\infty$, we give in Chapter 4 a criterion in terms of the distribution function of $(\lambda_n)_{n\geq 1}$.

3.5.12. THEOREM. *Let A be the generator of a strongly continuous group having a Riesz basis of eigenvectors, $A\varphi_n = -\lambda_n \varphi_n$, $n \geq 1$, and let $(\psi_n)_{n\geq 1}$ be its biorthogonal. Let further $\tau > 0$. The following assertions are equivalent.*

(1) There exists an operator B with $\operatorname{rank} B = 1$ *such that the system (A, B) is ENCO at time τ (that is $S(\tau)H = c(\tau)L^2(0, \tau; \mathcal{U})$).*

(2) $(e^{-\bar{\lambda}_n t})_{n\geq 1}$ is a Riesz sequence in $L^2(0, \tau)$ and

$$\sum_{n\geq 1} (|\operatorname{Re} \lambda_n| + 1) e^{-2 \operatorname{Re} \lambda_n \tau} < \infty. \tag{3.5.6}$$

(3) There is $\alpha > 0$ such that $(\lambda_n + \alpha)_{n\geq 1}$ satisfies the Carleson condition *(C) (see Section 4.2, Subsections C.3.2.17 or C.3.3.3(c) for the definition), and* (3.5.6) *holds.*

PROOF. By Theorem 3.5.5, the system (A, B) with $B : \mathbb{C} \longrightarrow H$, $B\lambda = \lambda y$, and y a fixed vector in H, is ENCO if and only if the sequence $\left(e^{-\bar{\lambda}_n t}(\psi_n, y)\right)_{n\geq 1}$ is a Riesz sequence and

$$\frac{2 \operatorname{Re} \lambda_n}{e^{2 \operatorname{Re} \lambda_n \tau} - 1} \asymp \frac{|(\psi_n, y)|^2}{\|\psi_n\|^2}, \quad n \geq 1.$$

Clearly the existence of $y \in H$ satisfying this relation is equivalent to (3.5.6). For the equivalence of (2) and (3) we refer the reader to Chapter 4, Subsection 4.5.1. □

3.5.13. COROLLARY. *There exists a time $\tau > 0$ and a rank one operator B such that the system (A, B) is ENCO at time τ if and only if $(\lambda_n)_{n\geq 1}$ satisfies the Carleson condition (C), and, after an increasing rearrangement of* $\operatorname{Re} \lambda_n$, *the condition*

$$\operatorname{Re} \lambda_n \geq \varepsilon \cdot \log n, \quad n > N,$$

where $\varepsilon > 0$.

Indeed, it is an exercise to verify that the last condition is equivalent to the existence of $\tau > 0$ such that (3.5.6) holds. $\square$

3.5.14. Conclusions about null controllability. The landscape of null controllability is richer and more various than that of exact controllability, even than that of renormalized controllability. In fact, there are more systems that are NCO.

If we still assume that the eigenvectors of A, $A\varphi_n = -\lambda_n\varphi_n$, $n \geq 1$, form a Riesz basis, then we may recapitulate the preceding results in the following way.

For *"big" control operators* B (invertible, or satisfying $\|B^*\psi_n\| \geq \varepsilon\|\psi_n\|$ and the condition that $(\mathcal{E}_n)_{n\geq 1}$ is a Riesz sequence) the system (A, B) is *always* NCO at every time (see Theorem 3.5.3 and Corollary 3.5.4; for the examples, see Section 1.5 as well as Subsections 3.1.1 and 3.1.2).

For *"small" control operators* B there are necessary and sufficient conditions for *ENCO*. Namely, the family $(\mathcal{E}_n)_{n\geq 1}$ is necessarily a Riesz sequence, which in case of a finite rank control B forces strong restrictions on the spectrum $(-\lambda_n)_{n\geq 1}$ such as the Carleson condition (C) (cf. Theorem 3.5.12 and Corollary 3.5.13) or being a finite union of Carleson sequences (cf. Subsection 3.8.4 and Theorem 3.8.5). If B is really "small" (belonging to an operator ideal) we get some sharp conditions on the growth of $\operatorname{Re}\lambda_n$ at infinity (which, in particular, *excludes the hyperbolic case*).

For *the "smallest" control operators* B ($\operatorname{rank} B = 1$), the necessary and sufficient spectral conditions for ENCO are the Carleson condition (C) and at least logarithmic growth of $\operatorname{Re}\lambda_n$ (see Theorem 3.5.12, Corollary 3.5.13 as well as Subsection 3.8.4 and Theorem 3.8.5). In particular, *parabolic spectra* (even $\lambda_n \in \mathbb{R}$) *with geometric growth* ($\lambda_{n+1}/\lambda_n \geq q > 1$) are admissible (see also Chapter 4). On the other hand, *classical spectra* of elliptic operators, given by $\lambda_n \asymp n^{2/N}$ (cf. Subsection 3.1.1), do not satisfy the Carleson condition and hence are not compatible with ENCO.

Therefore, parabolic systems are easier to dampen than hyperbolic ones. Recall that, on the contrary, renormalization works better for hyperbolic systems (see 3.4.6(10)).

3.6. Weak Controllability

It may occur that a system (A, B) is not controllable at all, even after a renormalization of the state space. Then it may be useful to consider a weaker type of controllability, actually we may be interested in dampening (null controlling) certain particularly important initial states. For instance, we could consider the resonances of the system, that is, the eigenvectors of A, $A\varphi_n = -\lambda_n\varphi_n$, $n \geq 1$.

3.6.1. DEFINITION. Let $(\varphi_n)_{n\geq 1}$ be the eigenvectors of the generator A of the system (A, B). The system is called *controllable for simple oscillations* if

$$\varphi_n \in c(\tau)L^2(0, \tau; \mathcal{U}), \quad n \geq 1.$$

The system is called *uniformly controllable for simple oscillations* if there is a constant $C > 0$ such that for every $n \geq 1$ there exists $u_n \in L^2(0, \tau; \mathcal{U})$ satisfying

$$\begin{aligned} c(\tau)u_n &= \varphi_n, \quad \text{and} \\ \|u_n\|_{L^2}^2 &\leq C(\|\varphi_n\|^2 + |\operatorname{Re} < A\varphi_n, \varphi_n > |). \end{aligned}$$

This means that with a suitable finite energy control we can dampen the states of pure oscillation $\varphi_n \in X$, $n \geq 1$.

3.6.2. THEOREM. *Suppose that the eigenvectors of A form a Riesz basis, and set as before $\mathcal{E}_n = e^{-\bar{\lambda}_n t} B^* \psi_n$, $n \geq 1$. Then the system is*

(1) controllable for simple oscillations if and only if $(\mathcal{E}_n)_{n\geq 1}$ is a minimal sequence (see 3.2 or A.5.1.1 for the definition),

(2) uniformly controllable for simple oscillations if and only if $(\mathcal{E}_n)_{n\geq 1}$ is a uniformly minimal sequence (see 3.2 or A.5.1.1 for the definition), and

$$\inf_{n\geq 1} \|B^*\psi_n\|/\|\psi_n\| > 0.$$

PROOF. (1) If $c(\tau)u_n = \varphi_n$, $n \geq 1$, for some $u_n \in L^2(0,\tau;\mathcal{U})$, then

$$\langle u_n, c(\tau)^*\psi_k\rangle = \langle c(\tau)u_n, \psi_k\rangle = \langle \varphi_n, \psi_k\rangle = \delta_{nk} \tag{3.6.1}$$

for every $n, k \geq 1$. This means that $\left(c(\tau)^*\psi_k\right)_{k\geq 1} = \left(e^{-\bar{\lambda}_k(\tau-t)}B^*\psi_k\right)_{k\geq 1}$ is a minimal sequence in $L^2(0,\tau;\mathcal{U})$. Then this will be also the case for $(\mathcal{E}_n)_{n\geq 1}$ since the operator J defined by $Jf(t) = f(\tau - t)$ is unitary on $L^2(0,\tau;\mathcal{U})$.

Conversely, if $\left(c(\tau)^*\psi_k\right)_{k\geq 1}$ is minimal, then the identity (3.6.1) shows that $c(\tau)u_n = \varphi_n$, $n \geq 1$, for every sequence $(u_n)_{n\geq 1}$ which is biorthogonal to $\left(c(\tau)^*\psi_k\right)_{k\geq 1}$.

(2) Again we have that $(\mathcal{E}_n)_{n\geq 1}$ is uniformly minimal if and only if $\left(c(\tau)^*\psi_k\right)_{k\geq 1}$ is, that is, if and only if there is a biorthogonal sequence $(u_n)_{n\geq 1}$ with $c(\tau)u_n = \varphi_n$, $n \geq 1$, such that $\sup_{n\geq 1}\|u_n\| \cdot \|c(\tau)^*\psi_n\| < \infty$ (see Section 3.2). Clearly for every biorthogonal sequence $(u_n)_{n\geq 1}$ with $c(\tau)u_n = \varphi_n$, $n \geq 1$, we have

$$\|u_n\| \cdot \|c(\tau)^*\psi_n\| \geq 1, \quad n \geq 1.$$

On the other hand, using $\sup_{n\geq 1}(-\operatorname{Re}\lambda_n) \leq \alpha_0(S) < \infty$, where $\alpha_0(S)$ is the growth bound, we get

$$\begin{aligned}
\|c(\tau)^*\psi_n\|^{-2} &= \|B^*\psi_n\|^{-2}\left(\frac{e^{-2\operatorname{Re}\lambda_n\tau} - 1}{-2\operatorname{Re}\lambda_n}\right)^{-1} \\
&\asymp \frac{1}{\|B^*\psi_n\|^2}(1 + |\operatorname{Re}\lambda_n|) \\
&\asymp \frac{\|\psi_n\|^2}{\|B^*\psi_n\|^2}\|\varphi_n\|^2(1 + |\operatorname{Re}\lambda_n|) \\
&= \frac{\|\psi_n\|^2}{\|B^*\psi_n\|^2}\left(\|\varphi_n\|^2 + |\operatorname{Re} < A\varphi_n, \varphi_n > |\right).
\end{aligned}$$

It is easy to verify that uniform controllability for simple oscillations is equivalent to uniform minimality of $(c(\tau)^*\psi_n)_{n\geq 1}$ (that is, of $(\mathcal{E}_n)_{n\geq 1}$) and $\|\psi_n\|^2/\|B^*\psi_n\|^2 \leq C$, $n \geq 1$. □

3.7. Squared (Energy) Optimization

Let (A, B) be a dynamical system, where A is a generator of a strongly continuous semigroup defined on X, $B \in L(\mathcal{U}, X)$ a control operator, and $c(\tau)$: $L^2(0,\tau;\mathcal{U}) \longrightarrow X$,

$$c(\tau)u = \int_0^\tau S(\tau - t)Bu(t)\,dt, \quad u \in L^2(0,\tau;\mathcal{U}),$$

the corresponding controllability map.

For a reachable state $a \in X$ we have

$$c(\tau)u = a, \quad a \in X,$$

and the norm $\|u\|^2_{L^2(0,\tau;\mathcal{U})}$ has the meaning of the energy of the control (at least in case $A \in L(X)$). One of the basic optimization problems related to control is to minimize this energy. This means that, given a state $a \in \operatorname{Range} c(\tau)$, we seek the least energy control

$$\inf\left\{\|u\|^2_{L^2(0,\tau;\mathcal{U})} : c(\tau)u = a\right\}.$$

In this section, we explain an algorithm to solve this problem provided the system is exactly controllable ($\operatorname{Range} c(\tau) = X$), or, if X and $\mathcal{U}$ are Hilbert spaces and the system is renormalized exactly controllable (cf. Section 3.4).

In fact, we consider this problem in a more general framework of the following optimization problem

$$\inf\left\{\|u\| : Tu = a\right\},$$

where $T \in L(Y, X)$ is a surjective operator from a Banach space Y onto another Banach space X. For control problems such as RECO we shall then set $Y = L^2(0, \tau; \mathcal{U})$ and $X = H(w_n)$ (with the notation of Section 3.4).

3.7.1. The general procedure. Let X, Y be Banach spaces such that Y and Y^* are *uniformly convex*, and let $T \in L(Y, X)$. In particular, every closed convex subset of Y or Y^* admits a unique vector of minimal norm, and for every $f \in Y^*$ there exists a unique vector $y \in Y$ with $\|y\| = 1$ and $\langle y, f\rangle = \|f\|$ (see Section 3.9 below for standard references for these properties and definitions).

Assume now that T is surjective, $TY = X$. Given $a \in X$, let $u_* \in Y$ be the *minimal norm solution* to the equation

$$Tu = a,$$

that is

$$Tu_* = a, \quad \|u_*\| = \inf\{\|u\| : Tu = a\}. \tag{3.7.1}$$

As is just mentioned, the uniform convexity implies the existence and the uniqueness of such a solution. The procedure to find a minimal solution is essentially based on the duality relation (Hahn-Banach)

$$\|u|E\| = \operatorname{dist}_Y(u, E_\perp) = \frac{1}{\operatorname{dist}_{Y^*}(\mathbf{0}, H \cap E)}.$$

Here, for $u \in Y$, we define $u|E$ as the linear form $\varphi \longmapsto \langle u, \varphi\rangle$, $\varphi \in E \subset Y^*$,

$$E_\perp = \{y \in Y : \langle y, \varphi\rangle = 0, \text{ for every } \varphi \in E\},$$

and

$$H = \{\varphi \in Y^* : \langle u, \varphi\rangle = 1\}.$$

However, what we really need are only the following two steps.
(1) Find the solution to the extremal problem

$$\inf\{\|T^* f\|_{Y^*} : f \in X^*, \langle a, f\rangle = 1\} =: \frac{1}{\lambda}.$$

In view of the assumptions on Y^* and T ($\|T^*f\| \geq \varepsilon\|f\|$, $f \in X^*$, for some $\varepsilon > 0$), there exists a unique vector $f_0 \in X^*$ such that

$$\langle a, f_0 \rangle = 1, \quad \|T^* f_0\|_{Y^*} = \frac{1}{\lambda}.$$

(2) Find the (unique) vector $u_0 \in Y$ such that

$$\|u_0\| = 1, \quad \langle u_0, T^* f_0 \rangle = \|T^* f_0\| = \frac{1}{\lambda}.$$

3.7.2. THEOREM. *With the preceding notation, the unique solution to the problem* (3.7.1) *is given by*

$$u_* = \lambda u_0.$$

PROOF. Let $u \in Y$ be such that $Tu = a$. Set $E_\perp = \operatorname{Ker} T$, $E = \operatorname{Range} T^*$. Then by the duality formula above and by the definition of λ, we have

$$\begin{aligned} \|u_*\| &= \inf\{\|v\| : a = Tu = Tv\} = \inf\{\|u - x\| : x \in \operatorname{Ker} T\} \\ &= \operatorname{dist}_Y(u, E_\perp) = \frac{1}{\operatorname{dist}_{Y^*}(\mathbf{0}, H \cap E)} = \lambda. \end{aligned}$$

Hence $\|u_* \lambda^{-1}\| = 1$. Moreover

$$\langle u_* \lambda^{-1}, T^* f_0 \rangle = \langle T u_* \lambda^{-1}, f_0 \rangle = \langle a \lambda^{-1}, f_0 \rangle = \frac{1}{\lambda}.$$

As u_0 is unique (cf. (2)) we get $u_* \lambda^{-1} = u_0$. □

3.7.3. COMMENTS. This elementary algorithm, using only basic Banach space properties, turns out to be efficient for many control problems, even in infinite dimensions. We refer the reader to Section 3.9 below for references to various examples and algorithms involving this procedure.

Let us return to control problems. By the recipe of Theorem 3.7.2, in order to find the optimal control function u_*, $c(\tau)u_* = a$ for a given state $a \in X$, we have to solve the following two extremal problems:

(1) Find

$$\inf\left\{\int_0^\tau \|B^* S^*(t) y\|^2 \, dt : y \in X^*, \langle a, y \rangle = 1\right\} = \frac{1}{\lambda^2},$$

and determine the corresponding extremal vector $y = y_0$.

(2) Find $u_0 \in L^2(0, \tau; \mathcal{U})$ such that $\|u_0\|_{L^2(0,\tau;\mathcal{U})} = 1$ and

$$\int_0^\tau \langle u_0(t), B^* S^*(t) y_0 \rangle \, dt = \frac{1}{\lambda}.$$

(3) The solution u_* to the extremal problem is then $u_* = \lambda u_0$.

3.7.4. Squared optimization in Hilbert spaces. Let Y be a Hilbert space (which will simply be $L^2(0, \tau; \mathcal{U})$ with a Hilbert space $\mathcal{U}$). Then it is clear that

$$u_* = P_{(\operatorname{Ker} T)^\perp} u = P_{\operatorname{Range} T^*} u$$

for every solution $Tu = a$.

If in addition the generator $A : H \longrightarrow H$ of the system (A, B) admits a minimal complete family of eigenvectors, $A\varphi_n = -\lambda_n \varphi_n$, such that its biorthogonal is also complete, then we get

$$\operatorname{Range} c(\tau)^* = \operatorname{span}(\mathcal{E}_n : n \geq 1)$$

provided the system is exactly controllable (at least up to a renormalization). Here, as before, $\mathcal{E}_n = e^{-\bar{\lambda}_n(\tau - t)} B^* \psi_n$, $n \geq 1$. Suppose now that $c(\tau)u = a$. For the smallest energy control we then get

$$u_* = P_{\text{Range}\, c(\tau)^*} u \sim \sum_{n \geq 1} \langle u, \mathcal{E}_n \rangle \mathcal{E}'_n,$$

where $(\mathcal{E}'_n)_{n \geq 1} \subset \text{Range}\, c(\tau)^*$ is the biorthogonal family associated to $(\mathcal{E}_n)_{n \geq 1}$. If $(\mathcal{E}_n)_{n \geq 1}$ is a basic sequence (in particular, a Riesz sequence), then the formal series above converges to the best control function u_*. To find u_*, it suffices to know the moments

$$(a, \psi_n)_H = (c(\tau)u, \psi_n)_H = (u_1, \mathcal{E}_n)_{L^2(0,\tau;\mathcal{U})}, \quad n \geq 1,$$

where $u_1(t) = u(\tau - t)$, $0 \leq t \leq \tau$. We have already mentioned this procedure as the *method of moments* (see Lemma 3.2.1 and Corollary 3.2.2).

3.8. Control at Time $\tau = \infty$ and Interpolation in H^2

As an illustration of the preceding rsults, we briefly consider the control problem on the infinite interval $[0, \infty)$. Let $c(\infty)$ be the controllability map

$$c(\infty)u = \int_0^\infty S(t)Bu(t)\, dt,$$

which is well-defined for an exponentially stable semigroup (see Lemma 1.4.2).

It is assumed throughout this section that the generator A possesses a Riesz basis of eigenvectors, $A\varphi_n = -\lambda_n \varphi_n$, $n \geq 1$ (with the corresponding biorthogonal $(\psi_n)_{n \geq 1}$). Then the exponential stability implies that

$$\inf_{n \geq 1} \text{Re}\, \lambda_n > 0. \tag{3.8.1}$$

Recall also that in this chapter we always assume $(A, B) \in ACO$.

3.8.1. Exact controllability and renormalization. It is easy to check that all the facts concerning ECO (see Section 3.3) and RECO (see Section 3.4) remain valid for $\tau = \infty$ (up to some obvious modifications). There are nevertheless two important differences. First, by (3.8.1), the spectral condition $\sup_{n \geq 1} |\text{Re}\, \lambda_n| < \infty$ has to be replaced by the following:

$$0 < c \leq \text{Re}\, \lambda_n \leq C, \quad n \geq 1, \tag{3.8.2}$$

where c, $C > 0$ are suitable constants. Secondly, the requirement that $(\mathcal{E}_n)_{n \geq 1} = (e^{-\bar{\lambda}_n t} B^* \psi_n)_{n \geq 1}$ is a Riesz sequence in $L^2(0, \infty; \mathcal{U})$ is much weaker than that on a finite interval (see Sections 4.2-4.5). In particular we have the following result.

3.8.2. THEOREM (ECO at $\tau = \infty$). *Suppose that the eigenvectors of the generator A form a Riesz sequence and $B \in L(\mathcal{U}, H)$.*

(1) The system (A, B) is ECO at time $\tau = \infty$ (which means $c(\infty)L^2(0, \infty; \mathcal{U}) = H$) if and only if the exponentials $(\mathcal{E}_n)_{n \geq 1}$ form a Riesz sequence in $L^2(0, \infty; \mathcal{U})$, the frequencies $(\lambda_n)_{n \geq 1}$ satisfy (3.8.2) *and* $\inf_{n \geq 1} \|B^* \psi_n\| / \|\psi_n\| > 0$.

(2) If the frequencies $(\lambda_n)_{n \geq 1}$ satisfy (3.8.2) *and the Carleson condition (C)*

$$|\lambda_n - \lambda_m| \geq \delta > 0, \quad n \neq m \tag{3.8.3}$$

for some $\delta > 0$, then (A, B) is exactly controllable at time $\tau = \infty$ for every control operator B satisfying $\|B^* \psi_n\| \geq \varepsilon \|\psi_n\|$, $n \geq 1$.

PROOF. For the assertion (1), see Theorem 3.3.2, and (2) will be clear from Section 4.2 below. □

3.8.3. THEOREM (GECO at $\tau = \infty$). *Under the same assumptions and notation as in Theorem 3.8.2, the following assertions hold.*

(1) If (A, B) is GECO at time $\tau = \infty$ (that is, if $\operatorname{Range} c(\infty) = H(w_n)$ *with $w_n = \|\psi_n\|/\|\mathcal{E}_n\|$, $n \geq 1$) and $N = \operatorname{rank} B < \infty$, then the spectrum $(-\lambda_n)_{n\geq 1}$ is a union of at most N Carleson sequences (that is, sequences satisfying* (3.8.3)*).*

(2) If $(\lambda_n)_{n\geq 1}$ is a Carleson sequence contained in a strip $\{z \in \mathbb{C} : 0 < \varepsilon \leq \operatorname{Re} z \leq c\}$, then (A, B) is GECO for arbitrary $B \in L(\mathcal{U}, H)$ such that $(A, B) \in ACO$.

PROOF. The reasoning of Theorem 3.4.3 carries over to this situation. Hence, GECO implies that $(\mathcal{E}_n)_{n\geq 1}$ is a Riesz sequence in $L^2(0, \infty; B^*H)$. Moreover, $\dim B^*H = N$, and Theorem 4.2.4 below finishes the proof.

(2) The Carleson condition implies that the family of exponentials $(e^{-\bar{\lambda}_n t})_{n\geq 1}$ is a Riesz sequence in $L^2(0, \infty)$ (see Theorem 4.2.2 and Remark 4.2.3(1)), and this is also true for $(\mathcal{E}_n)_{n\geq 1}$ (see the introduction to Chapter 4). Now, we can apply the arguments of Theorem 3.4.3 to obtain the result. □

3.8.4. Squared optimization and interpolation. We illustrate here the existing links between squared optimization (Section 3.7) and interpolation in the Hardy space H^2_+ of the upper half plane $\mathbb{C}_+ = \{z \in \mathbb{C} : \operatorname{Im} z > 0\}$ in the case of scalar control.

Let B be a rank-one operator, $B : \mathbb{C} \longrightarrow H$, $B\lambda = \lambda h$, where $h \in H$. Assume that the eigenvectors of the generator A, $A\varphi_n = -\lambda_n \varphi_n$, $n \geq 1$, form a minimal family such that the corresponding biorthogonal $(\psi_n)_{n\geq 1}$ is complete. Then the reachable states $a = c(\infty)u$, $u \in L^2(\mathbb{R}_+)$, are given by the equations

$$\begin{aligned}(a, \psi_n)_H &= (u, c(\infty)^* \psi_n)_{L^2} = \gamma_n (u, e^{-\bar{\lambda}_n t})_{L^2} \\ &= \gamma_n (u, e^{i\mu_n t})_{L^2},\end{aligned}$$

where $\gamma_n = (h, \psi_n)_H$, $\mu_n = i\bar{\lambda}_n$, $n \geq 1$. By (3.8.1) we have

$$\inf_{n\geq 1} \operatorname{Im} \mu_n > 0.$$

Recall also that if $a \in c(\infty)L^2(\mathbb{R}_+)$, then the least energy control u_*, $c(\infty)u_* = a$, belongs to $(\operatorname{Ker} c(\infty))^\perp = \overline{\operatorname{Range} c(\infty)^*}$.

If the family of exponentials $(e^{i\mu_n t})_{n\geq 1}$ is minimal in $L^2(\mathbb{R}_+)$, then the sequence $(\mu_n)_{n\geq 1}$ satisfies the Blaschke condition, and we obtain

$$\overline{\operatorname{Range} c(\infty)^*} = \operatorname{span}_{L^2(\mathbb{R}_+)}(e^{i\mu_n t} : n \geq 1) = \mathcal{F}\mathcal{K}_\Theta,$$

where $\mathcal{F}$ is the Fourier transform, $\mathcal{K}_\Theta = H^2_+ \ominus \Theta H^2_+$ and Θ is the corresponding Blaschke product associated to $(\mu_n)_{n\geq 1}$ (see Corollary A.6.4.3, Volume 1, and Chapter 4 below). The moments $(u, e^{i\mu_n t})_{L^2}$ are nothing but the values of the corresponding function $f = \mathcal{F}^{-1}u \in H^2_+$:

$$(u, e^{i\mu_n t})_{L^2} = \sqrt{2\pi} f(-\bar{\mu}_n), \quad n \geq 1.$$

We have the following result.

3.8.5. THEOREM. *Under the above assumptions, a state $a \in H$ is reachable if and only if there is a function $f \in H^2_+$ such that*

$$\sqrt{2\pi} f(-\overline{\mu}_n) = (a, \psi_n)_H \cdot \frac{1}{\gamma_n}, \quad n \geq 1, \tag{3.8.4}$$

where $\gamma_n = (h, \psi_n)_H$, $n \geq 1$.

If such a function exists, then the least energy control is given by $u_* = \mathcal{F} f_*$, where f_* is the unique function in $\mathcal{K}_\Theta$ satisfying the interpolation condition (3.8.4).

If $(\psi_n)_{n\geq 1}$ is a Riesz basis in H, then $\sum_{n\geq 1} |\gamma_n|^2 \|\varphi_n\|^2 < \infty$ and

$$c(\infty)L^2(\mathbb{R}_+) = \left\{ \sum_{n\geq 1} \gamma_n f(-\overline{\mu}_n)\varphi_n : f \in H^2_+, \sum_{n\geq 1} |\gamma_n|^2 \|\varphi_n\|^2 |f(-\overline{\mu}_n)|^2 < \infty \right\}.$$

Finally, if the exponentials $(e^{i\mu_n t})_{n\geq 1}$ also form a Riesz sequence (which is equivalent to $(\mu_n)_{n\geq 1}$ satisfying the Carleson condition, see Chapter 4), then

$$c(\infty)L^2(\mathbb{R}_+) = \left\{ \sum_{n\geq 1} a_n \varphi_n : \sum_{n\geq 1} |a_n|^2 \left(\|\varphi_n\|^2 + \frac{\operatorname{Im} \mu_n}{|\gamma_n|^2} \right) < \infty \right\}.$$

The last expression coincides with the renormalization of H given in Section 3.4. Notice that always $c(\infty)L^2(\mathbb{R}_+) \neq H$.

PROOF. Concerning the control, all has already been proved in Sections 3.4 and 3.7. It remains to cite the theorem on free interpolation in the Hardy space H^2_+ (see Section 4.2): if $(\mu_n)_{n\geq 1} \in (C)$ then

$$H^2_+ | M = \left\{ (a_n)_{n\geq 1} : \sum_{n\geq 1} |a_n|^2 \operatorname{Im} \mu_n < \infty \right\},$$

where $M = \{-\overline{\mu}_n\}_{n\geq 1}$ and $H^2_+ | M = \{ (f(-\overline{\mu}_n))_{n\geq 1} : f \in H^2_+ \}$. □

3.8.6. REMARK. In principle, this theorem remains valid for the case of control at a finite time (up to some obvious modifications), where we have to replace $L^2(\mathbb{R}_+)$ by $L^2(0, \tau)$, $e^{i\mu_n t}\chi_{\mathbb{R}_+}$ by $e^{i\mu_n t}\chi_{(0,\tau)}$, etc. The major difference concerns the bases of exponentials. In fact, the Carleson condition is no longer sufficient (but it remains necessary), see Section 4.5 for the corresponding criterion.

Finally, this last subsection, and in particular Theorem 3.8.5, emphasize the fundamental role of H^2_+-interpolation for control problems. In the case of multi-dimensional control, rank $B > 1$, one has to consider vector valued H^2_+-spaces (of dimension rank B), see Chapter 4 for details and Section 3.9 for references.

3.9. Notes and Remarks

General remarks. The case of semigroups which have a Riesz basis of eigenvectors is the most developped in the general theory. Such semigroups are actually similar to diagonal semigroups. For more information, we refer the reader to D. Russel [**Rus**], S. Dolecki and D. Russel [**DolR**], R. Curtain and H. Zwart [**CZ**], S. Avdonin and S. Ivanov [**AI2**] and finally A. Butkovskii [**Butk**].

The notion of renormalized exact controllability (RECO) introduced in Subsection 3.4.6 seems to be important for semigroups of this kind. Indeed, since exact controllability appears rather rarely, it is quite natural to adapt the state space to the behaviour of the dynamic system to regain exact controllability. It is curious to notice that a complete and elementary analysis of the renormalizations (as presented in 3.4.6) is possible without any additional assumptions.

Formal credits. For Section 3.1 we refer to the comments on Example 1.5 in Section 1.7. In particular, Example 3.1.2 originates from A. Butkovskii [**Butk**]. For asymptotics of the point spectrum of elliptic PDE see, for example, L. Hörmander [**Hör2**]. The same book is also an excellent source for spectral analysis of convolution operators mentioned in Subsection 3.1.4.

Theorem 3.3.1 is essentially contained in S. Avdonin and S. Ivanov [**AI2**] (and probably in other sources), whereas Theorem 3.3.2 seems to be new. A criterion for observability similar to our controllability criterion 3.3.2 is contained in D. Russel and G. Weiss [**RW**].

The notion of generalized exact controllability (GECO) was introduced in S. Avdonin and S. Ivanov [**AI2**]. We did not find the renormalized version of exact control (RECO) in the existing literature so that this notion appears to be new. Also, it seems that the crucial condition (3.4.1), $B\mathcal{U} \subset H(w_n)$, has never been analyzed/considered before. For the theory of symmetrically normed and quasi-normed operator ideals, and especially, for the Schatten–von Neumann ideals $\mathfrak{S}_p$ we refer to Part B of the book (Volume 1), Chapter B.7 (and also Chapters B.2 and B.8), to the book of A. Pietsch [**Pie2**], and to sources quoted in Chapters B.2, B.7 and B.8. For a quick reference, let us mention that a quasi-normed operator ideal is defined in the same way as a normed ideal (see Chapter B.7, Volume 1) but is equipped with a quasi-norm ($\|A + B\| \leq c(\|A\| + \|B\|)$) instead of a norm (in order to include such ideals as $\mathfrak{S}_p$ for $0 < p < 1$).

As mentioned in 3.4.6(9), the distribution of zeros of entire functions of exponential type is crucially important for applications of the Riesz bases approach to RECO. Basic references for this subtle matter are B. Levin [**Le1**], R. Boas [**Bo1**], N. Levinson [**LeN**]; see also L. Hörmander [**Hör2**], where an interesting extension of the well-known symetry property of zeros of Fourier transforms of functions with compact supports can be found. Some explanations of the physical-technical meaning of the "optical length" of a hyperbolic dynamical system can be found in S. Avdonin and S. Ivanov [**AI2**] and A. Butkovskii [**Butk**]. For the existence of this limit length for Riesz basis frequencies see S. Hruschev, N. Nikolski, and B. Pavlov [**HNP**] and N. Nikolski and S. Hruschev [**HrN**].

The notion of null controllability (NCO) is classical, and we refer to S. Avdonin and S. Ivanov [**AI2**] as well as to R. Curtain and H. Zwart [**CZ**] for an operator presentation. Exact null controllability (ENCO) is implicitely contained in A. Butkovskii [**Butk**] (and in other sources) and has been explicitely stated in S. Avdonin and S. Ivanov [**AI2**]. However, the presentation given in Section 3.5 seems to be new.

The weak controllabilities of Section 3.6 are introduced in S. Avdonin and S. Ivanov [**AI2**] but without the term $|\operatorname{Re}(Ax, x)|$ that appears in our definition of the energy of the state $x \in H$, see for example M. Birman and M. Solomyak [**BiS**] for this definition. For this reason, in order to prove an analog of our Theorem 3.6.2 (in the case $B = I$), the authors of [**AI2**] were constrained to change the standard notion of uniform minimality.

The optimization process presented in Section 3.7 seems to be folklore. For a special version of this process (and a lot of examples) we refer to A. Butkovskii

[**Butk**]. For general properties and applications of this "linear programming" duality approach see, for example, L. Kantorovich and G. Akilov [**KanA**], and for needed geometric background (as uniformly convex spaces, etc.) J. Lindenstrauss and L. Tzafriri [**LT**].

For other treatments of controllability at time $\tau = \infty$ (considered in Section 3.8), see P. Fuhrmann [**Fuh3**], C. Foiaş and A. Frazho [**FF**], J. Ball, I. Gohberg and L. Rodman [**BGR**] as well as R. Curtain and H. Zwart [**CZ**]. For interpolation problems for vector valued functions (mentioned in 3.8.5) we refer to C. Foiaş, A. Frazho, I. Gohberg and R. Kaashoek [**FFGK**], S. Avdonin and S. Ivanov [**AI2**] and N. Nikolski [**N19**].

CHAPTER 4

A Glance at Bases of Exponentials and of Reproducing Kernels

Following the preceding chapter, the geometric properties of the system of exponentials $(\mathcal{E}_n)_{n\geq 1}$,

$$\mathcal{E}_n(t) = e^{-\bar{\lambda}_n t} B^* \psi_n, \quad n \geq 1,$$

are crucially important for controllability properties of a dynamical system (A, B), at least in case the eigenvectors of the generator A form a Riesz basis.

Intending to study these properties, we can distinguish two extremal cases:

- when B^* is an isomorphism (onto its range), and/or
- when the control is of finite rank ($\operatorname{rank} B < \infty$), in particular $\operatorname{rank} B = 1$. In the latter case, we are concerned with a family of scalar exponentials

$$\mathcal{E}_n(t) = b_n e^{-\bar{\lambda}_n t} \quad n \geq 1$$

in $L^2(0, \tau)$. Here $b_n \in \mathbb{C}$ are suitable coefficients.

In the first case, if $(\varphi_n)_{n\geq 1}$, and hence also $(\psi_n)_{n\geq 1}$, is a Riesz basis of the state space, then there is an isomorphism of the control space $V : \mathcal{U} \longrightarrow \mathcal{U}$ that transforms the family $(B^*\psi_n)_{n\geq 1}$ into an orthogonal basis $(VB^*\psi_n)_{n\geq 1}$. Consequently, the isomorphism $I \otimes V : L^2(0, \tau; \mathcal{U}) \longrightarrow L^2(0, \tau; \mathcal{U})$ transforms the sequence $(\mathcal{E}_n)_{n\geq 1}$ into an orthogonal basis $(e^{-\bar{\lambda}_n t} VB^*\psi_n)_{n\geq 1}$ of its span.

Similarly, if we suppose that $(e^{-\bar{\lambda}_n t})_{n\geq 1}$ is a Riesz sequence, then there is an isomorphism $V' : L^2(0, \tau) \longrightarrow L^2(0, \tau)$ that turns this sequence into an orthogonal sequence $(V' e^{-\bar{\lambda}_n t})_{n\geq 1}$ in $L^2(0, \tau)$. Then the isomorphism $V' \otimes I : L^2(0, \tau; \mathcal{U}) \longrightarrow L^2(0, \tau; \mathcal{U})$ transforms $(\mathcal{E}_n)_{n\geq 1}$ into a Riesz sequence regardless of B. This example emphasizes the importance of geometric properties of scalar exponentials.

This chapter is devoted to an intensive study of these exponentials. Recall that we have repeatedly mentioned the links between the distribution of the frequencies $(\lambda_n)_{n\geq 1}$ (the eigenvalues of the generator A) and the behaviour of $\mathcal{E}_n$, $n \geq 1$.

We begin the chapter by recalling a classical approach to exponentials initiated by R. Paley and N. Wiener in 1934 and called the *perturbation method* (Section 4.1).

Next, in Section 4.2, we discuss bases of exponentials on an infinite interval, namely, in the space $L^2(\mathbb{R}_+)$, reducing this case to reproducing kernels of the Hardy space H^2, which have already been studied in Chapter C.3.

Our main theme, exponentials on a finite interval $I \subset \mathbb{R}$, starts in Section 4.3. Here we use B. Pavlov's *projection method* to reduce the Riesz bases problem on I to a similar problem on $\mathbb{R}_+$, at least for sets of frequencies arising from control theory.

After what we know from Part C, it is natural to expect that the right framework for the projection method is the problem of reproducing kernel bases in the model spaces $\mathcal{K}_\Theta$. This is explained in Section 4.4, where the main basis criteria are deduced from the Hankel–Toeplitz theory of Part B (Volume 1).

In Section 4.5, we return to exponentials in order to explain an explicit basis condition expressed in terms of the frequency distribution, and to the proof of "Kadets' 1/4 theorem" about uniform perturbations of the standard orthonormal exponential basis. In Section 4.6, we prove a theorem of N. Levinson giving an efficient sufficient condition for the completeness of the exponentials $(e^{i\lambda_n t})$ in $L^p(0, 2\pi)$. This condition turns out to be sufficient to yield the theorem of A. Ingham, which is the necessity counterpart of Kadets' theorem (another, simpler, proof of this theorem is given in Section 4.1).

In order to complete the quick picture of the geometry of exponentials and reproducing kernels given in this chapter, we finish our exposition by an account of the rest of the theory (Section 4.8).

4.1. Small Perturbations of Harmonic Frequencies

The only situation where the exponentials $(e^{i\mu_n t})_{n\in\mathbb{Z}}$ are orthogonal on $(0, \tau)$ is the "harmonic case", that is, when the frequencies μ_n are multiples of the principal frequency $2\pi/\tau$: $\mu_n = \frac{2\pi}{\tau}\cdot k_n$, $k_n \in \mathbb{Z}$. For instance, the classical orthonormal basis on $L^2(0, 2\pi)$ (with respect to normalized Lebesgue measure $dt/2\pi$) is $(e^{int})_{n\in\mathbb{Z}}$.

In this subsection, we recall some classical results about small perturbations $(e^{i\mu_n t})_{n\in\mathbb{Z}}$ of the harmonic basis $(e^{int})_{n\in\mathbb{Z}}$ in $L^2(0, 2\pi)$, namely, we consider the case when $|\mu_n - n| \le \delta$, $n \in \mathbb{Z}$. The simple translation $t \longmapsto t - a$ transforms the exponentials $e^{i\mu_n t}$, $\mu_n \in \mathbb{C}$, in $L^2(a, b)$ into $e^{i\mu_n(t-a)} = e^{i\mu_n t}a_n$ $(a_n \in \mathbb{C})$ in the space $L^2(0, b - a)$. This change of variable is obviously unitary, and as scalar factors are of no influence on the basis property, we can always replace the interval under consideration by another one of the same length.

4.1.1. THEOREM (R. Paley, N. Wiener, R. Duffin, J. Eachus). *Let $\mu_n \in \mathbb{C}$ be such that $|\mu_n - n| \le \delta < \frac{\log 2}{\pi}$ $(\simeq 0.22)$, $n \in \mathbb{Z}$. Then $(e^{i\mu_n t})_{n\in\mathbb{Z}}$ is a Riesz basis of $L^2(0, 2\pi)$.*

PROOF. The following statement is an obvious consequence of Theorem C.3.1.4.

Let V be an operator defined on $\mathcal{L}\mathrm{in}(f_n : n \ge 1)$ by $Vf_n = g_n$, $n \ge 1$. If V extends to an isomorphism $V : X \longrightarrow X$ then (f_n) and (g_n) are simultaneously unconditional bases or not (or Riesz bases in a Hilbert space setting).

Set $Ve^{inx} = e^{i\mu_n x}$, $n \in \mathbb{Z}$. Replacing $(0, 2\pi)$ by $(-\pi, \pi)$, we will show that V extends to an isomorphism on $L^2(-\pi, \pi)$. This will follow if we can show that $\|I - V\| < 1$.

Let $(a_n)_{n\in\mathbb{Z}}$ be a finitely supported sequence of complex numbers. Then

$$\begin{aligned}\Big\|(V-I)\Big(\sum_{n\in\mathbb{Z}} a_n e^{inx}\Big)\Big\| &= \Big\|\sum_{n\in\mathbb{Z}} a_n (V-I)e^{inx}\Big\| = \Big\|\sum_{n\in\mathbb{Z}} a_n (e^{i\mu_n x} - e^{inx})\Big\| \\ &= \Big\|\sum_{n\in\mathbb{Z}} a_n e^{inx} \sum_{k\geq 1} \frac{x^k (i(\mu_n - n))^k}{k!}\Big\| = \Big\|\sum_{k\geq 1} \frac{x^k}{k!} \sum_{n\in\mathbb{Z}} (i(\mu_n - n))^k a_n e^{inx}\Big\| \\ &\leq \sum_{k\geq 1} \frac{1}{k!} \Big\|x^k \sum_{n\in\mathbb{Z}} (i(\mu_n - n))^k a_n e^{inx}\Big\| \leq \sum_{k\geq 1} \frac{\pi^k}{k!} \Big\|\sum_{n\in\mathbb{Z}} i(\mu_n - n)^k a_n e^{inx}\Big\| \\ &= \sum_{k\geq 1} \frac{\pi^k}{k!} \Big(\sum_{n\in\mathbb{Z}} |\mu_n - n|^{2k} |a_n|^2\Big)^{1/2} \leq \sum_{k\geq 1} \frac{\delta^k}{k!} \pi^k \Big\|\sum_{n\in\mathbb{Z}} a_n e^{inx}\Big\| \\ &= (e^{\pi\delta} - 1) \Big\|\sum_{n\in\mathbb{Z}} a_n e^{inx}\Big\|.\end{aligned}$$

By assumption, we have $\pi\delta < \log 2$, and hence $e^{\pi\delta} - 1 < 1$. □

The question whether the bound $(\log 2)/\pi$ of the previous theorem is sharp has occupied many people. After several attempts, the following result was obtained.

4.1.2. THEOREM (A. Ingham, M. Kadets). *The following assertions are equivalent.*

(1) Let $\mu_n \in \mathbb{R}$ and suppose that $|n - \mu_n| \leq \delta$, $n \in \mathbb{Z}$. Then $(e^{i\mu_n t})_{n\in\mathbb{Z}}$ is a Riesz basis in $L^2(-\pi,\pi)$.

(2) $\delta < 1/4 = 0.25$.

The implication (1)$\Longrightarrow$(2) was proved by A. Ingham in 1937, whereas the converse remained open until M. Kadets' impact in 1964. We prove Kadets' implication in Subsection 4.5.7. An example which proves Ingham's implication is given below. Another proof of the same implication will be deduced from a completeness theorem by N. Levinson in Section 4.6. Some other results on perturbations of the orthonormal basis $(e^{int})_{n\in\mathbb{Z}}$ are outlined in Sections 4.7 and 4.8.

PROOF OF (1)$\Longrightarrow$(2). Let $\mu_0 = 0$ and $\mu_n = -\mu_{-n} = n - \frac{1}{4}$ for $n \geq 1$. We show that the family $(e^{i\mu_n t})_{n\in\mathbb{Z}}$ is not minimal. Indeed, it is clear that $L_+ = \mathrm{span}(e^{i\mu_n t} : n \geq 1) = e^{-it/4} e^{it} H^2$ and $L_- = \mathrm{span}(e^{i\mu_n t} : n \leq -1) = e^{it/4} H^2_-$. We check that $\mathrm{span}(L_+, L_-) = L^2$.

Indeed, if $f \in L^2$ and $f \perp e^{it/4} H^2_-$, then $f e^{-it/4} \in H^2$, and hence $f = e^{it/4} h$ with $h \in H^2$. If, in addition, $f \perp e^{3it/4} H^2$ then $e^{-it/2} h \perp H^2$, and hence $e^{-it/2} h \in H^2_-$. This yields $e^{-it} h^2 \in e^{-it} H^1_-$, and hence $h^2 \in H^1_-$. Since $h^2 \in H^1$, it follows that $h = 0$.

Therefore, $1 \in \mathrm{span}(e^{i\mu_n t} : n \neq 0)$, and the claim follows. □

4.2. Bases of Exponentials on the Half-Line

In this section we describe some results about bases of exponentials on the infinite interval $[0, \infty) = \mathbb{R}_+$ and their links with bases of reproducing kernels in the Hardy space H^2.

We begin with an elementary observation on frequencies of exponentials lying in $L^2(0,\infty) = L^2(\mathbb{R}_+)$. Namely,

$$e^{i\lambda t} \in L^2(\mathbb{R}_+) \Longleftrightarrow \mathrm{Im}\,\lambda > 0,$$

and

$$\int_0^\infty |e^{i\lambda t}|^2 dt = \int_0^\infty e^{-2\,\mathrm{Im}\,\lambda t} dt = \frac{1}{2\,\mathrm{Im}\,\lambda}.$$

The results of Chapter A.6 (Volume 1) allow us to turn exponentials in $L^2(\mathbb{R}_+)$ into reproducing kernels of the Hardy space $H^2(\mathbb{D})$. Namely, Theorem A.6.3.4 (change of variables) and Theorem A.6.3.5 (Paley–Wiener) show that the composition $\mathcal{F}U_2$ is a unitary map from $H^2(\mathbb{D})$ onto $L^2(\mathbb{R}_+)$ (see Section A.6.1, Volume 1, for the definition of U_2). Moreover, by exactly the same explicit computation as in A.8.3.1, we can verify the following lemma.

4.2.1. LEMMA. *Let $|\lambda| < 1$. Then*

$$\mathcal{F}U_2\Big(\frac{1}{1-\bar{\lambda}z}\Big) = c(\lambda)e^{-i\overline{\omega(\lambda)}t},$$

where

$$\omega(\lambda) = i\,\frac{1+\lambda}{1-\lambda}$$

is the standard conformal mapping from the unit disk onto the upper half plane, $c(\lambda) \in \mathbb{C}$ and $\mathrm{Im}\,\omega(\lambda) = \frac{1-|\lambda|^2}{|1-\lambda|^2} > 0$ (hence $\mathrm{Im}(-\overline{\omega(\lambda)}) > 0$). □

Then, as $\mathcal{F}U_2$ is unitary, all geometric Hilbert space properties (such as completeness, minimality, being a basis, an unconditional basis etc.) are the same for exponentials $e^{i\mu t}$, $\mu = -\overline{\omega(\lambda)}$, as for reproducing kernels $\frac{1}{1-\bar{\lambda}z}$.

Recall that the reproducing kernels k_λ, $|\lambda| < 1$, are those elements in H^2 satisfying

$$f(\lambda) = \langle f, k_\lambda\rangle, \quad f \in H^2.$$

From these simple observations it follows that a family of exponentials $(e^{i\mu t})_{\mu\in M}$ is never a basis of the whole space $L^2(0,\infty)$ (or equivalently, a family of reproducing kernels $(k_\lambda)_{\lambda\in\Lambda}$ is never a basis of H^2). Moreover, we have seen that a basis is uniformly minimal (Theorem A.5.1.6, Volume 1), but an easy exercise shows that if the family (of exponentials or reproducing kernels) is minimal, then it is incomplete.

The following theorem gives a complete description of Riesz sequences of exponentials on $\mathbb{R}_+$.

4.2.2. THEOREM. *Let $\Lambda \subset \mathbb{D}$, $\Lambda = (\lambda_n)_{n\geq 1}$, $\lambda_i \neq \lambda_j$, $i \neq j$. Let further $M = (\mu_n)_{n\geq 1}$ be the image of Λ with respect to the mapping ω of Lemma 4.2.1: $\mu_n = -\overline{\omega(\lambda_n)}$, $\mathrm{Im}\,\mu_n > 0$, $n \geq 1$. The following assertions are equivalent.*

(1) $(k_\lambda : \lambda \in \Lambda)$ is a Riesz sequence in H^2.

(2) $(e^{i\mu t} : \mu = -\overline{\omega(\lambda)}, \lambda \in \Lambda)$ is a Riesz sequence in $L^2(0,\infty)$.

(3) M is an interpolating sequence *for H^2_+, that is*

$$H^2_+|M = l^2\,(\mathrm{Im}\,\mu_n) = \Big\{(a_n)_{n\geq 1} : \sum_{n\geq 1} |a_n|^2\,\mathrm{Im}\,\mu_n < \infty\Big\}.$$

(4) Λ is an interpolating sequence for H^2, that is

$$H^2|\Lambda = l^2\left(1 - |\lambda_n|^2\right).$$

(5) Λ is an interpolating sequence for the space of bounded analytic functions on the unit disk:

$$H^\infty|\Lambda = l^\infty.$$

(6) Λ satisfies the Carleson condition*: there exists $\delta > 0$ such that*

$$(C) \qquad |B_n(\lambda_n)| \geq \delta > 0$$

for every $n \geq 1$, where $B = \prod_{n=1}^{\infty} b_{\lambda_n}$ is the Blaschke product associated to Λ, $b_{\lambda_n} = \frac{\lambda_n - z}{1-\bar{\lambda}_n z} \cdot \frac{|\lambda_n|}{\lambda_n}$, and $B_n = B/b_{\lambda_n}$.

(7) The sequence of reproducing kernels $(k_{\lambda_n})_{n\geq 1}$ is uniformly minimal *in H^2, that is, there exists $\varepsilon > 0$ such that*

$$\mathrm{dist}_{H^2}\left(\frac{k_{\lambda_n}}{\|k_{\lambda_n}\|} : \mathrm{span}(k_{\lambda_i} : i \neq n)\right) \geq \varepsilon$$

for every $n \geq 1$.

PROOF. All the results concerning reproducing kernels k_λ, $\lambda \in \Lambda$, are contained in Chapter C.3. The claims about the exponentials follow from Lemma 4.2.1. □

4.2.3. REMARK. **(1) The half-plane Carleson condition**. We can express the Carleson condition (C) in terms of the frequencies $(\mu_n)_{n\geq 1}$ by

$$(C) \qquad \prod_{k\neq n} \left|\frac{\mu_n - \mu_k}{\mu_n - \bar{\mu}_k}\right| \geq \delta > 0, \quad n \geq 1.$$

It can also be shown that in the case of exponentials $(e^{i\mu_n x})_{n\geq 1}$ of "hyperbolic type" (see Remark 3.3.3), that is,

$$0 < \inf_n(\mathrm{Im}\,\mu_n) \leq \sup_n(\mathrm{Im}\,\mu_n) < \infty,$$

the Carleson condition reduces to the existence of $\varepsilon > 0$ such that

$$|\mu_n - \mu_m| \geq \varepsilon, \quad n \neq m.$$

In other words, we have uniform separation in the euclidean metric. Take, for instance, a sequence $(i + \mu_n : n \in \mathbb{Z})$ with $|n - \mu_n| \leq \delta < 1/2$, $n \in \mathbb{Z}$. It obviously satisfies the Carleson condition, so $(e^{i\mu_n t}e^{-t} : n \in \mathbb{Z})$ is a Riesz sequence in $L^2(\mathbb{R}_+)$ (the reader should compare this result with the Ingham–Kadets theorem).

Let us comment a bit more on the links between the Carleson condition and the above separating condition. Recall that the *sparseness (separating)* condition (R) for the half-plane $\mathbb{C}_+$ means that

$$\Delta = \inf_{n\neq k}\left|\frac{\mu_n - \mu_k}{\mu_n - \bar{\mu}_k}\right| > 0$$

or, equivalently, that the disks

$$D(\mu_n, \varepsilon \cdot \mathrm{Im}\,\mu_n)$$

are pairwise disjoint for some $\varepsilon > 0$ (see C.3.3.4 for the conformally equivalent condition in the disk $\mathbb{D}$ and also Chapter 6 of Part A, Volume 1, for the change of variables from the unit disk to the half-plane). Therefore, in a band $0 < c \leq \mathrm{Im}\, z \leq C < \infty$, this is equivalent to the Euclidean separating condition mentioned above.

On the other hand, Carleson measures $\nu = \sum_{n\geq 1} \mathrm{Im}(\mu_n)\delta_{\mu_n}$ (see C.3.3.4 for the case of the disk) are characterized by

$$\nu D(x, r) \leq Cr, \quad \forall r > 0,\ \forall x \in \mathbb{R}.$$

It is clear that for band limited sequences (μ_n) as above, this condition follows from (R), and therefore (R) is equivalent to (C). Similarly, for $\mu_n = iy_n$, $y_n > 0$, where $(y_n)_{n\in\mathbb{Z}}$ is an increasing suquence of positive numbers, the condition (R) is equivalent to

$$y_{n+1} \geq qy_n \text{ (for } n \in \mathbb{Z}), \quad q = (1+\Delta)(1-\Delta)^{-1} > 1;$$

consequently, in this case, (R) implies (C) too (see also 4.7.2(b) below).

(2) Vector valued exponentials and reproducing kernels. We now turn to vector valued bases of exponentials, that is, to functions of the following type

$$\mathcal{E}_n(t) = e^{i\mu_n t}u_n, \quad t \in \mathbb{R}_+,$$

where $\mu_n \in \mathbb{C}_+$, $u_n \in \mathcal{U}$ and $\mathcal{U}$ is an auxiliary Hilbert space. It is easy to see that the above mentioned mappings U_2 and $\mathcal{F}$ carry over to isometries from the space of $\mathcal{U}$ valued square integrable functions $L^2(\mathbb{R}_+,\mathcal{U})$ onto the $\mathcal{U}$ valued Hardy space on $\mathbb{D}$, $H^2(\mathbb{D},\mathcal{U})$, and that $\mathcal{F}U_2$ then transforms the exponentials $\mathcal{E} = e^{i\mu t}\cdot u$ into the reproducing kernels $k_\lambda \cdot u$ $(u \in \mathcal{U})$ with $\mu = \overline{-\omega(\lambda)}$ (see Lemma 4.2.1).

The following two results are useful in control theory.

4.2.4. THEOREM. *Let* $\dim\mathcal{U} = N < \infty$, *and suppose that* $\mathcal{E}_n = e^{i\mu_n t}u_n$, $n \geq 1$, *is a Riesz sequence in* $L^2(\mathbb{R}_+,\mathcal{U})$. *Then* $(\mu_n)_{n\geq1}$ *is a union of at most* N *sequences satisfying the Carleson condition.*

See Section 4.8 for references and comments.

4.2.5. THEOREM. *Suppose again that* $\dim\mathcal{U} < \infty$ *and let* $\mathcal{E}_n = e^{i\mu_n t}u_n$, $n \geq 1$, *be a family of vector valued exponentials in* $L^2(\mathbb{R}_+,\mathcal{U})$. *The following assertions are equivalent.*

(1) $(\mathcal{E}_n)_{n\geq1}$ *is a Riesz sequence.*
(2) $(\mathcal{E}_n)_{n\geq1}$ *is uniformly minimal.*

The implication $(1) \Longrightarrow (2)$ is obvious, and $(2) \Longrightarrow (1)$ (*Treil's theorem*) is an analogue of the implications $(7) \Longrightarrow (6) \Longrightarrow (3)$ and $3 \Longleftrightarrow 2$ of Theorem 4.2.2.

4.3. Bases of Exponentials on Finite Intervals

We now examine the exponentials $(e^{i\mu_n t})_{n\in\mathbb{Z}}$ on a finite interval and start with some rather technical but important observations.

4.3.1. Technicalities. The following five facts are valid for both scalar- and vector valued exponentials. In order to simplify the notation, we restrict ourselves to the scalar case.

(1) Interval translation. As already mentioned, we may translate every interval (a,b) to $(0,b-a)$ or $(-c,c)$ with $c = (b-a)/2$, by the simple change of variable $t \longmapsto t+\alpha$.

(2) Frequency translation. The multiplication $f(x) \longmapsto e^{-\alpha x}f(x)$ is an isomorphism in $L^2(a,b)$, and thus, the families $(e^{i\mu_n t})_{n\in\mathbb{Z}}$ and $(e^{i(\mu_n+i\alpha)t})_{n\in\mathbb{Z}}$ have the same geometric Banach space properties such as being a basis, minimality, uniform minimality etc.

(3) Frequencies arising from control problems. In control theory, the frequencies $i\mu_n$ are the eigenvalues of the generator of a semigroup, and hence $\mathrm{Re}(i\mu_n) \leq$ const (see Subsections 2.2.1, 2.2.2), or simply $\mathrm{Im}\,\mu_n \geq -$const. Then, there is $\alpha \in \mathbb{R}$ such that

$$\mathrm{Im}(\mu_n + i\alpha) > 0, \quad n \geq 1,$$

or even

$$\operatorname{Im}(\mu_n + i\alpha) \ge \delta > 0, \quad n \ge 1.$$

(4) Exponentials and model spaces. Consider a family of exponentials $(e^{i\mu_n t})_{n\in\mathbb{Z}}$ in $L^2(0,\tau)$ such that $\operatorname{Im}\mu_n > 0$ for every n. Define the right translation operator $S_\tau : L^2(0,\infty) \longrightarrow L^2(0,\infty)$ by $S_\tau f(x) = f(x-\tau)$ if $x > \tau$ and $S_\tau f(x) = 0$ otherwise. We then have

$$\begin{aligned}\mathcal{F}^{-1}L^2(0,\tau) &= \mathcal{F}^{-1}(L^2(0,\infty) \ominus S_\tau L^2(0,\infty)) \\ &= H^2_+ \ominus e^{i\tau z}H^2_+ = U_2(H^2 \ominus \Theta H^2) = U_2\mathcal{K}_\Theta,\end{aligned}$$

where $\Theta = \exp(-\tau\,\frac{1+z}{1-z})$. This means that $L^2(0,\tau)$ is the isometric $\mathcal{F}U_2$-image of the model space

$$\mathcal{K}_\Theta := H^2 \ominus \Theta H^2.$$

Furthermore,

$$e^{i\mu_n t}\chi_{(0,\tau)} = P_{L^2(0,\tau)}e^{i\mu_n t}\chi_{\mathbb{R}_+},$$

where P_E is the orthogonal projection onto the subspace E, and hence

$$\begin{aligned}U_2^{-1}\mathcal{F}^{-1}e^{i\mu_n t}\chi_{(0,\tau)} &= U_2^{-1}\mathcal{F}^{-1}P_{L^2(0,\tau)}\mathcal{F}U_2(U_2^{-1}\mathcal{F}^{-1})e^{i\mu_n t}\chi_{\mathbb{R}_+} \\ &= P_{\mathcal{K}_\Theta}\frac{c}{1-\bar{\lambda}_n z},\end{aligned}$$

where, $c \in \mathbb{C}$ and $\mu_n = -\overline{\omega(\lambda_n)}$. This identity follows from Lemma 4.2.1. On the other hand, using the fact that k_λ is the reproducing kernel of H^2, we see that $k_\lambda^\Theta := \frac{1-\overline{\Theta(\lambda_n)}\Theta}{1-\bar{\lambda}_n z} \perp \Theta H^2$, and therefore

$$U_2^{-1}\mathcal{F}^{-1}e^{i\mu_n t}\chi_{(0,\tau)} = P_{\mathcal{K}_\Theta}\frac{c}{1-\bar{\lambda}_n z} = ck^\Theta_{\lambda_n}(z).$$

Recall that an analogue of the above computations has already been encountered in Section A.8.3 (Volume 1).

(5) Propagation of bases. *Let $\mu_n \in \mathbb{C}$, $n \ge 1$, be such that $\inf_{n\ge1}\operatorname{Im}\mu_n > -\infty$ and such that $(e^{i\mu_n t}\chi_{(0,\tau)})_{n\ge1}$ is a Riesz sequence in $L^2(0,\tau)$, $\tau > 0$. Then $(e^{i\mu_n t}\chi_{(0,\tau')})_{n\ge1}$ is likewise a Riesz sequence in $L^2(0,\tau')$ for every $\tau' > \tau$.*

Indeed, the assumption implies that

$$\begin{aligned}\Big\|\sum_{n\ge1} a_n e^{i\mu_n t}\Big\|^2_{L^2(0,\tau)} &\asymp \sum_{n\ge1}|a_n|^2\|e^{i\mu_n t}\|^2_{L^2(0,\tau)} = \sum_{n\ge1}|a_n|^2\frac{1-e^{-2\operatorname{Im}\mu_n\tau}}{2\operatorname{Im}\mu_n} \\ &\asymp \sum_{n\ge1}\frac{|a_n|^2}{|\operatorname{Im}\mu_n|+1}\end{aligned}$$

for every finitely supported sequence $(a_n)_{n\ge1}$, $a_n \in \mathbb{C}$. Note that this expression is independent of τ (but the constants involved in this expression do). Clearly $\|f\|^2_{L^2(0,\tau)} \le \|f\|^2_{L^2(0,\tau')} = \|f\|^2_{L^2(0,\tau)} + \|f\|^2_{L^2(\tau,\tau')}$ for every $f \in L^2(0,\tau')$. It thus only remains to estimate $\|\sum_{n\geq 1} a_n e^{i\mu_n t}\|^2_{L^2(\tau,\tau')}$. Without loss of generality we may assume that $\tau' \in]\tau, 2\tau]$ (and then proceed by induction). Then

$$\begin{aligned}\Big\|\sum_{n\geq 1} a_n e^{i\mu_n t}\Big\|^2_{L^2(\tau,\tau')} &- \int_\tau^{\tau'} \Big|\sum_{n\geq 1} a_n e^{i\mu_n t}\Big|^2 dt = \int_0^{\tau'-\tau} \Big|\sum_{n\geq 1} a_n e^{i\mu_n\tau} e^{i\mu_n t}\Big|^2 dt \\ &\leq \int_0^{\tau} \Big|\sum_{n\geq 1} a_n e^{i\mu_n\tau} e^{i\mu_n t}\Big|^2 dt \\ &\leq C_1 \sum_{n\geq 1} |a_n|^2 e^{-2\operatorname{Im}\mu_n \tau} \frac{1}{|\operatorname{Im}\mu_n| + 1} \\ &\leq C_2 \sum_{n\geq 1} |a_n|^2 \frac{1}{|\operatorname{Im}\mu_n| + 1},\end{aligned}$$

as required. □

4.3.2. Reduction to exponentials on $\mathbb{R}_+$. The preceding five observations (and certain others) enable us to deduce a criterion for exponentials to form a basis on a finite interval from such a criterion on the half-line $\mathbb{R}_+$. The following result is a first step in this direction. Later on, in Section 4.4, we use the same method to link bases of reproducing kernels of the space H^2 and its model subspaces.

4.3.3. LEMMA. *Let X, Y be Banach spaces, $T \in L(X, Y)$, and let $(x_n)_{n\geq 1} \subset X$ be a sequence satisfying $\|x_n\| \asymp \|Tx_n\|$, $n \geq 1$. If $(Tx_n)_{n\geq 1}$ is uniformly minimal, then $(x_n)_{n\geq 1}$ is also uniformly minimal.*

PROOF. This is an immediate consequence of the definition (see Section 3.2). □

4.3.4. EXAMPLE. Let $X = L^2(\mathbb{R}_+)$, $x_n = e^{i\mu_n t}\chi_{\mathbb{R}+}$, $(\operatorname{Im}\mu_n > 0)$, $Y = L^2(0,\tau)$ and $T = P_{L^2(0,\tau)}$. Then $\|x_n\|^2 = \frac{1}{2\operatorname{Im}\mu_n}$, and

$$\|P_{L^2(0,\tau)} e^{i\mu_n x}\|^2 = \int_0^\tau e^{-2\operatorname{Im}\mu_n x} dx = \frac{1 - e^{-2\operatorname{Im}\mu_n \tau}}{2\operatorname{Im}\mu_n}.$$

Hence

$$\|x_n\| \asymp \|P_{L^2(0,\tau)} x_n\|$$

if and only if

$$\inf_{n\geq 1}(\operatorname{Im}\mu_n) > 0. \tag{4.3.1}$$

4.3.5. COROLLARY. *If a family $\mathcal{E} = (e^{i\mu_n t}\chi_{(0,\tau)})_{n\geq 1}$ is uniformly minimal in $L^2(0,\tau)$, and if (4.3.1) holds, then $(\mu_n)_{n\geq 1} \in (C)$. In particular, if $\mathcal{E}$ is a Riesz sequence verifying (4.3.1), then $(\mu_n)_{n\geq 1} \in (C)$.*

This is an immediate consequence of Lemma 4.3.3, Example 4.3.4 and Theorem 4.2.2. □

4.3.6. THEOREM. *Let $(\mathcal{E}_n)_{n\geq 1} = (e^{i\mu_n t}\chi_{(0,\tau)})_{n\geq 1}$ be a sequence with frequencies satisfying (4.3.1). Then $(\mathcal{E}_n)$ is a Riesz sequence (respectively a Riesz basis) in $L^2(0,\tau)$ if and only if the following two conditions hold.*

(1) $(\mu_n)_{n\geq 1} \in (C)$,

(2) the orthogonal projection $P_{L^2(0,\tau)} : f \longrightarrow \chi_{(0,\tau)} f$ is an isomorphism of $\operatorname{span}_{L^2(\mathbb{R}_+)}(e^{i\mu_n t}\chi_{\mathbb{R}_+} : n \geq 1)$ *onto its range (respectively onto $L^2(0,\tau)$).*

PROOF. For the necessity, asssume that $(\mathcal{E}_n)_{n\geq1}$ is a Riesz sequence. By Corollary 4.3.5 we get that $(\mu_n)_{n\geq1} \in (C)$, and by Theorem 4.2.2 that $(x_n)_{n\geq1} = (e^{i\mu_n t}\chi_{\mathbb{R}_+})_{n\geq1}$ is also a Riesz sequence. By assumption, the projections $(\mathcal{E}_n)_{n\geq1} = (P_{L^2(0,\tau)}x_n)_{n\geq1}$ likewise form a Riesz sequence, and we get

$$\Big\|\sum_{n\geq1} a_n x_n\Big\|^2 \asymp \sum_{n\geq1} |a_n|^2\|x_n\|^2,$$

as well as

$$\Big\|\sum_{n\geq1} a_n \mathcal{E}_n\Big\|^2 \asymp \sum_{n\geq1} |a_n|^2\|\mathcal{E}_n\|^2$$

for every finite sequence of complex numbers $a_n \in \mathbb{C}$. Moreover, because of (4.3.1), we have

$$\|\mathcal{E}_n\| = \|P_{L^2(0,\tau)}x_n\| \asymp \|x_n\|, \quad n \geq 1,$$

and consequently, the projection $P_{L^2(0,\tau)}$ is an isomorphism of $\mathrm{span}(x_n : n \geq 1)$ onto $\mathrm{span}(\mathcal{E}_n : n \geq 1)$.

For the sufficiency, let $(\mu_n)_{n\geq1}$ be a sequence satisfying the Carleson condition (C). Theorem 4.2.2 shows that the sequence $(x_n)_{n\geq1}$ is a Riesz sequence, and consequently its isomorphic image, given by $\mathcal{E}_n = P_{L^2(0,\tau)}x_n$, $n \geq 1$, is also a Riesz sequence. □

4.3.7. REMARK. (1) As is already mentioned, the frequencies arising from control problems satisfy (4.3.1) (eventually after a suitable lifting, see 4.3.1(3)).

On the other hand, as is shown in Lemma 4.4.2 below, *the Carleson condition (C) is always necessary* for $(e^{i\mu_n t}\chi_{(0,\tau)})_{n\geq1}$ to be a Riesz sequence ($\mathrm{Im}\,\mu_n > 0$ is assumed).

(2) Note that the second condition of the above theorem is equivalent to

$$\int_0^\infty |f(t)|^2\,dt \leq C^2\int_0^\tau |f(t)|^2\,dt$$

for every finite linear combination $f = \sum a_n e^{i\mu_n t}\chi_{\mathbb{R}_+}$, $a_n \in \mathbb{C}$, where $C > 0$ is a constant not depending on f. The question is how to express this property in terms of the intrinsic data of the problem, that is, the frequencies $(\mu_n)_{n\geq1}$ and the length τ. It appears that this question is easier to handle in a more general framework that is more adapted to our problem. We return to the problem of exponential bases after an intermezzo on model spaces and reproducing kernels.

4.4. Bases of Reproducing Kernels in Model Spaces

It is interesting and important that the preceding results on exponential bases can be generalized almost automatically to model spaces $\mathcal{K}_\Theta$,

$$\mathcal{K}_\Theta = H^2 \ominus \Theta H^2,$$

where Θ is an arbitrary inner function (about inner functions, see A.1.4.2, Volume 1). Moreover, the framework of model spaces offers a more natural language for problems of exponential bases. Following 4.3.1(4), recall that a family of exponentials $(e^{i\mu_n t}\chi_{(0,\tau)})_{n\geq1}$ is an isometric image of a family of reproducing kernels

$(k_{\lambda_n}^{\Theta})_{n\geq 1}$, where

$$k_\lambda^\Theta(z) = \frac{1-\overline{\Theta(\lambda)}\Theta(z)}{(1-\bar{\lambda}z)},$$

and $\Theta(z) = \exp(-\tau\frac{1+z}{1-z})$. Having in mind the computations of 4.3.1(4), for a general inner function Θ, it is natural to consider the reproducing kernels

$$P_\Theta k_{\lambda_n} = k_{\lambda_n}^\Theta \in \mathcal{K}_\Theta,$$

where $P_\Theta = P_{\mathcal{K}_\Theta}$ and $\lambda_n \in \mathbb{D}$. A direct calculation yields

$$\|k_{\lambda_n}\|^2 = \frac{1}{1-|\lambda_n|^2}, \qquad \|k_{\lambda_n}^\Theta\|^2 = \frac{1-|\Theta(\lambda_n)|^2}{1-|\lambda_n|^2}.$$

Now, the first assertion of the next lemma is obvious, and the second one is contained in Section B.3.2 (Volume 1).

4.4.1. LEMMA. *(1) For a sequence $(\lambda_n)_{n\geq 1} \subset \mathbb{D} = \{z \in \mathbb{C} : |z| < 1\}$, the norms of k_{λ_n} and $k_{\lambda_n}^\Theta$ are equivalent, $\|k_{\lambda_n}\| \asymp \|k_{\lambda_n}^\Theta\|$, $n \geq 1$, if and only if*

$$\sup_{n\geq 1} |\Theta(\lambda_n)| < 1. \tag{4.4.1}$$

(2) If the sequence $(\lambda_n)_{n\geq 1}$ satisfies the Blaschke condition $\sum_{n\geq 1}(1-|\lambda_n|) < \infty$, then we have

$$\mathrm{span}_{H^2}(k_{\lambda_n} : n \geq 1) = \mathcal{K}_B = H^2 \ominus BH^2,$$

otherwise, $\mathrm{span}_{H^2}(k_{\lambda_n} : n \geq 1) = H^2$. Here, as before, B denotes the Blaschke product $B = \prod_{n\geq 1} b_{\lambda_n}$, $b_\lambda = \frac{\lambda - z}{1-\bar{\lambda}z}\frac{|\lambda|}{\lambda}$.

□

Now, we prove that the Carleson condition (C) is necessary for reproducing kernels to form a Riesz sequence. To this end, we employ the Gram matrices introduced in C.3.3.1.

4.4.2. LEMMA. *Let Θ be an inner function, and let $(k_{\lambda_n}^\Theta)_{n\geq 1}$ be a sequence of reproducing kernels in $\mathcal{K}_\Theta$ which is a Riesz sequence. Then $(\lambda_n)_{n\geq 1} \in (C)$.*

PROOF. Let $x_n = k_{\lambda_n}^\Theta / \|k_{\lambda_n}^\Theta\|$, and let

$$G = (\gamma_{i,j}(\{k_{\lambda_n}^\Theta\}))_{i,j\geq 1} = ((x_i, x_j))_{i,j\geq 1}$$

be the *Gram matrix* of the normalized sequence $(x_n)_{n\geq 1}$. We refer to C.3.3.1 for properties of Gram matrices. In particular, since $(x_n)_{n\geq 1}$ is a Riesz sequence, we obtain

$$\sup_{i\neq j} |\gamma_{i,j}(\{k_{\lambda_n}^\Theta\})| = \delta < 1.$$

On the other hand, we have

$$\begin{aligned}
\gamma_{i,j}(\{k_{\lambda_n}^\Theta\}) &= (x_i, x_j) = k_{\lambda_i}^\Theta(\lambda_j)/(\|k_{\lambda_i}^\Theta\| \cdot \|k_{\lambda_j}^\Theta\|) \\
&= \frac{(1-|\lambda_i|^2)^{1/2}(1-|\lambda_j|^2)^{1/2}}{1-\bar{\lambda}_i\lambda_j} \cdot \frac{1-\Theta(\lambda_j)\overline{\Theta(\lambda_i)}}{(1-|\Theta(\lambda_i)|^2)^{1/2}(1-|\Theta(\lambda_j)|^2)^{1/2}} \\
&= \gamma_{i,j}(\{k_{\lambda_n}\})/\gamma_{i,j}(\{k_{\Theta(\lambda_n)}\}).
\end{aligned}$$

Therefore, $|\gamma_{i,j}(\{k_{\lambda_n}\})| \leq |\gamma_{i,j}(\{k^{\Theta}_{\lambda_n}\})|$, and we get

$$\begin{aligned}\delta^2 &\geq \sup_{i\neq j}|\gamma_{i,j}(\{k_{\lambda_n}\})|^2 = \sup_{i\neq j}\frac{(1-|\lambda_i|^2)(1-|\lambda_j|^2)}{|1-\overline{\lambda}_i\lambda_j|^2}\\ &= \sup_{i\neq j}(1-|b_{\lambda_i}(\lambda_j)|^2),\end{aligned}$$

which means that $(\lambda_n)_{n\geq 1}$ is a sparse (separated) sequence, see C.3.3.4.

For the same reason, we have

$$\sum_{j\geq 1}|\gamma_{i,j}(\{k_{\lambda_n}\})|^2 \leq \sum_{j\geq 1}|\gamma_{i,j}(\{k^{\Theta}_{\lambda_n}\})|^2 \leq \|G\|$$

for every $i \geq 1$. This embedding property together with the separation property imply that $(\lambda_n)_{n\geq 1} \in (C)$; see also C.3.3.4. □

In the case when condition (4.4.1) is fulfilled, we can say more, exactly in the same way as in Theorem 4.3.6 above. The role that condition (4.4.1) plays for the bases problem is discussed in Subsection 4.4.9 below.

4.4.3. LEMMA. *Let Θ be an inner function, and assume that the sequence $\Lambda = (\lambda_n)_{n\geq 1}$ satisfies (4.4.1). Denote by B the Blaschke product associated with Λ. Then the sequence of reproducing kernels $(k^{\Theta}_{\lambda_n})_{n\geq 1}$ is a Riesz sequence (respectively a Riesz basis) in $\mathcal{K}_\Theta$ if and only if the following two conditions hold.*

(1) $(\lambda_n)_{n\geq 1} \in (C)$, and

(2) the restriction $P_\Theta|\mathcal{K}_B : \mathcal{K}_B \longrightarrow \mathcal{K}_\Theta$ is an isomorphism onto its range (respectively onto $\mathcal{K}_\Theta$).

PROOF. Clearly the arguments used in the proof of 4.3.6 carry over to this more general situation. □

The following lemma reduces the problem of invertibility of $P_\Theta|\mathcal{K}_B$ to invertibility of a Toeplitz operator.

4.4.4. LEMMA. *In the space $BH^2_- = H^2_- \oplus \mathcal{K}_B \subset L^2(\mathbb{T})$, the following identity holds*

$$\Theta J T_{\Theta\bar{B}} J\overline{B} = I_{H^2_-} \oplus (P_\Theta|\mathcal{K}_B),$$

where $P_\Theta = P_{\mathcal{K}_\Theta} = \Theta P_-\bar{\Theta}$ and $T_f g = P_+ fg$, $g \in H^2$, is the Toeplitz operator with symbol f. As before, we set $Jg = \bar{z}\bar{g}$, $g \in L^2(\mathbb{T})$. Here $P_+ = P_{H^2}$, and $P_- = P_{H^2_-}$ are the Riesz projections, and $H^2_- = L^2(\mathbb{T}) \ominus H^2$.

PROOF. This can be verified directly by using $J^2 = I$, $JH^2_+ = H^2_-$, $JH^2_- = H^2_+$ and $J\varphi J f = \bar{\varphi} f$ for every φ and f. □

4.4.5. COROLLARY. *The restriction $P_\Theta|\mathcal{K}_B$ is an isomorphism onto its range if and only if the Toeplitz operator $T_{\Theta\bar{B}}$ is an isomorphism onto its range, and $P_\Theta|\mathcal{K}_B$ is an isomorphism onto $\mathcal{K}_\Theta$ if and only if $T_{\Theta\bar{B}}$ is invertible.*

Indeed, this follows immediately from Lemma 4.4.4 by using the fact that multplication by B (and hence $\overline{B}$) and Θ, and the mapping J are isometries on $L^2(\mathbb{T})$. □

Now, making use of the Devinatz–Widom theorem B.4.3.1 (Volume 1), we exhibit several properties which are equivalent to the invertibility of $P_\Theta|\mathcal{K}_B$, and hence, to the basic sequence property for reproducing kernels.

4.4.6. THEOREM. *Let Θ be an inner function, and let $(\lambda_n)_{n\geq 1}$ be a sequence of distinct points of $\mathbb{D}$ satisfying condition* (4.4.1). *The following assertions are equivalent.*

(1) $(k^{\Theta}_{\lambda_n})_{n\geq 1}$ is a Riesz basis in $\mathcal{K}_{\Theta}$.

(2) $(\lambda_n)_{n\geq 1} \in (C)$, and $T_{\Theta\overline{B}}$ is invertible, where $B = \prod_n b_{\lambda_n}$ stands for the corresponding Blaschke product.

(3) $(\lambda_n)_{n\geq 1} \in (C)$, and $\operatorname{dist}(\Theta\overline{B}, H^\infty) < 1$, $\operatorname{dist}(B\overline{\Theta}, H^\infty) < 1$.

(4) $(\lambda_n)_{n\geq 1} \in (C)$, and there exists an outer function $h \in H^\infty$ such that $\|\Theta\overline{B} - h\|_\infty < 1$.

(5) $(\lambda_n)_{n\geq 1} \in (C)$, and there exist real valued bounded functions a, b and a constant $c \in \mathbb{R}$ such that $\Theta\overline{B} = e^{i(a+\tilde{b}+c)}$ and $\|a\|_\infty < \pi/2$.

(6) $(\lambda_n)_{n\geq 1} \in (C)$, and $\Theta\overline{B} = ge^{b+ia}$ where $g, 1/g \in H^\infty$, and a and b are real valued bounded functions with $\|a\|_\infty < \pi/2$.

(7) $(\lambda_n)_{n\geq 1} \in (C)$, and there exists a function h such that $h, 1/h \in H^\infty$ and $\Theta\overline{B}h$ is sectorial *(as defined in Corollary B.4.2.6, Volume 1).*

(8) $(\lambda_n)_{n\geq 1} \in (C)$, and there exists an outer function $h \in H^2$ such that $\Theta\overline{B} = \varepsilon h/\overline{h}$ and $|h|^2 \in (HS)$, where $|\varepsilon| = 1$ and (HS) means the Helson–Szegö condition from A.5.4.1 (Volume 1).

PROOF. This is a straightforward consequence of Lemma 4.4.3, Corollary 4.4.5, and Theorem B.4.3.1 (Volume 1). □

4.4.7. COROLLARY. *Let Θ be an inner function, and let $(\lambda_n)_{n\geq 1}$ be a sequence of distinct points of $\mathbb{D}$ satisfying condition* (4.4.1). *If $(\lambda_n)_{n\geq 1} \in (C)$ and $\|\Theta - B\|_\infty < 1$, then $(k^{\Theta}_{\lambda_n})_{n\geq 1}$ is a Riesz basis in $\mathcal{K}_{\Theta}$.* □

Similarly, we obtain a criterion for the Riesz sequence property, that is, for a sequence to be a Riesz basis in its span.

4.4.8. THEOREM. *Under the assumptions of Theorem 4.4.6, the following assertions are equivalent.*

(1) $(k^{\Theta}_{\lambda_n})_{n\geq 1}$ forms a Riesz sequence.

(2) $(\lambda_n)_{n\geq 1} \in (C)$, and $T_{\Theta\overline{B}}$ is left invertible.

(3) $(\lambda_n)_{n\geq 1} \in (C)$, and $\operatorname{dist}(\Theta\overline{B}, H^\infty) < 1$.

□

4.4.9. Comments on condition (4.4.1)**.** We denote the supremum introduced in (4.4.1) by $\alpha(\Theta, B)$ where B stands for the Blaschke product $B = \prod_n b_{\lambda_n}$, that is,

$$\alpha(\Theta, B) = \sup_{n\geq 1} |\Theta(\lambda_n)|.$$

By the maximum principle, $\alpha(\Theta, B) = 1$ if and only if $\overline{\lim}_n |\Theta(\lambda_n)| = 1$.

Throughout this subsection we always assume that $(\lambda_n)_{n\geq 1}$ is a sequence of distinct points.

(1) The spectrum and (4.4.1)**.** *It follows from the description of the spectrum $\sigma(\Theta)$, given in C.2.3.3, that condition* (4.4.1) *means that λ_n, $n \geq 1$, are "not far" from the spectrum. In any case, the relation $\lim_n \Theta(\lambda_n) = 0$ implies that* $\lim_n \operatorname{dist}(\lambda_n, \sigma(\Theta)) = 0$. □

(2) The spectrum and minimality. *If $(k^{\Theta}_{\lambda_n})_{n\geq 1}$ is a minimal family (and,* a fortiori, *if $(k^{\Theta}_{\lambda_n})_{n\geq 1}$ is a basis), then again* $\lim_n \operatorname{dist}(\lambda_n, \sigma(\Theta)) = 0$.

Indeed, assume that there exists a limit point λ of $(\lambda_n)_{n\geq 1}$ belonging to $\mathbb{T}\backslash\sigma(\Theta)$, and let $f \in \mathcal{K}_\Theta$ be such that $(f, k^{\Theta}_{\lambda_n}) = f(\lambda_n) = 0$ for $n \geq 2$. Since f is analytic in $\mathbb{D}\cup D(\lambda, \varepsilon)$ for a sufficiently small $\varepsilon > 0$ (see C.2.3.5), it follows from the uniqueness theorem that $f = 0$. Therefore, $(k^{\Theta}_{\lambda_n})_{n\geq 1}$ cannot be minimal. □

On the other hand, the basis property of $(k^{\Theta}_{\lambda_n})_{n\geq 1}$ does not imply $\lim_n \Theta(\lambda_n) = 0$. Indeed, if $\Theta = \exp(-a\frac{1-z}{1+z})$, then it can be shown that the function

$$B_\alpha = \frac{\Theta - \alpha}{1 - \overline{\alpha}\Theta}$$

(the *Frostman transform* of Θ) is an interpolating Blaschke product for every $\alpha \in \mathbb{D}\backslash\{0\}$ (that is, the zeros of B_α form an interpolating sequence), and $\|\Theta - B_\alpha\|_\infty < 1$, at least for $|\alpha| < 1/2$ (in fact, $\|\Theta - B_\alpha\|_\infty = 2|\alpha|$, since $|\Theta - B_\alpha| = |1 - \frac{1-\alpha\overline{\Theta}}{1-\overline{\alpha}\Theta}|$ on $\mathbb{T}$ and $z = 1 - \overline{\alpha\Theta(\xi)}$ ranges over the circle $1 + |\alpha|\cdot\mathbb{T}$). If $(\lambda_n)_{n\geq 1}$ are the zeros of B_α, then clearly (4.4.1) holds and also $(\lambda_n)_{n\geq 1} \in (C)$. Now, by Corollary 4.4.7, $(k^{\Theta}_{\lambda_n})_{n\geq 1}$ is a Riesz basis of $\mathcal{K}_\Theta$, and $\Theta(\lambda_n) = \alpha$ by construction for every $n \geq 1$. □

For another example see Section 4.5 below.

(3) The projection $P_\Theta|\mathcal{K}_B$ and $\alpha(\Theta, B)$. *If $(\lambda_n)_{n\geq 1} \in (C)$ and $P_\Theta|\mathcal{K}_B$ is left invertible (i.e., an isomorphism onto its range), then $(k^{\Theta}_{\lambda_n})_{n\geq 1}$ is a Riesz sequence. (Clear.)*

On the other hand, if $P_\Theta|\mathcal{K}_B$ is left invertible, then condition (4.4.1) *holds* automatically, *that is:* $\alpha(\Theta, B) < 1$.

Indeed, by virtue of Theorem B.4.3.1 (Volume 1), there exists $f \in H^\infty$ such that $\|\Theta - Bf\|_\infty < 1$, and hence $\sup_{n\geq 1} |\Theta(\lambda_n)| < 1$ (in fact, $\alpha(\Theta, B) \leq \operatorname{dist}(\Theta, BH^\infty)$). □

(4) The Carleson condition and $\alpha(\Theta, B)$. *Let $(\lambda_n)_{n\geq 1} \in (C)$, and let V be an orthogonalizer of the family $(k_{\lambda_n})_{n\geq 1}$, that is, an isomorphism such that $Vk_{\lambda_n} = e_n$, $n \geq 1$, form an orthogonal system. Then, if*

$$\|V\| \cdot \|V^{-1}\| \cdot \alpha(\Theta, B) < 1,$$

the restriction $P_\Theta|\mathcal{K}_B$ is left invertible, and hence, $(k^{\Theta}_{\lambda_n})_{n\geq 1}$ is a Riesz sequence.

In order to see this, let $f = \sum a_n k_{\lambda_n}$ be a finite linear combination, $Tf = \Theta \sum \overline{\Theta(\lambda_n)} a_n k_{\lambda_n}$. We have

$$\begin{aligned}
\|Tf\|^2 &= \left\|\sum \overline{\Theta(\lambda_n)} a_n k_{\lambda_n}\right\|^2 \leq \|V^{-1}\|^2 \cdot \left\|\sum \overline{\Theta(\lambda_n)} a_n e_n\right\|^2 \\
&\leq \|V^{-1}\|^2 \cdot \alpha(\Theta, B)^2 \sum |a_n|^2 \|e_n\|^2 \\
&= \|V^{-1}\|^2 \cdot \alpha(\Theta, B)^2 \left\|\sum a_n e_n\right\|^2 \\
&\leq \|V^{-1}\|^2 \cdot \alpha(\Theta, B)^2 \cdot \|V\|^2 \left\|\sum a_n k_{\lambda_n}\right\|^2 \\
&= \|V^{-1}\|^2 \cdot \|V\|^2 \cdot \alpha(\Theta, B)^2 \cdot \|f\|^2.
\end{aligned}$$

Since $P_\Theta f = f - Tf$, we get $\|P_\Theta f\| \geq \|f\| - \|V^{-1}\| \cdot \|V\| \cdot \alpha(\Theta, B) \cdot \|f\|$ for every f, and the result follows. □

(5) $\alpha(\Theta, B)$ **and larger spaces**. *Let* $(\lambda_n)_{n\geq 1} \in (C)$. *The following assertions are equivalent.*

i) $\alpha(\Theta, B) < 1$.

ii) There exists an integer $N \geq 1$ *such that* $P_{\Theta^N} | \mathcal{K}_B$ *is left invertible (and hence,* $P_{\Theta^m} | \mathcal{K}_B$ *is left invertible for every* $m \geq N$*).*

iii) There exists an integer $N \geq 1$ *such that* $\operatorname{dist}(\Theta^N, BH^\infty) < 1$ *(and hence,* $\operatorname{dist}(\Theta^m, BH^\infty) < 1$ *for every* $m \geq N$*).*

Indeed, ii) and iii) are equivalent by Theorem B.4.3.1 (Volume 1), and the rest follows from (3) and (4), since $\alpha(\Theta^m, B) = \alpha(\Theta, B)^m$. □

It is clear that in the case of a singular inner function Θ, the parameters N and m can take non-integer values. The same is true in the next claim (6).

(6) $\alpha(\Theta, B)$ **and bases in larger spaces**. *Let* $(\lambda_n)_{n\geq 1} \in (C)$ *and suppose that* $\alpha(\Theta, B) < 1$. *Then there exists an integer* $N \geq 1$ *such that* $(k_{\lambda_n}^{\Theta^N})_{n\geq 1}$ *is a Riesz sequence (and hence,* $(k_{\lambda_n}^{\Theta^m})_{n\geq 1}$ *is a Riesz sequence for every* $m \geq N$*).*

This is straightforward from (5) and Lemma 4.4.3. □

(7) Bases with $\lim_n \Theta(\lambda_n) = 0$. *Assume that* $\lim_n \Theta(\lambda_n) = 0$. *The following assertions are equivalent.*

i) There exists an integer $N \geq 1$ *such that* $(k_{\lambda_n}^{\Theta})_{n\geq N}$ *is a Riesz sequence.*

ii) $(\lambda_n)_{n\geq 1} \in (C)$.

Indeed, i) $\Rightarrow$ ii) by Lemma 4.4.2, and ii) $\Rightarrow$ i) by (4), since $\lim_N \alpha(\Theta, B_N) = 0$, where $B_N = \prod_{n\geq N} b_{\lambda_n}$. □

This property can be improved for inner functions Θ that are not Blaschke products as follows.

(8) Singular inner functions with $\lim_n \Theta(\lambda_n) = 0$. *Let* Θ *be an inner function which is* not *a Blaschke product, and suppose that* $\lim_n \Theta(\lambda_n) = 0$ *where* $(\lambda_n)_{n\geq 1} \in (C)$. *Then* $(k_{\lambda_n}^{\Theta})_{n\geq 1}$ *is a Riesz sequence and* $\dim(\mathcal{K}_\Theta \ominus \operatorname{span}(k_{\lambda_n}^{\Theta} : n \geq 1)) = \infty$.

Indeed, Property (7) can be applied to Θ and to any $\Theta_\alpha = BV^\alpha$, $0 < \alpha < 1$, where $\Theta = BV$ stands for the canonical factorization into a Blaschke product B and a non-trivial singular inner function V. We obtain that both $(k_{\lambda_n}^{\Theta})_{n\geq N}$ and $(k_{\lambda_n}^{\Theta_\alpha})_{n\geq N}$ are Riesz sequences for a suitable integer N. Since $\lim_n \Theta(\lambda_n) = 0$, we see that the norms $\|k_{\lambda_n}^{\Theta_\alpha}\|$ and $\|k_{\lambda_n}^{\Theta}\|$ are both comparable with $\|k_{\lambda_n}\|$, and hence, are comparable with each other.

On the other hand, $\mathcal{K}_{\Theta_\alpha} \subset \mathcal{K}_\Theta$, and hence, $k_{\lambda_n}^{\Theta_\alpha} = P_{\Theta_\alpha} k_{\lambda_n}^{\Theta}$ for $n \geq N$. The previous observations show that $P_{\Theta_\alpha} | \operatorname{span}(k_{\lambda_n}^{\Theta} : n \geq N)$ is an isomorphism onto its range: $P_{\Theta_\alpha} \operatorname{span}(k_{\lambda_n}^{\Theta} : n \geq N) = \operatorname{span}(k_{\lambda_n}^{\Theta_\alpha} : n \geq N) \subset \mathcal{K}_{\Theta_\alpha}$. Since

$$\dim \operatorname{Ker}(P_{\Theta_\alpha} | \mathcal{K}_\Theta) = \dim(\mathcal{K}_\Theta \ominus \mathcal{K}_{\Theta_\alpha}) = \infty,$$

it follows that

$$\dim(\mathcal{K}_\Theta \ominus \operatorname{span}(k_{\lambda_n}^{\Theta} : n \geq N)) = \infty.$$

Using Property (10) (below), we obtain that $(k_{\lambda_n}^{\Theta})_{n\geq 1}$ is a Riesz sequence. □

Is is worth mentioning that for a Blaschke product $\Theta = B$ this property fails. Indeed, let B be a Blaschke product whose zero-set $(\lambda_n)_{n\geq 2}$ satisfies the Carleson condition (C). Then $(k^\Theta_{\lambda_n})_{n\geq 2} = (k_{\lambda_n})_{n\geq 2}$ is a Riesz basis of $\mathcal{K}_\Theta$. Adding an arbitrary point $\lambda_1 \in \mathbb{D} \setminus (\lambda_n)_{n\geq 2}$ we obtain $\lim_n \Theta(\lambda_n) = 0$ and $(\lambda_n)_{n\geq 1} \in (C)$, but $(k^\Theta_{\lambda_n})_{n\geq 1}$ is not even minimal in $\mathcal{K}_\Theta = \operatorname{span}(k_{\lambda_n} : n \geq 2)$.

(9) Incompleteness and minimality properties. *Let Λ be a subset of $\mathbb{D}$ such that $\mathcal{K}_\Theta(\Lambda) = \operatorname{span}(k^\Theta_\lambda : \lambda \in \Lambda) \neq \mathcal{K}_\Theta$, and let $\mu \in \mathbb{D} \setminus \Lambda$. Then*

(a) *$k^\Theta_\mu \notin \mathcal{K}_\Theta(\Lambda)$; in particular, $(k^\Theta_\lambda : \lambda \in \Lambda)$ is a minimal family;*

(b) *$\mathcal{K}_\Theta((\Lambda\backslash\{\lambda\}) \cup \{\mu\}) \neq \mathcal{K}_\Theta$ for any $\lambda \in \Lambda$.*

Indeed, let $f \in \mathcal{K}_\Theta \ominus \mathcal{K}_\Theta(\Lambda)$, $f \neq 0$, and let $d(\zeta) \geq 0$ be its zero-multiplicity at a point $\zeta \in \mathbb{D}$. Then $f(\lambda) = (f, k^\Theta_\lambda) = 0$ for all $\lambda \in \Lambda$.

To check (a), we set

$$g = \left(b_\mu^{d(\mu)}(M_\Theta)\right)^* f = P_+ \bar{b}_\mu^{d(\mu)} f.$$

Clearly, $g \in \mathcal{K}_\Theta$. Moreover, $g = f / b_\mu^{d(\mu)}$, and hence, $g(\lambda) = (g, k^\Theta_\lambda) = 0$ for all $\lambda \in \Lambda$, but $(g, k^\Theta_\mu) = g(\mu) \neq 0$. The first claim follows. The minimality is now a consequence of $k^\Theta_\lambda \notin \mathcal{K}_\Theta(\Lambda\backslash\{\lambda\})$ for every $\lambda \in \Lambda$.

To prove (b), we set $g = f_0 \cdot \frac{z-\mu}{z-\lambda}$ where $f_0 = f/(z-\lambda)^{d(\lambda)-1}$. As before, $f_0 \in \mathcal{K}_\Theta$ and, still, $f_0(\lambda) = (f_0, k^\Theta_\lambda) = 0$ for all $\lambda \in \Lambda$; thus $f_0 \perp \mathcal{K}_\Theta(\Lambda)$. Moreover, $g = f_0 + f_0 \cdot \frac{\lambda-\mu}{z-\lambda} \in \mathcal{K}_\Theta$ (since $\mathcal{K}_\Theta = H^2 \cap \Theta H^2_-$), and, clearly, $g \neq 0$, but $g = 0$ on $(\Lambda\backslash\{\lambda\}) \cup \{\mu\}$. Hence, $g \perp \mathcal{K}_\Theta((\Lambda\backslash\{\lambda\}) \cup \{\mu\})$. □

(10) Completing a Riesz sequence up to a basis in $\mathcal{K}_\Theta$. *Let $(k^\Theta_\lambda)_{\lambda\in\Lambda}$ be a Riesz sequence, and let $\dim(\mathcal{K}_\Theta \ominus \mathcal{K}_\Theta(\Lambda)) = N$ with $0 < N \leq \infty$. Then, a finitely extended family $(k^\Theta_\lambda)_{\lambda\in\Lambda'}$ with $\Lambda' = \Lambda \cup \{\mu_1, ..., \mu_m\}$ is a Riesz sequence, whenever $m \leq N$ distinct points $\mu_i \in \mathbb{D} \setminus \Lambda$ are added ($m < \infty$). In particular, for $m = N < \infty$, we get a Riesz basis of the entire space $\mathcal{K}_\Theta$.*

Indeed, Property (9)(a) ensures that $\dim(\mathcal{K}_\Theta \ominus \mathcal{K}_\Theta(\Lambda \cup \{\mu_1\})) = N - 1$ for every $\mu_1 \in \mathbb{D} \setminus \Lambda$. It is also clear from (9)(a) that the extended family is a Riesz sequence. An induction completes the proof. □

4.5. Back to Exponentials

The key points for the study of bases of exponentials in L^2 on a finite interval are contained in Sections 4.3 and 4.4. We first draw some immediate consequences. We begin with some Riesz sequence properties of exponentials, and then continue with bases of the entire space $L^2(0,\tau)$. In most cases, we deal with families of exponentials $(e^{i\mu_n t}\chi_{(0,\tau)})_{n\geq 1}$ arising from control theory, that is, satisfying

$$\inf_{n\geq 1} \operatorname{Im} \mu_n > -\infty.$$

As usual, for more information and references we refer the reader to the end of the chapter, Sections 4.7 and 4.8.

4.5.1. Corollary. *Let $\tau > 0$, $\alpha > 0$ and let $(\mu_n)_{n\geq 1}$ be a sequence of complex numbers, $\mu_n \in \mathbb{C}$, such that $\inf_{n\geq 1} \operatorname{Im} \mu_n > -\alpha > -\infty$. Then the following assertions are equivalent.*

(1) The sequence $(e^{i\mu_n t}\chi_{(0,\tau)})_{n\geq 1}$ is a Riesz sequence in $L^2(0,\tau)$.

(2) $(\mu_n + i\alpha)_{n\geq 1} \in (C)$ *and* $\operatorname{dist}_{L^\infty(\mathbb{R})}(e^{i\tau t}\bar{B}, H^\infty_+) < 1$ *where* $H^\infty_+ = H^2_+ \cap L^\infty(\mathbb{R})$, $H^2_+ = H^2(\mathbb{C}_+)$ *is the Hardy space of the upper half plane, and* B *is the Blaschke product associated to* $(\mu_n + i\alpha)_{n\geq 1}$*:*

$$B(z) = \prod_{n\geq 1} \frac{z - (\mu_n + i\alpha)}{z - \overline{(\mu_n + i\alpha)}} \cdot \varepsilon_n.$$

Here ε_n, $n \geq 1$, *are the usual normalizing factors of modulus one:* $\varepsilon_n = 1$ *if* $\mu_n + i\alpha = i$ *and* $\varepsilon_n = (i - \overline{\mu}_n + i\alpha)/(i - \mu_n - i\alpha)$ *otherwise.*

To see this, it suffices to apply the results of Sections 4.3 and 4.4. □

The case when $\lim_n \operatorname{Im}\mu_n = +\infty$ is particularly interesting for control theory. In this case, the Carleson condition (C) already implies the distance condition $\operatorname{dist}(e^{i\tau t}\bar{B}, H^\infty_+) < 1$, see Remark 4.4.9(7) above. Moreover, since the function $\Theta = \exp(-\tau\frac{1+z}{1-z})$ is singular in $\mathbb{D}$, we can use points (4) and (6) of 4.4.9 in order to get the following corollaries.

4.5.2. Corollary. *Let* $\tau > 0$ *and let* $(\mu_n)_{n\geq 1}$ *be a sequence of complex numbers,* $\mu_n \in \mathbb{C}$, *verifying* $\lim_n \operatorname{Im}\mu_n = +\infty$. *Then the family* $(e^{i\mu_n t}\chi(0,\tau))_{n\geq 1}$ *is a Riesz sequence if and only if* $(\mu_n + i\alpha)_{n\geq 1} \in (C)$ *for a suitable* $\alpha > 0$. □

4.5.3. Corollary. *Let* $\mu_n \in \mathbb{C}$ *be such that* $\inf_{n\geq 1}\operatorname{Im}\mu_n > -\alpha > -\infty$. *Then there exists* $\tau > 0$ *such that* $(e^{i\mu_n t}\chi_{(0,\tau)})_{n\geq 1}$ *is a Riesz sequence in* $L^2(0,\tau)$ *if and only if* $(\mu_n + i\alpha)_{n\geq 1} \in (C)$. *Moreover, such a* τ *depends only on the constant* δ *of the Carleson condition (C).* □

4.5.4. A criterion for exponential bases using Devinatz–Widom and Helson–Szegö conditions. Now, we state a counterpart of Corollary 4.5.1 for bases of exponentials of the entire space $L^2(0,\tau)$. Of course, all seven characterizations of 4.4.6 may be used, but we restrict ourselves to the most useful ones (3), (5), and (8). The criteria obtained, in particular condition (3) of Corollary 4.5.5, are especially efficient for exponentials of *hyperbolic type*, that is, $\sup_{n\geq 1}|\operatorname{Im}\mu_n| < \infty$. We will justify this claim by two applications. Firstly, we explicitely express condition (3) of Corollary 4.5.5 in terms of the frequency distribution function that will be defined below. Secondly, this same condition allows us to deduce Kadets' 1/4-theorem (see Theorem 4.1.2).

4.5.5. Corollary. *Let* $\mu_n \in \mathbb{C}$ *be such that* $\inf_{n\geq 1}\operatorname{Im}\mu_n > -\alpha > -\infty$. *The following assertions are equivalent.*

(1) $(e^{i\mu_n t}\chi_{(0,\tau)})_{n\geq 1}$ *is a Riesz basis of* $L^2(0,\tau)$.

(2) $(\mu_n + i\alpha)_{n\geq 1} \in (C)$, *and* $\operatorname{dist}(e^{i\tau x}\overline{B}, H^\infty_+) < 1$, $\operatorname{dist}(Be^{-i\tau x}, H^\infty_+) < 1$, *where* B *stands for the Blaschke product in the upper half plane with zeros* $(\mu_n + i\alpha)_{n\geq 1}$, *and* $H^\infty_+ = H^\infty(\mathbb{C}_+)$.

(3) $(\mu_n + i\alpha)_{n\geq 1} \in (C)$, *and there exist a real valued function* $b \in L^\infty(\mathbb{R})$ *and a constant* $c \in \mathbb{R}$ *such that* $\|\tau x - \arg B(x) - \tilde{b} - c\|_\infty < \pi/2$.

(4) $(\mu_n + i\alpha)_{n\geq 1} \in (C)$, *and there exists an outer function* h *such that* $B(x) = \varepsilon\frac{h(x)}{\overline{h(x)}}e^{i\tau x}$ *and* $|h|^2 \in (HS)$, *where (HS) means the Helson–Szegö condition from A.5.4.1 (Volume 1) and* $|\varepsilon| = 1$.

The corollary follows from the results of Section 4.4 (see especially Theorem 4.4.6). Notice that in (3) and (4) we used the real line version of the Helson–Szegö condition since, in fact, it is invariant under a conformal change of variables. □

Now we are ready to state an explicit criterion for a sequence of exponentials to be a Riesz basis, at least for frequencies μ_n lying in a strip of finite width, that is, $\sup_n |\operatorname{Im}\mu_n| < \infty$. This will be given in terms of the following *frequency distribution function*

$$N_{(\mu_n)}(x) = \begin{cases} -\operatorname{card}\{\mu_n : \operatorname{Re}\mu_n \in [x, 0[\} & \text{if } x < 0, \\ \operatorname{card}\{\mu_n : \operatorname{Re}\mu_n \in [0, x]\} & \text{if } x \geq 0. \end{cases}$$

4.5.6. THEOREM. *Let $\mu_n \in \mathbb{R}$, $\tau > 0$. The following assertions are equivalent.*

(1) The sequence $\mathcal{E} = (e^{i\mu_n t}\chi_{(0,\tau)})_{n\geq 1}$ is a Riesz basis of the space $L^2(0,\tau)$.

(2) $\inf_{n\neq m} |\mu_n - \mu_m| > 0$, the function $a(x) = 2\pi N_{(\mu_n)}(x) - \tau x$, $x \in \mathbb{R}$, is in $BMO(\mathbb{R}) = \{u + \tilde{v} : u, v \in L^\infty(\mathbb{R})\}$ (see B.1.6.2, Volume 1), and there exists $y > 0$ such that the harmonic extension of a to the upper half plane C_+ satisfies

$$a(x + iy) = c + u(x) + \tilde{v}(x), \quad x \in \mathbb{R},$$

where $c \in \mathbb{R}$, $v \in L^\infty(\mathbb{R})$ and $\|u\|_{L^\infty(\mathbb{R})} < \pi/2$.

Moreover, each perturbation $(e^{i(\mu_n + i\alpha_n)t}\chi_{(0,\tau)})_{n\geq 1}$ with $\alpha_n \in \mathbb{R}$, $\sup_{n\geq 1} |\alpha_n| < \infty$, of a Riesz basis $\mathcal{E}$ is also a Riesz basis in $L^2(0,\tau)$.

In fact, this theorem is a corollary of a criterion given in terms of the so-called "generating functions", see 4.7.1(b) below. See also Section 4.8 for more comments. It is worth mentioning (cf. also Remark 4.2.3(1)) that the condition $\inf_{n\neq m} |\mu_n - \mu_m| > 0$ is equivalent to the Carleson condition (C) for the sequence $(\mu_n + i\alpha)_{n\geq 1}$, $\alpha > 0$.

4.5.7. Proof of Kadets' 1/4 theorem. Using condition (3) of Corollary 4.5.5 it is now easy to prove Kadets' 1/4-theorem 4.1.2. The following proof contains all the details except a few simple computations with Blaschke products

$$\prod_{n\geq 1} \left(\frac{x - \mu_n}{x - \bar{\mu}_n} \cdot c_n \right),$$

$|c_n| = 1$, $x \in \mathbb{R}$, on the real axis.

Suppose that $\delta < 1/4$.

(a) First of all, we consider harmonic frequencies, $\mu_n = n$, $n \in \mathbb{Z}$, and their translates $\mathbb{Z} + i\alpha$, $\alpha > 0$. For the Blaschke product B^α associated to such a sequence we have

$$B^\alpha(x) = \frac{\sin \pi(x - i\alpha)}{\sin \pi(x + i\alpha)}, \quad x \in \mathbb{R}.$$

One can easily show that

$$\arg B^\alpha(x) = 2\pi x + \pi + c_\alpha(x), \quad x \in \mathbb{R},$$

where $|c_\alpha(x)| \leq \text{const} \cdot e^{-\pi\alpha}$ for every $x \in \mathbb{R}$.

(b) Let now $\mu_n = n + \delta_n$, $|\delta_n| \leq \delta$ $(n \in \mathbb{Z})$ and $\mu_n + i\alpha$ its translates. Denote by B_1^α the corresponding Blaschke product. Then

$$\begin{aligned} A &:= |\arg B^\alpha(x) - \arg B_1^\alpha(x) - \text{const}| \\ &= \left| \sum_{n\in\mathbb{Z}} \left\{ \arg \frac{x - (n + i\alpha)}{x - (n - i\alpha)} - \arg \frac{x - (n + \delta_n + i\alpha)}{x - (n + \delta_n - i\alpha)} \right\} \right|. \end{aligned}$$

Notice that for a fixed $x \in \mathbb{R}$ this expression takes its maximal value (as a function of δ_n, $|\delta_n| \le \delta$) if all the terms of the sum are of the same sign and if $|\delta_n|$ is maximal, that is, $\delta_n = \delta$ (or $\delta_n = -\delta$). Hence

$$A \le \Big| \sum_{n\in\mathbb{Z}} \arg \frac{x-(n+i\alpha)}{x-(n-i\alpha)} - \arg \frac{x-(n+\delta+i\alpha)}{x-(n+\delta-i\alpha)} \Big|.$$

(c) The previous observation leads to the following estimate

$$\begin{aligned} |\arg B^\alpha(x) - \arg B_1^\alpha(x) - \text{const}\,| &\le |\arg B^\alpha(x) - \arg B^\alpha(x-\delta)| \\ &= |2\pi\delta + c_\alpha(x) - c_\alpha(x-\delta)| \\ &\le 2\pi\delta + \text{const}\cdot e^{-\pi\alpha}, \end{aligned}$$

where $x \in \mathbb{R}$ is arbitrary.

(d) Now, we apply the implication (3) $\Longrightarrow$ (1) of Corollary 4.5.5. By Remark 4.2.3(1) and as the sequence $(\mu_n)_{n\ge1}$ satisfies $|\mu_n - \mu_m| \ge 1/2$, we already have the Carleson condition (C). It remains to verify the Helson–Szegö condition for $\tau = 2\pi$ and $\Theta = e^{2\pi i x}$ (see 4.3.1(4) for the equality $L^2(0,2\pi) = \mathcal{F}U_2\mathcal{K}_\Theta$):

$$\begin{aligned} &\|2\pi x - \arg B_1^\alpha - (\text{const} - \pi)\|_\infty \\ &\quad\le \|2\pi x + \pi - \arg B^\alpha\|_\infty + \|\arg B^\alpha - \arg B_1^\alpha - \text{const}\,\|_\infty \\ &\quad\le \text{const}\cdot e^{-\pi\alpha} + 2\pi\delta + \text{const}\cdot e^{-\pi\alpha}. \end{aligned}$$

Using again Kadets' condition $\delta < 1/4$ we obtain for sufficiently large $\alpha > 0$ the estimate required by condition 4.5.5(3). □

4.5.8. Bases of exponentials and free interpolation. Here we briefly mention the links between bases of exponentials and reproducing kernels and free interpolation in model spaces. It was proved in Chapter C.3 (in a more general framework) that a sequence of reproducing kernels $(k^\Theta_{\lambda_n})$ is a Riesz sequence if and only if the trace space $\mathcal{K}_\Theta|\Lambda$ is an *ideal space*; here $\Lambda = (\lambda_n)_{n\ge1}$. Recall that this means that for all $f \in \mathcal{K}_\Theta$ and every function $a : \Lambda \longrightarrow \mathbb{C}$, $\lambda_n \longmapsto a_n$, verifying $|a_n| \le \text{const}\cdot|f(\lambda_n)|$, there exists $g \in \mathcal{K}_\Theta$ such that $a_n = g(\lambda_n)$, $n \ge 1$. It is then clear that $(k^\Theta_{\lambda_n})_{n\ge1}$ is a Riesz sequence if and only if the following description holds for the space $\mathcal{K}_\Theta|\Lambda$: $a \in \mathcal{K}_\Theta|\Lambda \Longleftrightarrow \sum_{n\ge1} |a_n|^2\, \|k^\Theta_{\lambda_n}\|^{-2} < \infty$. We shall say that a sequence Λ is an *interpolating sequence for* $\mathcal{K}_\Theta$ if $\mathcal{K}_\Theta|\Lambda$ is an ideal space (or if any other of the equivalent conditions above is fulfilled).

In the classical case of $\Theta = \exp(-\tau\frac{1+z}{1-z})$, and hence that of exponentials $(e^{i\mu_n x}\chi_{(0,\tau)})_{n\ge1}$ on $(0,\tau)$, we get free interpolation by entire functions of exponential type less than or equal to $\tau/2$ (namely by functions of the form $e^{-i\frac{\tau}{2}z}(\mathcal{F}^{-1}f)(z)$, $z \in \mathbb{C}$, where $f \in L^2(0,\tau)$); see Section 4.3 above and references and comments in Section 4.8. We also refer the reader to Section 3.7 for a discussion of the role of free interpolation in energy optimization (in control problems).

4.6. A Levinson Completeness Theorem

When speaking about the bases problem, it is impossible to avoid the classical completeness problem. This can be stated as follows. Given an interval $I \subset \mathbb{R}$ and an exponent p, $1 \le p \le \infty$, find a geometric criterion on a set $\mathcal{M} \subset \mathbb{C}$ for the family $(e^{i\mu x}\chi_I(x) : \mu \in \mathcal{M})$ to be complete in $L^p(I)$ (weak* complete for $p = \infty$), that is,

$$\text{span}_{L^p(I)}(e^{i\mu x}\chi_I : \mu \in \mathcal{M}) = L^p(I).$$

As in Section 4.3, completeness or incompleteness of a family $(e^{i\mu x}\chi_I : \mu \in \mathcal{M})$ does not depend on the location of I on the real line but on its length $|I| = a$, and on the exponent p as well. "Geometric" means that the desired criterion should be expressed directly in terms of distribution (or "density") properties of $\mathcal{M}$ over the complex plane $\mathbb{C}$. It should be noted in advance that such a criterion is not available at the moment (2001); see Section 4.8 below for some references on the present state of the problem.

Completeness is related to the bases problem in several ways. For instance, if $(e^{i\mu x}\chi_I : \mu \in \mathcal{M})$ is a basic sequence, then it is a basis of $L^p(I)$ if and only if it is complete in $L^p(I)$. Also, a family having a proper complete subfamily is not (even) minimal, and thus, is not a basis.

In this section we show a sharp sufficient completeness condition by N. Levinson for real frequencies $\mathcal{M} \subset \mathbb{R}$ (Theorem 4.6.1), and give yet another proof of the A. Ingham implication (1) $\Rightarrow$ (2) of Theorem 4.1.2. The Levinson condition is expressed in terms of the *symmetric distribution function* $n_{\mathcal{M}}$ of a subset $\mathcal{M} \subset \mathbb{R}$; namely,

$$n_{\mathcal{M}}(s) = \operatorname{card}\{\mu \in \mathcal{M} : |\mu| \leq s\}$$

for $s > 0$. Clearly, $n_{\mathcal{M}}(s) = N_{\mathcal{M}}(s) - N_{\mathcal{M}}(-s)$, where $N_{\mathcal{M}}(s)$ is the distribution function used in Theorem 4.5.6.

4.6.1. Theorem (N. Levinson, 1936). *Let $1 < p \leq \infty$, $I \subset \mathbb{R}$ an interval, and $\mathcal{M} \subset \mathbb{R}$ a set such that*

$$\int_1^t \frac{n_{\mathcal{M}}(s)}{s}\, ds > \frac{|I|}{\pi} t - \frac{1}{p}\log(t) - C \tag{4.6.1}$$

for some constant $C > 0$ and for all $t > 1$. Then the family $(e^{i\mu x}\chi_I : \mu \in \mathcal{M})$ is complete in $L^p(I)$ (weak complete if $p = \infty$).*

Proof. It is clear that we can deal with a centered interval $I = (-a/2, a/2)$ with $a = |I|$.

Let us assume that the conclusion is false. Then there exists $f \in L^{p'}(I)$ ($\frac{1}{p} + \frac{1}{p'} = 1$) such that $f \neq 0$ and $\int_I f(x)e^{i\mu x}\, dx = 0$ for $\mu \in \mathcal{M}$. In order to show that this is impossible, we consider the Fourier transform $\hat{f}$ of f:

$$\hat{f}(z) = \int_I f(x)e^{ixz}\, dx,$$

which is a well defined analytic function in $\mathbb{C}$ such that $\hat{f}(\mu) = 0$ for $\mu \in \mathcal{M}$. We obtain a contradiction by showing that

(a) on the one hand, since $f \in L^{p'}(I)$, we have $|\hat{f}(it)| = o(e^{a|t|/2}|t|^{-1/p})$ for $|t| \longrightarrow \infty$, $t \in \mathbb{R}$, and

(b) on the other hand, since $\hat{f}$ has "too many" real zeros in the sense of (4.6.1), we have

$$|\hat{f}(it)| \geq ce^{a|t|/2}|t|^{-1/p}$$

for a constant $c > 0$.

We carry out this plan in several steps.

(1) In order to prove (a), let $g \in L^{p'}(I)$. By Hölder's inequality,

$$\begin{aligned}|\hat{g}(z)| &\leq \int_I |g(x)|e^{-x\cdot \operatorname{Im}(z)}\,dx \leq \|g\|_{p'}\left(\int_{-a/2}^{a/2} e^{-px\cdot\operatorname{Im}(z)}\,dx\right)^{1/p}\\ &\leq \|g\|_{p'}\cdot|\operatorname{Im}(z)|^{-1/p}e^{a|\operatorname{Im}(z)|/2}\end{aligned}$$

for every $z \in \mathbb{C}$, and hence

$$\sup_{z\in\mathbb{C}}\left(|\hat{g}(z)|\cdot|\operatorname{Im}(z)|^{1/p}e^{-a|\operatorname{Im}(z)|/2}\right) \leq \|g\|_{p'}$$

for every $g \in L^{p'}(I)$. Similarly,

$$|\hat{g}(z)| \leq |\operatorname{Im}(z)|^{-1/p}e^{(a-\varepsilon)|\operatorname{Im}(z)|/2}\|g\|_{p'} \tag{4.6.2}$$

for every $g \in L^{p'}(I)$ with $\operatorname{supp}(g) \subset (\frac{-a+\varepsilon}{2}, \frac{a-\varepsilon}{2})$, and hence

$$\left(|\hat{g}(z)|\cdot|\operatorname{Im}(z)|^{1/p}e^{-a|\operatorname{Im}(z)|/2}\right) = o(1),$$

as $|\operatorname{Im}(z)| \longrightarrow \infty$ for all such functions g. Since $p' < \infty$, the set of those g is dense in $L^{p'}(I)$, and the result follows.

In order to prove the second statement (b), we need some preparations. Below, we denote by $c_1, c_2, ...$ constants depending only on $a, \mathcal{M}, f$.

(2) Without loss of generality, we may assume that $0 \notin \mathcal{M}$ and that $\hat{f}(0) \neq 0$; see 4.4.9, claim (9)b, for the proof.

(3) We show that

$$\int_1^r \frac{n_Z(s)}{s}\,ds \leq \frac{a}{\pi}r + c_4, \tag{4.6.3}$$

and that

$$n_{\mathcal{M}}(r) \leq c_1 r + c_2 \tag{4.6.4}$$

for all $r \geq 1$, where

$$Z = Z(\hat{f}) = \{z \in \mathbb{C} : \hat{f}(z) = 0\},$$

and n_Z stands for the corresponding distribution function (each zero is counted according to its multiplicity).

In order to estimate n_Z, we use the estimate of $|\hat{f}(r\xi)|$ from (1) and Jensen's inequality as stated in A.3.12.3(d) (Volume 1). We obtain from (4.6.2)

$$\begin{aligned}\int_1^r \frac{n_Z(s)}{s}\,ds &\leq \int_{\mathbb{T}} \log|\hat{f}(r\zeta)|\,dm(\zeta) + c_3 \leq \int_{\mathbb{T}} \frac{a}{2}|\operatorname{Im}(r\zeta)|\,dm(\zeta) + c_4\\ &= \frac{a}{4\pi}\int_0^{2\pi} r\,|\sin(t)|\,dt + c_4\\ &= \frac{a}{\pi}r + c_4,\end{aligned}$$

and hence we get (4.6.3). Moreover

$$n_Z(r)\log 2 = n_Z(r)\int_r^{2r}\frac{1}{s}\,ds \leq \int_r^{2r}\frac{n_Z(s)}{s}\,ds \leq \int_1^{2r}\frac{n_Z(s)}{s}\,ds \leq \frac{2a}{\pi}r + c_4.$$

Since

$$n_{\mathcal{M}}(s) \leq n_Z(s)$$

for every s, the claim follows.

(4) Now, we need the *Hadamard representation theorem* for entire functions of order not exceeding 1. Recall that a function $\varphi \in \mathrm{Hol}(\mathbb{C})$ is said to be of *order* $\leq \rho$, if $|\varphi(z)| = O(e^{|z|^{\rho+\varepsilon}})$ as $|z| \longrightarrow \infty$ for every $\varepsilon > 0$. The Hadamard theorem states (in particular) that every entire function of order ≤ 1 can be represented in the form

$$\varphi(z) = z^N e^{\alpha z+\beta} \prod_n \Big(1 - \frac{z}{\lambda_n}\Big) e^{z/\lambda_n},$$

where $\lambda_n \neq 0$, $n \geq 1$, stand for the zeros of φ, each repeated according to its multiplicity, and $\alpha, \beta \in \mathbb{C}$. Conversely, the product $\prod_n (1 - \frac{z}{\lambda_n}) e^{z/\lambda_n}$ converges to a function of order ≤ 1 if $n_{(\lambda_n)}(s) \leq c_1 s + c_2$ for $s > 0$. The quotient φ/ψ of two functions of order ≤ 1 is, if it is holomorphic, a function of order ≤ 1. All these facts are standard in the theory of entire functions; see, for instance, the textbook E. Titchmarsh [**Ti**], Theorems 8.2.4 and 8.4(IV).

In particular, the Fourier transform $\hat{f}$ is an entire function of order ≤ 1.

(5) Let

$$F(z) = \prod_n \Big(1 - \frac{z}{\mu_n}\Big) e^{z/\mu_n},$$

where $(\mu_n) = \mathcal{M}$ is an indexation of $\mathcal{M}$. We show that

$$\hat{f}(z) = e^{\alpha+\beta z} F(z)$$

for some constants α and β.

Indeed, by (4.6.4) and the Hadamard representation theorem F is an entire function of order ≤ 1, and so is $\varphi = \hat{f}/F$. Since $n_{Z(\varphi)}(s) = n_Z(s) - n_{\mathcal{M}}(s)$, we have, using (4.6.1),

$$\begin{aligned} \int_1^r \frac{n_{Z(\varphi)}(s)}{s}\, ds &= \int_1^r \frac{n_Z(s)}{s}\, ds - \int_1^r \frac{n_{\mathcal{M}}(s)}{s}\, ds \\ &\leq \frac{a}{\pi} r + c_4 - \Big(\frac{a}{\pi} r - \frac{1}{p}\log(r) - C\Big) \\ &= \frac{1}{p}\log(r) + c_5. \end{aligned}$$

Since $p > 1$ and $n_{Z(\varphi)}$ is monotone and integer valued, we get $n_{Z(\varphi)} \equiv 0$. From the Hadamard theorem we get $\varphi = e^{\alpha+\beta z}$, and the result follows.

(6) Let us show that $|F(it)| \geq c_6 e^{a|t|/2} |t|^{-1/p}$ for all $t \in \mathbb{R}$. Indeed,

$$\begin{aligned} 2 \cdot \log|F(it)| &= \sum_n \log\Big(1 + \frac{t^2}{\mu_n^2}\Big) = \int_0^\infty \log\Big(1 + \frac{t^2}{s^2}\Big)\, dn_{\mathcal{M}}(s) \\ &= \int_0^\infty \frac{4t^2 s}{(t^2+s^2)^2} \Big(\int_0^s \frac{n_{\mathcal{M}}(r)}{r}\, dr\Big)\, ds \end{aligned}$$

(the integrations by parts are easily justified by using the passage to the limit $\lim_{\varepsilon\to 0, A\to\infty} \int_\varepsilon^A$ and the fact that $n_{\mathcal{M}}(\varepsilon) = 0$ for some $\varepsilon > 0$ (which means that some neighbourhood of 0 is zero-free), and $n_{\mathcal{M}}(r) \le c_1 r + c_2$ (see (2) above)). Therefore,

$$\log|F(it)| \ \ge \ \int_0^\infty \frac{2t^2 s}{(t^2+s^2)^2}\Big(\frac{a}{\pi}s - \frac{1}{p}\log(s) - C\Big)\,ds$$

(change of variables $s = |t|y$)

$$= \ \int_0^\infty \frac{2y}{(1+y^2)^2}\Big(\frac{a}{\pi}|t|y - \frac{1}{p}\log|t| - \frac{1}{p}\log(y) - C\Big)\,dy.$$

Using $\int_0^\infty \frac{y^2}{(1+y^2)^2}\,dy = \frac{\pi}{4}$ (by residues) and $\int_0^\infty \frac{2y}{(1+y^2)^2}\,dy = 1$ we get the required estimate:

$$\log|F(it)| \ge \frac{a}{2}|t| - \frac{1}{p}\log|t| - c_7.$$

(7) Now, we are in a position to deduce a lower estimate for $|\hat{f}(it)|$ and to complete the proof of the theorem. Indeed, (5) implies that $|\hat{f}(it)| = e^{ct}|F(it)|$ for $t \in \mathbb{R}$, where $c \in \mathbb{R}$. Hence, (1) contradicts (6), either for $t \longrightarrow \infty$, or for $t \longrightarrow -\infty$. □

Now, we give another proof to the Ingham part of Theorem 4.1.2 (in fact, this is a generalization of that result).

4.6.2. COROLLARY. *Let $1 < p \le \infty$, and $\mu_0 = 0$, $-\mu_{-n} = \mu_n = n - \frac{1}{2p'}$ for $n \ge 1$. Then the family $(e^{i\mu_n x}\chi_{(-\pi,\pi)} : n \in \mathbb{Z})$ is not minimal in $L^p(-\pi,\pi)$; more precisely, the family $(e^{i\mu_n x}\chi_{(-\pi,\pi)} : n \in \mathbb{Z}\backslash\{0\})$ is complete in $L^p(-\pi,\pi)$ (weak* complete if $p = \infty$).*

Indeed, we simply verify condition (4.6.1) for $\mathcal{M} = (\mu_n)_{n\in\mathbb{Z}\backslash\{0\}}$. For this case,

$$n_{\mathcal{M}}(s) = \mathrm{card}\{\mu \in \mathcal{M} : |\mu| \le s\} = 2\cdot\max\{n \ge 1 : n - \frac{1}{2p'} \le s\} = 2\cdot[s + \frac{1}{2p'}].$$

Therefore,

$$\begin{aligned}
\int_1^r \frac{n_{\mathcal{M}}(s)}{s}\,ds \ &= \ c_1 + \int_{1-1/(2p')}^r \frac{n_{\mathcal{M}}(s)}{s}\,ds = c_1 + 2\int_{1-1/(2p')}^r \frac{[s+\frac{1}{2p'}]}{s}\,ds \\
&\ge \ c_1 + 2\sum_{k=1}^{[r-1+1/(2p')]} \int_{k-1/(2p')}^{k+1-1/(2p')} \frac{k}{s}\,ds \\
&= \ c_1 + 2\sum_{k=1}^{[r-1+1/(2p')]} k\cdot\log\frac{k+1-1/(2p')}{k-1/(2p')} \\
&\ge \ c_1 + 2\sum_{k=1}^{[r-1+1/(2p')]} k\cdot\Big\{\frac{1}{k-1/(2p')} - \frac{1}{2(k-1/(2p'))^2}\Big\} \\
&= \ c_1 + \sum\nolimits_1 - \sum\nolimits_2,
\end{aligned}$$

where we have used that $\log(1+x) > x - x^2/2$ for $x > 0$. For the first sum we have

$$\begin{aligned}\sum\nolimits_1 &= 2 \cdot \sum_{k=1}^{[r-1+1/(2p')]} \left\{1 + \frac{1}{2p'(k - 1/(2p'))}\right\} \\ &= 2 \cdot [r - 1 + \frac{1}{2p'}] + \frac{1}{p'} \sum_{k=1}^{[r-1+1/(2p')]} \frac{1}{k - 1/(2p')} \\ &= 2r + \frac{1}{p'} \log(r) + O(1),\end{aligned}$$

as $r \longrightarrow \infty$. For the second sum we have

$$\begin{aligned}\sum\nolimits_2 &= \sum_{k=1}^{[r-1+1/(2p')]} \frac{k}{(k - 1/(2p'))^2} \\ &= \sum_{k=1}^{[r-1+1/(2p')]} \frac{1}{k - 1/(2p')} + \sum_{k=1}^{[r-1+1/(2p')]} \frac{1}{2p'(k - 1/(2p'))^2} \\ &= \log(r) + O(1),\end{aligned}$$

and, finally,

$$\int_1^r \frac{n_{\mathcal{M}}(s)}{s} \, ds \geq 2r - \frac{1}{p} \log(r) + c_2.$$

The result follows. $\square$

Similarly, one can verify the following result on perturbations of the harmonic exponentials $(e^{inx}\chi_{(-\pi,\pi)} : n \in \mathbb{Z})$ in $L^p(-\pi, \pi)$ (which form, in fact, a basis of $L^p(-\pi, \pi)$ if $1 < p < \infty$, see A.5.7.3(f), Volume 1).

4.6.3. COROLLARY. *Let $\mathcal{M} = (\mu_n)_{n \in \mathbb{Z}} \subset \mathbb{R}$ and suppose that there exists $\delta > 0$ such that $|\mu_n| \leq |n| + \frac{1}{2p} + (\log |n|)^{-1-\delta}$ for all sufficiently large $|n|$. Then the family $(e^{i\mu_n x}\chi_{(-\pi,\pi)} : n \in \mathbb{Z})$ is complete in $L^p(-\pi, \pi)$ (weak* complete if $p = \infty$).* $\square$

4.6.4. REMARKS. **(1)** Recall that when considering (in)completeness of a system $(e^{i\mu_n x}\chi_I)$, we can replace any finite subset of frequencies $(\mu_n)_{n=1}^N$ by any other finite subset of frequencies $(\mu'_n)_{n=1}^N$, see (9)b of Subsection 4.4.9 above. In particular, in order to verify (4.6.1) for μ_n of Corollary 4.6.3, we can assume that the inequality holds for all $|n| \geq 2$, and $\mu_n = n$ for $|n| < 2$.

(2) In fact, Theorem 4.6.1 and Corollaries 4.6.2 and 4.6.3 are sharp. In particular, it can be shown that for every $\delta > 0$ there exists an incomplete system $(e^{i\mu_n x}\chi_{(-\pi,\pi)} : n \in \mathbb{Z})$ in $L^p(-\pi, \pi)$ such that $|\mu_n| \leq |n| + \frac{1}{2p} + \delta$ for $n \in \mathbb{Z}$; see Section 4.8 for references.

4.7. Exercises and Further Results

4.7.1. More about bases of exponentials on finite intervals. Let $\Lambda = (\lambda_n)_{n \in \mathbb{Z}}$ be a Blaschke sequence of pairwise distinct points in $\mathbb{C}_+ = \{z \in \mathbb{C} : \operatorname{Im} z > 0\}$ and $\Theta_a(z) = e^{iaz}$, where $a > 0$. The corresponding Blaschke product is denoted by

$$B = B_\Lambda = \prod_{n \in \mathbb{Z}} \varepsilon_n \frac{z - \lambda_n}{z - \overline{\lambda}_n},$$

where $|\varepsilon_n| = 1$. Most of the results presented in this subsection are due to S. Hruschev, N. Nikolski, and B. Pavlov (1981), [**HNP**].

(a) *A criterion in terms of the argument.* Let

$$a_\Lambda(x) = 2\int_0^x \sum_{n\in\mathbb{Z}} \frac{\mathrm{Im}(\lambda_n)}{|\lambda_n - t|^2}\,dt - ax, \qquad x \in \mathbb{R}.$$

(i) Show that $B(x)\overline{\Theta}_a(x) = \varepsilon \cdot e^{ia_\Lambda(x)}$ for $x \in \mathbb{R}$.

[*Hint:* use the formula $\int_0^x (\overline{\lambda} - t)^{-1}dt = \log(1 - \frac{x}{\overline{\lambda}})$ (the principal value), then take imaginary parts.]

(ii) Show that the family $(e^{i\lambda_n x})_{n\in\mathbb{Z}}$ forms a Riesz basis in $L^2(0,a)$ if $\Lambda \in (C)$ *and*

$$\mathrm{dist}_{L^\infty}(a_\Lambda, \mathcal{H}L^\infty + \mathbb{C}) < \frac{\pi}{2};$$

the converse is also true if $\inf_{n\geq 1} \mathrm{Im}\,\lambda_n > 0$.

(Here $\mathcal{H}$ stands for the harmonic conjugation (in the halfplane $\mathbb{C}_+$), $\mathcal{H}L^\infty = \{\mathcal{H}f : f \in L^\infty\}$, and $\mathrm{dist}_{L^\infty}(a_\Lambda, \mathcal{H}L^\infty + \mathbb{C}) = \inf\{\|a_\Lambda - h\|_\infty : h \in \mathcal{H}L^\infty\}$ ($\|f\|_\infty = \infty$ if $f \notin L^\infty$)).

[*Hint:* the "if" part follows from 4.4.6(5); for the "only if" part one needs to prove that the function a arising in 4.4.6(5) is automatically continuous under the above assumptions (see [**HNP**] for details); this implies that $a - a_\Lambda = \mathrm{const}$ and the result follows from 4.4.6(5).]

(b) *A criterion in terms of the generating function.* Let $F \in \mathrm{Hol}(\mathbb{C})$ be a function of *exponential type*, i.e., $|F(z)| \leq Me^{A|z|}$ for suitable $M > 0$ and $A > 0$. Recall that the function h_F,

$$h_F(\theta) = \overline{\lim}_{r\to\infty} \frac{\log|F(re^{i\theta})|}{r}, \qquad \theta \in [-\pi, \pi],$$

is the *indicator function* of F. Given $a > 0$, denote by E_a the set of all entire functions F with $h_F = H_a$, where

$$H_a(\theta) = \max(-a \cdot \sin\theta, 0).$$

(i) Let $(e^{i\lambda_n x})_{n\in\mathbb{Z}}$ be a complete minimal family in $L^2(0,a)$. Show that there exists a unique (up to a multiplicative constant) function $F = F_\Lambda \in E_a$ having the zero set $Z(F) = \Lambda = (\lambda_n)_{n\in\mathbb{Z}}$ (a simple zero at every λ_n).

F_Λ is called *generating function* for $(e^{i\lambda_n x})_{n\in\mathbb{Z}}$. (In fact, by the Hadamard factorization theorem, $F_\Lambda = e^{bz}\prod_{n\geq 1}(1 - \frac{z}{\lambda_n})e^{z/\lambda_n}$ for a suitable $b \in \mathbb{C}$).

[*Hint:* there exists a unique (up to a multiplicative constant) function $f \in L^2(0,a)$ such that $(e^{i\lambda_n x}, \overline{f}) = \delta_{n0}$ for $n \in \mathbb{Z}$; set

$$\hat{f}(z) = \int_0^a e^{izt} f(t)\,dt$$

and show that $Z(\hat{f}) = \Lambda\backslash\{\lambda_0\}$ (if $\hat{f}(\mu) = 0$ for $\mu \in \mathbb{C}\backslash\Lambda$, consider the function g,

$$g(x) = -ie^{-i\mu x}\int_0^x e^{i\mu t} f(t)\,dt, \qquad x \in [0,a],$$

and observe that $g \in L^2(0,a)$ and $\hat{g}(z) = \hat{f}(z)(z-\mu)^{-1}$, which contradicts the uniqueness of f);

set $F_\Lambda(z) = (1 - \frac{z}{\lambda_0})\hat{f}(z)$ and verify the other properties of F_Λ (for computing the indicator function of F_Λ consult B. Levin [**Le1**] or R. Boas [**Bo1**]).]

(ii) Assume that $\inf_{n\geq 1} \operatorname{Im}\lambda_n > 0$. Show that the family $(e^{i\lambda_n x})_{n\in\mathbb{Z}}$ is a Riesz basis in $L^2(0,a)$ if and only if $\Lambda \in (C)$ and there exists a generating function F_Λ satisfying

$$|F_\Lambda|^2|\mathbb{R} \in (HS),$$

where (HS) means the Helson–Szegö condition from A.5.4.1, Volume 1 (or, equivalently, $|F_\Lambda|^2|\mathbb{R}$ satisfies the Muckenhoupt condition (A_2), see A.5.7.2, Volume 1).

[*Hint:* use the necessary and sufficient condition 4.4.6(8) to be a Riesz basis (and the notation used there); it is equivalent to $B_\Lambda h = \bar{\varepsilon}\Theta_a\bar{h}$, where h is an outer function such that $|h|^2 \in (HS)$; this implies that $(z+i)^{-1}h \in H^2(\mathbb{C}_+)$, and hence $B_\Lambda h$ extends across the real line as a holomorphic function $z \longmapsto \bar{\varepsilon}\Theta_a(z)\bar{h}(\bar{z})$ in the lower half-plane $\mathbb{C}_-$; it is not hard to see that this extension F is an entire function, $F \in E_a$, $Z(F) = \Lambda$ and $|F|^2 = |h|^2 \in (HS)$; for details see [**HNP**].]

(iii)* In fact, $\mathcal{H}(\log|F_\Lambda(\cdot + iy)|^2) = -2a_{\Lambda+iy}(\cdot) + c$, where c is a constant and $\mathcal{H}$ stands for the harmonic conjugation.

(iv) *(B. Levin, V. Golovin, 1961)* Let F be an entire function of exponential type such that a) $F \in E_a$, where $a > 0$; b) there exists $y \in \mathbb{R}$ such that

$$0 < \inf_{x\in\mathbb{R}} |F(x+iy)| \leq \sup_{x\in\mathbb{R}} |F(x+iy)| < \infty;$$

c) F has simple zeros $\Lambda = (\lambda_n)_{n\in\mathbb{Z}}$ such that $\sup_{n\in\mathbb{Z}} |\operatorname{Im}\lambda_n| < \infty$; d) Λ is a separated (sparse) sequence, i.e., $\inf_{n\neq m} |\lambda_n - \lambda_m| > 0$. Show that $(e^{i\lambda_n x})_{n\in\mathbb{Z}}$ is a Riesz basis in $L^2(0,a)$.

A function F satisfying a), b) and c) is called *sine type function* (STF).

[*Hint:* for the case where $\inf_{n\geq 1} \operatorname{Im}(\lambda_n) > 0$, apply (ii); further, consult [**HNP**].]

(c) *Perturbations of frequencies.* (i) Let

$$P_z(t) = \pi^{-1}\frac{\operatorname{Im}(z)}{|z-t|^2}, \quad t \in \mathbb{R},$$

be the Poisson kernel at a point $z \in \mathbb{C}_+$, let $(\delta_n)_{n\in\mathbb{Z}}$ be a sequence of positive numbers, and let $(e^{i\lambda_n x})_{n\in\mathbb{Z}}$ be a Riesz basis in $L^2(0,a)$. Suppose that $|\lambda_n' - \lambda_n| \leq \delta_n$ for $n \in \mathbb{Z}$, and $\Lambda' = (\lambda_n')_{n\in\mathbb{Z}} \in (C)$ and

$$\inf_{y>0}\left\{\left\|2\pi\sum_{n\in\mathbb{Z}}\delta_n P_{\lambda_n+iy}\right\|_\infty + \operatorname{dist}_{L^\infty}(a_{\Lambda+iy}, \mathcal{H}L^\infty + \mathbb{C})\right\} < \frac{\pi}{2}.$$

Show that $(e^{i\lambda_n' x})_{n\in\mathbb{Z}}$ is still a Riesz basis in $L^2(0,a)$.

[*Hint:* assume (without loss of generality) that $\delta = \inf_{n\geq 1} \operatorname{Im}\lambda_n > 0$, and show that the condition $\|\sum_{n\in\mathbb{Z}}\delta_n P_{\lambda_n}\|_\infty < \infty$ implies that $a_\Lambda - a_{\tilde{\Lambda}} \in \mathcal{H}L^\infty + \mathbb{C}$ when considering vertical perturbations only, $\lambda_n' = \tilde{\lambda}_n$ with $\operatorname{Re}\tilde{\lambda}_n = \operatorname{Re}\lambda_n$, $|\tilde{\lambda}_n - \lambda_n| \leq \delta_n \leq \delta$ (use the definition of a_Λ);

for horizontal perturbations, $\lambda'_n = \lambda^*_n$ with $\operatorname{Im}\lambda^*_n = \operatorname{Im}(\lambda_n)$, show that

$$\int_0^x P_{\lambda_n}(t)dt - \int_0^x P_{\lambda^*_n}(t)\,dt = (\lambda^*_n - \lambda_n)\Big(P_{\lambda_n}(x) - P_{\lambda_n}(0)\Big)\Big(1 + O(\frac{1}{\operatorname{Im}(\lambda_n)})\Big),$$

and then make first horizontal and then vertical shifts to get

$$\begin{aligned}&\operatorname{dist}_{L^\infty}(a_{\Lambda'+iy}, \mathcal{H}L^\infty + \mathbb{C})\\ &\quad\le \operatorname{dist}_{L^\infty}(a_{\Lambda+iy}, \mathcal{H}L^\infty + \mathbb{C}) + \Big(1 + O(\frac{1}{y})\Big)\Big\|2\pi\sum_{n\in\mathbb{Z}}\delta_n P_{\lambda_n+iy}\Big\|_\infty;\end{aligned}$$

now, the result follows from (a)(ii).]

(ii) Let $(e^{i\lambda_n x})_{n\in\mathbb{Z}}$ be a Riesz basis in $L^2(0,a)$ and $\inf_{n\ge1}\operatorname{Im}\lambda_n > 0$. Show that there exists $\varepsilon > 0$ such that $(e^{i\lambda'_n x})_{n\in\mathbb{Z}}$ is still a Riesz basis in $L^2(0,a)$ provided that $|\lambda'_n - \lambda_n| \le \varepsilon$ for every $n \in \mathbb{Z}$.

[*Hint:* use (i) and (a).]

(iii)* Let $\Lambda = (\lambda_n)_{n\in\mathbb{Z}}$ and $\Lambda' = (\lambda'_n)_{n\in\mathbb{Z}}$ be complex sequences such that

$$\sup_{n\in\mathbb{Z}}|\operatorname{Im}(\lambda_n)| < \infty, \quad \sup_{n\in\mathbb{Z}}\Big|\operatorname{Im}(\lambda'_n)\Big| < \infty,$$
$$\text{and } \operatorname{Re}(\lambda_n) = \operatorname{Re}(\lambda'_n) \quad (\forall n \in \mathbb{Z}).$$

Show that $(e^{i\lambda_n x})_{n\in\mathbb{Z}}$ is a Riesz basis in $L^2(0,a)$ if and only if $(e^{i\lambda'_n x})_{n\in\mathbb{Z}}$ is.

(iv)* *(S. Avdonin, 1974)* Let $\Lambda = (\lambda_n)_{n\in\mathbb{Z}}$ be the zero set of a STF $F \in E_a$, where $a > 0$ (see (b) above), and let $\Lambda' = (\lambda'_n)_{n\in\mathbb{Z}}$ be a separated sequence of complex numbers such that

$$\sup_{n\in\mathbb{Z}}|\lambda_n - \lambda'_n| < \infty, \quad \overline{\lim}_{R\to\infty}\sup_{x\in\mathbb{R}}\frac{\Delta_x(R)}{2R} < \frac{1}{4}\cdot\frac{a}{2\pi},$$

where $\Delta_x(R) = \sum \operatorname{Re}(\lambda_k - \lambda'_k)$ and the sum is taken over all k with $|x - \operatorname{Re}\lambda_k| \le R$. Then $(e^{i\lambda'_n x})_{n\in\mathbb{Z}}$ is a Riesz basis in $L^2(0,a)$.

(d)* *Complementation up to a Riesz basis.*

(i) *(V. Vasyunin, 1981)* Let $\Lambda' = (\lambda'_n)$ be a Carleson interpolating sequence in $\mathbb{C}_+$ with $\lim_n \operatorname{Im}\lambda'_n = \infty$. Then, for every $a > 0$, there exists a sequence $\Lambda'' = (\lambda''_n)$ in $\mathbb{C}_+$ such that $(e^{i\lambda x} : \lambda \in \Lambda)$ is a Riesz basis in $L^2(0,a)$ where $\Lambda = \Lambda' \cup \Lambda''$.

(ii) *(K. Seip, 1995)* Let $\Lambda = \{\lambda_n = n(1 + |n|^{-1/2}) : n \in \mathbb{Z}\backslash\{0\}\}$. Then $(e^{i\lambda x} : \lambda \in \Lambda)$ is a Riesz sequence in $L^2(0,2\pi)$ but there exists no $\Lambda' \supset \Lambda$ such that $(e^{i\lambda x} : \lambda \in \Lambda')$ is a Riesz basis in $L^2(0,2\pi)$.

(e)* *(A. Minkin, 1991)* Let $a > 0$, $\Lambda \subset \mathbb{C}$ and $\Lambda_+ = \Lambda \cap (\mathbb{C}_+ \cup \mathbb{R})$, $\Lambda_- = \overline{\Lambda} \cap \mathbb{C}_+$ ($\overline{\Lambda}$ means complex conjugate). The following are equivalent.

1) $(e^{i\lambda x} : \lambda \in \Lambda)$ is a Riesz basis in $L^2(0,a)$.

2) $\Lambda_\pm + iy$ are Carleson interpolating sequences ($y > 0$), $\inf\{|\lambda - \lambda'| : \lambda, \lambda' \in \Lambda, \lambda \ne \lambda'\} > 0$ and $\operatorname{dist}(\Theta_a\overline{B}_y, H^\infty) < 1$, $\operatorname{dist}(\overline{\Theta}_a B_y, H^\infty) < 1$, where $B_y = B_{\Lambda_+ + iy}B_{\Lambda_- + iy}$.

3) $\Lambda_\pm + iy$ are Carleson interpolating sequences ($y > 0$), $\inf\{|\lambda - \lambda'| : \lambda, \lambda' \in \Lambda, \lambda \ne \lambda'\} > 0$, and there exists a generating function F_y for $(\Lambda_+ + iy) \cup (\Lambda_- + iy)$ (double points give multiple zeros) such that $|F_y||\mathbb{R} \in (HS)$. (Compare with (b) above).

4.7.2. Exponentials on $\mathbb{R}_+$ and complex powers on $(0,1)$. (a) *(C. Müntz; O. Szasz, 1916)* Let $Q \subset \{z \in \mathbb{C} : \operatorname{Re} z > 0\}$. Show that a family of powers $(x^q : q \in Q)$ is complete in the space $L^2((0,1); \frac{dx}{x})$ if and only if

$$\sum_{q \in Q} \frac{\operatorname{Re} q}{1+|q|^2} = \infty.$$

[*Hint:* make the change of variables $x = e^{-t}$, $t \in \mathbb{R}_+$, and use the Blaschke condition for the half-plane $\mathbb{C}_+$.]

(b) Let $Q = (q_n)_{n\in\mathbb{Z}}$ be a sequence of pairwise distinct points in $\{z \in \mathbb{C} : \operatorname{Re} z > 0\}$. Show that the family $(x^{q_n} : n \in \mathbb{Z})$ is a Riesz sequence in $L^2((0,1); \frac{dx}{x})$ if and only if $(iq_n)_{n\in\mathbb{Z}}$ satisfies the Carleson condition for $\mathbb{C}_+$.

In particular, for the case $0 < \inf(\operatorname{Re} q_n : n \in \mathbb{Z}) \le \sup(\operatorname{Re} q_n : n \in \mathbb{Z}) < \infty$, this is equivalent to

$$\inf_{n\neq m} |q_n - q_m| > 0,$$

and for the case $Q = (q_n)_{n\in\mathbb{Z}} \subset \mathbb{R}_+$, $q_n < q_{n+1}$, to

$$\inf_{n\in\mathbb{Z}} \frac{q_{n+1}}{q_n} > 1.$$

[*Hint:* make the change of variables $x = e^{-t}$, $t \in \mathbb{R}_+$, and use the Carleson condition for $\mathbb{C}_+$.]

(c)* *(V.I. Gurarii and V. Matsaev, 1966)* Let $Q = (q_n)_{n\in\mathbb{Z}} \subset \mathbb{R}_+$, $q_n < q_{n+1}$, and $1 < p < \infty$. The following are equivalent.

1) $(x^{q_n} : n \in \mathbb{Z})$ is a uniformly minimal sequence in $L^p((0,1); \frac{dx}{x})$.

2) $(x^{q_n}/\|x^{q_n}\|_p : n \in \mathbb{Z})$ is equivalent to the standard basis in $l^p(\mathbb{Z})$ (and hence, it is an unconditional basis sequence in $L^p((0,1); \frac{dx}{x})$).

3) $(iq_n)_{n\in\mathbb{Z}}$ is a Carleson interpolating sequence in $\mathbb{C}_+$, that is,

$$\inf_{n\in\mathbb{Z}} \frac{q_{n+1}}{q_n} > 1.$$

(d)* *Asymptotically orthogonal families of exponentials.* Let $\Lambda = (\lambda_n)_{n\ge1} \subset \mathbb{C}_+$ be a sequence of distinct points.

(i) *(A. Volberg, 1982)* The following are equivalent.

1) $(e^{i\lambda_n x}\chi_{\mathbb{R}_+})_{n\ge1}$ is an *asymptotically orthogonal Riesz sequence* in $L^2(\mathbb{R}_+)$, which means that

$$G(X) = I + K,$$

where $K \in \mathfrak{S}_\infty$ and $G(X)$ stands for the Gram matrix of the normalized sequence $X = (x_n)_{n\ge1} = (e^{i\lambda_n x}/\|e^{i\lambda_n x}\|_{L^2(\mathbb{R}_+)})_{n\ge1}$ (see C.3.3.1).

2) $V = U + K$ for a unitary operator U and $K \in \mathfrak{S}_\infty$, where V stands for an orthogonalizer of X (see C.3.3.1).

3) Λ is a *vanishing Carleson sequence*, that is,

$$\lim_n |B_{\lambda_n}(\lambda_n)| = 1,$$

where $B_{\lambda_n} = B/b_{\lambda_n}$ (see C.3.3.3 for the notation).

(ii) *(E. Fricain, 1999)* 1) If $(e^{i\lambda_n x}\chi_{\mathbb{R}_+})_{n\geq 1}$ is an asymptotically orthogonal Riesz sequence in $L^2(\mathbb{R}_+)$ and $\sup_{n\geq 1}|(\lambda'_n - \lambda_n)(\lambda'_n - \overline{\lambda}_n)^{-1}| < 1$, then $(e^{i\lambda'_n x}\chi_{\mathbb{R}_+})_{n\geq 1}$ is.

2) If $a > 0$ and $(e^{i\lambda_n x}\chi_{(0,a)})_{n\geq 1}$ is an asymptotically orthogonal Riesz sequence in $L^2(0,a)$, so is $(e^{i\lambda_n x}\chi_{\mathbb{R}_+})_{n\geq 1}$ in $L^2(\mathbb{R}_+)$.

3) If $(e^{i\lambda_n x}\chi_{\mathbb{R}_+})_{n\geq 1}$ is an asymptotically orthogonal Riesz sequence in $L^2(\mathbb{R}_+)$ and $\lim_n \operatorname{Im}\lambda_n = \infty$, so is $(e^{i\lambda_n x}\chi_{(0,a)})_{n\geq 1}$ in $L^2(0,a)$ for every $a > 0$.

(e)* *Interior compact families of exponentials (P. Koosis, 1957).* Let $\Lambda = (\lambda_n)_{n\geq 1} \subset \mathbb{C}_+$ be a sequence of distinct points and

$$L^2_\Lambda = \operatorname{span}_{L^2(\mathbb{R}_+)}\Big(e^{i\lambda_n x}\chi_{\mathbb{R}_+} : n \geq 1\Big).$$

The following are equivalent.

1) The restriction operator $f \longmapsto \chi_{(a,\infty)}f$ is compact from L^2_Λ to $L^2(a,\infty)$ for every $a > 0$.

2) The Hankel operator $H_{\overline{B}\Theta_a} = P_-\overline{B}\Theta_a : H^2(\mathbb{C}_+) \longrightarrow H^2_-(\mathbb{C}_+)$ is compact for every $a > 0$, where $\Theta_\alpha = e^{iax}$.

3) $\lim_n \operatorname{Im}\lambda_n = \infty$ and $\lim_{|x|\longrightarrow\infty}\sum_{n\geq 1}\frac{\operatorname{Im}(\lambda_n)}{|\lambda_n - x|^2} = 0$.

4.7.3. More about bases of reproducing kernels. Let Θ be an inner function, and let $\Lambda = (\lambda_n)_{n\geq 1}$ be a sequence of distinct points in $\mathbb{D}$.

(a) *(S. Hruschev, N. Nikolski, and B. Pavlov, 1981)*

(i) Assume that $(k^\Theta_{\lambda_n})_{n\geq 1}$ is a Riesz sequence. Show that

$$\sum_{n\geq 1}\frac{1-|\lambda_n|^2}{1-|\Theta(\lambda_n)|^2} < \infty;$$

in particular, if Θ is a singular inner function and μ is a singular measure associated with Θ, then

$$\sum_{n\geq 1}\Big(\operatorname{dist}(\lambda_n, \operatorname{supp}(\mu))\Big)^2 \leq \mu(\mathbb{T})\sum_{n\geq 1}\left(\int_{\mathbb{T}}\frac{d\mu(\zeta)}{|\zeta-\lambda_n|^2}\right)^{-1} < \infty.$$

[*Hint:* for every $f \in \mathcal{K}_\Theta$,

$$\sum_{n\geq 1}\frac{1-|\lambda_n|^2}{1-|\Theta(\lambda_n)|^2}|f(\lambda_n)|^2 = \sum_{n\geq 1}\left|\left(f, \frac{k^\Theta_{\lambda_n}}{\|k^\Theta_{\lambda_n}\|}\right)\right|^2 \leq c\|f\|^2;$$

then, take $f = P_\Theta 1 = 1 - \overline{\Theta(0)}\Theta$ and use $|f(\lambda_n)| \geq 1 - |\Theta(0)| > 0$;

for the special case of a singular inner function Θ, we have

$$1-|\Theta(\lambda_n)|^2 = 1 - \exp\Big(-2\int_{\mathbb{T}}\frac{1-|\lambda_n|^2}{|\zeta-\lambda_n|^2}\,d\mu(\zeta)\Big) \leq 2\int_{\mathbb{T}}\frac{1-|\lambda_n|^2}{|\zeta-\lambda_n|^2}\,d\mu(\zeta).]$$

(ii) Show that the following assertions are equivalent.

1) $(k^\Theta_{\lambda_n})_{n\geq 1}$ is a Riesz basis in $\mathcal{K}_\Theta$ (respectively, a Riesz sequence).

2) $\Lambda \in (C)$, and there exists a function $G \in H^2$ such that the operator $P_\Theta T_{\overline{G}}$ can be extended up to an isomorphism from $\mathcal{K}_B$ onto $\mathcal{K}_\Theta$.

(Here, $T_{\overline{G}}f = P_+\overline{G}f$ stands for a Toeplitz operator which is well-defined on the dense subset $\mathcal{L}\text{in}((1-\overline{\lambda}z)^{-1} : \lambda \in \Lambda)$ of $\mathcal{K}_B$).

[*Hint:* for 1) $\Rightarrow$ 2), get $\Lambda \in (C)$ from Lemma 4.4.2; now, using (i), find a function $G \in H^2$ such that $G(\lambda_n) = (1-|\Theta(\lambda_n)|^2)^{-1/2}$ for every $n \geq 1$, and observe that

$$P_\Theta T_{\overline{G}} k_{\lambda_n} = (1-|\Theta(\lambda_n)|^2)^{-1/2} P_\Theta k_{\lambda_n} = (1-|\Theta(\lambda_n)|^2)^{-1/2} k_{\lambda_n}^\Theta,$$

where $k_{\lambda_n} = (1-\overline{\lambda}_n z)^{-1}$; therefore, $P_\Theta T_{\overline{G}}$ is an isomorphism since it transforms a Riesz basis into a Riesz basis (sequence) and $\|k_{\lambda_n}\| = \|(1-|\Theta(\lambda_n)|^2)^{-1/2} k_{\lambda_n}^\Theta\|$ for $n \geq 1$.]

(b) *Orthogonal bases of reproducing kernels and D. Clark's measures.* Let $\Theta = B_\Lambda S$ be an inner function factorized into its Blaschke product and singular part, and let

$$s_\Theta = \left\{ \lambda \in \mathbb{T} : \sum_{n\geq 1} \frac{1-|\lambda_n|^2}{|\lambda-\lambda_n|^2} + \int_{\mathbb{T}} \frac{d\mu(\zeta)}{|\lambda-\zeta|^2} < \infty \right\}.$$

(i)* *(P. Ahern and D. Clark, 1970)* Let $\lambda \in \mathbb{T}$. The following are equivalent.

1) There exists a number c such that $(1-\overline{c}\Theta)(1-\overline{\lambda}z)^{-1} \in H^2$.

2) $\lambda \in s_\Theta$.

Moreover, for $\lambda \in s_\Theta$, the limit $\lim_{r\longrightarrow 1} \Theta(r\lambda) = \Theta(\lambda)$ exists and $\Theta(\lambda) = c$; therefore

$$\lim_{r\to 1} k_{r\lambda}^\Theta = k_\lambda^\Theta \in \mathcal{K}_\Theta.$$

(ii) *(D. Clark (1972), D. Georgijevic (1979))* Let $\lambda, \lambda' \in s_\Theta$, $\lambda \neq \lambda'$ and $\Theta(\lambda) = \Theta(\lambda') = c$, where $|c| = 1$. Show that $(k_\lambda^\Theta, k_{\lambda'}^\Theta) = 0$.

[*Hint:* a straightforward computation using (i).]

(iii)* Let $\alpha \in \mathbb{T}$ and σ_α be the probability (Herglotz) measure (see A.3.9.2, Volume 1) representing the harmonic function

$$\operatorname{Re}\left(\frac{\alpha+\Theta(z)}{\alpha-\Theta(z)}\right) = \frac{1-|\Theta(z)|^2}{|\alpha-\Theta(z)|^2} = \int_{\mathbb{T}} \frac{1-|z|^2}{|\zeta-z|^2} d\sigma_\alpha(\zeta).$$

The measures σ_α (called *Clark measures* of Θ) are pairwise singular and singular with respect to the Lebesgue measure m. The restriction map $j : f \longrightarrow f|\operatorname{supp}(\sigma_\alpha)$ (well defined on reproducing kernels k_λ^Θ, $|\lambda| < 1$) extends to a unitary mapping

$$\mathcal{K}_\Theta \longrightarrow L^2(\sigma_\alpha)$$

(D. Clark, 1972). Moreover, every function $f \in \mathcal{K}_\Theta$ has nontangential limits σ_α-a.e. and

$$(jf)(\zeta) = \lim_{r\longrightarrow 1} f(r\zeta)$$

σ_α-a.e. *(A. Poltoratski, 1993).* The measures σ_α satisfy $\int_{\mathbb{T}} \sigma_\alpha \, dm(\alpha) = m$ (weak* convergent integral) *(A. Aleksandrov, 1987).*

(iv) *(D. Clark, 1972)* Let $Z_\alpha : L^2(\sigma_\alpha) \longrightarrow L^2(\sigma_\alpha)$ be the unitary operator defined by $Z_\alpha f = zf$, $f \in L^2(\sigma_\alpha)$, where σ_α is a Clark measure associated to Θ. Show that $\operatorname{rank}(j^{-1}Z_\alpha j - M_\Theta) \leq 1$, where j is the unitary operator defined in (iii) above.

In fact, if $\Theta(0) = 0$, $j^{-1}Z_\alpha j = M_\Theta + \alpha(\cdot, \overline{z}\Theta)$.

[*Hint:* take $f \in \mathcal{K}_\Theta$ such that $f(0) = (f, P_\Theta 1) = 0$ and use (iii) to check that $j^{-1} Z_\alpha^* j f = M_\Theta^* f$.]

(v)* *(D. Clark, 1972)* The space $\mathcal{K}_\Theta$ has an orthogonal basis of reproducing kernels $(k_\lambda^\Theta : \lambda \in \Lambda)$, $\Lambda \subset \mathbb{T}$, if and only if there exists a purely atomic Clark measure σ_α.

For example, if the set $\mathbb{T} \backslash s_\Theta$ is at most countable then the family

$$\{k_\lambda^\Theta : \lambda \in s_\Theta, \Theta(\lambda) = \alpha\}$$

is an orthogonal basis in $\mathcal{K}_\Theta$ for every α, $\alpha \in \mathbb{T}$; in particular, this is the case if $\Theta = \Theta_\mu$ is a singular function such that $\text{supp}(\mu)$ is at most countable.

(c) *Existence of bases of reproducing kernels (S. Hruschev, N. Nikolski, and B. Pavlov, 1981)*

(i) Let Θ be an inner function having a purely atomic Clark measure σ_α. Show that there exists a Riesz basis of reproducing kernels $(k_{\lambda_n}^\Theta)_{n \geq 1}$, $\lambda_n \in \mathbb{D}$, in $\mathcal{K}_\Theta$.

[*Hint:* use the (trivial) sufficiency part of (b)(v) to get an orthogonal basis $\{k_{\mu_n}^\Theta : \mu_n \in s_\Theta,\ \Theta(\mu_n) = \alpha, n \geq 1\}$ in $\mathcal{K}_\Theta$; then, set $\lambda_n = r_n \mu_n$ with r_n, $0 < r_n < 1$ sufficiently close to 1 so that

$$\sum_{n \geq 1} \left\| \frac{k_{\mu_n}^\Theta}{\|k_{\mu_n}^\Theta\|} - \frac{k_{\lambda_n}^\Theta}{\|k_{\lambda_n}^\Theta\|} \right\|^2 < 1;$$

using A.5.7.1(f), Volume 1 (see also 4.4.1 above), complete the proof.]

(ii)* There exists a Carleson interpolating sequence $\Lambda \subset \mathbb{D}$ such that $s_B = \emptyset$, where $B = B_\Lambda$ is the Blaschke product whose zero set is Λ; therefore $\mathcal{K}_B$ has a Riesz basis of reproducing kernels $\{k_\lambda^\Theta = k_\lambda : \lambda \in \Lambda\}$ but contains no orthogonal bases of reproducing kernels.

(iii) Show that for every inner function Θ there exists a Blaschke product $B = B_\Lambda$ having a Carleson interpolating sequence as zero set Λ such that the space $\mathcal{K}_{\Theta B} = \mathcal{K}_\Theta \oplus \Theta \mathcal{K}_B$ has a Riesz basis of reproducing kernels. (Observe that $\mathcal{K}_B$ has such a basis too).

[*Hint:* using P. Jones result [**J3**] (see A.3.13, Volume 1, for the statement), find Carleson interpolating Blaschke products B and C such that $\|\Theta - C/B\|_\infty = \|\Theta B - C\| < 1$; then use 4.4.7.]

(d)* *(E. Fricain, 1999)* Let $(k_{\lambda_n}^\Theta)_{n \geq 1}$ be a Riesz sequence in $\mathcal{K}_\Theta$ such that $\sup_{n \geq 1} |\Theta(\lambda_n)| < 1$, let $\delta = \inf_{n \geq 1} |B_{\lambda_n}(\lambda_n)| > 0$ be the Carleson constant of $(\lambda_n)_{n \geq 1}$ and $\gamma = \text{dist}(\Theta \overline{B}, H^\infty)$. Then every sequence $(k_{\lambda_n'}^\Theta)_{n \geq 1}$ satisfying

$$\sup_{n \geq 1} |b_{\lambda_n}(\lambda_n')| < \frac{\delta^6}{8} \cdot \frac{1 - \gamma}{1 + \gamma}$$

is a Riesz sequence in $\mathcal{K}_\Theta$.

4.7.4. Reproducing kernels and the Schur–Nevanlinna parameters. Let Θ be an inner function and let $B = \prod_{n\ge1} b_{\lambda_n}$ be the Blaschke product corresponding to a sequence $\Lambda = (\lambda_n)_{n\ge1}$, $\lambda_n \in \mathbb{D}$, $\sum_{n\ge1}(1-|\lambda_n|) < \infty$. Define the *Schur–Nevanlinna functions* Θ_n and the *Schur–Nevanlinna parameters* γ_n by the formulas

$$\begin{array}{ll} \Theta_1 = \Theta, & \gamma_1 = \gamma_1(\Lambda) = \Theta_1(\lambda_1), \\ \Theta_2 = (b_{\gamma_1} \circ \Theta_1)/b_{\lambda_1}, & \gamma_2 = \gamma_2(\Lambda) = \Theta_2(\lambda_2), \\ \dots, & \dots \\ \Theta_{n+1} = (b_{\gamma_n} \circ \Theta_n)/b_{\lambda_n}, & \gamma_{n+1} = \gamma_{n+1}(\Lambda) = \Theta_{n+1}(\lambda_{n+1}), \\ \dots, & \dots . \end{array}$$

It is clear that $|\gamma_n| \le 1$, and Schwarz' lemma implies that $|\gamma_n| = 1$ if and only if Θ is a Blaschke product of degree $n-1$ (and then $\gamma_k = 0$ for $k > n$). Below, it is supposed that Θ *is not a finite Blaschke product.* Therefore, $|\gamma_n(\Lambda)| < 1$ for every n, $n \ge 1$.

The following results are due to *I. Boricheva (1995, 1999).*

(a)* Let $\sum_{n\ge1} |\gamma_n(\Lambda)| < \infty$. Then

$$\operatorname{dist}(\Theta\overline{B}, H^\infty)^2 \le 1 - \prod_{n\ge1} \frac{1-|\gamma_n|}{1+|\gamma_n|} \le 1 - \exp\left(-\sum_{n\ge1} \frac{2|\gamma_n|}{1-|\gamma_n|}\right) < 1.$$

Therefore, if in addition $\Lambda \in (C)$, then the family $(k^\Theta_{\lambda_j})_{j\ge1}$ is a Riesz sequence in $\mathcal{K}_\Theta$ (see Theorem 4.4.8).

(b)* Let $\sigma = \{\lambda_n : n \ge 1\}$ and let d be the divisor of $(\lambda_n)_{n\ge1}$, i.e., $d(\lambda) = \operatorname{card}\{n : \lambda = \lambda_n\}$, and let $\sigma_\lambda = \{\lambda, \lambda, ..., \lambda\}$, $d(\lambda)$ times.

(i) If $d(\lambda) > 0$, then $\gamma_1(\sigma_\lambda) = \Theta(\lambda)$.

(ii) If $\sup\{|\gamma_n(\sigma_\lambda)| : 1 \le n \le d(\lambda), \lambda \in \sigma\} < 1$ and $\sup\{\sum_{n=1}^{d(\lambda)} |\gamma_n(\sigma_\lambda)| : \lambda \in \sigma\} < \infty$, then

$$\sup_{\lambda\in\sigma} \operatorname{dist}(\Theta\overline{b}_\lambda^{d(\lambda)}, H^\infty) < 1.$$

(iii) If $\sup_{\lambda\in\sigma} \operatorname{dist}(\Theta\overline{b}_\lambda^{d(\lambda)}, H^\infty) < 1$, then $\sup\{\sum_{n=1}^{d(\lambda)} |\gamma_n(\sigma_\lambda)|^2 : \lambda \in \sigma\} < \infty$.

(iv) If $\sup_{\lambda\in\sigma} d(\lambda) < \infty$, then $\sup_{\lambda\in\sigma} \operatorname{dist}(\Theta\overline{b}_\lambda^{d(\lambda)}, H^\infty) < 1$ if and only if

$$\sup\{|\gamma_n(\sigma_\lambda)| : 1 \le n \le d(\lambda), \lambda \in \sigma\} < 1.$$

If, in addition, Θ is a singular inner function, then the latter is equivalent to $\sup_{\lambda\in\sigma} |\Theta(\lambda)| < 1$.

(c)* Let $G = G(\{k^\Theta_{\lambda_j}\}_{j=1}^n)$ be the Gram matrix of reproducing kernels (see C.3.3.1). Then

$$\det(G) = \prod_{j=1}^n \frac{|B_{j-1}(\lambda_j)|}{1-|\lambda_j|} \cdot \prod_{1\le j<l\le n} \frac{|1-\overline{\gamma}_j\Theta_j(\lambda_l)|}{1-|\gamma_j|^2},$$

where $B_i = \prod_{l=1}^i b_{\lambda_l}$ for $i \ge 1$, and $B_0 = 1$.

(d)* Let θ_j, $1 \le j \le n$, be the Schur–Nevanlinna functions associated with Θ and the finite sequence $\lambda_1, ..., \lambda_{k-1}, \lambda_{k+1}, ..., \lambda_n, \lambda_k$, and let

$$\Pi_n(\lambda_k) = \prod_{j=2}^{n} \frac{1 - |\theta_j(\lambda_k|^2}{1 - |\theta_j(\lambda_k) b_{\lambda_{j-1}}(\lambda_j)|^2}.$$

(i) The following distance formula holds.

$$\operatorname{dist}^2_{H^2}\left(\frac{k^\Theta_{\lambda_k}}{\|k^\Theta_{\lambda_k}\|}, \mathcal{L}\mathrm{in}(k^\Theta_{\lambda_j} : 1 \le j \le n, j \ne k)\right) = \left|\frac{B_n}{b_{\lambda_k}}(\lambda_k)\right|^2 \Pi_n(\lambda_k).$$

[Hint: apply (c) and C.3.3.1(g)(ii).]

(ii) The following are equivalent.

1) The family $(k^\Theta_{\lambda_j})_{j\ge1}$ is uniformly minimal.

2) $\Lambda = (\lambda_j)_{j\ge1} \in (C)$ and

$$\inf_{k\ge1} \Pi_\infty(\lambda_k) > 0,$$

where Π_∞ is constructed as in (i) for the sequence $\lambda_1, ..., \lambda_{k-1}, \lambda_{k+1}, ...$.

(e)* Assume that $\Lambda = (\lambda_n)_{n\ge1}$ is a sequence of distinct points. The following are equivalent.

1) The family $(k^\Theta_{\lambda_j})_{j\ge1}$ is *not* minimal.

2) $\mathrm{span}(k^\Theta_{\lambda_j} : j \ne k) = \mathcal{K}_\Theta$ for every k, $k \ge 1$.

3) $\Theta\overline{B} + H^\infty$ is an extreme point in L^∞/H^∞.

4.7.5. Frames. A sequence of vectors $(x_n)_n$ in a Hilbert space H is called a *frame* if there exist positive constants c, C such that

$$c^2\|x\|^2 \le \sum_n |(x, x_n)|^2 \le C^2\|x\|^2$$

for every $x \in H$. For a RKHS H on a set $\Omega \subset \mathbb{C}$, a subset $\Lambda = (\lambda_n)_n \subset \Omega$ is a *sampling set* if the reproducing kernels $x_n = k_{\lambda_n}$ form a frame.

(a) *(E. Whittacker (1915), V. Kotelnikov (1933))* Let PW_τ be the *Paley–Wiener space* of entire functions of exponential type $\le \tau$ whose restriction to $\mathbb{R}$ belongs to $L^2(\mathbb{R})$ endowed with the norm induced by $L^2(\mathbb{R})$. Show that

1) $(\mathrm{Sinc}(\frac{\tau x}{\pi} - n))_{n\in\mathbb{Z}}$ is an orthonormal basis in PW_τ and

$$f = \sum_{n\in\mathbb{Z}} f\left(\frac{n\pi}{\tau}\right) \cdot \mathrm{Sinc}\left(\frac{\tau x}{\pi} - n\right)$$

for every $f \in PW_\tau$, where

$$\mathrm{Sinc}(t) = \frac{\sin(\pi t)}{\pi t}, \quad t \in \mathbb{C};$$

2) $\Lambda = \frac{\pi}{\tau}\mathbb{Z}$ is a sampling set for PW_τ; in fact,

$$\|f\|^2 = \frac{\pi}{\tau}\sum_{n\in\mathbb{Z}} \left|f\left(\frac{n\pi}{\tau}\right)\right|^2$$

for every $f \in PW_\tau$.

[*Hint:* by Paley–Wiener's theorem A.6.6.2, Volume 1, $PW_\tau = \mathcal{F}L^2(-\tau,\tau)$; write $f = \mathcal{F}g$, where $g \in L^2(-\tau,\tau)$, expand in Fourier series

$$g = \sum_{n\in\mathbb{Z}} \sqrt{\frac{\pi}{\tau}} \cdot f\Big(\frac{n\pi}{\tau}\Big) \cdot \frac{1}{\sqrt{2\tau}} e^{in\pi x/\tau}$$

and observe that $\mathcal{F}\Big(\frac{1}{\sqrt{2\tau}} e^{in\pi x/\tau}\chi_{(-\tau,\tau)}\Big) = \sqrt{\frac{\tau}{\pi}} \cdot \mathrm{Sinc}(\frac{\tau x}{\pi} - n)$.]

(b) *(K. Seip, 1995)* (i) Let $\Lambda \subset \mathbb{R}$. Show that the embedding

$$\sum_{\lambda\in\Lambda} |f(\lambda)|^2 \le M^2\|f\|^2, \quad \forall f \in PW_\tau$$

holds if and only if the set Λ is a finite union of sparse (separated) sets, i.e., sets satisfying $|\lambda - \lambda'| \ge \delta > 0$ for $\lambda \ne \lambda'$.

[*Hint:* for the "if" part, use $\mathcal{F}L^2(0,2\tau) = \mathcal{K}_{\Theta_{2\tau}} \subset H^2(\mathbb{C}_+)$, 3.3.4(f) and the remark after Theorem 4.5.6 (applied to $\Lambda + i\tau$).]

(ii)* Every sampling set $\Lambda \subset \mathbb{R}$ for PW_τ contains a separated sampling subset.

(c)* *(K. Seip, 1995)* Let $\Lambda \subset \mathbb{R}$; the *lower Beurling density* of Λ is defined by

$$D^-(\Lambda) = \sup\nolimits_{\Lambda'\subset\Lambda} d^-(\Lambda'),$$

where $d^-(\Lambda') = \lim_{r\to\infty}(r^{-1}\inf\{\mathrm{card}(\Lambda' \cap (a,b)) : a,b\in\mathbb{R}, b-a=r\})$ and Λ' runs over all separated subsets of Λ.

(i) If $(e^{i\lambda x}\chi_{(-\pi,\pi)} : \lambda \in \Lambda)$ is a frame in $L^2(-\pi,\pi)$ then Λ is a finite union of separated sets and $D^-(\Lambda) \ge 1$.

(ii) If Λ is a finite union of separated sets and $D^-(\Lambda) > 1$ then $(e^{i\lambda x}\chi_{(-\pi,\pi)} : \lambda \in \Lambda)$ is a frame in $L^2(-\pi,\pi)$.

(d)* *(K. Seip, 1995)* Let $\Lambda = \{n(1-|n|^{-1/2}) : n \in \mathbb{Z}\backslash\{0\}\}$. Then $(e^{i\lambda x}\chi_{(-\pi,\pi)} : \lambda \in \Lambda)$ is a frame in $L^2(-\pi,\pi)$ but there exists no $\Lambda' \subset \Lambda$ such that $(e^{i\lambda x}\chi_{(-\pi,\pi)} : \lambda \in \Lambda')$ is a Riesz basis in $L^2(-\pi,\pi)$.

4.7.6. More about completeness of exponentials and reproducing kernels. Let Θ be an inner function, and let $\Lambda = (\lambda_n)_{n>1}$ be a sequence of distinct points in $\mathbb{D}$.

(a) Show that $\mathrm{span}(k^\Theta_{\lambda_n} : n \ge 1) = \mathcal{K}_\Theta$ if and only if either $\sum_{n\ge1}(1-|\lambda_n|) = \infty$, or $\mathrm{Ker}\, T_{B\overline{\Theta}} = \{0\}$, where $B = B_\Lambda$ is a Blaschke product having the zero set Λ.

[*Hint:* observe that $\mathrm{span}(k^\Theta_{\lambda_n} : n \ge 1) = \mathrm{clos}(P_\Theta \mathcal{K}_B)$, and hence $\mathrm{span}(k^\Theta_{\lambda_n} : n \ge 1) = \mathcal{K}_\Theta$ if and only if $\mathrm{Ker}(P_B|\mathcal{K}_\Theta) = \{0\}$; by Lemma 4.4.4, the latter equality is equivalent to $\mathrm{Ker}\, T_{B\overline{\Theta}} = \{0\}$.]

(b) *(E. Fricain, 1999)* Suppose $\mathrm{span}(k^\Theta_{\lambda_n} : n \ge 1) = \mathcal{K}_\Theta$ and $\sum_{n\ge1}|b_{\lambda_n}(\lambda'_n)| < \infty$. Show that $\mathrm{span}(k^\Theta_{\lambda'_n} : n \ge 1) = \mathcal{K}_\Theta$.

[*Hint:* assume that there exists $f \in \mathcal{K}_\Theta$, $f \ne 0$, such that $f(\lambda'_n) = 0$, $n \ge 1$; setting $f_0 = f$ and

$$f_n = \frac{b_{\lambda'_n} - b_{\lambda'_n}(\lambda_n)}{b_{\lambda'_n}} f_{n-1}, \quad n \ge 1,$$

observe that $\|f_n - f_{n-1}\| = |b_{\lambda'_n}(\lambda_n)| \cdot \|f_{n-1}\|$, and hence

$$(1 - |b_{\lambda'_n}(\lambda_n)|)\|f_{n-1}\| \le \|f_n\| \le (1 + |b_{\lambda'_n}(\lambda_n)|)\|f_{n-1}\|,$$

$$\|f\| \cdot \prod_{n=1}^{m}(1 - |b_{\lambda'_n}(\lambda_n)|) \le \|f_m\| \le \|f\| \cdot \prod_{n=1}^{m}(1 + |b_{\lambda'_n}(\lambda_n)|);$$

now, since $\|f_m\|$ are bounded from above and from below, it follows that

$$\|f_{n+p} - f_n\| \le \sum_{k=1}^{p} \|f_{n+k} - f_{n+k-1}\| \le c\|f\| \sum_{k=1}^{p} |b_{\lambda'_{n+k}}(\lambda_{n+k})|$$

and hence the sequence $(f_n)_{n\ge 1}$ is norm convergent; clearly, $0 \ne g := \lim_n f_n \in \mathcal{K}_\Theta$ and $g(\lambda_n) = 0$ for every n, $n \ge 1$; this contradiction completes the proof.]

(c) *(W. Alexander and R. Redheffer, 1967)* Let $a > 0$; $\lambda_n, \lambda'_n \in \mathbb{C}$ and $\mathrm{span}(e^{i\lambda_n x}\chi_{(0,a)} : n \ge 1) = L^2(0,a)$ and

$$\sum_{n\ge 1} \frac{|\lambda_n - \lambda'_n|}{1 + |\operatorname{Im}\lambda_n| + |\operatorname{Im}\lambda'_n|} < \infty.$$

Show that $\mathrm{span}(e^{i\lambda'_n x}\chi_{(0,a)} : n \ge 1) = L^2(0,a)$.

[*Hint (for the case $\lambda_n, \lambda'_n \in \mathbb{C}_+$):* pass to $\lambda_n + i$, $\lambda'_n + i$ (without loss of generality) and apply (a) for the case $\Theta_a = e^{iaz}$.]

(d)* *Completeness of biorthogonal sequences.*

(i) *(E. Fricain, 1995)* Let Θ be an inner function in $\mathbb{C}_+$ such that $\Theta' \in L^\infty(\mathbb{R})$, and let $\lambda_n \in \mathbb{C}_+$ be such that $(k^\Theta_{\lambda_n} : n \ge 1)$ is a complete minimal sequence in $\mathcal{K}_\Theta$. Then the sequence biorthogonal to $(k^\Theta_{\lambda_n} : n \ge 1)$ in $\mathcal{K}_\Theta$ is also complete.

(ii) *(R. Young (1981), Yu. Lyubarskii (1996))* Let $a > 0$ and $\lambda_n \in \mathbb{C}$ be such that the family $(e^{i\lambda_n x}\chi_{(0,a)} : n \ge 1)$ is minimal and complete in $L^2(0,a)$. Then the sequence biorthogonal to $(e^{i\lambda_n x}\chi_{(0,a)} : n \ge 1)$ is also complete in $L^2(0,a)$.

(e)* *The Beurling–Malliavin formula for the completeness radius.*

Let $d : \mathbb{C} \longrightarrow \mathbb{Z}_+$ be a divisor. The *completeness radius* of d is

$$\rho(d) = \sup\left\{\tau : \mathrm{span}(x^k e^{i\lambda x}\chi_{(-\tau,\tau)} : 0 \le k < d(\lambda), \lambda \in \mathbb{C}) = L^2(-\tau,\tau)\right\};$$

by definition, $\rho\{\emptyset\} = 0$. The case of pure exponentials $(e^{i\lambda x} : g \in \Lambda)$ corresponds to $d = \chi_\Lambda$, and we write

$$\rho(d) = \rho(\Lambda).$$

(i) Let $\sum_{\lambda\in\mathbb{C}\setminus\{0\}} d(\lambda)|\operatorname{Im}\frac{1}{\lambda}| = \infty$. Show that $\rho(d) = \infty$.

[*Hint:* if $\Lambda = \{\lambda : d(\lambda) > 0\}$ has a finite accumulation point then $\hat{f}(\lambda) = 0$, $\lambda \in \Lambda$, implies $f = 0$, where $f \in L^2(I)$ (see 4.7.1(b) for the notation); if Λ is a sequence tending to ∞, and, say, $\sum_{\operatorname{Im}\lambda>0} d(\lambda)|\operatorname{Im}\frac{1}{\lambda}| = \infty$ then $(x^k e^{i\lambda x}\chi_{\mathbb{R}_+} : 0 \le k < d(\lambda), \operatorname{Im}\lambda > 0)$ is complete in $L^2(\mathbb{R}_+)$ (the Blaschke condition and the Paley–Wiener theorem A.6.3.5, Volume 1), and hence $\mathrm{span}(x^k e^{i\lambda x}\chi_{(0,a)} : 0 \le k < d(\lambda), \lambda \in \mathbb{C}) = L^2(0,a)$ for every $a > 0$.]

(ii) *(A. Beurling and P. Malliavin, 1967)* Let $\sum_{\lambda\in\mathbb{C}\backslash\{0\}} d(\lambda)|\operatorname{Im}\frac{1}{\lambda}| < \infty$. Define

$$d_*(x) = \sum_{1/x=\operatorname{Re}(1/\lambda)} d(\lambda)$$

for $x \in \mathbb{R}\backslash\{0\}$ and $d_*(0) = 0$, $d_*|(\mathbb{C}\backslash\mathbb{R}) = 0$. Then $\rho(d) = \rho(d_*)$.

In particular, if $\Lambda \subset \mathbb{C}$ such that any circle $\operatorname{Re}(1/z) = 1/\lambda_*$, where $\lambda_* \in \mathbb{R}\backslash\{0\}$, contains at most one point of Λ and $\Lambda_* = \{|\lambda|^2/\operatorname{Re}\lambda : \lambda \in \Lambda, \operatorname{Re}\lambda \neq 0\}$, then $\Lambda_* \subset \mathbb{R}$ and $\rho(\Lambda) = \rho(\Lambda_*)$.

(iii) *(A. Beurling and P. Malliavin, 1967)* Let d be a divisor, $\operatorname{supp}(d) \subset \mathbb{R}$, and

$$N_d(x) = \sum_{0\le y\le x} d(y) \text{ for } x \ge 0, \quad N_d(x) = -\sum_{x\le y\le 0} d(y) \text{ for } x < 0$$

be the counting (distribution) function of d (compare with N_Λ as defined in Theorem 4.5.6), and assume that there exists $R \ge 0$ such that

$$\int_{\mathbb{R}} \frac{|N_\Lambda(x) - Rx|}{1+x^2} dx < \infty.$$

Then $\rho(\Lambda) = \pi R$.

(iv) *(A. Beurling and P. Malliavin, 1967)* Let d be a divisor, $\operatorname{supp}(d) \subset \mathbb{R}$, and for a given $a > 0$, let M_a be the class of continuous piecewise differentiable functions φ satisfying $0 \le \varphi'(x) \le a$ at points where the derivative exists. Let

$$B(d) = \inf\left\{a > 0 : \exists\varphi \in M_a \text{ s.t. } \int_{\mathbb{R}} \frac{|N_d(x) - \varphi(x)|}{1+x^2} dx < \infty\right\}.$$

Then

$$\rho(d) = \pi B(d).$$

Consequently, if d and d' are two real divisors and

$$\int_{\mathbb{R}} \frac{|N_d(x) - N_{d'}(x)|}{1+x^2} dx < \infty,$$

then $\rho(d) = \rho(d')$.

(v) *(R. Redheffer, 1967)* Let d be a divisor and let $\Lambda = (\lambda_n)_n$ be the corresponding sequence of complex numbers, where each $\lambda \in \mathbb{C}\backslash\{0\}$ is repeated $d(\lambda)$ times. Let

$$C(d) = \inf\left\{c > 0 : \sum_n \left|\frac{1}{\lambda_n} - \frac{c}{m_n}\right| < \infty, m_n \text{ are distinct integers}\right\},$$

where $\inf\{\emptyset\} = \infty$. Then $\rho(d) = \pi C(d)$.

Consequently, $\sum_n \left|\frac{1}{\lambda_n} - \frac{1}{\lambda'_n}\right| < \infty$ implies that $\rho(\Lambda) = \rho(\Lambda')$.

4.8. Notes and Remarks

Bases of exponentials. The problem of expanding a given function f in a series of exponentials with given frequencies λ_n

$$f = \sum_n a_n e^{i\lambda_n z}$$

is one of the basic problems of harmonic analysis. The domain and the meaning of the convergence depend on the origin of such a series and may vary considerably. For example, remember the Euler rule for the general solution of a differential equation $p(\frac{d}{dx})f = 0$ and its generalizations by J. Delsart and L. Schwartz for mean periodic functions satisfying a distributional convolution equation $S * f = 0$, in one or in several variables. In particular, as mentioned in 3.1.4 above, the set of frequencies $\Lambda = (\lambda_n)$ appears as the zero set of the characteristic equation $\hat{S}(z) = 0$ (as common zeros of the corresponding ideal for the case of a system of equations). We refer to general surveys where further references to the immense literature can be found: V. Havin [**Hav**], G. Mackey [**Mac**], N. Nikolski [**N8**].

The interest in exponentials on intervals of the real line $\mathbb{R}$, primary, with real frequencies and then with complexe ones, was stimulated by problems arising from signal processing and filtering theory. First systematic studies were initiated by N. Wiener and R. Paley, see [**Wi1**], [**Wi4**], [**PW**] and P. Masani [**Mas**]. Two fundamental ideas were developed: first, to consider frequencies $\Lambda = (\lambda_n)$ that are sufficiently close to the classical harmonic grid $\Lambda_a = \frac{2\pi}{a}\mathbb{Z}$ generating a periodic analysis, and secondly, to derive properties of $(e^{i\lambda_n z} : \lambda_n \in \Lambda)$ from analytic properties of the "generating function" F_Λ,

$$F_\Lambda(z) = e^{bz} \prod_n (1 - \frac{z}{\lambda_n}) e^{z/\lambda_n},$$

which plays the role of the characteristic function $\hat{S}$ mentined above (really, N. Wiener did not use the term "generating function"); see Chapters 6 and 7 of [**PW**].

Important impacts to this field were made by N. Levinson [**LeN**] and B. Levin [**Le2**]. In the latter paper, the concepts of generating functions and sine type functions were introduced, see 4.7.1(b) above for the definitions. Many interesting results, partially represented in Section 4.7, were obtained by these methods. The next advancement of analytic techniques was proposed by B. Pavlov [**P3**] by using Hilbert space geometry of subspaces generated by exponentials and introducing such substantial tools of harmonic analysis as Hardy classes and Muckenhoupt weights. These techniques were transformed into the language of weighted Riesz projections (S. Hruschev [**Hr**]) and Toeplitz and Hankel operators (N. Nikolski [**N16**], [**N19**]); see S. Hruschev, N. Nikolski and B. Pavlov [**HNP**] for a synthetic exposition. This geometric-analytic approach also works for vector valued exponentials, see S. Avdonin and S. Ivanov [**AI2**] for further references.

Also, nonharmonic Fourier and Dirichlet series were systematically treated in R. Young [**Y1**], J. Higgins [**Hig1**], S. Mandelbrojt [**Man1**], [**Man3**]. For the analysis of series of exponentials originated from convolution equations and partial differential equations we refer to L. Schwartz [**Schw1**], [**Schw3**], L. Ehrenpreis [**Ehr**], J.-P. Kahane [**Kah1**], A. Leontiev [**Leo1**], [**Leo2**], L. Hörmander [**Hör2**], A. Sedletskii [**Sed**].

Bases of reproducing kernels. The problem of convergence or summation of series

$$\sum_n a_n k_{\lambda_n}$$

of reproducing kernels of a RKHS H on a set Ω is a dual form of attempts to describe the restriction spaces $H|\Lambda$ onto (discrete) subsets Λ of Ω. For a few general facts about this duality we refer to N. Nikolski [**N19**], [**N8**], [**N15**]. In a sense, bases of reproducing kernels of model spaces, i.e., $(k^\Theta_{\lambda_n})$ for an inner function Θ, form a natural framework even for the particular case of exponentials, where $\Theta = e^{iaz}$. Also, there is a number of applied problems (in Regge scattering), where the problem of bases $(k^\Theta_{\lambda_n})$ appears in all its generality (in an equivalent form of joint bases of eigenvectors $(\Theta B/(\lambda_n - z))$ of a model operator $M_{\Theta B}$ and those, (k_{λ_n}), of its adjoint $M^*_{\Theta B}$), see [**HNP**], N. Nikolski and S. Hruschev [**HrN**] for details. The same type of bases is useful for some problems arising in the framework of L. de Branges spaces of entire functions, see L. de Branges [**dB2**] and A. Baranov [**Bara**].

It is worth mentioning that there is no analog of the generating function method for exploring the geometry of reproducing kernels $(k^\Theta_{\lambda_n})$ for a general inner function Θ. It was replaced in N. Nikolski [**N16**] by an analysis of the corresponding Toeplitz operators $T_{\overline{\Theta}B}$ and $T_{\Theta\overline{B}}$. As for exponentials, all depends on the behaviour of the argument $\arg(\overline{\Theta}B)$ on the unit circle (or on the line) but the roles of Θ (i.e., of the space $\mathcal{K}_\Theta$) and B (i.e., of the frequency set) are explicitely separated. For more about properties of kernels k^Θ_λ see [**N19**], [**HNP**], and E. Fricain [**Fri2**], as well as the comments in the *Formal credits* below.

Frames, signal processing, embedding theorems and equivalent norms. The utility of frames and bases for economical signal transmissions was understood by the 1930's due to E. Whittacker [**Whi1**] and V. Kotelnikov [**Kot**]. Later on, it was rediscovered by C. Shannon [**Shan**], see also C. Shannon and W. Weaver [**ShanW**]. For the meaning of frames, bases and interpolation formulas in electrical engineering we refer to Ya. Hurgin and V. Yakovlev [**HY**], J. Higgins [**Hig2**], R. Marks [**Mark**]. For instance, for the space PW_τ (see 4.7.5 above), the sampling property of $\Lambda = (\lambda_n)$, means, roughly speaking, the stability of the transmission of a signal whose spectrum is located in $[-\tau, \tau]$ making use of its sample values $(f(\lambda_n))$ only. For a survey on frames theory see O. Christensen [**Chr**], where Riesz bases are also presented but in a rather incomplete way.

It is clear from the definition (see 4.7.5 above) that the concept of frames is closely related to embedding theorems of Carleson type as presented in C.3.2.23, C.3.3.3, C.3.3.4, C.3.4 and A.5.7.2 (Volume 1). However, theorems of this type become technically much more complicated when the embeddings are restricted to the model spaces $\mathcal{K}_\Theta$, (e.g., $\mathcal{K}_\Theta \subset L^2(\mu)$) than they are for the case of the entire Hardy spaces (e.g., $H^2 \subset L^2(\mu)$). Although the problem of describing those measures μ in $\operatorname{clos}\mathbb{D}$ for which $\mathcal{K}^p_\Theta \subset L^p(\mu)$ is still open in its full generality (2001), many interesting results are obtained, especially by W. Cohn, A. Aleksandrov, S. Treil, A. Volberg, K. Dyakonov, X. Zhong. In particular, for the so-called "one-component" inner functions Θ (the level set $\{z : |\Theta(z)| < \varepsilon\}$ is connected for some

$0 < \varepsilon < 1$), the embedding follows from the reproducing kernel test (RKT)

$$\int_{\overline{\mathbb{D}}} |k_\lambda^\Theta(z)|^2 \, d\mu(z) \le C^2 \|k_\lambda^\Theta\|_2^2 = C^2 \frac{1 - |\Theta(\lambda)|^2}{1 - |\lambda|^2}, \quad \lambda \in \mathbb{D},$$

(W. Cohn, [**Cohn**]), but for general Θ this is not the case (A. Volberg, [**V6**]). We refer to A. Aleksandrov [**Aleks8**], [**Aleks7**] and K. Dyakonov [**Dya3**] for recent results, surveys and further references. For the special case of the spaces PW_τ (see 4.7.5 above), corresponding to $\Theta = \Theta_\tau = e^{2i\tau z}$, much more is known due to B. Paneyah, V. Lin, V. Logvinenko and Yu. Sereda, B. Levin, V. Katsnel'son, E. Gorin, V. Havin and B. Jöricke, A. Volberg. We refer to V. Havin and B. Jöricke [**HJ**] and A. Volberg [**V2**] for advanced results, surveys and further references.

The completeness problem. The problem of describing subsets $\Lambda \subset \mathbb{C}$ such that

$$\operatorname{span}_{L^p(I)} \left(e^{i\lambda x} \chi_I : \lambda \in \Lambda \right) = L^p(I)$$

for a given interval $I \subset \mathbb{R}$ is still open in spite the enormous efforts of several generations of analysts and numerous impressive achievements. By duality, this is nothing but the problem of the uniqueness sets for the space of Fourier transforms $\mathcal{F}L^{p'}(I)$, where $\frac{1}{p'} + \frac{1}{p} = 1$. Of course, the natural framework for this problem is to consider an arbitrary Borel measure μ in $\mathbb{C}$ and to try to describe those divisors $d : \mathbb{C} \longrightarrow \mathbb{Z}_+ = \mathbb{Z} \cap [0, \infty)$ for which $z^k e^{i\lambda z} \in L^p(\mu)$ for $0 \le k < d(\lambda)$ and

$$\operatorname{span}_{L^p(\mu)} \left(z^k e^{i\lambda z} : 0 \le k < d(\lambda), \lambda \in \mathbb{C} \right) = L^p(\mu).$$

In this setting, the problem includes the problem of completeness of polynomials considered in Sections A.4.8 and A.4.9 (Volume 1). For the Hilbert space case, the reproducing kernel analogs are also of great interest: given a RKHS H on a set Ω, describe $\Lambda \subset \Omega$ such that $H = \operatorname{span}(k_\lambda : \lambda \in \Lambda)$. In the case $H \subset \operatorname{Hol}(\Omega)$, the problem should include multiplicities,

$$\operatorname{span}\left(\frac{d^j}{d\overline{\lambda}^j} k_\lambda : 0 \le j < d(\lambda), \lambda \in \Omega\right) = H.$$

There is a vast literature on completeness of exponentials, and a more or less complete survey of the field would require tens of pages. We limit ourselves to listing a few monographs and fundamental survey papers: R. Paley and N. Wiener [**PW**], N. Levinson [**LeN**], L. Schwartz [**Schw1**], [**Schw2**], A. Akhiezer [**Akh3**], B. Levin [**Le1**], S. Mandelbrojt [**Man1**], [**Man2**], R. Young [**Y1**], J. Higgins [**Hig1**], A. Beurling and P. Malliavin [**BeuM2**], R. Redheffer [**Red3**], P. Koosis [**Ko5**], [**Ko6**], [**Ko1**], V. Havin and B. Jöricke [**HJ**]. The most significant results are due to N. Levinson [**LeN**] and A. Beurling and P. Malliavin [**BeuM2**]. Theorem 4.6.1 is one of the simplest sample of a series of results by Levinson [**LeN**], but for its proof several typical ingredients of the field are required: distribution (counting) functions of sequences, Jensen or Carleman formulas for zeros of entire functions, infinite Hadamard products, delicate asymptotic properties of Fourier transforms. The more for the proof of Beurling–Malliavin's completeness radius formula [**BeuM2**] quoted in 4.7.6(e). The main ingredients of the proof are the so-called "small multiplier" theorem by the same authors [**BeuM1**] and a careful analysis of "densities" of measures on $\mathbb{R}$. P. Koosis's book [**Ko6**] is especially devoted to a presentation of these results. Many interesting papers deal with interpretations and simplifications of some points of the proof, see J.-P. Kahane [**Kah3**], J. Korevaar [**Kore**], R. Redheffer [**Red3**], I. Krasichkov–Ternovskii [**Kra1**].

The nature of the completeness problem for reproducing kernels depends on the corresponding RKHS and may vary considerably, even if we restrict ourselves to the case of holomorphic spaces H on an open set Ω such that $H \subset \mathrm{Hol}(\Omega)$. In this case, the completeness is nothing but the uniqueness property: $(f \in H$, $0 = (f, k_\lambda) = f(\lambda)$, $\lambda \in \Lambda) \Rightarrow f = 0$. Every book on complex analysis contains material on uniqueness theorems for one or the other (Hilbert) space of holomorphic functions. Therefore, it is no matter to try to give some general review of the field. However, the function-operator flavour of kernels k_λ^Θ of the model spaces is shared with some other spaces such as, for instance, weighted Bergman spaces and their duals. For this reason we quote here the books M. Djrbashian and F. Shamoyan [**DjSh**] and H. Hedenmalm, B. Korenblum and K. Zhu [**HKZ**].

Very few facts are known about completeness of the families $(k_\lambda^\Theta : \lambda \in \Lambda)$ in the spaces $\mathcal{K}_\Theta^p = H^p \cap \Theta H_-^p$ beyond the exponential case $\Theta = e^{i\tau z}$. Following 4.7.6(a), the completeness problem is reduced to kernels of Toeplitz operators $\mathrm{Ker}\, T_{B\overline{\Theta}}$; see references and comments about these kernels in Section B.4.8 (Volume 1). Also, some information on completeness of families $(k_\lambda^\Theta : \lambda \in \Lambda)$ in the spaces $\mathcal{K}_\Theta^p$ can be found in N. Nikolski [**N19**], S. Hruschev, N. Nikolski and B. Pavlov [**HNP**], N. Nikolski and S. Hruschev,[**HrN**] and E. Fricain [**Fri2**].

Formal credits. Theorem 4.1.1 is proved in R. Paley and N. Wiener [**PW**] for $\delta < \pi^{-2}$ instead of $\delta < \log 2/\pi$; the latter improvement of this sufficient condition, as well as the proof presented in the text, can be found in R. Duffin and J. Eachus [**DuE**]. For Theorem 4.1.2, the proofs of both the implication (1) $\Rightarrow$ (2) (A. Ingham [**In**]) and the implication (2) $\Rightarrow$ (1) (M. Kadec [**Kad**]) differ from the original ones; the second one, given in 4.5.7, follows V. Vasyunin and is taken from N. Nikolski [**N19**]. Ingham's proof is cited in [**LeN**] and [**HNP**]; it consists in showing that a function which is orthogonal to $e^{i\mu_n x}$ *should be* $(\cos(x/2))^{-1/2}$, but it is not in $L^2(-\pi, \pi)$.

In Section 4.2 we follow [**N19**]. Theorem 4.2.4 is proved by N. Nikolski and B. Pavlov (1970, without giving the sharp value N of Carleson series) and V. Vasyunin (1976), see [**N19**] for a discussion and original references. The essential part of Theorem 4.2.5 is due to S. Treil [**Tre4**], [**Tre9**].

Section 4.3, as a whole, follows [**N19**] and [**HNP**]. See the paragraph "Bases of exponentials" above for a few general remarks on the projection method employed there. It is clear from 4.3.1(3) that the Carleson condition $(\mu_n + iy)_{n \geq 1} \in (C)$ holds for sufficiently big $y > 0$ for any uniformly minimal family $(e^{i\mu_n x}\chi_{(0,a)} : n \geq 1)$ in $L^2(0, a)$ satisfying

$$\inf_n \mathrm{Im}\, \mu_n > -\infty.$$

Lemma 4.4.2 is proved in N. Nikolski [**N19**]; it is strengthened in 4.7.4(d)(ii) (I. Boricheva [**Bori2**]). Lemma 4.4.4 is from [**N19**], [**N18**]; Theorems 4.4.6 and 4.4.8 are from N. Nikolski [**N16**], see also [**N19**], [**HNP**].

The comment 4.4.9(2) can be compared with 4.7.3(a); the properties 4.4.9(3)-4.4.9(10) improve several propositions from [**N19**], [**HNP**] and some others papers.

Section 4.5 is written following N. Nikolski [**N19**] and S. Hruschev, N. Nikolski and B. Pavlov[**HNP**]. In particular, Theorem 4.5.6 is proved in [**HNP**]; it gives a criterion for being a Riesz basis expressed in the language of the "signed" counting (distribution) function $N_{(\mu_n)}$ traditional for this field (see N. Levinson [**LeN**]). The proof of Kadec's theorem 4.1.2, as is mentioned, is taken from [**N19**] (following an idea of V. Vasyunin); yet another proof is given in [**HNP**].

The proof of the Levinson completeness theorem 4.6.1 employs the symmetric counting function $n_{\mathcal{M}}$. It is adapted from N. Levinson [**LeN**] and keeps a flavour of the classical theory of entire functions. Other presentations of completeness theorems can be found in B. Levin [**Le1**], R. Boas [**Bo1**], P. Koosis [**Ko5**], V. Havin and B. Jöricke [**HJ**]. Corollary 4.6.3 and Remarks 4.6.4 can also be found in [**LeN**].

The criteria for being a Riesz basis in $L^2(0,a)$, as given in 4.7.1(a) and 4.7.1(b), complete the picture traced in Theorem 4.5.6 (and, in fact, their proofs should preceed the latter theorem). It should be noted that 4.7.1(b) comes directly from B. Pavlov's original paper [**P3**], but 4.7.1(a) is found in [**HNP**]. Sine type functions are introduced in B. Levin [**Le2**]; later on, they proved to be a useful and flexible tool in harmonic analysis. Several references for other results in this direction can be found in [**HNP**], [**N19**], [**HrN**]. The final criterion 4.7.1(e) for Riesz bases of exponentials is due to A. Minkin [**Min1**], [**Min2**]. Its proof is based on the same techniques as explained above.

Stability properties of bases or complete families of exponentials represent a classical subject in harmonic analysis starting from Paley–Wiener's result 4.1.1. See also the completeness results presented in 4.7.6. The proofs of 4.7.1(c) (i)-(iv) can be found in [**HNP**]. The original reference for S. Avdonin's useful theorem 4.7.1(c) (iv) is [**Av**].

The complementation and elimination (or, deficiency and excess) problems for families of exponentials or, more generally, reproducing kernels were raised in N. Nikolski [**N19**] and S. Hruschev, N. Nikolski, and B. Pavlov [**HNP**]. Namely, the problem is whether a given Riesz sequence $(k_\lambda^\Theta : \lambda \in \Lambda)$ (respectively, a frame) can be completed up to a Riesz basis $(k_\lambda^\Theta : \lambda \in \Lambda \cup \Lambda')$ in $\mathcal{K}_\Theta$ (respectively, can be reduced to a Riesz basis $(k_\lambda^\Theta : \lambda \in \Lambda \backslash \Lambda')$ in $\mathcal{K}_\Theta$). First results (due to S. Vinogradov and V. Vasyunin) are contained in [**HNP**] (as, for example, Theorem 4.7.1(d) (i)), but more complete ones appeared 15 years later in K. Seip [**Sei3**] (see also [**Sei1**], [**Sei2**]). In particular, K. Seip proved that most of the known sufficient conditions for being a Riesz sequence (or a frame) which are based on various density criteria lead to families of exponentials complementable (respectively, reducible) to a basis in $\mathcal{K}_\Theta$. However, in general, this is not the case, see the examples 4.7.1(d)(ii) and 4.7.5(d). Up to now, no criterion of complementability or reducibility (in the above sens) is known.

Subsection 4.7.2 deals with exponentials in $L^2(\mathbb{R}_+)$ and, therefore, contains complementary results to the main criterion given in Chapter C.3: $(e^{i\mu_n x}\chi_{\mathbb{R}_+} : n \geq 1)$ is a Riesz sequence in $L^2(\mathbb{R}_+)$ if and only if $(\mu_n)_{n\geq 1} \in (C)$. Asymptotically orthogonal families $(e^{i\mu_n x}\chi_{\mathbb{R}_+} : n \geq 1)$, as presented in 4.7.2(d), were studied by A. Volberg [**V3**]. The vanishing Carleson condition from 4.7.2(d) (i) has already appeared in C.3.3.5(b) (iv); for some interesting properties of such sequences see C. Sundberg and Th. Wolff [**SW**]. The stability properties of the above asymptotic orthogonality as claimed in 4.7.2(d) (ii) are from E. Fricain [**Fri2**], [**Fri3**]. Another kind of asymptotic sparseness of frequencies, presented in 4.7.2(e), is studied by P. Koosis [**Ko1**] and P. Lax [**Lax1**]. Under the condition of 4.7.2(e), Toeplitz operators

$T_{\overline{B}\Theta_a}$, responsible for the basis property of $(e^{i\lambda_n x}\chi_{(0,a)} : n \geq 1)$ (see Theorems 4.4.6 and 4.4.8), are unitary modulo compact operators, which means that exponentials restricted to finite intervals are "compact close" to full exponentials $(e^{i\lambda_n x}\chi_{\mathbb{R}_+} : n \geq 1)$.

The results of 4.7.2(a) and (b) on completeness and bases properties of families of powers are obtained by simple changes of variable from the corresponding results for exponentials. Of course, the original methods of C. Müntz [**Mün**] and O. Szasz [**Sza**] (presented in C.3.3.1(i)) were different. Actually, similar theorems hold for every $L^p(0,1)$ space, $1 \leq p < \infty$ and for the space $C[0,1]$; for these and other generalizations see N. Akhiezer [**Akh3**] and R. DeVore and G. Lorentz [**DeVL**]. The unconditional bases theorem of V.I. Gurarii and V. Matsaev, 4.7.2(c), is from [**GuV**].

Point 4.7.3(a) contains a few known properties (taken from S. Hruschev, N. Nikolski and B. Pavlov [**HNP**]) of bases of reproducing kernels $(k^{\Theta}_{\lambda_n})_{n\geq 1}$ in a general position, that is, without the hypothesis $\alpha(\Theta, B) = \sup_{n\geq 1} |\Theta(\lambda_n)| < 1$, which is sometimes rather inconvenient. These few facts should be compared with the comments in 4.4.9 and with the role played by the Schur–Nevanlinna parameters $\gamma_n(\Lambda)$ which generalize $\alpha(\Theta, B)$, see 4.7.4. The stability property 4.7.3(d) is from E. Fricain [**Fri1**], [**Fri2**]; in fact, it generalizes and strengthens 4.7.1(c) (ii) and (iii), but unlike the Ingham–Kadec theorem 4.1.2, it is most likely that the constant given in 4.7.3(d) is not sharp.

The essential part of Subsection 4.7.3 is occupied by the so-called *Clark bases*, i.e., *orthogonal* bases of reproducing kernels $(k^{\Theta}_{\lambda_n})_{n\geq 1}$. Such an orthogonality is possible only if λ_n, $n \geq 1$, belong to a subset $s_\Theta \subset \mathbb{T}$ as described in 4.7.3(b). The basic theorems are (iii), (iv) and (v) of 4.7.3(b). All these facts are due to D. Clark [**Cl2**], excepting that, actually, D. Clark proved the existence of the limits $\lim_{r\to 1} f(r\zeta)$ σ_α-a.e. for f from a dense subset of $\mathcal{K}_\Theta$ only. A. Poltoratski [**Pol1**] proved it for all $f \in \mathcal{K}_\Theta$. Poltoratski also obtained many other properties of Clark measures (in several subsequent papers) and found applications of these techniques to perturbation theory of unitary and selfadjoint operators as well; see [**Pol2**] for an account of this field and further references. For a vector valued generalization of these techniques see J. Ball and A. Lubin [**BaL**]. Formally speaking, 4.7.3(b) (i) is from P. Ahern and D. Clark [**AhC**]; 4.7.3(b) (ii) is from D. Clark [**Cl2**] and (independently) D. Georgijevic [**Geo**]; 4.7.3(b) (iii) is due to D. Clark [**Cl2**], A. Poltoratski [**Pol1**] and A. Aleksandrov [**Aleks4**], and 4.7.3(b) (iv) and (v) are proved in D. Clark [**Cl2**].

The existence of bases of reproducing kernels is an interesting unsolved problem; it is discussed in 4.7.3(c). If an inner function Θ is "sufficiently singular", i.e. if there exists an atomic Clark measure σ_α (as for singular functions $\Theta = \Theta_\mu$ with singular measures having an at most countable support $\text{supp}(\mu)$), then such a basis exists (4.7.3(c), [**HNP**]). Points (ii) and (iii), taken from N. Nikolski [**N19**] and [**HNP**], show possible approaches to the general case of the problem.

Point 4.7.3(d) is commented above.

Subsection 4.7.4 is taken from I. Boricheva [**Bori1**], [**Bori2**]. For general information and classical results on the Schur–Nevanlinna procedure we refer to J. Garnett [**Gar**] and M. Bakonyi and T. Constantinescu [**BaC**]; the latter also contains some applications to operators and system theory (see also C. Foiaş and A.

Frazho [**FF**], D. Alpay [**Alp**]). In fact, I. Schur [**Sch2**], motivated by Caratheodory–Féjer and Herglotz interpolation problems and moments problems, considered only the case, where $\lambda_1 = \lambda_2 = ... = 0$. The subsequent history of the Schur algorithm and references can be found in sources quoted above; see also B.3.4.2 and Section B.3.5 (Volume 1). Among other results of 4.7.4, it is worth mentioning point (d) (ii) which, in particular, states that uniform minimality of $(k_{\lambda_n}^{\Theta})_{n\geq 1}$ *always* implies that $(\lambda_n)_{n\geq 1} \in (C)$.

Some general remarks on Subsection 4.7.5 are made above. It is worth mentioning that the proof of the sampling property 2) from 4.7.5(a) follows Kotelnikov's original proof [**Kot**] excepting that he did not know the Plancherel theorem but used the Fourier inversion formula subjected to the Dirichlet sufficient conditions. (It is curious to notice that it was G. Hardy who observed as late as in 1941 that this sampling formula is, actually, an orthogonal expansion). On the contrary, E. Whittacker [**Whi1**], [**Whi2**] did not impose precise restrictions on the expanded functions requiring simply that f "has no rapid oscillations". In practice, sampling nodes are not uniformly distributed (for example, the eye retina has an inhomogeneous distribution of receptors), and in order to get frame decompositions of the corresponding signals you need to treat general families of nodes. The results presented in 4.7.5 (b), (c) and (d) are from K. Seip [**Sei3**] where some other papers are quoted. See also the sources on frame theory quoted above, the book of L. Aizenberg [**Aiz**], and recent papers by K. Seip [**Sei1**], [**Sei2**], Yu. Lyubarskii and K. Seip [**LyS2**], Yu. Lyubarskii and A. Rashkovskii [**LyR**], K. Florens, Yu. Lyubarskii, and K. Seip [**FLS**].

Completeness of exponentials and reproducing kernels is already commented above, including the links between Toeplitz operators and completeness as presented in 4.7.6(a). The theorem on stability of completeness from 4.7.6(b) is taken from E. Fricain [**Fri1**], [**Fri2**]. Results on stability of completeness for exponentials are numerous, and even in our short list, points 4.7.6 (c), (e)(iv) and (e)(v) contain such results. For the result 4.7.6(c) by W. Alexander and R. Redheffer as well as for Theorem 4.7.6(e)(v) by R. Redheffer, we refer to R. Redheffer [**Red3**] which gives a complete survey on the completeness problem until 1977. In a more general framework, stability of completeness, as well as its companion - approximation with constrained coefficients of approximating aggregates, was studied from long ago. For instance, V.I. Gurarii and M. Melitidi [**GuM**] proved that for every complete sequence $(x_n)_{n\geq 1}$ in a Banach space X, there exists a sequence of positive numbers $(\varepsilon_n)_{n\geq 1}$ such that

$$\Big(y_n \in X, \|y_n - x_n\| < \varepsilon_n\Big) \quad \Rightarrow \quad \operatorname{span}_X(y_n : n \geq 1) = X.$$

For surveys on stable and constrained approximations see S. Khavinson [**Kha3**], V. Eiderman [**Eid1**], [**Eid2**], R. DeVore and G. Lorentz [**DeVL**].

Another particular but important subject is traced in 4.7.6(d). The completeness of the biorthogonal of a complete minimal family is not an automatic consequence of the definition, and moreover, fails very often. See comments and references about hereditarily complete sequences in A.5.6, A.5.7.1 and A.5.8 (Volume 1). The result of 4.7.6(d) (ii) is for R. Young [**Y2**] and (independently, and with a different proof) Yu. Lyubarskii [**Lyu**]. E. Fricain's result 4.7.6(d) (i) is from [**Fri1**], [**Fri2**]. The question whether the biorthogonal of a complete minimal sequence of reproducing kernels $(k_{\lambda_n}^{\Theta})_{n\geq 1}$ is always complete in $\mathcal{K}_{\Theta}$ seems to be open. Recall that for $(k_{\lambda_n})_{n\geq 1}$ in $\mathcal{K}_B$ this is always the case (see C.3.3.3(a) (iii), and C.3.3.3(f) for another proof), and hence the same is true for exponentials $(e^{i\lambda_n x}\chi_{\mathbb{R}_+} : n \geq 1)$ on the half-line $\mathbb{R}_+$.

The principal result of Subsection 4.7.6 is the Beurling and Malliavin formula for the completeness radius. It is largely commented above. It is worth adding that the frequency transformation $\lambda \longmapsto x$, $1/x = \mathrm{Re}(1/\lambda)$, which does not disturb the completeness of exponentials and reduces the problem to the case of real frequencies (see 4.7.6(e) (ii)), was already introduced by N. Levinson [**LeN**]. Also, various concepts of "asymptotic density" of a real sequence were studied in connection with the completeness of exponentials on finite intervals. Any exposition of the subject should pay attention to this technical problem. The densities $B(d)$ (from 4.7.6(e) (iv)) and $C(d)$ (from 4.7.6(e) (v)), were introduced by J.-P. Kahane [**Kah3**] and R. Redheffer [**Red1**], [**Red2**], respectively. Both appeared as an interpretation and an attempt to simplify the basic geometric density $D(d)$ of Beurling and Malliavin for which we refer to [**BeuM2**] or to any of the sources mentioned in this paragraph or above in "*Completeness problem*".

Several topics related to the geometry of exponentials and reproducing kernels are omitted in this short survey. It is worth mentioning two of them which are very closely related to problems of operator theory and control: (1) bases and completeness problems for Mittag–Leffler functions (and other generalizations of exponentials), and (2) vector valued exponentials and reproducing kernels of (vector valued) model spaces $\mathcal{K}_\Theta$. Without attempting to give exhausting references we mention several publications. For the problem (1), see M. Djrbashian [**Djr**], N. Nikolski and S. Hruschev [**HrN**], G. Gubreev [**Gub1**] (with a good survey of the field). For problem (2), we refer to S. Ivanov and B. Pavlov [**IP**], S. Avdonin and S. Ivanov [**AI2**], [**AI1**], G. Gubreev [**Gub2**], G. Gubreev and E. Olefir [**GubO**], S. Treil [**Tre9**], S. Treil and A. Volberg [**TV3**], [**TV2**], A. Volberg [**V5**]. In particular, in [**AI2**], [**AI1**] and [**GubO**] matrix valued generating entire functions are introduced and studied (see 4.7.1(b) for the scalar case); however, the corresponding vector Muckenhoupt condition concerns mostly diagonal matrices. An essentially non commutative Muckenhoupt condition and its applications to weighted expansions of vector valued functions is given in [**Tre9**], [**TV4**], [**TV3**]; see also comments and references in Section A.5.8 (Volume 1).

CHAPTER 5

A Brief Introduction to H^∞ Control

In this chapter, we consider an approach to control systems by means of the so-called transfer functions and a method of controlling at time ∞. Hankel operator techniques will be essentially used.

We then briefly study the notion of sensitivity of a control system as well as minimization problems with the help of feedback compensators. We finish with the problem of robust stabilization and give some comments on finite dimensional systems.

5.1. Input-Output Maps and Transfer Functions

In this section we introduce the basic input-output maps in the state domain as well as in the frequency domain and present the *realization problem* for transfer functions.

5.1.1. The input-output map. Consider the control system

$$\begin{cases} x' & = Ax + Bu \\ y & = Cx + Du \\ x(0) & = \mathbf{0}, \end{cases} \tag{5.1.1}$$

where $B \in L(\mathcal{U}, X)$, $C \in L(X, Y)$, $D \in L(\mathcal{U}, Y)$; $X, Y, \mathcal{U}$ are Hilbert spaces and A is the generator of a strongly continuous semigroup $S(\cdot)$. The system (5.1.1) will be refered to as the system (A, B, C, D). We first recall some terminology from the first chapter of this part. The functions $t \longmapsto u(t) \in \mathcal{U}$, $t \longmapsto x(t) \in X$ and $t \longmapsto y(t) \in Y$ are respectively the *input*, *state* and *output function.* In what follows we assume that $x_0 = 0$. We have already seen (see (1.1.2)), that

$$x(t) = c(\tau)u = \int_0^t S(t-r)Bu(r)dr, \quad t \geq 0,$$

and consequently

$$y(t) = \int_0^t CS(t-r)Bu(r)dr + Du(t), \quad t \geq 0.$$

We introduce some additional definitions. The mapping $u \longmapsto x$ is called *input-state map*, and $u \longmapsto y$ is the *input-output map.*

5.1.2. THEOREM. *Let $S(\cdot)$ be an exponentially stable semigroup (see Section 1.4). Then the input-state and the input-output maps are continuous as applications from $L^2(\mathbb{R}_+, \mathcal{U})$ to $L^2(\mathbb{R}_+, X)$, respectively from $L^2(\mathbb{R}_+, \mathcal{U})$ to $L^2(\mathbb{R}_+, Y)$.*

PROOF. We extend all functions defined on $\mathbb{R}_+$ by zero on $\mathbb{R}_-$. Then we can rewrite

$$x = (\chi_{\mathbb{R}_+} \cdot S(\cdot)) * Bu$$

and

$$y = (\chi_{\mathbb{R}_+} \cdot CS(\cdot)) * Bu + Du,$$

where the convolution of an operator valued function f and a vector valued function g is defined in the standard way by

$$(f * g)(t) = \int_{\mathbb{R}} f(t-x)g(x)\,dx, \quad t \in \mathbb{R}.$$

As $\chi_{\mathbb{R}_+} S(\cdot)$ and $\chi_{\mathbb{R}_+} CS(\cdot)$ are (respectively $L(X)$ and $L(X,Y)$ valued) integrable functions, we get

$$\begin{aligned} \|x\|_{L^2(\mathbb{R}_+,X)} &\le \|\chi_{\mathbb{R}_+} S(\cdot)\|_{L^1} \cdot \|B\| \cdot \|u\|_{L^2(\mathbb{R}_+,\mathcal{U})}, \\ \|y\|_{L^2(\mathbb{R}_+,Y)} &\le \left(\|\chi_{\mathbb{R}_+} S(\cdot)\|_{L^1} \cdot \|C\| \cdot \|B\| + \|D\|\right) \cdot \|u\|_{L^2(\mathbb{R}_+,\mathcal{U})}, \end{aligned} \tag{5.1.2}$$

which completes the proof. □

5.1.3. REMARK. It can be added that for $u \in L^2(\mathbb{R}_+, \mathcal{U})$ the solution x is, in fact, bounded on $\mathbb{R}_+$:

$$\begin{aligned} \|x(t)\| &\le \left(\int_0^t \|S(t-r)\|^2\,dr\right)^{1/2} \cdot \|B\| \cdot \left(\int_0^t \|u(r)\|_{\mathcal{U}}^2\,dr\right)^{1/2} \\ &\le \text{const} \cdot \|u\|_{L^2(\mathbb{R}_+,\mathcal{U})}. \end{aligned}$$

Here we have used the exponential stability of $S(\cdot)$.

5.1.4. The Laplace transform and the transfer function. As a consequence of Theorem 5.1.2 we can observe that the vector valued Laplace transform

$$\hat{f}(\lambda) = \int_{\mathbb{R}_+} e^{-\lambda t} f(t)\,dt,$$

is well-defined for every $\lambda \in \mathbb{C}$ with $\operatorname{Re}\lambda > 0$ for the three functions $f = u, x, y$ (as soon as the control is supposed to be of finite energy, $u \in L^2(\mathbb{R}_+, \mathcal{U})$, see also (5.1.2)). Moreover, the transformed functions $\hat{u}$, $\hat{x}$ and $\hat{y}$ are analytic in

$$\mathbb{C}^+ = \{\lambda : \operatorname{Re}\lambda > 0\},$$

and by the Paley–Wiener theorem A.6.3.5 (Volume 1), they belong respectively to the Hardy classes $H^2(\mathbb{C}^+, E)$ with $E = \mathcal{U}, X, Y$. We have the following result.

5.1.5. THEOREM. *Let $S(\cdot)$ be an exponentially stable semigroup. Then the Laplace transform of the input-output map $u \longmapsto y$, $u \in L^2(\mathbb{R}_+, \mathcal{U})$, is a multiplication operator*

$$\hat{u} \longmapsto \hat{y} = T\hat{u}, \quad \hat{u} \in H^2(\mathbb{C}^+, \mathcal{U}),$$

by a bounded analytic function on $\mathbb{C}^+$, $T \in H^\infty(\mathbb{C}^+, L(\mathcal{U}, Y))$, and T is given by

$$T(\lambda) = D + C(\lambda I - A)^{-1}B, \quad \operatorname{Re}\lambda > 0.$$

PROOF. By the preceding remarks, it is clear that the system (5.1.1) is equivalent to the following identities

$$\begin{aligned} \lambda\hat{x}(\lambda) &= A\hat{x}(\lambda) + B\hat{u}(\lambda), \quad \operatorname{Re}\lambda > 0, \\ \hat{y}(\lambda) &= C\hat{x}(\lambda) + D\hat{u}(\lambda), \quad \operatorname{Re}\lambda > 0. \end{aligned}$$

On the other hand, as $S(\cdot)$ is exponentially stable, the resolvent $(\lambda I - A)^{-1}$ exists for every $\lambda \in \mathbb{C}^+$, and we get

$$\hat{x}(\lambda) = (\lambda I - A)^{-1} B\hat{u}(\lambda),$$

which implies that

$$\hat{y}(\lambda) = T(\lambda)\hat{u}(\lambda), \quad \lambda \in \mathbb{C}^+.$$

Moreover, the Paley–Wiener theorem states that the Laplace transform (exactly as the Fourier transform) is an isometry (up to a constant) between the vector valued function spaces $L^2(\mathbb{R}_+, E)$ and $H^2(\mathbb{C}^+, E)$. Here, E is supposed to be a Hilbert space. Consequently, by Theorem 5.1.2, the application $\hat{u} \longmapsto T\hat{u}$ is continuous from $H^2(\mathbb{C}^+, \mathcal{U})$ into $H^2(\mathbb{C}^+, Y)$. By a standard argument we conclude that $\sup_{\lambda\in\mathbb{C}^+} \|T(\lambda)\| < \infty$, but we include the following alternative proof of this claim based on the exponential stability of $S(\cdot)$. Indeed,

$$\begin{aligned} \|(\lambda I - A)^{-1}\| &= \left\| \int_{\mathbb{R}_+} S(t)e^{-\lambda t}\, dt \right\| \le \int_{\mathbb{R}_+} \|S(t)\| e^{-\operatorname{Re}\lambda\cdot t}\, dt \\ &\le C \int_{\mathbb{R}_+} e^{-\alpha t} e^{-\operatorname{Re}\lambda \cdot t}\, dt \\ &\le C\alpha^{-1} \end{aligned}$$

for every $\lambda \in \mathbb{C}^+$. Here $\alpha = \alpha(S(\cdot))$ is the growth bound of $S(\cdot)$ (see 1.4 and 2.2.1). □

5.1.6. DEFINITION. A *stable* system means a system with exponentially stable semigroup. The function T defined in the preceding theorem is called *transfer function* of a stable system (A, B, C, D), and we say that $\hat{u}, \hat{x}, \hat{y} = T\hat{u}$ is the *representation in the frequency domain* of the system.

5.1.7. REMARK. The proof of Theorem 5.1.5 shows that the stability of $S(\cdot)$ implies the boundedness of the resolvent $\sup_{\lambda\in\overline{\mathbb{C}^+}} \|(\lambda I - A)^{-1}\| < \infty$. In fact, it is easy to see that the converse is also true, that is,

if $(\lambda I - A)^{-1}$ *exists for all* λ, $\operatorname{Re}\lambda \ge 0$, *and* $\sup_{\lambda\in\overline{\mathbb{C}^+}} \|(\lambda I - A)^{-1}\| < \infty$, *then the semigroup* $S(\cdot)$ *is exponentially stable.*

Indeed, since $S(\cdot)$ is strongly continuous, the growth bound $\alpha = \alpha(S(\cdot))$ is finite, $\alpha < \infty$ (see Subsection 2.2.1). For $c > \alpha$, we get $e^{-ct}S(t) \in L^2(\mathbb{R}_+, L(X))$, and the Paley–Wiener theorem shows that

$$\lambda \longmapsto ((\lambda + c)I - A)^{-1} x = \int_0^\infty e^{-(\lambda+c)t} S(t)x\, dt$$

belongs to $H^2(\mathbb{C}^+, X)$ for every $x \in X$. Because of the *Hilbert identity*

$$(\lambda I - A)^{-1} ((\lambda + c)I - A)^{-1} x = \frac{1}{c}\left((\lambda I - A)^{-1} - ((\lambda + c)I - A)^{-1} \right) x$$

and our assumption, the function $\lambda \longmapsto (\lambda I - A)^{-1}x$, $\operatorname{Re}\lambda > 0$, also belongs to $H^2(\mathbb{C}^+, X)$. As $(\lambda I - A)^{-1}x = \widehat{S(t)}x$, we can again apply the Paley–Wiener theorem to get $S(\cdot)x \in L^2(\mathbb{R}_+, X)$, $x \in X$, and then Theorem 1.4.3 completes the proof. □

The problem of finding an intrinsic description of transfer functions is known as the *realization problem.*

5.1.8. DEFINITION. A function $F \in H^\infty(\mathbb{C}^+, L(\mathcal{U}, Y))$ is called *realizable* if there exists a stable system (A, B, C, D) such that the associated transfer function coincides with F:

$$F = T = T(A, B, C, D).$$

For information, we state the following result; for other realization theorems see Section 5.6 below.

5.1.9. THEOREM (The realization theorem). *Let $\mathcal{U}, Y$ be Hilbert spaces of finite dimension and $F \in H^\infty(\mathbb{C}^+, L(\mathcal{U}, Y))$. Then F is realizable with an $A \in L(X)$,* $\dim X < \infty$, *if and only if F is a rational function holomorphic in $\overline{\mathbb{C}^+} \cup \{\infty\}$. Among all realizations of F, $F = T(A, B, C, D)$, those with minimal possible dimension* $\dim X$ *are characterized by the condition that (A, B) and (A^*, C^*) are controllable.*

5.1.10. Discrete-time systems. A theory similar to that of Subsections 5.1.1-5.1.9 exists also for discrete time signals:

$$\begin{cases} x_{n+1} &= Ax_n + Bu_n, \quad n \geq 0 \\ y_n &= Cx_n + Du_n, \quad n \geq 0. \end{cases} \tag{5.1.3}$$

In this case, we still consider finite energy signals $u = (u_n)_{n\geq 0} \in l^2(\mathbb{Z}_+, \mathcal{U})$, $x \in l^2(\mathbb{Z}_+, X)$ and $y \in l^2(\mathbb{Z}_+, Y)$, and the obvious discrete analog of the Laplace transform

$$\hat{u}(\lambda) = \sum_{n\geq 0} u_n \lambda^{-n}, \quad |\lambda| > 1.$$

Summing up the equations in (5.1.3) we obtain

$$\begin{aligned} \lambda\hat{x}(\lambda) &= A\hat{x}(\lambda) + B\hat{u}(\lambda) + \lambda x_0, \quad |\lambda| \geq 1, \\ \hat{y}(\lambda) &= C\hat{x}(\lambda) + D\hat{u}(\lambda), \quad |\lambda| \geq 1. \end{aligned}$$

Then

$$\hat{y}(\lambda) = T(\lambda)\hat{u}(\lambda), \quad |\lambda| \geq 1,$$

where $T(\lambda) = D + C(\lambda I - A)^{-1}B$, $|\lambda| \geq 1$, is the associated *transfer function.*

Actually, the theory of discrete-time systems is even easier than its continuous counterpart.

5.2. Noise Minimization, Feedback Control, and Sensitivity

This section is devoted to some basic notions of H^∞-control, namely we discuss uniform noise minimization at the output of a dynamical system by using regulators (compensators).

5.2.1. Feedback compensators. Consider an exponentially stable system (A, B, C, D) and its representation in the frequency domain

$$\hat{y} = T\hat{u},$$

where $T(\lambda) = D + C(\lambda I - A)^{-1}B$, $\mathrm{Re}\,\lambda > 0$, is the transfer function. For the sake of simplicity, we assume that the control space $\mathcal{U}$ and the *observation space* Y are Hilbert spaces of the same dimension, and we can thus identify them $\mathcal{U} = Y = E$.

We represent the action of T by a diagram (see Figure 5.2.1), where all functions are considered in the frequency domain, and to simplify the notation, we have omitted the "hat" over the functions. Further, we consider a perturbation $\tilde{y}$ of the output signal y by a kind of random noise v (see Figure 5.2.2). Actually, we

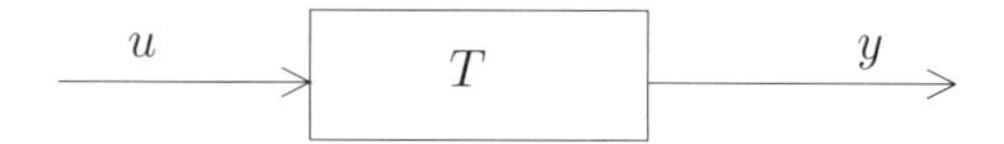

FIGURE 5.2.1

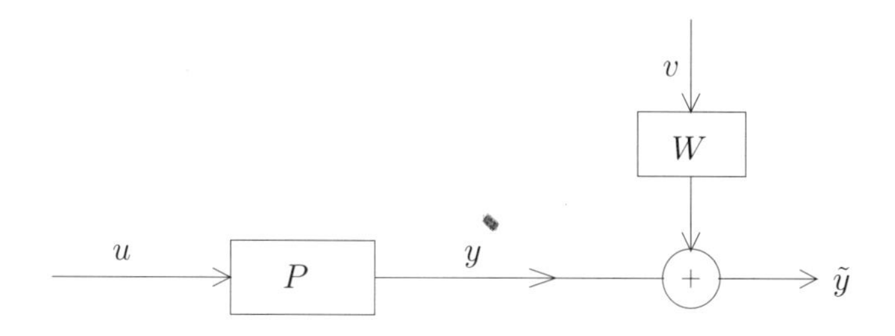

FIGURE 5.2.2

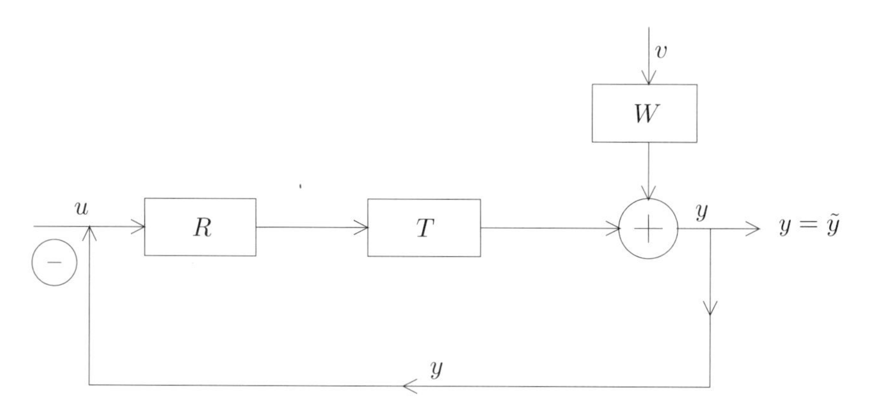

FIGURE 5.2.3

assume that v, in turn, passes through a noise transformer W and belongs to H^2 as the Laplace transform of a function on $\mathbb{R}_+$. In order to reduce the noise at the output, engineers introduce the so-called *regulators* or *compensators of negative feedback type*. These are devices $\mathcal{R}$ of convolution type in the state domain that turn into multipliers R in the frequency domain with $R \in H^\infty(E', E)$. Here E' is some auxiliary Hilbert space. Figure 5.2.3 illustrates how such a modified system works. The equation of the circuit taken at point y is as follows

$$TR(u - y) + Wv = y,$$

or

$$TRu + Wv = (I + TR)y.$$

Assuming that $I + TR$ is invertible, we obtain at the output

$$y = (I + TR)^{-1}TRu + (I + TR)^{-1}Wv.$$

Now, the problem is to find a compensator $\mathcal{R}$ that *minimizes the output noise energy*

$$\|(I + TR)^{-1}Wv\|_2 \tag{5.2.1}$$

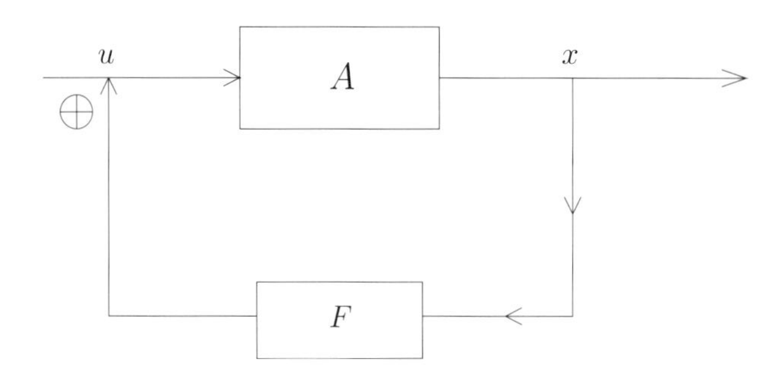

FIGURE 5.2.4

uniformly in $v \in H^2$, $\|v\|_2 \leq 1$. Mathematically this means that we seek R_{min} such that

$$\mu := \|(I + TR_{min})^{-1}W\|_\infty = \inf \|(I + TR)^{-1}W\|_\infty. \tag{5.2.2}$$

The infimum is taken over all compensators R that keep the system *internally stable*, which means that all inner and outer arrows in Figure 5.2.3 are of finite energy. It is known that this is equivalent to

$$R(I + TR)^{-1} \in H^\infty(L(E)).$$

(Observe that this is well coherent with preceding statements about stability; see, for example, 5.1.2 and 5.1.5.)

5.2.2. DEFINITION. The operator $(I+TR)^{-1}$ is called *sensitivity* of the system, and the minimization problem (5.2.2) is refered to as the *sensitivity problem*.

Denoting

$$F = R(I + TR)^{-1},$$

where, by the above comment, $F \in H^\infty(L(E))$, we get

$$TF = TR(I + TR)^{-1} = I - (I + TR)^{-1},$$

and hence

$$(I + TR)^{-1} = I - TF.$$

The sensitivity problem (5.2.2) can thus be rewritten in the following way

$$\mu = \inf_{F\in H^\infty} \|(I - TF)W\|_\infty = \inf_{F\in H^\infty} \|W - TFW\|_\infty. \tag{5.2.3}$$

5.2.3. COROLLARY. *If F is a solution of* (5.2.3), *then* $R = F(I - TF)^{-1} = (I - FT)^{-1}F$ *is a* minimizing compensator.

Here, minimizing is in the sense of (5.2.1). The second expression for R is from $F(I + TR) = R$, or $F = (I - FT)R$. □

5.2.4. REMARK (Another inclusion of the compensator). Let us return to the space-time domain (in place of the amplitude-frquency domain considered above). There, feedback compensators are plugged into the circuit as shown in Figure 5.2.4.

The equation of the circuit at the point x is the following

$$x' = Ax + B(u + Fx),$$

or

$$x' = (A + BF)x + Bu.$$

This makes it possible (at least theoretically) to control certain spectral properties of the new generator $A + BF$ by a suitable choice of the compensator F. See also Subsections 1.4.4 and Section 5.5.

5.2.5. Minimizing the sensitivity by using Hankel operators. In order to solve the sensitivity problem, we need to recall some facts about canonical factorization and Hankel operators. In particular, the canonical inner-outer factorization (see C.1.2.1) of a function $\Theta \in H^\infty(L(E, E'))$, $\Theta \not\equiv \mathbf{0}$, is the unique decomposition $\Theta = \Theta_{inn}\Theta_{out}$ of Θ into an inner function $\Theta_{inn} \in H^\infty(L(F, E'))$, that is, the boundary values $\Theta_{inn}(\xi)$ are isometries a.e. on $\mathbb{T}$, and an outer function $\Theta_{out} \in H^\infty(L(E, F))$, that is

$$\text{clos}_{H^2(E)}\, \Theta_{out} \cdot H^2(E) = H^2(F),$$

and also $\Theta = \Theta_{*out}\Theta_{*inn}$, where $\Theta_{*inn} \in H^\infty(L(E, F))$ is $*$-inner, which means that $(\Theta_{*inn}(\bar{z}))^*$ is inner, and $\Theta_{*out} \in H^\infty(L(F, E'))$ is $*$-*outer*, which means that $(\Theta_{*out}(\bar{z}))^*$ is outer.

Recall also that a Hankel operator (see B.1.6.1, Volume 1) with operator valued symbol $f \in L^\infty(L(E', E))$ is defined by

$$H_f x := P_- f x, \quad x \in H^2(E'),$$

where P_- is the orthogonal projection of $L^2(\mathbb{T}, E)$ onto $H^2_-(E) = L^2(\mathbb{T}, E) \ominus H^2(E)$. The Nehari theorem B.1.6.1 (Volume 1) states that

$$\|H_f\| = \text{dist}_{L^\infty(L(E',E))}\left(f, H^\infty(L(E', E))\right).$$

5.2.6. THEOREM. *Let $T \in H^\infty(L(E))$, $W \in H^\infty(L(E', E))$. Suppose that $T = T_{inn}T_{out}$ with T_{out} invertible ($T_{out}{}^{-1} \in H^\infty(L(E))$) and $T_{inn} \in H^\infty(L(E))$ inner and $*$-inner, and $W = W_{*out}W_{*inn}$ with W_{*out} invertible ($W_{*out}{}^{-1} \in H^\infty(L(E))$). Then the infimum of the sensitivity problem* (5.2.2) *associated to the transfer function T is given by*

$$\mu = \|H_{T^*_{inn}W_{*out}}\| = \text{dist}_{L^\infty(L(E))}\left(T^*_{inn}W_{*out}, H^\infty(L(E))\right).$$

PROOF. Following (5.2.3), we have

$$\mu = \inf\left\{\|W + TFW\|_\infty : F \in H^\infty(L(E))\right\}.$$

Moreover, $T_{out}H^\infty(L(E)) = H^\infty(L(E))$ and $H^\infty(L(E))W_{*out} = H^\infty(L(E))$ by assumption, and also $\|AW_{*inn}\|_\infty = \|A\|_\infty$ for arbitrary $A \in L^\infty(L(E))$. Hence

$$\begin{aligned}
\mu &= \inf\left\{\|W_{*out} + T_{inn}gW_{*out}\|_\infty : g \in H^\infty(L(E))\right\} \\
&= \inf\left\{\|W_{*out} + T_{inn}f\|_\infty : f \in H^\infty(L(E))\right\} \\
&= \inf\left\{\|T^*_{inn}W_{*out} + f\|_\infty : f \in H^\infty(L(E))\right\} \\
&= \|H_{T^*_{inn}W_{*out}}\|.
\end{aligned}$$

□

5.2.7. REMARK. An important special case for the applications occurs when T and W are rational operator valued functions. Then there is an explicit algorithm that enables us to construct the best approximation F and hence the compensator R. See comments in Section 5.7 below.

5.3. Remarks on Robust Stabilization

It happens often that the system that we wish to control or stabilize is known only up to a certain error. Then the transfer function is given by

$$\tilde{T} = T + \Delta.$$

This is the case of *additive uncertainty*. Suppose that the operator Δ satisfies

$$\|W_1 \Delta W_2\|_\infty < \varepsilon,$$

where $\varepsilon > 0$ is fixed, and W_1, W_2 are $H^\infty(L(E))$-functions for which $W_1^{-1}, W_2^{-1} \in H^\infty(L(E))$; these are called *weights*.

The system represented in Figure 5.2.3 is called *robustly stable* if there is $\varepsilon > 0$ such that $(R, T + \Delta)$ is *internally stable* for every Δ with $\|W_1 \Delta W_2\|_\infty < \varepsilon$. Recall that a system with feedback compensator is internally stable if all applications in Figure 5.2.3 are stable (and hence given by H^∞-functions). For information about robust stability and robust stabilization we refer to sources given in Section 5.7 below.

5.4. Scattering Type Input-Output, and Hankel Operators

In this section, we describe another approach to dynamical systems, in a sense, "à la" scattering theory.

5.4.1. Control at $-\infty$ and observation at ∞. To begin with, we make the simple observation that under the assumption of exponential stability, the *controllability map*

$$c(\tau)u = \int_0^\tau S(\tau - t)Bu(t)\,dt,$$

and the *observability operator*

$$\mathfrak{o}(\tau)u := \int_0^\tau CS(\tau - t)Bu(t)\,dt$$

extend in an obvious way to finite energy input signals u defined on the whole real line. In fact, we can set

$$\begin{aligned} c(\tau)u &= \int_{-\infty}^\tau S(\tau - t)Bu(t)dt, \\ \mathfrak{o}(\tau)u &= \int_{-\infty}^\tau CS(\tau - t)Bu(t)dt \end{aligned}$$

for $u \in L^2(\mathbb{R}, \mathcal{U})$. Now, we consider only control signals u which stop at time $t = 0$, that is, input functions $u \in L^2(\mathbb{R}, \mathcal{U})$ with $u(t) = \mathbf{0}$, $t > 0$, and we will study the reaction of the system at moments $\tau > 0$, that is the values $c(\tau)u$ and $\mathfrak{o}(\tau)u$ for $\tau > 0$.

For this situation we get the following two scattering-type operators:

$$c_\infty : L^2(\mathbb{R}_-, \mathcal{U}) \longrightarrow L^2(\mathbb{R}_+, X),$$

$$(c_\infty u)(\tau) = \int_{\mathbb{R}_-} S(\tau - t)Bu(t)\,dt, \quad \tau > 0,$$

and

$$\begin{aligned}\mathfrak{o}_\infty &: L^2(\mathbb{R}_-,\mathcal{U}) \longrightarrow L^2(\mathbb{R}_+,Y),\\ (\mathfrak{o}_\infty u)(\tau) &= \textstyle\int_{\mathbb{R}_-} CS(\tau - t)Bu(t)\,dt, \quad \tau > 0.\end{aligned}$$

The change of variables $t \longmapsto -t$ turns the above operators into the following ones (we will use the same symbols):

$$\begin{aligned}c_\infty &: L^2(\mathbb{R}_+,\mathcal{U}) \longrightarrow L^2(\mathbb{R}_+,X),\\ (c_\infty u)(r) &= \int_{\mathbb{R}_+} N_1(r+t)u(t)dt, \quad r \geq 0,\end{aligned}$$

where $N_1(t) = S(t)B$ $(t \geq 0)$ and

$$\begin{aligned}\mathfrak{o}_\infty &: L^2(\mathbb{R}_+,\mathcal{U}) \longrightarrow L^2(\mathbb{R}_+,Y),\\ (\mathfrak{o}_\infty u)(r) &= \int_{\mathbb{R}_+} N_2(r+t)u(t)dt, \quad r \geq 0,\end{aligned}$$

where $N_2(t) = CS(t)B$ $(t \geq 0)$.

Recall from Definition B.4.6.5, Volume 1 (where the scalar case is considered), that an integral operator T,

$$\begin{aligned}T &: L^2(\mathbb{R}_+,\mathcal{U}) \longrightarrow L^2(\mathbb{R}_+,X),\\ (Tf)(r) &= \int_{\mathbb{R}_+} k(r,t)f(t)dt, \quad r \geq 0,\end{aligned}$$

is called a *Hankel operator with integral kernel* N if there exists a ($L(\mathcal{U}, X)$ valued) function N such that

$$k(r,t) = N(r+t), \quad r,t \geq 0.$$

In this case we write $T = \Gamma_N$.

5.4.2. COROLLARY. *The scattering-type controllability and observability maps, c_∞ and $\mathfrak{o}_\infty$, are Hankel operators. If moreover B and C are of rank one, $B \in L(\mathbb{C}, X)$, $B\lambda = \lambda b$, $\lambda \in \mathbb{C}$, $b \in X$ and $C \in L(X, \mathbb{C})$, $Cx =< x, c >$ where $x \in X$, $c \in X^*$, then the observability operator $\mathfrak{o}_\infty$ is a scalar Hankel operator Γ_{N_2} with kernel*

$$N_2(r+t) =< S(r+t)b, c >, \quad r,t \geq 0.$$

□

5.4.3. Scattering type input-output maps for discrete time. The same results, as in the continuous case, hold for discrete time systems. In this case, exponential stability of a system

$$\begin{cases} x_{n+1} &= Ax_n + Bu_n, \quad n \geq 0,\\ y_n &= Cx_n + Du_n, \quad n \geq 0,\end{cases}$$

simply means that $\sigma(A) \subset \mathbb{D} = \{z \in \mathbb{C} : |z| < 1\}$. As for the continuous case, we suppose now that $u_n = 0$ for $n > 0$ and perform the change of variables $n \longmapsto -n$.

This leads to

$$\begin{aligned}(c_\infty u)(m) &= \sum_{n\geq 0} A^{n+m}Bu_n, \quad m\geq 0,\\ (\mathfrak{o}_\infty u)(m) &= \sum_{n\geq 0} CA^{n+m}Bu_n, \quad m\geq 0.\end{aligned}$$

With the same notations as in Subsection 5.4.1 we get

$$c_\infty = \Gamma_{N_1} : l^2(\mathbb{Z}_+,\mathcal{U}) \longrightarrow l^2(\mathbb{Z}_+, X),$$

where

$$N_1(m) = A^m B, \quad m \geq 0,$$

and

$$\mathfrak{o}_\infty = \Gamma_{N_2} : l^2(\mathbb{Z}_+,\mathcal{U}) \longrightarrow l^2(\mathbb{Z}_+, Y),$$

where

$$N_2(m) = CA^m B, \quad m \geq 0.$$

In the case of scalar control and observation, Corollary 5.4.2 carries over to

$$\begin{aligned}\mathfrak{o}_\infty &= \Gamma_{N_2} : l^2(\mathbb{Z}_+) \longrightarrow l^2(\mathbb{Z}_+),\\ N_2(m) &=< A^m b, c >, \quad m\geq 0.\end{aligned}$$

It should be mentioned that the scattering type operators c_∞, $\mathfrak{o}_\infty$ and the original controllability and observability operators $c(\cdot)$, $\mathfrak{o}(\cdot)$ are in a certain sense complementary to each other. Namely, in the initial notation (before the time inversion $n \longmapsto -n$), we had

$$\begin{aligned}c(m)u &= \sum_{n=0}^{m} A^{m-n}Bu_n, \quad m\geq 0,\\ \mathfrak{o}(m)u &= \sum_{n=0}^{m} CA^{m-n}Bu_n, \quad m\geq 0,\end{aligned}$$

and

$$\begin{aligned}(c_\infty u)(m) &= \sum_{n\leq 0} A^{m-n}Bu_n, \quad m\geq 0,\\ (\mathfrak{o}_\infty u)(m) &= \sum_{n\leq 0} CA^{m-n}Bu_n, \quad m\geq 0.\end{aligned}$$

Define a matrix on $\mathbb{Z}\times\mathbb{Z}$ by

$$a_{m,n} = \begin{cases} A^{m-n}B, & \text{if } m\geq n,\\ \mathbf{0}, & \text{if } m<n.\end{cases}$$

Then $(a_{m,n})$ is a triangular Toeplitz matrix (of analytic type). Indeed, $a_{m,n} = \gamma_{m-n}$, $m,n\in\mathbb{Z}$, where the sequence $(\gamma_n)_{n\in\mathbb{Z}}$ satisfies $\gamma_n = \mathbf{0}$, $n<0$. Then the relation between $c(\cdot)$ and c_∞ can be illustrated as in Figure 5.4.1 (we get the same scheme for $\mathfrak{o}(\cdot)$ and $\mathfrak{o}_\infty$ by applying C to each $a_{m,n} = \gamma_{m-n}$). Of course, this figure is similar to that opening the Part B of the book (Volume 1).

We will use the techniques of Hankel operators to solve some problems of "scattering-type" control.

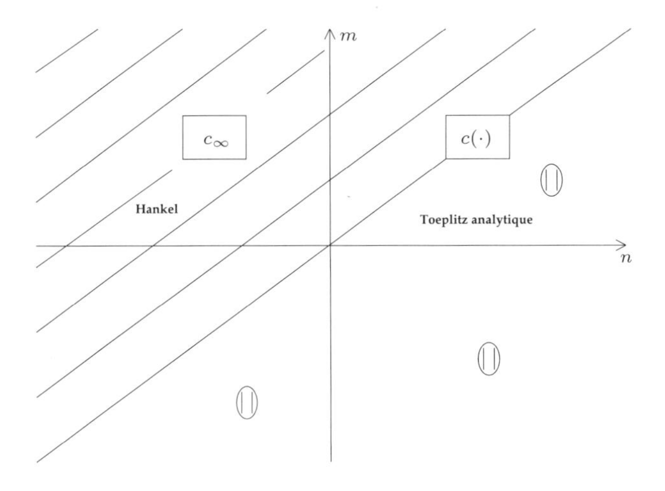

FIGURE 5.4.1

5.4.4. Generalized exact controllability and observability. As in the case of standard control theory, the main purpose in scattering type control problems is the description of the ranges of c_∞ and $\mathfrak{o}_\infty$. We only consider the discrete-time case, the continuous one being basically similar. We say that the (discrete-time, scattering type) system is *exactly controllable* if

$$c_\infty l^2(\mathbb{Z}_+,\mathcal{U}) = l^2(\mathbb{Z}_+, X),$$

and *exactly observable* if

$$\mathfrak{o}_\infty l^2(\mathbb{Z}_+,\mathcal{U}) = l^2(\mathbb{Z}_+, Y).$$

The following well-known lemma states that in contrast to classical control, exact controllability can never happen for the scattering type control.

5.4.5. LEMMA. *Let*

$$\Gamma_N : l^2(\mathbb{Z}_+,\mathcal{U}) \longrightarrow l^2(\mathbb{Z}_+, X)$$

be a bounded Hankel operator with kernel N. Then $\operatorname{Range}\Gamma_N \neq l^2(\mathbb{Z}_+, X)$. *Also,*

$$\inf\{\|\Gamma_N x\| : x \in l^2(\mathbb{Z}_+,\mathcal{U}), \|x\| = 1\} = 0. \tag{5.4.1}$$

PROOF. Clearly ${\Gamma_N}^* = \Gamma_{N^*}$ where $N^*(m) = N(m)^*$, $m \geq 0$. By Lemma 1.3.12, it is sufficient to show (5.4.1) (with N replaced by N^*). Let $\delta_n u^* = (\delta_{nm} u^*)_{m\geq 0}$ for $u^* \in \mathcal{U}^*$, $\|u^*\| = 1$. Then we have $\|\delta_n u^*\|_{l^2(\mathbb{Z}_+,\mathcal{U}^*)} = 1$, but $\|\Gamma_{N^*}\delta_n u^*\|^2_{l^2(\mathbb{Z}_+,X^*)} = \sum_{m\geq n} \|N^*(m)u\|^2_{X^*}$ tends to zero as $n \to \infty$. □

In view of the preceding lemma, we introduce another notion of exact controllability (respectively, observability), in a sense close to the original one. Namely, we say that the system is *generalized exactly controllable* if

$$\operatorname{Range} c_\infty \text{ is a closed subspace of } l^2(\mathbb{Z}_+, X),$$

respectively *generalized exactly observable* if

$$\operatorname{Range} \mathfrak{o}_\infty \text{ is a closed subspace of } l^2(\mathbb{Z}_+, Y).$$

Before giving the criteria for such a controllability (respectively, observability), we recall the links between Hankel operators with kernel:

$$\Gamma_N : l^2(\mathbb{Z}_+,\mathcal{U}) \longrightarrow l^2(\mathbb{Z}_+, X),$$

and Hankel operators with symbol:

$$H_f : H^2(\mathcal{U}) \longrightarrow H^2_-(X).$$

Namely, Nehari's theorem B.1.6.1 (Volume 1) establishes a one-to-one correspondence between these. More precisely the operator Γ_N is unitarily equivalent to JH_f (where $Jf = \bar{z}f(\bar{z})$, $z \in \mathbb{T}$) if and only if

$$\hat{f}(-m-1) = N(m), \quad m \geq 0,$$

and the symbol f may always be chosen in $L^\infty(T, L(\mathcal{U}, X))$. This relation explains why the following theorem provides the desired criterion for generalized exact controllability and observability of scattering type.

5.4.6. THEOREM (P. Fuhrmann, 1976). *Let $\mathcal{U}, X$ be finite-dimensional Hilbert spaces, and let*

$$H_f : H^2(\mathcal{U}) \longrightarrow H^2_-(X)$$

be a bounded Hankel operator, $f \in L^\infty(\mathbb{T}, L(\mathcal{U}, X))$. Then $\operatorname{Range} H_f$ *is a closed subspace of $H^2_-(X)$ if and only if there is an inner function $\Theta \in H^\infty(X, \mathcal{V})$, a function $\varphi \in H^\infty(\mathcal{U}, \mathcal{V})$ and a constant $\varepsilon > 0$ such that*

$$f = \Theta^*\varphi, \qquad (5.4.2)$$
$$\|\Theta(z)^*v\| + \|\varphi(z)^*v\| \geq \varepsilon\|v\|$$

for every $v \in \mathcal{V}$ (here $\mathcal{V}$ is an auxiliary Hilbert space).

PROOF. We restrict ourselves to the one-dimensional case $\dim \mathcal{U} = \dim X = 1$. The following steps contain all essential points for the proof. Some minor details are left to the readers. It is easy to see that $\operatorname{Range} H_f$ is closed if and only if zero is an isolated point of the spectrum of the self-adjoint operator $H_f^*H_f$. Hence, if $\operatorname{Range} H_f$ is closed, then Lemma 5.4.5 implies that $\operatorname{Ker} H_f \neq \{\mathbf{0}\}$. Recall from B.2.4.2 (Volume 1) that $\operatorname{Ker} H_f$ is a z-invariant subspace, and thus Beurling's theorem A.1.3.2 (Volume 1) gives the existence of an inner function $\Theta \in H^\infty$ such that $\operatorname{Ker} H_f = \Theta \cdot H^2$. In particular, this shows that $f\Theta \in H^\infty$, from which we deduce that $f = \varphi\bar{\Theta}$ for some $\varphi \in H^\infty$. It is easy to see that the inner part φ_{inn} and Θ are coprime. The reader can now verify (cf. also Section 4.4) that $H_f(H^2) = V\varphi(M_\Theta)\mathcal{K}_\Theta$ for some isometry V (in fact, V is just the multiplication operator $f \longmapsto \overline{\Theta}f$), and $\varphi(M_\Theta)$ is defined on the model space $\mathcal{K}_\Theta$, as before, by $\varphi(M_\Theta)x = P_\Theta\varphi x$, $x \in \mathcal{K}_\Theta$.

As $\operatorname{Ker}\varphi(M_\Theta) = \{\mathbf{0}\}$ and $\operatorname{Range}\varphi(M_\Theta)$ is closed, the operator $\varphi(M_\Theta)$ is invertible (first show that it is left invertible), and hence, by the commutant lifting theorem C.2.4.3, there exist functions $f, g \in H^\infty$ such that

$$\varphi f + \Theta g \equiv 1.$$

Clearly, this implies (5.4.2) in the special situation $\dim \mathcal{U} = \dim X = 1$.

Conversely, assume that (5.4.2) holds. Using the Carleson corona theorem C.3.2.10 we find functions $f, g \in H^\infty$ such that $\varphi f + \Theta g \equiv 1$ (for the general case of arbitrary finite dimension of X and $\mathcal{U}$ we have to use the *Fuhrmann–Vasyunin theorem*, see Section B.9.2, Volume 1). The last identity implies that $\operatorname{Range} H_f = \operatorname{Range} V\varphi(M_\Theta)$ is closed. □

5.5. Remarks on Finite Dimensional Systems

In order to control ordinary differential equations, it is often sufficient to consider finite-dimensional systems: $\dim X < \infty$, $\dim \mathcal{U} < \infty$ and $\dim Y < \infty$. In this case, the theory simplifies considerably. Here, we restrict ourselves to a list of results using the standard notation for dynamical systems:

$$\begin{cases} x' &= Ax + Bu, \\ y &= Cx + Du, \end{cases}$$

or

$$x' = (A + BF)x + Bu,$$

if a linear feedback operator $F : X \longrightarrow \mathcal{U}$ is included.

5.5.1. Controllability criteria. The system (A, B) is controllable if and only if

$$\mathcal{L}\text{in}(A^k B\mathcal{U} : 0 \le k \le n) = X,$$

where $n = \dim X$.

5.5.2. Controllability of feedback stabilized systems. If (A, B) is controllable, then $(A + BF, B)$ is likewise controllable for an arbitrary $F : X \longrightarrow \mathcal{U}$.

5.5.3. Attaining the prescribed spectrum by a feedback compensator. *The system (A, B) is controllable if and only if for every set of $n(= \dim X)$ distinct points $\Lambda = \{\lambda_1, ..., \lambda_n\} \subset \mathbb{C}$ there exists $F : X \longrightarrow \mathcal{U}$ such that $\sigma(A + BF) = \Lambda$.*

PROOF. Let us consider the key case $\Lambda \cap \sigma(A) = \emptyset$. If the system (A, B) is controllable, then $\mathcal{L}\text{in}\left((\lambda_i I - A)^{-1} B\mathcal{U} : \lambda_i \in \Lambda\right) = X$. Hence, if $(x_i)_{i=1}^n$ is a basis of X, then there exist $y_i \in \mathcal{U}$, $i = 1, \ldots, n$, such that $x_i = (\lambda_i I - A)^{-1} B y_i$, $1 \le i \le n$. Setting now

$$Fx_i := y_i, \quad 1 \le i \le n,$$

we get the required compensator. Indeed, we have $x_i = (\lambda_i I - A)^{-1} BFx_i$, and hence

$$(\lambda_i I - A)x_i - BFx_i = \mathbf{0}, \quad 1 \le i \le n;$$

in other words, x_i, $1 \le i \le n$, are the eigenvectors of $A + BF$ associated to the eigenvalues λ_i.

The converse part is also elementary. □

5.5.4. Rank one controllability of stabilized systems. Let (A, B) be a controllable system and let $b \in X$, $b \ne \mathbf{0}$. Set $\bar{b}\lambda = \lambda b$, $\lambda \in \mathbb{C}$. Then there exists $F : X \longrightarrow \mathcal{U}$ such that $(A + BF, \bar{b})$ is controllable (this means that b is a cyclic vector of $A + BF$).

5.5.5. Typical and robust character of controllability. The set $\{(A, B),$ such that (A, B) is controllable$\}$ is open and everywhere dense in $L(X) \times L(\mathcal{U}, X)$; moreover it is of complete Lebesgue measure.

5.6. Exercises and Further Results

5.6.1. Discrete-time realization theorems. Let $\mathcal{U}$, Y be Hilbert spaces, and let T be an $L(\mathcal{U}, Y)$ valued function defined and holomorphic in a neighbourhood of $z = 0$. It is *realizable* if there exists a discrete-time dynamical system $\Sigma = (A, B, C, D)$ such that

$$T(z) = T_\Sigma(1/z)$$

for $|z| < r$, where $T_\Sigma(\lambda) = D + C(\lambda I - A)^{-1}B$ is the transfer function of Σ (see 5.1.10 for the definition). The following results are due to *J. Helton (1974)*.

(a) Let T be as above. Show that the following are equivalent.

1) T is realizable with a finite energy (internally stable) system Σ.
2) $T \in H^\infty(L(\mathcal{U}, Y))$.

In fact, if T is realizable as in 1) then the realization can be chosen controllable and completely observable.

[*Hint:* the implication 1)$\Rightarrow$ 2) is contained in the computations of Subsection 5.1.10; for the converse, set

$A = S^* : H^2(Y) \longrightarrow H^2(Y)$, where S^* is the backward shift operator,
$C : H^2(Y) \longrightarrow Y$, where $Cf = f(0)$,
$D : \mathcal{U} \longrightarrow Y$, where $D = T(0)$, and
$B : \mathcal{U} \longrightarrow H^2(Y)$, where $B = (S^*\Gamma_T)|\mathcal{U}$ is a restriction of the Hankel operator $S^*\Gamma_{S^*} = \Gamma_{S^*T} = JH_{S^*T} : H^2(\mathcal{U}) \longrightarrow H^2(Y)$ onto the constants (see Subsection 5.4.4 for the notation); then

$$\begin{aligned} T_\Sigma(1/z)u &= Du + zC(I - zA)^{-1}Bu = T(0)u + \sum_{n\geq 0} z^{n+1}CA^nBu \\ &= T(0)u + \sum_{n\geq 0} z^{n+1}(S^{*(n+1)}\Gamma_T u)(0) = T(0)u + \sum_{n\geq 0} z^{n+1}\hat{T}(n+1)u \\ &= T(z)u \end{aligned}$$

for every $u \in \mathcal{U}$; therefore, T is realizable;

next, notice that the controllability operator $c : l^2(\mathcal{U}) \longrightarrow H^2(Y)$ is unitarily equivalent to the Hankel operator Γ_{S^*T}:

$$c((u_n)_{n\geq 0}) = \sum_{n\geq 0} A^nBu_n = \sum_{n\geq 0} S^{*(n+1)}\Gamma_T u_n = \sum_{n\geq 0} S^*\Gamma_T S^n u_n = S^*\Gamma_T(\sum_{n\geq 0} S^n u_n),$$

and hence, the subspace of reachable states is

$$\operatorname{clos}(c(l^2(\mathcal{U}))) = \operatorname{clos}(S^*\Gamma_T(H^2(\mathcal{U})));$$

therefore, in order to get a controllable realization, replace the state space $H^2(Y)$ by $\operatorname{Range}(\Gamma_T) = \operatorname{clos}(\Gamma_{S^*T}H^2(\mathcal{U}))$; the adjoint to the observability operator is

$$\mathfrak{o}^*((y_n)_{n\geq 0}) = \sum_{n\geq 0} A^{*n}C^*y_n = \sum_{n\geq 0} S^n y_n,$$

and hence it coincides with the canonical isomorphism of $l^2(Y)$ onto $H^2(Y)$.]

(b) Let T be as above. Show that T is realizable with a controllable and observable system Σ.

[*Hint:* apply (a) to the function $T(\rho z)$ with sufficiently small $\rho > 0$ to get a controllable and observable system $\Sigma = (A, B, C, D)$ such that $T(\rho z) = T_\Sigma(1/z)$; then show that $T(z) = T_{\Sigma'}(1/z)$ for $\Sigma = (\rho^{-1}A, B, C, D)$.]

5.6.2. Output normal realizations of Hankel operators: the discrete-time case. The following results are due to *A. Megretskii, V. Peller, and S. Treil (1995)*.

(a) Let $R \in L(H)$, $R^* = R$ and $\operatorname{Ker} R = \{0\}$, and let $q \in H$, $\|q\| \leq 1$.

(i) Show that $R^2 - (\cdot, Rq)Rq \geq \mathbb{O}$.

(ii) Let $\Delta \in L(H)$ be such that $\Delta^2 = R^2 - (\cdot, Rq)Rq$. Show that there exists a unique operator $A \in L(H)$ such that $\Delta = AR$, $\|A\| \leq 1$.

[*Hint:* use $\Delta^2 \leq R^2$ and 1.3.12 (or B.1.4.5, Volume 1) to show the existence of A; the uniqueness follows from $\operatorname{Ker} R = \{0\}$.]

(b) Let R and A be as in (a), let $\Sigma = (A, B, C, 0)$ be a discrete-time dynamic system

$$\begin{aligned} x_{n+1} &= Ax_n + Bu_n, \\ y_n &= Cx_n, \end{aligned}$$

with one-dimensional control operator $B \in L(\mathbb{C} \longrightarrow H)$, $Bz = zRq$ ($z \in \mathbb{C}$) and one-dimensional output $C \in L(H \longrightarrow \mathbb{C})$, $Cx = (x, q)$ (for $x \in H$).

Suppose that A is *asymptotically stable*, i.e.,

$$\lim_{n \longrightarrow \infty} \left\| A^n x \right\| = 0, \quad x \in H.$$

Show that

(i) the system Σ is *output normal*, that is $L_{\mathfrak{o}} = I$, where $L_{\mathfrak{o}}$ is the *observability Gramian*,

$$L_{\mathfrak{o}} = \sum_{n \geq 0} (A^*)^n C^* C A^n;$$

(ii) R is unitarily equivalent to $\Gamma_N | (\operatorname{Ker} \Gamma_N)^\perp$, where Γ_N is the observability (Hankel) operator

$$\mathfrak{o}_\infty = \Gamma_N : l^2 \longrightarrow l^2$$

with the matrix $N = ((A^{m+n}Rq, q))_{m,n \geq 0}$ (see 5.4.3 for the notation);

(iii) $\operatorname{Ker} \Gamma_N = \{0\}$ if and only if $\|q\| = 1$ and $q \notin R(H)$.

[*Hint:* for (i), use $\Delta = AR$, and observe that $R^2 = RA^*AR + (\cdot, Rq)Rq$; then, using that $\operatorname{clos} R(H) = H$, get $I = A^*A + (\cdot, q)q$, and hence

$$\|x\|^2 = \|Ax\|^2 + |(x, q)|^2$$

for every $x \in H$; iterating and using the asymptotic stability of A, obtain that

$$\|x\|^2 = \sum_{n \geq 0} |(A^n x, q)|^2, \quad \forall x \in H,$$

which means that $L_{\mathfrak{o}} = I$;

for (ii), consider the isometry $V : H \longrightarrow l^2$ defined by $Vx = ((A^n x, q))_{n \geq 0}$; since $V^* e_n = A^{*n} q$, where $(e_n)_{n \geq 0}$ is the standard orthonormal basis of l^2, we have:

$$(VRV^* e_n, e_m) = (RA^{*n}q, A^{*m}q) = (A^m RA^{*n}q, q) = (A^{n+m}Rq, q)$$

(where it is used that $\Delta = AR = RA^*$); therefore (ii) holds with $\operatorname{Ker}\Gamma = \operatorname{Ker} V^*$; for (iii), we refer to A. Megretskii, V. Peller and S. Treil [**MPT**].]

(c) *Outline of the proof of Theorem B.9.1.12 (Volume 1).* Let Γ be a selfadjoint operator Γ with scalar spectral measure $\mu = \mu_\Gamma$ and let $d = d_\Gamma$ be its *spectral multiplicity function* satisfying the following conditions (condition (2) of the *Megretskii–Peller–Treil theorem B.9.1.12*):

(a) $0 \in \operatorname{supp}(\mu)$ and $d(0) \in \{0, \infty\}$ if $d(0)$ is well-defined, i.e., if $\mu(\{0\}) > 0$;
(b) $d(x) - 2 \le d(-x) \le d(x) + 2$ μ_a a.e.;
(c) $d(x) - 1 \le d(-x) \le d(x) + 1$ μ_s-a.e. .

Then there exists a Hankel operator Γ_N unitarily equivalent to Γ.

[*Hint:* 1) set $R = \Gamma|(\operatorname{Ker}\Gamma)^\perp$;

2) show that there exists $q \in (\operatorname{Ker}\Gamma)^\perp$, $\|q\| \le 1$, and a selfadjoint operator Δ satisfying $\Delta^2 = R^2 - (\cdot, Rq)Rq$ and such that the operator A defined in (a) is asymptotically stable; ensure that q can be chosen in such a way that $\operatorname{Ker}\Gamma = \{0\} \Leftrightarrow (\|q\| = 1, q \notin \operatorname{Range}(R))$; (for this part of the proof we refer to [**MPT**]);

3) apply (b)(ii).]

5.7. Notes and Remarks

Classical facts about transfer functions and feedback controllers as presented in Sections 5.1 through 5.3 can be found in any book on control theory. Our short exposition is based on R. Curtain and H. Zwart [**CZ**], P. Fuhrmann [**Fuh3**], C. Foiaş and A. Frazho [**FF**], B. Francis [**Fr**] and J. Ball, I. Gohberg, and L. Rodman [**BGR**].

In connection with 5.1.5 and 5.1.7 see also the tauberian theorems from Subsection 1.6.1.

The realization theorems 5.1.9 and 5.6.1 can be found in P. Fuhrmann [**Fuh3**], as well as in other sources mentioned above and below. The original reference for 5.6.1 is J. Helton [**Helt1**]. For algorithms mentioned in 5.2.7 see C. Foiaş and A. Frazho [**FF**].

In Section 5.4, we mostly follow [**Fuh3**]. In fact, the scheme presented here is nothing but the filtering of Chapter A.7 (Volume 1). The fundamental fact is that an input-output operator of a stationary system is a Hankel operator. This is another natural place to raise the inverse problem for Hankel operators (see Chapter B.9, Volume 1): how to give an unitarily invariant description of Hilbert space operators which can be input-output operators of finite energy systems? Selfadjoint operators which are unitarily equivalent to Hankel operators are described in a unitarily invariant language by A. Megretskii, V. Peller, and S. Treil [**MPT**]; see B.9.1.12 and B.9.4 (Volume 1) for the statement and a discussion, and 5.6.2 above for a partial proof. The omitted (most difficult) part of the proof is based on a test for asymptotic stability by B. Sz.-Nagy–C. Foiaş (see C.1.5.4(f) and D.1.7) and on the *balanced realization method* developed by R. Ober [**Ob**], R. Ober and S. Montgomery-Smith [**ObM**], D. Salamon [**Sal**], and N. Young [**You**]. A discrete-time dynamic system $\Sigma = (A, B, C, D)$ is called *balanced* if it is controllable, observable and the controllability and observability Gramians exist and coincide, i.e., $L_c = L_\mathfrak{o}$, where

$$L_\mathfrak{o} = \sum_{n \ge 0} A^{*n} C^* C A^n, \quad L_c = \sum_{n \ge 0} A^n B B^* A^{*n}.$$

For every bounded Hankel operator Γ there exists a balanced system such that the associated "scattering controllability operator" $c_\infty = \Gamma_{N_1}$ (see 5.4.3 for the definitions) is unitarily equivalent to Γ (N. Young [**You**]). Balanced realizations of Hankel operators are also used in S. Treil's theorem that describes moduli $|\Gamma|$ of Hankel operators, see Chapter B.9 (Volume 1).

Theorem 5.4.6 on Hankel operators with closed range is mentioned by D. Clark [**Cl2**] (for the scalar case) and developed by P. Fuhrmann [**Fuh2**]; see also J. Helton [**Helt1**].

The theory presented in Subsection 5.6.2 is from A. Megretskii, V. Peller and S. Treil [**MPT**], see comments above.

Bibliography

[Ab1] E.V. Abakumov, *Inverse spectral problem for finite rank Hankel operators*, Zapiski Nauchn. Semin. POMI, 217 (1994), 5-15 (Russian); English transl.: J. Math. Sci. (New York), 85 (1997), no. 2, 1759–1766.

[Ab2] ——— *Cyclicity and approximation by lacunary power series*, Michigan Math. J., 42 (1995), 277–299.

[Ab3] ——— *Essais sur les opérateurs de Hankel et la capacité d'approximation des séries lacunaires*, Thesis, Univ. Bordeaux, 1994.

[AbB] E.V. Abakumov and A.A. Borichev, *Shift invariant subspaces with arbitrary indices in weighted ℓ^p_A spaces*, to appear in J. Funct. Analysis.

[AbrKr] M.B. Abrahamse and T.L.Kriete, *The spectral multiplicity of a multiplication operator*, Indiana Univ. Math. J., 22 (1972/73), 845–857.

[AAK1] V.M. Adamyan, D.Z. Arov and M.G. Krein, *On infinite Hankel matrices and the generalized problems of Carathéodory–Fejer and F. Riesz*, Funkcional. Analiz i Prilozhen., 2 (1968), 1-19 (Russian); English transl.: Functional. Anal. Appl. 2 (1968), 1–18.

[AAK2] ——— *On bounded operators commuting with a contraction of class C_{00} having non-unitarity rank 1*, Funkcional. Analiz i Prilozhen., 3 (1969), 86-87 (Russian); English transl.: Functional. Anal. Appl. 3 (1969), 242–243.

[AAK3] ——— *Infinite Hankel block matrices and related problems of extension*, Izvestia Akad. Nauk Armyan. SSR Ser. Mat., 6 (1971), 87–112 (Russian); English transl.: Amer. Math. Soc., Translat., II. Ser. 111 (1978) 133–156.

[AAK4] ——— *Analytic properties of Schmidt pairs of a Hankel operator and the generalized Schur-Takagi problem*, Matem. Sbornik, 86 (128) (1971), 33–73 (Russian); English transl.: Math. USSR Sbornik, 15 (1971), 31–73.

[Ag1] J. Agler, *The Arveson extension theorem and coanalytic models*, Integral Eq. and Oper. Theory, 5 (1982), 608–631.

[Ag2] ——— *Some interpolation theorems of Nevanlinna-Pick type*, preprint, 1989.

[Ag3] ——— *Nevanlinna-Pick interpolation on Sobolev space*, Proc. Amer. Math. Soc. 108 (1990), 341–351.

[Ag4] ——— *On the representation of certain holomorphic functions defined on a polydisc*, pp. 47–66 in: Operator Theory: Adv. and Appl., 48, Birkhäuser–Verlag, Basel, 1990.

[AY] J. Agler and N.J. Young, *A converse to a theorem of Adamyan, Arov, and Krein*, J. Amer. Math. Soc. 12 (1999), no. 2, 305–333.

[Agm] S. Agmon, *Sur un problème de translations*, C.R. Acad. Sci. Paris, 229 (1949), no. 11, 540–542.

[AhC] P.R. Ahern and D.N. Clark, *Radial limits and invariant subspaces*, Amer. J. Math., 92 (1970), no. 2, 332–342.

[Air] G.M. Airapetyan, *Multiple interpolation and the property of being a basis for certain biorthogonal systems of rational functions in the Hardy H^p classes*, Izvestia Akad. Nauk SSSR, Ser. Math., 12 (1977), 262–277 (Russian).

[Aiz] L.A. Aizenberg, *Carleman's formulas in complex analysis. First applications*, Novossibirsk, Nauka (Sibirskoe Otdelenie), 1990 (Russian); English transl.: Carleman's formulas in complex analysis, theory and applications; revised translation of the 1990 Russian original, Mathematics and its Applications, 244, Kluwer Academic Publishers Group, Dordrecht, 1993.

[AizAD] L.A. Aizenberg, A. Aytuna, and P. Djakov, *An abstract approach to Bohr's phenomenon*, Proc. Amer. Math. Soc., 128 (2000), no. 9, 2611–2619.

[Akh1] N.I. AKHIEZER [AHIEZER], *On a proposition of A.N. Kolmogorov and a proposition of M.G. Krein*, Doklady Akad. Nauk SSSR, 50 (1945) (Russian).

[Akh2] ——— *The classical moment problem*, Moscow, GIFML, 1961 (Russian); English transl.: Olivier & Boyd, Edinburgh–London, 1965.

[Akh3] ——— *Lectures on approximation theory*, 2nd edition, Nauka, Moscow (Russian); German transl.: *Vorlesungen über Approximationstheorie*, Klaus Fiedler, Mathematische Lehrbücher, Band II Akademie-Verlag, Berlin, 1967.

[Akh4] ——— *On the weighted approximation of continuous functions by polynomials on the entire real axis*, Uspehi Matem. Nauk, 11 (1956), no. 4, 3–43 (Russian); English transl.: AMS Transl. Series, 22 Ser. 2 (1962), 95–137.

[AG] N.I. AKHIEZER [AHIEZER] AND I.M. GLAZMAN, *Theory of linear operators in Hilbert space*, 2nd edition, Nauka, Moscow, 1966 (Russian); German transl.: Berlin, Akademie–Verlag, 1960; English transl.: Frederick Ungar, New York, 1961.

[AK] N.I. AKHIEZER [AHIEZER] AND M. KREIN, *Some questions in the theory of moments*, Nauchn.-Tech. Izd., Kharkov, 1938 (Russian); English transl.: Transl. Math. Monographs, vol. 2, Amer. Math. Soc., Providence, RI, 1962.

[Aleks1] A.B. ALEKSANDROV, *Discrete measures with compact nonwhere dense support that are orthogonal to the rational functions*, Zapiski Nauchn. Semin. Math. Inst. Steklov (LOMI), 73 (1977), 7–15 (Russian); English transl.: J. Soviet Math., 34 (1986), no. 6, 2023–2028.

[Aleks2] ——— *Invariant subspaces of the backward shift operator in the space H^p $(p \in (0,1))$*, Zapiski Nauchn. Semin. LOMI, 92 (1979), 7–29 (Russian).

[Aleks3] ——— *Invariant subspaces of the shift operators. An axiomatic approach*, Zapiski Nauchn. Semin. Steklov Math. Institute (LOMI), 113 (1981), 7–26 (Russian); English transl.: J. Sov. Math., 22 (1983), 1695–1708.

[Aleks4] ——— *Multiplicity of boundary values of inner functions*, Izvestia Akad. Nauk Armian. SSR, Mat., 22 (1987), no. 5, 490–503 (Russian).

[Aleks5] ——— *Inner functions and related spaces of pseudocontinuable functions*, Zapiski Nauchn. Semin. Steklov Math. Institute (LOMI), 170 (1989), 7–33 (Russian); English transl.: J. Soviet Math., 63 (1993), no. 2, 115–129.

[Aleks6] ——— *Lacunary series and pseudocontinuations*, Zap. Nauchn. Sem. S.-Peterburg. Otdel. Mat. Inst. Steklov. (POMI) 232 (1996), Issled. po Linein. Oper. i Teor. Funktsii. 24,16–32 (Russian); translation in J. Math. Sci. (New York) 92 (1998), no. 1, 3550–3559.

[Aleks7] ——— *On embedding theorems for coinvariant subspaces of the shift operator, II*, Zapiski Nauchn. Semin. Steklov Math. Institute (St. Petersburg), 262 (1999), 5–48 (Russian).

[Aleks8] ——— *On embedding theorems for coinvariant subspaces of the shift operator, I*, pp. 45–64 in: Operator Theory: Adv. and Appl., 113 (The S.A. Vinogradov Memorial Volume), Basel, Birkhäuser, 2000.

[AP] A.B. ALEKSANDROV AND V.V. PELLER, *Hankel operators and similarity to a contraction*, Int. Math. Res. Notices, (1996) no. 6, 263–275.

[ARR] A. ALEMAN, S. RICHTER AND W. ROSS, *Pseudocontinuations and the backward shift*, Indiana Univ. Math. J., 47 (1998), 223–276.

[ARS] A. ALEMAN, S. RICHTER, C. SUNDBERG, *Beurling's theorem for the Bergman space*, Acta Math. 177 (1996), 275–310.

[All] G.R. ALLAN, *On one-sided inverses in Banach algebras of holomorphic vector-valued functions*, J. London Math. Soc., 42 (1967), 463–470.

[Alp] D. ALPAY, *Algorithme de Schur, espaces à noyau reproduisant et théorie des systhèmes*, Panoramas et Synthèses, vol. 6, Soc. Math. France, Paris, 1998.

[Am1] E. AMAR, *Ensembles d'interpolation sur le spectre d'une algèbre d'opérateurs*, Thèse, Dept. Math., Université Paris XI, Orsay, 1977.

[AB] E. AMAR AND A. BONAMI, *Mesures de Carleson d'ordre α et solutions au bord de l'équation $\bar\partial$*, Bull. Soc. Math. France, 107 (1979), 23–48.

[Ander] J.M. ANDERSON, *Bloch functions: the basic theory*, pp. 1–19 in: Operators and function theory, ed. S. Power, Reidel Publ., Dordrecht, 1985.

[And1] T. ANDO, *On a pair of commutative contractions*, Acta Sci. Math., 24 (1963), 88–90.

[And2] ——— *De Branges spaces and analytic operator functions*, Lect. Notes, Sapporo University, 1990.

[And3] ——— *Majorization and inequalities in matrix theory*, Linear Algebra Appl., 199 (1994), 17–67.

[ABFP] C. Apostol, H. Bercovici, C. Foiaş and C. Pearcy, *Invariant subspaces, dilation theory, and the structure of the predual of a dual algebra, I*, J. Funct. Anal., 63 (1985), 369–404.

[AFHV] C. Apostol, L.A. Fialkow, D.A. Herrero and D. Voiculescu, *Approximation of Hilbert space operators, Vol. II*, Res. Notes Math. 102, Pitman, Boston, 1984.

[ArG] N.U. Arakelian and P.M. Gauthier, *Propagation of smallness and uniqueness for harmonic and holomorphic functions*, J. Contemp. Math. Anal. (Nat. Acad. Sci. Armenia) 30 (1995), no. 4, 2–24.

[AFP1] J. Arazy, S. Fischer and J. Peetre, *Möbius invariant function spaces*, J. Reine Angew. Math., 363 (1985), 110–145.

[AFP2] ——— *Hankel operators on weighted Bergman spaces*, Amer. J. Math., 110 (1988), 989–1054.

[Are1] W. Arendt, *Spectral bound and exponential growth: survey and an open problem*, pp. 397–401 in: Ulmer Seminare Funktionalanalysis und Differentialgleichungen, Heft 1, Universität Ulm, 1996.

[ArBa] W. Arendt and C.J.K. Batty, *Tauberian theorems and stability of one-parameter semigroups*, Trans. Amer. Math. Soc., 306 (1988), 837–852.

[ArN] W. Arendt and N. Nikolski, *Vector-valued holomorphic functions revisited*, Math. Zeit., 234 (2000), no. 4, 777–805.

[Aro] N. Aronszajn, *Theory of reproducing kernels*, Trans. Amer. Math. Soc., 68 (1950), 337–404.

[ACS] R. Arosena, M. Cotlar and C. Sadosky, *Weighted inequalities in L^2 and lifting property*, Math. Anal. and Appl., Adv. in Math. Suppl. Stud., 7A (1981), 95–128.

[ArGh] Gr. Arsene and A. Gheondea, *Completing matrix contractions*, J. Operator Theory, 7 (1982), 179–189.

[Arv1] W.B. Arveson, *Subalgebras of C^*-algebras. I*, Acta Math., 123 (1969), no. 3–4, 141–224.

[Arv2] ——— *Subalgebras of C^*-algebras.II*, Acta Math., 128 (1972), no. 3–4, 271–308.

[Arv3] ——— *Interpolation problems in nest algebras*, J. Funct. Anal., 20 (1975), 208–233.

[Arv4] ——— *Ten lectures on operator algebras*, CBMS series 55, Amer. Math. Soc., Providence, 1983.

[Arv5] ——— *Subalgebras of C^*-algebras. III: multivariable operator theory*, preprint, Berkeley, 1999.

[At] F.V. Atkinson, *Discrete and continuous boundary problems*, Academic Press, N.Y., 1964.

[Atz] A. Atzmon, *Multilinear mappings and estimates of multiplicity*, Integral Eq. Operator Theory, 10 (1987), 1–16.

[Au] B. Aupetit, *Propriétés spectrales des algèbres de Banach*, Lect. Notes Math., 735, Springer, Heidelberg, 1979.

[Av] S.A. Avdonin *Towards the question on Riesz bases of exponentials in L^2*, Vestnik Leningradskogo Universiteta, ser. matem., mehan. i astron., 13 (1974), 5–12 (Russian).

[AI1] S.A. Avdonin and S.A. Ivanov, *Controllability of systems with distributed parameters and families of exponentials*, UMK VO, Kiev, 1989 (Russian).

[AI2] ——— *Families of exponentials*, Cambridge Univ. Press, Cambridge, 1995.

[Ave] R.A. Avetisyan, *Inequalities of Bernstein type for derivatives of meromorphic functions, and approximation by meromorphic functions on the real line*, Izvestia National. Akad. Nauk Armenii, Matematika, 28 (1993), no. 6, 14–29 (Russian); English transl.: J. Contemp. Math. Anal., 28 (1993), no. 6, 11–24.

[Ax1] Sh. Axler, *Factorization of L^∞ functions*, Ann. Math., 106 (1977), 567–572.

[Ax2] ——— *Bergman spaces and their operators*, in: Surveys of some recent results in operator theory, Vol. 1 (eds. J.B. Conway and B.B. Morrel), Pitman Res. Notes Math. Series 171, 1–50, (1988).

[AChS] Sh. Axler, A.S.-Y. Chang and D. Sarason, *Products of Toeplitz operators*, Integral Eq. Oper. Theory, 1 (1978), 285–309.

[AZ] Sh. Axler and D. Zheng, *The Berezin transform on the Toeplitz algebra*, Studia Math., 127 (1998), 113–136.

[AzCl] E.A. Azoff and K.F. Clancey, *Spectral multiplicity for direct integrals of normal operators*, J. Operator Theory, 3 (1980), no. 2, 213–235.

[Ba] K.I. Babenko, *On harmonic conjugate functions*, Dokl. Akad. Nauk SSSR, 62 (1948), 157–160 (Russian).

[Bach] G.F. Bachelis *On the upper and lower majorant properties in $L^p(G)$*, Quart. J. Math. Oxford, (2)24 (1973), 119–128.

[BadPau] C. Badea and V.I. Paulsen *Schur multipliers and operator-valued Foguel–Hankel operators*, Preprint (2000), Univ. Lille-1 (badea@agat.univ-lille1.fr).

[Baez] L. Baez-Duarte, *On Beurling's real variable reformulation of the Riemann hypothesis*, Adv. Math. 101 (1993), no. 1, 10–30.

[BaS] F. Bagemihl and W. Seidel, *A problem concerning cluster sets of analytic functions*, Math. Z., 62 (1955), 99–110.

[BaC] M. Bakonyi and T. Constantinescu, *Schur's algorithm and several applications*, Pitman Res. Notes Math. 261, Longman, Harlow, and Wiley, New York, 1992.

[Bal] M.Balazard, *Completeness problems and the Riemann hypothesis: an annotated bibliography*, manuscript, Univ. de Bordeaux, 2000.

[BS] M. Balazard et E. Saias, *Notes sur la fonction ζ de Riemann, 1–3*, manuscripts, Univ. de Bordeaux, 1997.

[Ball] J.A. Ball, *Models for noncontractions*, J. Math. Anal. Appl., 52 (1975), 235–254.

[BGR] J.A. Ball, I. Gohberg, and L. Rodman, *Interpolation of rational matrix functions*, (Operator theory: Advances and applications, 45), Basel, Birkhäuser, 1990.

[BaH] J.A. Ball and J.W. Helton, *A Beurling-Lax theorem for the Lie group $U(m,n)$ which contains most classical interpolation theory*, J. Oper. Theory, 9 (1983), 107–142.

[BaL] J.A. Ball and A. Lubin, *On a class of contractive perturbations of restricted shifts*, Pacific J. Math., 63 (1976), no. 2, 309–323.

[Ban1] S. Banach, *Über einige Eigenschaften der lacunären trigonometrischen Reihen*, Studia Math., 2 (1930), 207–220.

[Ban2] ——— *Théorie des opérations linéaires*, Monografie Matematyczne, Warszawa, 1932.

[Bara] A. Baranov, *Differentiation in the de Branges spaces and embedding theorems*, to appear.

[BaLe] L. Baratchart and J. Leblond, *Hardy approximation to L^p functions on subsets of the circle with $1 \le p < \infty$*, Constructive Approx., 14 (1998), 41–56.

[BLP] L. Baratchart, J. Leblond and J.R. Partington, *Hardy approximation to L^∞ functions on subsets of the circle*, Constructive Approx., 12 (1996), 423–436.

[BKö] K. Barbey and H. König, *Abstract analytic function theory and Hardy algebras*, Lect. Notes Math., 593, Springer–Verlag, Berlin, 1977.

[Bar1] N.K. Bari, *On bases in a Hilbert space*, Doklady Akad. Nauk SSSR, 54 (1946), 383–386 (Russian).

[Bar2] ——— *Biorthogonal systems and bases in a Hilbert space*, Uchenye zapiski MGU (Moscow State Univ.), no. 148, vol. 4 (1951), 69–107 (Russian).

[BSch] K.F. Barth and W.J. Schneider, *An asymptotic analog of the F. and M. Riesz radial uniqueness theorem*, Proc. Amer. Math. Soc. 22 (1969), no. 1, 53–54.

[Bat] C.J.K. Batty, *Asymptotic behaviour of semigroups of operators*, pp. 35–52 in: Banach Center Publications, vol. 30 (ed. J. Zemanek), Warszawa, 1994.

[BY] C.J.K. Batty and S.B. Yeates, *Weighted and local stability of semigroups of operators*, Math. Proc. Cambridge Philos. Soc. 129 (2000), no. 1, 85–98.

[Bau] H. Baumgärtel, *Analytic perturbation theory for matrices and operators* Birkhäuser Verlag, Basel, 1985.

[Bax] G. Baxter, *A norm inequality for a finite-section Wiener-Hopf equation*, Illinois J. Math. 7 (1963), 97–103.

[BeaBu] K. Beatrous and J. Burbea, *Reproducing kernels and interpolation of holomorphic functions*, pp. 25-46 in: *Complex Analysis, Functional Analysis and Approximation Theory* (Ed., J.Mujica), 1986.

[Beau] B. Beauzamy, *Introduction to operator theory and invariant subspaces*, North-Holland, 1988.

[BN] N. Benamara and N.K. Nikolski, *Resolvent test for similarity to a normal operator*, Proc. London Math. Soc., (3) 78 (1999), 585–626.

[Be] H. BERCOVICI, *Operator theory and arithmetic in H^∞*, Math. Surveys and Monographs, no. 26, Amer. Math. Soc., Providence, RI, 1988.

[BF] H. BERCOVICI AND C.FOIAŞ, *A real variable restatement of Riemann's hypothesis*, Israel J. Math. 48 (1984), 57–68.

[BFP] H. BERCOVICI, C. FOIAŞ AND C. PEARCY, *Dual algebras with applications to invariant subspaces and dilation theory*, CBMS series 56, Amer. Math. Soc., Providence, RI, 1985.

[Ber] C. BERENSTEIN (ED.), *Complex analysis*, Lect. Notes Math., 1275, 1276, 1277, Berlin etc., Springer–Verlag, 1987.

[BD] C. BERENSTEIN AND M. DOSTAL, *Analytically uniform spaces and their applications to convolution equations*, Lect. Notes Math., 256, Berlin etc., Springer–Verlag, 1972.

[BT] C. BERENSTEIN AND B.A. TAYLOR, *A new look at interpolation theory for entire functions of one variable*, Adv. in Math., 33 (1979), 109–143.

[BeChR] C. BERG, J.P.R. CHRISTENSEN AND P. RESSEL, *Harmonic analysis on semigroups. Theory of positive definite and related functions*, Springer-Verlag, NY-Heidelberg, 1984.

[B.I] I.D. BERG, *An extension of the Weyl-von Neumann theorem to normal operators*, Trans. Amer. Math. Soc. 160 (1971), 365–371.

[Berge] C.A. BERGER, *Normal dilations*, Thesis, Cornell University, 1963; Dissert. Abstracts, 1964, 24 (1964), no. 7, p. 2918.

[BergeSh] C.A. BERGER AND B.I. SHAW, *Intertwining, analytic structure, and the trace norm estimate*, Lect. Notes Math., 345 (1973), 1–6.

[BLo] J. BERGH AND J. LÖFSTRÖM, *Interpolation spaces. An introduction*, Grundlehren der Mathematischen Wissenschaften, no. 223, Springer–Verlag, Berlin–New York, 1976.

[Berg1] S. BERGMAN, *Über die Entwicklung der harmonischen Funktionen der Ebene und des Raumes nach Orthogonalfunktionen*, Math. Ann., 86 (1922), 238–271.

[Berg2] ——— *The kernel function and conformal mapping*, Amer. Math. Soc. Math. Surveys, vol. 5, Providence, RI, 1950.

[BerA] A. BERNARD, *Algèbres quotients d'algèbres uniformes*, C.R. Acad. Sci. Paris, Sér. A, 272 (1971), 1101–1104.

[Bern1] S.N. BERNSTEIN, *Le problème de l'approximation des fonctions continues sur tout l'axe réel et l'une de ses applications*, Bull. Soc. Math. France, 52 (1924), 399–410.

[Bern2] ——— *Leçons sur ler propriétés extremales et la meilleure approximation des fonctions analytiques d'une variable réelle*, Gauthier–Villars, Paris, 1926.

[Bes] O.V. BESOV, *On a family of function spaces. Embedding and extension theorems*, Doklady Akad. Nauk SSSR, 126 (1959), 1163–1165.

[Beu1] A. BEURLING, *Sur les fonctions limites quasianalytiques des fractions rationnelles*, pp. 199–210 in: 8 Scand. Math. Congr. Stockholm, 1934.

[Beu2] ——— *On two problems concerning linear transformations in Hilbert space*, Acta Math. 81 (1949), no. 1–2, 79–93.

[Beu3] ——— *A closure problem related to the Riemann ζ-function*, Proc. Nat. Acad. USA, 41 (1955), no. 5, 312–314.

[Beu4] ——— *A critical topology in harmonic analysis on semigroups*, Acta Math. 112 (1964), 215–228.

[BeuM1] A. BEURLING AND P. MALLIAVIN, *On Fourier transforms of measures with compact support*, Acta Math., 107 (1962), 291–309.

[BeuM2] ——— *On the closure of characters and the zeros of entire functions*, Acta Math., 118 (1967), 79–93.

[Bha] R. BHATIA, *Matrix analysis*, Springer–Verlag, NY, 1997.

[BhaH] R. BHATIA AND A.R. HOLBROOK, *A softer, stronger Lidskii theorem*, Proc. Indian Acad. Sci. (Math. Sci.), 99 (1989), no. 1, 75–83.

[Bir] G.D. BIRKHOFF *Démonstration d'un théorème élémentaire sur les fonctions entières*, C.R. Acad. Paris, 189 (1929), 473–475.

[BiS] M.S. BIRMAN AND M.Z. SOLOMYAK, *Spectral theory of selfadjoint operators in Hilbert space*, Leningrad. Univ., Leningrad, 1980 (Russian); English transl.: Mathematics and its Applications (Soviet Series), D. Reidel Publishing Co., Dordrecht, 1987.

[BiBr] E. BISHOP AND D. BRIDGES, *Constructive analysis*, Springer Verlag, Heidelberg, 1985.

[Bl1] O. BLASCO, *Boundary values of functions in vector valued Hardy spaces and geometry of Banach spaces*, J. Funct. Anal., 78 (1988), 346–364.

[Bl2] ——— *Vector-valued Hardy inequalities and B-convexity*, Ark. Mat., 38 (2000), 21-36.

[Bl3] ——— *Remarks on vector-valued $BMOA$ and vector-valued multipliers*, preprint (2001), to appear in *Positivity*.

[BJH] P. Bloomfeld, N.P. Jewell, and E. Hayashi, *Characterization of completely non-deterministic stochastic processes*, Pacific J. Math., 107 (1983), 307–317.

[Bo1] R.P. Boas, *Entire functions*, N.Y., Academic Press, 1954.

[Bo2] ——— *Majorant problems for Fourier series*, J. Anal. Math., 10 (1962), 253–271.

[Boch] S. Bochner, *Harmonic analysis and the theory of probability*, Univ. Calif. Press, 1955.

[BPh] S. Bochner and R.S. Phillips, *Absolutely convergent Fourier expansions for non-commutative normed rings*, Ann. of Math., (2) 43 (1942), 409–418.

[Boh] H. Bohr, *A theorem concerning power series*, Proc. London Math. Soc., (2) 13 (1914), 1–5.

[BoB] A. Bonami and J. Bruna, *On truncations of Hankel and Toeplitz operators*, Publicacions Matematiques (Barcelona), 43 (1999), 235–250.

[BoD] F.F. Bonsall and J. Duncan, *Complete normed algebras*, Berlin etc., Springer–Verlag, 1973.

[BoGi] F.F. Bonsall and T.A. Gillespie, *Hankel operators with PC symbols and the space $H^\infty + PC$*, Proc. Royal Soc. Edinburgh, 89 A (1981), 17–24.

[BP] F.F. Bonsall and S.C. Power, *A proof of Hartman's theorem on compact Hankel operators*, Math. Proc. Camb. Phil. Soc., 78 (1975), 447–450.

[Boo] G. Boole, *On the comparison of transcendents, with certain applications to the theory of definite integrals*, Phil. Trans. Royal Soc., 147 (1857), 745-803.

[Bore] E. Borel, *Remarques sur la Note de M. Wolff*, C.R. Acad. Sci. Paris, 173 (1921), 1056–1057.

[Bor1] A. Borichev, *The generalized Fourier transformation, Titchmarsh's theorem and asymptotically holomorphic functions*, Algebra i Analiz, 1 (1989), no. 4, 17–53; English transl.: Leningrad Math. J., 1 (1990), 825–857.

[Bor2] ——— *Beurling algebras and the generalized Fourier transform*, Proc. London Math. Soc., (3) 73 (1996), 431–480.

[Bor3] ——— *Estimates from below and cyclicity in Begman-type spaces*, Internat. Math. Research Notes (1996), no. 12, 603–611.

[Bor4] ——— *Invariant subspaces of given index in Banach spaces of analytic functions*, Journ. reine und angew. Math., 505 (1998), 23–44.

[Bor5] ——— *On the closure of polynomials in weighted spaces of functions on the real line*, Indiana Univ. Math. J., 50 (2001), 829–846.

[BH1] A. Borichev and H. Hedenmalm, *Completeness of translates in weighted spaces on the half-line*, Acta Math., 174 (1995), no. 1, 1–84.

[BH2] ——— *Harmonic functions of maximal growth: invertibility and cyclicity in Bergman spaces*, J. Amer. Math. Soc., 10 (1997), 761–796.

[BHV] A. Borichev, H. Hedenmalm and A. Volberg, *Invertibility, cyclicity and subspaces of big index in large Bergman spaces*, manuscript, 2000.

[BoS1] A. Borichev and M. Sodin, *The Hamburger moment problem and weighted polynomial approximation on discrete subsets of the real line*, J. Analyse Math., 76 (1998), 219–264.

[BoS2] ——— *Weighted polynomial approximation and the Hamburger moment problem*, Acta Univ. Upsaliensis, 64 (1999), 110–122.

[BoS3] ——— *Krein's entire functions and the Bernstein approximation problem*, Illinois J. Math., 45 (2001), 167–185.

[Bori1] I.A. Boricheva, *An application of the Schur method to the free interpolation problems in the model space*, Algebra i Analiz, 7 (1995), no. 4, 44–64 (Russian); English Transl.: St.Petersburg Math. J., 7 (1996), no. 4, 543–560.

[Bori2] ——— *Geometric properties of projections of reproducing kernels on z^*-invariant subspaces of H^2*, J. Funct. Anal., 161 (1999), 397–417.

[BoriD] I.A. Boricheva and E.M. Dyn'kin, *A non-classical free interpolation problem*, Algebra i Analiz, 4:5 (1992), 45-90 (Rusian); English transl.: St.Petersburg Math. J., 4:5 (1993).

[Bö] A. Böttcher, *Toeplitz operators with piecewise continuous symbols — a neverending story?*, Jahresber. Deutsch. Math.-Verein. 97 (1995), no. 4, 115–129.

[BöG] A. Böttcher and S.M. Grudsky, *Toeplitz operators with discontinuous symbols — phenomena beyond piecewise continuity*, In: Operator theory: Advances and applications, 90 (1996), Basel, Birkhäuser, 55–118.

[BöG2] ______ *Toeplitz matrices, asymptotic linear algebra and functional analysis*, Hindustan Book Agency, New Delhi, 2000.

[BöK] A. Böttcher and Yu.I. Karlovich, *Carleson curves, Muckenhoupt weights, and Toeplitz operators*, Birkhäuser Verlag, Basel, 1997.

[BöSe] A. Böttcher and M. Seybold, *Discrete one-dimensional zero-order pseudodifferential operators on spaces with Muchenhoupt weights*, Algebra i Analiz, 13:2 (2001); (Reprinted in St. Petersburg Math. J., 13:2 (2002)).

[BSi1] A. Böttcher and B. Silbermann, *Invertibility and asymptotics of Toeplitz matrices*, Akademie–Verlag, Berlin, 1983.

[BSi2] ______ *Analysis of Toeplitz operators*, Akademie–Verlag, Berlin, 1989, and Springer–Verlag, 1990.

[BSi3] ______ *Introduction to large truncated Toeplitz matrices*, New York etc., Springer–Verlag, 1999.

[Bour] N. Bourbaki, *Théories spectrales*, Eléments de mathématique, Fasc. XXXII, Hermann, Paris, 1967.

[BouSh] P.S. Bourdon and J.H. Shapiro, *Cyclic phenomena for composition operators*, Mem. Amer. Math. Soc. 125 (1997), no. 596,

[Bou1] A. Bourgain, *Some remarks on Banach spaces in which martigale difference sequences are unconditional*, Ark. Mat., 21 (1983), no. 2, 163–168.

[Bou2] ______ *Bilinear forms on H^∞ and bounded bianalytic functions*, Trans. Amer. Math. Soc., 286 (1984), no. 1, 313–337.

[Bou3] ______ *On the similarity problem for polynomially bounded operators on Hilbert space*, Israel J. Math., 54 (1986), 227–241.

[Boz] M. Bozejko, *Positive-definite kernels, length functions on groups and a noncommutative von Neumann inequality*, Studia Math., 95 (1989), 107–118.

[Br] J. Bram, *Subnormal operators*, Duke Math. J., 22 (1955), no. 1, 75–94.

[BR] O. Bratteli and D. Robinson, *Operator algebras and quantum statistical mechanics*, New York, Springer-Verlag, 1981.

[Bre] S. Brehmer, *Über vertauschbare Kontraktionen des Hilbertschen Raumes*, Acta Sci. Math. Szeged, 22 (1961), 106–111.

[Bro1] M.S. Brodskii, *On a problem of I.M. Gelfand*, Uspehi Matem. Nauk, 12 (1957), no. 2, 129–132 (Russian).

[Bro2] ______ *On unicellularity of the integration operator and a Titchmarsh theorem*, Uspehi Matem. Nauk, 20 (1965), no. 5, 189–192 (Russian).

[Bro3] ______ *Triangular and Jordan representation of linear operators*, Moscow, Nauka, 1969 (Russian); English transl.: Amer. Math. Soc., Providence, 1971.

[BrL] M.S. Brodskii and M.S. Livshic, *Spectral analysis of non-selfadjoint operators and intermediate systems*, Uspehi Matem. Nauk, 13 (1958), no. 1, 3–85 (Russian).

[Brown] A. Brown, *A version of multiplicity theory*, pp. 129–160 in: Topics in operator theory (ed. C. Pearcy), Math. Surveys series, 13, Amer. Math. Soc., Providence, 1974.

[BDF] L.G. Brown, R.G. Douglas, and P.A. Fillmore, *Unitary equivalence modulo the compact operators and extensions of C^*-algebras*, Lect. Notes Math., Springer–Verlag, 345 (1973), 58–128.

[BrH] A. Brown and P. Halmos, *Algebraic properties of Toeplitz operators*, J. Reine Angew. Math., 213 (1963), 89–102.

[BSZ] A. Brown, A. Shields, and K. Zeller, *On absolutely convergent exponential sums*, Trans. Amer. Math. Soc. 96 (1960), no. 1, 162–183.

[Bu1] A.V. Bukhvalov, *Continuity of operators in spaces of measurable vector-valued functions with applications to the study of Sobolev spaces and spaces of analytic functions in the vector-valued case*, Doklady Akad. Nauk SSSR, 246 (1979), no. 3, 524–528 (Russian); English transl.: Sov. Math., Dokl., 20 (1979), 480–484.

[Bu2] ______ *Application of methods of the theory of order-bounded operators to the theory of operators in L^p-spaces*, Uspehi Matem. Nauk, 38 (1983), no. 6, 37–83 (Russian); English transl.: Russ. Math. Surv. 38 (1983), no. 6, 43–98.

[BuD] A.V. BUKHVALOV AND A.A. DANILEVICH, *Boundary properties of analytic and harmonic functions with values in Banach spaces*, Matem. Zametki, 31 (1982), no. 2, 203–214 (Russian); English transl.: Math. Notes, 31 (1983), no. 1–2, 104–110.

[Bun] J.W. BUNCE, *Models for n-tuples of non-commuting operators*, J. Funct. Anal., 57 (1984), 21–30.

[Burk] D.L. BURKHOLDER, *A geometric condition that implies the existence of certain singular integrals of Banach-space-valued functions*, pp. 270–286 in: Conference on Harmonic Analysis in Honor of Antony Zygmund, eds. W. Beckner, A. Calderon, R. Fefferman, and P. Jones, Wadsworth Int. Math. Series, 1983.

[Bur] L. BURLANDO, *Continuity of spectrum and spectral radius in Banach algebras*, pp. 53–100 in: Banach Center Publications, vol. 30 (ed. J. Zemanek), Warszawa, 1994.

[Butk] A.G. BUTKOVSKII, *Methods of control of distributed parameter systems*, Nauka, Moscow, 1995 (Russian); English Transl.: *Theory of optimal control of distributed parameter systems*, American Elsevier, 1969.

[But] J.R. BUTZ, *s-numbers of Hankel matrices*, J. Funct. Anal., 15 (1974), 297–305.

[Can] D.G. CANTOR, *Power series with integral coefficients*, Bull. Amer. Math. Soc., 69 (1963), 362-366.

[Cara] S.R. CARADUS, *Universal operators and invariant subspaces*, Proc. Amer. Math. Soc., 23 (1969), no. 3, 526–527.

[Car1] T. CARLEMAN, *Zur Theorie der linearen Integralgleichungen*, Math. Zeit., 9 (1921), 196–217.

[Car2] ______ *Sur les séries* $\sum \frac{A_n}{z-z_n}$, C.R. Acad. Sci. Paris, 174 (1922), 1056–1057.

[Car3] ______ *Les fonctions quasi analytiques*, Gauthier-Villars, Paris, 1926.

[C1] L. CARLESON, *Representations of continuous functions*, Math. Zeit., 66 (1957), 447–451.

[C2] ______ *An interpolation problem for bounded analytic functions*, Amer. J. Math., 80 (1958), 921–930.

[C3] ______ *Interpolations by bounded analytic functions and the corona problem*, Ann. of Math., 76 (1962), 547–559.

[C4] ______ *On convergence and growth of partial sums of Fourier series*, Acta Math., 116 (1966), 135–157.

[Cha] B.L. CHALMERS, *Some interpolation problems in Hilbert spaces*, Michigan Math. J., 18 (1971), 41–50.

[ChL] C. CHICONE AND YU. LATUSHKIN, *Evolution semigroups in Dynamical systems and differential equations*, Math. Surveys and Monographs, 70, Amer. Math. Soc., Providence, 1999.

[Chr] O. CHRISTENSEN, *Frames, Riesz bases, and discrete Gabor/wavelet expansions*, Bull. Amer. Math. Soc., 38 (2001), no. 3, 273–291.

[CiRos] J.A. CIMA AND W.T. ROSS, *The backward shift on the Hardy space*, Amer. Math. Soc. Surveys and Monographs, vol. 79, Providence, 2000.

[CiSt] J.A. CIMA AND M. STESSIN, *On the recovery of analytic functions*, Canad. J. Math., 48 (2) (1996), 288–301.

[Clan] K. CLANCEY, *Seminormal operators*, Lect. Notes Math., 742, Springer–Verlag, 1979.

[CG] K. CLANCEY AND I. GOHBERG, *Factorization of matrix functions and singular integral operators*, Operator theory: Advances and applications, 3, Basel, Birkhäuser, 1981.

[Cl1] D.N. CLARK, *On commuting contractions*, J. Math. Anal. Appl., 32 (1970), 590–596.

[Cl2] ______ *One dimensional perturbations of restricted shifts*, J. Analyse Math., 25 (1972), 169–191.

[Cl3] ______ *On Toeplitz operators with loops, II*, J. Oper. Theory, 7 (1982), 109–123.

[Cle] PH. CLÉMENT, H. HEIJMANS, S. ANGENENT, C. VAN DUIJN, B. DE PAGTER (EDS.), *One-parameter semigroups*, CWI Monographs, 5. North-Holland Publishing Co., Amsterdam–New York, 1987.

[Cob1] L.A. COBURN, *Weyl's theorem for nonnormal operators*, Mich. Math. J., 13 (1966), 285–288.

[Cob2] ______ *The C^*-algebra generated by an isometry.I*, Bull. Amer. Math. Soc., 73 (1967), 722–726.

[Coh] P. COHEN, *A note on constructive methods in Banach algebras*, Proc. Amer. Math. Soc., 12:1 (1961), 159-164.

[Cohn] W.S. Cohn, *Carleson measures for functions orthogonal to invariant subspaces*, Pacific J. Math., 103 (1982), no. 2, 367–384.

[CV] W.S. Cohn and I.E. Verbitskii, *Factorization of tent spaces and Hankel operators*, J. Funct. Anal., 175 (2000), no. 2, 308–329.

[CM] R.R. Coifman and Y. Meyer, *Ondelettes et opérateurs. III. Opérateurs multilinéaires*, Paris, Hermann, 1990.

[CR] R.R. Coifman and R. Rochberg, *Representation theorems for holomorphic and harmonic functions in L^p*, pp. 11–66 in: Representation theorems for Hardy classes, Astérisque 77, Soc. Math. France, Paris, 1980.

[CRW] R.R. Coifman, R. Rochberg, and G. Weiss, *Factorization theorems for Hardy spaces in several variables*, Ann. Math., 103 (1976), 611–635.

[CoiW] R.R. Coifman and G. Weiss, *Transference methods in analysis*, CBMS regional conf. series in math., 31, Amer. Math. Soc., Providence, RI, 1977.

[CLW] B.J. Cole, K. Lewis, and J. Wermer, *A characterization of Pick bodies*, J. London Math. Soc. (2), 48 (1993), 316-328.

[CW1] B.J. Cole and J. Wermer, *Pick interpolation, von Neumann inequalities, and hyperconvex sets*, In: Complex potential theory (Montreal, PQ, 1993), 89–129, NATO Adv. Sci. Inst. Ser. C Math. Phys. Sci., 439, Kluwer Acad. Publ., Dordrecht, 1994.

[CW2] ——— *Ando's theorem and sums of squares*, Indiana Univ. Math. J., 48 (2000), no. 3, 767–791.

[CL] E.F. Collingwood and A.J. Lohwater, *The theory of cluster sets*, Cambridge Univ. Press, 1966.

[CF] I. Colojoara and C. Foiaş, *The theory of generalized spectral operators*, Gordon and Breach, N.Y., 1968.

[Con] J.B. Conway, *The Theory of Subnormal Operators*, Math. Surveys and monographs 36, Amer. Math. Soc., Providence, RI, 1991.

[CY] J.B. Conway and Liming Yang, *Some open problems in the theory of subnormal operators*, pp. 201–209 in: Holomorphic Spaces, eds. Sh. Axler, J. McCarthy, and D. Sarason, MSRI Publications, 33, Cambridge Univ. Press, 1998.

[Cot1] M. Cotlar, *A combinatorial inequality and its application to L^2 spaces*, Rev. Math. Cuyana, 1 (1955), 41–55.

[CS1] M. Cotlar and C. Sadosky, *On the Helson–Szegö theorem and a related class of modified Toeplitz kernels*, pp. 383–407 in: Harmonic analysis in Euclidean spaces (Williamstowm, MA, 1978), Part 1, eds. G. Weiss, and S. Wainger, Proc. Symp. Pure Math. 35, Amer. Math. Soc., Providence, 1979.

[CS2] ——— *On some L^p versions the Helson–Szegö theorem*, pp. 306–317 in: Conference on harmonic analysis in honor of Antoni Zygmund, Vol. I, II (Chicago, Ill., 1981), Wadsworth Math. Ser., Wadsworth, Belmont, Calif., 1983.

[CS3] ——— *Generalized Toeplitz kernels, stationarity and harmonizability*, J. Analyse Math., 44 (1984/85), 117–133.

[CS4] ——— *A lifting theorem for subordinated invariant kernels*, J. Funct. Anal., 67 (1986), 345–359.

[CS5] ——— *Toeplitz liftings of Hankel forms*, pp. 22–43 in: Function spaces and applications (Lund, 1986), Lect. Notes Math., 1302, Springer, Berlin and New York, 1988.

[CS6] ——— *Weakly positive matrix measures, generalized Toeplitz forms, and their applications to Hankel and Hilbert transform operators*, pp. 93–120 in: Operator Theory: Adv. and Appl. (Basel, Birkhäuser), vol. 58, 1992.

[CS7] ——— *Abstract, weighted, and multidimensional Adamyan-Arov-Krein theorems, and the singular numbers of Sarason commutators*, Integr. Eq. Oper. Theory, 17 (1993), 169–201.

[CS8] ——— *Nehari and Nevanlinna-Pick problems and holomorphic extensions in the polydisk in terms of restricted BMO*, J. Funct. Anal., 124 (1994), no. 1, 205–210.

[CD] M.J. Cowen and R.G. Douglas, *Operators possessing an open set of eigenvalues*, Colloq. Math. Soc. Janos Bolyai, 323–341; Amsterdam, North-Holland, 1980.

[CrD] M. Crabb and A.M. Davie, *Von Neumann's inequality for Hilbert space operators*, Bull. London Math. Soc., 7 (1975), 49–50.

[CPr] R.F. Curtain and A.J. Pritchard, *Infinite-dimensional linear systems theory*, Lect. Notes CIS, vol. 8, Springer–Verlag, Berlin, 1978.

[CZ] R.F. Curtain and H. Zwart, *An introduction to infinite-dimensional linear systems theory*, Springer–Verlag, Heidelberg, 1995.

[Cur] R. Curto, *An operator theoretic approach to truncated moment problem*, pp. 75-104 in: *Banach Center Publications*, Vol.38 (Eds. J. Janas, F.H. Szafraniec and J. Zemanek), Warszawa, 1997.

[Dan] V.W. Daniel, *Convolution operators on Lebesgue spaces at the half-line*, Trans. Amer. Math. Soc., 164 (1972), 479–488.

[Dat] R. Datko, *Extending a theorem of A.M. Liapounov to Hilbert Space*, J. Math. Anal. Appl., 32 (1970), 610–616.

[Da] G. David, *Wavelets and singular integrals on curves and surfaces*, Lect. Notes Math., 1465, Berlin, Springer–Verlag, 1991.

[Dav] K.R. Davidson, *Nest algebras. Triangular forms for operator algebras on Hilbert space*, Pitman Res. Notes Math. Series, 191, Longman Sci&Tech, UK, 1988.

[DP] K.R. Davidson and V.I. Paulsen, *Polynomially bounded operators*, J. Reine Angew. Math., 487 (1997), 153–170.

[DPi1] K.R. Davidson and D.R. Pitts, *Nevanlinna-Pick interpolation for non-commutative analytic Toeplitz algebras*, Integral Eq. Oper. Theory, 31 (1998), no. 3, 321–337.

[DPi2] ——— *The algebraic structure of non-commutative analytic Toeplitz algebras*, Math. Ann. 311 (1998), no. 2, 275–303.

[Davies] E.B. Davies, *Spectral theory and differential operators*, Cambridge Univ. Press, Cambridge, 1995.

[Davi] Ch. Davis, *J-unitary dilation of a general operator*, Acta Sci. Math. Szeged, 31 (1970), 75–86.

[DF] Ch. Davis and C. Foiaş, *Operators with bounded characteristic functions and their J-unitary dilations*, Acta Sci. Math. Szeged, 32 (1971), 127–139.

[dB1] L. de Branges, *The Bernstein problem*, Proc. Amer. Math. Soc., 10 (1959), 825–832.

[dB2] ——— *Hilbert spaces of entire functions*, Prentice Hall, Englewood Cliffs, N.J., 1968.

[dB3] ——— *Square summable power series*, Unpublished manuscript, 400 pp., 1984.

[dB4] ——— *Underlying concepts in the proof of the Bieberbach conjecture*, Proc. Internat. Congr. Math. at Berkeley 1986, Berkeley, 25–42.

[dB5] ——— *A conjecture which implies the Riemann hypothesis*, J. Funct. Anal., 121 (1994), 117–184.

[dB6] ——— *A proof of the Riemann hypothesis*, preprint, Purdue University, Lafayette, 2001.

[dB7] ——— *Invariant subspaces and the Stone-Weierstrass theorem*, preprint, Purdue University, Lafayette, 2000.

[dBR1] L. de Branges and J. Rovnyak, *Square summable power series*, Holt, Rinehart and Winston, N.Y., 1966.

[dBR2] ——— *Canonical models in quantum scattering theory*, pp. 295–392 in: Perturbation theory and its applications in quantum mechanics, Madison, 1965, ed. C.H.Wilcox, Wiley, N.Y., 1966

[dBS] L. de Branges and L. Schulman, *Perturbations of unitary transformations*, J. Math. Anal. and Appl., 23 (1968), no. 2, 294–326.

[Del] R. Delaubenfels, *Similarity to a contraction, for power bounded operators with finite peripheral spectrum*, Trans. Amer. Math. Soc., 350 (1998), no. 8, 3169–3191.

[dLR] K. de Leeuw and W. Rudin, *Extreme points and extreme problems in H_1*, Pacific J. Math., 8 (1958), 467–485.

[Den] A. Denjoy, *Sur les séries de fractions rationnelles*, Bull. Soc. Math. France, 52 (1924), 418–434.

[Dev1] A. Devinatz, *The factorization of operator valued functions*, Ann. of Math., (2) 73 (1961), 458–495.

[Dev2] ——— *Toeplitz operators on H^2 spaces*, Trans. Amer. Math. Soc., 112 (1964), 304–317.

[Dev3] ——— *On Wiener-Hopf operators*, in Functional Analysis (B.R.Gelbaum, ed.), Proc. Conf., Irvine, CA, 1966, pp.81–118; Washington D.C., Thompson Book Co., 1967.

[DevSh] A. Devinatz and M. Shinbrot, *General Wiener-Hopf operators*, Trans. Amer. Math. Soc., 145 (1969), 467-494.

[DeVL] R.A. DeVore and G.G. Lorentz, *Constructive approximation*, Springer–Verlag, Heidelberg–Berlin, 1993.

[Dij] A. DIJKSMA, *Almost commutant lifting in a Krein space setting*, Analysis Conference in honor of C. Foiaş, October 18-19, 2000, Amsterdam.

[Dix] J. DIXMIER, *Les moyennes invariantes dans les semi-groupes et leurs applications*, Acta Sci. Math. Szeged, 12 (1950), 213–227.

[Djr] M.M. DJRBASHIAN [DZHRBASHYAN], *Integral transforms and representations of functions of complex variable*, Nauka, Moscow, 1966 (Russian).

[DjSh] M.M. DJRBASHIAN AND F.A. SHAMOYAN, *Topics in theory of A^p_α spaces*, Teubner–Verlag, Leipzig, 1988.

[DolR] S. DOLECKI AND D. RUSSEL, *A general theory of observation and control*, SIAM J. Control and Optim., 15 (1977), 185–220.

[Don1] W.F. DONOGHUE, *The lattice of invariant subspaces of a completely continuous quasinilpotent transformation*, Pacif. J. Math., 7 (1957), no. 2, 1031–1035.

[Don2] ______ *Monotone matrix functions and analytic continuation*, N.Y., Springer–Verlag, 1974.

[Dou1] R.G. DOUGLAS, *On factoring positive operator functions*, J. Math. Mech., 16 (1966), 119–126.

[Dou2] ______ *On majorization, factorization and range inclusion of operators in Hilbert space*, Proc. Amer. Math. Soc., 17 (1966), 413–415.

[Dou3] ______ *Structure theory for operators, I.*, J. Reine Angew. Math., 232 (1968), 180–193.

[Dou4] ______ *On the operator equations $S^*XT = X$ and related topics*, Acta Sci. Math. Szeged, 30 (1969), no. 1–2, 19–32.

[Dou5] ______ *Banach algebra techniques in operator theory*, N.Y., Academic Press, 1972.

[Dou6] ______ *Banach algebra techniques in the theory of Toeplitz operators*, CBMS series, 15, Amer. Math. Soc., Providence, RI, 1973.

[Dou7] ______ *Canonical models*, pp. 163–218 in: Topics in operator theory (ed., C. Pearcy), Math. Surveys series, 13, Amer. Math. Soc., Providence, 1974.

[DR] R.G. DOUGLAS AND W. RUDIN, *Approximation by inner functions*, Pasif. J. Math., 31 (1969), 313–320.

[DSar] R.G. DOUGLAS AND D. SARASON, *Fredholm Toeplitz operators*, Proc. Amer. Math. Soc., 26 (1970), 117–120.

[DSS] R.G. DOUGLAS, H.S. SHAPIRO, AND A.L. SHIELDS, *Cyclic vectors and invariant subspaces of the backward shift operator*, Ann. Inst. Fourier, 20 (1970), 37–76.

[DN] L.N. DOVBYSH AND N.K. NIKOLSKI, *Two ways to avoid the hereditary completeness*, Zapiski Nauchn. Semin. LOMI, 65 (1976), 183–188 (Russian); English transl.: J. Sov. Math., 16 (1981), 1175–1179.

[DNS] L.N. DOVBYSH, N.K. NIKOLSKI, AND V.N. SUDAKOV, *How "good" can a non-hereditarily complete family be?*, Zapiski Nauchn. Semin. LOMI, 73 (1977), 52–69 (Russian); English transl.: J. Soviet Math., 34 (1986), 2050–2060.

[Dow] H.R. DOWSON, *Spectral theory of linear operators*, Academic Press, London New York, 1978.

[DFT] J.C. DOYLE, B.A. FRANCIS AND A.R. TANNENBAUM, *Feedback control theory*, McMillan, 1982.

[Dru1] S. DRURY, *A generalization of von Neumann's inequality to the complex ball*, Proc. Amer. Math. Soc., 68 (1978), no. 3, 300–304.

[Dru2] ______ *Remarks on von Neumann's inequality*, pp. 14–32 in: Banach spaces, harmonic analysis, and probability theory, Proceedings 1980–1981 (eds. R. Blai and S. Sydney), Lect. Notes Math., 995, Springer–Verlag, Heidelberg, 1983.

[DuE] R.J. DUFFIN AND J.J. EACHUS, *Some notes on an expansion theorem of Paley and Wiener*, Bull. Amer. Math. Soc., 48 (1942), no. 12, 850–855.

[DS1] N. DUNFORD AND J. SCHWARTZ, *Linear operators. Part 1. General theory*, Wiley–Interscience, New York, 1958.

[DS2] ______ *Linear operators. Part 2. Spectral theory. Self-adjoint operators in Hilbert space*, Wiley–Interscience, New York, 1963.

[DS3] ______ *Linear operators. Part 3. Spectral operators*, Wiley–Interscience, New York, 1971.

[Du1] P. DUREN, *Theory of H^p spaces.*, Academic Press, New York, 1970.

[Du2] ______ *Bergman spaces*, AMS, 2001.

[DRSh] P.L. DUREN, B.W. ROMBERG, AND A.L. SHIELDS, *Linear functionals on H^p spaces with $0 < p < 1$*, J. Reine Angew. Math., 238 (1969), 32–60.

[Dya1] K.M. DYAKONOV, *Generalized Hardy inequalities and pseudocontinuable functions*, Ark. Mat., 34 (1996), 231-244.

[Dya2] ——— *Kernels of Toeplitz operators via Bourgain's factorization theorem*, J. Funct. Anal., 170 (2000), 93–106.

[Dya3] ——— *Continuous and compact embeddings between star-invariant subspaces*, pp. 65–76 in: Operator Theory: Adv. and Appl., 113 (The S.A. Vinogradov Memorial Volume), Basel, Birkhäuser, 2000.

[Dym1] H. DYM, *J contractive matrix functions, reproducing kernel Hilbert spaces and interpolation*, Amer. Math. Soc., Providence, R.I., 1989.

[Dym2] ——— *Book review: The commutant lifting approach to interpolation problems*, by C. Foiaş and A. Frazho [FF], Bull. Amer. Math. Soc., 31:1 (1994), 125-140.

[Dym3] ——— *A Basic Interpolation Problem*, pp. 381–424 in: Holomorphic Spaces, eds. Sh. Axler, J. McCarthy, and D. Sarason, MSRI Publications, 33, Cambridge Univ. Press, 1998.

[DyMc] H. DYM AND H.P. MCKEAN, *Gaussian processes, function theory and the inverse spectral problem*, NY, Academic Press, 1976.

[Dyn1] E.M. DYN'KIN, *An operator calculus based on the Cauchy–Green formula, and quasianalyticity of the classes $\mathcal{D}(h)$*, Zapiski Nauchnyh Seminarov LOMI, 19 (1970), 221–226 (Russian); English transl.: Seminars in Math., V.A.Steklov Math. Inst., Leningrad, vol. 19, 128–131, Consultant Bureau, N.Y.-London, 1972.

[Dyn2] ——— *An operator calculus based on the Cauchy-Green formula*, Zapiski Nauchnyh Seminarov LOMI, 30 (1972), 33–39 (Russian); English transl.: J. Soviet Math., 4 (1975), no. 2 (1976) 329–334, Consultant Bureau, N.Y., 1976.

[Dyn3] ——— *Methods of the Theory of Singular Integrals: Hilbert Transform and Calderon-Zygmund Theory*, pp. 197–292 in: Itogi Nauki i Techniki, Contemp. Problems Math. (Russian Math. Encyclopaedia), 15, VINITI, Moscow, 1987 (Russian); English transl.: *Encyclopaedia of Math. Sci.*, vol. 15, 167–260, Berlin etc., Springer–Verlag, 1991.

[Dyn4] ——— *The pseudoanalytic extension*, J. Analyse Math., 60 (1993), 45–70.

[Dyn5] ——— *Inequalities for rational functions*, J. Approx. Theory, 91 (1997), 349-367.

[DO] E.M. DYN'KIN AND B.P. OSILENKER, *Weighted estimates of singular integrals and their applications*, in Itogi Nauki i Techniki Ser. Math. Anal., vol. 21, 42–129, VINITI, Moscow, 1983 (Russian); English transl.: J. Soviet Math., 30 (1985), 2094–2154.

[Ed.H] H.M. EDWARDS, *Riemann's Zeta function*, Academic Press, New York, 1974.

[Ed] R.E. EDWARDS, *Fourier series. A modern introduction*, vol. 1&2, Springer Verlag, Heidelberg, 1982.

[EG] R.E. EDWARDS AND G.I. GAUDRY, *Littlewood–Paley and multiplier theory*, Berlin etc., Springer–Verlag, 1977.

[Ehr] L. EHRENPREIS, *Fourier analysis in several variables*, Wiley–Interscience, New York, 1970.

[Eid1] V.YA. EIDERMAN, *On a sum of values of a function of a certain class along a sequence of points*, Izvestia VUZ'ov. Mathematika, 1 (1992), 89–97 (Russian); English transl.: Russ. Math., 36 (1992), no. 1, 87–95.

[Eid2] ——— *Metric characteristics of exceptional sets and unicity theorems in function theory*, Steklov Math. Inst. St.Petersburg, 2nd Doctor Thesis, 1999, 192 pp (Russian).

[ElR] O. EL-FALLAH AND T. RANSFORD, *Extremal growth of powers of operators satisfying conditions of Reiss-Ritt type*, Preprint, Univ. Laval, Québec, Canada, 2000.

[ENZ] O. EL-FALLAH, N.K. NIKOLSKI AND M. ZARRABI, *Resolvent estimates in Beurling-Sobolev algebras*, Algebra i Analiz, 6 (1998), 1–80 (Russian); English transl.: St.Petersburg Math. J., 6 (1999), 1–69.

[Enf] P. ENFLO, *A counterexample to the approximation property in Banach spaces*, Acta Math., 130 (1973), 309–317.

[ESZ] J. ESTERLE, E. STROUSE, AND F. ZOUAKIA, *Stabilité asymptotique de certains semigroupes d'opérateurs et ideaux primaires de $L^1(\mathbb{R}^+)$*, J. Operator Theory, 28 (1992), 203–227.

[EV] J. ESTERLE AND A. VOLBERG, *Sous-espaces invariants par translations bilatérales de certains espaces de Hilbert de suites quasi-analytiquement pondérées*, C.R. Acad. Sci. Paris, Sér.I, 326 (1998), 295–300.

[Ex] R. EXEL, *Hankel matrices over right ordered amenable groups*, Canad. Math. Bull., 33 (1990), no. 4, 404–415.

[FP] L.D. FADDEEV AND B.S. PAVLOV, *Scattering theory and automorphic functions*, Zapiski Nauchn. Semin. LOMI 27 (1972), 161–193 (Russian); English transl.: J. Sov. Math., 3 (1975), 522–548.

[Fa] P. FATOU, *Séries trigonométriques et séries de Taylor*, Acta Math., 30 (1906), 335–400.

[FS] C. FEFFERMAN AND E.M. STEIN, *H^p spaces of several variables*, Acta Math., 129 (1972), no. 3–4, 137–193.

[Fe] A. FEINTUCH, *Robust control theory in Hilbert spaces*, Ben-Gurion Univ. of the Negev, Lect. Notes, 1995.

[GF] A.I. FELDMAN AND I. GOHBERG, *see* I. Gohberg and A.I. Feldman.

[Fej] L. FEJÉR, *Ueber trigonometrische Polynome*, J. reine angew. Math., 146 (1916), 53–82.

[Fer] S.H. FERGUSON, *Polynomially bounded operators and Ext groups*, Proc. Amer. Math. Soc., 124 (1996), no. 9, 2779–2785.

[F] J.-P. FERRIER, *Spectral theory and complex analysis*, North-Holland, Amsterdam N.Y., 1973.

[Fil] P.A. FILLMORE, *Notes on operator theory*, Van Nostrand Reinhold Co, N.Y. etc, 1970.

[FLS] K.M. FLORENS, YU.I. LYUBARSKII AND K. SEIP, *A direct interpolation method for irregular sampling*, Preprint, Trondheim University, 2000.

[Fo1] S.R. FOGUEL, *Normal operators of finite multiplicity*, Comm. Pure Appl. Math., 11 (1958), no. 3, 297–313.

[Fo2] ——— *A counterexample to a problem of Sz.-Nagy*, Proc. Amer. Math. Soc., 15 (1964), 788–790.

[Foi1] C. FOIAŞ, *Sur certains théorèmes de J. von Neumann concernant les ensembles spectraux*, Acta Sci. Math. Szeged, 18 (1957), 15–20.

[Foi2] ——— *Unele aplicatii ale multimilor spectrale. I. Masura armonica-spectrala*, Studii si cercetari. mat. Acad. RPR, 10 (1959), no. 2, 365–401 (Rumanian).

[Foi3] ——— *Some applications of structural models for operators on Hilbert spaces*, Actes Congrès Int. Math. (Nice, 1970), vol. 2, pp. 433–440, Gauthier-Villars, Paris, 1971.

[Foi4] ——— *On the scalar parts of a decomposable operator*, Rev. Roumaine Math. Pures Appl., 17 (1972), 1181–1198.

[FF] C. FOIAŞ AND A.E. FRAZHO, *The commutant lifting approach to interpolation problems*, Operator theory: Advances and applications, 44, Basel, Birkhäuser, 1990.

[FFGK] C. FOIAŞ, A.E. FRAZHO, I. GOHBERG AND M.A. KAASHOEK, *Metric constrained interpolation, commutant lifting and systems interpolation*, Basel, Birkhäuser, 1998.

[FW] C. FOIAŞ AND J. WILLIAMS, *On a class of polynomially bounded operators*, preprint circa 1980, unpublished.

[FoS] Y. FOURÈS AND I.E. SEGAL, *Causality and analyticity*, Trans. Amer. Math. Soc., 78 (1955), 385–405.

[Fr] B. FRANCIS, *A course in H^∞ control theory*, Lect. Notes CIS, 88, Springer–Verlag, 1987.

[FR1] R. FRANKFURT AND J. ROVNYAK, *Finite convolution operators*, J. Funct. Analysis, 49 (1975), no. 2, 347–374.

[FR2] ——— *Recent results and unsolved problems on finite convolution operators*, pp. 133–150 in: Linear spaces and approximation (eds. P. Butzer and B. Sz.-Nagy), ISNM vol. 40, Birkhäuser Verlag, Basel, 1978.

[Fra] A.E. FRAZHO, *Models for noncommuting operators*, J. Funct. Anal., 48 (1982), 1–11.

[Fri1] E. FRICAIN, *Uniqueness theorems for analytic vector valued functions*, Zapiski Nauchn. Semin. POMI, 247 (1997), 242–267; J. Soviet Math., 101 (2000), no. 3, 3193–3210.

[Fri2] ——— *Propriétés géométriques des suites de noyaux reproduisants dans les espaces modèles*, Thèse, Bordeaux, 1999.

[Fri3] ——— *Bases of reproducing kernels in model spaces*, J. Operator Theory, to appear.

[Frie] K.O. FRIEDRICHS, *On certain inequalities and characteristic value problems for analytic functions and for functions of two variables*, Trans. Amer. Math. Soc., 41 (1937), 321–364.

[Fro] O. FROSTMAN, *Potrntiel d'équilibre et capacité des ensembles avec quelques applications à la théotie des fonctions*, Medd. Lunds Univ. Mat. Sem., 3 (1935), 1–118.

[Fug] B. FUGLEDE, *A commutativity theorem for normal operators*, Proc. Amer. Math. Soc., 36 (1950), 35–40.

[Fuh1] P.A. FUHRMANN, *On the corona theorem and its applications to spectral problems in Hilbert space*, Trans. Amer. Math. Soc., 132 (1968), 55–66.

[Fuh2] ——— *Exact controllability and observability and realization theory in Hilbert space*, J. Math. Anal. Appl., 53 (1976), 377–392.

[Fuh3] ——— *Linear systems and operators in Hilbert space*, McGraw Hill Int., NY, 1981.

[Ful] W. FULTON, *Eigenvalues, invariant factors, highest weights, and Schubert calculus*, Bull. Amer. Math. Soc., 37:3 (2000), 209-249.

[Gab] R.M. GABRIEL, *Some results concerning the integrals of moduli of regular functions along curves of certain types*, Proc. London Math. Soc., 28 (1928), 121–127.

[Gai] D. GAIER, *Vorlesungen über Approximation im Komplexen*, Birkhäuser, 1980.

[Gama1] M.F. GAMAL, *Weak generators of the measure algebra and unicellularity of convolution operators*, Zapiski Nauchn. Semin. POMI, 232 (1996), 73-85 (Russian); English transl.: J. Soviet Math., 92 (1998), no. 1.

[Gama2] ——— *Invariant subspaces and generators of the commutant of a model operator*, Thesis, Steklov Inst. Math., St.Petersburg, 1997 (Russian).

[Gama3] ——— *Quasisimilarity, pseudosimilarity, and lattices of invariant subspaces of weak contractions*, Abstracts of Tenth Summer St. Petersburg Meeting in Math. Analysis, p. 14, Euler Int. Math. Inst., August 2001.

[Gam] T.W. GAMELIN, *Uniform Algebras*, Englewood Cliffs N.J., Prentice Hall, 1969.

[Gan] F.R. GANTMACHER, *The theory of matrices*, 2nd edition, Moscow, Nauka, 1966 (Russian); English transl. of the 1st edition: N.Y., Chelsea, 1960.

[GCRF] J. GARCIA-CUERVA AND J.L. RUBIO DE FRANCIA, *Weighted norm inequalities and related topics*, Math. Studies 116, North Holland, 1985.

[Gar] J.B. GARNETT, *Bounded analytic functions*, Academic Press, New York, 1981.

[GarN] J.B. GARNETT AND A. NICOLAU, *Interpolating Blaschke products generate H^∞*, Pacific J. Math., 173 (1996), no. 2, 501–510.

[GHVid] E.M. GAVURINA, V.P. HAVIN [KHAVIN] AND I.V. VIDENSKII, *Analogs of the interpolation formula of Carleman–Krylov–Goluzin*, pp. 359–388 in: Operator Theory and Function Theory, 12, Leningrad Gosud. Universitet, Leningrad, 1983 (Russian).

[Gea] L. GEARHART, *Spectral theory for contraction semigroups on Hilbert space*, Trans. Amer. Math. Soc., 236 (1978), 385–394.

[Gel1] I.M. GELFAND, *A problem*, Uspehi Matem. Nauk, 5 (1938), 233 (Russian).

[Gel2] ——— *A remark on N.K. Bari's paper "Biorthogonal systems and bases in a Hilbert space"*, Uchenye Zapiski MGU (Moscow State Univ.), 4 (1951), no. 148, 224–225 (Russian).

[GN] I.M. GELFAND AND M.A. NAIMARK, *On inclusion of a normed ring in the ring of operators on a Hilbert space*, Matem. Sbornik, 12 (1943), 197–213 (Russian); English transl. in: Gelfand I.M., *Collected papers. I.*, Springer–Verlag, 1987.

[GRS] I.M. GELFAND, D.A. RAIKOV AND G.E. SHILOV, *Commutative normed rings*, Moscow, Fizmatgiz, 1960 (Russian); English transl.: NY, Chelsea 1964.

[Geo] D. GEORGIJEVIC, *Bases orthogonales dans les espaces $H^p(e)$ et H^p*, C.R. Acad. Sci. Paris, Ser. A, 289 (1979), 73–74.

[Gersh] S.A. GERSHGORIN, *Über die Abgrenzung der Eigenwerte einer Matrix*, Izvestia Acad. Sci. SSSR, Fiz.-Mat. (1931), 749–754.

[GNPTV] A. GILLESPIE, F. NAZAROV, S. POTT, S. TREIL AND A. VOLBERG, *Logarithmic growth for weighted Hilbert transform and vector Hankel operators*, Preprint, Michigan State Univ., 1998; to appear in Algebra i Analiz (St.Petersburg Math. J.).

[GiN] J.I. GINSBERG AND D.J. NEWMAN, *Generators of certain radical algebras*, J. Approx. Theory, 3 (1970), no. 3, 229–235.

[GS] YU.P. GINZBURG AND L.V. SHEVCHUK, *On the Potapov theory of multiplicative representations*, pp. 28–47 in: Operator Theory: Adv. Appl., 72, Birkhäuser, Basel, 1994.

[Goh1] I. GOHBERG, *On an application of the theory of normed rings to singular integral equations*, Uspehi Matem. Nauk, 7 (1952), no. 2, 149–156 (Russian).

[GF] I. GOHBERG AND A.I. FELDMAN, *Convolution equations and projection methods for their solutions*, RIO of Moldavian Academy, Kishinev, and Nauka, Moscow, 1971 (Russian); English transl.: Amer. Math. Soc., Providence, R.I., 1974.

[GGK] I. GOHBERG, S. GOLDBERG AND M.A. KAASHOEK, *Classes of linear operators*, Vol. I and II, Birkhäuser, Basel, 1990, 1993.

[GK1] I. GOHBERG AND M.G. KREIN, *Introduction to the theory of linear non-selfadjoint operators on Hilbert space*, Nauka, Moscow, 1965 (Russian); English transl.: Amer. Math. Soc., Providence, R.I., 1969.

[GK2] ——— *The theory of Volterra operators in Hilbert space and applications*, Nauka, Moscow, 1967 (Russian); English transl.: Amer. Math. Soc., Providence, R.I., 1970.

[GK3] ——— *On a description of contractions similar to unitaries*, Funktsional. Analiz i Prilozhen., 1 (1967), no. 1, 38–60 (Russian).

[GKru1] I. GOHBERG AND N.YA. KRUPNIK, *On the algebra generated by Toeplitz matrices*, Funktsional. Analiz i Prilozhen., 3 (1969), no. 2, 46–56 (Russian).

[GKru2] ——— *One-dimensional linear singular integral equations*, "Shtiintsa", Kishinev, 1973 (Russian); English transl.: Vol. I and II, Birkhäuser Verlag, Basel, 1992.

[GM] I. GOHBERG AND A. MARKUS, *On certain inequalities between eigenvalues and matrix entries of linear operators*, Izvestia Acad. Sci. Mold. SSR, 5 (1962), 103–108 (Russian).

[Go] G.M. GOLUZIN, *Geometric theory of functions of a complex variable*, Moscow, Nauka, 1966 (Russian); English transl.: Amer. Math. Soc., Providence, R.I., 1969.

[GoK] G.M. GOLUZIN AND V.I. KRYLOV, *Generalized Carleman formula and its applications to analytic continuation of functions*, Matem. Sbornik, 40 (1933), no. 2, 144–149 (Russian).

[Gon] A.A. GONCHAR, *Quasianalytic continuation of analytic functions across a Jordan arc*, Doklady Akad. Nauk SSSR, 166 (1966), 1028–1031 (Russian); English transl.: Soviet Math. Doklady, 7 (1966), 213–216.

[GHVin] E.A. GORIN, S.V. HRUSCHEV [KHRUSCHEV] AND S.A. VINOGRADOV, *Interpolation in H^∞ along P. Jones' lines*, Zapiski Nauchn. Semin. LOMI, 113 (1981), 212–214 (Russian); English transl.: J. Soviet Math., 22 (1983), 1838–1839.

[GorK] E.A. GORIN AND M.I. KARAHANIAN, *An asymptotic version of the Fuglede-Putnam theorem for commutators of elements of a Banach algebra*, Matem. Zametki, 22 (1977), no. 2, 179–188 (Russian). English transl.: Soviet Math. Notes 22 (1977), no. 1–2, 591–596 (1978).

[GMcG] C.C. GRAHAM AND O.C. MCGEHEE, *Essays in Commutative Harmonic Analysis*, New York – Heidelberg, Springer–Verlag, 1979.

[GSz] U. GRENANDER AND G. SZEGÖ, *Toeplitz forms and their applications*, Univ. of California Press, Berkeley, 1958.

[GrN] M.B. GRIBOV AND N.NIKOLSKI, *Invariant subspaces and rational approximation*, Zapiski Nauchnyh Semin. LOMI, 92 (1979), 103–114 (Russian).

[Gri] L.D. GRIGORYAN, *Estimates of the norm of holomorphic components of meromorphic functions in domains with a smooth boundary*, Mat. Sbornik, 100(142):1 (1976), 156-164 (Russian); English transl.: Math. USSR Sbornik, 29 (1976).

[Gr-E] K.-G. GROSSE-ERDMANN, *Universal families and hypercyclic operators*, Bull. Amer. Math. Soc., 36 (1999), no. 3, 345–381.

[Gub1] G.M. GUBREEV, *Spectral analysis of biorthogonal expansions generated by Muckenhoupt weights*, Zapiski Nauchnyh Seminarov LOMI, 190 (1991), 34–80 (Russian); English transl.: J. Math. Sci. 71 (1994), no. 1, 2192–2221.

[Gub2] ——— *Unconditional bases of Hilbert spaces composed of values of vector valued entire functions of exponential type*, Funkzion. Analiz i Prilozhen., 33 (1999), no. 1, 62–65 (Russian); English transl.: Funct. Anal. Appl. 33 (1999), no. 1, 52–55

[GubO] G.M. GUBREEV AND E.I. OLEFIR, *Unconditional bases of certain families of functions, the matrix Muckehoupt condition, and Carleson series in the spectrum*, Zapiski Nauchnyh Seminarov POMI (St.Petersburg), 262 (1999), 90–126 (Russian).

[GuM] V.I. GURARII AND M.A. MELITIDI, *Stability of completeness of sequences in Banach spaces*, Bull. Acad. Sci. Polon., ser. math., astron. et phys., 18:9 (1970), 533-536 (Russian).

[GuV] V.I. GURARII AND V.I. MATSAEV [MACAEV], *Lacunary power sequences in spaces C and L^p*, Izvestia Akad. Nauk SSSR, SEr. Mat., 30 (1966), 3–14 (Russian).

[Gur] V.P. GURARII, *Group theoretical methods in commutative harmonic analysis*, Itogi Nauki i Techniki, Contemp. Problems Math. (Russian Math. Encyclopaedia), 25, VINITI, Moscow, 1988 (Russian); English transl.: Encyclopaedia of Math. Sci., vol. 25, Berlin etc., Springer–Verlag, 1997.

[Had] J. HADAMARD, *Leçons sur la propagation des ondes*, Hermann, Paris, 1903.

[Hall] T. HALL, *Sur l'approximation polynômiale des fonctions d'une variable réelle*, pp. 367–369 in: IX-ème Congrès Math. Scand. (1938), Helsingfors, 1939.

[Hal1] P. HALMOS, *Normal dilations and extensions of operators*, Summa Brasil. Math., 2 (1960), 125–134.

[Hal2] ——— *A Hilbert space problem book*, Springer–Verlag, Berlin etc., 1967.

[Hal3] ——— *Ten problems in Hilbert space*, Bull. Amer. Math. Soc., 76 (1970), 887-933.

[Ham1] H. HAMBURGER, *Über eine Erweiterung des Stieltjesschen Momentproblems*, I, II, Math. Ann., 81, 82 (1920, 1921).

[Ham2] ——— *Über die Zerlegung des Hilbertschen Raumes durch vollstetige lineare Transformationen*, Math. Nachr., 4 (1951), 56–69.

[HaL1] G.H. HARDY AND J.E. LITTLEWOOD, *Some new properties of Fourier constants*, Math. Ann., 97 (1926), 159–209.

[HaL2] ——— *A new proof of a theorem on rearrangements*, J. London Math. Soc., 23 (1948), 163-168.

[HLP1] G.H. HARDY, J.E. LITTLEWOOD, AND G. POLYA, *Some simple inequalities satified by convex functions*, Messenger of Math., 58 (1929), 145–152.

[HLP2] ——— *Inequalities*, Cambridge, 1934.

[HaWr] G.H. HARDY AND E.M. WRIGHT, *An introduction to the theory of numbers*, Fourth Ed., Clarendon Press, Oxford, 1962.

[Har] P. HARTMAN, *On completely continuous Hankel matrices*, Proc. Amer. Math. Soc., 9 (1958), 862–866.

[HW1] P. HARTMAN AND A. WINTNER, *On the spectra of Toeplitz's matrices*, Amer. J. Math., 72 (1950), 359–366.

[HW2] ——— *The spectra of Toeplitz's matrices*, Amer. J. Math., 76 (1954), 867–882.

[Hart1] A. HARTMANN, *Une approche de l'interpolation libre généralisée par la théorie des opérateurs et caractérisation des traces* $H^p|\Lambda$, J. Operator Theory, 35 (1996), 281–316.

[Hart2] ——— *Free interpolation in Hardy–Orlicz spaces*, Studia Math., (2) 135 (1999), 179–190.

[HSr] M. HASUMI AND T. SRINIVASAN, *Invariant subspaces of continuous functions*, Canad. J. Math., 17 (1965), no. 4, 643–651.

[Hau] F. HAUSDORFF, *Mengenlehre*, 3. Aufl. Berlin-Leipzig, W. de Gruyter, 1935.

[Hav] V.P. HAVIN [KHAVIN], *Methods and structure of commutative harmonic analysis*, pp. 1–112 in: Itogi Nauki i Techniki, Contemp. Problems Math. (Russian Math. Encyclopaedia), 15 (ed. N. Nikolski), VINITI, Moscow, 1987 (Russian); English transl.: Encyclopaedia of Math. Sci., 15, Berlin etc., Springer–Verlag, 1991.

[HJ] V.P. HAVIN [KHAVIN] AND B. JÖRICKE, *The uncertainty principle in harmonic analysis*, Berlin etc., Springer–Verlag, 1974.

[HN1] V.P. HAVIN [KHAVIN] AND N.K. NIKOLSKI, *V.I. Smirnov's results in complex analysis and their subsequent developments*, pp. 111–145 in: V.I. Smirnov, Selected papers. Complex Analysis, Mathematical Diffraction Theory, Leningrad State Univ. Publishers, 1988 (Russian).

[HN2] V.P. HAVIN [KHAVIN] AND N.K. NIKOLSKI, *Stanislav Aleksandrovich Vinogradov, his life and mathematics*, pp. 1–18 in: Complex Analysis, operator theory, and related topics: S.A. Vinogradov Memorial Volume (eds. V. Havin and N. Nikolski) Operator Theory: Adv. and Appl., 113, Basel, Birkhäuser–Verlag, 2000.

[HVin] V.P. HAVIN [KHAVIN] AND S.A. VINOGRADOV, *Free interpolation in* H^∞ *and in some other function classes, I and II*, Zapiski Nauchn. Semin. LOMI, 47 (1974), 15–54, 56 (1976), 12–58 (Russian); English transl.: J. Soviet Math., 9 (1978), 137–171, 14 (1980), 1027–1065.

[HWo] V.P. HAVIN [KHAVIN] AND H. WOLF, *Poisson kernel is the only approximative identity asymptotically multiplicative on* H^∞, Zapiski Nauchn. Seminarov LOMI, 170 (1989), 82–89 (Russian).

[Hay] E. HAYASHI, *The kernel of a Toeplitz operator*, Integral Eq. Oper. Theory, 9 (1986), 588–591.

[Ha] W.K. HAYMAN, *Identity theorems for functions of bounded characteristic*, J. London Math. Soc., (2) 58 (1998), 127–140.

[HKZ] H. HEDENMALM, B. KORENBLUM AND K. ZHU, *Theory of Bergman spaces*, N.-Y.-Heidelberg, etc., Springer–Verlag, 2000.

[HeV] H. HEDENMALM, A. VOLBERG, *Zero-free invariant subspaces in weighted Bergman spaces with critical topology*, manuscript, 1996.

[Hei] E. HEINZ, *Ein v. Neumannscher Satz über beschränkte Operatoren im Hilbertschen Raum*, Göttinger Nachr. (1952), 5–6.

[Hel] H. HELSON, *Lectures on invariant subspaces*, N.Y., Academic Press, 1964.

[HL1] H. HELSON AND D. LOWDENSLAGER, *Invariant subspaces*, pp. 251–262 in: Proc. Intern. Symp. Linear Spaces, Jerusalem, Pergamon Press, Oxford, 1961.

[HL2] ——— *Prediction theory and Fourier series in several variables, II*, Acta Math., 106 (1961), 175–213.

[HSa] H. HELSON AND D. SARASON, *Past and future*, Math. Scand., 21 (1967), 5–16.

[HS] H. HELSON AND G. SZEGÖ, *A problem of prediction theory*, Ann. Mat. Pur. Appl., 51 (1960), 107–138.

[Helt1] J.W. HELTON, *Discrete time systems, operator models and scattering theory*, J. Funct. Anal., 16 (1974), 15–38.

[Helt2] ——— *Operator theory, analytic functions, matrices and electrical engineering*, CBMS conf. series in Math., 68, AMS, Providence, RI, 1987.

[HM] J.W. HELTON AND O. MERINO, *Classical control using H^∞ methods*, Univ. of California, San Diego, 1994.

[Her] G. HERGLOTZ, *Über Potenzreihen mit positiven reellen Teil im Einheitskreise*, Berichte Verh. Kgl.-sächs. Gesellsch. Wiss. Leipzig, Math.-Phys. Klasse, 63 (1911), 501–511.

[Herr1] D.A. HERRERO, *On multicyclic operators*, Integral Equat. and Operator Theory, 1 (1978), 57–102.

[Herr2] ——— *Approximation of Hilbert space operators*, Vol.I, Res. Notes Math. 72, Pitman, Boston, 1982.

[HerrMcD] D.A. HERRERO AND J. MC DONALD, *On multicyclic operators and the Vasyunin—Nikolski discotheca*, Integral Equat. and Operator Theory, 6 (1983), 206–223.

[HTW] D.A. HERRERO, TH.J. TAYLOR, AND Z.Y. WANG, *Variation of the point spectrum under compact perturbations*, pp. 113–158 in: Operator Theory: Adv. and Appl. 32, Birkhäuser–Verlag, Basel, 1988.

[Hig1] J.R. HIGGINS, *Completeness and basis properties of sets of special functions*, Cambridge Univ. Press, Cambridge, 1977.

[Hig2] ——— *Five short stories about the cardinal series*, Bull. Amer. Math. Soc., 12 (1985), 45–89.

[Hil1] D. HILBERT, *Grundzüge einer allgemeinen Theorie der linearen Integralgleichungen*, I–VI, Nachr. Akad. Wiss. Göttingen, Math.-Phys. Kl., (1904), 49–91; (1905), 213–259; (1905), 307–338; (1906), 157–227; (1906), 439–480; (1910), 355–417.

[Hil2] ——— *Gründzüge einer allgemeinen Theorie der linearen Integralgleichungen*, Leipzig, 1912.

[HPh] E. HILLE AND R.S. PHILLIPS, *Functional analysis and semi-groups*, Amer. Math. Soc. Coll. Publ. 31, Providence, 1957.

[HT] E. HILLE AND J.D. TAMARKIN, *On the absolute integrability of Fourier transforms*, Fund. Math., 25 (1935), 329–352.

[Hir] R.A. HIRSCHFELD, *On polynomials in several Hilbert space operators*, Math. Zeit., 127 (1972), 224–234.

[Hi] D. HITT, *Invariant subspaces of H^2 of the annulus*, Pacific J. Math., 134 (1988), 101–120.

[Hof] K. HOFFMAN, *Banach spaces of analytic functions*, Prentice–Hall, Englewood Cliffs, N.J., 1962.

[Hol] J.A. HOLBROOK, *Spectral variation of normal matrices*, Linear Alg. Appl., 174 (1992), 131–141.

[WH] E. HOPF AND N. WIENER, *see* N. Wiener and E. Hopf.

[Hör1] L. Hörmander, *Estimates for translation invariant operators in L^p spaces*, Acta Math., 104 (1960), 93–140.

[Hör2] ______ *The analysis of partial differential operators*, vol. 1 and 2, Springer–Verlag, Heidelberg, 1983.

[HoJ] R.A. Horn and C.R. Johnson, *Topics in matrix analysis*, Cambridge Univ. Press, NY, 1990.

[How1] J.S. Howland, *Trace class Hankel operators*, Quart. J. Math., Oxford, (2) 22 (1971), 147–159.

[How2] ______ *On a theorem of Gearhart*, Integral Equations and Operator Theory, 7 (1984), 138–142.

[How3] ______ *Spectral theory of operators of Hankel type, I.*, Indiana Univ. Math. J., 41 (1992), no. 2, 409–426.

[How4] ______ *Spectral theory of operators of Hankel type, II.*, Indiana Univ. Math. J., 41 (1992), no. 2, 427–434.

[Hr] S.V. Hruschev [Khruschev], *Perturbation theorems for bases of exponentials and the Muckenhoupt condition*, Doklady Akad. Nauk SSSR, 247 (1979), no. 1, 44–48 (Russian); English transl.: Soviet Math. Dokl., 20 (1979), no. 4, 665–669.

[HrN] S.V. Hruschev [Khruschev] and N.K. Nikolski, *A function model and some problems in the spectral theory of functions*, Trudy Mat. Inst. Steklova, 176 (1987), 97–210 (Russian); English transl.: Proc. Steklov Inst. Math. (1988), no. 3, 101–214.

[HNP] S.V. Hruschev [Khruschev], N.K. Nikolski and B.S. Pavlov, *Unconditional bases of exponentials and reproducing kernels*, Lect. Notes Math., 864 (1981), 214–335.

[HP1] S.V. Hruscev [Khruschev] and V.V. Peller, *Hankel operators, best approximations and stationary Gaussian processes*, Uspehi Mat. Nauk, 37 (1982), no. 1, 53–124 (Russian); English transl.: Russian Math. Surveys, 37 (1982), no. 1, 61–144.

[HP2] ______ *Moduli of Hankel operators, past and future*, pp. 92–97 in: Linear and Complex analysis problem book. 199 research problems (eds. V. Havin, S. Hruschev and N. Nikolski), Lect. Notes Math. 1043, Springer–Verlag, Berlin, 1984.

[HP3] ______ *Hankel operators of Schatten-von Neumann class and their applications to stationary processes and best approximations*, published as Appendix 5 in N. Nikolski [**N19**].

[HV] S.V. Hruschev and S.A. Vinogradov, *Free interpolation in the space of uniformly convergent taylor series*, Lect. Notes Math., 864 (1981), 171–213.

[HMW] R. Hunt, B. Muckenhoupt and R.L. Wheeden, *Weighted norm inequalities for the conjugate function and Hilbert transform*, Trans. Amer. Math. Soc., 176 (1973), 227–251.

[HY] Ya.I. Hurgin [Khurgin] and V.P. Yakovlev, *Compactly supported functions in physics and engineering*, Nauka, Moscow, 1971 (Russian).

[IR] I.A. Ibragimov and Yu.A. Rozanov, *Gaussian stochastic processes*, Moscow, Nauka, 1970 (Russian); English transl.: N.Y.–Heidelberg etc., Springer–Verlag, 1978.

[In] A.E. Ingham, *A note on Fourier transforms*, J. London Math. Soc., 9 (1934), 29–32.

[Io] I.S. Iokhvidov, *Hankel and Toeplitz matrices and forms. Algebraic theory*, Moscow, Nauka, 1974 (Russian); English transl.: Birkhuser, Boston, Mass., 1982.

[IK] Sh.-I. Izumi and T. Kawata, *Quasi-analytic class and closure of $\{t^n\}$ in the interval* $(-\infty, \infty)$, Tôhoku Math. J., 43 (1937), 267–273.

[IP] S.A. Ivanov and B.S. Pavlov, *Vector valued systems of exponentials and zeros of entire matrix functions*, Vestnik Leningrad University, 1980, no. 1, 25–31 (Russian); English transl.: Vestn. Leningr. Univ. Math., 13 (1981), 31–38.

[JaPa] B. Jacob and J.R. Partington, *The Weiss conjecture on admissibility of observation operators for contraction semigroups*, Preprint, 2000.

[JPR] S. Janson, J. Peetre, and R. Rochberg, *Hankel forms and the Fock space*, Revista Mat. Iberoamer., 3 (1987), 61–138.

[J1] P. Jones, *Extension theorems for BMO*, Indiana Univ. Math. J., 29 (1979), 41–66.

[J2] ______ *L^∞ estimates for the $\bar\partial$-problem in a half-plane*, Acta Math., 150 (1980), 137–152.

[J3] ______ *Ratios of interpolating Blaschke products*, Pacific J. Math., 95 (1981), no. 2, 311–321.

[JZ] V.I. JUDOVICH AND V.P. ZAKHARYUTA, *The general form of a linear functional in H'_p*, Uspekhi Matem. Nauk, 19 (1964), no. 2, 139–142 (Russian).

[Ju] G. JULIA, *Sur la représentation analytique des opérateurs bornés ou fermés de l'espace hilbertien*, C.R. Acad. Sci. Paris, 219 (1944), 225–227.

[Kac.I] I.S. KAC, *Inclusion of Hamburger's power moment problem in the spectral theory of the canonical systems*, Zapiski Nauchn. Semin. POMI (Steklov Math. Inst., St.Petersburg), vol.262 (1999), 147-171 (Russian).

[Kac.IKr] I.S. KAC AND M.G. KREIN, *On spectral functions of strings*, in: Appendix 2 to the Russian translation of the book F. Atkinson [At], pp. 648-737 (Russian), Izdat. Mir, Moscow, 1968.

[Kac] M. KAC, *Can one hear the shape of a drum?*, Amer. Math. Monthly, 73 (1966), no. 4, part II, 1–24.

[KSt] S. KACZMARZ AND H. STEINHAUS, *Theorie der Orthogonalreihen*, Monografje matematyczne, Warszawa–Lwow, 1935.

[Kad] M.I. KADEC [KADETS], *The exact value of the Paley–Wiener constant*, Doklady Akad. Nauk SSSR, 155 (1964), no. 6, 1253–1254 (Russian); English transl.: Soviet Math. Doklady, 5 (1964), no. 2, 559–561.

[Kah1] J.-P. KAHANE, *Lectures on mean periodic functions*, Tata Institute Fund. Research, Bombay, 1959.

[Kah2] ——— *Séries de Fourier absolument convergentes*, Springer–Verlag, Berlin, etc., 1970.

[Kah3] ——— *Travaux de Beurling et Malliavin*, Séminaire Bourbaki, 7 (1995), no. 225, 27–39.

[Kai] T. KAILATH, *Linear Systems*, Prentice Hall, 1980.

[Kal] G.K. KALISCH, *A functional analysis proof of Titchmarch's theorem on convolution*, J. Math. Anal. and Appl., 5 (1962), no. 2, 176–183.

[KFA] R.E. KALMAN, P.L. FALB, AND M.A. ARBIB, *Topics in mathematical system theory*, McGraw–Hill, 1969.

[KanA] L.V. KANTOROVICH AND G.P. AKILOV, *Functional analysis*, 2nd edition, Nauka, Moscow, 1977 (Russian); English trans.: Pergamon Press, Oxford–Elmsford, N.Y., 1982.

[Kap1] V.V. KAPUSTIN, *Reflexivity of operators: general methods and a criterion for almost isometric contractions*. Algebra i Analiz, 4 (1992), no. 2, 141–160 (Russian); English transl.: St.Petersburg Math. J., 4 (1993), no. 2.

[Kap2] ——— *Function calculus for almost isometric operators*, Zapiski Nauchn. Seminarov POMI, 217 (1994), 59–73 (Russian); English transl.: J. Soviet Math.

[KLi1] V.V. KAPUSTIN AND A.V. LIPIN, *Operator algebras and lattices of invariant subspaces. I*, Zapiski Nauchn. Seminarov LOMI (Leningrad), 178 (1989), 23–56 (Russian); English transl.: J. Soviet Math., 61 (1992), no. 2, 1963–1981.

[KLi2] ——— *Operator algebras and lattices of invariant subspaces. II*, Zapiski Nauchn. Seminarov LOMI (Leningrad), 190 (1991), 110–147 (Russian); English transl.: J. Math. Sci., 71 (1994), no. 1, 2240–2262.

[KasS] B.S. KASHIN AND A.A. SAAKYAN, *Orthogonal series*, "Nauka", Moscow, 1984 (Russian).

[Kat] T. KATO, *Perturbation theory for linear operators*, Springer–Verlag, Berlin etc., 1966.

[KKY] V.E. KATSNEL'SON [KACNEL'SON], A.Y. KHEIFETS AND P.M. YUDITSKII, *An abstract interpolation problem and the theory of extensions of isometric operators*, pp. 83–96 in: Operators in function spaces and problems in function theory, ed. V.A. Marchenko, Kiev, Naukova Dumka, 1987 (Russian); English transl.: Topics in interpolation theory, ed. H. Dym et al., Oper. Theory Adv. Appl. 95, Basel, Birkhäuser, 1997, pp. 283–298.

[KM] V.E. KATSNEL'SON [KACNEL'SON] AND V.I. MATSAEV [MACAEV], *Spectral sets for operators in a Banach space and estimates of functions of finite-dimensional operators*, Teoria Funckcii, Funkcional. Analiz i Prilozhenia (Kharkov), 1966, no. 3, 3–10 (Russian).

[Ka] Y. KATZNELSON, *An introduction to harmonic analysis*, New York, Dover, 1976.

[KaTz] Y. KATZNELSON AND L. TZAFRIRI, *On power bounded operators*, J. Funct. Anal., 68 (1986), 313–328.

[Ke] M.V. KELDYSH, *Sur l'approximation en moyenne par polynômes des fonctions d'une variable complexe*, Matem. Sbornik, 16 (58) (1945), 1–20.

[KZ] K. KELLAY AND M. ZARRABI, *Normality, non-quasianalyticity and invariant subspaces*, J. Operator Theory, to appear.

[KLS] D. KHAVINSON, T.L. LANCE, AND M.I. STESSIN, *Wandering property in the Hardy space*, Mich. Math. J., 44 (1997), no. 3, 597–606.

[Kha1] S.YA. KHAVINSON [S.JA. HAVINSON], *Extremum problems for functions satisfying supplementary restrictions inside the region and an application to problems of approximation*, Doklady Akad. Nauk SSSR, 135 (1960), 29–32 (Russian); English transl.: Soviet Math. Dokl., 1 (1960), 1263–1266.

[Kha2] ______ *Some approximation theorems involving the magnitude of the coefficients of the approximating functions*, Doklady Akad. Nauk SSSR, 196:6 (1971) (Russian); English transl.: Soviet Math. Dokl., 12:1 (1971), 366-370.

[Kha3] ______ *On complete systems in Banach spaces*, Izvestia Akad. Nauk Armen. SSR, Mat., 20 (1985), no. 2, 89–110 (Russian).

[Kh] A. KHEIFETS, *The Abstarct Interpolation Problem and applications*, pp. 351–380 in: Holomorphic Spaces, eds. Sh. Axler, J. McCarthy, and D. Sarason, MSRI Publications, 33, Cambridge Univ. Press, 1998.

[Kis1] S.V. KISLYAKOV, *On projections onto the set of Hankel matrices*, Zapiski Nauchnyh Sem. LOMI (Steklov Inst., Leningrad), 126 (1983), 109-116 (Russian); English transl.: J. Soviet Math., 27 (1984), 2495–2500.

[Kis2] ______ *Classical themes of Fourier analysis*, in Itogi Nauki i Techniki, Contemp. Problems Math. (Russian Math. Encyclopaedia), vol. 15 (eds. V. Khavin and N. Nikolski), 135–196, VINITI, Moscow, 1987 (Russian); English transl.: Encyclopaedia of Math. Sci., vol. 15, 113–166, Springer–Verlag, Berlin etc., 1991.

[Kis3] ______ *Exceptional sets in harmonic analysis*, pp. 199–227 in: Itogi Nauki i Techniki, Contemp. Problems Math. (Russian Math. Encyclopaedia), 42 (eds. V. Khavin and N. Nikolski), VINITI, Moscow, 1989 (Russian); English transl.: Encyclopaedia of Math. Sci., vol. 42, 195–222, Berlin etc., Springer–Verlag, 1992.

[Kle] I. KLEMEŠ, *Finite Toeplitz matrices and sharp Littlewood conjectures*, Algebra i Analiz, 13:1 (2001), 39-59 (Reprinted in: St. Petersburg Math. J., 13:1 (2002)).

[K1] A.N. KOLMOGOROV, *Sur les fonctions harmoniques conjuguées et les séries de Fourier*, Fund. Math., 7 (1925), 24–29.

[K2] ______ *Stationary sequences in Hilbert space*, Bull. Moscow Univ. Math. 2 (1941), no. 6, 1–40 (Russian).

[K3] ______ *Une série de Fourier-Lebesgue divergente presque partout*, Fundamenta Math., 4 (1923), 324–338.

[K4] ______ *Une série de Fourier-Lebesgue divergente partout*, C.R. Acad. Sci. Paris, 183 (1926), 1327–1329.

[Kö] H. KÖNIG, *Eigenvalue distribution of compact operators*, Oper. Theory: Adv. and Appl. 16, Birkhäuser-Verlag, Basel, 1986.

[Kon] S.V. KONYAGIN, *On the Littlewood problem*, Izvestia Akad. Nauk SSSR, ser. mat., 45:2 (1981), 243-265 (Russian).

[Ko1] P. KOOSIS, *Interior compact spaces of functions on a half-line*, Comm. Pure Appl. Math., 10 (1957), no. 4, 583–615.

[Ko2] ______ *L'approximation pondérée par des polynômes et par des sommes exponentielles imaginaires*, Ann. Sci. Ecole Norm. Sup., 81 (1964), no. 4, 387–408.

[Ko3] ______ *Moyennes quadratiques pondérées de fonctions périodiques et de leur conjuguées harmoniques*, C.R. Acad. Sci. Paris, Ser. A, 291 (1980), 255–257.

[Ko4] ______ *Introduction to H^p spaces*, Cambridge Univ. Press, Cambridge etc., 1980.

[Ko5] ______ *The logarithmic integral, Vol.I&II*, Cambridge Univ. Press, 1988,1992.

[Ko6] ______ *Leçons sur le théorème de Beurling et Malliavin*, Les Publications CRM, Montréal, 1996.

[Ko7] ______ *Carleson's interpolation theorem deduced from a result of Pick*, pp. 151–162 in: Complex analysis, operators, and related topics, Operator Theory: Adv. Appl., 113, Birkhauser, Basel, 2000.

[SzK*] A. KORANYI AND B. SZŐKEFALVI-NAGY, *see* B. Szőkefalvi-Nagy and A. Koranyi.

[Kor1] B. KORENBLUM, *Closed ideals of the ring A^n*, Funkt. Analyz i ego prilozh., 6 (1972), no. 3, 38–52 (Russian).

[Kor2] ______ *A Beurling type theorem*, Acta Math., 138 (1977), 265–293.

[Kor3] ______ *Cyclic vectors in some spaces of analytic functions*, Bull. Amer. Math. Soc., 5 (1981), 317–318.

[Kore] J. Korevaar, *Zero distribution of entire functions and spanning radius for a set of complex exponentials*, pp. 293–312 in: Aspects of contemporary complex analysis (Proc. NATO Adv. Study Inst., Univ. Durham, Durham, 1979), Academic Press, London–New York, 1980.

[Kot] V.A. Kotelnikov, *Conducting capacity of the "ether" and a wire in electrical communications*, in: Vsesojuznyi Energeticheskii Komitet, Materials to the First All-Union Congress on Problems of the Technical Reconstruction of Communications and Developments of the Weak Current Engineering. Izdat. of the Communication Administration of RKKA, Moscow, 1933 (Russian).

[KT] G. Köthe und O. Toeplitz, *Lineare Räume mit unendlich vielen Koordinaten und Ringe unendlicher Matrizen*, J. Reine Angew. Math., 171 (1934), 251–270.

[KP] I.V. Kovalishina and V.P. Potapov, *An indefinite metric in the Nevanlinna–Pick problem*, Doklady AN Armyan. SSR, 59 (1974), 17–22 (Russian).

[Kra1] I.F. Krasichkov-Ternovskii, *An interpretaion of the Beurling–Malliavin theorem on the radius of completeness*, Matem. Sbornik, 180 (1989), no. 3, 397–423 (Russian); English transl.: Math. USSR Sbornik, 66 (1990), no. 2, 405–429.

[LKMV] N. Kravitsky, M.S. Livshic, A.S. Markus and V. Vinnikov, *see* M.S. Livshic, N. Kravitsky, A.S. Markus and V. Vinnikov.

[Kr1] M.G. Krein, *On a generalization of some investigations of G. Szegö, V. Smirnov and A. Kolmogorov*, Doklady Akad. Nauk SSSR, 46 (1945), no. 3, 95–98 (Russian).

[Kr2] ———, *On a generalization of investigations of Stieltjes*, Doklady Akad. Nauk SSSR, 87:6 (1952), 881–884 (Russian).

[Kr3] ——— *The theory of self-adjoint extensions of semi-bounded Hermitian operators and applications, I&II*, Mat. Sbornik, 20 (1947), 431-495; 21 (1947), 365–404 (Russian).

[Kr4] ——— *Integral equations on a half-line with kernel depending upon the difference of the arguments*, Uspehi Matem. Nauk, 13 (1958), no. 5, 3–120 (Russian); English transl.: Amer. Math. Soc. Transl., (2) 22 (1962), 163–288.

[Kr5] ——— *Introduction to the geometry of indefinite J-spaces and to the theory of operators in those spaces*, pp. 15–92 in: Vtoraia Letniaya Matematicheskaia Shkola, Part I, Kiev, Naukova Dumka, 1965 (Russian); English transl.: Amer. Math. Soc. Transl., (2) 93 (1970), 103–176.

[Kr6] ——— *Analytical problems and results in theory of linear operators on a Hilbert space*, pp. 189–216 in: Proc. Int. Math. Congress 1966, Moscow, Mir, 1968 (Russian).

[KN] M.G. Krein and A.A. Nudel'man, *The Markov moment problem and extremal problems*, Moscow, Nauka, 1973 (Russian); English transl.: Transl. Math. Monographs, Vol. 50, Amer. Math. Soc., Providence, R.I., 1977.

[KR] M.G. Krein and P.G. Rehtman, *On the problem of Nevanlinna–Pick*, Trudi Odes'kogo Derzh. Univ. Mat., 2 (1938), 63–68 (Russian).

[KPS] S.G. Krein, Yu.I. Petunin and E.M. Semenov, *Interpolation of linear operators*, Moscow, Nauka, 1978 (Russian); English transl.: Amer. Math. Soc. Providence, RI, 1982.

[Kre] H.-O. Kreiss, *Über die Stabilitätsdefinition für Differenzengleichungen die partielle Differentialgleichungen approximieren*, Nord. Tidskr. Inf. (BIT), 2 (1962), 153–181.

[Kri1] T.L. Kriete, *Complete non-selfadjointness of almost selfadjoint operators*, Pacific J. Math., 42 (1972), 413–437.

[Kri2] ——— *Splitting and boundary behavior in certain H^2 spaces*, pp. 80–86 in: Linear and Complex Analysis problem Book 3, vol. II, eds. V.P. Havin and N.K. Nikolski, Lect. Notes Math., 1574 (1994).

[Kro] L. Kronecker, *Zur Theorie der Elimination einer Variablen aus zwei algebraischen Gleichungen*, Königl. Preuss. Akad. Wiss. (Berlin) (1881), 535–600; see also pp. 113–192 in: Leopold Kronecker's Werke II, Chelsea Publishing Co., New York 1968.

[Kru] N.Ya. Krupnik, *Banach algebras with symbol and singular integral operators*, Shtiintsa, Kishinev, 1984 (Russian); English transl.: Birkhäuser–Verlag, Basel, 1987.

[Kry] V.I. Krylov, *On functions regular in a half-plane*, Matem. Sbornik, 6 (48) (1939), 95–138 (Russian); English transl.: Amer. Math. Soc. Transl. (2) 32 (1963), 37–81.

[Kup] S. Kupin, *Linear resolvent growth test for similarity of a weak contraction to a normal operator*, Arkiv Math., 39 (2001), no. 1, 95–119.

[KuT] S. KUPIN AND S. TREIL, *Linear resolvent growth of a weak contraction does not imply its similarity to a normal operator*, Illinois J. Math., to appear.

[Ku] S.T. KURODA, *On a theorem of Weyl–von Neumann*, Proc. Japan Acad. 34 (1958), 11–15.

[LTh] M. LACEY AND C. THIELE, *L^p estimates on the bilinear Hilbert transform*, Proc. Nat. Acad. Sci. USA, 94 (1997), no. 1, 33–35.

[Lar] R. LARSEN, *An introduction to the theory of multipliers*, Springer–Verlag, N.Y.–Heidelberg, 1971.

[Lav] M.A. LAVRENTIEV, *Towards the theory of conformal mappings*, pp. 129–245, in Proc. Fiz.-Mat. Steklov Institute, Mathematics, vol V (1934), Izd. Akad. Nauk SSSR (Russian).

[Lax1] P.D. LAX, *Remarks on the preceeding paper*, Comm. Pure Appl. Math., 10 (1957), no. 4, 617–622.

[Lax2] ______ *Translation invariant spaces*, Acta Math., 101 (1959), 163–178.

[LPh1] P.D. LAX AND R.S. PHILLIPS, *Scattering theory*, AP, N.Y.–London, 1967.

[LPh2] ______ *Scattering theory for automorphic functions*, Princeton Univ. Press, Princeton, N.J., 1976.

[LM] N.A. LEBEDEV AND I.M. MILIN, *On an inequality*, Vestnik Leningrad. Universiteta, Ser. mat., meh., astron., 19 (1965), no. 4, 157–158 (Russian).

[Leb] A. LEBOW, *On von Neumann's theory of spectral sets*, J. Math. Anal. and Appl., 7 (1963), no. 1, 64–90.

[Leg] R. LEGGETT, *On the invariant subspace structure of compact dissipative operators*, Indiana Univ. Math. J., 22 (1973), no. 10, 919–928.

[Leh] O. LEHTO, *On the first boundary value problem for functions harmonic in the unit circle*, Ann. Acad. Sci. Fenn., Ser. AI, 210 (1955), 1–26.

[Leo1] A.F. LEONTIEV, *Exponential series*, Nauka, Moscow, 1976 (Russian).

[Leo2] ______ *Sequences of polynomials of exponentials*, Nauka, Moscow, 1980 (Russian).

[Le1] B.JA. LEVIN, *Distribution of zeros of entire functions*, Moscow, GITTL, 1956 (Russian); English transl.: Providence, RI, Amer. Math. Soc., 1980.

[Le2] ______ *On bases of exponentials in L^2*, Zapiski Mat. Otdel. Fiz.-Mat. Fac. Harkov Univ. and Harkov Mat. Obschestva, Ser. 4, 27 (1961), 39–48 (Russian).

[LeN] N. LEVINSON, *Gap and density theorems*, Amer. Math. Soc. Coll. Publ., XXVI, 1940.

[Lew] S. LEWIN, *Über einige mit der Konvergenz im Mittel verbundenen Eigenschaften von Funktionalfolgen*, Math. Zeitschr., 32 (1930), no. 4.

[LinR] P. LIN AND R. ROCHBERG, *The essential norm of Hankel operator on the Bergman space* Integral Equations Operator Theory 17 (1993), no. 3, 361–372.

[Lin] V.YA. LIN, *Holomorphic fiberings and multivalued functions of elements of a Banach algebra*, Funkt. Analiz i Prilozh., 7 (1973), no. 2, 43–51 (Russian); English transl.: Funct. Anal. Appl. 7 (1973), 122–128.

[Lind] E. LINDELÖF, *Sur un principe général de l'analyse et ces applications à la théorie de la représentation conforme*, Acta Soc. Sci. Fenn., 46:4 (1915).

[LT] J. LINDENSTRAUSS AND L. TZAFRIRI, *Classical Banach spaces, vol. I&II*, Springer–Verlag, Berlin etc., 1977, 1979.

[Lio] J.-L. LIONS, *Optimal control of systems described by partial differential equations*, Springer–Verlag, Heidelberg, 1971.

[LiP] J.-L. LIONS AND J. PEETRE, *Sur une classe d'espaces d'interpolation*, Inst. Hautes Etudes Sci. Publ. Math., 19 (1964), 5–68.

[Lit] J.E. LITTLEWOOD, *On inequalities in the theory of functions*, Proc. London Math. Soc., 23 (1925), 481–519.

[LS] G.S. LITVINCHUK AND I. SPITKOVSKY, *Factorization of measurable matrix functions*, Berlin, Akademie–Verlag, and Basel, Birkhäuser–Verlag, 1987.

[Liv1] M.S. LIVSHIC, *On an application of the theory of Hermitian operators to the generalized moment problem*, Dokl. AN SSSR, 44 (1944) (Russian).

[Liv2] ______ *On a class of linear operators on Hilbert space*, Matem. Sbornik, 19 (1946), 239–260 (Russian).

[Liv3] ______ *Operators, Oscillations, Waves. Open Systems*, Nauka, Moscow, 1966 (Russian); English transl.: Transl. Math. Monographs 34, Amer. Math. Soc., Providence, 1973.

[LKMV] M.S. Livshic, N. Kravitsky, A.S. Markus and V. Vinnikov, *Theory of commuting nonselfadjoint operators*, Kluwer Publ., Dordrecht, 1995.

[LNS] A.I. Loginov, M.A. Naimark and V.S. Shulman, *Nonselfadjoint operator algebras on a Hilbert space*, Itogi Nauki, VINITI, Moscow, Math. Anal. 12 (1974), 413–465 (Russian); Engl. transl.: J. Soviet. Math., 5 (1976), 250–278.

[Lor] E.R. Lorch, *Bicontinuous linear transformations in certain vector spaces*, Bull. Amer. Math. Soc., 45 (1939), 564–569.

[Low] D. Lowdenslager, *On factoring matrix valued functions*, Ann. of Math., (2) 78 (1963), 450–454.

[Lu] D.H. Luecking, *Characterizations of certain classes of Hankel operators on the Bergman space of the unit disk*, J. Funct. Anal., 110 (1992), no. 2, 247–271.

[Lus] N.N. Lusin, *Integral and trigonometric series*, Moscow, 1915 (Russian); 2nd edition 1951, Izd. Akad. Nauk SSSR (Russian).

[LPr] N.N. Lusin and I.I. Privalov, *Sur l'unicité et la multiplicité des fonctions analytiques*, Ann. Sci. Ecole Norm. Sup., (3) 42 (1925), 143–191.

[LZ] W.A.J. Luxembourg and A.C. Zaanen, *Riesz spaces. I.*, North–Holland, 1971.

[Ly] Yu.I. Lyubarskii, *Completeness of a biorthogonal family*, manuscript, Bordeuax, 1997.

[LyR] Yu.I. Lyubarskii and A. Rashkovskii, *Complete interpolating sequences for Fourier transforms supported by convex symmetric polygons*, Preprint, Trondheim University, 2000.

[LyS1] Yu.I. Lyubarskii and K. Seip, *A uniqueness theorem for bounded analytic functions*, Bull. London Math. Soc. (2) 58 (1997), 127–140.

[LyS2] ——— *Weighted Paley–Wiener spaces*, Preprint, Trondheim University, 2000.

[Lyu] Yu.I. Lyubich, *Functional analysis*, Itogi nauki i tehniki. Fundamental'nye napravlenia, vol. 19 (ed. N.K.Nikolski), VINITI Publ., Moscow, 1988 (Russian); English transl.: Encyclopaedia of Math. Sciences, vol. 19, Berlin etc., Springer–Verlag, 1992.

[LyuV] Yu.I. Lyubich and Vũ Quôc Phóng, *Asymptotic stability of linear differential equations on Banach spaces*, Studia Math., 88 (1988), 37–42.

[Mac] G.W. Mackey, *Harmonic analysis as the exploitation of symetry — a historical survey*, Rice Univ. Studies, 64 (1978), no. 2–3, 73–228.

[McL] G.R. MacLane, *Asymptotic values of holomorphic functions*, Rice Univ. Studies Monographs Math., Houston, 1963.

[Mag] W. Magnus, *On the spectrum of Hilbert's matrix*, Amer. J. Math., 72 (1950), 699–704.

[MakB] B.M. Makarov, *On the moment problem in certain function spaces*, Doklady Akad. Nauk SSSR, 127 (1959), 957–960 (Russian).

[MV1] N.G. Makarov and V.I. Vasyunin, *A model for noncontractions and stability of the continuous spectrum*, Lect. Notes Math., 864 (1981), 365–412.

[MV2] ——— *On quasi-similarity of model contractions with non-equal defects*, Zapiski Nauchn. Seminarov LOMI (Leningrad), 149 (1986), 24–37 (Russian); English transl.: J. Soviet Math., 42 (1988), no. 2, 1550–1561.

[Man1] S. Mandelbrojt, *Séries adhérentes. Régularisation des suites. Applications*, Paris, Gauthier–Villars, 1952.

[Man2] ——— *Closure theorems and composition theorems*, Izdat. Inostr. Literatury, Moscow, 1962 (Russian).

[Man3] ——— *Séries de Dirichlet*, Gauthier–Villars, Paris, 1969.

[Marc] J. Marcinkiewicz, *Sur l'interpolation d'opérations*, C.R. Acad. Sci. Paris, 208 (1939), 1272–1273.

[MZ] J. Marcinkiewicz and A. Zygmund, *Quelques inégalités pour les opérateurs linéaires*, Fund. Math., 32 (1939), 115–121.

[MM] M. Marcus and H. Minc, *A survey of matrix theory and matrix inequalities*, Allyn and Bacon, Boston, 1964.

[Mark] R.J. Marks II, *Introduction to Shannon sampling and interpolation*, Springer Verlag, New York, 1991.

[Ma1] A.S. Markus, *Some completeness criteria for the system of root vectors of a linear operator on a Banach space*, Matem. Sbornik, 70 (1966), 526–561 (Russian); English transl. in American Mathematical Society Translations, (2) 85, Twelve papers on functional analysis and geometry, American Mathematical Society, Providence, R.I., 1969.

[Ma2] ______ *The spectral synthesis problem for operators with the point spectrum*, Izvestia AN SSSR, ser. matem., 34 (1970), no. 3, 662–688 (Russian); English transl.: Math. USSR Izvestia, 4 (1970), 670–696.

[MaS] A.S. MARKUS AND A.A. SEMENTSUL, *A version of Bochner-Phillips-Allan's theorem*, Izvestia Akad. Nauk Moldavian SSR, Ser. Phys. Techn. and Mathem. Sci., 2 (1989), 14–18 (Russian).

[MAT] R.A. MARTINEZ–AVENDAÑO AND S.R. TREIL, *An inverse spectral problem for Hankel operators*, to appear in J. Operator Theory.

[MO] A.W. MARSHALL AND I. OLKIN, *Inequalities: theory of majorization and its applications*, Academic Press, NY, 1979.

[Mar] D.E. MARSHALL, *Blaschke products generate H^∞*, Bull. Amer. Math. Soc., 82 (1976), 494–496.

[Mas] P. MASANI, *Wiener's contribution to generalized harmonic analysis, prediction theory and filter theory*, Bull. Amer. Math. Soc. 72 (1966), no. 1, part II, 73–125.

[MW1] P.R. MASANI AND N. WIENER, *The prediction theory of multivariate stochastic processes*, Acta Math., 98 (1957), 111–150, 99 (1958), 93–137.

[MW2] ______ *On bivariate stationary processes and the factorization of matrix valued functions*, Teoria Veroyant. i Primenenia, 4 (1959), no. 3, 322–331.

[Masl] V.P. MASLOV, *Operator methods*, Nauka, Moscow, 1973 (Russian); French transl.: Mir, Moscow, 1987.

[Mats] V.I. MATSAEV [MACAEV], *On a class of completely continuous operators*, Doklady Akad. Nauk SSSR, 139 (1961), no. 3, 548–551 (Russian); English transl.: Sov. Math., Dokl., 2 (1961), 972-975.

[Mat] R.F. MATVEEV, *On the regularity of discrete time multivariate stochastic processes*, Doklady Akad. Nauk SSSR, 126 (1959), no. 5, 713–715 (Russian).

[McE] B. MCENNIS, *Dilations of holomorphic semigroups*, J. Operator Theory, 23 (1990), 21–42.

[McGPS] O.C. MCGEHEE, L. PIGNO, AND B. SMITH, *Hardy's inequality and the L^1-norm of exponential sums*, Ann. Math., 113 (1981), 613–618.

[Meg] A. MEGRETSKII, *A quasinilpotent Hankel operator*, Algebra i Analiz, 2 (1990), no. 4, 201–212 (Russian); English transl.: Leningrad Math. J., 2 (1991), no. 4, 879–889.

[MPT] A. MEGRETSKII, V. PELLER AND S. TREIL, *The inverse spectral problem for self-adjoint Hankel operators*, Acta Math., 174 (1995), no. 2, 241–309.

[MeM] R. MENNICKEN AND M. MÖLLER, *Nonselfadjoint boundary eigenvalue problems*, manuscript, 2000.

[Mer1] S.N. MERGELYAN, *On an integral related to analytic functions*, Izvestia AN SSSR, Seria Matem., 4 (1951), 395–400 (Russian).

[Mer2] ______ *Weighted approximation by polynomials*, Uspehi Matem. Nauk, 11 (1956), no. 5, 107–152 (Russian); English transl.: AMS Transl. Series, Ser. 2, 10 (1958), 59–106.

[Mey] Y. MEYER, *Ondelettes et opérateurs. II. Opérateurs de Calderon–Zygmund*, Paris, Hermann, 1990.

[Mi1] S.G. MIKHLIN, *Composition of singular integrals*, Doklady Acad. Nauk SSSR, 2 (II) (1936), no. 1 (87), 3–6 (Russian).

[Mi2] ______ *Singular integral equations*, Uspehi Mat. Nauk, 3 (1936), 29–112 (Russian).

[Mi3] ______ *Multivariate singular integrals and integral equations*, GIFML, Moscow, 1936 (Russian); English transl.: Pergamon Press, Oxford, 1965.

[Mi4] ______ *Numerical realization of variational processes*, Nauka, Moscow, 1936 (Russian).

[Min1] A.M. MINKIN, *Reflection of exponents and unconditional bases of exponentials*, Algebra i Analiz, 3 (1991), no. 5, 109–134 (Russian); English transl.: St. Petersburg Math. J., 3 (1992), no. 5, 1043–1068.

[Min2] ______ *Reflection of indices and unconditional bases of exponentials*, manuscript, 1997.

[Ml] W. MLAK, *Dilations of Hilbert space operators (general theory)*, Dissertationes Math., 153 (1978).

[Moe] C. MOELLER, *On the spectra of some translation invariant spaces*, J. Math. Anal. Appl., 4 (1962), 267–296.

[Mol] C. MOLLER, *General properties of the characteristic matrix in the theory of elementary particles. I*, Danske Vid. Selsk. Mat.-Fys. Medd., 23 (1945), no. 1, 1–48.

[Mu] B. MUCKENHOUPT, *Weighted norm inequalities for the Hardy maximal function*, Trans. Amer. Math. Soc., 165 (1972), 207–226.

[Muh] P.S. MUHLY, *Compact operators in the commutant of a contraction*, J. Funct. Anal., 8 (1971), 197–224.

[Mün] C.H. MÜNTZ, *Über den Approximationssatz von Weierstrass*, pp. 303–312 in: H.A. Schwarz Festschrift; Ch. 11:345, Berlin, 1914.

[Mur] G.J. MURPHY, *C^*-algebras and operator theory*, Boston etc., Academic Press, 1990.

[Nab1] S.N. NABOKO, *Absolutely continuous spectrum of a nondissipative operator, and the function model; I and II*, Zapiski Nauchn. Seminarov LOMI, 65 (1976), 90–102, 73 (1977), 118–135 (Russian); English transl.: J. Soviet Math., 16 (1981), no. 3, 1109–1118, 34 (1986), 2090–2101.

[Nab2] ______ *The function model of perturbation theory and its applications to scattering theory*, Trudy Math. Inst. Steklova, 147 (1980), 86–114, Moscow (Russian); English transl.: Proc. Steklov Math. Inst., 147 (1981), 85-116.

[Nab3] ______ *Conditions for similarity to unitary and selfadjoint operators*, Funktsional'nyi Analiz i ego Prilozhenia, 18 (1984), no. 1, 16–27 (Russian); English transl.: Funct. Anal. Appl., 18 (1984), 13–22.

[Nag] R. NAGEL (ED.), *One parameter semigroups of positive operators*, Lect. Notes Math., 1184 (1984), Springer–Verlag, Berlin.

[SzNF*] C. FOIAŞ AND B. SZŐKEFALVI-NAGY, *see* B. Szőkefalvi-Nagy and C. Foiaş.

[Nai1] M.A. NAIMARK, *Spectral functions of a symmetric operator*, Izvestia Acad. Nauk SSSR, ser. matem., 4 (1940), no. 3, 277–318 (Russian).

[Nai2] ______ *On a representation of additive operator valued set functions*, Doklady Acad. Nauk SSSR, 41 (1943), 373–375 (Russian).

[Nai3] ______ *Positive definite operator functions on a commutative group*, Izvestia Acad. Nauk SSSR, ser. matem., 7 (1943), 237–244.

[Nai4] ______ *Normed rings*, Nauka, Moscow (2nd edition, 1968) (Russian); English transl.: Wolters–Noordhoff Publishing, Groningen, 1972.

[LNS] M.A. NAIMARK, A.I. LOGINOV AND V.S. SHULMAN, *see* A.I. Loginov, M.A. Naimark and V.S. Shulman.

[Nak1] T. NAKAZI, *Commuting dilations and uniform algebras*, Canadian J. Math., XLII (1990), no. 5, 776–789.

[Nak2] ______ *ρ-dilations and hypo-Dirichlet algebras*, Acta Sci. Math. Szeged, 56 (1992), 175–181.

[NY] T. NAKAZI AND T. YAMAMOTO, *Some singular integral operators and Helson–Szegö measures*, J. Funct. Analysis, 88 (1990), no. 2, 366–384.

[Nat] I.P. NATANSON, *On an infinite system of linear equations*, Izvestia Fiz.-Mat. Obschestva Kazan, (3) 7 (1934/35), 97–98 (Russian; German summary).

[Naz] F.L. NAZAROV, *Local estimates of exponential polynomials and their applications to inequalities of uncertainty principle type*, Algebra i Analiz, 5 (1993), no. 4, 3–66 (Russian); English transl.: St.Petersburg Math. J., 5 (1994), no. 4, 663–717.

[NPTV] F. NAZAROV, G. PISIER, S. TREIL AND A. VOLBERG, *Sharp estimates in vector Carleson imbedding theorem and for vector paraproducts*, to appear in J. Reine angew. Math.

[NT] F. NAZAROV, S. TREIL, *The weighted norm inequalities for Hilbert transform are now trivial*, C.R. Acad. Sci. Paris, Sér. I, 323 (1996), 717–722.

[NTV1] F. NAZAROV, S. TREIL AND A. VOLBERG, *Cauchy integral and Calderon–Zygmund operators on nonhomogeneous spaces*, Int. Math. Research Notices, 15 (1997), 703–726.

[NTV2] ______ *Counterexample to the infinite dimensional Carleson embedding theorem*, C.R. Acad. Sci. Paris, Sér. I, 325 (1997), 383–388.

[Neh] Z. NEHARI, *On bounded bilinear forms*, Ann. of Math., 65 (1957), 153–162.

[Nel] E. NELSON, *The distinguished boundary of the unit operator ball*, Proc. Amer. Math. Soc., 12 (1961), 994–995.

[vN1] J. VON NEUMANN, *Zur Algebra der Funktionaloperatoren und Theorie der normalen Operatoren*, Math. Annalen, 102 (1929), 370–427.

[vN2] ______ *Charakterisierung des Spektrums eines Integraloperators*, Act. Sci. et Ind., Paris, 1935.

[vN3] ——— *Some matrix inequalities and metrization of matrix space*, Zapiski Tomskogo Universiteta, 1 (1937), 286–300; also published in "Collected works", vol.IV, pp.205–219, Pergamon Press, Oxford, 1962.

[vN4] ——— *On rings of operators, Reduction theory*, Ann. of Math., 50 (1949), 401–485.

[vN5] ——— *Eine Spektraltheorie für allgemeine Operatoren eines unitären Raumes*, Math. Nachr., 4 (1951), 258–281.

[NS] J. VON NEUMANN AND R. SCHATTEN, *The cross-space of linear transformations, I, II and III*, Ann. of Math. (2), 47 (1946), 73–84; 47 (1946), 608–630; 49 (1948), 557–582.

[Ne] F. NEVANLINNA AND R. NEVANLINNA, *Über die Eigenschaften analytischer Functionen in der Umgebung einer singulären Stelle oder Linie*, Acta Soc. Sci. Fenn., 50 (1922), no. 5, 1–46.

[Nev.O1] O. NEVANLINNA, *On the growth of the resolvent operators for power bounded operators*, pp. 247–264 in: Banach Center Publications, 38 (eds. J. Janas, F.H. Szafraniec and J. Zemanek), Warszawa, 1997.

[Nev.O2] O. NEVANLINNA, *Meromorphic resolvents and power bounded operators*, International Linear Algebra Year (Toulouse, 1995), BIT 36 (1996), no. 3, 531–541.

[Nev1] R. NEVANLINNA, *Über beschränkte analytische Funktionen*, Ann. Acad. Sci. Fenn., 32 (1929), no. 7.

[Nev2] ——— *Eindeutige analytische Funktionen*, 2nd ed., Springer–Verlag, 1953.

[New] J.D. NEWBURGH, *The variation of spectra*, Duke Math. J., 18 (1951), 165–176.

[Newm] D.J. NEWMAN, *The closure of translates in l^p*, Amer. J. Math., 86 (1964), no. 3, 651–667.

[NPT] A. NICOLAU, J. PAU AND P.J. THOMAS, *Smallness sets for bounded holomorphic functions*, J. Anal. Math. 82 (2000), 119–148.

[Nie] J.I. NIETO, *Sur le théorème d'interpolation de J. Lions et J. Peetre*, Can. Math. Bull., 14 (1971), 373–376.

[N.L] L.N. NIKOLSKAIA, *A stability criterion for the point spectrum of a linear operator*, Matematicheskie Zametki, 18 (1975), 601–607 (Russian); English transl.: Math. Notes, 18 (1975), 946–949.

[N1] N.K. NIKOLSKI [NIKOL'SKII], *On invariant subspaces of unitary operators*, Vestnik Leningrad. Univ, Matem. Mehanika Astronom., 21 (1966), 36–43 (Russian).

[N2] ——— *On spaces and algebras of Toeplitz matrices acting on l^p*, Sibirskii Mat. Zhurnal, 7 (1966), no. 1, 146–158 (Russian); English transl.: Sib. Math., J. 7, (1966), 118–126.

[N3] ——— *Invariant subspaces of the shift operator in some sequence spaces*, PhD Thesis, Leningrad University (Russian), 1966.

[N4] ——— *Complete extensions of Volterra operators*, Izvestia Akad. Nauk SSSR, ser. matem., 33 (1969), no. 6, 1349–1355 (Russian); English transl.: Math. URSS – Izvestia, 3 (1969), no. 6, 1271–1276.

[N5] ——— *On perturbations of the spectra of unitary operators*, Matem. Zametki, 5 (1969), no. 3, 341–350 (Russian); English transl.: Math. Notes, 5 (1969), 207–211.

[N6] ——— *The multiple shift with the simple spectrum*, Zapiski Nauchn. Seminarov LOMI, 19 (1970), 227–236 (Russian); English transl.: Seminars Math. Steklov Inst. Leningrad, 19 (1970), 132–136.

[N7] ——— *Five problems on invariant subspaces*, Zapiski Nauchn. Seminarov LOMI, 23 (1971), 115–127 (Russian); English transl.: J. Sov. Math., 2 (1974), 441–450.

[N8] ——— *Invariant subspaces in operator theory and function theory*, pp. 199–412 in: Itogi Nauki (VINITI, Moscow), Math. Anal., 12 (1974) (Russian); Engl. transl.: J. Soviet. Math., 5 (1976), 129–249.

[N9] ——— *Selected problems of weighted approximation and spectral analysis*, Trudy Math. Inst. Steklova, 120 (1974), Moscow (Russian); English transl.: Proc. Steklov Math. Inst., 120 (1974), AMS, Providence, 1976.

[N10] ——— *Lectures on the shift operator. IV.*, Zapiski Nauchn. Seminarov LOMI, 65 (1976), 103–132 (Russian); English transl.: J. Soviet Math., 16 (1981), no. 3, 1118–1139.

[N11] ——— *The present state of the problem of spectral analysis and synthesis*, pp. 240–282 in: Proc. First Summer School on Linear Operators and Functional Spaces, Novosibirsk, 1975; Nauka, Sibirsk. Otdel., Novosibirsk, 1977 (Russian); English transl.: pp. 240–282 in: American Mathematical Society Translations (2), 124, Fifteen papers on functional analysis, American Mathematical Society, Providence, RI, 1984.

[N12] ——— *A tauberian theorem on the spectral radius*, Sibir. Mat. Zh., 18 (1977), no. 6, 1367–1372 (Russian); English transl.: Sib. Math. J., 18 (1978), 969–973.

[N13] ——— *Two problems on spectral synthesis*, Zapiski Nauchn. Seminarov LOMI, 81 (1978), 139–141 (Russian); English transl.: J. Soviet Math., 26 (1984), 2185–2186; reprinted in Lect. Notes Math., 1043 (1984), 378–381.

[N14] ——— *Bases of invariant subspaces and operator interpolation*, Trudy Math. Inst. Steklova, 130 (1978), 50–123 (Russian); English transl.: Proc. Steklov Math. Inst., 130 (1979), no. 4, 55–132.

[N15] ——— *What problems do spectral theory and complex analysis solve for each other?*, pp. 631–638 in: Proc. Internat. Congr. Math. (Helsinki, 1978), Vol. 2, Acad. Sci. Fenn., Helsinki, 1980.

[N16] ——— *Bases of exponentials and values of reproducing kernels*, Doklady Akad. Nauk SSSR, 252 (1980), no. 6, 1316–1320 (Russian); English transl.: Soviet Math. Dokl., 21 (1980), no. 3, 937–941.

[N17] ——— *Methods for calculations of the spectral multiplicity of orthogonal sums*, Zapiski Nauchn. Seminarov LOMI, 126 (1983), 150–158 (Russian); English transl.: J. Soviet Math. 27 (1984), 2521–2526.

[N18] ——— *Ha-plitz operators: a survey of some recent results*, pp. 87–138 in: Operators and function theory, Proc. NATO ASI at Lancaster, 1984 (ed. S. Power), Dordrecht, Reidel Publ. Co., 1985.

[N19] ——— *Treatise on the shift operator*, Springer-Verlag, Berlin etc., 1986.

[N20] ——— *Interpolation libre dans l'espace de Hardy*, C.R. Acad. Sci. Paris, Sér. I, 304 (1987), no. 15, 451–454.

[N21] ——— *A few notes on generalized free interpolation*, U.U.D.M. Report 1988:3, Uppsala.

[N22] ——— *Multicyclicity phenomenon. I. An introduction and maxi-formulas*, pp. 9–57 in: Toeplitz Operators and Spectral Function Theory, Operator Theory: Adv. and Appl., 42, Birkhaüser, 1989.

[N23] ——— *Distance formulae and invariant subspaces, with an application to localization of zeros of the Riemann ζ-function*, Ann. Inst. Fourier, 45 (1995), no. 1, 143–159.

[N24] ——— *Yngve Domar's forty years in harmonic analysis*, Acta Univ. Upsaliensis, 58 (1995), 45–78.

[N25] ——— *Featured Review of the paper by A. Borichev and H. Hedenmalm "Completeness of translates in weighted spaces on the half-line" (Acta Math., 174 (1995), no. 1, 1–84)*, Math. Rev., 96f:43003.

[N26] ——— *In search of the invisible spectrum*, Ann. Inst. Fourier, 49 (1999), no. 6, 1925–1998.

[N27] ——— *Remarks concerning completeness of translates in function spaces*, J. Approx. Theory, 98 (1999), 303–315.

[HN*] N.K. Nikolski and V.P. Havin [Khavin], *see* V.P. Havin [Khavin] and N.K. Nikolski.

[HrN] N.K. Nikolski and S.V. Hruschev [Khruschev], *see* S.V. Hruschev [Khruschev] and N.K. Nikolski.

[NiP1] N.K. Nikolski and B.S. Pavlov, *Expansions in characteristic vectors of non-unitary operators and the characteristic function*, Zap. Nauchn. Semin. Leningr. Otd. Mat. Inst. Steklov, 11 (1968), 150–203 (Russian); English transl.: Semin. Math., V.A.Steklov Math. Inst., Leningr., 11 (1968), 54–72.

[NiP2] ——— *Bases of eigenvectors of completely nonunitary contractions and the characteristic function*, Izv. Akad. Nauk SSSR, Ser. Mat., 34 (1970), 90–133 (Russian); English transl.: Math. USSR Izv., 4 (1970), 91–134 (1970).

[NTr] N. Nikolski and S. Treil, *Linear resolvent growth of rank one perturbation of a unitary operator does not imply its similarity to a normal operator*, to appear in J. d'Analyse Math.

[NVa1] N.K. Nikolski and V.I. Vasyunin, *Control subspaces of minimal dimension. Elementary introduction. Discotheca*, Zapiski Nauchnyh Semin. LOMI, 113 (1981), 41–75 (Russian).

[NVa2] ______ *Control subspaces of minimal dimension, unitary and model operators*, J. Operator Theory, 10 (1983), 307–330.

[NVa3] ______ *Control subspaces of minimal dimension and rootvectors*, Integral Equat. Oper. Theory, 6 (1983), no. 2, 274–311.

[NVa4] ______ *Notes on two function models*, pp. 113–141 in: The Bieberbach conjecture, Proc. Symp. on the occasion of the proof, Amer. Math. Soc. Math. Surveys and Monographs, 21 (1986) (A. Baernstein, D. Drasin, P. Duren and A. Marden, eds.), Providence, R.I.

[NVa5] ______ *A unified approach to function models, and the transcription problem*, pp. 405–434 in: The Gohberg anniversary collection (Calgary, 1988), vol. 2, eds. H.Dym et al., Operator Theory: Adv. Appl., 41, Birkhäuser, Basel, 1989.

[NVa6] N.K. Nikolski and V.I. Vasyunin, *Quasiorthogonal decompositions with respect to complementary metrics, and estimates of univalent functions*, Algebra i Analyz, 2 (1990), no. 4, 1–81 (Russian); English transl.: Leningrad Math. J., 2 (1991), no. 4, 691–764.

[NVa7] ______ *Operator valued measures and coefficients of univalent functions*, Algebra i Analyz, 3 (1991), no. 6, 1–75 (Russian); English transl.: Leningrad Math. J., 3 (1992), no. 6, 1199–1270.

[NVa8] ______ *Elements of spectral theory in terms of the free function model. Part I: basic constructions*, pp. 211–302 in: Holomorphic Spaces, eds. Sh. Axler, J. McCarthy and D. Sarason, MSRI Publications, 33, Cambridge Univ. Press, 1998.

[NV] N.K. Nikolski and A.L. Volberg, *Tangential and approximate free interpolation*, pp. 277–299 in: Analysis and partial differential equations, ed. C. Sadosky, N.Y., Marcel Dekker, 1990.

[S.Nik] S.M. Nikol'skii, *Approximation of functions of several variables, and embedding theorems*, Nauka, Moscow, 1969 (Russian); English transl.: Springer–Verlag, Berlin etc., 1975.

[Nor] E.A. Nordgren, *Compact operators in the algebra generated by essentially unitary C_0 operators*, Proc. Amer. Math. Soc., 51 (1975), 159–162.

[Nos] K. Noshiro, *Cluster sets*, Springer–Verlag, Berlin–Heidelberg, 1960.

[Nud] A.A. Nudel'man, *On a new problem of moment type*, Doklady AN SSSR, 233 (1977), 792–795 (Russian); English transl.: Soviet Math. Dokl., 18 (1977), 507–510.

[Ny] B. Nyman, *On the one-dimensional translation group and semi-group in certain function spaces*, Thesis, Uppsala Univ., 1950.

[Ob] R.J. Ober, *A note on a system theoretic approach to a conjecture by Peller–Khruschev: the general case*, IMA J. Math. Control and Inform., 7 (1990), 35–45.

[ObM] R.J. Ober and S. Montgomery-Smith, *Bilinear transformation of infinite-dimensional state-space systems and balanced realizations of non-rational transfer functions*, SIAM J. Control and Optimization, 28 (1990), 438–465.

[O'N] M.D. O'Neill, *The convex hull of the interpolating Blaschke products generates H^∞*, Michigan Math. J., 44 (1997), no. 3, 419–434.

[Orl] W. Orlicz, *Über unbedingte Konvergenz in Funktionenräumen*, Studia Math., 2 (1933), 41–47.

[Ost] A. Ostrovski, *Sur quelques applications des fonctions convexes et concaves au sens de I. Schur*, J. Math. Pures Appl., 31 (1952), 253–292.

[Pag1] L.B. Page, *Bounded and compact vectorial Hankel operators*, Trans. Amer. Math. Soc., 150 (1970), 529–539.

[Pag2] ______ *Applications of the Sz.-Nagy and Foiaş lifting theorem*, Indiana Univ. Math. J., 20:2 (1970), 135–145.

[Pal] R.E.A.C. Paley, *On the lacunary coefficients of power series*, Annals of Math., 34 (1933), 615–616.

[PW] R.E.A.C. Paley and N. Wiener, *Fourier transforms in the complex domain*, Amer. Math. Soc. Colloquium Publ., 19, Providence, RI, 1934.

[Pa] A. Papoulis, *Signal analysis*, Auckland etc., McGraw-Hill, 1984.

[Paro] M. Parodi, *La localisation des valeurs caractéristiques des matrices et ses applications*, Gauthier–Villars, Paris, 1959.

[Par1] S. Parrott, *Unitary dilations for commuting contractions*, Pacific Math. J., 34 (1970), no. 2, 481–490.

[Par2] ——— *On a quotient norm and the Sz.-Nagy–Foiaş lifting theorem*, J. Funct. Anal., 30 (1978), 311–328.

[Part] J.R. PARTINGTON, *An introduction to Hankel operators*, Cambridge, Cambridge Univ. Press, 1988.

[PartW] J.R. PARTINGTON AND G. WEISS, *Admissible observation operators for the right shift semigroups*, preprint, 1999.

[Pat] D.I. PATIL, *Recapturing H^2-function on a polydisc*, Trans. Amer. Math. Soc., 188 (1974), 97–103.

[PauT] J. PAU AND P. THOMAS, *Decrease of bounded holomorphic functions along discrete sets*, Prépublication N° 217, 14pp., Université Toulouse III, 2001.

[Pau] V.I. PAULSEN, *Completely bounded maps and dilations*, Pitman Res. Notes Math., 146, New York, Longman, 1986.

[P1] B.S. PAVLOV, *Dilation theory and spectral anlysis of nonselfadjoint differential operators*, pp. 3–69 in: Operator theory in linear spaces, Proc. 7th Winter School, Drogobych, 1974 (ed. B. Mityagin), Central Econom. Institute, Moscow, 1975 (Russian); English transl.: Amer. Math. Soc. Transl. (2), 115 (1981). 103–142.

[P2] ——— *On conditions for separation of the spectral components of a dissipative operator*, Izvestia Akad. Nauk SSSR, Ser. Matem., 39 (1975), 123–148 (Russian); English transl.: Math. USSR Izvesyia, 9 (1976), 113–137.

[P3] ——— *Bases of exponentials and the Muckenhoupt condition*, Doklady Akad. Nauk SSSR, 247 (1979), no. 1, 37–40 (Russian); English transl.: Soviet Math. Dokl., 20 (1979), no. 4, 655–659.

[FP] B.S. PAVLOV AND L.D. FADDEEV, *see* L.D. Faddeev and B.S. Pavlov.

[Paz] A. PAZY, *Semigroups of linear operators and applications to partial differential equations*, Springer–Verlag, Heidelberg, 1983.

[Pea] C. PEARCY, *Some recent developments in operator theory*, CBMS series, 36, Amer. Math. Soc., Providence, RI, 1978.

[PSh] C. PEARCY AND A.L. SHIELDS, *A survey of the Lomonosov technique in the theory of invariant subspaces*, pp. 219–229 in: Topics in operator theory (ed. C. Pearcy), Math. Surveys series, 13, Amer. Math. Soc., Providence, 1974.

[Pee] J. PEETRE, *New thoughts on Besov spaces*, Duke University Math. Series 1, Math. Dept., Duke Univ., Durham, NC, 1976.

[Pek1] A.A. PEKARSKII, *Inequalities of Bernstein type for derivatives of rational functions, and inverse theorems of rational approximation*, Matem. Sbornik, 124 (166) (1984), no. 4, 571–588 (Russian); English transl.: Math. USSR Sbornik, 52 (1985), no. 2, 557–574.

[Pek2] ——— *Estimates of derivatives of rational functions in $L_p(-1,1)$*, Matem. Zametki, 39 (1986), no. 3, 388–394; English transl.: Math. Notes, 39 (1986), no. 3–4, 212–216.

[Pek3] ——— *Norm comparison for rational functions in the Bloch space and the Carathéodory-Fejér space*, Algebra i Analiz, 11:4 (1999), 139-150 (Russian); English transl.: St. Petersburg Math. J., 11:4 (2000).

[Pel1] A. PELCZYNSKI, *On universality of some Banach spaces*, Vestnik Len. Gos. Universiteta, 13 (1962), no. 3, 22–29 (Russian).

[Pel2] ——— *Banach spaces of analytic functions and absolutely summing operators*, CBMS regional conf. series math., 30, Amer. Math. Soc., Providence, RI, 1977.

[Pel] V.V. PELLER, *Hankel operators of class $\mathfrak{S}_p$ and their applications (rational approximation, Gaussian processes, the problem of majorization of operators)*, Matem. Sbornik, 113 (155) (1980), no. 4, 538–581 (Russian); English transl.: Math. USSR Sbornik, 41 (1982), 443–479.

[Pe2] ——— *An analogue of von Neumann's inequality, isometric dilation of contractions, and approximation by isometries in spaces of measurable functions*, Trudy Inst. Steklova, 155 (1981), 103–150 (Russian); English transl.: Proc. Steklov Inst. Math. (1983), 101–145.

[Pe3] ——— *Estimates of functions of power bounded operators on Hilbert spaces*, J. Operator Theory, 7 (1982), 341–372.

[Pe4] ——— *Vectorial Hankel operators and related operators of the Schatten-von Neumann class $\mathfrak{S}_p$*, Int. Equat. Operator Theory, 5 (1982), 244-272.

[Pe5] ——— *A description of Hankel operators of the class $\mathfrak{S}_p$ for $p > 0$, investigation of the rate of rational approximation and other applications*, Matem. Sbornik 122 (164)

(1983), no. 4, 481–510 (Russian); English transl.: Math. USSR Sbornik, 50 (1985), 465–494.

[Pe6] ——— *Estimates of functions of Hilbert space operators, similarity to a contraction and related function algebras*, pp. 199–204 in: Linear and Complex Analysis problem Book, Lect. Notes Math., 1043 (eds. V.P. Havin, S.V. Hruscev and N.K. Nikolski), Springer–Verlag, Berlin, 1984.

[Pe7] ——— *Nuclear Hankel operators acting between H^p spaces*, pp. 213–220 in: Operator Theory: Adv. Appl., 14, Birkhäuser Verlag, Basel, 1984.

[Pe8] ——— *Spectrum, similarity, and invariant subspaces of Toeplitz operators*, Izvestia Akad. Nauk SSSR, ser. matem., 50 (1986), no. 4, 776–787 (Russian); English transl.: Math. USSR Izvestya, 29 (1987), 133–144.

[Pe9] ——— *Hankel operators and multivariate stationary processes*, pp. 357-371 in: *Operator Theory: Operator Algebras and Applications, NH, 1988*, Proc. Sympos. Pure Math., 51, Part 1, Amer. Math. Soc., Providence, RI, 1990.

[Pe10] ——— *An excursion into the theory of Hankel operators*, pp. 65–120 in: Holomorphic Spaces, eds. Sh. Axler, J. McCarthy and D. Sarason, MSRI Publications, 33, Cambridge Univ. Press, 1998.

[HP*] V.V. Peller and S.V. Hruschev [Khruschev], *see* S.V. Hruschev [Khruschev] and V.V. Peller.

[Peterm] S. Petermichl, *Dyadic shift and a logarithmic estimate for Hankel operator with matrix symbol*, C.R. Acad. Sci. Paris, Ser. I Math., 330:6 (2000), 455–460.

[PetW] S. Petermichl and J. Wittwer, *An estimate for weighted Hilbert transform via Bellman functions*, Preprint, 16 pp., Mich. State Univ., 2000.

[Pet] K.E. Petersen, *Brownian motion, Hardy spaces and bounded mean oscillation*, Cambridge University Press, 1977.

[Pic] G. Pick, *Über die Beschränkungen analytischer Funktionen, welche durch vorgegebene Funktionswerte bewirkt werden*, Math. Ann., 77 (1916), 7–23.

[Pie1] A. Pietsch, *Eigenvalues and s-numbers*, Akad. Verlagstellenschaft Geest&Portig K.-G., Leipzig, 1987.

[Pie2] ——— *Operator ideals*, Berlin, Math. Monographien Bd. 16, Deutscher Verlag Wissensch., 1978.

[Pi1] G. Pisier, *Similarity problems and completely bounded maps*, Lect. Notes Math., 1618, Berlin–NY, Springer–Verlag 1995; 2nd expanded edition, 2001.

[Pi2] ——— *A polynomially bounded operator on Hilbert space which is not similar to a contraction*, J. Amer. Math. Soc., 10 (1997), 351–369.

[Pl1] A.I. Plessner, *Über das Verhalten analytischer Funktionen am Rande ihres Definitionsbereichs*, J. Reine Angew. Math., 158 (1927), 219–227.

[Pl2] ——— *On semi-unitary operators*, Doklady Akad. Nauk SSSR, 25 (1939), 708–710 (Russian).

[Pol1] A.G. Poltoratski, *On the boundary behavior of pseudocontinuable functions*, Algebra i Analiz, 5 (1993), no. 2, 189–210 (Russian); English transl.: St.Petersburg Math. J., 5 (1994), 389–406.

[Pol2] ——— *Integral representation and uniqueness sets for star-invariant subspaces*, to appear in: Proc. of IWOTA 2000, Operator Theory: Adv. Appl., 129, Birkhäuser, Basel, 2001.

[PSz] G. Polya and G. Szegö, *Aufgaben und Lehrsätze aus der Analysis, Vol. 1,2*, Berlin, Springer, 1925; English transl.: Berlin and N. Y., Springer–Verlag, 1972.

[Pom] Ch. Pommerenke, *Univalent functions*, Vandenhoeck and Göttingen Ruprecht, 1975.

[Pon] L.S. Pontryagin, *Hermitian operators in spaces with an idefinite metric*, Izvestia AN SSSR, Seria Matem., 8 (1944), no. 1, 243–280 (Russian).

[Pop1] G. Popescu, *von Neumann inequality for $(\mathcal{B}(H)^n)_1$*, Math. Scand., 68 (1991), 292–304.

[Pop2] ——— *Noncommutative disc algebras and their representations*, Proc. Amer. Math. Soc., 124 (1996), 2137–2148.

[Pot] V.P. Potapov, *On the multiplicative structure of J-nonexpanding matrux functions*, Trudy Moskov. Mat. Obschestva, 4 (1955), 125–236 (Russian); English transl.: Amer. Math. Soc. Transl. (2), 15 (1960), 131–243.

[Po1] S.C. Power, *Hankel operators on Hilbert space*, Bull. London Math. Soc., 12 (1980), 422–442.

[Po2] ——— *Hankel operators on Hilbert space*, Pitman Res. Notes Math., 64, Boston etc., Pitman, 1982.

[Pr1] I.I. PRIVALOV, *Cauchy integral*, Saratov University, 1919 (Russian).

[Pr2] ——— *Boundary properties of analytic functions*, 2nd edition, Moscow, 1950 (Russian); German transl.: Berlin, Deutscher Verlag, 1956.

[PV] V. PTÁK AND P. VRBOVÁ, *Operators of Toeplitz and Hankel type*, Acta Sci. Math. Szeged, 52 (1988), 117–140.

[Pu] M. PUTINAR, *Generalized eigenfunction expansions and spectral decompositions* pp. 265–286 in: *Banach Center Publications*, Vol. 38 (Eds., J. Janas, F.H. Szafraniec and J. Zemanek), Warszawa, 1997.

[PuV] M. PUTINAR AND F.-H. VASILESCU, *Solving moment problems by dimensional extension*, Ann. of Math., 149 (1999), 1087–1107.

[Put1] C.R. PUTNAM, *On normal operators in Hilbert space*, Amer. J. Math., 73 (1951), 357–362.

[Put2] ——— *Commutation properties of Hilbert space operators and related topics*, Springer Verlag, Berlin etc., 1967.

[Put3] ——— *An inequality for the area of hyponormal spectra*, Math. Zeitschr., 28 (1971), 473–477.

[Q] P. QUIGGIN, *For which reproducing kernel Hilbert spaces is Pick's theorem true?*, Integral Eq. Oper. Theory, 16 (1993), 244–266.

[RR] H. RADJAVI AND P. ROSENTHAL, *Invariant subspaces*, Berlin etc., Springer–Verlag, 1973.

[Red1] R.M. REDHEFFER, *A note on completeness*, Notices Amer. Math. Soc., 16 (1967), 830.

[Red2] ——— *Two consequences of the Beurling–Malliavin theory*, Proc. Amer. Math. Soc., 36 (1972), no. 1, 116–122.

[Red3] ——— *Completeness of sets of complex exponentials*, Adv. Math., 24 (1977), 1–62.

[RS] M. REED AND B. SIMON, *Methods of modern mathematical physics, vol.1–4*, Academic Press, N.Y., 1972–1979.

[Re] E. REICH, *On non-Hermitian Toeplitz matrices*, Math. Scand., 10 (1962), 145–152.

[Rei] H. REINHARD, *Eléments de mathématiques du signal, Tome 1–2*, Paris, Dunod, 1995.

[Rick] C.E. RICKART, *General theory of Banach algebras*, Van Nostrand, 1960.

[Ric] W.J. RICKER, *The spectral theorem: a historical viewpoint*, pp. 365–393 in: Ulmer Seminare, Funktionalanalysis und Differentialgleichungen, Heft 4 (eds. W. Arendt, W. Balser, W. Kratz, U. Stadtmüller), Ulm, 1999.

[Rie1] F. RIESZ, *Sur certains systèmes singuliers d'équations intégrales*, Ann. Ecole Normale Sup. (3), 28 (1911), 33–62.

[Rie2] ——— *Über die Randwerte einer analytische Funktion*, Math. Z., 18 (1923), 87–95.

[RieR] F. RIESZ AND M. RIESZ, *Über die Randwerte einer analytische Funktion*, pp. 27–44 in: Quatrième Congrès des Math. Scand., Stockholm, 1916.

[RSzN] F. RIESZ AND B. SZŐKEFALVI-NAGY, *Leçons d'analyse fonctionnelle*, Troisème édition, Szeged, Akadémiai Kiado, 1955.

[RieM1] M. RIESZ, *Sur le problème des moments. I; II; III*, Ark. för Mat. Astr. Fys., 16 (1922); 17 (1923).

[RieM2] ——— *Sur les fonctions congjugées et les séries de Fourier*, C.R. Acad. Sci. Paris, Ser. A–B, 178 (1924), 1464–1467.

[RieM3] ——— *Sur les maxima des formes bilinéaires et sur les fonctionnelles linéaires*, Acta Math., 49 (1926), 465–497.

[RieM4] ——— *Sur les fonctions conjuguées*, Math. Zeit., 27 (1927), 218–244.

[Ro1] R. ROCHBERG, *Trace ideal criteria for Hankel operators and commutators*, Indiana Univ. Math. J., 31 (1982), no. 6, 913–925.

[Ro2] ——— *Decomposition theorems for Bergman spaces and their applications*, pp. 225–278 in: Operators and function theory, ed. S.Power, Reidel Publ., Dordrecht, 1985.

[Ro3] ——— *Higher-order Hankel forms and commutators*, pp. 155–178 in: Holomorphic Spaces, eds. Sh. Axler, J. McCarthy and D. Sarason, MSRI Publications, 33, Cambridge Univ. Press, 1998.

[Rol] S. ROLEWICZ, *On orbits of elements*, Studia Math., 33 (1969), 17–22.

[Ros1] M. ROSENBLUM, *On the Hilbert matrix, I*, Proc. Amer. Math. Soc., 9 (1958), 137–140.

[Ros2] ——— *On the Hilbert matrix, II*, Proc. Amer. Math. Soc., 9 (1958), 581–585.

[Ros3] ——— *On a theorem of Fuglede and Putnam*, J. London Math. Soc., 33 (1958), 376–377.

[Ros4] ——— *Summability of Fourier series in $L^p(\mu)$*, Trans. Amer. Math. Soc., 105 (1962), no. 1, 32–42.

[Ros5] ——— *Self-adjoint Toeplitz operators*, Summer Institute of Spectral Theory and Statistical Mechanics 1965, Brookhaven National Laboratory, Upton, NY, 1966 (MR 34 #4084).

[Ros6] ——— *A corona theorem for countably many functions*, Integral Equations and Operator Theory, 3 (1980), 125–137.

[RoR] M. Rosenblum and J. Rovnyak, *Hardy classes and operator theory*, Oxford Univ. Press, 1985.

[RosSha] W.T. Ross and H.S. Shapiro, *Notes on generalized analytic continuation*, manuscript, 143pp., 2001.

[Rota] G.C. Rota, *On models for linear operators*, Comm. Pure Appl. Math., 13 (1960), no. 3, 468–472.

[Roz1] Yu.A. Rozanov, *Spectral theory of multi-dimensional stationary random processes with discrete time*, Uspehi Matem. Nauk, 13 (1958), no. 2, 93–142 (Russian); English transl.: Selected Transl. in Math. Stat. and Probability, 1 (1961), 253–306.

[Roz2] ——— *Linear extrapolation of discrete time multivariate stochastic processes of rank* 1, Doklady Akad. Nauk SSSR, 125 (1959), no. 2, 277–280.

[Roz3] ——— *Stationary stochastic processes*, Moscow, Fizmatgiz, 1963 (Russian); English transl.: San Francisco etc., Golden–Day 1967.

[Ru1] W. Rudin, *The radial variation of analytic function*, Duke Math. J., 22 (1955), 235–242.

[Ru2] ——— *Boundary values of continuous functions*, Proc. Amer. Math. Soc., 7 (1956), 808–811.

[Ru3] ——— *Fourier analysis on groups*, Wiley&Sons, N.Y., 1962.

[Ru4] ——— *Real and complex analysis*, McGraw–Hill Book Company, N.Y. etc., 1966.

[Ru5] ——— *Function theory in polydiscs*, New York, Benjamin, 1969.

[Ru6] ——— *Functional Analysis*, McGraw–Hill Book Company, N.Y. etc., 1973.

[RuVin] S.E. Rukshin and S.A. Vinogradov, *On free interpolation of germs of analytic functions in Hardy classes*, Zapiski Nauchn. Semin. LOMI, 107 (1982), 36–45 (Russian); English transl.: J. Soviet Math., 36 (1987), no. 3, 319–325.

[Rus] D.L. Russel, *Controlability theory for linear partial differential equations : recent progress and open problems*, SIAM Review, 20 (1978), 639–739.

[RW] D.L. Russel and G. Weiss, *A general necessary condition for exact observability*, SIAM J. Control and Optim, 32 (1994), 1–23.

[Sad1] C. Sadosky, *Some applications of majorized Toeplitz kernels*, pp. 581–626 in: Topics in Modern Harmonic Analysis, Proc. Seminar Torino and Milano (May–June 1982), Vol. II, Inst. Naz. Alta Matematica F.Severi, Roma, 1983.

[Sad2] ——— *Liftings of kernels shift-invariant in scattering theory*, pp. 303–336 in: Holomorphic Spaces, eds. Sh. Axler, J. McCarthy and D. Sarason, MSRI Publications, 33, Cambridge Univ. Press, 1998.

[Sai] S. Saitoh, *Theory of reproducing kernels and its applications*, Pitman Res. Notes Math. Series 189, Longman, 1988.

[Sakh] L.A. Sakhnovich, *Spectral analysis of Volterra operators and inverse problems*, Doklady Akad. Nauk SSSR, 115 (1957), no. 4, 666–669.

[Sal] D. Salamon, *Realization theory in Hilbert space*, Math. Systems Theory, 21 (1989), 147–164.

[S1] D. Sarason, *The H^p spaces on an annulus*, Mem. Amer. Math. Soc., 56 (1965).

[S2] ——— *On spectral sets having connected complement*, Acta Sci. Math. Szeged, 26 (1965), 289–299.

[S3] ——— *Weak-star generators of H^∞*, Pacific J. Math., 17 (1966), 519–528.

[S4] ——— *Generalized interpolation in H^∞*, Trans. Amer. Math. Soc., 127 (1967), no. 2, 179–203.

[S5] ——— *Weak-star density of polynomials*, J. Reine Angew. Math., 252 (1972), 1–15.

[S6] ——— *On products of Toeplitz operators*, Acta Sci. Math. Szeged, 35 (1973), 7–12.

[S7] ——— *Functions of vanishing mean oscillation*, Trans. Amer. Math. Soc., 207 (1975), 391–405.

[S8] ——— *Nearly invariant subspaces for the backward shift*, pp. 481–493 in: Operator theory: Adv. and Appl., 35, Basel, Birkhäuser, 1988.

[S9] ——— *Sub-Hardy Hilbert spaces in the unit disk*, Univ. of Arkansas Lect Notes, 10, N.Y. etc., John Wiley&Sons, 1994.

[Scha] H.H. SCHAEFER, *Topological vector spaces*, Springer–Verlag, Berlin, 1991.

[Schm1] E. SCHMIDT, *Entwicklung willkürlicher Funktionen nach Systemen vorgeschriebener*, Math. Annalen, 63 (1907), 433–476.

[Schm2] ——— *Über die Auflösung linearer Gleichungen mit unendlich vielen Unbekannten*, Rend. Circolo Mat. Palermo, 25 (1908), 53–77.

[Schmi] H. SCHMIDT, Jahresber. deutsch. Math. Verein., Section 2, 43 (1933), 6–7.

[Schu] J. SCHUBERT, *The corona theorem as an operator theorem*, Proc. Amer. Math. Soc., 69 (1978), 73–76.

[Sch1] I. SCHUR, *Bemerkungen zur Theorie der beschränkten Bilinearformen mit unendlich vielen Veränderlichen*, J. reine angew. Math., 140 (1911), 1–28.

[Sch2] ——— *Über die Potenzreihen, die im Innern des Einheitskreises beschränkt sind, I&II*, J. Reine Angew. Math., 147 (1917), 205–232; 148 (1918), 122–145 (German); English transl.: Operator Theory: Adv. Appl., 18, Birkhäuser Verlag, Basel, 1986.

[Schw1] L. SCHWARTZ, *Etude des sommes d'exponentielles réelles*, Hermann, Paris, 1943.

[Schw2] ——— *Approximation d'une fonction quelconque par des sommes d'exponentielles imaginaires*, Ann. Fac. Sci. Univ. Toulouse (4), 6, (1943). 111–176.

[Schw3] ——— *Théorie générale des fonctions moyenne-périodiques*, Ann. Math., 48 (1947), no. 4, 857–927.

[Sed] A.M. SEDLETSKII, *Biorthogonal expansions of functions in series of exponentials on intervals of the real line*, Uspehi Mat. Nauk, 37 (1982), no. 5, 51–95 (Russian); English transl.: Russian Math. Surveys, 37 (1983), no. 5, 57–108.

[Sei1] K. SEIP, *Regular sets of sampling and interpolation for weighted Bergman spaces*, Proc. Amer. Math. Soc., 117 (1993), no. 1, 213–220.

[Sei2] ——— *Beurling type density theorems in the unit disk*, Invent. math., 113 (1993), 21–39.

[Sei3] ——— *On the connection between exponential bases and certain related sequences in* $L^2(-\pi,\pi)$, J. Funct. Anal., 130 (1995), no. 1, 131–160.

[Se] D. SELEGUE, *A* C^**-algebra extension of the Szegö trace formula*, Talk given at the GPOTS, Arizona State Univ., Tempe, 1996.

[Sem] S. SEMMES, *Trace ideal criteria for Hankel operators, and applications to Besov classes*, Integral Equations Operator Theory, 7 (1984), no. 2, 241–281.

[Sham] F.A. SHAMOYAN, *Weak invertibility in some spaces of analytic functions*, Akad. Nauk Armyan. SSR Dokl., 74 (1982), no. 4, 157–161 (Russian).

[Shan] C.E. SHANNON, *A mathematical theory of communications*, Bell System Tech. J., 27 (1948), 379–423, 623–656.

[ShanW] C.E. SHANNON AND W. WEAVER, *The mathematical theory of communication*, Urbana, Illinois, 1949.

[Sha1] H.S. SHAPIRO, *Weakly invertible elements in certain function spaces, and generators in* l_1, Mich. Math. J., 11 (1964), 161–165.

[Sha2] ——— *A class of singular functions*, J. Canad. Math., 20 (1968), no. 6, 1425–1431.

[Sha3] ——— *Generalized analytic continuation*, pp. 151–163 in: Symposia on theoretical physics and mathematics, vol. 8 (Madras), Plenum Press, New York, 1968.

[Sha4] ——— *Some function-theoretic problems motivated by the study of Banach algebras*, pp. 95–113 in: *Proc. NRL Conference on Classical Function Theory*, ed. F.Gross, Washington, D.C., NRL, 1970.

[ShaS] H.S. SHAPIRO AND A.L. SHIELDS, *On some interpolation problems for analytic functions*, Amer. J. Math., 83 (1961), 513–532.

[Shap] J.H. SHAPIRO, *Composition operators and classical function theory*, Springer–Verlag, 1993.

[Sh1] A.L. SHIELDS, *Weighted shift operators and analytic function theory*, pp. 49–128 in: Topics in operator theory, ed. C. Pearcy, Math. Surveys 13, Amer. Math. Soc., Providence, 1974.

[Sh2] ——— *On Möbius bounded operators*, Acta Sci. Math. Szeged, 40 (1978), 371–374.

[Sh3] ——— *Cyclic vectors in Banach spaces of analytic functions*, pp. 315–349 in: Operators and function theory, ed. S.C. Power, NATO ASI Series 153, Dordrecht etc., Reidel, 1985.

[Shim1] S.M. SHIMORIN, *Wold-type decompositions and wandering subspaces for operators close to isometries*, J. Reine Angew. Math., 531 (2001), 147-189.

[Shim2] ——— *Approximate spectral synthesis in the Bergman space*, Duke Math. J., 101 (2000), no. 1, 1–39.

[Shi] N.A. SHIROKOV, *Analytic functions smooth up to the boundary*, Lect. Notes Math., 1312, Springer–Verlag, 1988.

[ShV] N.A. SHIROKOV AND I.V. VIDENSKII, *An extremal problem in the Wiener algebra*, Algebra i Analyz, 11 (1999), no. 6, 122–138 (Russian); English transl.: St. Petersburg Math. J., 11 (2000), no. 6, 1035–1049.

[ST] J. SHOHAT AND J. TAMARKIN, *The problem of moments*, Math. Surveys I, Amer. Math. Soc., Providence, RI, 1943.

[Sib1] R. SIBILËV, *Théorèmes d'unicité pour les séries de Wolff–Denjoy, et des opérateurs normaux*, Thèse, Université de Bordeaux, 1995.

[Sib2] ——— *A uniqueness theorem for Wolff–Denjoy series*, Algebra i Analiz, 7 (1995), no. 1, 170–199 (Russian); English transl.: St. Petersburg Math. J., 7 (1996), no. 1, 145–168.

[Si1] B. SIMON, *Trace ideals and their applications*, Lect. Notes London Math. Soc., Cambridge Univ. Press, Cambridge, 1979.

[Si2] ——— *Pointwise domination of matrices and comparison of $\mathfrak{S}_p$ norms*, Pacif. J. Math., 97 (1981), no. 2, 471–475.

[Sim1] I.B. SIMONENKO, *Riemann boundary problem with measurable coefficients*, Doklady Akad. Nauk SSSR, 135 (1960), 538–541 (Russian); English transl.: Soviet Math. Dokl., 1 (1960), 1295–1298.

[Sim2] ——— *A new general method for studying linear operator equations of the type of singular integral equations*, Doklady Akad. Nauk SSSR, 158 (1964), 790-793 (Russian); English transl.: Soviet Math. Dokl., 5 (1964), 1323–1326.

[Simo] A. SIMONIC, *An extension of Lomonosov's techniques to non-compact operators*, Trans. Amer. Math. Soc., 348 (1996), no. 3, 975–995.

[Sin] I. SINGER, *Bases in Banach spaces. I&II*, Springer–Verlag, Heidelberg, 1970, 1981.

[Sm1] V.I. SMIRNOV, *Sur la théorie des polynômes orthogonaux à une variable complexe*, J. Leningrad Fiz.-Mat. Obsch., 2 (1928), no. 1, 155–179.

[Sm2] ——— *Sur les valeurs limites des fonctions régulières à l'intérieur d'un circle*, J. Leningrad Fiz.-Mat. Obsch., 2 (1928), no. 2, 22–37.

[Sm3] ——— *Sur les formules de Cauchy et Green et quelques problèmes qui s'y rattachent*, Izvestia AN SSSR, ser. fiz.-mat., 3 (1932), 338–372.

[So1] B.M. SOLOMYAK, *Spectral multiplicity of analytic Toeplitz operators*, Doklady Akad. Nauk SSSR, 286 (1986), no. 6, 1308–1311 (Russian); English transl.: Soviet Math. Dokl., 33 (1986), no. 1, 286–290.

[So2] ——— *A functional model for dissipative operators. A coordinate-free approach*, Zapiski Nauchn. Seminarov LOMI, 178 (1989), 57–91 (Russian); English transl.: J. Soviet Math. 61 (1992), no. 2, 1981–2002.

[So3] ——— *Scattering theory for almost unitary operators and the functional model*, Zapiski Nauchn. Seminarov LOMI, 178 (1989), 92–119 (Russian); English transl.: J. Soviet Math. 61 (1992), no. 2, 2002–2020.

[SoV1] B.M. SOLOMYAK AND A.L. VOLBERG, *Multiplicity of analytic Toeplitz operators*, pp. 87–192 in: Toeplitz Operators and Spectral Function Theory (ed. N. Nikolski), Operator Theory: Adv. and Appl., 42, Birkhäuser, 1989.

[SoV2] ——— *Operator of multiplication by an analytic matrix valued function*, pp. 193–208 in: Toeplitz Operators and Spectral Function Theory (ed. N. Nikolski), Operator Theory: Adv. and Appl., 42, Birkhäuser, 1989.

[Sp] I.M. SPITKOVSKY, *Singular integral operators with PC symbols on the spaces with general weights*, J. Funct. Anal., 105 (1992), 129–143.

[Ste] S.B. STECHKIN, *On bilinear forms*, Dokl. Akad. Nauk SSSR, 71 (1950), no. 2, 237–240 (Russian).

[St1] E. STEIN, *Singular integrals and differentiability properties of functions*, Princeton Univ. Press, Princeton, N.J., 1970.

[St2] ——— *Harmonic analysis*, Princeton Univ. Press, Princeton, N.J., 1993.

[StW] E.M. STEIN AND G. WEISS, *Introduction to analysis on Euclidean spaces*, Princeton Univ. Press, Princeton, NJ, 1971.

[Sti] W.F. STINESPRING, *Positive functions on C^*-algebras*, Proc. Amer. Math. Soc., 6 (1955), 211–216.

[StWa] J.C. STRIKWERDA AND B.A. WADE, *A survey of the Kreiss matrix theorem for power bounded families of matrices and its extensions*, pp. 339–360 in: Banach Center Publications, 38 (eds. J. Janas, F.H. Szafraniec, and J. Zemanek), Warszawa, 1997.

[Stro] K. STROETHOFF, *The Berezin transform and operators on spaces of analytic functions*, pp. 361-380 in: *Banach Center Publications*, Vol.38 (Eds. J. Janas, F.H. Szafraniec and J. Zemanek), Warszawa, 1997.

[Str] E. STROUSE, *Finite rank intermediate Hankel operators*, Archiv Math., 67 (1996), 142–149.

[SV] I. SUCIU AND I. VALUSESCU, *Factorization theorems and prediction theory*, Rev. Roumaine Math. Pures Appl., 23 (1978), 1393–1423.

[SW] C. SUNDBERG AND TH. WOLFF, *Interpolating sequences for QA_B*, Trans. Amer. Math. Soc., 276 (1983), no. 2, 551–581.

[Sza] O. SZASZ, *Über die Approximation stetiger Funktionen durch lineare Aggregate von Potenzen*, Math. Ann., 77 (1916), 482–496.

[Sz1] G. SZEGÖ, *Beiträge zur Theorie der Toeplitzschen Formen, I*, Math. Zeit., 6 (1920), Heft 3/4, 167–202.

[Sz2] ——— *Über die Entwicklung einer analytischen Funktion nach den Polynomen eines Orthogonalsystems*, Math. Ann., 82 (1921), 188–212.

[Sz3] ——— *Über die Randwerte einer analytischen Funktion*, Math. Ann., 84 (1921), Heft 3/4, 232–244.

[Sz4] ——— *Orthogonal polynomials*, Amer. Math. Soc. Colloquium Publ., vol. XXIII, AMS, N.Y., 1959.

[SzN1] B. SZŐKEFALVI-NAGY, *On uniformly bounded linear transformations in Hilbert space*, Acta Sci. Math. Szeged, 11 (1947), 152–157.

[SzN2] ——— *A moment problem for self-adjoint operators*, Acta Math. Acad. Sci. Hungarica, 3 (1952), 285–293.

[SzN3] ——— *Sur les contractions de l'espace de Hilbert*, Acta Sci. Math. Szeged, 15 (1953), 87–92.

[SzN4] ——— *Completely continuous operators with uniformly bounded iterates*, Publ. Math. Inst. Hung. Acad. Sci., 4 (1959), 89–92.

[SzN5] ——— *Unitary dilations of Hilbert space operators and related topics*, CBMS conf. series, 19, Amer. Math. Soc., Providence, 1974.

[SzN6] ——— *A problem on operator valued bounded analytic function*, Zapiski Nauchn. Semin. LOMI (Steklov Inst. Math., Leningrad), 81 (1978), p. 99; see also J. Soviet Math., 26 (1984), no. 5.

[SzNF1] B. SZŐKEFALVI-NAGY AND C. FOIAŞ, *Modèles fonctionnels des contractions de l'espace de Hilbert. La fonction caracréristique*, C.R. Acad. Sci. Paris, 256 (1963), 3236–3239.

[SzNF2] ——— *Propriétés des fonctions caracréristiques, modèles fonctionnels et une classification des contractions*, C.R. Acad. Sci. Paris, 258 (1963), 3413–3415.

[SzNF3] ——— *Sur les contractions de l'espace de Hilbert. X. Contractions similaires à des transformations unitaires*, Acta Sci. Math. Szeged, 26 (1965), 79–91.

[SzNF4] ——— *Analyse harmonique des opérateurs de l'espace de Hilbert*, Akadémiai Kiado, Budapest, 1967; English transl.: *Harmonic analysis of operators on Hilbert space*, North Holland, New York, 1970.

[SzNF5] ——— *On the structure of intertwining operators*, Acta Sci. Math. Szeged, 35 (1973), 225–254.

[SzNF6] ——— *On contractions similar to isometries and Toeplitz operators*, Ann. Acad. Sci. Fenn., A I, 2 (1976), 553–564.

[SzNF7] ——— *Contractions without cyclic vectors*, Proc. Amer. Math. Soc., 87 (1983), no. 4, 671–674.

[SzK1] B. Szőkefalvi-Nagy and A. Koranyi, *Relations d'un problème de Nevanlinna et Pick avec la théorie des opérateurs de l'espace hilbertien*, Acta Math. Acad. Sci. Hungar., 7 (1957), 295–302.

[SzK2] ——— *Operatortheoretische Behandlung und Verallgemeinerung eines Problemkreises in der komplexen Funktionentheorie*, Acta Math., 100 (1958), 171–202.

[Tch] P.L. Tchebyshev, *On mean values*, Matem. Sbornik, vol. 2 (1867) (Russian).

[T] J.E. Thomson, *Approximation in the mean by polynomials*, Ann. of Math. (2) 133 (1991), no. 3, 477–507.

[Ti] E.C. Titchmarsh, *The theory of functions*, Oxford, 1939.

[Tit] ——— *The theory of the Riemann ζ-function*, Oxford, 1951.

[Toe] O. Toeplitz, *Zur Theorie der quadratischen und bilinearen Formen von unendlichvielen Veränderlichen*, Math. Ann., 70 (1911), 351–376.

[To1] V. Tolokonnikov, *Estimates in Carleson's corona theorem and finitely generated ideals in the algebra H^∞*, Funkzionalnyi Analyz i Prilozh., 14 (1980), 85–86 (Russian); English transl.: Funct. Anal. Appl., 14 (1980), no. 4, 320–321.

[To2] ——— *Estimates in Carleson's corona theorem. Ideals of the algebra H^∞, the problem of Szőkefalvi-Nagy*, Zapiski Nauchn. Semin. LOMI, 113 (1981), 178–198 (Russian); English transl.: J. Soviet Math., 22 (1983), 1814–1828.

[To3] ——— *The corona theorem in algebras of bounded analytic functions*, A paper deposed at VINITI (Moscow), No 251-84 Dep. (1984), 1–61 (Russian): English transl.: Amer. Math. Soc. Transl. (2), 149 (1991), 61–95.

[To4] ——— *Hankel and Toeplitz operators on Hardy spaces*, Zapiski Nauchn. Semin. LOMI, 141 (1985), 165–175 (Russian); English transl.: J. Soviet Math., 37 (1987), 1359–1364.

[Tom] Yu. Tomilov, *Resolvent approach to stability of operator semigroups*, J. Operator Theory, to appear.

[Tre1] S.R. Treil, *An operator theory approach to weighted norm inequalities for singular integrals*, Zapiski Nauchn. Semin. LOMI, 135 (1984), 150-174 (Russian).

[Tre2] ——— *The Adamyan–Arov–Krein theorem: a vector version*, Zapiski Nauchn. Semin. LOMI, 141 (1985), 56–71 (Russian); English transl.: J. Sov. Math., 37 (1987), 1297–1306.

[Tre3] ——— *Moduli of Hankel operators and the V.V. Peller–S.V. Khruschev problem*, Dokl. Akad. Nauk SSSR, 283 (1985), no. 5, 1095–1099 (Russian); English transl.: Soviet Math. Dokl., 32 (1985), 293–297.

[Tre4] ——— *A spacially compact system of eigenvectors forms a Riesz basis if it is uniformly minimal*, Doklady Akad. Nauk SSR, 288 (1986), 308–312 (Russian); English transl.: Soviet Math. Dokl., 33 (1986), 675–679.

[Tre5] ——— *Extreme points of the unit ball of the operator Hardy space $H^\infty(E \longrightarrow E)$*, Zapiski Nauchn. Semin. LOMI, 149 (1986), 160–164 (Russian); English transl.: J. Soviet Math., 42 (1988), no. 2, 1653–1656.

[Tre6] ——— *The invertibility of a Toeplitz operator does not imply its invertibility by the projection method*, Dokl. Akad. Nauk SSSR, 292 (1987), no. 3, 563–567 (Russian); English transl.: Soviet Math. Dokl., 35 (1987), no. 1, 103–107.

[Tre7] ——— *The resolvent of a Toeplitz operator may have an arbitrary growth*, Zapiski Nauchn. Semin. LOMI, 157 (1987), 175–177 (Russian); English transl.: J. Soviet Math., 44 (1989), no. 6, 868–869.

[Tre8] ——— *Angles between coinvariant subspaces and an operator valued corona problem. A question of Szőkefalvi-Nagy*, Doklady Akad. Nauk SSR, 302 (1988), no. 5, 1063–1068 (Russian); English transl.: Soviet Math. Dokl., 38 (1989), no. 2, 394–399.

[Tre9] ——— *Geometric methods in spectral theory of vector valued function: some recent results*, pp. 209–280 in: Toeplitz Operators and Spectral Function Theory (ed., N. Nikolski), Operator Theory: Adv. and Appl., 42, Basel, Birkhäuser–Verlag, 1989.

[Tre10] ——— *Hankel operators, imbedding theorems, and bases of invariant subspaces of the multiple shift*, Algebra i Analiz, 1 (1989), no. 6, 200–234 (Russian); English transl.: Leningrad Math. J., 1 (1990), no. 6, 1515–1548.

[Tre11] ——— *An inverse spectral problem for moduli of Hankel operators, and balanced realizations*, Algebra i Analyz, 2 (1990), no. 2, 158–182 (Russian); English transl.: Leningrad Math. J., 2 (1991), no. 2, 353–375.

[Tre12] ______ *The stable rank of the algebra H^∞ equals* 1, J. Functional Analysis, 109 (1992), no. 1, 130–154.

[Tre13] ______ *On the uniqueness of the best approximation by rational functions (Schur-Takagi problem*, manuscript, 2000.

[Tre14] ______ *Estimates in the corona theorem and ideals of H^∞: a problem of T. Wolff*, to appear in J. Analyse Math.

[TVa] S.R. Treil and V.I. Vasyunin, *An inverse spectral problem for the modulus of a Hankel operators*, Algebra i Analyz, 1 (1989), no. 4, 54–66; (Russian); English transl.: Leningrad Math. J., 1 (1990), no. 4, 859–870.

[TV1] S. Treil and A. Volberg, *A fixed point approach to Nehari's problem and its applications*, pp. 165–186 in: Toeplitz operators and related topics, The Harold Widom anniversary volume, Workshop on Toeplitz and Wiener-Hopf operators, Santa Cruz, CA, USA, September 20–22, 1992, ed. E.L. Basor, Operator Theory: Adv. and Appl., 71, Basel, Birkhäuser, 1994.

[TV2] ______ *Wavelets and the angle between past and future*, J. Funct. Analysis, 143 (1997), no. 2, 269–308.

[TV3] ______ *Continuous frame decomposition and a vector Hunt-Muckenhoupt-Wheeden theorem*, Ark. Mat., 35 (1997), 363–386.

[TV4] ______ *A simple proof of the Hunt–Muckenhoupt–Wheeden theorem*, Preprint, Mich. State Univ., 1997.

[TV5] ______ *Completely regular multivariate stationary processes and Muckenhoupt condition*, Pacif. J. Math., 190 (1999), no. 2, 361–382.

[Tren] T.T. Trent *A new estimate for the vector valued corona problem*, to appear in J. Funct. Analysis.

[Tr] H. Triebel, *Spaces of Besov–Hardy–Sobolev type*, Leipzig, Teubner Verlag, 1978.

[Ts] O.D. Tsereteli, *Metric properties of conjugate functions*, pp. 18–57 in: Itogi Nauki i Techniki Sovrem. Probl. Mat., 7, VINITI, Moscow, 1975 (Russian); English transl.: J. Soviet Math. 7 (1977), 309–414.

[Tum] G.Ts. Tumarkin, *A description of a class of functions permitting approximation by fractions with preassigned poles*, Izvestia Akad. Nauk Armyan. SSR, Ser. Fiz. Mat., 1 (1966), no. 2, 89–105 (Russian).

[Us] I.V. Ušakova [Ushakova], *A uniqueness theorem for holomorphic functions bounded in the unit circle*, Doklady Akad. Nauk SSSR, 130 (1960), 29–32 (Russian); English transl.: Soviet Math. Doklady, 1 (1960), 19–22.

[VanC1] J.A. Van Casteren, *Operators similar to unitary or selfadjoint ones*, Pacific J. Math., 104 (1983), no. 1, 241–255.

[VanC2] ______ *Boundedness properties of resolvents and semigroups of operators*, pp. 59–74 in: Banach Center Publications, 38 (eds. J. Janas, F.H. Szafraniec, and J. Zemanek), Warszawa, 1997.

[VanN] J.M.A.M. Van Neerven, *Asymptotic behaviour of semigroups of linear operators*, Operator Theory: Adv. Appl., 88, Birkhäuser Verlag, Basel, 1996.

[Var1] N.Th. Varopoulos, *Ensembles pics et ensembles d'interpolation pour les algèbres uniformes*, C.R. Acad. Sci. Paris, Sér. A, 272 (1971), 866–867.

[Var2] ______ *Some remarks on Q-algebras*, Ann. Inst. Fourier, 22 (1972), 1–11.

[Var3] ______ *Sur une inégalité de von Neumann*, C.R. Acad. Sci. Paris, ser. A-B, 277 (1973), A19–A22.

[Vasil] F.-H. Vasilescu, *Analytic functional calculus*, Editura Academiei (Bucharest) and Reidel Publ. Co (Dordrecht), 1982.

[Vas1] V.I. Vasyunin, *Unconditionally convergent spectral decompositions and nonclassical interpolation*, Doklady Akad. Nauk SSSR, 227 (1976), no. 1, 11–14 (Russian); English transl.: Soviet Math. Dokl., 17 (1976), no. 2, 309–313.

[Vas2] ______ *Construction of the function model of B. Sz.-Nagy and C. Foiaş*, Zapiski Nauchn. Semin. LOMI, 73 (1977), 6–23 (Russian); English transl.: J. Soviet Math., 34 (1986), no. 6, 2028–2033.

[Vas3] ______ *Unconditionally convergent spectral decompositions and interpolation problems*, Trudy Math. Inst. Steklova, 130 (1978), 5–49 (Russian); English transl.: Proc. Steklov Math. Inst. (1979), no. 4, 1–53.

[Vas4] ______ *Traces of bounded analytic functions on finite unions of Carleson sets*, Zapiski Nauchn. Semin. LOMI, 126 (1983), 31–34 (Russian); English transl.: J. Soviet Math., 27 (1984), no. 1, 2448–2450.

[Vas5] ______ *Formula for multiplicity of contractions with finite defect indices*, pp. 281–304 in: Toeplitz Operators and Spectral Function Theory (ed., N. Nikolski), Operator Theory: Adv. and Appl., 42, Birkhäuser, 1989.

[Vas6] ______ *On a biorthogonal function system related to the Riemann hypothesis*, Algebra i Analiz, 7 (1995), no. 3, 113–135 (Russian); English transl.: St. Petersburg Math. J., 7 (1996), no. 3, 405–419.

[MV*] V.I. Vasyunin and N.G. Makarov, *see* N.G. Makarov and V.I. Vasyunin.

[NVa*] V.I. Vasyunin and N.K. Nikolski, *see* N.K. Nikolski and V.I. Vasyunin.

[TVa] V.I. Vasyunin and S.R. Treil, *see* S.R. Treil and V.I. Vasyunin.

[GHVid] I.V. Videnskii, E.M. Gavurina and V.P. Havin [Khavin], *see* E.M. Gavurina, V.P. Havin [Khavin] and I.V. Videnskii.

[ShV] I.V. Videnskii and N.A. Shirokov, *see* N.A. Shirokov and I.V. Videnskii.

[Vid] V.S. Videnskii, *On normally increasing functions*, Uspehi Matem. Nauk, 9 (1954), no. 2, 212–213.

[Vinn] V. Vinnikov, *Commuting operators and function theory on a riemann surface*, pp. 445–476 in: Holomorphic Spaces, eds. Sh. Axler, J. McCarthy and D. Sarason, MSRI Publications, 33, Cambridge Univ. Press, 1998.

[Vin1] S.A. Vinogradov, *On interpolation and zeros of power series with coefficients in l^p*, Doklady AN SSSR, 160 (1965), no. 2, 262–266 (Russian); English transl.: Soviet Math. Dokl., 6 (1965), 57–60.

[Vin2] ______ *Interpolation theorems of Banach-Rudin-Carleson type, and norm estimates of embeddings of certain classes of analytic functions*, Zapiski Nauchn. Semin. LOMI, 19 (1970), 6–54 (Russian); English transl.: Seminars in Mathematics, V.A. Steklov Math. Inst., Leningrad, 19 (1972), 1–28.

[Vin3] ______ *Properties of multipliers of Cauchy-Stiltjes integrals, and some factorization problems for analytic functions*, pp. 5–39 in: Function theory and functional analysis, Proc. Seventh Winter School, Drogobych, 1974 (ed. B.S. Mityagin), C.E.M.I. AN SSSR, Moscow, 1976 (Russian); English transl.: Transl. Amer. Math. Soc. (2), 115 (1980), 1–32.

[Vin4] ______ *A refinement of Kolmogorov's theorem on the conjugate function and interpolation properties of uniformly convergent power series*, Trudy Mat. Inst. Steklova, 155 (1983), 7–40 (Russian); English transl.: Proc. Steklov Inst. Math. 155 (1983), 3–37.

[Vin5] ______ *Some remarks on free interpolation by bounded and slowly growing analytic functions*, Zapiski Nauchn. Semin. LOMI, 126 (1983), 35–46 (Russian); English transl.: J. Soviet Math. 27 (1984), 2450–2458.

[GHVin] S.A. Vinogradov, E.A. Gorin and S.V. Hruschev [Khruschev], *see* E.A. Gorin, S.V. Hruschev [Khruschev] and S.A. Vinogradov.

[HVin] S.A. Vinogradov and V.P. Havin [Khavin], *see* V.P. Havin [Khavin] and S.A. Vinogradov.

[RuVin] S.A. Vinogradov and S.E. Rukshin, *see* S.E. Rukshin and S.A. Vinogradov.

[Vit1] P. Vitse, *A tensor product approach to the operator corona problem*, to appear in Journal Operator Theory.

[Vit2] ______ *Smooth operators in the commutant*, Preprint, Université Laval, Québec, 2001.

[Voi] D. Voiculescu, *A note on quasitriangularity and trace class self-commutators*, Acta Sci. Math. Szeged, 42 (1980), 195–201.

[V1] A.L. Volberg, *The simultaneous approximation by polynomials on the circle and in the interior of the disc*, Zapiski Nauchn. Semin. LOMI, 92 (1979), 60–84 (Russian).

[V2] ______ *Thin and thick families of rational functions*, Lect. Notes Math., 864 (1981), 440–480.

[V3] ______ *Two remarks concerning the theorem of S. Axler, S.-Y. A. Chang and D. Sarason*, J. Operator Theory, 7 (1982), 209–218.

[V4] ______ *Weighted density of polynomials on the line for strongly nonsymmetric weights*, Zapiski Nauchn. Semin. LOMI, 126 (1983), 47–54 (Russian); English transl.: J. Soviet Math., 27 (1984), 2458–2462.

[V5] ——— *Matrix A_p weights via S-functions*, J. Amer. Math. Soc., 10 (1997), no. 2, 445–466.

[V6] ——— *Carleson measures for $\mathcal{K}_\Theta$ cannot be checked on reproducing kernels*, Address to the IWOTA conference, June 13–16, 2000, Bordeaux.

[Vu] Q.Ph. Vũ [Vũ, Quôc Phóng], *Theorems of Katznelson–Tzafriri type for semigroups of operators*, J. Funct. Anal., 103 (1992), 74–84.

[Wa] J.-K. Wang, *Note on a theorem of Nehari on Hankel forms*, Proc. Amer. Math. Soc., 24 (1970), 103–105.

[Wei] G. Weiss, *L^p-stability of a linear semigroup on a Hilbert space implies exponential stability*, J. Diff. Equations, 76 (1988), 269–285.

[We1] J. Wermer, *On invariant subspaces of normal operators*, Proc. Amer. Math. Soc., 3 (1952), 270–277.

[We2] ——— *On restrictions of operators*, Proc. Amer. Math. Soc., 4 (1953), no. 6, 860–865.

[Wey1] H. Weyl, *Über beschränkte quadratische Formen, deren Differenz vollstetig ist*, Rend. Circolo Mat. Palermo, 27 (1909), 373–392.

[Wey2] ——— *Inequalities between the two kinds of eigenvalues of a linear transformation*, Proc. Acad. Sci. USA, 35 (1949), 408–411.

[Whi1] E.T. Whittacker, *On the functions, which are represented by expansions of the interpolation theory*, Proc. Roy. Soc. Edinburgh, 35 (1915), 181.

[Whi2] ——— *Fourier theory of cardinal series*, 1935.

[Wid1] H. Widom, *Inversion of Toeplitz matrices. II*, Illinois J. Math., 4 (1960), 88–99.

[Wid2] ——— *Inversion of Toeplitz matrices. III*, Notices Amer. Math. Soc., 7 (1960), p. 63.

[Wid3] ——— *Hankel matrices*, Trans. Amer. Math. Soc., 127 (1966), 179–203.

[Wi1] N. Wiener, *On the closure of certain assemblages of trigonometric functions*, Proc. Nat. Acad. Sci. USA, 13 (1927), 27.

[Wi2] ——— *Generalized harmonic analysis*, Acta Math., 55 (1930), 117-258.

[Wi3] ——— *The Fourier integral and certain of its applications*, New York, Cambridge Univ. Press, 1933.

[Wi4] ——— *Extrapolation, interpolation, and smoothing of stationary time series. With engineering applications*, Cambridge, Mass., MIT Press; New York, Wiley&Sons, 1949.

[WH] N. Wiener and E. Hopf, *Über eine Klasse singulärer Integralgleichungen*, S.-B. Preuss Akad. Wiss. Berlin, Phys.-Math. Kl., 30/32 (1931), 696–706.

[MW*] N. Wiener and P.R. Masani, *see* P.R. Masani and N. Wiener.

[Wog1] W.R. Wogen, *On some operators with cyclic vectors*, Indiana Univ. Math. J., 27 (1978), no. 1, 163–171.

[Wog2] ——— *On cyclicity of commutants*, Integral Equat. and Operator Theory, 5 (1982), 141–143.

[Woj] P. Wojtaszczyk, *Banach spaces for analysts*, Cambridge Univ. Press, 1991.

[Wol] H. Wold, *A study in the analysis of stationary time series*, Almquist och Wiksell, Uppsala, 1938.

[HWo] H. Wolf and V.P. Havin [Khavin], *see* V.P. Havin [Khavin] and H. Wolf.

[Wolf] J. Wolff, *Sur les séries $\sum \frac{A_n}{z-z_n}$*, C.R. Acad. Sci. Paris, 173 (1921), 1056–1059, 1327–1328.

[Wolff1] Th.H. Wolff, *Counterexamples to two variants of the Helson-Szegö theorem*, manuscript, 1980.

[Wolff2] ——— *Two algebras of bounded functions*, Duke Math. J., 49 (1982), 321–328.

[Wolff3] ——— *A refinement of the corona theorem*, pp. 399–400 in: *Linear and complex analysis problem book. 199 research problems* (Eds., N. Nikolski, V. Havin and S. Hruschev), Lect. Notes Math., 1043, Springer Verlag, 1984.

[Wo] W.M. Wonham, *Linear multivariable control: a geometric approach*, Springer–Verlag, Heidelberg, 1979.

[Wr] V. Wrobel, *Analytic functions into Banach spaces and a new characterization for isomorphic embeddings*, Proc. Amer. Math. Soc., 85 (1982), 539-543.

[Wu] Zh. Wu, *Function theory and operator theory on the Dirischlet space*, pp. 179–199 in: Holomorphic Spaces, eds. Sh. Axler, J. McCarthy and D. Sarason, MSRI Publications, 33, Cambridge Univ. Press, 1998.

[Xia] D. Xia, *Spectral theory of hyponormal operators*, Operator Theory: Advances and Appl., 10, Birkhäuser–Verlag, 1983.

[Yaf] D.R. YAFAEV, *Mathematical scattering theory*, St. Petersburg University, St. Petersburg, 1994 (Russian); English transl.: Transl. Math. Monographs series, 105, Amer. Math. Soc., Providence, RI, 1992.

[Yak1] D.V. YAKUBOVICH, *Riemann surface models of Toeplitz operators*, pp. 305–415 in: Operator theory: Advances and applications, 42, Basel, Birkhäuser, 1989.

[Yak2] ——— *Spectral multiplicity of Toeplitz operators with smooth symbols*, Amer. J. Math., 115 (1993), no. 6, 1335–1346.

[Yak3] ——— *Subnormal operators of finite type, I and II.*, Revista Mat. Iberoamericana, 14 (1998), no. 1, 95–115, 14 (1998), no. 3, 623–681.

[You] N.J. YOUNG, *Balanced realizations in infinite demensions*, pp. 449–471 in: Operator Theory: Adv. Appl., 19, Birkhäuser Verlag, Basel, 1976.

[Y1] R.M. YOUNG, *An introduction to nonharmonic Fourier series*, London, Academic Press, 1980.

[Y2] ——— *On complete biorthogonal systems*, Proc. Amer. Math. Soc., 83 (1981), no. 3, 537–540.

[Zab] J. ZABCZYK, *A note on C_0-semigroups*, Bull. Acad. Polon. Sci. Sér. Math., 23 (1975), 895–898.

[JZ] V.P. ZAKHARYUTA AND V.I. JUDOVICH, *see* V.I. Judovich and V.P. Zakharyuta.

[Zas] V.N. ZASUKHIN, *On the theory of multivariate stationary processes*, Doklady Akad. Nauk SSSR, 33 (1941), 435–437 (Russian).

[Zhu] K.H. ZHU, *Operator theory in function spaces*, Monographs and Textbooks in Pure and Appl. Math., 139, Marcel Dekker Inc, N.Y., 1990.

[Z] A. ZYGMUND, *Trigonometric series, Vol. I and II.*, Cambridge University Press, 1959.

Author Index

The indexes to this volume give references to the entire book; reference I-$\alpha\beta\gamma$ means page $\alpha\beta\gamma$ of Volume 1.

Subject Index

The indexes to this volume give references to the entire book; reference I-$\alpha\beta\gamma$ means page $\alpha\beta\gamma$ of Volume 1.

Symbol Index

The indexes to this volume give references to the entire book; reference I-$\alpha\beta\gamma$ means page $\alpha\beta\gamma$ of Volume 1.

Errata to Volume 1

The following misprints have been detected after the impression of the first volume.

1) On page 58_7, in the displayed formula, a λ should be replaced by ζ in the following way

$$C(f,\zeta) = \{\lambda \in \overline{\mathbb{C}} : \lambda = \lim_n f(z_n), \text{ where } z_n \longrightarrow \zeta \text{ nontangentially}\}.$$

2) On pages 114 (Exercise A.5.7.2(k)), 132_{22}, 140_{17} as well as in the Author index on page 430 and in the bibliography on page 446: *Wittwer* instead of *Witter.*

3) On page 131^{10}: it should be $c\log(n)$ instead of $c(\log(n))^{1/2}$.

4) On page 186_{17}: H_2 instead of H^2 (in the index of the identity operator).

5) On page 270_5: D.4.7, D.4.8 instead of D.5.7.

6) On page 374^8 and 374^{15}: *Bergh* instead of *Berg* (as well as in the Author index on page 401).

7) On page 375^3 and 375^6, the correct references are: *P. Duren, B. Romberg, and A. Shields for 8.7.1(f) and 8.7.1(i) for the case p=1.*

8) On page 401, a 'v' is missing in Arov's name in the references [AKK1]–[AAK4] (corrected in Volume 2).

Selected Titles in This Series

(Continued from the front of this publication)

61 **W. Norrie Everitt and Lawrence Markus,** Boundary value problems and symplectic algebra for ordinary differential and quasi-differential operators, 1999

60 **Iain Raeburn and Dana P. Williams,** Morita equivalence and continuous-trace C^*-algebras, 1998

59 **Paul Howard and Jean E. Rubin,** Consequences of the axiom of choice, 1998

58 **Pavel I. Etingof, Igor B. Frenkel, and Alexander A. Kirillov, Jr.,** Lectures on representation theory and Knizhnik-Zamolodchikov equations, 1998

57 **Marc Levine,** Mixed motives, 1998

56 **Leonid I. Korogodski and Yan S. Soibelman,** Algebras of functions on quantum groups: Part I, 1998

55 **J. Scott Carter and Masahico Saito,** Knotted surfaces and their diagrams, 1998

54 **Casper Goffman, Togo Nishiura, and Daniel Waterman,** Homeomorphisms in analysis, 1997

53 **Andreas Kriegl and Peter W. Michor,** The convenient setting of global analysis, 1997

52 **V. A. Kozlov, V. G. Maz′ya, and J. Rossmann,** Elliptic boundary value problems in domains with point singularities, 1997

51 **Jan Malý and William P. Ziemer,** Fine regularity of solutions of elliptic partial differential equations, 1997

50 **Jon Aaronson,** An introduction to infinite ergodic theory, 1997

49 **R. E. Showalter,** Monotone operators in Banach space and nonlinear partial differential equations, 1997

48 **Paul-Jean Cahen and Jean-Luc Chabert,** Integer-valued polynomials, 1997

47 **A. D. Elmendorf, I. Kriz, M. A. Mandell, and J. P. May (with an appendix by M. Cole),** Rings, modules, and algebras in stable homotopy theory, 1997

46 **Stephen Lipscomb,** Symmetric inverse semigroups, 1996

45 **George M. Bergman and Adam O. Hausknecht,** Cogroups and co-rings in categories of associative rings, 1996

44 **J. Amorós, M. Burger, K. Corlette, D. Kotschick, and D. Toledo,** Fundamental groups of compact Kähler manifolds, 1996

43 **James E. Humphreys,** Conjugacy classes in semisimple algebraic groups, 1995

42 **Ralph Freese, Jaroslav Ježek, and J. B. Nation,** Free lattices, 1995

41 **Hal L. Smith,** Monotone dynamical systems: an introduction to the theory of competitive and cooperative systems, 1995

40.4 **Daniel Gorenstein, Richard Lyons, and Ronald Solomon,** The classification of the finite simple groups, number 4, 1999

40.3 **Daniel Gorenstein, Richard Lyons, and Ronald Solomon,** The classification of the finite simple groups, number 3, 1998

40.2 **Daniel Gorenstein, Richard Lyons, and Ronald Solomon,** The classification of the finite simple groups, number 2, 1995

40.1 **Daniel Gorenstein, Richard Lyons, and Ronald Solomon,** The classification of the finite simple groups, number 1, 1994

39 **Sigurdur Helgason,** Geometric analysis on symmetric spaces, 1994

38 **Guy David and Stephen Semmes,** Analysis of and on uniformly rectifiable sets, 1993

37 **Leonard Lewin, Editor,** Structural properties of polylogarithms, 1991

36 **John B. Conway,** The theory of subnormal operators, 1991

35 **Shreeram S. Abhyankar,** Algebraic geometry for scientists and engineers, 1990

For a complete list of titles in this series, visit the AMS Bookstore at **www.ams.org/bookstore/**.